High Temperature Processing of Milk and Milk Products

High Temperature Processing of Milk and Milk Products

Hilton C. Deeth

School of Agriculture and Food Sciences,
The University of Queensland,
Brisbane, Australia

Michael J. Lewis

Department of Food and Nutritional Sciences,
University of Reading,
Reading, UK

WILEY Blackwell

This edition first published 2017

Registered Office
John Wiley & Sons Ltd, The Atrium, Southern Gate, Chichester, West Sussex, PO19 8SQ, UK

Editorial Offices
111 River Street, Hoboken, NJ 07030, USA
9600 Garsington Road, Oxford, OX4 2DQ, UK
The Atrium, Southern Gate, Chichester, West Sussex, PO19 8SQ, UK

For details of our global editorial offices, customer services, and more information about Wiley products visit us at www.wiley.com.

Wiley also publishes its books in a variety of electronic formats and by print-on-demand. Some content that appears in standard print versions of this book may not be available in other formats.

Library of Congress Cataloging-in-Publication Data

[9781118460504]

Cover Design: Wiley
Cover image: Leong Thian Fu / EyeEm/Gettyimages

Set in 10/12pt Warnock by SPi Global, Pondicherry, India

10 9 8 7 6 5 4 3 2 1

Contents

About the Authors

Hilton C. Deeth

Hilton Deeth grew up on a dairy farm in Australia which engendered in him a love of all things dairy from an early age. After completing a science degree and a PhD in organic chemistry at the University of Queensland, he worked as a research food scientist in the Queensland Department of Primary Industries (QDPI) for 23 years. His areas of research included quality aspects of milk, butter and cheese, specialising in lipase and lipolysis, and flavour problems associated with milk fat. He also initiated and led a seafood research group which became the major seafood group in Australia. At the time of leaving QDPI to take up an academic position at the University of Queensland in 1995, he was Manager of Food Research and Development, responsible for a range of projects on dairy, seafood, meat, and fruits and vegetables.

At the University of Queensland, he taught dairy science, seafood science, emerging food technologies and food product development. He also supervised research projects on various dairy and seafood topics and was advisor for more than 30 PhD and research Masters students, as well as several coursework Masters students. His main dairy research interests, and the topics of his students' theses, were UHT processing and products, quality aspects of milk, yogurt and milk powders, and new processing technologies.

In 1996, he established a specialist Centre for UHT Processing and Products for the Australian dairy industry at the University of Queensland and directed the Centre until it was merged with four other specialist dairy centres in Australia to form Dairy Innovation Australia Ltd (DIAL) in 2008. The research initiated in the UHT Centre was continued in a DIAL-funded Food Science Research Program at the University of Queensland which he managed until his retirement in 2011. He has published 150 research papers and reviews, and 28 book chapters. A large proportion of these are on topics directly related to this book and hence the book contains numerous references to them.

Since retiring as Emeritus Professor of Food Science, he has remained involved in dairy science and technology as a consultant assisting dairy companies with product and process development, and trouble shooting, and also providing technical training in UHT processing and other dairy topics for companies in Australia and other countries. This book also draws on the knowledge and experience gained in these industry involvements.

Michael J. Lewis

Mike Lewis worked for over 38 years at the University of Reading as a Lecturer and Senior Lecturer in the School of Chemistry, Food and Pharmacy before retiring in Sept 2011. He was educated at the University of Birmingham in the Department of Chemical Engineering where he gained a BSc, MSc (Biol Eng) and PhD. Over the last 40 years, he has acquired considerable expertise in many topics related to food science and technology, including physical properties of foods, food processing operations, milk and milk processing, heat treatment, evaporation, drying and membrane technology. He has an extensive publication record in these areas, with over 80 refereed papers and over 20 book chapters and three books. In addition, he was actively involved in maintaining the University pilot plant and generating considerable income from industry and research funders. In the context of this book he has worked with a UHT pilot plant since 1976, when Reading University acquired an APV Junior UHT plant. Since then it has been used for teaching, product development and research and has earned the University in excess of £350,000 from outside work. He has helped many companies with product and process development and staff training activities in different countries.

He supervised over 30 PhD students and over 150 BSc and MSc project students. His research activities have focused on minerals in milk and their interactions with proteins, especially with regard to calcium and also magnesium and their role in casein micelle stability. Stability aspects that have been studied include ethanol stability, heat coagulation, involving heat coagulation times, stability to in-container sterilization, UHT sterilization, involving fouling of heat exchangers and deposit formation and fouling of UF membranes. His most recent work involves developing procedures for measuring pH and ionic calcium at high temperatures, to better understand their role in heat stability of milk and the effects of chelating agents on these parameters related to calcium fortification, calcium removal and stabilizer addition. Much of his research has been conducted in close partnership with the food and dairy food industries and he has recently completed four sets of workshops for Dairy Innovation Australia Ltd (DIAL) on UHT processing and several for major multi-national producers of milk products. He remains research active and is a regular reviewer for a number of the major food and dairy journals. He loves teaching and sharing his knowledge and wishes to continue doing this.

Preface

This book has arisen from a productive period of collaboration between the authors. They first met when Hilton visited the University of Reading in 1995, on a fact-finding mission for setting up a UHT Centre in Australia. In 2003, he returned to the UK to spend a short sabbatical period at Reading University and shortly afterwards Mike spent time at the UHT Centre at the University of Queensland. This led to funding by Dairy Innovation Australia Ltd (DIAL) for a PhD student, who studied at both Universities and helped to cement a fruitful partnership and long lasting friendship.

In 2000, Mike produced a book in collaboration with Neil Heppell on continuous thermal processing (Lewis & Heppell, 2000). This arose from a suggestion by the publisher for a revision of Harold Burton's book on UHT processing of milk (Burton, 1988), which was published in 1988 and which was then out of print. Harold dedicated that book to all those who worked on UHT processing and aseptic filling at the National Institute for Research in Dairying (NIRD) between 1948 and 1985 and particularly those in the Process Engineering Group, of which he was Head for much of the time. During that period Harold's group carried out much of the fundamental work on understanding the safety and quality of UHT milk and his name was known worldwide. This was the major publication in this area in that era and should still be consulted by anybody involved with UHT processing. Harold retired in 1985. Mike's relationship with Harold extended for over 20 years and is described in the preface to the Lewis and Heppell book.

The Lewis and Heppell book concentrated on continuous processing and aimed to expand the range of food products that were featured beyond milk products. This aim was simple to state but more difficult in practice to achieve, as the majority of publications dealt with milk and milk-based products. Today, the commercial reality is that the range of heat-treated, particularly sterilised, products available to the consumer is much wider, although important technical information on matters such as formulations and processing conditions is less readily available in the public domain. This book also incorporated pasteurisation and heat treatments designed to further extend the shelf-life of pasteurised products and also acidic products such as fruit juices. In fact, pasteurised products are more widespread than sterilised products in many countries.

In this volume we have aimed to produce a book that gives a clear explanation of the principles involved in high-temperature heat treatment processes. The main emphasis throughout is on product safety and quality. To fully understand these issues involves integrating a number of important scientific disciplines covering the physical aspects of foods, the transfer of energy and the effects of heat on the chemical, biochemical and sensory characteristics and the problems inherent in dealing with biological raw materials. Thus there is a section which describes the basic physical properties of the products that are to be heat treated. We have aimed to provide a good balance between the engineering aspects and the chemical, biochemical, microbiological and sensory issues which have to be considered to produce foods which are both safe and of a high quality. One of the innovations is a better understanding of factors affecting heat stability and the role of pH and ionic calcium and the interesting relationship between them, along with suggestions for measuring these parameters at sterilisation conditions. Another is the use of temperature-time profiles for assessing the microbiological and chemical effects of a given process.

The book covers microbiological issues, other thermal processes such as pasteurisation, extended shelf-life (ESL) and in-container sterilisation, UHT processing conditions and characterisation of processes, engineering aspects, heat stability, fouling and cleaning, changes during storage, quality assurance procedures, alternative technologies, shelf-stable products other than sterilised cow's milk, products that can be manufactured from UHT milk and analytical procedures. We have devoted a chapter specifically to products other than white cow's milk to reflect the increasing importance of these products.

Some products that are UHT processed are considerably more viscous than milk or cream are, and some of these contain discrete particles, deliberately added and not present as sediment. Thus it covers situations where streamline flow conditions are likely to prevail, as well as the thornier problem of heat-treating products containing particles, ideally ensuring uniform heating of the solid and liquid phases. One observation is that there are still relatively few UHT products in this category, although this may change as Chinese consumers like drinks containing particulates. The Lewis and Heppell book still remains worth consulting in this area.

There is a great deal of interest and research activity in alternative technologies and processes for pasteurising and sterilising foods and these have been addressed in this book. However, these technologies have to compete against heat treatment, which is a very effective, convenient and energy-efficient method of processing foods. In fact, the application of heat in HTST pasteurisation and UHT sterilisation are two well established processes. Nevertheless, alternative technologies are finding applications, mainly in niche areas and these aspects are discussed. In most cases, they add a considerable processing cost to the product.

The layout of this volume should help the reader who wishes to explore specific topics in depth. We have taken care to ensure that the book is well cross-referenced and indexed, which will help the reader who wishes to browse. Perhaps a novelty is the chapter on analytical procedures which can be used to further understand some of the issues involved in UHT processing and products. A recent excellent publication on analytical procedures for milk and milk products contains only three indexed references to UHT milk. We hope this chapter goes some way to redressing this imbalance. One interesting challenge regarding analytical methods is to identify applications for some of the powerful instrumental techniques which are now available. Some of these, such as proteomics and molecular-based microbiological techniques, are now well established and represent quantum leaps in milk analysis.

In the final chapter (Chapter 12) we have outlined several aspects on which we believe there is currently insufficient information and require further research. These have been identified through our research and consultancy activities and confirmed during the preparation of this book. In that chapter we have also collated several key references to books, book chapters and review articles which can be consulted for further information on specific topics.

We hope that this book will be stimulating to undergraduate and postgraduate students of food science and technology, as well as industry biotechnologists, food technologists and engineers who are involved or interested in heating and cooling milk and other products of a biological nature.

We believe that it will provide a useful reference source for the food industry and provide a focus for gaining a better understanding of the factors influencing safety and quality of heat-treated products. We are confident that a major strength of the book is the combination of theoretical knowledge derived from the considerable research output in the subject area with our practical experience of heat processing. We have tried to make our explanations as clear as possible, especially when interpreting results from those articles where it was unclear what was really intended.

There have been many other constraints and competing pressures in meeting the publisher's deadlines. However, as we have both recently retired, issues such as teaching, quality audits, research assessment exercises, enjoyable teaching activities, research priorities and University administration are no longer excuses.

Finally, returning to Harold Burton, in brewing technology, there is a process known as "Burtonising" the water, to ensure that water used for beer production has a composition similar to that found in Burton-on-Trent, one of the great brewing centres in the UK. It could well be argued, considering his enormous contribution to the subject area, that the term Burtonising the milk, should be the synonymous with UHT treatment. In support of this, it is especially noteworthy that Harold Burton's 1988 book on UHT processing has recently been reprinted by Springer, in its original form.

List of Abbreviations

A.	*Anoxybacillus*
AQL	Acceptable quality limit
ATP	Adenosine triphosphate
B.	*Bacillus*
BCA	Bicinchoninic acid
C.	*Cronobacter, Coxiella*
CAR	Carboxen
CCP	Colloidal calcium phosphate
Cl.	*Clostridium*
CMC	Carboxymethyl cellulose
cP	Centipoise
cSt	Centistoke (Unit of kinematic viscosity)
DEFT	Direct epifluorescent technique
DSHP	Disodium hydrogen phosphate
DVB	Divinylbenzene
E401	Sodium alginate
E407	Carrageenan
E410	Locust bean gum
E412	Guar gum
E451	Sodium & potassium triphosphates
E466	Cellulose gum, carboxymethyl cellulose
E471	Mono & diglycerides of fatty acids
EDTA	Ethylenediaminetetraacetic acid
EGTA	Ethylene glycol tetraacetic acid
ELISA	Enzyme-linked immunosorbent assay
ESL	Extended shelf-life
FCM	Flow cytometry
FDNB	1-Fluoro-2,4-dinitrobenzene
FFA	Free fatty acids
FID	Flame ionization detection
FITC	Fluorescein isothiocyanate
FOS	Fructo-oligosaccharide
FPD	Freezing point depression
FTIR	Fourier transform infrared
G.	*Geobacillus*

GC	Gas chromatography
GOS	Galacto-oligosaccharide
HCA	Hierarchical cluster analysis
HCT	Heat coagulation time
HMF	Hydroxymethylfurfural
HPCD	High pressure carbon dioxide
HPH	High pressure homogenisation
HPLC	High performance liquid chromatography
HPP	High pressure processing
HTST	High-Temperature, Short-Time
ICP	Inductively coupled plasma
IDF	International Dairy Federation
L.	*Listeria*
LA	Lactic acid
LAL	Lysinoalanine
LMTD	Logarithmic mean temperature difference
LPS	Lactoperoxidase system
LRM	Lactose-reduced milk
LTI	Low-temperature inactivation
MCC	Microcrystalline cellulose
MF	Microfiltration
MPC	Milk protein concentrate
MSNF	Milk solids non-fat
MWCO	Molecular weight cut-off
NCN	Non-casein nitrogen
NMR	Nuclear magnetic resonance
NPN	Non-protein nitrogen
OHTC	Overall heat transfer coefficient
OPA	Ortho-phthalaldehyde
PAGE	Polyacrylamide gel electrophoresis
PATP	Pressure-assisted thermal processing
PATS	Pressure-assisted thermal sterilisation
PCA	Principal component analysis
PDMS	Polydimethylsiloxane
PEF	Pulsed electric field
PFPD	Pulsed flame photometric detector
PHE	Plate heat exchanger
PPC	Post-processing contamination or post-pasteurisation contamination
Ps.	*Pseudomonas*
RDA	Recommended daily allowance
RDI	Recommended daily intake
RP	Reversed phase
RSMP	Reconstituted skim milk powder
SCC	Somatic cell count
SDHP	Sodium dihydrogen phosphate
SDS	Sodium dodecylsulfate
SFR	Sterility failure rate

SGE	Starch gel electrophoresis
SHMP	Sodium hexametaphosphate
SMP	Skim milk powder
SPME	Solid phase microextraction
St.	*Staphylococcus*
Str.	*Streptococcus*
TA	Titratable acidity
TBA	Thiobarbituric acid
TCA	Trichloroacetic acid
TNBS	Trinitrobenzenesulfonic acid
TSC	Trisodium citrate
TVC	Total viable count
UHT	Ultra-high temperature
UTP	Uniform trans-membrane pressure
WPC	Whey protein concentrate

1

History and Scope of the Book

1.1 Setting the Scene

Bovine milk is the main source of milk in the world today. Table 1.1 illustrates some production data for the leading bovine milk producing countries in the world. The first column shows total milk production, whereas the second shows milk production expressed as per head of population. Thus countries like New Zealand and Ireland (see footnote) produce large quantities per capita, whereas countries such as China, although positioned in the top five milk producers in the world, are most probably not producing sufficient milk for their increasing populations who are developing a taste for milk and milk products. USA and Brazil are also large producers of bovine milk. Much of the milk in Brazil is consumed as liquid milk with a fair proportion being UHT processed.

It is very exciting time to be writing a book on high-temperature processing, particularly ultra-high temperature (UHT) processing. UHT is a continuous process and as such is applicable to any product that can be pumped through a heat exchanger and then aseptically packaged, although the vast majority of products are either milk or milk-based. UHT milk and milk products are now global commodities and are being transported large distances to all parts of the world. In a number of traditional milk-drinking countries, for example, UK, Greece and Australia, pasteurised milk is still the milk of preference and the cooked flavour that is associated with UHT and sterilised milk is given as a major reason for maintaining this status quo (Perkins & Deeth, 2011). In contrast, in some other countries much more UHT milk is consumed than pasteurised milk. For example, in France, Belgium and Portugal, more than 90% of all liquid milk purchased is UHT-treated, whereas in the UK, Norway, Sweden, Australia and New Zealand, it is less than 10%. Similar variations are also found in other parts of the world, with less than 5% of UHT milk being consumed in India and USA but over 60% in Vietnam and China. In other words, availability and also preferences for pasteurised or sterilised milk vary from country to country. Some examples for Europe and other parts of the world are given in Table 1.2.

Recently, there has been a substantial increase in UHT capacity in all parts of the world. In part, this is to supply the increased demand for UHT milk from China. It is also predicted that there will be an increased demand from Africa and other parts of South East Asia. Since UHT milk does not require refrigeration and has a long shelf-life, it provides a very convenient way of providing good quality milk to large populations in remote areas, without the need for the expensive cold chain infrastructure. UHT milk is

High Temperature Processing of Milk and Milk Products, First Edition. Hilton C. Deeth and Michael J. Lewis.

Table 1.1 Leading producers of bovine milk in 2012, with populations and production per head of population.

	Milk production, 2012 (billion L)	Population (billion)	Per capita consumption (L/person)
United States of America	90.9	0.318	286
India	54.0	1.244	43.4
China	37.8	1.364	27.7
Brazil	32.3	0.204	158
Russian Federation	31.6	0.146	216
Germany	30.5	0.081	377
France	24.0	0.066	364
New Zealand	20.0	0.0046	4350
Turkey	16.0	0.078	205
United Kingdom	13.9	0.065	214
World	**620.3**	**7.25**	**85.6**

from: http://dairy.ahdb.org.uk/market-information/supply-production/milk-production/world-milk-production/#.VzxQVHn2aUk and world population figures

Table 1.2 Percentage of drinking milk which is UHT processed in various European countries and worldwide.

Europe	
Greece	0.9
Norway	5.3
UK	8.4
Austria	20.3
Germany	66.1
France	95.5
Spain	95.7
Belgium	96.7
Worldwide	
US	2
India	3
Australia	11
Japan	11
Malaysia	28
China	32
Thailand	46
Vietnam	62

Information from Wikipedia and Datamonitor (China has the largest forecast growth increase in UHT milk consumption over the period 2012 to 2020. India also has a high projected growth rate but is starting from a much lower base level).

now transported to China and other parts of South East Asia from countries such as Australia, New Zealand and even longer distances from USA and several countries in Europe. Both large multinational conglomerates and much smaller companies are engaged in these activities.

The demand for UHT milk is increasing worldwide. It has been estimated that the compound annual growth rate for UHT milk in the world between 2013 and 2019 will be 12.5%, with the global market reaching USD 137.6 billion in 2019 (Persistance Market Research, 2014). In locations where fresh milk is not available, UHT milk can be produced from milk powder. Also milk demand is increasing in locations where there has previously been no strong culture of drinking milk; there is a continuing investment in UHT capability in various parts of the world to meet this demand.

Demand for UHT milk is not the only factor that is changing in relation to the market for milk and milk products. The variety of milk-based beverages is constantly expanding. In the early days of UHT processing, only white milk and some cream products were processed. The variety in milk drinks has since mushroomed and now includes flavoured milk and products containing additives offering health benefits, derived either from naturally occurring components in the milk or non-milk components, such as plant extracts, fruit juices and other substances such as melatonin and dietary fibre (see Tables 1.3 and 1.4). There are also many products of non-dairy origin; these are covered in more detail in Chapter 9.

Whatever type of UHT product is being produced, a key consideration is to ensure that the formulation has good heat stability. The first consideration is a knowledge of the chemical composition of raw milk which is complex and subject to day-to-day and seasonal variation, as illustrated by data on a bulk milk supply collected in the UK over 15 months (Chen *et al.*, 2014) (see Table 1.5). Secondly, it is crucial to understand how different additives, for example, fruit essences, flavours, mineral salts, stabilizers and emulsifiers will influence heat stability in order to ensure that fouling of the UHT plant and sediment formation in the treated product are minimized. This has been one of the authors' main areas of research and an aim of this book is to share our experiences dealing with these topics. Similar issues arise with some non-bovine milk products, such as goat's and camel's milk, which have poorer heat stability than cow's milk and need to be stabilized to be suitable for UHT processing. Historically, pH was considered to be a very important determinant of heat stability of milk, but now the role of both pH and ionic calcium and their interrelationship is better understood, as is how they change when milk is heated to 140 °C and then cooled; these issues are discussed in Chapter 6.

The first and overriding objective is to make UHT products safe to drink by ensuring that they are adequately sterilised and that they will not cause outbreaks of food poisoning. The most heat resistant pathogen is *Clostridium botulinum.* It is noteworthy that raw milk is not considered to be a source of this pathogen and incidents of botulinum have not been attributed to liquid milk and only very rarely to milk products. However, raw milk may contain some bacterial spores that are more heat resistant than *Cl. botulinum* and ensuring that these are inactivated during the UHT process will ensure the UHT milk is free of *Cl. botulinum*, even if the bacterium may have inadvertently found its way into a formulated milk product from other sources. In fact, some recent work using a probabilistic assessment model predicted that contamination of a UHT product with *Cl. botulinum* might arise only once in 367 years (Pujol *et al.*, 2015). The release of product containing the thermophilic spore-former *Geobacillus stearothermophilus* was

Table 1.3 Some drinking milk products available commercially or being developed.

Milk types
Full-cream, skim, semi-skim – HTST, ESL, UHT, sterilised
Flavoured
Lactose-reduced
Carbonated
Goat's, sheep, buffalo's, horse and camel's milk
Microfiltered
Breakfast milks
A2 milk
Yogurt drink
Pet milk
Soy, almond, oat and other plant "milks"
Additives/fortifiers
Calcium and other minerals
Vitamins
Plant sterols and stannols
Omega-3, conjugated linoleic acid (CLA)
Microparticulated whey protein
Milk bioactive peptides
Dietary fibre (e.g., β-glucan, inulin)
Melatonin
Polyphenols
Oligosaccharides

calculated to be much higher than this, but this is not a food pathogen and will only be problematic where the temperature of the products during storage is allowed to reach >50 °C, such as in hot climates.

The ideal UHT milk product should be free of environmental contaminants and also be commercially sterile. This is the combined responsibility of the milk producer, the milk processor and the packaging technologist. However, this is by no means the end of the process because UHT milk will then be expected to be acceptable to the consumer and have a "best before date" of at least six months (Rysstad & Kolstad, 2006). There are sound scientific explanations why six months is a reasonable period and problems may be encountered if this is extended. Although it is possible to eliminate microbial activity, it is not possible to prevent chemical and physical reactions taking place; in some circumstances, enzymatic reactions such as proteolysis and lipolysis may also be encountered. Thus, there is in place a dynamic situation in UHT milk during storage, where its active components are reacting or interacting and, as a result, some of its important quality attributes are also changing. The rate at which these changes take place is

Table 1.4 Some high-temperature-processed milk products from different countries.

Product	Country	Brand Name	Format	Other comments
On-the-go snack	USA	Dynamoo	UHT, 8 fl oz, boxes	
Red-bean flavoured milk	Taiwan	Acacia Lover	UHT, 250 mL bottle	
Fruit and milk drink	France	Danoo Mon fruit prefere	1 L re-sealable carton	Acidic product
Milk shake	USA	Cold Stone Milk Shake	UHT, 12 fl oz plastic bottle	
Nutritious weight loss shake	UK	USlim	UHT, 250 mL plastic bottle	
Dairy based functional drink	Latvia	Lakto	100 g plastic pots	P24, digest, acidic product
Coffee milk	Finland	Kahvi Maito, Valio	1 L cartons	Claimed to foam well
Milk with real fruit pieces	China	Meng niu	UHT cartons	Available since 2007
Milk with cereal grain	China	Yi Li	UHT cartons	Thai rice and Euro wheat
Milk with oat cereals	China	Meng niu	UHT cartons, "Miao Dian"	
Black cereal milk	China	Guanxi Huangshi Dairy Co ltd	UHT cartons	Black sesame, rice and beans
Milk and peanut protein			UHT cartons	
Breakfast milks	Australia UK UK	Up and Go (sanitarium) Weetabix The Fuel Station	UHT cartons and plastic bottles Tetra Prisma	Many flavoured varieties 2014 Caffe Latte
High protein drinks (whey based)	UK	Upbeat (Good Whey Company)	ESL, plastic bottles	Microparticulated whey protein (8%)
Milk based recovery and build drink		Maxi-Nutrition	UHT, plastic bottles	9% protein, fruit flavoured products
Infant and follow-on formulations	UK	SMA, first infant milk	Sterilised, glass bottle	Protein, 1.3%;fat 3.6%, C/H 7.3%
	UK	Aptamil, first milk	UHT, 100 mL bottle	Contains GOS, FOS
	UK	Aptamil, toddler milk	UHT, 200 mL bottle	Contains GOS, FOS

influenced by storage temperature. Within the life-span of a carton of UHT milk, it may be stored at temperatures from -10 °C to over 50 °C. For example, during transportation from UK to China it may go through fluctuating temperatures as it passes from the UK through the Gulf states and across the equator. Furthermore, large countries such as China, Australia and USA have several climatic zones, and ambient temperature may

Table 1.5 Composition of bulk raw milk from one farm collected over 15 months.

Composition/properties	Mean (n=25)	Range	Seasonal variation
pH	6.79	6.73–6.87	SP > SM and A; W > A
Ca^{2+} (mM)	2.05	1.68–2.55	NS
Total solids (%)	12.78	12.31–13.31	A > SM
Protein (%)	3.29	2.89–3.56	SP > SM and A
Total casein (%)	2.36	2.08–2.52	SP > SM and A
Fat (%)	4.08	3.62–4.77	A > SP, SM and W
Lactose (%)	4.59	4.52–4.69	NS
Ash (%)	0.71	0.53–1.03	NS
Total Ca (mM)	29.29	24.53–31.53	NS
Total Mg (mM)	5.11	4.21–5.81	NS
Total citrate (mM)	9.04	8.22–10.09	NS
Total P (mM)	27.52	22.58–33.57	NS
Urea (%)	0.0237	0.016–0.033	NS
SCC ('000)	155	65–357	W > SP, SM and A

Source: Chen *et al.*, 2014. Reproduced with permission of Elsevier.
(SP=Spring; SM=Summer; A=Autumn; W=Winter; NS=Non-significant difference ($p > 0.05$) (from Chen *et al.*, 2014)

extend over a wide range, from below 0 °C to above 50 °C. Also, individual milk cartons from the same specific production batch may have had totally different temperature storage histories by the end of six months. The expectations are that each one of these individual cartons will be still acceptable to the consumer. In our opinion, this is a lot to expect from the product and there is no doubt that the number of complaints will increase from stored products where the best-before period exceeds six months. In fact, the expected best-before time period is now creeping up to nine months or even one year, which is posing some new challenges for the UHT milk producer.

Also, the consumer is becoming more discerning. For example, it is reported that the Chinese consumer spends more time than any other reading food labels. There are a number of things any consumer might notice which could result in their making a complaint. On pouring the product, any physical defects such as fat separation, gelation or sediment will be obvious. The more inquisitive consumer will also notice what is left in the carton after the contents have been removed. Sediment and fat may be left in the carton and, if observed, may be a source of complaint, although this is unlikely to be a safety issue. The colour of the milk may also give cause for concern, especially if it is browner that expected. On tasting, any physical defects which change the mouthfeel might be noticed, for example, increased viscosity or presence of sediment giving rise to a powdery or gritty mouthfeel. Finally, its flavour must be acceptable and not be too cooked, oxidized or lipolytically rancid, as well as free of any other off-flavours and taints. Overall, expecting UHT milk to have a best-before period of longer than six months under all possible conditions is taking it out of its comfort zones. Consumers do allow some leeway for product imperfections such as a small fat layer on the milk but this cannot be pushed too far. One anecdote is that some consumers are prepared to accept a degree of fat separation, as this indicates that fat is actually present in the product.

1.2 Scope of the Book

This book aims to integrate the scientific information arising from several disciplines that needs to be considered in order to ensure that UHT and other highly heat-treated products are both safe and acceptable to the consumer. UHT processing requires an understanding of aspects of fluid flow and heat transfer, and a detailed knowledge of the properties of the food being processed and of the mechanisms of the various changes that occur during processing and storage. This includes knowledge of its chemical composition, the enzymes that are present and its microbial flora as well as an awareness of possible environmental contaminants.

When any food is subjected to UHT treatment, a large number of heat-induced reactions take place, which, if properly understood and controlled, ensure that the food is safe and that it will have a good appearance and taste for up to six months and probably for considerably longer. The material in this book is derived from the scientific literature related to UHT processing and the personal insights from two practitioners who have spent much of their working lives involved with UHT products and processes.

In order to put UHT processing and products in perspective in dairy processing, an overview of the heat treatments of milk is initially given (see Chapters 2 and 3). Furthermore, the microbiological aspects of these heat treatments and their associated products are provided (Chapter 4) as it has to be remembered that the basic reason why heat treatments are carried out is to destroy micro-organisms. A good understanding of the microbiological aspects is therefore fundamental to ensuring the safety and quality of the products. In these chapters, extended shelf-life (ESL) processing, a sub-UHT heat treatment, is covered in some detail because of the growing demand for ESL products.

1.3 Reasons for Heating Foods

In addition to inactivating micro-organisms, both pathogenic and spoilage, foods are heated to inactivate enzymes, as foods may change and become unacceptable due to reactions catalysed by enzymes. Milk contains about 60 indigenous enzymes (Fox, 2003), some of which, such as lipases and proteases, may cause flavour changes, whereas in fruit, browning may occur as a result of polyphenol oxidase activity. The process of heating a food may also induce physical changes and chemical reactions, such as starch gelatinisation, protein denaturation and Maillard browning, which in turn affect the sensory characteristics, such as colour, flavour and texture, either beneficially or adversely. For example, during the manufacture of canned evaporated milk, forewarming of milk prior to evaporation is essential for preventing gelation and thickening during the subsequent evaporation and canning steps. Heat treatment is also crucial in yogurt manufacture to achieve the required final texture in the product. However, such heating processes may result in loss of important nutrients, although these losses can be reduced by controlling the heating conditions.

Thermal processes vary considerably in their severity, ranging from mild processes such as thermisation and pasteurisation, intermediate processes such as used for ESL milk, through to more severe processes such as UHT and in-container sterilisation processes (see Chapters 2 and 3). The severity of the process affects both the shelf-life and quality characteristics of the product.

A UHT process contains heating, holding and cooling stages. After the product has been heated to the desired temperature, it is held for a short period of time to inactivate the microorganisms before being finally cooled and packaged under aseptic conditions. Continuous processes provide scope for energy savings, whereby the hot fluid is used to heat the incoming fluid; this is known as heat regeneration and saves both heating and cooling costs (see Chapter 5)

A wide range of products are heat-treated, ranging from low-viscosity fluids such as milk and fruit juices, through to highly viscous fluids. The process is more complicated when particles are present, as it becomes necessary to ensure that both the liquid and solid phases are adequately and, if possible, equally heated. A secondary issue is keeping the particulates suspended during storage, especially in transparent containers. The presence of dissolved air in either of the phases becomes a problem as air becomes less soluble as temperature increases and will come out of solution. Air is a poor heat-transfer fluid in comparison to steam and hence its presence affects the rate of heating of the food. For this reason, deaeration is sometimes used.

1.4 Brief History of Sterilisation Processes

Food sterilisation in sealed containers is often attributed to the pioneering work of Nicholas Appert. However, Cowell (1994, 1995) reported that investigations on heating foods in sealed containers were documented and took place earlier than this. He describes the commercialisation of the canning process in East London at the turn of the nineteenth century, which included the contributions not only of Nicholas Appert, but Peter Durand, Bryan Donkin, John Gamble and Phillipe de Girard. It is both noteworthy and worrying that bacteria which are the causative agents of food poisoning and spoilage were not understood until considerably later in the nineteenth century, through the work of Pasteur. He confirmed that the many food fermentations which were spoiling foods were not spontaneous but caused by microbial metabolism. He also discovered that both yeasts and *Acetobacter* could be destroyed by relatively mild heat treatments at about 55 °C. According to Wilbey (1993), Pasteur's work on producing beer, wine and vinegar laid the foundations for hygienic processing and the recognition of the public health implications of hygiene and heat treatment.

Early sterilisation processes were essentially of a batch nature and the food was heated in the container. Batch processing still has an important role in food processing operations and provides the small-scale food producer with a cheap and flexible means of heat-treating foods. The steps involved in a batch sterilisation process are shown in Figure 1.1. Continuous sterilisers had been patented and constructed and were able to heat milk to temperatures of 130-140 °C before the end of the nineteenth century, again well before the benefits of the process were understood. Hostettler (1972) recalls that in 1893, a continuous-flow heating apparatus with an output of up to 5000 L/h had been constructed which could heat milk to 125 °C, with a holding time of up to 6 min.

Figure 1.1 The batch canning process (from Jackson & Shinn, 1979).

Around 1909, a number of patents were registered which involved contacting milk with jets of hot air, gases and steam. Aseptically canned milk was produced in 1921 and a steam injection system was developed in 1927 by Grindrod in USA. However, the major initiatives leading to commercialisation of the UHT process began in the late 1940s, through the development of concentric-tube sterilisers and the uperisation steam-into-milk UHT system, which was developed in conjunction with the Dole aseptic canning system. UHT milk was not commercially available in the 1940s and early 1950s, as evidenced by the absence of information in both Cronshaw (1947) and Davis (1955). During the first half of the twentieth century, investigations took place side-by-side into in-container sterilisation and UHT processing, but the unsolved difficulty of filling the sterilised milk, without recontamination, into containers caused the interest in continuous processes to wane, so sterilisation of milk in sealed containers retained its dominance at this time. It is also noteworthy that many of these early investigations involved direct heating and the only mention of UHT-type processing in Davis (1955) was to uperisation, a steam injection process. In fact the marketing of uperised UHT milk in cans was first practised in Switzerland in 1953, with milk heated by steam injection at 150 °C for 2.4 s and flash cooled. The dominance of sterilised milk around that time is also illustrated in Davis (1955).

As mentioned, early commercial aseptic filling machines filled milk into metal cans, which were usually sterilised by superheated steam, which could be used at atmospheric pressure and avoided condensation and wet cans. A shelf-life of 4 to 6 months was claimed for the product.

The main developments in getting UHT milk to the market place occurred between the early 1960s and 1972 and were rapid. A major development was the use of hydrogen peroxide to sterilise the packaging material. Typical conditions then were 17% w/v solution, with a wetting agent. Hydrogen peroxide was evaporated off with hot air at about 180 °C. Equipment using this procedure was first commercialised in 1961 and from this point availability of UHT products started to accelerate.

Regulations permitting UHT milk in the UK were introduced in 1965. In 1968 UHT milk was introduced in Germany and in 1969 it commanded less than 2% of the liquid milk market. Its success there is illustrated by the fact that now over 90% of milk consumed is UHT treated. In Australia, the first successful UHT operation commenced in 1968 although an earlier installation ceased operation after a few years due in part to technical difficulties such as age gelation (Zadow, 1998).

In 1970, Hsu published the first book on UHT processing of dairy products and this was followed in 1972 by the first International Dairy Federation (IDF) monograph on UHT milk and a revised version in 1981. These publications catalogued most of the technical challenges that had been recognized and investigated in order to produce sterile milk of long shelf-life by means of a continuous-flow process involving heating at a high temperature for a short time, followed by aseptic packaging. By that time it had become well accepted as a method for heat treatment of milk for consumption.

A more detailed account of the early development of UHT processing, before it was properly commercialized is given by IDF (1972). The history of the continuous sterilisation process has also been reviewed by Burton (1988).

It is interesting that in the early 1970s there was no clear statement about how long UHT milk should keep. However, it was probably quite short because of the numerous challenges in UHT processing and the lack of a good understanding of the technology and its effects on product quality. An indication of this was given by Singh and Patel (1988) who reported that the shelf-life of UHT milk in India was only 15 days although the expected shelf-life was three months. They identified numerous aspects of the UHT process which required attention to improve the shelf-life including the initial bacterial content of raw milk, selection of suitable time – temperature conditions, problems related to heat-resistant proteases, sedimentation and deposit formation, and problems with the packaging system; these would have been similar to those encountered by the early UHT processors. With the developments in technology and a better understanding of the key determinants of shelf-life, together with market demands, it is not uncommon for the "best-before" period to be now set at nine months, and more recently 12 months, as discussed above.

At this point it is instructive to state two descriptions of a UHT treatment from the latest EU regulations (Hickey, 2009): "Continuous flow at a high temperature for a short time with not less than 135 °C for a suitable holding time such that there are no viable spores capable of growing in the treated product when kept in an aseptic container at ambient temperature" and "Sufficient to ensure that the products remain microbiologically stable after incubating at 15 days at 30 °C in closed containers, or 7 days at 55 °C in closed containers, or after any other method that demonstrates that appropriate heat treatment has been applied." The EU regulations no longer state what level of microbial activity would constitute microbial sterility after these incubation periods, whereas previous regulations stipulated it to be less than 100 cfu/mL, which seems to be a reasonable standard. This was illustrated by Quratulain and Saeed (2004) who found two brands of commercial UHT milk had mesophile counts of 75 and 96 cfu/mL after storage for 40 days; they commented that the milk met the "requirements of the standard". The current Australia and New Zealand Food Standards match the EU regulation and state that UHT milk and cream "should comply with a test for commercial sterility" (FSANZ, 2011).

In conclusion, it is worthwhile considering what factors have changed over the past 15 years since the publication of the Lewis and Heppell (2000) book. The basic processing technology and heat exchanger configurations have changed little although improvements continue to be made. There is now more recognition of the roles of the heating and cooling profiles. This has led to a wider use of the concept of bacterial and chemical indices (see Chapter 3) for characterising the process and understanding the effects of different processing conditions on the quality of the products.

The processing run times that can be achieved have increased considerably. It is now claimed that it is possible to obtain runs of 40 h. The main way of achieving this has been to include a protein stabilisation tube. One explanation is that this does not eliminate fouling but it causes the fouled deposit to accumulate in the protein stabilisation tube, which is away from areas where its build-up may be more critical.

The control and instrumentation has improved and information on when the plant needs to be cleaned and also when cleaning has been completed is more readily available. One possible consequence of longer run times is that the cleaning times may be longer, although this has not been reported to be the case. Also, a lot more information is now available to UHT process operators to provide them with a better understanding of the performance of the heat exchanger.

There have been other more subtle changes, such as improvements in homogeniser valve design, which should lead to an improvement in emulsion stability. This is crucially important as the "best-before" date for many products is now nine months or twelve months.

The product range continues to expand and there is now more emphasis on environmental considerations; for example, how much water and energy is used and how much waste is generated. One of the advantages that UHT processing offers is that the product does not need to be refrigerated during transportation or storage, although refrigeration or some form of temperature control may be beneficial in hot climatic conditions.

It has been difficult selecting a concise title for this book to reflect its entire content. However, we have chosen high temperature processing of milk and milk products. One reason for this is the dominance of white milk and other milk-based products in the

Table 1.6 Volumes of liquid dairy and dairy-like products sold worldwide in 2015 (Source: Reproduced with permission of Tetra Pak Compass).

Product	Volume (billion litres)
White milk	216.8
Baby and toddler milk	19.6
Flavoured milk	18.4
Soy milk	17.8
Drinking yogurt	9.4
*RNGS milk	8.3
Dairy cream	4.0
Sweetened condensed milk	2.5
Buttermilk	2.3
Evaporated milk	1.6

* RNGS is rice, nuts, grains and seeds products

global dairy products market, as shown in Table 1.6. Almost all of the beverages listed are subjected to thermal processing of some kind and many of them to UHT processing. Non-dairy products such as the rice, nuts, grains and seeds (RNGS) products are making inroads into the nutritive beverage market. These and most of the other products listed in Table 1.6 are discussed in Chapter 9 while some emerging technologies which have potential for processing these products are covered in Chapter 10.

References

Burton, H. (1988) *Ultra High Temperature Processing of Milk and Milk Products.* Elsevier Applied Science, London.

Chen, B., Lewis, M.J. & Grandison, A.S. (2014) Effect of seasonal variation on the composition and properties of raw milk destined for processing in the UK. *Food Chemistry* **158**, 216–223.

Cowell, N.D. (1994) *An Investigation of early Methods of Food Preservation by Heat.* PhD thesis, University of Reading, Reading, UK.

Cowell, N.D. (1995) Who invented the tin can? A new candidate. *Food Technology* **49**(12), 61–64.

Cronshaw, H.B. (1947) *Dairy Information.* Dairy Industries Ltd, London.

Davis, J.G. (1955) *A Dictionary of Dairying,* 2nd edition. Leonard Hill, London.

Fox P.F. (2003) Indigenous enzymes in milk. In: *Advanced Dairy Chemistry* Volume 1 *Proteins.* 3rd edn. pp. 467–471. Kluwer Academic/Plenum Press, New York.

FSANZ. (2001) User guide to Standard 1.6.1 – Microbiological limits for food with additional guideline criteria. https://www.foodstandards.gov.au/code/userguide/documents/Micro_0801.pdf. (accessed September 2015)

Hickey, M. (2009) Current legislation of market milks. In: *Milk Processing and Quality Management.* (ed. A.Y. Tamime), pp. 101–138, Wiley Blackwell, Oxford.

Hostettler, H. (1972) History of the development of UHT [ultra-high temperature] processes. In: *Monograph on UHT Milk.* IDF Annual Bulletin Part V, pp. 169–174. International Dairy Federation, Brussels.
Hsu, D.S. (1970) *Ultra-high-temperature (U.H.T) Processing and Aseptic Packaging (A.P.) of Dairy Products.* Damana Tech Inc, New York
IDF (1972) *Monograph on UHT milk.* IDF Annual Bulletin, Part V. International Dairy Federation, Brussels.
IDF (1981) *New Monograph on UHT Milk.* IDF Doc. 133, International Dairy Federation, Brussels.
Jackson J.M. & Shinn, B.M, (eds) (1979) *Fundamentals of Food Canning Technology, Canned Foods,* AVI, Westport.
Lewis, M.J. & Heppell, N. (2000) *Continuous Thermal Processing of Foods: Pasteurization and UHT Sterilization.* Aspen Publishers, Gaithersburg, MD.
Perkins, M.L. & Deeth, H.C. (2001) A survey of Australia consumers' attitude towards UHT milk. *Australian Journal of Dairy Technology* **56**, 28–34.
Persistance Market Research (2014) Global UHT milk market will reach USD 137.7 billion in 2019. http://www.persistencemarketresearch.com/mediarelease/uht-milk-market.asp. (accessed September 2015).
Pujol, L., Albert, L., Magras, C., Johnson, N.B. & Membré. J.-M. (2015) Probabilistic exposure assessment model to estimate aseptic-UHT product failure rate. *International Journal of Food Microbiology* **192**, 124–141.
Quratulain, S. & Saeed, A. (2004) Storage effect on bacteria and enzymes in UHT buffalo milk at 37 °C. *Milchwissenschaft* **59**, 605–608.
Rysstad G. & Kolstad J. (2006) Extended shelf-life milk - advances in technology. *International Journal of Dairy Technology* **59**, 85–96.
Singh, R.R.B. & Patil, G.R. (1988) UHT processing of milk under Indian conditions. *Indian Dairyman* **40**(2), 85–91.
Wilbey, R.A. (1993) Pasteurization of foods: Principles of pasteurization: In: *Encyclopedia of Food Science, Food Technology and Nutrition.* (eds R Macrae, R.K. Robinson & M.J. Sadler), pp. 3437–3441, Academic Press, London, UK.
Zadow, J.G. (1998) The development of UHT processing in Australia. *Australian Journal of Dairy Technology* **53**, 195–198.

2

Heat Treatments of Milk – Thermisation and Pasteurisation

2.1 Introduction

This chapter explains the important principles and procedures for producing heat-treated milk which is safe and of high quality. It includes information gained by the authors through their combined experiences of teaching, pilot plant work, research and troubleshooting.

Raw (or untreated) milk consumption has fallen considerably worldwide over the past 30 years and in some areas it is now illegal to sell raw milk for direct human consumption. In the UK, raw milk consumption now accounts for less than 0.1% of liquid milk consumption and in many countries, for example, Scotland and Australia, its sale is prohibited. Consequently most milk for consumption is now heat treated.

The two main treatments are pasteurisation and sterilisation, with treatments somewhere between these for extended shelf-life (ESL) products. The main aims of heat treatment of raw milk are to reduce the microbial population, both pathogenic and spoilage, to inactivate enzymes and to minimise chemical reactions and physical changes during storage. Such heating may also alter the sensory characteristics of the milk, for example, its overall appearance, colour, flavour and texture, as well as its nutritional value, but will make it safe for consumption and improve its keeping quality.

Most of the milk destined for conversion to dairy products is also heat treated at some point, an exception being those cheeses which are made from raw milk. Such processes include thermisation, which is milder than pasteurisation and used for extending the storage time of raw milk, preheating or forewarming applied to milks prior to evaporation and powder production, a high pasteurisation treatment used in yogurt manufacture, as well as pasteurisation, ESL treatment and sterilisation. The most common heat treatments, in order of increasing severity, are thermisation, pasteurisation, ESL treatment, ultra-high temperature (UHT) treatment and in-container sterilisation. The heating conditions for these are summarized in Table 2.1 together with their bactericidal effects and their effects on selected enzymes. These treatments and their effects on milk are reviewed in this chapter. As high-temperature treatments are the focus of this book, an overview only is given of sterilisation treatments, especially UHT processing, here as these are covered in detail in subsequent chapters.

High Temperature Processing of Milk and Milk Products, First Edition. Hilton C. Deeth and Michael J. Lewis.

Table 2.1 Heat treatments used for milk (in increasing order of severity).

Heat treatments	Temperature–time conditions	Bactericidal effect	Effect on selected enzymes
Thermisation	57-68 °C for 5-20 s	Destroys most non-spore-forming psychrotrophic spoilage bacteria	Does not inactivate milk alkaline phosphatase, lipase, lactoperoxidase, plasmin or bacterial proteases/lipases
Pasteurisation	63 °C for 30 min; 65 °C for 15 min (batch) 72-82 °C for 15-30 s[1] (continuous, HTST)	Destroys non-spore-forming pathogens and psychrotrophic spoilage bacteria	Inactivates milk alkaline phosphatase and lipase but not lactoperoxidase, plasmin or bacterial proteases/lipases
ESL (extended shelf-life) processing	123-145 °C for <1-5 s	Destroys all non-spore-forming bacteria and most psychrotrophic and mesophilic spores	Inactivates milk alkaline phosphatase, lipase and lactoperoxidase but not plasmin or bacterial proteases/lipases
UHT (ultra high temperature) processing	138-145 °C for 1-10 s	Destroys all non-spore-forming bacteria and all spores except highly heat-resistant spores	Inactivates milk alkaline phosphatase, lipase, lactoperoxidase; and most plasmin but not all bacterial proteases/lipases
In-container sterilisation	115-120 °C for 10-30 min (conventional) 125 °C for 4 min (e.g., Shaka™ technology)	Destroys all non-spore-forming bacteria and all spores except highly heat-resistant spores	Inactivates virtually all enzymes

[1] 72 °C for 15 s are the regulated minimum conditions in most countries

2.2 Thermisation

Thermisation is the mildest heat treatment given to milk. It is used to improve the keeping quality of raw milk when it is necessary for the milk to be held chilled for some time before being further processed. Thermised milk is subsequently used for other heat-treated milk or converted into various milk products. The aim of thermisation is to reduce the growth of psychrotrophic bacteria which may release heat-resistant proteases and lipases into the milk if allowed to reach high levels. These enzymes will not be totally inactivated during subsequent heat treatments and may give rise to off-flavours in processed milk or in subsequently manufactured cheese or milk powders. Conditions used for thermisation are 57 to 68 °C for 5-20 s, followed by refrigeration. Humbert *et al.* (1985) recommended 65 °C for 20 s as these were the minimum conditions for extending the shelf-life by four days at 4 °C. According to IDF (1984), thermised raw milk can be stored at a maximum of 8 °C for up to 3 days. Similarly, Stadhouders (1982) found that thermisation at 64-68 °C for 10 s extended the shelf-life by 3 days at 4-7 °C. Thermised milk is phosphatase-positive which distinguishes it from pasteurised milk, which is phosphatase-negative. Thermisation causes virtually no whey protein denaturation, does not affect the milk's heat stability as measured by the heat coagulation time at 130 °C (Coghill *et al.*, 1982) and reduces lipase activity by about 50% (Humbert *et al.*, 1985). While thermisation reduces psychrotrophic bacterial growth, it may activate and initiate germination of bacterial spores and accelerate the build-up of the thermoduric bacterium *Streptococcus thermophilus* in the regeneration section of the pasteuriser (Stadhouders, 1982).

2.3 Pasteurisation

2.3.1 Introduction

"Pasteurisation of milk represents one of the singularly successful contributions to the safety of foods of animal origin" (Holsinger *et al.*, 1997). Pasteurisation was first practised on wine, prior to 1857 and slightly later on beer. In terms of milk processing, the history of pasteurisation between 1857 and the end of that century came chiefly from the medical profession interested in infant feeding. The first commercial positive holder pasteurisation system for milk was introduced in Germany in 1895 and in the USA in 1907. A most important principle was recognised as early as 1895 that an effective pasteurisation process "will destroy all disease germs" and "a thoroughly satisfactory product can only be secured where a definite quantity of milk is heated for a definite period of time at a definite temperature. Then too, an apparatus to be efficient must be arranged so that the milk will be uniformly heated throughout the whole mass. Only when all particles of milk are actually raised to the proper temperature for the requisite length of time is the pasteurisation process complete" (Cronshaw, 1947). This remains the main guiding principle underpinning current heat treatment regulations for ensuring a successful pasteurisation process. The description of pasteurisation given by the IDF (1986) remains very appropriate: "a process applied with the aim of avoiding public health hazards arising from pathogenic microorganisms associated with milk, by heat treatment which is consistent with minimal chemical, physical and organoleptic changes in the product". This implies that pasteurised milk should be little different to raw milk in terms of its sensory characteristics and nutrient content (Deeth, 2005).

Pasteurisation of milk is now universally accepted, although it did meet with resistance when first introduced (Satin, 1996). There are still devotees who prefer to drink raw milk and many artisan cheesemakers do not use pasteurisation. Pasteurisation is now mostly performed as a continuous process, which is known as the high-temperature, short-time (HTST) process. This allows it to benefit from economies of scale with capacities of modern HTST units of up to 50,000 L/h. These units operate at high heat regeneration efficiencies (>95%) and are capable of long run times of up to 20 h before cleaning is required. A recently opened dairy in the UK has a capacity of pasteurising 1.3 billion litres of milk each year, which is about 10% of all the raw milk produced in the UK. Pasteurised milk does require refrigeration to ensure a long shelf-life, which incurs substantial energy requirements. In many countries it remains the preferred option to UHT milk, for example, UK, Scandinavia, Greece, USA, Australia and New Zealand.

The conditions used in pasteurisation are designed to inactivate the most heat-resistant, non-spore-forming pathogenic bacteria in milk, *Mycobacterium tuberculosis* and *Coxiella burnetii*. According to Codex Alimentarius (2003), pasteurisation is designed to achieve at least a 5-log reduction of *C. burnetii* in whole milk. It therefore results in very substantial reduction in populations of pathogens that might be present in raw milk with the exception of the spore-former *Bacillus cereus* of which some strains can be toxigenic (Juffs & Deeth, 2007) (see Section 3.2.2.2.1) for more information on *B. cereus*).

There are now some alternative non-thermal processes which have been developed to replace or augment thermal pasteurisation. These include microfiltration, high pressure processing, high pressure homogenisation, pulsed electric field technology, and UV and gamma irradiation (Deeth *et al.*, 2013). These are discussed in Chapter 10. However, a major factor preventing these alternative technologies gaining widespread acceptance is that thermal processes, especially pasteurisation, are firmly established and accepted as being capable of producing safe, high quality and highly nutritious foods in large volumes and at relatively low processing costs. Many of the alternative technologies also face regulatory hurdles. To date, they have not been able to compete in terms of scale of operation, length of processing runs and energy efficiency for high-volume products like milk.

2.3.2 Historical Background

To chart the developments that have taken place with milk pasteurisation, it is interesting to note what was known about the process 50 to 60 years ago, by reference to publications such as Cronshaw (1947) and Davis (1955) which are well worth consulting. Halfway through the twentieth century (~1950), batch pasteurisation was still widely used but the principles of HTST processes were well established. Continuous pasteurisers were available, processing, on average, just under 10,000 L/h. As mentioned in Section 2.3.1, it was well recognised that every element of milk being pasteurised needed to be sufficiently heat-treated. Although from the start pasteurised milk had to satisfy a plate count requirement of less than 10^5 cfu/mL, in the 1940s it became evident from emphasis on keeping quality that these plate count standards had shortcomings. From 1946, the official test for pasteurisation efficacy became the phosphatase test and, as an indicator of milk keeping quality, the methylene blue reduction test was used. The phosphatase test remains in use throughout the world but the methylene blue test is seldom used now. The methylene blue test is a simple way of providing a rough estimate of the bacterial state of a milk sample. Although it is much less used now, it was recently reported being used for assessing the microbial quality of UHT milks imported into Iraq (Al-Shamary & Abdalali, 2011). Traditionally, it has been used more for assessing the bacteriological quality of raw milk.

When pasteurised milk was first introduced, its keeping quality was poor and its shelf-life was short. Household refrigeration was not widespread and usually milk was stored in the larder. A satisfactory keeping quality meant that it would remain sweet and palatable for 24 h after delivery to the consumer and up to 48 h if the consumer was lucky. If milk with a longer shelf-life was required, the only alternative was milk which had been sterilised in the bottle, with its strong cooked flavour and brown colour. UHT milk was not then available. Even in the 1960s, the choice of milk products was limited (UK Milk Marketing Boards, 1964). There was hardly any mention of skim milk. In the UK, 69% of milk produced went to liquid sales, 31% to manufacture, 6.2% was consumed raw, 18% went into condensed milk, and less than 2.6% was used for other products; fermented products such as yogurt received no mention. No breakdown was provided of what proportion of milk was pasteurised or sterilised and, at this juncture, the heat treatment regulations for UHT milk had just been introduced. Thus considerable commercial interest arose in UHT milk between 1950 and 1965 (see Section 3.4.2).

HTST continuous processes were developed between 1920 and 1927 and for some time the ability of this process to produce safe milk was questioned. The importance of flow control and temperature control was known and it was appreciated that there was a distribution of residence times. Scales of operation were fairly substantial; Davis (1955) quotes HTST plants between 50 and 5,000 gal/h, although the most favoured were about 2,000 gal/h. (note that Imperial units were widely used: 1 gal/h = 4.54 L/h). Run times were cited as being up to 5 h. Milk was cooled to below 43 °F (5 °C) for distribution after pasteurisation, and brine cooling was popular. Energy regeneration up to 72% was achieved and Davis (1955) reported that 75% of liquid milk was processed (pasteurised) using HTST methods. Gaskets were a problem on the early equipment. Milk was not often homogenised, as a visible cream line was a popular feature. Where homogenisation was used, the pasteuriser was run at a slightly higher temperature. Scale formation was also mentioned as being a problem, most likely occurring when poorer quality milk was being processed. If temperatures were not well controlled, a cooked flavour may have resulted and/or the cream line been diminished. Time – temperature conditions which induce a cooked flavour and result in loss of cream line were well known. According to Cronshaw (1947), momentary heating at 169-172 °F (76.1-77.8 °C) or 30 min hold at 158-162 °F (70-72.2 °C) would cause the cooked flavour to appear.

The role of pasteurisation in inactivating *M. tuberculosis* was well established. A key development was in 1927, when North and Park established a wide range of temperature – time conditions to inactivate tubercle bacilli (Cronshaw, 1947). These experiments were performed by heating milk heavily infected with tubercle bacilli at different conditions and injecting them into guinea pigs. A selection of conditions where negative results were found, that is, those where the animals survived, were: 212 °F (100 °C) for 10 s; 160 °F (71.1 °C) for 20 s; 140 °F (60 °C) for 10 min and 130 °F (54.4 °C) for 60 min.

The phosphatase test was in widespread use as an index of correct heat treatment of milk, in particular to ensure that no milk was under-treated. It was developed from pioneering work reported by Kay and Graham (1935) and was based upon the finding that the naturally occurring alkaline phosphatase in milk had similar inactivation kinetics to the inactivation of *M. tuberculosis.* It is interesting that about 70 years later the bacterium *Mycobacterium avium* subsp. *paratuberculosis* (MAP) became of concern to the dairy community (Griffiths, 2006). One procedure recommended to ensure its destruction was to increase the pasteurisation holding time from 15 s to 25 s (Grant *et al.*, 2005). Hickey (2009) pointed out that while this recommendation has been adopted widely by the UK industry, and supported by many retailers, it is a recommendation that is voluntary and is not a legal requirement for HTST pasteurisation, which still remains at 72 °C for 15 s. MAP is discussed in more detail in Section 4.4.1. Further developments were made in the classification of tests for evaluating the pasteurisation process; these included tests for raw milk quality (the platform test); pasteurisability (survival of thermodurics); efficiency of pasteurisation (pathogens and phosphatase); recontamination (thermophilic and coliform bacteria and the methylene blue test); and general bacterial quality, including organisms surviving pasteurisation plus contaminating organisms (plate count).

It was also recognized that it would be more difficult to inactivate microorganisms in situations where clumping of bacteria occurred, although this is not discussed much now. The role of thermoduric and thermophilic microorganisms was recognised and it was fully appreciated that some microorganisms would survive pasteurisation. It is noteworthy that the role of thermoduric bacteria has started to be questioned again (Gleeson *et al.,* 2013); this is discussed in more detail in Chapter 4. Maintaining the cream line was important as most milk was packaged in glass bottles where the cream line was clearly visible. In fact, taking the temperature up to about 78 °C was one method of losing the cream line. Odour and taste were also important quality characteristics. The role of post-pasteurisation contamination (PPC) was recognised, although this became more fully appreciated once pasteurised milk was stored in domestic refrigerators. Davis (1955) reported that when pasteurised milk soured or deteriorated rapidly it was almost invariably due to post-pasteurisation contamination. The situation today is very similar.

A number of installations were introduced for batch pasteurising milk sealed in bottles. Although the keeping quality was comparable to that of HTST pasteurised milk (Davis, 1955), there were some major technical problems and costs were considered to be higher. Consequently, this innovation was relatively short-lived.

"In considering the history of pasteurisation, it is important to remember that, although scientists everywhere agreed fairly closely on the necessary degree of heat treatment, the process itself was loosely (less well) controlled in commercial practice. Milk was frequently either over-heated or under-heated so that it either gave a cooked flavour or was found to contain viable tuberculosis bacteria. In addition, pasteurised milk was often so badly contaminated by unsterile plant, that its keeping quality was decreased" (Davis, 1955).

Several changes have influenced heat treatment of milk over the last 50 years. Some of these are:

- A much wider variety of milk products is available, including skim, semi-skim, flavoured, lactose-reduced, calcium-fortified and a range of speciality milk products with added nutritional and health benefits.
- Milk from species other than cows is more widely available and in the UK goat's milk has increased in popularity. An interesting phenomenon that both authors have encountered is that pasteurised goat's milk has a better keeping quality than pasteurised cow's milk; this still remains a curiosity.
- Scales of operation have increased, with dairies handling upward of 5 million litres of milk a day, most of which is heat-treated in some way.
- Considerable advances have been made in understanding the role of raw milk quality and the role of PPC in keeping quality.
- Domestic refrigeration is much more widely available and the cold chain, involving refrigerated transport and storage systems, has improved. The role of low temperatures in extending shelf-life is better understood.
- With improvement in refrigeration, there has emerged a better understanding of the role of psychrotrophic bacteria, as raw milk remains refrigerated for longer periods prior to pasteurisation and pasteurised milk remains acceptable for longer.
- Homogenisation is now widespread.
- There is a wider variety of packaging options.

- Much less milk is sold in glass bottles; it was 95% in 1975 but is now less than 5% in the UK.
- There is a demand for extended-shelf-life products.
- Environmental issues have become more important in terms of reducing energy and water use, reducing product waste, minimising effluent, reducing detergent usage and minimising the carbon footprint.

2.3.3 Pasteurisation Equipment

2.3.3.1 Holder or Batch Heating

Cronshaw (1947) and Davis (1955) both provide excellent descriptions of equipment for the holder or batch process – individual vessels (heated internally) and externally heated systems with one or more holding tanks. These processes are more labour-intensive than continuous processes and involve filling, heating, holding, cooling, emptying and cleaning. Temperatures attained are between 63 and 65 °C for 15-30 min. They are still used, particularly by small-scale producers who require flexibility and the ability to treat relatively small volumes of a wide variety of products. They are relatively time-consuming and heating and cooling times are considerable; the total time for one batch may be up to 2 h. The time required to reach the pasteurisation temperature can be determined from the following equation:

$$t = \frac{Mc}{AU} \ln\left(\frac{\theta_h - \theta_i}{\theta_h - \theta_f}\right)$$

t = heating time (s) c = specific heat ($J\,kg^{-1}\,K^{-1}$)
M = mass batch (kg) A = surface area (m^2)
U = overall heat transfer coefficient ($W\,m^{-2}\,K^{-1}$): θ = temperature, i, initial: f, final; h, heating medium temperatures (see Section 5.2.1.8.3.3).

The dimensionless temperature ratio represents the ratio of the initial temperature driving force to that of the final approach temperature. The same dimensionless ratio can be used to evaluate cooling times, which tend to be longer than heating times, because of the limitations of chilled water temperature and hence a lower approach temperature. Cooling times can be shortened by using glycol systems, but this adds to the complexity. These factors have been discussed in more detail by Lewis (1990). One major advantage of the batch system is its flexibility, that is, it is easy to change from one product to another. Also, provided the product is well mixed, there is no distribution of residence times (see Section 5.2.1.8.4).

An interesting question is whether HTST pasteurisation produces a better quality product that the holder process. Yale in 1933 (cited in Cronshaw, 1947) concluded that one method of pasteurisation is as good as the other when sound methods are used and when conditions are comparable. The authors are unaware of anything of late to contradict this, although most pasteurised milk is now produced by the HTST process. Homogenisation just prior to or after pasteurisation is simple in a continuous flow system. However, it is more difficult to link homogenisation with batch pasteurisation as the time delay between homogenisation and when the milk reaches pasteurisation temperature can result in an unacceptable amount of lipolysis (Deeth, 2002). However, this problem can be largely overcome by homogenising the milk at ~60 °C.

2.3.3.2 Continuous Heating

HTST pasteurisation permits the use of continuous processing, regeneration of energy and long run times. The main types of indirect heat exchanger for milk are the plate heat exchanger and the tubular heat exchanger. Plate heat exchangers (PHE) are most widely used for pasteurisation of milk, cream and ice-cream mix. They have a high overall heat transfer coefficient (OHTC) and are generally more compact than tubular heat exchangers. Their main limitation is pressure, with an upper limit of about 2 MPa. The normal gap width between the plates is between 2.5 and 5 mm but wider gaps are available for viscous liquids to prevent large pressure drops. In general, PHEs are the cheapest option and the one most widely used for low viscosity fluids. Maintenance costs may be higher than for tubular heat exchangers, as gaskets may need replacing and the integrity of the plates also needs evaluating regularly as pin-holes may appear in the plates of older heat exchangers. This may lead to pasteurised milk being recontaminated, for example, if such plates are in the regeneration section, a cracked or leaking plate may allow raw milk to contaminate already pasteurised milk. They are also more prone to fouling, but this is a more serious problem in UHT processing (see Section 6.2.2).

Tubular heat exchangers have a lower OHTC than plates and generally occupy a larger space. They have slower heating and cooling rates with a longer transit time through the heat exchanger. In general, they have fewer seals and provide a smoother flow passage for the fluid. A variety of tube designs are available to suit different product characteristics. Most tubular plants use a multi-tube design. They can withstand higher pressures than PHEs. Although they are still susceptible to fouling, high pumping pressures can be used to overcome the flow restrictions. Tubular heat exchangers give longer processing times than PHEs with viscous materials and with products which are more susceptible to fouling. Thus they may be used with more viscous milk-based desserts. They are also widely used in UHT processing of milk and milk products.

The viscosity of the product is a major factor that affects the choice of the most appropriate heat exchanger and the selection of pumps. Viscosity will influence the pressure drop causing a problem in the cooling section and when phase transitions such as coagulation or crystallization take place. For more viscous products or products containing particulates, for example, starch-based desserts or yogurts with fruit pieces, a scraped-surface heat exchanger may be required. Viscosity data for a range of milk products at different temperatures were presented by Kessler (1981).

One of the main advantages of continuous systems over batch systems is that energy can be recovered in terms of heat regeneration. The layout for a typical regeneration section is shown in Figure 2.1. The hot fluid can be used to heat the incoming fluid, thereby saving on heating and cooling costs. Regeneration efficiencies over 90% can be obtained.

In terms of the temperatures at different locations, the regeneration efficiency (RE) is given by:

$$RE = \frac{\theta_2 - \theta_1}{\theta_3 - \theta_1} x100$$

θ_1 = inlet temperature; θ_2 = temperature after regeneration; θ_3 = final temperature

Although higher regeneration efficiency results in considerable savings in energy, it necessitates the use of higher surface areas, resulting from the lower temperature driving force, and a slightly higher capital cost for the heat exchanger. This also means that the heating and cooling rates are slower, and the transit times longer, which may affect product quality.

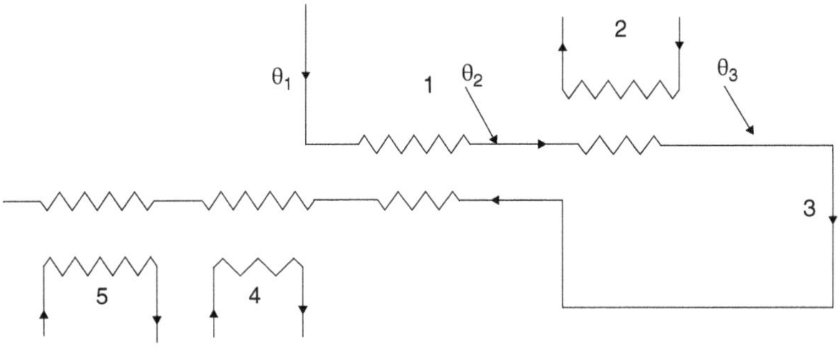

Figure 2.1 Heat exchanger sections for HTST pasteuriser: 1 regeneration; 2, Hot water section; 3, Holding tube; 4 mains water cooling; 5 chilled water cooling. (Source: Lewis, 1994. Reproduced with permission of Elsevier.)

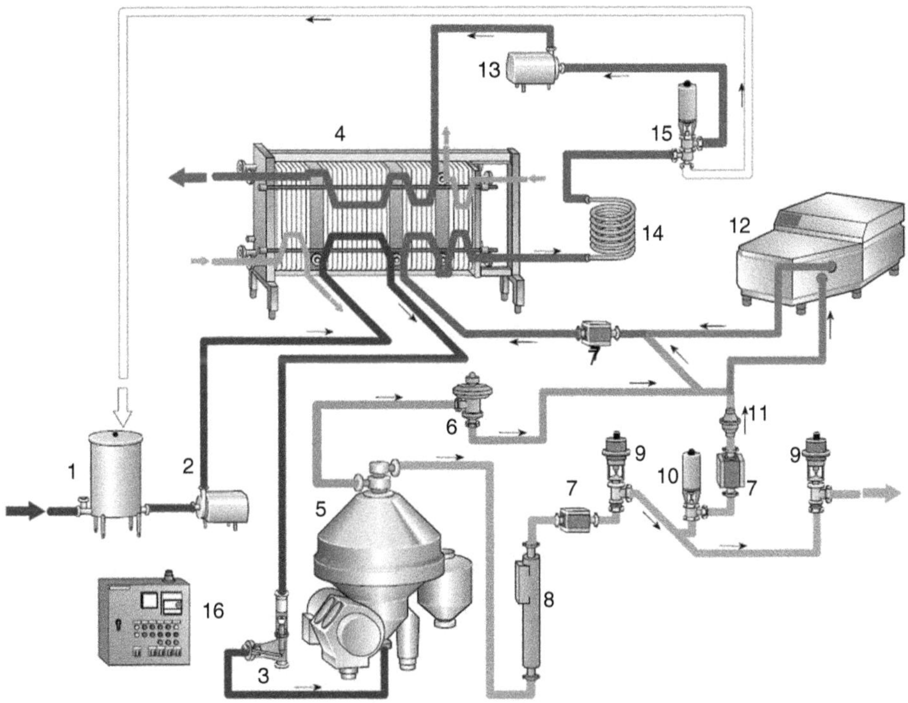

Figure 2.2 Production line for pasteurised milk with partial homogenisation. 1 Balance tank; 2 Product feed pump; 3 Flow controller; 4 Plate heat exchanger; 5 Separator; 6 Constant pressure valve; 7 Flow transmitter; 8 Density transmitter; 9 Regulating valve; 10 Shut-off valve; 11 Check valve; 12 Homogeniser; 13 Booster pump; 14 Holding tube; 15 Flow diversion valve; 16 Process control. (Source: Reproduced with permission of Tetra Pak.)

For milk containing substantial fat and for various cream products, homogenisation must be incorporated to prevent fat separation. However, as homogenisation of raw milk is a very effective way of initiating lipolysis (Deeth & Fitz-Gerald, 2006), it must be carried out immediately before or after pasteurisation, which inactivates the native lipase.

The layout of a typical HTST pasteuriser and its ancillary services is shown in Figure 2.2. The holding time is controlled either by using a positive displacement pump or by a centrifugal pump linked to a flow controller, and the temperature is usually controlled and recorded. Note that a booster pump can be incorporated to ensure that the pasteurised milk is at a higher pressure than the raw milk in the regeneration section to eliminate PPC in this section. A flow diversion valve diverts under-processed fluid back to the feed tank. In continuous processing operations there will be a distribution of residence times, and it is vital to ensure that the minimum residence time, that is, the time for the fastest element of the fluid to pass through the holding tube, is greater than the stipulated time in order to avoid under-processing. In a fully developed turbulent flow, the minimum residence time is about 0.83 × average residence time (t_{av}). This will usually be the situation for milk, but it may be different for more viscous fluids. In this situation, the minimum residence time will only be 0.5 × t_{av} and the distribution of residence times will be much wider (see Section 5.2.1.8.4).

Since most HTST pasteurisers are of the plate type, the plates themselves should be regularly tested for pinhole leaks, as discussed earlier. Consideration should be given to ensuring that if leaks do occur, they do so in a safe fashion, that is, pasteurised milk is not contaminated with cooling water or raw milk in the regeneration section. This can be achieved by making sure that the pressure on the milk side (downstream of the holding tube) is higher than on the water side, or on the raw milk side in the regeneration section. The control instrumentation, diversion valves and other valves should be checked regularly.

2.3.4 Process Characterisation

A number of parameters have been used to characterise heat treatment processes in order to allow comparisons of different temperature – time profiles. Two which have universal application are D-value and z-value. A parameter used specifically for pasteurisation is the pasteurisation unit (PU).

2.3.4.1 D-value

The time required at a particular temperature to reduce the bacterial population by 90%, that is, 1 log cycle or decimal reduction.

The number of decimal reductions is equal to the heating time divided by the decimal reduction time (*D*-value). For example, if the original bacterial count is 10^3 cfu/mL, the remaining count after various heat treatments will be as shown in Table 2.2.

Table 2.2 Remaining bacterial counts after heat treatments causing one to nine decimal reductions. Initial count = 10^3/mL.

		Remaining count	
Log reduction	**Percentage reduction**	**Per mL**	**Per L**
1D	90	100	100,000
2D	99	10	10,000
3D	99.9	1	1,000
6D	99.9999	0.001	1
9D	99.9999999	0.000001	0.001

The practical inference from this table is that the higher the initial count, the higher will be the count of remaining bacteria after heat treatment.

2.3.4.2 z-value

The change in temperature required to produce a tenfold change in the decimal reduction time (D-value).

The z-value is therefore the slope of the semi-logarithmic curve of D-value versus temperature.

The reported z-values for a range of pathogenic non-sporeforming bacteria that can contaminate milk are shown in Table 2.3 assembled from Juffs and Deeth (2007). Lovett *et al.* (1982) quoted the expected range of these z-values to be 5.56 ± 1.1 °C for heat treatments in the range 54.4 to 71.1 °C.

These z-values are lower than those for destruction of bacterial spores which, in turn, are considerably lower than those of chemical reactions (see Table 3.1). The practical significance of this is that the lower the *z*-value, the greater will be the effect of raising the temperature of processing. This is highly significant for UHT processing as it is the basis of one of the two fundamental principles on which UHT technology is based (see Section 3.4.2).

2.3.4.3 Pasteurisation Unit (PU)

One PU results from heating at a temperature of 60 °C (140 °F) for 1 min. The equivalent z-value is 10 °C (18 °F), which is high for vegetative bacteria (see Table 2.2 for z-values of

Table 2.3 z-values (°C) of selected pathogenic vegetative bacteria.

Bacterium	z-value
Brucella abortus	5.3, 4.7-4.8, 4.3-4.8, 4.4-5.5
Campylobacter jejuni	5.1, 4.94-5.6, 7.0, 8.0
Coxiella burnetii	4.4-5.5
Enterobacter (now *Cronobacter*) *sakazakii*	5.82, 3.1, 3.6, 5.7
Escherichia coli	5.07, 4.61, 3.2-3.4, 4.72
E. coli 0157:H7	1.8, 3.1
Listeria innocua	4.8-5.9
L. monocytogenes	6.1-7.4
Mycobacterium tuberculosis	4.4-5.6
M. bovis	4.8, 4.8, 4.9, 5.2
M. avium subsp. *paratuberculosis*	7.11, 8.6 (mean of 5 strains)
Salmonella Typhimurium	5.3
Staphylococcus aureus	6.04, 5.1, 4.4-6.7, 9.46, 4.83, 4.5
Streptococcus pyogenes	4.4-6.7
Yersinia enterocolitica	5.78, 5.22, 5.11

Data extracted from Juffs and Deeth (2007)

vegetative pathogenic bacteria). Thus, the number of pasteurisation units for a heating temperature (T,°C) and heating time (t, min) is given by:

$$PU = 10^{(T-60)/10)}.t$$

Thus, a temperature of 63 °C for 30 min would have a value of approximately 60 PU (Wilbey, 1993), whereas HTST conditions (72 °C/15 s; originally 161 °F or 71.7 °C/15 s) would give only 3.96 PU. One might expect the values to be similar, and the discrepancy probably arises from the large z-value. Perhaps a lesson to be learnt is that it may not be meaningful to extrapolate this to continuous pasteurisation processes.

2.3.4.4 p*

Another parameter, introduced by Kessler, is p*. This is based on a reference temperature of 72 °C and a z-value of 8 °C. Processing conditions of 72 °C for 15 s are designed to provide a safe pasteurisation process for milk and are given an arbitrary p* of 1.

It can be calculated from:

$$p^* = \frac{10^{\frac{T-72}{z}}}{15} t$$

T = temperature (°C); t = time (s)

Figure 2.3 shows the time-temperature combinations that correspond to a p* value of 1 (normal pasteurisation) as well as other p* values (0.1 to 10).

This simplified equation ignores the contribution of the heating and cooling section. Both these factors provide an additional measure of safety (in terms of inactivating pathogens) but they may not help to extend shelf-life; they are further discussed by Kessler in IDF (1986). Knowledge of the heating and cooling profiles will enable their contribution to be determined. The procedure for this is, from the temperature – time profile, to plot p against time and determine the area under the curve. Alternatively, the activation energy (285 kJ/mol) can be used. As mentioned earlier, it is important to check the minimum residence time and that this exceeds the residence time required by regulations. Dye injection methods can be used to check this. It is also important to calibrate temperature probes at regular intervals. Note from earlier discussions that subjecting milk to higher p* than 1 will ensure that the milk is safe but it will not necessarily extend its shelf-life (see also Section 2.3.5).

It is worth reiterating that these different pasteurisation parameters make use of different z-values: If the holder process (63 °C/30 min) is considered to be equivalent to HTST conditions (72 °C/15 s), in terms of microbial inactivation, this would give those microorganisms a z-value of about 4.3 °C. It is probably not to be recommended to extrapolate PU from a batch to a continuous process, or p* from continuous to batch processes. The z values used are:

Comparing holder/HTST process	4.3
p*	8
PU	10

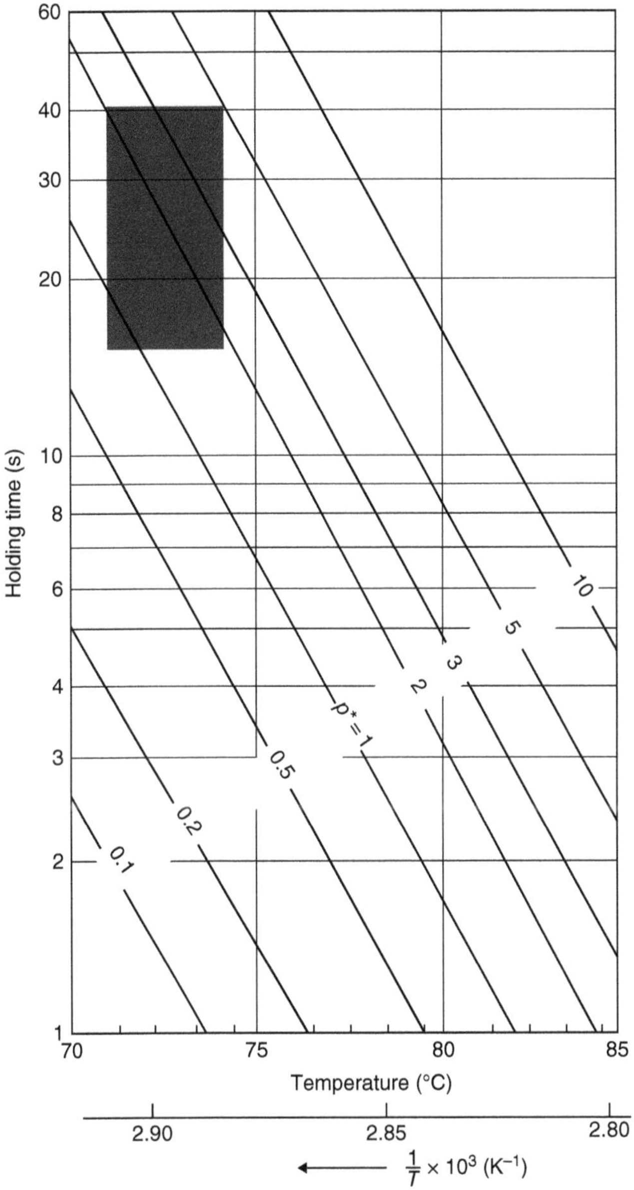

Figure 2.3 Temperature–time combinations required to give p* values in the range of 0.1 to 10. (Source: Kessler, 1989. Reproduced with permission of Elsevier.)

2.3.5 Processing Conditions

While the specified minimal pasteurisation conditions for milk are generally 72 °C for 15 s, in some situations they are exceeded. From the above discussion, it is not clear what the rationale is for selecting higher processing temperatures. It may be that they are being used to overcome problems related to poor quality milk. This is further discussed in the following section.

However, pasteurisation conditions do vary from one country to another. In the USA, a wide range of conditions are used including 63 °C for 30 min, 77 °C for 15 s, 90 °C for 0.5 s and 100 °C for 0.01 s. Similarly in Australia, a range of conditions are used for commercial pasteurisation of milk. Of 10 batch pasteurisation plants surveyed, the conditions ranged from 63 °C for 30 min to 80 °C for 5 min while of 61 HTST plants, most used conditions in the range 72-82 °C for 15 to 30 s; no holding times were <15 s (Juffs & Deeth, 2007).

Normal HTST pasteurisation conditions for milk are a minimum of 72 °C for a minimum of 15 s. As many processes are more severe than this, it is worthwhile considering the positive and negative effects associated with these more severe processing conditions. One surprising finding is that increasing the severity for pasteurisation of milk reduces its keeping quality at low temperature. This has been identified by Kessler and Horak (1984), Schroder and Bland (1984), Schmidt *et al.* (1989), Gomez Barroso (1997) and Barrett *et al.* (1998). It seems counterintuitive as it could be expected that a more severe heating process (e.g., at ~80-90 °C) would result in an improvement in keeping quality. Because these higher pasteurisation heat treatments do not result in milks with better keeping quality, a relatively new class of milk known as ESL milk which is mostly produced by heating at >120 °C has been developed and is commercially produced in several countries. ESL technology is described in Section 3.3.

The earliest explanation for the poorer keeping quality of milks subjected to these higher pasteurisation heat treatments was that the more severe conditions caused activation of the spores and that their subsequent germination and growth reduced the keeping quality. This is possible as it is known that such conditions activate *B. cereus* spores, some of which can grow at low temperatures (see Section 4.2.2.2.1). Another explanation is that there is reduced competition from other spoilage bacteria which would be destroyed at the harsher processing conditions. A third explanation is that the natural antibacterial lactoperoxidase system (LPS) in milk is inactivated. The LPS system involves the enzyme lactoperoxidase (LP), hydrogen peroxide and thiocyanate, all of which are present in raw milk. The oxidation products, for example, hypothiocyanite, exhibit strong anti-microbial activity by oxidising sulfydryl groups of bacterial cell walls (Reiter and Harnulv, 1982). The LPS system can be further activated in raw milk by small additions of thiocyanate and hydrogen peroxide and can be used to keep raw milk longer in countries where refrigeration is not widespread (IDF, 1988). LP inactivation is very temperature-sensitive with z-values of about 4 °C. Marks *et al.* (2001) showed that pasteurisation at 72 °C for 15 s resulted in active LPS remaining in the milk which greatly increased the keeping quality of milks inoculated with *Pseudomonas aeruginosa*, *Staphylococcus aureus* and *Str. thermophilus*, when compared to heating at 80 °C for 15 s.

The LPS also appeared to affect the shelf-life of pasteurised milk manufactured from raw milk which had been kept for 1 to 7 days at refrigeration conditions prior to pasteurisation (Ravanis & Lewis, 1995). The longest shelf-life was found for milk that was pasteurised after 3 or 4 days' storage and not for milk stored for 1 day. It was found that lactoperoxidase activity changed during storage of raw milk and was usually higher later in its storage period.

2.3.6 Changes During Pasteurisation

2.3.6.1 Microbiological Aspects

Raw milk from healthy animals has a very low microbial count, but it easily becomes contaminated with spoilage and sometimes pathogenic microorganisms. These need to be inactivated and this is readily achieved by heat treatment. From a milk processor's standpoint, raw milk composition and its microbial loading will vary from day to day.

The primary reason for pasteurisation is to destroy the small numbers of pathogenic microorganisms in raw milk which may be picked up from the farm environment. These include non-spore-forming bacteria such as *St. aureus, Campylobacter jejuni, Salmonella* spp, *Escherichia coli* including *E. coli* O157:H7, *Yerisinia enterocolitica, Listeria monocytogenes, C. burnetii* and *M. tuberculosis,* together with spore-formers such as *Bacillus cereus* and *Clostridium* spp. An excellent review of pathogenic bacteria in raw milk was published by Griffiths (2010). The non-spore-forming pathogens can be effectively controlled by pasteurisation; according to Codex Alimentarius (2003), pasteurisation achieves a 5-log reduction of *C. burnetii*, so inactivation of the above non-spore-forming pathogenic bacteria will be greater. A summary of heat resistance data for various pathogens was provided by Lewis and Heppell (2000).

The secondary but very important microbiological reason for pasteurisation is to destroy spoilage organisms, particularly psychrotrophic bacteria which can grow and cause spoilage during storage of milk at low temperatures. *Pseudomonas* species are the major psychrotrophic bacteria in cold-stored raw milk but several other psychrotrophs also occur (see Section 4.2.1). Since these bacteria increase in numbers over time, raw milk should be processed as quickly as possible. The commercial reality is that some raw milk is pasteurised within 24 h of milking, but some may be up to one week old before it is pasteurised. In addition, it is important for the raw milk to be maintained at a low temperature, preferably ≤4 °C. In countries where it is not possible to refrigerate raw milk, its keeping quality can be extended by activation of the naturally occurring milk lactoperoxidase system. This has been demonstrated by Fweja *et al.* (2008) and discussed in recent reviews by the FAO/WHO (2006).

While pasteurisation destroys the pathogenic and most of the spoilage non-spore-forming bacteria in raw milk, some thermoduric bacteria remain. A count of the microbial flora in pasteurised milk determined soon after processing provides a measure of these thermoduric bacteria. High thermoduric counts are sometimes found in raw milk and these give rise to high counts in freshly pasteurised milk. Milk processors occasionally experience spikes in thermoduric counts, which are generally spasmodic and short lived. These most probably arise from brief lapses in hygiene during milking, poor temperature control or unfavourable climatic conditions. Thermoduric bacteria play a minor role in spoilage of pasteurised milk.

The thermodurics include non-spore-formers and spores of spore-formers which are more heat-resistant than the non-spore-formers; the spores all survive 80 °C for 30 min and some survive 100 °C for 30 min. Some non-spore-formers such as coryneforms also survive 80 °C for 30 min. The main non-spore-forming thermoduric genera are *Microbacterium, Micrococcus, Enterococcus, Lactobacillus, Corynebacterium* and *Streptococcus* and the main spore-forming genera are *Bacillus, Geobacillus, Paenibacillus* and *Clostridium.* While the thermoduric non-spore-forming bacteria are not pathogenic, some sporefomers are. Of particular interest are *Clostridium* species and some strains of *B. cereus.* These are further discussed in Section 4.2.2.

2.3.6.2 Enzyme Inactivation

Milk contains an abundance of indigenous milk enzymes. These have been extensively reviewed in Fox and McSweeney (2003) and Kelly *et al.* (2006); the heat stabilities of several of them were reviewed by Griffiths (1986) and Andrews *et al.* (1987). Some of these enzymes have particular relevance to the heat treatments discussed in this book. These include alkaline phosphatase, lactoperoxidase, lipase and plasmin (alkaline protease).

The role of alkaline phosphatase has been discussed earlier in this chapter. The original phosphatase test for assessing the adequacy of pasteurisation was based upon the reaction of phosphatase with disodium phenyl phosphate. It was claimed to be able to detect the presence of about 0.2% raw milk (addition) in pasteurised milk, as well as under-processing, for example, heating at 62 °C instead of 63 °C for 30 min or 70 °C rather than 72 °C for 15 s. Since then, more automated tests based on fluorescence measurement (e.g., the Fluorophos® ALP Test System) have increased the sensitivity of the method, being able to detect the presence of 0.006% added raw milk. This is a very useful quality assurance test procedure as it detects low levels of post-pasteurisation contamination by raw milk, and helps to minimise the incidence of pathogens in pasteurised milk.

Griffiths (1986) determined the heat resistance of several indigenous milk enzymes. Alkaline phosphatase was the most heat labile of those measured ($D_{69.8} = 15$ s; $z = 5.1$ °C), compared to lactoperoxidase ($D_{70} = 940$ s; $z = 5.4$ °C). Acid phosphatase was much more heat resistant (about 100 fold) than alkaline phosphatase. Some discrepancies were also noticed between data obtained from capillary tube experiments and those obtained from HTST conditions using plate heat exchangers. Lactoperoxidase activity was thought to provide a useful indicator of over-processing. Activities determined on a PHE for 15 s were generally lower than those obtained from the capillary tube method. Using a PHE, lactoperoxidase activity was almost all destroyed at 78 °C for 15 s and completely destroyed at 80 °C for 5 s. The enzyme appeared sensitive to temperatures above 75 °C, with a z-value of 5.4 °C. Therefore, if milk is processed at or slightly above the specified minimum heating conditions of 72 °C for 15 s, it should test negative for alkaline phosphatase but positive for lactoperoxidase.

Ribonuclease was found to be more heat resistant than lactoperoxidase. Again there were discrepancies between laboratory studies and PHE studies. No loss of activity was observed at 80 °C for 15 s (lab), whereas a 40% loss of activity was found when using a PHE at 80 °C for 15 s.

Andrews *et al.* (1987) determined the following percentage retentions of enzyme activities in milk samples heated for precisely 15 s at 72 °C in glass capillary tubes; acid phosphatase, >95%; α-D-mannosidase, 98%; xanthine oxidase, 78%; γ-glutamyl transpeptidase, 75%; α-L-fucosidase, 26%; N-acetyl-β-D-glucosaminidase, 19%; and lipoprotein lipase, 1%. It was recommended that N-acetyl-β-D-glucosamiindase could be used for more detailed studies between 65 and 75 °C and γ-glutamyl transpeptidase (GGTP) between 70 and 80 °C. N-acetyl-β-D-glucosaminidase (NAGase) has been used as an alternative to somatic cell counts as a measure of mastitic infection (Kitchen *et al.*, 1978).

Patel and Wilbey (1994) recommend measuring GGTP activity for assessing the severity of HTST heat treatments above the minimum for whole milk, skim milk, sweetened milks, creams and ice-cream mixes. There was a good correlation between the reduction in GGTP activity, destruction of streptococci and water activity (Lewis, 1994).

Zehetner *et al.* (1996) investigated endogenous milk enzymes as indicators of heat treatment and concluded that α-fucosidase and phosphohexose isomerase were useful for thermisation, phosphodiesterase and γ-glutamyltransferase were suitable for pasteurisation and α-mannosidase was an appropriate indicator for high-temperature-pasteurised milk.

Lipolysis by indigenous lipoprotein lipase produces free fatty acids which give rise to soapy, rancid off-flavours and reduce the foaming capacity of milk, an important consideration for making cappuccino coffee. This topic has been reviewed by Deeth (2006) and Deeth and Fitz-Gerald (2006). Lipolysis can be initiated on the farm or in the factory if raw milk is subjected to mechanical actions such as agitation or homogenisation which disrupt the milk fat globule membrane and allow access of the lipase to the fat. For example, this can happen when mixing flavoured milks or other similar products, using raw milk, at temperatures of about 50 °C before pasteurisation (Fitz-Gerald, 1974). Fortunately, pasteurisation destroys virtually all of the natural lipase in milk. If this did not happen, all pasteurised (homogenised) milk would have a high level of free fatty acids and taste rancid (Pearse, 1993).

Plasmin is an indigenous alkaline protease in milk. It is very heat-resistant and survives pasteurisation and even some higher-temperature treatments such as UHT processing. However, plasmin-related problems are not reported in pasteurised milk; this is probably due to the short shelf-life of pasteurised milk and the very low activity of plasmin at low temperature. Problems related to residual plasmin activity are more serious in UHT milk. The plasmin system in milk and its effects in UHT milk are addressed in Chapters 6 and 7.

Enzymes may also arise from psychrotrophic bacteria. Many of these are very heat-resistant and survive pasteurisation. However, it is unlikely that residual bacterial lipases and proteases will cause problems in pasteurised milk because of its relatively short shelf-life, the refrigerated storage conditions and the very low activity of the residual protease. In general, however, it is best to avoid using aged milk for pasteurisation because of the risk of its containing bacterial enzymes but also because of its higher microbial count, higher acidity (lower pH), reduced heat stability, and greater likelihood of having off-flavours.

2.3.6.3 Other Changes

There are some other important changes that take place during pasteurisation. As far as chemical reactions are concerned, pasteurisation can be considered to be a mild process. Between 5 and 15% of the whey protein is denatured in milk. This is not sufficient to significantly release levels of volatile sulfur compounds which cause the development of cooked flavour as occurs with higher temperature treatments (see Section 6.1.6.1). Whey protein denaturation is higher during pasteurisation of skim milk concentrates produced by ultrafiltration, increasing with the increase in the concentration factor (Guney, 1989). There is some suggestion that the holder process may cause slightly more whey protein denaturation than the HTST process (Painter and Bradley, 1996). Pasteurisation results in little change in the renneting properties of milk and little association of whey proteins with casein; as a result, good quality cheddar cheese can be produced from pasteurised milk and the majority of milk for cheesemaking is subjected to pasteurisation. No dephosphorylation and no significant reduction in pH and ionic calcium occur during pasteurisation and there is very little effect on the heat-sensitive water-soluble vitamins. Overall, pasteurisation results in little change in texture, flavour and colour, compared to raw milk (Deeth, 1986).

Wilson (1942) reported that it was clear that the majority of people are unable to distinguish between raw and pasteurised milk. Also, the difference in taste between different raw milks appears to be as great as or greater than the difference between raw and pasteurised milks. There is no evidence to suggest that this observation has changed over the past 65 years. However, as the sale of raw milk for human consumption is now prohibited in most countries, comparison of the flavour of raw and pasteurised milk is generally not possible. Nursten (1995) reported that pasteurisation barely alters the flavour of milk and that the volatile flavours responsible for cooked flavour are negligible.

2.3.7 Changes During Storage

2.3.7.1 Changes Due to Post-Pasteurisation Contamination (PPC)

PPC with Gram-negative psychrotrophic bacteria is the major cause of spoilage of pasteurised milk and is a very important determinant of the keeping quality of milk. Muir (1996a,b) describes how this became widely recognised for both milk and cream in the early 1980s, although Davis (1955) had drawn attention to this much earlier. PPC encompasses the recontamination of the product anywhere downstream of the end of the holding tube. It can occur in the regeneration or cooling sections, in storage tanks and in the final packaging of the product, due to poor hygienic practices. It is much reduced by ensuring that all internal plant surfaces in contact with the product are heated at 95 °C for 30 min before processing is commenced. It can only be completely eliminated by sterilising and employing aseptic techniques downstream of the holding tube. One of the main safety concerns is recontamination of the product with pathogens from raw milk; this could occur via by-passing of the holding tube by a number of possible routes, including pinhole leaks in plates. The presence in pasteurised products of high counts of microorganisms (e.g., coliforms and pseudomonads) which should be inactivated by pasteurisation is indicative of PPC. A number of different tests which can be used to determine the extent of the problem are catalogued in IDF (1993). In practical situations where the keeping quality of milk starts to deteriorate or is below expectations, the most likely explanation is an increase in PPC and this should be the first factor to be investigated. Bintsis *et al.* (2008) pointed out that the test for Enterobacteriaceae instead of coliforms is a more sensitive test for post-pasteurisation contamination, since the test detects all of the heat-sensitive, non-spore-forming, Gram-negative rods and provides good evidence that contamination has occurred. In this case the media used for the test must contain glucose instead of lactose.

2.3.7.1.1 Ropiness – a PPC Issue

In 2016, an interesting post-processing contamination of pasteurised milk resulted in the formation of ropy milk in the UK. The most striking feature was the production of thin strings when the milk was poured or when a spoon was placed on the surface and pulled away. The ropiness could be dispersed by mild agitation. We had not encountered this before and we later found that many experienced dairy colleagues had also never seen it. It had apparently been common when pasteurisation was first introduced but is much less common now. Davis (1955) reported that these defects occurred occasionally in pasteurised milk stored about a week and Hubbell Jr. and Collins (1962) reported that a few dairy plants in California had occasionally encountered ropiness in pasteurised milk. However, these authors commented that improved equipment, upgraded cleaning methods, and better sanitizing procedures had reduced its incidence to a low level.

The milk also appeared to be more viscous and it had a distinct fruity aroma. Its pH was 6.50, compared to 6.69 for unspoilt pasteurised milk and its FPD was 535 m°C, compared to 513 m°C for unspoilt milk. However, its viscosity was lower than that of unspoilt milk, at 2.4 cP, compared to 2.6 cP. It was unstable in 70% alcohol, whereas unspoilt milk was stable in 80% alcohol. On centrifugation (3000 G for 30 min), there was no noticeable separation in the ropy milk and no noticeable difference to centrifuged unspoilt milk. As for its taste, it was still sweet, but slightly acidic, fruity with a slightly bitter after taste.

The best account of ropy milk was found in Davis (1955). One pertinent observation from that source was that ropy milk was of bacterial origin and the microorganisms involved would be inactivated by efficient pasteurisation. A small amount (1 mL) was added to 20 mL unspoilt milk and incubated at 5 °C and 20 °C to see if ropiness could be induced. The inoculated milk samples showed signs of ropiness after 24 h incubation at 5 °C and at 20 °C. The pH had fallen slightly and a fruity aroma was not apparent but the milk produced strings on pouring. This suggested that ropiness was bacterial in origin and that the offending bacteria grew well at 5 °C. Davis (1955) also mentioned that low temperatures also favour the formation of the slimy material by the bacteria. Thus, it was possible to produce ropy milk by inoculating a small amount into unspoilt milk, within 24 h in the refrigerator. It was also shown that it was inactivated by pasteurisation (63 °C for 30 min) and is therefore most likely to be a post-processing contaminant. In this case there were fewer than 20 consumer complaints about the product, so it was assumed that it was a very low level post-processing contamination, which was very quickly rectified. Nevertheless, it also interesting when something which is rarely seen, reappears. Coincidentally, another case of ropy milk was found in an ESL chocolate milk about one month later, from a completely different location in the UK.

In these recent cases, the bacteria responsible for the defect were not identified; however, several pyschrotropic bacteria have been reported to cause ropiness. These include *Alcaligenes* (Gainor & Wegemer, 1954, Samaras *et al.*, 2003, Morton & Barrett, 1982), *Escherichia intermedia* (Marth *et al.*, 1964), *Ps. aeruginosa, Klebsiella oxytoca* and others (Cheung & Westoff, 1983), *Acinetobacter* spp (Morton & Barrett, 1982) and *Bacillus aerogenes* and *B. lactis viscous* (Davis, 1955). The rope consists of long chains of bacteria, held together by their capsules. The capsules are polysaccharides which one report identified as levans which are polymers of fructose (Wegemer & Gainor, 1954).

2.3.7.2 Other Changes

It is instructive to outline the changes which lead to spoilage of pasteurised milk. These are illustrated by the results of Gomez Barosso (1997) who compared the shelf-life (days) for two batches of milk which were pasteurised at 72 °C for 15 s and 80 °C for 15 s and stored at 8 °C. The shelf-life was assessed by the aroma, tendency to form a clot on boiling, ethanol stability, titratable acidity, pH and dissolved oxygen, and the results were compared with those corresponding to a total viable count (TVC) of 10^7 cfu/mL. The results are summarised in Table 2.4. It is apparent that even when the TVC reaches 10^7 cfu/mL, the milk may still be acceptable by other criteria.

Table 2.4 Tests used to measure the shelf-life of milk processed at 72 °C and 80 °C for 15 s and stored at 8 °C.

	Shelf-life (days at 8 °C)			
	Processed at 72 °C/15 s		Processed at 80 °C/15 s	
Tests and spoilage criteria used	Trial 1	Trial 2	Trial 1	Trial 2
Total viable count > 10^7 cfu/mL	13	15	13	16
Unfresh aroma	17	20	14	17
Clot on boiling	19	22	19	21
Stability in 68% alcohol	17	20	15	18
Titratable acidity increase > 0.4% lactic acid	16	20	15	18
pH fall below 6.5	15	18	13	16
Dip in dissolved oxygen level	14	17	13	15

(from Gomez Barroso, 1997)

2.3.8 Pasteurisation of Other Milk-Based Products

As well as providing safe market milk, pasteurisation is crucial in the manufacture of cream, cheese, ice-cream and powdered milk to ensure that they are free of pathogenic microorganisms. In the UK, the minimum temperature – time conditions for cream are 72 °C for 15 s and for ice cream mix 79 °C for 15 s. Conditions for cream products are more severe in some countries. One major difference between milk and creams or ice-cream mix is the higher viscosity of the latter products, which results in lower Reynolds numbers, which may mean that the flow of product through the pasteuriser is less turbulent or even streamline. A transition from turbulent to streamline flow alters the residence time distribution and significantly reduces the minimum residence time achieved. Also, the water activity of ice-cream mix will be marginally lower than that of milk and the higher fat content might offer a protective effect toward microbial inactivation (Enright *et al.*, 1961; Champagne *et al.*, 1994).

Table 2.5 shows the heat treatment equivalents to pasteurisation for various types of milk and dairy products published by the Australia New Zealand Dairy Authorities Committee (ANZDAC) in Guideline Validation and Verification of Heat Treatment Equipment and Processes (FSANZ, 2009). It illustrates the higher temperatures recommended for products with larger particles and higher fat or total solids contents for each holding time.

The conditions used for pre-heating milk for powder production vary according to the powder being produced. These are summarized in Table 2.6. Skim milk powders are classified according to the concentration of undenatured whey proteins, which reflects the severity of the pre-heating conditions. Milk for whole milk powder production receives a pre-heat treatment similar to that used for medium-heat skim milk powder in order to generate free sulfydryl groups which act as antioxidants to protect the fat from oxidation during storage of the powder.

Table 2.5 Heat treatment equivalents to pasteurisation for various types of milk and dairy products.

Minimum holding time	Milk with <10% fat (no added sweeteners)			Dairy products with >10% fat and/or added sweeteners and concentrated dairy products with >15% total solids			Ice-cream mix
	Particle diameter (μm)						
	<200	200 - <500	500- <1000	<200	200 - <500	500- <1000	<1000
	Minimum temperature (°C)						
5 s	75.7	76.5	79.0	78.5	79.3	81.8	na
10 s	73.3	73.7	74.6	76.1	76.5	77.4	85.5
15 s	72.0	72.1	72.7	74.8	74.9	75.5	79.5
30 s	70.7	70.8	70.9	73.5	73.6	73.7	na
1 min	69.4	69.4	69.5	72.2	72.2	72.3	na
5 min	66.4	66.4	66.4	69.2	69.2	69.2	na
10 min	65.1	65.1	65.1	67.9	67.9	67.9	74.0
15 min	64.4	64.3	64.3	67.1	67.1	67.1	na
20 min	63.8	63.8	63.8	66.6	66.6	66.6	69.0
30 min	63.0	63.0	63.0	65.8	65.8	65.8	na

(based on data from FSANZ, 2009)
na = not available

Table 2.6 Pre-heat treatments used in the manufacture of various milk powders (in increasing order of severity).

Milk powder	Temperature–time conditions	Undenatured whey protein (g/L)
Low-heat skim milk powder (SMP)	70 °C for 15 s	>6.0
Medium-heat SMP; whole milk powder	85 °C for 1 min; 90 °C for 30 s; 105 °C for 30 s	1.5-6.0
High-heat SMP	90 °C for 5 min; 120 °C for 1 min; 135 °C for 30 s	<1.5
High-high-heat SMP	>120 °C for >40 s	<<1.5

(adapted from Kelly *et al.*, 2012)

References

Al-Shamary, A. & Abdalali, N. (2011) Detection of microbial load in imported UHT milk in Baghdad markets. *Al-Anbar Journal of Veterinary Science* **4**(2), 103–107. [http://www.iasj.net/iasj?func=fulltext&aId=40997]

Andrews, A.T., Anderson, M. & Goodenough, P.W. (1987) A study of the heat stabilities of a number of indigenous milk enzymes. *Journal of Dairy Research* **54**, 237–246.

Barrett, N., Grandison, A. S. & Lewis, M. J. (1998) Contribution of lactoperoxidase to the keeping quality of pasteurized milk. *Journal of Dairy Research* **66**, 73–80.

Bintsis. T., Angelides A.S. & Psoni, L. (2008) Modern Laboratory Practices- Analysis of Dairy Products. In: *Advanced Dairy Science and Technology*. (eds T.J. Britz & R.K. Robinson), Blackwell Publishing Ltd, Oxford.

Champagne, C.P., Laing, R.R., Roy, D., Mafu, A.A. & Griffiths, M.W. (1994) Psychrotrophs in dairy-products - their effects and their control. *Critical Reviews in Food Science and Nutrition* **34**(1), 1–30.

Cheung, B.A. & Westhoff, D.C. (1983) Isolation and identification of ropy bacteria in raw-milk. *Journal of Dairy Science* **66**(9), 1825–1834.

Codex Alimentarius (2003) Proposed Draft Code of Hygienic Practice for Milk and Milk Products at Step 5 (ALINORM 03/13A, Appendix III).

Coghill, D.M., Mutzelburg, I.D. & Birch, S.J. (1982) Effect of thermization on the bacteriological and chemical quality of milk. *Australian Journal of Dairy Technology* **37**(2), 48–50.

Cronshaw. H.B. (1947) *Dairy Information*. Dairy Industries Ltd, London.

Davis, J.G. (1955) *A Dictionary of Dairying*. 2nd edn, Leonard Hill, London.

Deeth, H.C. & Fitz-Gerald, C.H. (2006) Lipolytic enzymes and hydrolytic rancidity. *Advanced Dairy Chemistry, 2 Lipids*, (eds P.F. Fox and P. McSweeney) pp. 481–556, Springer, New York.

Deeth, H.C. (1986) The appearance, texture, flavour and defects of pasteurised milk. In: *Monograph on Pasteurized Milk*. IDF Bulletin 200, pp. 22–26. International Dairy Federation, Brussels.

Deeth, H.C. (2002) Lipolysis in milk and dairy products – A research journey. *Food Australia* **54**, 433–436.

Deeth, H.C. (2005) Lipoprotein lipase and lipolysis in milk. *International Dairy Journal* **16**, 555–562.

Deeth, H.C., Datta, N & Versteeg, C. (2013) Non thermal technologies In *Advances in Dairy Ingredients* (eds G. Smithers & M.A. Augustin). pp 161–216. Wiley-Blackwell Publishing, Ames, Iowa.

Enright, J.B. (1961) Pasteurization of cream, chocolate milk and ice cream mixes containing organism of Q fever. *Journal of Milk and Food Technology* **24**(11), 351–355.

FAO/WHO (2006) *Benefits and Potential Risks of the Lactoperoxidase system of Raw Mil Preservation*. Report of an FAO/WHO technical meeting, FAO Headquarters, Rome, Italy, 28 November - 2 December, 2005.

Fitz-Gerald, C.H. (1974) Milk lipase activation by agitation – Influence of temperature. *Australian Journal of Dairy Technology* **29**, 28–32.

FSANZ (2009) *Primary Production and Processing Standard for Dairy Products A Guide to Standard 4.2.4 Primary Production and Processing Standard for Dairy Products Part 3: Dairy Processing*. Appendix 2. p. 37.

Fox P.F. & McSweeney P.L.H. (eds) (2003) *Advanced Dairy Chemistry*, Volume 1 *Proteins*. 3rd edn. pp. 467–603. Kluwer Academic/Plenum Press, New York.

Fweja, L.W.T., Lewis, M.J. & Grandison, A.S. (2008) Challenge testing the lactoperoxidase system against a range of bacteria using different activation agents. *Journal of Dairy Science* **91**, 2566–2574.

Gainor, C. & Wegemer, D. E. (1954) Studies on a psychrophilic bacterium causing ropiness in milk .1. Morphological and physiological considerations. *Applied Microbiology* **2**(2), 95–97.

Gleeson, D., O'Connell, A.O. & Jordan, K. (2013) Review of potential sources and control of thermoduric bacteria in bulk-tank milk. *Irish Journal of Agricultural and Food Research* **52**, 217–227.

Gomez Barroso, B. (1997) *Effect of Raw Milk Quality on Keeping Quality of Pasteurized Milk*, MSc thesis, The University of Reading, Reading, UK.

Grant, I.R., Rowe, M.T., Williams, A.G. & Muir, D.D. (2005) Efficacy of various pasteurization time/temperature conditions in combination with homogenization on the inactivation of *Mycobacterium avium* subsp. *paratuberculosis* in milk. *Applied and Environmental Microbiology* **71**, 2853–2861.

Griffiths, M.W. (1986) Use of milk enzymes and indices of heat treatments. *Journal of Food Protection* **49**, 696–705.

Griffiths, M.W. (2010) The microbiological safety of raw milk. In: *Improving the Safety and Quality of Milk*. (ed. M.W. Griffiths M W), pp. 27–63. Woodhead Publishing, Cambridge.

Griffiths, M.W. (2006) *Mycobacterium paratuberculosis*. In: *Emerging Foodborne Pathogens*, (eds Y. Motarjemi & M. Adams). pp. 522–556. Woodhead Publishing, Cambridge, U.K.

Guney, P. (1989) *Improving Pasteurized and Extended Shelf Life Milk*. PhD thesis, University of Giessen, Germany.

Hickey, M. (2009), Current legislation of market milks. In: *Milk Processing and Quality Management*. (ed. A.Y. Tamine), pp. 101–138, SDT, Wiley-Blackwell, Chichester, UK.

Holsinger, V.H., Rajkowski, K.T. & Stabel, J.R. (1997) Milk pasteurisation and safety: a brief history and update. *Revue Scientifique et Technique de L'Office International des Epizooties* **16**(2), 441–451.

Hubbell, Jr. B.E. & Collins, E.B. (1962) Ropiness in milk. Pyschrotrophic bacteria and California milk quality. *California Agriculture* **10**, 14–15.

Humbert, E.S., Campbell, J.N., Blankenagel, G. & Gebreegziabher, A. (1985) Extended storage of raw-milk. 2. The role of thermization. *Canadian Institute of Food Science and Technology Journal* **18**(4), 302–305.

IDF (1984) *The Thermization of Milk*. IDF Bulletin 182. International Dairy Federation, Brussels.

IDF (1986) *Monograph on Pasteurised Milk. IDF* Bulletin 200. International Dairy Federation, Brussels.

IDF (1988) *Code of Practice for the Preservation of Milk by the Lactoperoxidase System*. IDF Bulletin 234. International Dairy Federation, Brussels.

IDF (1993) *Catalogue of Tests for the Detection of PPC of Milk*. IDF Bulletin 281. International Dairy Federation, Brussels.

Juffs, H. & Deeth, H.C. (2007) *Scientific Evaluation of Pasteurisation for Pathogen Reduction in Milk and Milk Products*. FSANZ, Canberra (ISBN 978-0-642-34555-4).

Kay, H. & Graham, W. (1935) The phosphatase test for pasteurised milk. *Journal of Dairy Research* **6**, 191–203.

Kelly, A., Datta, N. & Deeth, H.C. (2012) Thermal processing of dairy products. In: *Thermal Food Processing: New Technologies and Quality Issues* 2nd Edn (ed. D-W. Sun), pp. 273–307. CRC Press, Taylor & Francis Group, Boca Raton, FL.

Kelly, A., McSweeney, P., Farkye, N., Jost, R. & Deeth, H. (eds) (2006) Proceedings of the first IDF symposium on indigenous enzymes in milk. *International Dairy Journal* **16**(6), 499–716.

Kessler, H.G. & Horak, F.P. (1984) Effect of heat treatment and storage conditions on keeping quality of pasteurized milk. *Milchwissenschaft* **39**, 451–454.

Kessler, H.G. (1981) *Food Engineering and Dairy Technology*, Verlag A Kessler, Freising, Germany.

Kessler, H.G. (1989) Effect of thermal processing on milk. In: *Food Preservation – 5.* (ed. S. Thorne) pp. 80–129, Elsevier Applied Science, London.

Kitchen, B.J., Middleton, G. & Salmon, M. (1978) Bovine milk N-acetyl-B-D-glucosaminidase and its significance in the detection of abnormal udder secretions. *Journal of Dairy Research* **45**, 15–25.

Lewis, M.J. (1990) *Physical Properties of Foods and Food Processing Systems*, Woodhead Publishers, Cambridge, UK.

Lewis, M.J. (1994) Advances in the heat treatment of milk. In: *Modern Dairy Technology*, Volume 1. (ed. R.K. Robinson), pp. 1–60. Elsevier Applied Science, London.

Lewis, M.J. & Heppell, N. (2000) *Continuous Thermal Processing of Foods: Pasteurization and UHT Sterilisation*, Aspen Publishers, Gaithersburg, MD.

Lovett, J., Bradshaw, J. G. & Peeler, J. T. (1982) Thermal inactivation of *Yersinia enterocolitica* in milk. *Applied and Environmental Microbiology* **44**(2), 517–519.

Marks, N.E., Grandison, A.S. & Lewis, M.J. (2001) Challenge testing of the lactoperoxidase system in pasteurized milk. *Journal of Applied Microbiology* **91**, 735–741.

Marth, E.H., Ingold, D.L. & Hussong, R.V. (1964) Ropiness in milk caused by strain of *Escherichia intermedia*. *Journal of Dairy Science* **47**(11), 1265–1267.

Morton, D.J. & Barrett, E.L. (1982) Gram-negative respiratory bacteria which cause ropy milk constitute a distinct cluster within the genus *Acinetobacter*. *Current Microbiology* **7**(2), 107–112.

Muir, D.D. (1996a) The shelf life of dairy products: 1 Factors influencing raw milk and fresh products. *Journal of the Society of Dairy Technology* **49**, 24–32.

Muir, D.D, (1996b) The shelf life of dairy products: 2 Raw milk and fresh products. *Journal of the Society of Dairy Technology* **49**, 44–48.

Nursten, H. (1995) Heat induced changes in the flavour of milk. In: *Heat Induced Changes in Milk.* 2nd edn, pp. 308–317. IDF Document S.I. 9501. International Dairy Federation, Brussels.

Painter, C.J. & Bradley, R.L. (1996) Residual alkaline phosphatase activity in milks subjected to various time/temperature treatments, In: *Heat Treatments and Alternative Methods*. pp. 396–402. IDF Document 9602, International Dairy Federation, Brussels.

Patel, S.S. & Wilbey, R.A. (1994) Thermal inactivation of γ-glutamyltranspeptidase and *Enterococcus faecium* in milk-based systems. *Journal of Dairy Research* **61**, 263–270.

Pearse, M.A. (1993) Pasteurization of liquid products. In: *Encyclopedia of Food Science, Food Technology & Nutrition* pp. 3441–3450, Academic Press, London.

Ravanis, S. and Lewis, M.J. (1995) Observations on the effect of raw milk quality on the keeping quality of pasteurised milk. *Letters in Applied Microbiology* **20**, 164–167.
Reiter, B. & Harnulv, B.G. (1982) The preservation of refrigerated and uncooled milk by its natural lactoperoxidase system. *Dairy Industries International* **47**, 13–19.
Samaras, F. I., Kehagias, C., Arkoudelos, J. S. & Bocaris, M. I. (2003) Investigation on ropiness development by isolates of the genera *Lactobacillus, Alcaligenes* and Feta cheese starter cultures. *Food Microbiology* **20**(5), 503–509.
Satin, M. (1996) *Food Irradiation - A Guidebook*, Technomic Publishing Co Ltd, Lancaster.
Schmidt, D., Cromie, S.J. & Dommett, T.W. (1989) Effects of pasteurisation and storage conditions on the shelf life and sensory quality of aseptically packaged milk. *Australian Journal of Dairy Technology* **44**, 19–24.
Schroder, M. A. & Bland, M.A. (1984) Effect of pasteurisation temperature on keeping quality of whole milk *Journal of Dairy Research* **51**, 569–578.
Stadhouders, J. (1982) Cooling and thermization as a means to extend the keeping quality of raw milk. *Kieler Milchwirtschaftliche Forschungsberichte* **34**(1), 19–28.
UK Milk Marketing Boards (1964) *Dairy Facts and Figures, 1964.* The Federation of United Kingdom Milk Marketing Boards, London.
Wegemer, D.E. & Gainor, C. (1954) Studies on a psychrophilic bacterium causing ropiness in milk. 2. Chemical nature of the capsular polysaccharide. *Applied Microbiology* **2**(2), 97–99.
Wilbey, A. (1993) Pasteurization of foods. a) Principles of pasteurization. In: *Encyclopedia of Food Science, Food Technology & Nutrition.* pp. 3437 -3441, Academic Press, London.
Wilson, G.S. (1942) *Pasteurisation of Milk.* Arnold, London.
Yale (1933), cited in Cronshaw (1947), pp. 620–621.
Zehetner, G., Bareuther, C., Henle, T. & Klostermeyer, H. (1996) Endogenous milk enzymes as indicators for low-heat treatment of milk. In: *Heat Treatments and Alternative Methods.* IDF Document S.I. 9602, pp. 403–408. International Dairy Federation, Brussels.

3

Heat Treatments of Milk – ESL, UHT and in-Container Sterilisation

3.1 Introduction

This chapter provides information on the three high-temperature processes used for milk and other dairy products: Extended shelf-life (ESL) processing and the two sterilisation processes, UHT processing and in-container sterilisation. ESL processing is dealt with in some detail here, as, apart from the discussion of the microbiological aspects of ESL milk in Chapter 4, it is not covered elsewhere in the book.

As UHT is the main focus of this book, this chapter provides an introduction only to that topic and puts it into perspective relative to the other two processes. Further information on UHT processing and products is provided in subsequent chapters. An overview of these three heat treatments, together with thermisation and pasteurisation discussed in Chapter 2, was given in Table 2.1.

3.2 Some Important Definitions

In addition to the D-value and z-value defined in Sections 2.3.4.1 and 2.3.4.2, there are other important parameters which are most applicable to the high-temperature treatments discussed in this chapter.

3.2.1 Q_{10}

The increase in rate of a reaction with a temperature change of 10 °C. The reaction can be bacterial destruction or chemical change.

Like z-value, Q_{10} refers to the change in the rate of a reaction, either bactericidal or chemical, with a change in temperature. While the z-value is given in degrees Celsius, the Q_{10} value is a dimensionless number. They are related as follows:

$$z(°C) = 10 / \log Q_{10} \; or \; Q_{10} = 10^{(10/z)}$$

Both Q_{10} and z-values can be considered to be constant only over a limited range of temperatures (Burton, 1988). A comparison of the magnitude of the Q_{10} and z-values for some chemical and bactericidal effects is given in Table 3.1.

A case study illustrating the benefit of a knowledge of the relevant Q_{10} values is as follows. A processor is experiencing contamination in UHT milk due to heat-resistant

High Temperature Processing of Milk and Milk Products, First Edition. Hilton C. Deeth and Michael J. Lewis.

Table 3.1 Typical Q_{10} and z-values for bacterial destruction and chemical changes.

Bacterial destruction			Chemical change		
Bacteria	**Q_{10}**	**z-value (°C)**	**Chemical reaction**	**Q_{10}**	**z-value (°C)**
Mesophilic spores	8.9-9.6	10.2-10.7	Browning	2.5	25
Thermophilic spores	8.8-10.0	10-10.6	Whey protein denaturation 130-150 °C	2.2	30
G. stearothermophilus	8.5-31.1	6.7-9.4	Thiamine loss	2.1	30
B. subtilis	8.6-17.8	6.6-10.7	Lactulose formation	2.6	25
Clostridium spp	3.65-10	10-17.8	Plasmin inactivation	2.5	25

Based on data in Burton (1988)

bacterial spores. The processor is reluctant to increase the sterilisation temperature to destroy the spores for fear of causing excessive browning in the product. Based on the Q_{10} values shown in Table 3.1 (10 for bactericidal effect and 2.5 for browning), the effect of increasing the sterilisation temperature on the bacterial count will be of the order of 3.5 to 4 times the effect on browning. The effect can also be estimated using the kinetics of the browning reaction (see Chapter 6), and assuming a z-value for inactivation of thermophilic spores of 10.5. Accordingly, increasing the sterilisation temperature (in a direct processing plant [see Chapter 5]) from 146 to 149 °C will approximately double the bactericidal effect but only increase browning by about one third. Therefore the processor would be assured that a slight rise in sterilisation temperature would have a marked effect on spores but a relatively minor effect on browning and other chemical reactions.

3.2.2 Bacterial Indices, B* and F_0

These indices provide a measure of the bactericidal effect of a heat process. The major one of relevance to UHT processing is B*, whereas the most applicable to in-container sterilisation is F_0. Sometimes F_0 is also used for UHT processes but is not the most appropriate as the reference temperature for F_0 is 121.1 °C (or 250 °F), whereas for B* it is 135 °C. Hence B* is a measure of the bactericidal effect of a heat treatment compared to a treatment at 135 °C. A process with B* = 1 causes a 9-decimal reduction of thermophilic spores and is equivalent to holding the product at 135 °C for 10.1 s. This assumes a z-value of 10.5 °C. B* can be computed from the following equation:

$$B^* = \int 10^{((T-135)/10.5)} * dt / 10.1$$

where T is the processing temperature and t is time.

As a general rule, a UHT process should have a B* of at least 1. In practice, however, most commercial UHT plants operate at much higher B* values. For example, Tran *et al.* (2008) calculated the B* values of 22 commercial UHT plants and found they ranged from around 2 to 20.

The calculations of B* of a UHT plant assume plug flow through the plant. In other words, all particles in the product experience exactly the same heat treatment. This is not realistic as plug flow is never attained; commercial plants operate under turbulent conditions which minimize the residence time distribution of the particles in the liquid being processed but does not eliminate it. Hence account should taken of the residence time distribution in calculating B* by adjusting it to apply to the fastest moving particle. While the nature of the flow in plants varies, a reasonable adjusted B* value (B^*_{adj}) is calculated as $B^*_{adj} = B^*/1.2$.

The index F_0 is widely used for in-container sterilisation processes such as canning. It is defined as the equivalent time in minutes at 121.1 °C of a heat process, assuming a z-value of 10 °C. It can be expressed as follows:

$$F_0 = \int 10^{((T-121.1)/10)} * dt$$

where T is the processing temperature and t is time

F_0 is based on the destruction of spores of the anaerobic pathogen *Clostridium botulinum*. While it is a measure of the bacterial lethality of the heat treatment, it is more appropriate to heating at temperatures around 120 °C than around 140 °C. However, for comparative purposes, a process with a B* of 1 has approximately the same bactericidal effect as one with an F_0 of 3. Hence, a UHT processes should have an $F_0 \geq 3$ to ensure bacteriological safety.

The B* and F_0 values should be used as a guide only for the bactericidal effect of a process. The z-values assumed, 10.5 for B* and 10.0 for F_0, can never be entirely applicable for a particular processing situation as they vary considerably for different species and strains of bacteria or spores as is evident from the range of reported values summarized in Table 3.1.

3.2.3 Chemical Index, C*

C* provides a measure of the effect of a heat treatment on the chemical components of a product, for example, milk in the context of this book. The reference temperature is 135 °C and assumes a z-value of 31.4. A C* of 1 indicates a process which reduces the concentration of thiamine by 3% and is equivalent to holding the product at 135 °C for 30.5 s. C* can be computed by the equation:

$$C^* = \int 10^{((T-135)/31.4)} * dt / 30.5$$

where T is the processing temperature and t is time.

As a general rule, a UHT plant should have a C* of no greater than 1. However, most UHT plants, particularly indirect heating plants, operate at much higher C* values. Tran *et al.* (2008) found the range of C* for 22 commercial plants to be 0.4 to 3. Processes with high C* values give rise to products with greater amounts of chemical change, considerable cooked flavour but improved storage stability (see Chapter 7).

Although C* is based on the kinetics of destruction of thiamine, and it is used as a general chemical index of heat treatment, that is, the higher the C* the greater the change in the chemical components of milk, it needs to be borne in mind that not all changes in the chemical components of milk follow the same kinetics as that of thiamine destruction. For example, heating at 90 °C for 30 seconds, a common UHT preheat treatment, destroys less that 0.1% of thiamine but denatures ~75% of β-lactoglobulin and inactivates the native milk protease, plasmin (Newstead *et al.*, 2006).

The choice of thiamine on which to base the chemical index C* is interesting, given the fact that thiamine is present at very low levels in milk (~40 µg/L) and is seldom measured. However, it does serve as a useful theoretical construct for describing the heat input into a system.

Sometimes F_0 values are used to indicate the severity of heat treatments in relation to its chemical effect such as production of cooked flavour. This is clearly a misuse of this index and can be quite misleading. C* should be used for this purpose. Major reasons for this misuse of F_0 is that technologists are familiar with the use of F_0 to describe in-container sterilisation and there is no commonly used chemical index for this process equivalent to C* for UHT sterilisation.

3.3 Extended Shelf-Life (ESL) Milk Processing

ESL milk refers to milk which has a refrigerated shelf-life longer than that of pasteurised milk. The "use-by" period for pasteurised milk varies between countries but is in the range of 5-21 days. The desired shelf-life of ESL is >30 days and preferably up to 60 days or longer. The two major approaches to producing ESL milk are by heating in the range of 123-145 °C for <1-5 s, and microfiltration using a 0.8-1.4 µm pore size membrane combined with an HTST heat treatment. Other approaches to extending shelf-life include the use of an antibacterial agent such as nisin in association with a thermal treatment, use of multiple heat treatments such as tyndallisation, and combining thermal and non-thermal technologies such as pulsed electric field technology (Sepulveda *et al.*, 2005) (see Sections 3.3.1 to 3.3.6).

ESL milk products are now widely produced and, in some countries such as Austria and parts of Germany, they now constitute a significant share of the drinking milk market. ESL milk also provides a means of supplying "fresh" milk to countries which have no fresh milk supply.

Some consumers may be suspicious of a chilled product which has a long shelf-life. In this regard, the flavour of ESL milk is most important for consumer acceptance.

3.3.1 ESL Milk by Thermal Treatment

Heat treatment for production of ESL milk lies somewhere between pasteurisation and sterilisation. It is not clear exactly what heat treatment ESL milk should be subjected to, since there are no specific temperature – time requirements for this category in most countries. An exception is USA where an ESL milk, designated "ultrapasteurised", is specified to be produced by a heat treatment at ≥138 °C for ≥2 s. In a review on ESL milk, Rysstad and Kolstad (2006) used the following definition: *ESL products are products that have been treated in a manner to reduce the microbial count beyond normal pasteurisation, packaged under extreme hygienic conditions, and which have a defined prolonged shelf life under refrigeration conditions.*

ESL milk, unlike UHT milk, has to be stored refrigerated. It is expected to have a longer shelf-life than pasteurised milk but not as long as UHT milk. Thus, thermally produced ESL milk should include any milk which is given a more severe heat treatment than normal pasteurisation conditions but less severe than sterilisation conditions. One proposal is that it should give a negative test for lactoperoxidase to distinguish it from pasteurised milk. In former EU regulations (EU, 2005), such a milk would need to be labelled as "high pasteurised", but this legal restriction no longer applies; however, it remains an appropriate means of distinguishing between pasteurised and thermally produced ESL milk.

Reported commercial processing conditions for ESL are mostly in the range 123-127 °C for 1-5 s (Mayr *et al.*, 2004a,b; Kaufmann & Kulozik, 2008; Kaufmann *et al.*, 2010; Lorenzen *et al.*, 2011; Buckenhüskes, 2015). Higher temperatures can also be used. As mentioned above, US regulations define a process "ultrapasteurisation" for heating milk at ≥138 °C for ≥2 s, whereas Henyon (1999) reported that ESL milk processing in Canada is at 131-138 °C for 2 s, usually using steam infusion. Even higher temperatures for a shorter time, 130-145 °C for <1 s, also using steam infusion, have also been advocated (Fredsted *et al.*, 1996; Rysstad & Kolstad, 2006). These higher-temperature treatments may seem severe for ESL milk, but they are still sub-UHT treatments in terms of their sporicidal effects, that is, B* values are <1.

ESL milks produced at lower temperatures are also marketed. In South Africa, most ESL milk is processed at 94 –100 °C (Mugadza & Buys, 2015). This product is similar to the ESL milks reported by Manji (2000) which were processed by direct steam heating at 89 to 100 °C.

The heat treatment for ESL milk can be by either direct (steam infusion or steam injection) or indirect (using plate or tubular heat exchangers) heating. The direct processes are favoured as a high bacterial kill can be achieved with minimal chemical damage. Rysstad and Kolstad (2006) favoured steam infusion over steam injection on theoretical grounds but a significant difference has not been demonstrated empirically. These authors also made a theoretical comparison of a direct and an indirect process with equal bactericidal effect, that is, equal F_0/B* values (see Section 3.2.2): the direct treatment at 135 °C for 0.5 s and the indirect (with plate heat exchanger which heats faster than tubular systems) at 127 °C for 1 s. The difference in chemical effect was given by denaturation of β-Lg which was 13.6% and 83.5% respectively. Therefore, in order to produce ESL milk with few, if any, psychrotrophic spore-formers (the only organisms which can cause spoilage in aseptically packaged ESL milk containing no non-spore-formers) and retain a flavour similar to that of pasteurised milk, but much less cooked than UHT milk, use of direct heating at high temperatures (≥135 °C) for very short holding times (<1 s) followed by aseptic packaging is favoured. This will result in a high-quality product with a long shelf-life.

Mayer *et al.* (2010) investigated acid-soluble β-lactoglobulin (β-Lg) and furosine as relevant indicators of the heat load applied to ESL milk products. Liquid milk samples, including 7 raw, 33 pasteurised, 71 ESL and 17 UHT, were obtained from retail outlets in Austria and analysed. Only 45% of the analysed samples designated as ESL milk showed furosine contents <40 mg/100 g protein and acid-soluble β-Lg contents >1800 mg/L milk, a proposed limit for ESL milk (Note: an undenatured β-Lg content of 1800 mg/L represents 45-50% denaturation). A further 55% of the analysed ESL milk samples had low acid-soluble β-Lg (<500 mg/L; >85% denaturation) and high furosine contents (>40 mg/100 g protein), levels comparable to those in UHT milk and indicative of excessive heat treatment for ESL milk. In a previous paper, Mayer *et al.* (2009) analysed retail milk samples and found the acid-soluble β-Lg contents of pasteurised, ESL and UHT milk samples were 3000, 140-3680 and 212 mg/L, respectively. They identified good ESL milk as having furosine levels of 11.6 mg/100 g protein and 2729 mg/L β-Lg and bad ESL milk as having 71.3 mg/100 g protein furosine and 246 mg/L β-Lg. Fredsted *et al.* (1996) reported β-Lg levels for Pure-Lac™ ESL (produced with steam infusion), pasteurised and raw milk as 2572, 2877 and 3291 mg/L, respectively. This shows a 22% denaturation of β-Lg in the Pure-Lac™ ESL milk; Rysstad and Kolstad (2006) reported a 13.6% denaturation in a comparable ESL milk processed at 135 °C for 0.5 s, whereas Huijs *et al.* (2004) reported ~20-25% denaturation of β-Lg in milk treated with Innovative Steam Injection (ISI) heating at 150-180 °C for <0.1 s. All Pure-Lac™ ESL, pasteurised and raw milk samples analysed by Fredsted *et al.* (1996) had lactulose levels of <40 mg/L. An important finding of Huijs *et al.* (2004) was that, although plasmin was not inactivated by ISI treatment, storage of the ISI-treated milk at 7 °C (as for ESL milks) caused no development of bitterness up to 28 days but storage at 20 °C (as for UHT milks) caused considerable bitterness within a few weeks. The inference from these results is that plasmin exhibits little or no activity at refrigeration temperature.

An expectation is that ESL processing should provide a longer shelf-life than pasteurisation without the production of a distinct cooked flavour. Many consumers do not like the cooked flavour (also referred to as heated, boiled or cabbage [see Sections 6.1.6 and 7.1.3]) that is associated with milk that has been severely heat treated, such as UHT and in-container sterilised milk. Therefore, it is important to minimize the intensity of this flavour in ESL milk. Cooked/sulfurous flavour begins to appear in heated milks when denaturation of β-lactoglobulin reaches ~60% (Gaafar, 1987). Therefore, appropriate heating conditions should be chosen in order to minimize denaturation of β-lactoglobulin and development of this flavour. This is exemplified by the Pure-Lac™ ESL milk discussed above processed at up to 145 °C for <1 s. Direct heating processes, either injection or infusion processes, are ideal for achieving these conditions and causing minimal flavour change.

Based on the spread of data from their investigation of commercial ESL milks, Mayer *et al.* (2010) concluded that there was an urgent need for an EU regulation to define legal limits for the tolerable heat load of milk classified as ESL. Given the usefulness of predicting the extent of chemical changes occurring during heat treatment, as demonstrated in this book (Chapters 5 and 6), it is suggested that if such limits are to be set, they should be based on the chemical index, C*, denaturation of β-lactoglobulin (a major cause of production of volatile sulphur compounds and the cooked flavour), lactulose production or furosine levels. The limits should not be set according to bacteriological indices such as B* or F_0.

The microbiological aspects of ESL milk are discussed in Section 4.4.2. In summary, the shelf-life of ESL milks produced thermally and not packaged aseptically is generally limited by growth of post-processing bacterial contaminants. In contrast to the PPCs in pasteurised milks, the reported contaminants are mostly Gram-positive organisms. Such contamination can be avoided if the milk is packaged aseptically. In terms of processing, the aim should be to heat milk so that all non-spore-forming bacteria and almost all spores are inactivated. Of particular importance is inactivation of psychrotrophic spores, especially those of *B. cereus*, some strains of which are pathogenic and can grow at ≤7 °C. This can be best achieved using direct heat treatments at high temperatures (≥134 °C) for short times which are bactericidally effective without causing excessive flavour change (Blake *et al.*, 1995).

Based on the above discussion it is apparent that the optimum heating conditions for producing ESL milk should be based on the need for elimination of psychrotrophic spores and the need to minimize cooked flavour. We suggest that the former be based on the data of Blake *et al.* (1995) which showed a treatment at ≥134 °C for 4 s was required. This is in line with the conclusion of Bergere and Cerf (1992) that milk processed at ≥134 °C should not contain *B. cereus* spores (see also Section 4.2.2.2.1). If it is assumed that the z-value for psychrotrophic spore inactivation is the same as that for thermophilic spores, 10.5 °C, a heat treatment of 134 °C for 4 s would have a B* of 0.32, considerably less than 1.0, the minimum for UHT processing (see Section 3.4.2). This is shown in Figure 3.1 in which "ESL line 1" joins points with a B* of 0.32 and is placed to the left of the B* = 1 line. Interestingly, the minimum heating conditions used for producing ultrapasteurised milk in USA, 138 °C for 2 s, has a B* of 0.4 which would lie just to the right of ESL line 1.

To accommodate the second condition, we suggest the heat input should be less than that which theoretically causes ~50% denaturation of β-lactoglobulin according to the kinetics reported by Lyster (1980). This is consistent with the finding of Gaafar (1987) that sulfurous flavours become noticeable when denaturation of β-lactoglobulin reaches ~60% and the proposed limit of 1800 mg/L of undenatured β-lactoglobulin (Mayer *et al.*, 2010). The temperature–time conditions to achieve 50% denaturation β-lactoglobulin are shown by the line labelled thus in Figure 3.1. Hence, points along the ESL line below the 50% β-lactoglobulin denaturation line represent the optimum conditions for producing ESL milk to ensure destruction of psychrotrophic spores and

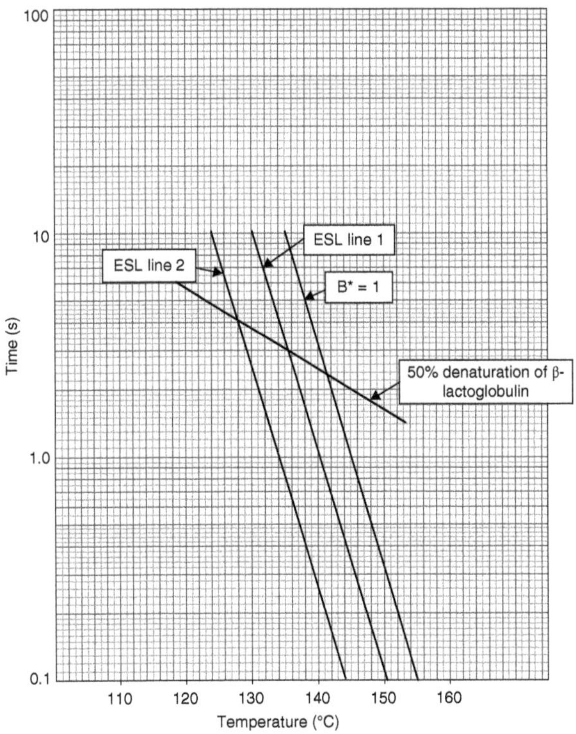

Figure 3.1 Temperature – time combinations for ESL processing of milk. (line "B* = 1" represents 9-log reduction of thermophilic spores, the heating conditions for the lower limit for UHT processing; "ESL line 1" represents conditions which would inactivate psychrotrophic spores; "ESL line 2" represents conditions equivalent to those commonly used commercially for producing ESL milk).

minimal flavour change. ESL line 2 in Figure 3.1 shows temperature – time conditions equivalent to 127 for 5 s and joins points with a B* of ~0.09.

Points in the zone between ESL lines 1 and 2 and below the 50% denaturation β-lactoglobulin line (the 50% line) represent reasonable conditions for commercial production of ESL milk, with the nearer the conditions are to ESL line 1, the more likely that spores of psychrotrophic spore-formers will be destroyed. Furthermore, points in this zone furthest from the 50% line will have the freshest flavour. Points in the zone between ESL lines 1 and 2 and above the 50% line will have similar bactericidal effects to those below the 50% line but will have more cooked flavours. Points in the zones between ESL line 1 and the B* = 1 line will have the greatest bactericidal stability but milk processed by conditions in these zones will still require refrigerated storage to ensure its stability.

The temperature–time combinations used to construct the lines shown in Figure 3.1 are based on a direct heating process where the heat-up and cool-down sections contribute very little to the overall heat load. Furthermore, the successful adoption of these conditions for long shelf-life ESL milk would occur only if there was no post processing contamination and packaged aseptically (see Section 3.3.1.1).

3.3.1.1 ESL Milk by Thermal Treatment Plus Aseptic Packaging

Most ESL milk to date has been packaged under very clean but not aseptic conditions. The commercial success of the product is evidence of the effectiveness of this approach in producing milk with extended shelf-life. However, the major cause of spoilage of ESL milk is post-heat-treatment bacterial contamination. When this is eliminated by packaging the milk aseptically, a considerable extension in shelf-life can be achieved. This has been demonstrated in several studies. Schmidt *et al.* (1989), using pasteurisation conditions of 72-74 °C for 15 s with aseptic packaging, obtained a shelf-life of up to 49 days when the milk was stored at 3 °C. Similarly, Brown *et al.* (1984) obtained a shelf-life of 35 days at 4 °C for milk pasteurised at 72 °C for 15 s combined with aseptic packaging. While we do not recommend these low-temperature heat treatments for commercial ESL milk production, these studies illustrate the marked improvement in shelf-life which is possible when post-pasteurisation contamination is excluded.

The heat treatments in the above studies would not have inactivated thermoduric bacteria, either spore-forming or non-spore-forming. Higher temperatures are required to achieve this. Wirjantoro and Lewis (1996) found that milk heated to 115 °C for 2 s and packaged under sterile conditions had a total plate count of <10 cfu/mL after 28 days' storage at 10 °C; its shelf-life at 10 °C would have been considerably longer than 28 days and even longer at lower temperatures. Ranjith (2000) heated cream at 115, 117.5, 120, 122.5 and 125 °C for 1 s with aseptic packaging and found that cream processed at 120-125 °C had counts of only 15-40 cfu/mL after 41 days' storage at 7 °C while cream processed at ≤117.5 °C had counts of >3 × 10^6 cfu/mL after 13 days at 7. Milk and cream treated at 120-125 °C for 1 s and packaged aseptically had shelf-lives of >37 and >49 days respectively when stored at 7 °C. Blake *et al.* (1995) used direct UHT heating at 120-140 ° for 4 s with aseptic packaging and obtained >60 days' shelf-life as judged by a formal taste panel. Continued sensory testing of the milk samples by a small informal panel found the milks still acceptable after 240 days. It is therefore apparent that very long refrigerated shelf-lives can be achieved with aseptic packaging and, provided the heat treatment is sufficient to inactivate any psychrotrophic pathogenic spore-formers, such as *B. cereus*, the safety of the ESL milk can also be assured.

3.3.2 ESL Milk by Microfiltration Plus HTST Heat Treatment

The second approach to producing ESL using microfiltration has been used commercially for several years in some countries. Some milk which is microfiltered and also HTST-treated may be sold as pasteurised milk in some countries (e.g., UK) or ESL in others (e.g., Austria). The process which is described in Chapter 10 involves separation of milk into skim and cream, microfiltration of the skim, high-temperature treatment of the cream at 100-120 °C, recombining the microfiltered skim and heat-treated cream, and HTST pasteurisation of the recombined mix. The pasteurisation step is a legal requirement in most countries.

Information on factors limiting the shelf life of ESL milk produced in this way (MF-ESL) is limited. Schmidt *et al.* (2012) analysed MF-ESL milk at different stages of the production process and during storage at 4 °C, 8 °C and 10 °C and evaluated the changes in bacterial counts, microbial diversity and enzymatic quality. Microfiltration decreased the microbial loads by 5–6 logs to <1 cfu/mL. However, bacterial counts at the end of shelf-life were extremely variable and ranged between <1 and 10^8 cfu/mL. Some (8%) of all samples had counts higher than 10^6 cfu/mL. The main groups of spoilage bacteria were Gram-negative bacteria (*Acinetobacter, Chryseobacterium, Psychrobacter, Sphingomonas*) and the spore-formers *Paenibacillus* and *B. cereus,* while other spore-formers and *Microbacterium* spp. did not reach spoilage levels. *Paenibacillus* spp. and *B. cereus* apparently influenced enzymatic spoilage, as indicated by increased free fatty acid production, pH 4.6-soluble peptide fractions and off-flavours. In some cases, enzymatic spoilage was observed although microbial counts were well below 10^6 cfu/mL. Also, 13 *B. cereus* isolates were characterized for their toxin profiles and psychrotolerance; Hbl, nhe, and cytK toxin genes were detected in 10, 13 and four isolates respectively and three isolates were psychrotolerant. Generally, large discrepancies in microbial loads and biodiversity were observed at the end of shelf life, even among packages of the same production batch. It was suggested that such unexpected differences may be due to very low cell counts after ESL treatment, causing stochastic variations of initial species distributions in individual packages. This would result in the development of significantly different bacterial populations during cold storage, including the occasional development of high numbers of pathogenic species such as *B. cereus* or *Acinetobacter.* This may be the major downside in the production of MF-ESL milk.

Lorenzen *et al.* (2011) compared MF-ESL milks with thermally (both direct and indirect) processed ESL milks produced by 17 German companies. They also compared the ESL milks with raw, HTST pasteurised and UHT milks in some analyses. Their data for acid-soluble β-Lg (averages in mg/L) were: raw, 4111; HTST, 4100; MF-ESL, 3817; direct ESL, 2084, indirect ESL, 507; and UHT 153, whereas their data for furosine (range in mg/100 g protein) were: HTST, 5.5-7.2; MF-ESL, 8.2-12.7; direct and indirect ESL, 11.1-22.6; and UHT, 145.7. The greater heat load indicated for the MF-ESL milk over the HTST pasteurised milks reflects the high-temperature treatment given to the cream in the MF-ESL process. They also found that the HTST-pasteurised and the MF-ESL milks were lactoperoxidase-positive but the thermally processed ESL and UHT milks were lactoperoxidase-negative. The authors concluded that the different classes of milk could be clearly differentiated on the basis of their furosine and acid soluble whey protein contents.

3.3.3 ESL Milk by Thermal Treatment Plus Bactofugation

Bactofugation is an alternative non-thermal treatment to microfiltration for removing bacterial cells, including spores, from milk. It is currently used for reducing the levels of spores, particularly *Clostridium tyrobutyricum* from cheese milk (see Section 10.6). It also has potential in the production of ESL milk.

Mugadza and Buys (2015) reported the use of bactofugation with thermal pasteurisation at 72-73 °C to produce ESL milk. The milk which has also been introduced onto the South African market is cold-filled under aseptic conditions. Bactofugation (see Section 10.6) reduces the bacterial load by about 2 logs so the ESL milk will have a lower number of spores compared to pasteurised milk produced without bactofugation and would be expected to have a correspondingly longer shelf-life. Such an ESL milk would be lactoperoxidase positive.

3.3.4 ESL Milk by Thermal Treatment Plus an Antibacterial Agent

A third approach is to use small additions of an antibacterial agent, such as the bacteriocin, nisin. The addition of small amounts of nisin (40 IU/mL) has been found to be effective in reducing microbial growth following heat treatment at 72 °C for 15 s or 90 °C for 15 s. The treatments were particularly effective at inhibiting *Lactobacillus* at both temperatures. Heating at 117 °C for 2 s with 150 IU/mL nisin was even more effective at inhibiting bacterial growth. Such milks have been successfully stored for over 150 days at 30 °C with only very low levels of spoilage (Wirjantoro *et al.*, 2001). Local regulations would need to be checked to establish whether nisin is a permitted additive in milk and milk-based beverages. It is worth noting that nisin is not effective against Gram-negative contaminants, such as pseudomonads (Phillips *et al.*, 1983; Wirjantoro *et al.*, 2001). This is very significant since most of the PPC in thermally produced ESL milks have been shown to be Gram-positive. This is discussed in Chapter 4. By contrast, several Gram-negative bacteria have been isolated from MF-ESL milks (Schmidt *et al.*, 2012).

3.3.5 ESL Milk by Thermal Treatment Plus a Non-Thermal Technology Treatment

The use of aseptic packaging after pasteurisation discussed in Section 3.3.1.1 eliminates post processing microbial contamination. Another approach to extending the shelf-life of pasteurised milk is to subject the heated milk to a non-thermal technology to inactivate remaining bacteria. This contrasts with the use of microfiltration and bactofugation which remove bacteria. The approach has been shown to be effective using three different technologies: ultra-violet (UV) irradiation, pulsed electric field (PEF) technology and gamma-irradiation (see also Chapter 10).

3.3.5.1 UV irradiation

The Panel on Dietetic Products, Nutrition and Allergies of the European Food Safety Association has concluded that UV-treated pasteurised milk is safe for the general public except for infants less than 1-year old (EFSA, 2016). Data submitted to the panel stated that the shelf-life of pasteurised milk was increased from 12 to 21 days by the UV treatment. A further advantage of the UV treatment was that the vitamin D3 level was increased in fat-containing milk (to 0.5–3.2 μg/100 g in whole milk and to 0.1–1.5 μg/100 g in semi-skimmed milk) through conversion of 7-dehydrocholesterol (more details on UV irradiation are provided in Section 10.7).

3.3.5.2 Pulsed Electric Field (PEF) Technology

Using PEF treatment (35 kV/cm, total treatment time 11.5 μs at 65 °C) of milk immediately after HTST pasteurisation, Sepulveda *et al.* (2005) obtained an extension of shelf-life of 60 days at 4 °C. When the PEF treatment was performed on the pasteurised milk 8 days after storage at 4 °C, a shelf-life of 78 days was achieved (more details on PEF are provided in Section 10.4).

3.3.5.3 Gamma-Irradiation

Gamma-irradiation has not been used for the commercial processing of milk, largely because of the adverse effects of irradiation on the flavour of milk, particularly at the high doses, >10 kGy, necessary to effect sterilisation. However, Sadoun *et al.* (1991) reported that pasteurised milk irradiated at room temperature with a dose of 0.25 kGy had an extended shelf-life without a change in organoleptic properties. The bacterial count after 18 days was $<10^4$ cfu/mL, whereas the count in a control milk after 10 days was $>10^7$ cfu/mL. More recently, de Oliveira Silva *et al.* (2015) irradiated raw whole milk at 1, 2 and 3 kGy and found that during storage at 4 °C for 60 days, the bacterial counts remained constant at around 2.5, 1.5 and 1.2 cfu/mL for the 1, 2 and 3 kGy doses, respectively. The results suggest that irradiation at 1 or even 2 kGy causes little adverse change in sensory properties and could technically be used to produce milk with an extended refrigerated shelf-life of ≥60 days. Consumer opposition to food irradiation is likely to prevent its commercial use on milk for some time (more details on gamma irradiation are provided in Section 10.8).

3.3.6 ESL Milk by Multiple Thermal Treatments

Another approach which has been investigated but not commercialised is tyndallisation. This involves successive heat treatments in order to inactivate spores. According to Wilbey (2002), Tyndall in 1877 suggested that if a medium was heated at 100 °C for 3 min on three successive days, first the vegetative cells would be killed and the spores would germinate and then be killed on either the second or third days. In practice, such triple heat treatments are rarely used, and the process suffers from the unpredictability of the spore germination process.

Mesquita *et al.* (1998) investigated three tyndallisation processes on goat's milk sealed in glass bottles: immersion in boiling water for 30, 45 and 60 min on three successive days with ambient storage between treatments. Milk treated for 30, 45 and 60 min showed 4, 1 and 0%, respectively, "cumulative microbial defects" during storage at room temperature for up to 90 days. When the milks were initially inoculated with 20,000 *Bacillus subtilis* spores/mL, "cumulative microbial defects" occurred in 59% and 46% of the 30 and 45 min samples, respectively, but none occurred in the 60 min samples. All samples had a strong heated flavour with the flavour scores being lowest for the 60 min samples. This work demonstrates that a tyndallisation process can be effective but is unlikely to be commercially acceptable because of the long treatment and storage times, and the unfavourable heated flavour generated. A three-stage process was also developed for sterilisation of cheese whey (Ghaly & El-Taweel, 1995). This involved heating the whey to 70 °C for 45 min, cooling in an ice-bath for 30 min, warming to room temperature, holding at that temperature for 24 h to allow the spores to germinate, and then repeating the process two more times.

A variation of the tyndallisation process is a double heat treatment process. Brown *et al.* (1979) studied the effect of two heat treatments of 80 °C for 10 min with various inter-treatment periods from 1 to 24 h at 30 °C on reconstituted skim milk inoculated with either a strain of *B. cereus*, a strain of *B. subtilus* or a mixture of natural spore-forming bacteria from pasteurised milk. There was little effect on the spore counts and the total count increased, especially for the samples held for 24 h between heat treatments. Possibly, any benefit was negated by the inter-treatment incubation at 30 °C during which time the germinated spores from the first heat activation and Gram-positive thermoduric vegetative bacteria such as coryneforms would have grown. Furthermore, heating at 80 °C for 10 min does not inactivate all non-spore-forming bacteria which could multiply during storage at 30 °C.

Guirguis *et al.* (1983) found that *Bacillus* spores were optimally activated at 115 °C for 1 s and suggested a modified tyndallisation process after the work of Brown *et al.* (1979) to extend the shelf-life of fresh dairy products. The process involves an initial heat treatment of 115 °C for 1 s followed by an HTST heat treatment after allowing a period of germination. Stewart (1975) reported that "double pasteurisation" of cream with an inter-heating interval of 12-18 h during which the cream was held cold, was effective in extending the shelf-life of the cream. This was attributed to activation and germination of spores during the first heat treatment and their inactivation during the second. Clearly, further investigation of multiple heat treatment processes is required before they could be adopted commercially for ESL products.

3.4 Sterilisation

3.4.1 Introduction

Sterilisation of milk, which allows it to be kept without refrigeration for several months, became a commercial proposition in 1894. Milk can either be sterilised in bottles or other sealed containers, or by continuous UHT processing followed by aseptic packaging, which is the main topic of this book. Very good accounts of the procedures for producing in-container sterilised milk, and problems associated with it have been provided by Cronshaw (1947), Davis (1955) and IDF (1981).

Sterilisation processes aim to destroy all bacteria but this is not possible because of the logarithmic nature of the destruction (see Chapter 4). Hence even in a correctly managed sterilisation operation, a small number of bacterial cells will always remain; however, they are unlikely to multiply under the normal conditions of storage. For this reason UHT and in-container sterilised products are referred to as "commercially sterile".

From a safety standpoint, the main concern is inactivation of the most heat-resistant pathogenic spore, namely *Cl. botulinum*. Since milk is a low-acid food (pH >4.5), the main criterion is to achieve a 12-log reduction of *Cl. botulinum*. This occurs when a product is heated at 121 °C for 3 min, at its slowest heating point (the so-called botulinum cook). The microbial severity of an in-container sterilisation process is traditionally expressed in terms of its F_0 value. This is evaluated by taking into account the contributions of the heating, holding and cooling periods to the total lethality. It is expressed in terms of time (min) at 121 °C and provides a useful means of comparing

Table 3.2 F_0 values that have been successfully used for canned products on the UK market.

Product	Can size(s)	F_0 values
Baby foods	Baby food	3–5
Beans in tomato sauce	All	4–6
Peas in brine	Up to A2	6
	A2-A10	6–8
Carrots	All	3–4
Green beans in brine	Up to A2	4–6
	A2-A10	6–8
Celery	A2	3–4
Mushrooms in brine	A1	8–10
Mushrooms in butter	Up to A1	6–8
Meats in gravy	All	12–15
Sliced meat in gravy	Ovals	10
Meat pies	Tapered, flat	10
Sausages in fat	Up to 1 lb	4–6
Frankfurters in brine	Up to 16Z	3–4
Curries, meats, and vegetables	Up to 16Z	8–12
Poultry and game, whole in brine	A2.5–A10	15–18
Chicken fillets in jelly	Up to 16 oz	6–10
"Sterile" ham	1, 2 lb	3–4
Herrings in tomato	Ovals	6–8
Meat soups	Up to 16Z	10
Tomato soup, not cream of	All	3
Cream soups	A1–16Z	4–5
	Up to A10	6–10
Milk puddings	Up to 16Z	4–10
Cream	4–6 oz	3–4

(from Brennan *et al.* (1976), with permission)

different sterilisation procedures. The minimum F_0 value for any low-acid food should be 3. Some recommended values for products on the UK market are shown in Table 3.2.

Cl. botulinum is not associated with or normally found in raw milk. However, other Clostridia are ubiquitous members of the dairy ecosystem and a relatively small proportion may be transmitted to milk; these should not be confused with *Cl. botulinum*. Nevertheless, total counts of Clostridia in raw milk are usually low (Lewis, 2013). More common spore-formers are *Bacillus* and *Bacillus*-like species of which some, such as *Geobacillus stearothermophilus* (formerly *B. stearothermophilus*) and *Bacillus sporothermodurans* (Hammer *et al.*, 1996) form highly heat-resistant spores, which are not destroyed by a process with an F_0 of 3. These bacteria may cause spoilage, but they are not pathogenic. Thus, a minimum "botulinum cook" will produce a product which is safe, but not necessarily sterile. For foods which may contain highly heat-resistant spores, a heat treatment achieving two or more decimal reductions is recommended, corresponding to an F_0 value of 8. The aim should be to reduce the spoilage rate to less than 1 in every 10,000 containers. Further information on spore-formers is given in Chapter 4.

3.4.2 UHT Processing

3.4.2.1 Introduction

Processing of milks at high temperature in a continuous-flow heater was first developed in 1893 when milk was indirectly heated at 125 °C for 6 min. This was followed by the invention of direct steam heating in 1912 to obtain temperatures of 130-140 °C with a shorter holding time and to avoid the burn-on encountered with the indirect system (Hostettler, 1972). However, UHT processing did not become practical before the 1940s and early 1950s when it was combined with aseptic filling into pre-sterilised cans. This process, although used commercially, was expensive and led to the development of aseptic packaging of UHT processed milk into paperboard cartons. This commenced in Switzerland in 1961 (Robertson, 2013). In principle, modern UHT processing is similar to that used in the 1960s but with refinements in the heating technologies and associated process controls, and the development of packaging alternatives such as plastic bottles and pouches (see Chapter 5).

UHT processing of milk combined with aseptic packaging was introduced to produce a shelf-stable product with much less chemical change compared with in-container sterilised milk. UHT milk may have a shelf-life of up to 12 months, although in practice it is usually consumed much earlier than this. In countries where it commands a small segment of the liquid milk market, it is often used as a stand-by product, to be used when pasteurised milk is not available. In other countries it is the major type of milk available and it is used regularly. In the former situation, UHT milk may need to be stable over a long period of time (6-12 months), whereas in the latter case, the desired shelf life may be $\leq$3 months.

UHT treatment is normally in the range 138 to 145 °C in combination with appropriate holding times (1-10 s) necessary to achieve "commercial sterility" (Burton, 1988; Lewis & Heppell, 2000). In practice, the products are checked for sterility by incubating at 55 °C for 7 days and at 30 °C for 15 days, and testing for bacterial growth after incubation (see Chapter 8).

There are two basic principles of UHT processing which distinguish it from in-container sterilisation.

3.4.2.2 UHT Principle 1

For the same bacterial destruction, a high-temperature–short-time treatment results in less chemical change than a low-temperature–long-time treatment.

This is due to the different Q_{10} values for chemical reactions and bacterial destruction discussed in Section 3.2.1. For the treatment temperatures used in sterilisation treatments the only relevant bactericidal effect is the sporicidal effect. Table 3.3 illustrates the change in relative chemical effect with temperature, for the same sporicidal effect, for treatment temperatures between 115 and 150 °C which encompass the conditions

Table 3.3 Relative chemical effect at various temperatures for the same sporicidal effect (based on Q_{10} for thermophilic spore destruction of 10 and for chemical change of 2.5).

Temperature (°C)	Time of heating for same sporicidal effect	Chemical change for same sporicidal effect
115	1	100
125	0.1	25
135	0.01	6.25
145	0.001	1.56
150	0.0001	0.26

for both in-container sterilisation and UHT processing. The figures are based on typical Q_{10} values for chemical reactions of 2.5 and for thermophilic bacterial spore inactivation of 10. They show that the extent of chemical change at 150 °C is less than 1% of that at 115 °C. Given this quite dramatic effect, the question arises as to the optimum combination of temperature and time. A partial answer to this lies partly in the second principle of UHT processing below (Section 3.4.3.2).

This first principle is very significant for milk as it explains why UHT milk has less-intense heated flavour and brown discolouration than in-container sterilised milk. It also explains why UHT milk produced by a direct heating process which rapidly heats milk to the sterilisation temperature, holds it for a short time and then rapidly cools it, has less "cooked" flavour than milk produced by an indirect process which slowly heats the product up to and cools it down from the sterilisation temperature (Datta *et al.*, 2002) (see Chapter 5). It also explains why heating milk rapidly to a very high temperature (150-180 °C) and holding it for a very short time, even <1 second (Van Asselt *et al.*, 2008), produces a sterile product with little flavour impairment. It is also the principle used in Section 3.3.1 in determining optimum conditions for ESL processing.

3.4.2.3 UHT Principle 2

Minimum times and temperatures are dictated by the need to inactivate thermophilic bacterial spores while the maximum times and temperatures are determined by the need to minimize undesirable chemical alterations.

3.4.2.3.1 The UHT Zone

This is illustrated graphically in Figure 3.2 which shows a shaded area containing feasible temperature–time combinations for UHT processing. It is bounded by a lower line corresponding to a 9-log reduction of thermophilic spores (the $B^* = 1$ line) and an upper line representing 3% reduction in thiamine (the $C^* = 1$ line).

The specification of a heat treatment with a $B^* = 1$ as the desired lower limit is now generally accepted and has replaced a previous specification for a less-intense treatment aimed at a 9-log reduction of mesophilic spores. In some countries, the lower limit for UHT processing is defined in terms of the lethality index, F_0, more commonly used for canned foods; an F_0 of 3 min is considered approximately equal to a B^* of 1.

The discovery of extremely heat-resistant mesophilic sporeformers, such as *B. sporothermodurans*, in UHT milk (IDF, 2000), prompted the recommendation of a more

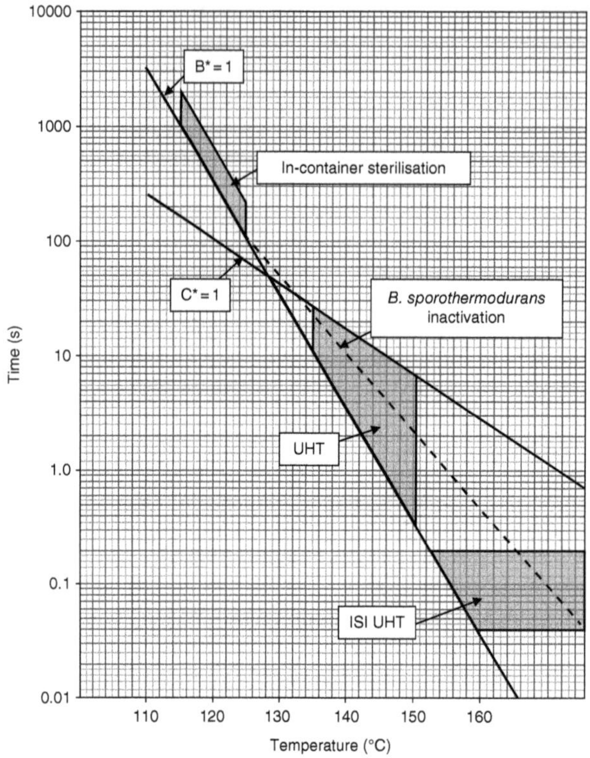

Figure 3.2 Temperature–time combinations for sterilisation of milk. (These combinations are for the holding conditions only. The "C* = 1" line represents 3% destruction of thiamine, the "B* = 1" line represents 9-log reduction of thermophilic spores and the dotted line represents 9-log reduction of *B. sporothermodurans*. ISI = Innovative Steam Injection).

severe lower heat treatment limit. This is illustrated in Figure 3.2 by a dotted line above the B* = 1 line representing a 9-log reduction of such highly heat-resistant spores. Most commercial UHT plants operate at a B* >1 in a region above this dotted line (see Section 3.2.2).

The desirable upper limit of temperature–time combinations for UHT processing of milk is determined by the maximum acceptable amount of chemical change to the components of milk. In Figure 3.2, this is defined as a heat treatment with a chemical index, C*, of 1.

The boundary at the left of the UHT shaded area is at temperatures below which the holding time necessary to ensure adequate sterilisation is practically excessive. Most UHT processors would not use temperatures lower than 135 °C. At this temperature, the reference temperature for B* calculations, a holding time of 10.1 seconds is required to achieve a B* of 1 (Kessler, 1981). It is instructive to note that some regulations set the lower limit for UHT sterilisation conditions at 135 °C for 1 s; this is equivalent to a B* of ~0.1 which is clearly inadequate to produce a safe shelf-stable product.

The boundary to the right of the UHT zone is arbitarily set here at 155 °C as heating to temperatures higher require the capability of the equipment to have a very short (but measurable for validation purposes) holding time to prevent excessive chemical damage. Most commercial UHT plants do not currently have this capability, the exception being some direct plants configured to have a very short holding time (<1 s). In practice, use of such high temperatures rarely occurs, although future developments are likely to be in this direction in order to provide high B* values with low C* values. An example of a very high temperature technology is the Innovative Steam Injection (ISI) process in which the product is rapidly heated to as high as 180 °C for ≤0.2 s (Van Asselt *et al.*, 2008). This is shown in Figure 3.2 as the ISI zone.

A feature of UHT processing is the range of temperature–time profiles that can be used. These profiles include heat-up and cool-down sections whose shapes vary considerably; however, one feature which is common to all UHT plants is a high-temperature holding section, a tube through which the product flows at the highest temperature reached in the plant, for a nominal time of 2-25 s (Tran *et al.*, 2008), commonly about 4-6 s. Although the sections of the plant immediately before and after the holding tube can contribute substantially to both B* and C*, particularly in indirect plants, validation of UHT plants is often based entirely on the holding tube conditions. This is the basis of regulatory requirements in some countries, for example, the lower limit of 135 °C for 1 s, mentioned above, refers to the conditions of the holding tube only.

While the holding temperatures and times are widely cited to describe UHT processes, the heat-up and cool-down parts of the process can and often do make major contributions to both B* and C*. The exceptions to this are systems involving very rapid heating up to, and cooling down from, the sterilisation temperature, for example, direct steam heating systems (see Chapter 5); for these systems, the heat-up and cool-down sections contribute little to either B* or C*. In contrast, in indirect heating systems, the contribution of the heat input into the product during the slower heat-up to and cool-down sections is substantial. Any heat input at temperatures over ~75 °C and ~90 °C affects C* and B* values, respectively. For 22 commercial UHT plants (17 indirect, 5 direct), Tran *et al.* (2008) calculated that the heat-up and cool-down sections of the indirect and direct plants contributed an average of 53% and 17%, respectively, to the overall B* of the plants, and 76 and 39%, respectively, to the overall C* values. These data show how much B* and C* are underestimated by considering only the holding temperatures and times in UHT plants, particularly indirect plants. Major reasons for not taking the heat-up and cool-down sections into consideration are, firstly, that the detailed temperature–time profile of the whole plant is seldom known and secondly, when the profile is known, it is not easy to compute the contributions to B* and C* from these sections. However, this can be facilitated by computer programs such as Excel (Browning *et al.*, 2001; Tran *et al.*, 2008). Matlab (Dinca, 2004) and commercially available programs such as *NIZO Premia*® (Smit *et al.*, 2001), when the temperature–time profile of a plant is available. For new installations, the equipment manufacturer is able to supply the profile and the relevant B*(or F_0) values and a chemical heat index such as C* or lactulose values (see Chapter 6 for discussion on chemical heat indices). For older plants, detailed physical measurements of the temperatures and times for each section of the plant need to be made and a temperature–time profile developed as described by Tran *et al.* (2008). Plants are often modified after they are installed by adding or subtracting sections of piping and/or altering temperatures. Therefore, before calculations are performed using a temperature–profile it is advisable to check that it represents the current configuration.

3.4.3 In-Container Sterilisation

3.4.3.1 Conventional Retort Processes

Foods have been sterilised in sealed containers for over 200 years. Milk was originally sterilised in glass bottles sealed with a crown cork but plastic bottles have been used more recently (Ashton & Romney, 1981). Milk sterilisation really developed after 1930 with the advent of the crown cork, which helped with the mechanisation of the bottle filling process, and the reuse of bottles. In general, the basic principles have remained the same.

The main aim is to inactivate heat-resistant spores, thereby producing a product which is "commercially sterile", with an extended shelf-life. Some drawbacks of in-container sterilisation processes are that the product heats and cools relatively slowly, and that temperatures are limited by the internal pressure generated. However, many dairy products are still produced in this way worldwide, including sterilised milk, flavoured milk, evaporated milk, and canned desserts, such as custards, rice puddings and other cereal desserts.

The manufacturing procedure used today for sterilised milk and milk products is similar to that used over 50 years ago. Milk is clarified using a centrifuge. A specialised centrifuge, for example, a Bactofuge®, is especially useful, as it not only removes foreign matter but also removes greater than 99% (2-log reduction) of the spores. The milk is pre-heated using similar equipment to that used for pasteurisation. It is then homogenised at 63–82 °C, for example, at a single stage pressure of about 20 MPa or double stage at about 17 and 3.5 MPa. It is then filled into glass bottles between 74 and 80 °C under conditions which give minimal frothing, and sealed using a crown cork. Plastic bottles are sealed at a lower temperature of 54-55 °C. Care should be taken to avoid conditions in balance tanks which may be conducive to growth of thermophiles. Ashton and Romney (1981) cite sterilisation processing conditions of 110 to 116 °C for 20-30 min, depending upon the extent of cooked flavour and colour preferred by the consumer. Batch or continuous retorting processes may be used (Davis, 1955). Other processing details are outlined by Ashton and Romney (1981); these include more detail on continuous retorts, such as hydrostatic or rotary valve sealed sterilisers, which are capable of higher temperatures and shorter times (132–140 °C for 12 min), and the use of steam for glass bottles or steam/air mixtures for plastic bottles.

In general, the heating conditions for in-container sterilisation are in the range of temperature–time combinations from ~115 °C for 2000 s to 125 °C for 200 s as shown by the designated shaded area in Figure 3.2. An F_0 of 3, which is equivalent to 3 min at 121.1 °C (121 °C is used in this book), is the lower limit for sterilisation and forms the lower boundary of the sterilisation zone. It represents a 12-log reduction for *Cl. botulinum*.

Excessive sediment and coagulation of sterilised milk is rarely reported, so heat stability does not appear to be a problem. It is noteworthy that gelation of sterilised milk during storage is seldom reported, which is in contrast to the occurrence of "age gelation" in UHT milk (see Section 7.2.2). Chen *et al.* (2015) studied the heat stability of sterilised milk by measuring sediment formation as well as the effects of small additions of stabiliser and calcium chloride (see Section 6.2.1). All the control samples showed good heat stability, whereas small additions of the stabilisers disodium hydrogen phosphate (DSHP) and trisodium citrate (TSC) resulted in more sediment being formed and the product being noticeably browner.

Davis (1955) pointed out that raw milk to be used for sterilisation should not be contaminated with bacterial spores. Today, this still remains an important control variable. Sweet curdling was the chief bacterial fault in sterilised milk, due to highly heat-resistant spores of *B. subtilis* and *B. cereus*. Bacterial growth was found to produce other taints, such as carbolic, bad (e.g. oxidised) or cardboard taints. Ashton and Romney (1981) reported that the failure level of well produced sterilised milks is of the order of 1 in 1000 units, although it may be higher in situations where there are large numbers of thermotolerant spores in the raw material or other contamination arising in the process. The microbiology of sterilised milk is discussed in more detail in Section 4.4.4.

A recognised test for ensuring adequate sterilisation is the turbidity test, developed by Aschaffenburg (1950) (see Section 11.2.33). This test measures undenatured whey protein. Although not directly measuring microbial activity, it was based on the assumption that if whey protein was completely denatured, this would indicate that microbial activity was also eliminated and thus the milk was adequately sterilised. This could be likened to the use of the phosphatase test for ensuring milk is adequately pasteurised. Heat treatment regulations in some countries may require sterilised milk to show a negative turbidity result, that is, it should be heated at such a temperature for such a time as to fully denature the whey proteins. In principle, UHT heating of milk should result in some undenatured whey protein, but milk produced by UHT processes with extensive heating and cooling profiles also give a negative turbidity test result. In fact, most UHT milk samples in the UK show a negative result. Thus, the turbidity test is not useful for distinguishing UHT milk from sterilised milk. Methods for distinguishing between sterilised and UHT milk have been discussed in detail by Burton (1988). This is not easy because of the wide range of temperature–time conditions that are in use for both processes. There is also no pressing need to distinguish between the two processes, as there was in the early days of UHT processing. Measurement of lactulose and furosine which is useful for distinguishing between UHT and in-container sterilised milk is discussed in Chapter 6.

Some sterilised milk is made by a combination heating process. This involves the production of milk under UHT conditions, for example, 137 °C for 4 s, which is then filled into bottles which are then sealed and passed through a conventional retorting process. Usually the retorting period is much reduced, but it should be sufficient to ensure a negative turbidity result. In terms of determining the sterilisation effect for the whole process, the critical point is to ensure that the milk does not become recontaminated in the intermediate filling process, especially with bacterial spores. This process was found to reduce the incidence of spoilage caused by spore survivors (Ashton & Romney, 1981).

Sterilised milk has a rich creamy appearance, perhaps helped by Maillard browning components, a distinct cooked flavour (described variously as rich, nutty, caramelised), which, once acquired, makes other heat-treated products taste insipid. It is considerably browner than raw milk and UHT milk, the extent of browning depending upon the severity of the heat treatment. The Maillard browning reaction is affected by pH and hence some of the variation that is found will be influenced by the pH of the milk, as it may vary considerably (see Section 7.2.5). Without doubt, this reaction contributes to product quality in terms of its colour and flavour, although not everybody finds sterilised milk to be as palatable as pasteurised or UHT milk.

Sterilised milk is still produced in quantity in some countries, with much of it now packaged in retortable disposable plastic bottles with a metal foil cap, rather than in returnable glass bottles. To minimize light-induced and oxidative changes in the milk during storage, the bottles should not allow the milk to be affected by light or allow transmission of oxygen into the milk.

3.4.3.2 Alternative Retort Processes

As indicated above, heat transfer rates in retorts can be improved in continuous retorts, such as hydrostatic and rotary retorts. It can also be improved by agitating the product during the sterilisation procedure. An example of a process based on this principle is the Shaka® retort process (Walden & Emanuel, 2010) which the manufacturers claim "blurs the lines between UHT/aseptic packaging and retort sterilisation". According to Jacob and Walden (2013), the retort agitates the packaged food using a 15 cm double-stroke reciprocating movement at speeds > 100 double strokes/min. The intense horizontal back and forth movement is equivalent to as high as 2-3 g of force. The agitation creates a more uniform heat distribution within the package than with conventional retorts and as a consequence sterilises food 5-10 times faster, depending on the package and food. Faster processing through this technology is claimed to improve the taste, texture, appearance, and nutritional quality of shelf-stable food, as a result of the faster heating and cooling rates.

Any type of food which is not too viscous and which will move within the container when agitated will sterilise quickly under this process. Soups, sauces, ready meals, dips, spreads, baby food and pet foods are all possible candidates. A Shaka® overpressure retort equipped with a suitable retort basket to contain the packages can sterilise cans, pouches, cartons, glass jars and bottles, small foodservice containers, and plastic cups, trays, and bottles. As mentioned, agitation is necessary for products such as rice pudding and custard to obtain a uniform consistency throughout.

Temperature—time profiles of a conventional retort and a Shaka® retort are shown in Figure 3.3. These have been drawn to approximate those shown in an equipment

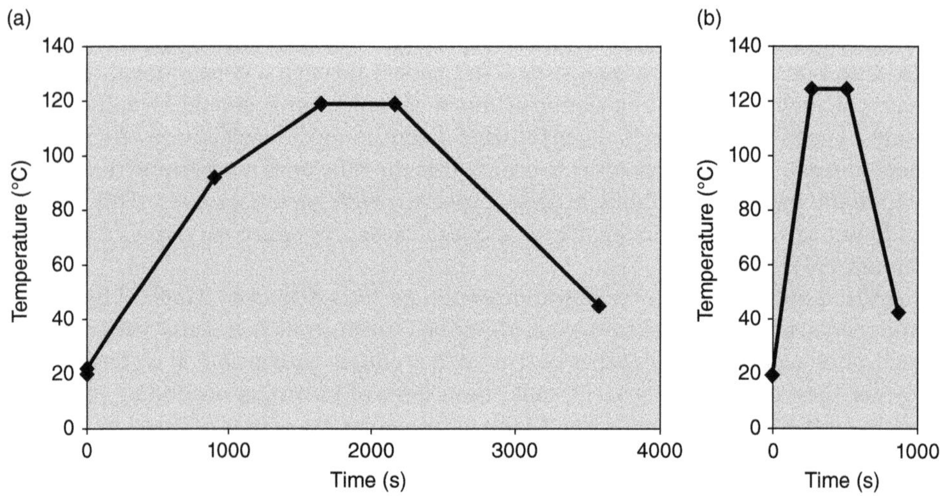

Figure 3.3 Temperature–time profiles of (a) a conventional retort and (b) an alternative, for example, Shaka®, retort.

brochure (http://www.steriflow.com/en/solutions/shaka) to illustrate the differences between the two processes. Calculations based on an Excel program (Browning *et al.*, 2001; Tran *et al.*, 2008) using these temperature–time profiles show that the Shaka® retort has a F_0 of 11.1, B* of 3.0 and a C* of 5.3. In comparison, the conventional retort has a F_0 of 7.3, B* of 2.1 and a C* 11.3 (the indices F_0, B* and C* are discussed in Sections 3.2.2 and 3.2.3). Therefore the Shaka® retort, although having higher bacteriological indices, F_0 and B*, has a much lower C* indicating it would cause considerably less change in colour, flavour and nutrient value than the conventional retort; this is consistent with the manufacturers' claims.

References

Aschaffenburg, R. (1950) A simple test for sterilised milk. *Journal of the Society of Dairy Technology* **3**, 236–237.

Ashton, T.R. & Romney, A.J.D. (1981) In-container sterilisation. In: *Factors affecting the Keeping Quality of Heat Treated Milk*, IDF Document 130. pp. 55–70. International Dairy Federation, Brussels. pp. 55–70.

Bergere, J.-L. & Cerf, O. (1992) Heat resistance of *Bacillus cereus* spores. In: *Bacillus cereus in Milk and Milk Products.* IDF Bulletin 275, pp. 23–25. International Dairy Federation, Brussels.

Blake, M.R., Weimer, B.C., McMahon, D.J. & Savello, P.A. (1995) Sensory and microbial quality of milk processed for extended shelf life by direct steam injection. *Journal of Food Protection* **58**, 1007–1013.

Brennan, J.G., Butters, J.R., Cowell, N.D. & Lilly, A.E.V. (1976) *Food Engineering Operations.* 2nd edn, Applied Science Publishers Ltd., London.

Brown, J.V., Wiles, R. & Prentice, G.A. (1979) The effect of a modified tyndallization process upon the spore-forming bacteria of milk and cream. *Journal of Society of Dairy Technology* **32**, 109–112.

Brown, J.V., Ranjith, H.M.P. & Prentice, G.A. (1984) Comparative shelf-lives of skimmed, semi-skimmed and whole milks. *Journal of the Society of Dairy Technology* **37**(4), 132–135.

Browning, E., Lewis, M. & MacDougall, D. (2001) Predicting safety and quality parameters for UHT-processed milks. *International Journal of Dairy Technology* **54**, 111–120.

Buckenhuskes, H.J. (2015) ESL milk production. DLG-Expert Report 4/4-14. (http://www.dlg.org/fileadmin/downloads/food/Expertenwissen/Ernaehrung/e_2014_4_Expertenwissen_ESL.pdf).

Burton, H. (1988) *Ultra High Temperature Processing of Milk and Milk Products.* Elsevier Applied Science Publishers, New York.

Chen, B.Y., Grandison, A.S. & Lewis, M.J. (2015) Effect of seasonal variation on some physical properties and heat stability of milk subjected to ultra-high temperature and in-container sterilisation. *Food Chemistry* **181**, 227–234.

Cronshaw, H.B. (1947) *Dairy Information*, Dairy Industries Ltd, London.

Datta, N., Elliott, A.J., Perkins M.L. & Deeth, H.C. (2002) Ultra-High Temperature (UHT) treatment of milk: Comparison of direct and indirect modes of heating. *Australian Journal of Dairy Technology* **57**, 211–227.

Davis, J.G. (1955) *A Dictionary of Dairying*, 2nd edn. Leonard Hill, London.

de Oliveira Silva, A. C., de Oliveira, L. A. T., De Jesus, E. F. O., Cortez, M. A. S., Alves, C. C. C., Monteiro, M. L. G. & Conte, C. A. (2015) Effect of gamma irradiation on the bacteriological and sensory analysis of raw whole milk under refrigeration. *Journal of Food Processing and Preservation* **39**(6), 2404–2411.

Dinca, E. (2004) *Modelling the Effects on Milk of High-Temperature Heat and Subsequent Storage*. M.Phil thesis, University of Queensland, Brisbane, Australia.

EFSA (2016) Safety of UV-treated milk as a novel food pursuant to Regulation (EC) No 258/97. Report of Panel on Dietetic Products, Nutrition and Allergies. *EFSA Journal* **14**(1), 1–14.

EU (2005) Commission Regulation (EC) No. 2074/2005.

Fredsted L.B., Rysstad G. & Eie T. (1996) Pure-Lac™: The new milk with protected freshness and extended shelf life. In: *Heat Treatments and Alternative Methods*. IDF Document 9602, pp. 104–125. International Dairy Federation, Brussels.

Gaafar, A.M.M. (1987) *Investigation into the Cooked Flavour in Heat-Treated Milk*. PhD thesis, University of Reading, Reading, UK.

Ghaly, A.E. & El-Taweel, A.A. (1995) Effect of lactose concentration on batch-production of ethanol from cheese whey using *Candida pseudotropicalis*. *Transactions of the ASAE (American Society of Agricultural Engineers)* **38**(4), 1113–1120.

Guirguis, A.H., Griffiths, M.W. & Muir, D.D. (1983) Spore-forming bacteria in milk .1. Optimization of heat-treatment for activation of spores of *Bacillus* species. *Milchwissenschaft* **38**(11), 641–644.

Hammer, P., Lembke, F., Suhren, G. & Heeschen, W. (1996) Characterisation of heat resistant mesophilic *Bacillus* species affecting the quality of UHT milk, *Heat Treatments and Alternative Methods*, pp. 9–16. IDF Document SI 9602, International Dairy Federation, Brussels.

Henyon, D.K. (1999) Extended shelf-life milks in North America: a perspective. *International Journal of Dairy Technology* **52**, 95–101.

Hostettler, H. (1972) History of the development of UHT processes. In: *IDF Monograph on UHT milk*. IDF Document 68, pp.169–174. International Dairy Federation, Brussels.

Huijs, G., Van Asselt, A. J., Verdurmen, R. E. M. & De Jong, P. (2004) High speed milk. *Dairy Industries International* **69**, 30–32.

IDF (1981) *New Monograph on UHT Milk*. IDF Document 133, International Dairy Federation, Brussels.

IDF (2000) *Bacillus sporothermodurans* - a *Bacillus* forming highly heat-resistant spores. IDF Bulletin 357, pp. 3–27. International Dairy Federation, Brussels.

Jacob. G. & Walden, R. (2013) Shaka retorts shake up the shelf-stable food paradigm. http://www.retorts.com/white-papers/shaka-retorts-shake-up-the-shelf-stable-food-paradigm/ (Accessed August 2015).

Kaufmann, V. & Kulozik U. (2008) Processing factors influencing quality and stability of extended shelf life (ESL) milk. *DMZ, Lebensmittelindustrie und Milchwirtschaft* **129**, 28–30.

Kaufmann, V., Scherer S. & Kulozik U. (2010) Procedures for the prolongation of the storability of consumable milk and its material changes: ESL milk. *Journal Fur Verbraucherschutz Und Lebensmittelsicherheit-Journal of Consumer Protection and Food Safety* **5**(1), 59–64.

Kessler, H.G. (1981) *Food Engineering and Dairy Technology*, Verlag A Kessler, Freising, Germany.

Lewis, M. (2013) The mythical *Clostridium botulinum*. *Food Management Today* Nov/Dec, 12-17.
Lewis, M.J. & Heppell, N. (2000) *Continuous Thermal Processing of Foods: Pasteurization and UHT Sterilisation*. Aspen Publishers, Gaithersburg, MD.
Lorenzen, P.C., Clawin-Raedecker I., Einhoff K., Hammer P., Hartmann R., Hoffmann W., Martin D., Molkentin J., Walte H.G. & Devrese M. (2011) A survey of the quality of extended shelf life (ESL) milk in relation to HTST and UHT milk. *International Journal of Dairy Technology* **64**, 166–178.
Lyster, R.L.J. (1970) The denaturation of α-lactalbumin and β-lactoglobulin in heated milk. *Journal of Dairy Research* **37**, 233–243.
Manji, B. (2000) Regulatory perspective related to extended shelf life products. In: *Proceedings of the Conference on Extended Shelf Life Milk*, Dresden, September, 2000, International Dairy Federation, Brussels.
Mayer, H.K., Raba, B., Meier, J. & Schmid, A. (2009) Extended shelf life (ESL) milk - milk with excessive heat load? *European Dairy Magazine* **7**, 18–22.
Mayer, H.K., Raba, B., Meier, J. & Schmid, A. (2010) RP-HPLC analysis of furosine and acid-soluble β-lactoglobulin to assess the heat load of extended shelf life milk samples in Austria. *Dairy Science and Technololgy* **90**, 413–428.
Mayr, R., Gutser, K., Busse, M. & Seiler, H. (2004a) Gram-positive non-spore-forming bacteria are frequent spoilage organisms of German retail ESL (Extended Shelf Life) milk. *Milchwissenschaft* **59**, 262–266.
Mayr, R., Gutser, K., Busse, M. & Seiler, H. (2004b) Indigenous aerobic spore-formers in high heat treated (127 °C, 5 s) German ESL (Extended Shelf Life) milk. *Milchwissenschaft* **59**, 143–146.
Mesquita, J., Teixeira, M. & Brandao, S. (1998) Tindalization of goats' milk in glass bottles. *Journal of Dairy Science* **81**(Suppl 1), Abs 80.
Mugadza D. & Buys E. (2014) Diversity of spore former and non-spore former bacteria in Extended Shelf Life (ESL) milk. South African Society of Dairy Technology 2014 Dairy Symposium. http://www.sasdt.co.za/wp-content/uploads/2015/04/Session-8-Speaker-26-Mugadza1.pdf (Accessed August 2015).
Newstead D.F., Paterson, G., Anema, S. G., Coker, C.J. & Wewala, A.R. (2006) Plasmin activity in direct-steam-injection UHT-processed reconstituted milk: effects of preheat treatment. *International Dairy Journal* **16**, 573–579.
Phillips, J.D., Griffiths, M.W. & Muir, D.D. (1983) Effect of nisin on the shelf-life of pasteurized double cream. *Journal of the Society of Dairy Technology* **36**, 17–21.
Ranjith, H.M.P. (2000) High temperature pasteurization. In: *Proceedings of the Conference on Extended Shelf Life Milk*, Dresden, September, 2000, International Dairy Federation, Brussels.
Robertson, G.L. (2013) Aseptic packaging of foods. In: *Food Packaging: Principles and Practice*. 3rd edn, pp. 367–382. CRC Press, Boca Raton.
Rysstad, G. & Kolstad, J. (2006) Extended shelf life milk - advances in technology. *International Journal of Dairy Technology* **59**, 85–96.
Sadoun, D., Couvercelle, C., Strasser, A., Egler, A. & Hasselmann, C. (1991) Low-dose irradiation of liquid milk. *Milchwissenschaft* **46**(5), 295–299.
Schmidt, D., Cromie, S. J. & Dommett, T.W. (1989) Effects of pasteurisation and storage conditions on the shelf life and sensory quality of aseptically packaged milk. *Australian Journal of Dairy Technology* **44**, 19–24.

Schmidt, V.S., Kaufmann,V., Kulozik, U., Scherer, S. & Wenning, M. (2012) Microbial biodiversity, quality and shelf life of microfiltered and pasteurized extended shelf life (ESL) milk from Germany, Austria and Switzerland. *International Journal of Food Microbiology* **154**(1-2), 1–9.

Sepulveda, D.R., Gongora-Nieto, M.M., Guerrero, J.A. & Barbosa-Canovas, G.V. (2005) Production of extended-shelf life milk by processing pasteurized milk with pulsed electric fields. *Journal of Food Engineering* **67**(1-2), 81–86.

Smit, F., De Jong, P., Straatsma, J. & Verschueren, M. (2001) [NIZO Premia as knowledge management tool for industry] *Voedingsmiddelentechnologie* **34**, 23–26.

Stewart, D.B. (1975) Factors influencing the incidence of *B. cereus* spores in milk. *Journal of the Society of Dairy Technology* **28**(2), 80–85.

Tran, H., Datta, N., Lewis, M.J. & Deeth, H.C. (2008) Processing parameters and predicted product properties of industrial UHT milk processing plants in Australia. *International Dairy Journal* **18**, 939–944.

van Asselt, A. J., Sweere, A. P. J., Rollema, H. S. & de Jong, P. (2008) Extreme high-temperature treatment of milk with respect to plasmin inactivation. *International Dairy Journal* **18**, 531–538.

Walden, R. & Emanuel J. (2010) Developments in in-container retort technology: the Zinetec Shaka® process. In: *Case Studies in Novel Food Processing Technologies: Innovations in Processing, Packaging, and Predictive Modelling* **197**, 389–406.

Wilbey, R.A. (2002) Microbiology of cream and butter. In: *Dairy Microbiology Handbook – The Microbiology of Milk and Milk Products*, 3rd edn (ed. R.K. Robinson) pp. 123–174, John Wiley and Sons, New York.

Wirjantoro, T.I. & Lewis, M.J. (1996) Effect of nisin and high temperature pasteurisation on the shelf-life of whole milk. *Journal Society Dairy Technology* **4**, 99–102.

Wirjantoro, T.I., Lewis, M.J., Grandison, A.S., Williams, G.C. & Delves-Broughton, J. (2001) The effect of nisin on the keeping quality of reduced heat treated (RHT) milks, *Journal Food Protection* **64**, 213–219.

4

Microbiological Aspects

4.1 Introduction

Milk with its high water content, abundant nutrients and a pH of 6.6–6.8 is an excellent medium for growth of micro-organisms. Energy is available from the major components, lactose, milk fat and protein, as well as minor components such as citrate and non-protein nitrogen compounds including peptides, amino acids, ammonia and urea. The micro-organisms in milk can be pathogenic or spoilage organisms. The former are of most concern from a public health perspective and are the major reason why milks are heat treated before consumption. However, spoilage organisms in milk are a major focus for the dairy industry as they cause deterioration during storage and reduce the shelf-life of the product. Spoilage is caused by the action of enzymes such as proteases, peptidases, lipases, glycosidases and oxidases produced by the micro-organisms which degrade the protein, fat and lactose to generate compounds suitable for growth of the organisms. These changes result in production of compounds which cause off-flavours and changes in the physical characteristics of the milk (Frank & Hanssan, 2003).

4.2 Bacteria in Raw Milk

Raw milk initially has a predominance of mesophilic Gram-positive lactic acid bacteria and, if the milk is not refrigerated, these bacteria will readily sour the milk. However, these organisms do not grow or grow very slowly under refrigeration conditions and so at the time of processing the raw milk usually contains mostly Gram-negative psychrotrophic micro-organisms, which have grown during refrigerated storage. They are predominantly *Pseudomonas* species but other genera such as *Flavobacterium, Achromobacter, Aeromonas, Alcaligenes* and *Chromobacterium* may also be present (Muir, 1996a). These are non-spore-forming bacteria and are destroyed by heat treatments such as high-temperature, short-time (HTST) pasteurisation. Milk also contains thermoduric bacteria which survive pasteurisation; these include non-spore-formers such as coryneforms, micrococci and some streptococci, as well as spore-formers such as *Bacillus, Geobacillus, Paenibacillus* and *Clostridium* species. Thermoduric non-spore-forming bacteria grow slowly at low temperature but can contribute to spoilage of heat-processed milk. Ultra high-temperature (UHT) and in-container sterilisation processes are designed to destroy all non-spore-forming bacteria and most spore-forming organisms with only some of the most heat-resistant *Bacillus* and

High Temperature Processing of Milk and Milk Products, First Edition. Hilton C. Deeth and Michael J. Lewis.

Geobacillus species surviving UHT treatments. *Clostridium* species, which are the major target of food sterilisation processes such as canning because of their pathogenic nature, are less heat-resistant than *Bacillus* species and do not survive UHT and in-container sterilisation treatments.

As discussed in Chapter 2, raw milk may also contain low levels of a range of pathogenic bacteria and the primary purpose of pasteurisation is to inactivate these bacteria and render the milk safe for human consumption. They include *Staphylococcus aureus, Campylobacter jejuni, Salmonella* spp, *Escherichia coli* including *E. coli* O157:H7, *Yerisinia enterocolitica, Listeria monocytogenes, Coxiella burnetii* and *Mycobacterium tuberculosis.* Recently, however, the presence of another potential pathogen, *Mycobacterium avium* ssp. *paratuberculosis* (MAP), has raised some concern. It is the causative agent of Johne's disease in animals and some reports have indicated that, when present in milk, it may survive pasteurization and cause Crohn's disease in humans (Greenstein, 2003). The situation regarding MAP was reviewed by Griffiths (2006); an overview is included here.

MAP levels found in raw milk appear to be low, but there is no real indication of true levels because of the decontamination procedures used during its analysis to remove the other bacteria in raw milk and its extremely slow growth rate in the laboratory. MAP levels in milk subjected to pasteurisation, if present, are low but there are many inconsistencies in the experimental results (Grant *et al.*, 2001; Hammer *et al.*, 1998).

Using the holder process (63 °C/30 min), most investigators found some survivors after pasteurisation, but the level of inoculum was much higher than would be found in raw milk. The D_{63} values quoted were 2.7–2.9 min, which would give a high level of inactivation (>10 log reductions) during batch pasteurization, and would provide a more than adequate process. Most other results have suggested that the holder process was not as efficient as this. Tails were also found in the survivor curves, which implied the presence of a more heat-resistant sub-population, though this could be an artifact. Results from HTST studies are also inconsistent and suggest great variability in the heat resistance data. Sung and Collins (1998) suggested a D_{71} value of 11.67 s and a z-value of 7.11 °C for pooled data of clinical strains of MAP. A review by the UK Food Standards Agency (2002), reported that values of D_{72} for MAP of 12 to 14 s have been widely cited within the industry. According to this, normal HTST conditions would only achieve about a 1.3-log reduction, which would mean that all samples inoculated with 100 cfu/mL MAP would contain surviving MAP after pasteurisation. However, results from milks inoculated with 10^7 and 10^4 cfu/mL indicated that about 20% and 40% of samples contained no viable MAP after HTST treatment, which suggested at least 7- and 4-log reductions, respectively, in these samples. More recently, Foddai *et al.* (2010) reported that the mean D_{63} and D_{72} values for four MAP strains were 81.8, 9.8, and 4.2 s, respectively, yielding a mean z-value of 6.9 °C. Complete inactivation of 10^6 to 10^7 cfu/mL of MAP in milk was not observed for any of the time–temperature combinations studied; 5.2- to 6.6-$\log_{10}$ reductions in numbers were achieved depending on the temperature and time.

It has also been suggested that MAP inactivation is not temperature-sensitive. However, experiments at 75, 78, 80, 85 and 90 °C showed that the survival rates appeared to be higher after heat treatments at 80 °C than at 75 °C and 78 °C. At first sight this is unexpected but it could demonstrate that MAP is inhibited by an active lactoperoxidase system, which would be inactivated at 80 °C. This apparent lack of temperature dependence is unusual in a bacterium and is worthy of further investigation, as is any protective effect that may be conferred by the lactoperoxidase system (Marks *et al.*, 2001).

Results from surveys on raw milks and pasteurised milks are also inconclusive in that MAP was found in 2% of both raw and pasteurised milk samples tested (Food Standards Agency, 2002). This again would suggest that pasteurisation is having no significant effect. Although the heat resistance data generated to date for MAP is inconclusive, Foddai *et al.* (2010) using a modified phage amplification technique have shown that a significant number of decimal reductions of MAP can be achieved by pasteurisation. This methodology allowed results to be obtained in 24 h, compared to up to 60 days using conventional plating techniques.

Information on MAP has been published by the IDF (1998, 2001). In the UK it has been recommended that HTST pasteurization conditions should be increased to from 72 °C for 15 s to 72 °C for 25 s as part of a strategy for controlling MAP in cow's milk (Food Standards Agency, 2002). Hickey (2009) pointed out that while this recommendation has been widely adopted by the UK industry, and supported by many retailers, it is a recommendation that is voluntary and is not a legal requirement for HTST heat treatment, which remains at 72 °C for 15 s. In a short time, some of the rapid and sensitive new methods will determine the wisdom of this advice.

In circumstances where there is no refrigeration, the keeping quality of raw milk can be extended by stimulating the natural milk lactoperoxidase system (FAO, 2005). It has the ability to exert a bacteriostatic effect on spoilage organisms in milk including both Gram-negative and Gram-positive (Haddadin *et al.*, 1996). The requirements for the lactoperoxidase system to be active, apart from the enzyme itself, are a source of hydrogen peroxide and an electron donor such as thiocyanate or iodide. The standard electron donor is thiocyanate. Note that both thiocycnate and iodide are present in small amounts in raw milk. Sachets are available which can be added to milk to provide the necessary reagents. The results of some trials performed by Fweja *et al.* (2007) in Tanzania are shown in Table 4.1. They show that activating the lactoperoxidase system extends the shelf-life of raw milk at the three different storage temperatures of 4, 19 and 30 °C. Using iodide as an activator was more effective than thiocyanate. We are not aware of any research in which LPS-activated raw milk has been used for UHT or ESL

Table 4.1 Effect of activation of the lactoperoxidase system on the shelf-life of raw milk.

		Shelf-life (h)				
Trial	**Temp. (°C)**	**Raw milk**	**10 ppm SCN^-; 80 ppm H_2O_2**	**10 ppm SCN^-; 80 ppm percarbonate**	**10 ppm I^-; 80 ppm H_2O_2**	**10 ppm I^-; 80 ppm percarbonate**
1	4	192	384	300	360	444
	19	25	40	35	55	76
	30	18	25	21	43	50
2	4	264	432	480	576	768
	19	28	48	48	75	100
	30	7	24	22	31	50

Data from Fweja *et al.*, (2007).

milk and whether such activation confers any positive or negative effects on the quality of the processed milk.

Two types of bacteria in raw milk which are of most concern in regard to the keeping quality of high-temperature processed milks such as Extended Shelf-Life (ESL) and UHT-treated milk are the psychrotrophic bacteria, chiefly pseudomonads, and *Bacillus* and *Bacillus*-like spore-formers. The former are of concern because they produce heat-resistant enzymes, such as proteases and lipases, during growth in the raw milk prior to heat treatment. These enzymes can remain active in milk after UHT treatment and cause breakdown of proteins and lipids in the UHT milk during storage (Muir, 1996b). These enzymes are of concern because of their heat resistance.

4.2.1 Non-Spore-Forming Psychrotrophic Bacteria and their Heat-Resistant Enzymes

Pseudomonas species, the major non-spore-forming psychrotrophic bacteria in raw and pasteurised milk, are motile, Gram-negative rods, which can grow at refrigeration temperatures although their optimum growth temperatures are between 25 and 30 °C (Muir, 1996a). They have short generation times; for some species isolated from raw milk they were shown to be 5.5 to 10.5 h at 3 to 5 °C (Suhren, 1989). About 50% of *Pseudomonas* species are of the fluorescent type, characterised by the production of a diffusible pigment (pyoverdin) during growth (Muir, 1996a).

The psychrotrophic bacteria vary considerably in their propensity to elaborate proteases and lipases and hence the metabolic processes of these bacteria are more important than their total numbers (Cousin, 1980; Haryani *et al.,* 2003). Species of *Pseudomonas,* particularly *Ps. fluorescens* and *Ps. fragi,* are the most important producers of heat- resistant enzymes (Shelley *et al.*, 1987; Champagne *et al.* 1994; Muir, 1996a). Once produced in raw milk, the heat-resistant proteases and lipases can remain in milk after UHT processing and cause proteolysis and lipolysis, respectively, during storage of the UHT milk. Up to 50% of activity of these enzymes can remain after UHT treatment (Marshall, 1996; Muir, 1996c). Given that UHT milk is stored at room temperature, often for several months, even traces of these enzymes can cause significant spoilage. In fact it has been estimated that for UHT milk to be stable for at least 4 months at room temperature, it should contain less than ~0.3 ng/mL of protease (Mitchell & Ewings, 1985). Button *et al.* (2011) showed that as little as 0.0003% of a cell-free supernatant of a *Ps. fluorescens* culture (grown to ~10^6 cells/mL in sterile milk) when added aseptically to UHT milk could cause detectable proteolysis during storage at room temperature.

Therefore the microbial quality of raw milk used for UHT processing is very important for the quality of the final product. As a rule of thumb, raw milk destined for UHT processing should have a total count of less than 10^6 cfu/mL, preferably less than 10^5 cfu/mL. However, the total bacterial count is not always a good guide to the risk of contamination from extracellular enzymes as different bacteria have different propensities to produce these enzymes. Significant quantities of enzymes can be produced in milk with bacterial counts as low as 10^5/mL while some milks with counts of 10^7/mL do not contain significant quantities (Haryani *et al.,* 2003).

The requirement for good quality raw milk also applies to the manufacture of ESL milk. Excessive levels of bacterial enzymes may remain in the milk after the heat treatment which is not as intense as UHT treatment. However, ESL milks are stored under refrigeration and for a much shorter time than UHT milk and hence the risk of bitter or rancid flavours resulting from heat-resistant bacterial proteases and lipases, respectively, is less than in UHT milk which is stored at ambient temperature.

According to Teh *et al.* (2014), enzymes produced by bacteria in biofilms attached to the surface of raw milk tankers can also affect the quality of UHT milk. They demonstrated that more proteolysis occurred in UHT milk made from raw milk which had been in contact with the milk tanker biofilms than in control milks which had not been exposed to the biofilms.

4.2.2 Spore-Forming Bacteria

4.2.2.1 Non-Pathogenic Spore-Formers

The spore-forming bacteria, particularly *Bacillus* and *Bacillus*-like species, are the major targets of UHT and in-container sterilisation heat treatments. They are Gram-positive, motile, spore-forming, rod-shaped organisms. Their vegetative forms are comparatively easily destroyed by heat but their spores are heat-resistant, with some being able to withstand sterilisation treatments. Spore-formers can be categorised as psychrotrophic, mesophilic or thermophilic according to the temperatures at which they can grow.

Psychrotrophic spore-formers are able to grow at ≤7 °C (Meer *et al.*, 1991) although their optimum temperature for growth is 20-30 °C. These are of particular relevance in refrigerated milk such as ESL. They include strains of *B. circulans, B. cereus* and *Paenibacillus* species. Meer *et al.* (1993) isolated 12 different psychrotrophic *Bacillus* species from milk, including *B. licheniformis* which is generally considered to mesophilic or thermophilic. Their generation and lag times at refrigeration temperatures, 2–7 °C, are longer than those of *Pseudomonas* species (McKellar, 1989). In general, the heat stability of the spores of psychrotrophic spore-formers is lower than that of mesophilic and thermoduric spore-formers discussed below, although there are exceptions, for example, some *Paenibacillus* species and some strains of *B. cereus*. Furthermore, these organisms can be important in cold climates (Blake *et al.*, 1995); this is particularly significant for ESL milk (see Section 4.4.2). Psychrotrophic *Bacillus* species produce proteinases, lipases and phospholipases which have comparable heat stabilities to the corresponding enzymes from *Pseudomonas* species (Meer *et al.*, 1991; Sorhaug & Stepaniak, 1997).

Paenibacillus, which was formerly part of the genus *Bacillus,* has emerged in recent years as an organism of concern to the dairy industry. It has been found in silage and feed concentrates and may enter the milk from these sources. It has been reported to be a major spoilage bacterium in pasteurized milk in the USA (Fromm & Boor, 2004; Ranieri & Boor, 2009) and can cause bitterness due to its production of proteases (Martin *et al.*, 2011). Scheldeman *et al.* (2004) reported *Paenibacillus* spores in UHT milk, suggesting its resistance to UHT processing. Curiously, *Paenibacillus* species have a broad range of growth temperatures from ~5 to 55 °C. Their growth in pasteurised milk is evidence of their psychrotolerance but Scheldeman *et al.* (2004) reported they could grow at temperatures as high as 55 °C. The optimum growth temperatures of *Paenibacillus* species range from 28 to 42 °C (Bosshard *et al.*, 2002). Therefore, *Paenibacillus* can be psychrotrophic, mesophilic or thermophilic and produce very heat-resistant spores. This is an unusual combination of properties and an unfortunate one for the dairy industry as it indicates some strains of these organisms can survive all common milk heat treatments and grow at both refrigeration and ambient temperatures.

The spore-forming organisms which have attracted most attention in UHT processing from a spoilage point of view are the thermophilic and mesophilic *Bacillus* and *Bacillus*-like bacteria. The thermophilic spore-formers of interest are the obligate thermophiles such as *Geobacillus*, chiefly *G. stearothermophilus*, and some *Bacillis* species, whereas the mesophiles of interest are *Bacillus* species, chiefly *B. subtilis*, *B. sporothermodurans* and *B. cereus*. Some *Bacillus* species are facultative thermophiles and are able to grow at both thermophilic and mesophilic temperatures. These include *B. licheniformis*, *B. coagulans*. *B pumilus* and *B. subtilis* (Intaraphan, 2001; Scheldeman *et al.*, 2005; Burgess *et al.*, 2014). The thermophiles that produce highly heat-resistant spores, for example, *G. stearothermophilus*, have optimum growth temperatures of >50 °C and hence will not grow in UHT milk unless it is severely temperature abused. On the other hand, mesophiles have optimum growth temperatures of 20 to 40 °C (Vasavada & Cousin, 1993) and hence can grow at room temperature and cause spoilage. Some mesophilic spore-formers, notably *B. sporothermodurans*, produce highly heat-resistant spores and pose a significant problem for UHT processors.

Spore counts in raw milk rarely exceed 10^3 cfu/mL. Low levels, up to 10^2 cfu/mL, are common in milk from grazing cows, such as those Australia and New Zealand, but higher in milk from cows that are housed (Scheldemann *et al.*, 2005). Hence they can be higher in winter than in summer where cows are housed during winter. Cook and Sandeman (2000) surveyed the spores in bulk milk from farms in two regions of Australia and found mean mesophilic spore counts of 7 cfu/mL and 73 cfu/mL for the two regions. The authors commented that these counts were lower than those in four other reports which were 10-1132 cfu/mL, 400-760 cfu/mL, 52-404 cfu/mL and (a year-round mean of) 240 cfu/mL. However, Bramley and McKinnon (1990) reported that spore counts can reach 5 x 10^3 cfu/mL and McGuiggan *et al.* (2002) reported even higher levels, with mesophilic spore counts averaging 7.6 x 10^3/mL, with occasional counts of over 2.4 x 10^5/mL; pyschrotrophic spore counts were very low, with a maximum of 3.5 cfu/mL and thermophilic spores were slightly higher with a maximum of 54 cfu/mL. Spores are mainly derived from surfaces of teats through contact with soil, dust, mud, mature, water and bedding materials (Cook & Sandemans, 2000). Feeding silage also leads to higher spore levels.

Spore-forming bacteria in raw milk are predominantly *Bacillus* and *Bacillus*-like species (Jay, 1996). The most common *Bacillus* spores isolated from teat surfaces are *B. licheniformis*, *B. subtilis* and *B. pumilis* with lower numbers of *B. cereus*, *B. firmus* and *B. circulans*. Of the *Bacillus* species, *B. licheniformis*, *B. cereus*, *B. subtilis* and *B. megaterium* are the most commonly isolated from raw milk (Shehata *et al.*, 1983). Other spore-formers in raw milk are *Geobacillus*, chiefly *G. stearothermophilus*, and *Paenibacillus*. Very heat-resistant spores, such as *G. stearothermophilus* and *B. sporothermodurans* are usually only a small proportion of the total.

Bacteria which produce highly heat-resistant spores are attracting increasing attention due to the increased international trade in UHT milk with the associated transport across tropical climatic zones and the likely storage of such milks at high ambient temperatures. It appears that these bacteria are continuing to be detected and identified and this is expected to continue. The discoveries of *B. sporothermodurans* (Huemer *et al.*, 1998; Scheldeman *et al.*, 2006), *Anoxybacillus flavithermus* (see Section 4.4.3.2) (Rueckert *et al.*, 2004), *Paenibacillus* spp. (Scheldeman *et al.* (2004), and *B. thermoamylovorans* and *B. thermolactis* (Coorevits *et al.*, 2011) are good examples. A survey of

Table 4.2a Summary of highly heat-resistant sporeformers[1] isolated from raw milk (Scheldeman *et al.*, 2005).

Spore-former	% of isolates from raw milk
Aneuribacillus	1.2
Bacillus barbaricus	1.2
B. fortii	1.2
B. licheniformis	22.3
B. pallidus	15.1
B. smithii	1.2
B. subtilis	1.2
Other *Bacillus*	9.6
Brevibacillus	10.8
Paenibacillus	10.2
Virgibacillus	9.0
Ureibacillus	6.6

Table 4.2b Summary of the major highly heat-resistant sporeformers[1] isolated from raw milk and dairy farm environments (Scheldeman *et al.*, 2005).

>10% of total isolates	>4% of total isolates
Bacillus pallidus	*B. subtilis* group
B. licheniforms	*B. farraginis*
	Brevibacillus agri
	Other *Brevibacillus*
	Geobacillus spp.
	B. smithii

[1] Defined as surviving heat treatment of 100 °C for 30 min

raw milk and dairy farm environments in Belgium indicates a wide range of highly heat-resistant spore-formers (defined as surviving heat treatment of 100 °C for 30 min) which have potential for spoilage of UHT milk (Scheldeman *et al.*, 2005). A summary of these organisms isolated from raw milk is shown in Table 4.2a and a summary of the main organisms isolated from animal feed and milking equipment is given in Table 4.2b.

4.2.2.2 Pathogenic Spore-Formers

From a public health perspective, *Cl. botulinum* and *B. cereus* are considered the most pertinent microbial hazards in UHT products (Pujol *et al.*, 2015). *Clostridium* species are often present in raw milk at low levels. However, in some countries they are at higher levels in winter because cows are housed and the milk can become contaminated from spore-containing bedding materials and silage (Ridgway, 1958; Bramley & McKinnon,

1990). Under normal circumstances, *Cl. botulinum* in milk and dairy products poses little risk to human health as it has a low probability of being present and would seldom encounter suitable conditions (e.g., anaerobic) for growth. Only two outbreaks of botulism attributable to consumption of dairy products, mostly cheese, have been reported in recent times (Lindström *et al.*, 2010). Most *Clostridium* spores are less heat-resistant than *Bacillus* spores and are highly unlikely to survive the heating conditions used in UHT processing which are designed to destroy heat-resistant *Bacillus* spores. However, the thermophile *Cl. thermosaccharolyticum* has been isolated from in-container sterilised milk, indicating its thermal resistance (Langevel, 1967).

4.2.2.2.1 *Bacillus cereus*

B. cereus is a concern because some strains are pathogenic, producing diarrhoeagenic or emetic toxins, and hence deserves special consideration. It is a widespread contaminant of milk and milk products, including infant formula (Becker *et al.*, 1994; Larsen & Jørgensen, 1999). It also occurs in many other foods such as rice which are of significance to milk products (Becker *et al.*, 1994). It has some closely related species; in fact, the *B. cereus* group comprises seven species: *B. cereus, B. anthracis, B. thuringiensis, B. mycoides, B. pseudomycoides, B. weihenstephanensis* and *B. cytotoxicus* which are difficult to distinguish phenotypically (Markland *et al.*, 2013). It can therefore be assumed that in most cases where "*B. cereus*" has been isolated from milk and milk-based products, it could include other members of the *B. cereus* group. Using 16S rRNA gene sequence data of dairy isolates, Ivy *et al.* (2012) identified *B. cereus, B. weihenstephanensis,* and *B. mycoides* from the *B. cereus* group.

Several surveys of the incidence of this organism in raw and pasteurised milk and milk-based products have been conducted with the percentage of *B. cereus*-positive samples ranging from very low to 100% (van Netten *et al.*, 1990; Champagne *et al.*, 1994; Frank, 1997). Its incidence in raw and pasteurised milk is commonly 20 to 60%. Griffiths and Phillips (1990) reported that three studies found it to be the main psychrotrophic spore-former in raw milk and others have found it to be the dominant psychrotrophic spore-former in pasteurised milk (e.g., Franklin, 1969; Coghill and Juffs, 1979). In general the counts of *B. cereus* in raw milk are low, <100/mL (Ahmed *et al.*, 1983). Becker *et al.* (1994) examined 261 samples of infant formulae from 17 countries and found 54% contained *B. cereus* with counts of <1 to 600 cfu/g; most had counts of <100 cfu/g. However, Larsen and Jørgensen (1997) found 56% of pasteurised milk samples and 29% of raw milk samples that had been held at 7 °C for 8 days contained higher counts of *B. cereus*, in the range 10 to 3 x 10^5 cfu/mL; 45% of the positive samples exceeded 10^3 cfu/mL.

Like other spore-formers, *B. cereus* can enter milk from numerous sources. On farm, the main sources are water, cows' udders and teat surfaces, dust, soil and milkstone deposits on farm bulk tanks and pumps, pipelines, gaskets and processing equipment in factories (Stewart, 1975; Meer *et al.*, 1991). Stewart (1975) stated that the cleaning and sterilizing systems used on equipment are not very effective in eliminating spores and may even activate spores of *B. cereus.* In the factory, fouling deposits and stainless steel surfaces to which the spores can readily attach (Simmonds *et al.*, 2003) are the main sources (Te Giffel *et al.*, 1996; Salustiano *et al.*, 2009). Franklin (1970) reported a case of a very heat-resistant *B. cereus* spore which contaminated UHT cream (processed at 140 °C for 2 s) being traced to an upstream homogeniser. When attached to stainless steel, spores show enhanced resistance to cleaning solutions (Te Giffel *et al.*, 1997) and enhanced heat resistance; Simmonds *et al.* (2003) found an average increase of 205% in D_{90} values of three *B. cereus* strains when attached to stainless steel compared with those of planktonic cells.

Several authors have referred to a seasonal effect on the incidence of aerobic spores, and *B. cereus* spores in particular, in milk. However, the reports are inconsistent. Where cows are housed during the winter, high spore counts are often encountered and attributed to contamination from bedding and fodder. However, Stewart (1975) and Bartoszewiez *et al.* (2008) observed higher counts during spring and summer. Stewart (1975) considered this may be due to the greater amount of dust during summer. Slaghuis *et al.* (1997) reported that the milk from cows grazing pasture during summer contained more *B. cereus* than milk from cows that were housed and fed conserved feed; this suggests that pasture may be another reservoir of *B. cereus* spores.

The major interest in *B. cereus* is due to the production of enterotoxins by several strains and its potential to cause two types of illness, diarrhoeal and emetic syndromes. The two types of toxins are very different. The diarrhoeagenic toxins are proteins with molecular weights in the range 38,000-46,000. They are produced by actively growing cells and are thermolabile, being inactivated by heating at 56 °C for 30 min. Two of three forms of the diarrhoeagenic toxins are believed to cause food poisoning in humans. In contrast, the emetic toxin is a cyclic peptide, cereulide, with a molecular weight of 1,200, which is extremely resistant to heat, surviving heating at 126 °C for 90 min. The diarrhoeagenic toxins are believed to be produced during vegetative growth of *B. cereus* in the small intestine while there is evidence that the emetic toxin is produced by *B. cereus* growing in the food before it is consumed (Granum & Lund, 1997). Psychrotrophic strains of *B cereus* growing at low temperature do not produce the emetic toxin but may produce the diarrhoeagenic toxins, albeit slowly and in low concentration (EFSA, 2005).

Despite its widespread presence in milk and milk-based products, *B. cereus* has been implicated in very few cases of illness; however, one outbreak in The Netherlands in which pasteurised milk was implicated involved 280 patients (van Netten *et al.*, 1990). Several authors have sounded a warning of the potential for this organism to cause disease (Te Giffel *et al.*, 1997; Notermans *et al.*, 1997) although it has been suggested that toxin-producing strains of *B. cereus* in milk and milk products are unlikely to cause food poisoning because their production of toxin, even at high counts, is very low (Agata *et al.*, 2002; Jooste & Anelich, 2008). Another reason why *B. cereus* rarely causes food poisoning is because it produces an intense bitter flavour making the contaminated products organoleptically unacceptable before they become toxic (Lewis, 1999).

Te Giffel *et al.* (1997) tested 37 isolates from pasteurised milk and found that, based on PCR, 28 produced enterotoxin. Strains which fermented lactose produced more enterotoxin than strains which did not. Van Netten *et al.* (1990) found 25% of psychrotrophic *B. cereus* isolates from pasteurised milk were enterotoxin positive. Of 38 *B. cereus* isolates recovered from milk-based infant formulae, Rowan and Anderson (1998) found 1 (2.6%), 2 (5.3%) and 6 (15.8%) grew well at 4, 6 and 8 °C, respectively, and were enterotoxigenic. However, enterotoxin was only detected when the count of *B. cereus* had reached 2.3×10^5 cfu/mL. Notermans *et al.* (1997) stated that $\geq 10^5$ cfu/mL of toxigenic *B. cereus* in pasteurised milk is generally considered to be a health hazard. They estimated that such numbers could be present in 7% of milk in the Netherlands at the time of consumption. However, epidemiological evidence does not indicate that *B. cereus* in milk causes diseases to anywhere near this extent and hence the dose–response relationship needs to be revisited. Since the growth rate and enterotoxin production of *B. cereus* is low at 4 °C, several authors have concluded that milk or milk-based products stored at or below this temperature present a very low risk of becoming toxic (Rowan & Anderson, 1998; EFSA, 2005), provided products are not stored for unduly long periods of time, for example, >20 days (Juffs & Deeth, 2007). Since ESL milk is designed to have a shelf-life of at least 30 days, it is possible for *B. cereus* to reach high counts by the end of its shelf-life if it is not destroyed by the heat process or the ESL milk is not packaged aseptically and is contaminated with the organism after the heat treatment. This is a good reason for producing ESL milk at high temperatures for short times such as ≥134 °C for ≤4 s (see Chapter 3).

B. cereus can grow at a range of temperatures but the optimum growth temperature is generally in the range 30-37 °C. The maximum temperature for most strains is 45-50 °C. However, some strains are capable of growing at low temperatures; these are of most concern for products such as pasteurised and ESL milks. They are termed psychrotrophic (able to grow at 7 °C) or psychrotolerant (able to grow at 4 °C but not at 43 °C). The species name, *B. weihenstephanensis*, has been used for this sub-group of *B. cereus* (Markland *et al.*, 2013). In surveys, the percentage of strains capable of growing at 7 °C have varied. For example, Te Giffel *et al.* (1995) isolated 766 *B. cereus* strains from farm environments and raw milk and found the percentage of isolates capable of growing at 7 °C was 40% and 30% respectively. Similarly in a survey of pasteurised milk samples, Te Giffel *et al.* (1997) found that 53% of 106 isolates tested were psychrotrophic. However, in a survey of milk from a fluid milk processing plant and a milk powder plant, these authors found only 6% of isolates from the first and none from the second plant were psychrotrophic. An interesting phenomenon observed by Mayr *et al.* (2004b) was that *B. cereus* (and three other spore-formers) grew at 8 °C after culturing at 30 °C but had not previously grown in milk at 10 °C.

Some strains are capable of growing at temperatures <7 °C but both the number of strains capable of growing at these temperatures and their growth rates decrease considerably as the temperature is decreased (Coghill & Juffs, 1979). For example, Rowan and Anderson (1998) reported that of 38 psychrotrophic *B. cereus* isolates from milk-based infant formulae, 1, 4 and 16 isolates showed growth after 15 days at 4, 6 and 8 °C, respectively. Dufeu and Leesement (1974) reported average generation times for four strains to be 1.3 h at 30 °C, 9.1 h at 8 °C and 54 h at 3 °C. This compares favourably with 9.4 to 75 h (average 8.2 h) at 7 °C reported by Dufrenne *et al.* (1995). In general it has been shown that storage of milk and milk products at ≤4 °C would severely limit the risk of growth of *B. cereus* and, as indicated above, would also prevent enterotoxin production (EFSA, 2005). This was demonstrated by Odumeru *et al.* (1997) who found 43 of 112 (38.4%) pasteurized milk samples that had been incubated at 10 °C until their expiry dates contained *Bacillus* diarrhoeal enterotoxin but none of the 112 samples had detectable enterotoxin when incubated at 4 °C.

As well as being potentially pathogenic, *B. cereus* can cause substantial spoilage. It produces protease which causes sweet curdling (Overcast & Atmaram, 1974) and bitterness. It also produces phospholipase C (sometimes referred to as lecithinase) which degrades phospholipids of the milk fat globule membrane and causes fat globule coalescence, or chemical churning, resulting in defects such as bitty cream (Stone & Rowlands, 1952). The phospholipase C from *B. cereus* is specific for glycerophospholipids, with a preference for phosphatidyl choline which largely resides in the outside (aqueous side) of the milk fat globule membrane (Deeth, 1997). The bitty cream problem largely disappeared when homogenisation was introduced to evenly disperse the fat throughout the milk.

Pujol *et al.* (2015) developed a probabilistic exposure assessment model to estimate product failure rate of UHT milk associated with *Cl. botulinum, B. cereus* and *G. stearothermophilus.* From the model they determined the highest Sterility Failure Rate (SFR) would be due to *G. stearothermophilus* and the lowest to *Cl. botulinum.* They attributed the high SFR due to *G. stearothermophilus* to its high heat stability and ability to survive the UHT heat treatment. By contrast, the SFR due to *B. cereus* was attributed to the possibility of airborne recontamination of the aseptic tank (49%) and to post-sterilisation packaging contamination (33%). Interestingly, the ability to survive the UHT heat treatment was not a major consideration for *B. cereus*; as indicated above this is generally, but not entirely, valid.

Because of the food safety aspects of *B. cereus* discussed above, it is instructive to consider the heat resistance of its spores; this is particularly relevant to ESL milk (see Section 4.4.2). While the spores of many strains are not very heat-resistant, there is actually a wide range of heat resistance amongst the spores of *B. cereus* isolates with some strains being highly heat resistant. For example, Mikolajcik (1970) reported that *B. cereus* (and *B. licheniformis*) produced the most heat-resistant spores in milk, and Intaraphan (2001) identified *B. cereus* (with *B. licheniformis* and *G. stearothermophilus*) among heat-resistant isolates from milk and farm environments selected on the basis of surviving a heat treatment of 100 °C for 30 min. Franklin (1970) and Vyletelova *et al.* (2002) showed that some *B. cereus* spores survive UHT treatment. The range of heat resistance is illustrated by the following D-values at 90 to 121 °C of *B. cereus* spores reported in 21 different studies and summarised by Bergere and Cerf (1992):

- $D_{90°C}$ = 3.6-10.8 min;
- $D_{95°C}$ = 0.5-20.2 min;
- $D_{100°C}$ = 0.6-27 min;
- $D_{105°C}$ = 11.2 min;
- $D_{110°C}$ = 11.5 min;
- $D_{121°C}$ = 0.03-2.35 min; and
- $D_{135°C}$ = >4 min

A report by Dufrenne *et al.* (1995) gave the range of D_{90}-values for spores of 11 *B. cereus* isolates as 2.2-9.2 min but one other strain had a D_{90}-value of >100 min. The z-values ranged from 6.7 to 13.8 °C, with most in the range 8-11 °C, in the Bergere and Cerf (1992) report and 8.5 °C in Dufrenne *et al.* (1995). Van Asselt and Zweitering (2006) collated 465 datapoints from 12 publications and determined the mean D_{120}-values of *B. cereus* spores to be 0.041 min and the upper 95% prediction interval D_{120}-value to be 0.52 min; the z-value was 12.9 °C. These data again show the range of heat resistance of these spores. Further data are given in Table 4.3; unfortunately, specific data for the spores of psychrotrophic *B. cereus* do not appear to be available.

In about half of the studies reviewed by Bergere and Cerf (1992), the heat inactivation curves were not linear; two showed shoulders and 10 showed tails. The importance of this is exemplified in a very heat-resistant *B. cereus* isolate from UHT cream which originated from an upstream homogeniser; although the majority of the spores were destroyed at 95-100 °C, a resistant fraction of ~1 in 10^5-10^6 survived heating at 135 °C for 4 h (Franklin, 1970). This small resistant fraction was apparently sufficient to cause contamination of the UHT cream.

Table 4.3 D-values and z-values for some spores relevant to high-temperature heat treatment of milk and milk products.

	D-value (s)								
Bacteria	**94-95 °C**	**99-100 °C**	**103-105 °C**	**110 °C**	**115 °C**	**120-121 °C**	**135 °C**	**z-value (°C)**	**Reference**
B. licheniformis	121-708	67-304	31-202	16-110	7-97	2-28	0-11		1
		171-246						6.4-7.8	2
B. subtilis	107-196	58-128	17-70	6-44	3-23	0-14	0-0		1
						0.001-139		6.5-32.1	4
B. cereus	107-166	58-138	16-45	4-25	2-11	0-5	0-0		1
	30-1212	162-186	672	690		1.8-141	>2400		2
	81	36-1520	4-7					6.7- 13.8	7
		13-21						8.5	8
B. sphaericus	200-414	79-119	19-43	8-11	4-5	0-4	0-0		1
	264	95							2
B. megaterium	286-328	117-137	44-66	20-26	5-17	0-3	0-0		1
	318	130						9.4	2
B. coagulans	238	76	25	10	4	0	0		1
	414	118							2
						18-100		6.0-35.9	4
B. circulans								11.5	2
B. pumilus	216	50	16	4	0	0	0		1
	242	53							2
	84							9.7	3
G. stearothermophilus[a]						5-562		5.6-8.8	4
						191			
							0.9	9.1	5
B. sporothermodurans						150	3.4-7.9[b]	13.1-14.2	5
							7.5 (4.7)[b]	14.0	6
Paenibacillus spp		14-103							9
B. thermoamylovorans						114		12.2	10

[a] reported as *B. stearothermophilus*;
[b] determined at 140 °C
References: 1, Janštová & Lukášová, 2001; 2, Mikolajcik, 1970; 3, Meer *et al.*, 1991; 4, Burton, 1988, collated from several reports; 5, Huemer *et al.*, 1998; 6, Scheldeman *et al.*, 2006; 7, Bergere & Cerf, 1992; 8, Wescott *et al.*, 1995; 9, Scheldeman *et al.*, (2004); 10, Berendsen *et al.*, 2015.

Bradshaw *et al.* (1975) isolated a *B. cereus* strain from a blown can of soup whose spores had D-values as follows: at 115.6 °C, 11.4 min; at 121.1 °C, 2.3 min; at 126.7 °C: 0.3 min; and at 129.4 °C, 0.24 min. The z-value was 7.9 °C. They also isolated a *B. cereus* strain from spoiled canned soup which had a D-value at 121.1 °C of 0.03 min and a z-value of 9.9 °C. They commented that D- and z-values similar to those of this isolate are most commonly reported for *B. cereus* but there have been other reports of very heat-resistant strains akin to their heat-resistant isolate. These are in addition to the anomalous stain of Franklin (1970) which had a very small population of a very heat resistant spores. No such heat-resistant tails were observed by Bradshaw *et al.* (1975).

Spores of *B. cereus* have to germinate and the resulting vegetative cells grow before they can cause spoilage or produce toxin. Often spores need to be activated, for example by heat treatment, before they can geminate. *B. cereus* spores are able to geminate without preliminary heat treatment but the rate of germination and the proportion of spores that germinate are higher when the spores are subjected to a heat treatment (Bergere, 1992). *B. cereus* spores can be activated by heating them in milk at temperatures in the range 65-95 °C for various times. The literature varies in regard to the optimum activation conditions, e.g., 65-75 °C (Coghill & Juffs, 1979), 74 °C for 10 s (Stadhouders & Hup, 1980), 80 °C for 15 s (Overcast & Atmaram, 1974), >80 °C (Wilkinson & Davies, 1973), 85 °C for 2 min (Stadhouders *et al.*, 1980), 95 °C for 15 s (Griffiths & Phillips, 1990) and 115 °C for 1 s (Guirguis *et al.*, 1983). The germination medium as well as the temperature is significant. Wilkinson and Davies (1973) reported that milk heated at 65 to 75 °C for 15 s provided the best medium for germination while Stadhouders *et al.* (1980) found that milk heated at 94 °C for 10 s was a better germination and growth medium than HTST-pasteurised milk.

A complication with germination of *B. cereus* spores is that they exist as both slow-germinating and fast-germinating, with the slow-germinating spores requiring a more intense heat-activation treatment than the fast-germinating spores (Stadhouders *et al.*, 1980). Heating fast-germinating spores of *B. cereus* in milk at 65 °C for 2 min or 72 °C for 10 s causes almost total germination at 20 °C in 24 h while heating the slow-germinating spores at 85-90 °C for 2 min resulted in the same level of germination. It has been shown that a germinant or germination factor is produced in milk by heat treatment (Wilkinson & Davies, 1973; Becker *et al.*, 1994). Therefore pasteurisation conditions are sufficient to cause germination of the fast-germinating spores of *B. cereus*, but not the slow-germinating spores (Labots & Hup, 1964; Bergere, 1992). Fast-germinating spores may be activated by ESL heat treatments although specific reports of this effect have not been located. These fast-germinating spores of *B. cereus* originate from soil, manure and fodder, whereas slow-germinating spores seem to come from milking equipment (Bergere, 1992).

Temperature has a major effect on the growth of vegetative cells following activation/germination. *B. cereus* spores may germinate at low temperatures but not show growth at these temperatures for months (Stewart, 1975). The abilities of *B. cereus* strains to germinate and to grow at low temperatures are not necessarily correlated. For example, Anderson Borge *et al.*, (2001) showed that, in a mixture of 11 mesophilic and psychrotolerant strains, the psychrotolerant strains exhibited both the highest and the lowest germination rates in milk at 7 and 10 °C.

In summary, *B. cereus* is a spore-former of some concern to the dairy industry. It is widespread in the environment and a common contaminant of raw and heat-treated milk. It belongs to a group which contains seven closely related species and hence its precise identification in the reported studies cannot be assured. A major issue for the dairy industry is the wide range of growth temperatures amongst its strains. Most strains are mesophilic but some are psychrotrophic, which means that they can grow in products such as ESL milk during refrigerated storage. The strains vary considerably in heat resistance also; most are readily inactivated by UHT processing but some very heat-resistant strains have been reported. *B. cereus* is often classed as a pathogen because some strains produce toxins; however, very little toxin is produced at low temperatures.

4.3 Heat Inactivation of Bacteria

Theoretically, during the thermal inactivation of bacteria, including bacterial spores, at a particular temperature, the number of bacteria decreases logarithmically with the time of heating as shown in Figure 4.1. While the inactivation of some bacteria follows, or approximately follows, this linear pattern, the shape of the thermal death curve for some bacteria is convex while for others it is concave as shown in Figure 4.2 (Moats, 1971; Lewis & Heppell, 2000). Convex thermal death curves occur when there is a lag phase in the inactivation while concave curves have been attributed to heterogeneity of the population with regard to susceptibility to heat. A further departure from linearity of the curve is when tailing occurs at the end of the graph, that is, when only a small number of bacteria remain. This has been attributed to the presence of a small population of very heat-resistant cells.

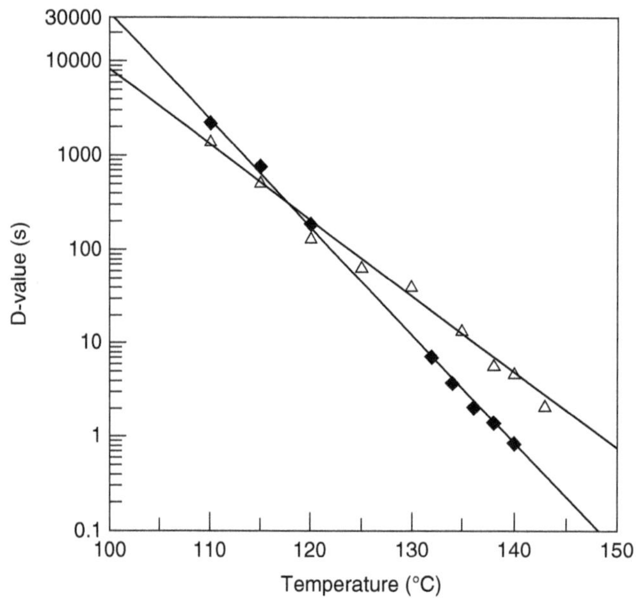

Figure 4.1 Thermal death curves of spores of G. stearothermophlilus (Δ) and B. sporothermodurans (◆). (Source: Huemer *et al.*, 1998. Reproduced with permission of Elsevier.)

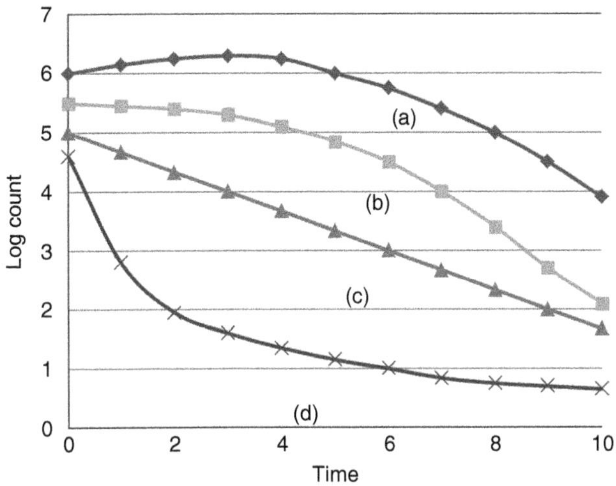

Figure 4.2 Typical types of bacterial thermal death curves. (a) activation shoulder, (b) concave downward, (c) linear, (d) concave upward.

The logarithmic thermal death of spores has certain practical consequences. Firstly, theoretically, the number of bacteria cannot be reduced to zero although the numbers may be extremely low. Secondly, the severity of the heat treatment required to reduce the numbers to a particular level depends on the initial bacterial count; for this reason, it is important for the quality of the final heat-treated product that the number of bacteria in the raw material be kept as low as possible. This can be illustrated in the following example. If the initial number of bacterial cells, whether vegetative or spores, is 10^3/mL, the count remaining after:

- One decimal reduction (90%) is 10^2/mL;
- Two decimal reductions (99%) is 10/mL;
- Three decimal reductions (99.9%) is 1/mL;
- Five decimal reductions (99.999%) is 0.01/mL (target reduction for pasteurisation)
- 6 decimal reductions (99.9999%) is 10^{-3}/mL or 1/L; and
- 9 decimal reductions (99.9999999%) is 10^{-6}/ml or 1/1000 L (target reduction of UHT processing)

Since the aim of UHT processing is to reduce the number of thermophilic spores by 9 decimal reductions, if the initial spore count in the raw milk is 10^4/mL rather than 10^3/mL as in the example, the UHT milk will contain 10/1000 L which would be excessive. However, it needs to be kept in mind that a UHT process designed to cause a 9-log reduction in thermophilic spores, a B* of 1, will have a much greater effect on the majority of the spores present.

The thermal resistance of bacterial cells including spores can be described by the decimal reduction time, the D-value, and the z-value (see Section 2.3.4). Some examples of D-values and z-values of spores are given in Table 4.3 (see Table 2.2 for z-values of non-spore-forming pathogens). These should be taken as indicative only as they can vary considerably between different strains of the same species. The heat resistance can also change considerably when the spores are produced in the laboratory compared with when they are in the environment from which they were isolated. For example, when Huemer *et al.* (1998) prepared a suspension of spores of *B. sporothermodurans* after 10 culture passages, they found the D-value at 140 °C to be decreased from 5.0 s to 1.6 s for a spore suspension prepared from the original stock culture. Furthermore, the heat resistance can vary according to the medium in which the spores are heated. Reported D-values have been obtained for spores suspended in milk, water and buffer and hence are not entirely comparable. The z-value ranges listed in Table 4.3 include some high values, i.e. >30 °C. These high values are exceptions as most reported values for spores of relevance to high-heat treated milk are <14 °C, with most in the range of 8-11 °C. Another source of variation in reported D- and z-values is the method used to obtain the thermal inactivation data. Common methods use capillary tubes, continuous flow systems and UHT pilot plants (Burton 1988). Wescott *et al.* (1995) found higher D-values for *B. cereus* spores when obtained in a batch system than when obtained in continuous system. The z-values obtained by the batch and continuous systems were the same, 8.5 °C. They also established that non-linear survivor curves with pronounced tailing were observed when thermal inactivation studies were conducted in a batch system but were always linear when a continuous flow system was used.

4.4 Microflora in Processed Milks

4.4.1 Pasteurised Milk

HTST pasteurisation of milk at a minimum of 72 °C for 15 s is the major commercial heat treatment of milk throughout the world. Its effect on the bacteria in milk has been reviewed in several publications (e.g., Lewis & Deeth, 2009, Touch & Deeth, 2009; Kelly *et al.*, 2012) and only an overview is included here.

HTST pasteurisation is designed to destroy pathogenic organisms in milk with the minimum conditions being devised to kill the most heat-resistant (non-spore-forming) pathogenic bacteria, *Coxiella burnetii* and *Mycobacterium tuberculosis.* In general, it kills the majority of psychrotrophic non-spore-forming bacteria but some thermoduric non-spore-formers such as coryneforms, micrococci and some streptococci such as *Str. thermophilus* are resistant. In addition, most bacterial spores survive HTST treatment. The surviving thermoduric bacteria do not grow or grow slowly at refrigeration temperatures and are mostly non-pathogenic. Exceptions are the psychrotrophic spore-formers such as *B. cereus* and *B. circulans* (Cromie *et al.*, 1989); some strains of *B. cereus* are pathogenic (see Section 4.2.2.2.1).

Pasteurised milk has a shelf-life of 5-21 days, depending on processing, packaging and storage conditions. The shelf-life is determined largely by the growth of post-pasteurisation contamination by Gram-negative bacteria such as *Pseudomonas* species which is greatly influenced by the temperature of storage. Craven and Macauley (1992a) surveyed the bacteria in pasteurised milk from three processors over a 15-month period and found that 87% and 83% were *Pseudomonas* species in milk stored at 4 °C and 7 °C, respectively, until the counts reached $\geq 10^7$ cfu/mL. Of these, the two major species were *Ps. fluorescens* and *Ps. fragi,* with low levels of *Ps. putida* and *Ps. maltophilia.* Griffiths and Phillips (1986) found *Ps. fluorescens, Ps. putida* and *Ps. cepacia* to be the most common bacteria in pasteurised milk after storage at 6 °C. In another investigation,

Table 4.4 Bacterial types in pasteurized milk stored at 4.4 °C to 19.6 °C until counts reached $10^{7.5}$ cfu/mL (data from Griffiths & Phillips, 1988).

	Percentage isolates from pasteurised milk stored at (°C)						
Bacterial group	**4.4**	**6.5**	**8.5**	**10.4**	**12.5**	**15.5**	**19.6**
Pseudomonas	92.1	65.2	56.6	55.3	35.8	14.7	20.7
Enterobacteriaceae	2.9	12.2	16.1	8.6	27.9	42.6	27.5
Other Gram-negatives	1.1	3.7	4.1	5.9	4.5	6.8	3.2
Bacillus	0.1	3.7	3.9	7.2	8.5	10.2	15.4
Other Gram-positives	3.8	15.2	19.3	23.0	23.3	25.8	34.2

bacteria in pasteurised milk at spoilage were *Ps. fluorescens, Ps. putida, Ps. cepacia* and *Ps. aureofaciens* (Deeth *et al.*, 2002).

In contrast, the major spoilage organism in pasteurised milk in New York State was reported to be a spore-former, *Paenibacillus* (Fromm & Boor, 2004; Ranieri & Boor, 2009). While *Paenibacillus* has been isolated from milk in several studies, its dominance in pasteurised milk had not been previously reported. In another study, in HTST pasteurised milk which was packaged aseptically and had an extended shelf-life, the main spoilage organism was the spore-former, *B. circulans* (Cromie *et al.*, 1989).

Temperature of storage of pasteurised milk has a major effect on its shelf-life. As an indication of the effect, raising the temperature from 4 °C to 10 °C reduces the shelf-life by at least 50%. The storage temperature also affects the nature of the bacteria. This was demonstrated by Griffiths and Phillips (1988) who showed that, at low temperatures, spoilage was mainly due to *Pseudomonas* species while at temperatures >10 °C, Enterobacteriaceae and Gram-positive bacteria became important in spoilage. A selection of the data from Griffiths and Phillips (1988) is shown in Table 4.4.

A further influence on shelf-life is the milk composition. For example, pasteurised skim milk has been reported to have a shorter shelf-life than full-fat milk; this has been attributed to the greater susceptibility of skim milk to proteolytic degradation (Deeth *et al.*, 2002). Griffiths and Phillips (1986) observed low-fat milk to have a shorter shelf-life than full-fat milk.

Some spoilage organisms in pasteurised milk have been found to be processor-specific indicating that a reservoir of certain bacteria develops in plants and continually contaminates products (Griffiths & Phillips, 1988; Craven & Macauley, 1992b). Griffiths and Phillips (1986) found different spoilage microflora between processors at the genus level while Craven and Macauley (1992b) found differences at the species and biovar level of *Pseudomonas.* While Craven and Macauley (1992b) found a predominance of *Pseudomonas* in the spoilage bacteria of milk of three different processors, faster spoilage occurred in the milk from one processor which contained very proteolytic and lipolytic strains of *Pseudonomas*, particularly *Ps. fragi.* The authors concluded that the shelf-life of the pasteurised milk was closely related to the efficiency of cleaning and standard of hygiene of the processor. They suggested that the ability of the highly proteolytic and lipolytic strains of *Pseudomonas* responsible for spoilage in their study to degrade proteins and lipids may contribute to their survival in the milk processing environment. This demonstrates the need for processor-specific remedial action to reduce bacterial contamination in pasteurised milk. Constant monitoring of the post-pasteurisation contamination level with sensitive test methods such as the Psychrofast test (Craven *et al.*, 1994) or the pre-incubation test of Griffiths and Phillips (1986) and strict attention to the cleanliness of surfaces with which the product comes in contact after heat treatment will ensure low contamination levels and maximise the shelf-life of the product. Heating plant surfaces at 95 °C for 30 min prior to processing also helps to minimise contamination arising from the plant.

Listeria monocytogenes is a significant human pathogen and has been linked to outbreaks of listeriosis arising from consumption of pasteurised milk. Of the seven species of the genus *Listeria*, it is the only one which is pathogenic. Other species include *L. innocua*, *L. seeligeri*, *L. ivanovii* and *L. welshimeri* (Lund, 1990). It is now generally agreed that *L. monocytogenes* is destroyed by pasteurisation although some earlier reports had questioned this (Bradshaw *et al.*, 1991; Prentice, 1994). However, it would survive a thermisation treatment of 62 °C for 15 s (Juffs & Deeth, 2007). Therefore incidences of the organism being present in pasteurised milk have been attributed to post-pasteurisation contamination, for example, due to leakage of raw milk into the pasteurised product (Lund, 1990). It is psychrotrophic, being able to grow at temperatures as low as 0.5 °C, and is therefore ideally suited to grow in pasteurised milk (Lund, 1990).

In 2007, an outbreak of listeriosis in Boston, USA, was attributed to consumption of flavoured milk. The milk was found to be pasteurized adequately but the *L. monocytogenes* was a post-pasteurisation contamination (Cumming *et al.*, 2008). The organism is known to occur widely in the environment, where it can survive for long periods. Major environmental sources in dairy factories are enclosed areas such as milk meters and hollow rollers on conveyers where it can readily form biofilms (Tompkin *et al.*, 2002). It was reported that the processing plant associated with the Boston outbreak had no environmental monitoring program for *Listeria*.

A post-pasteurisation contaminant in milk powder manufacture is *Cronobacter sakazakii*, previously known as *Enterobacter sakazakii* (Iversen *et al.*, 2008). It has also been isolated from UHT milk (see Section 4.4.3). This is an emerging food pathogen which has become a concern to the dairy industry. It is described as a ubiquitous organism having been isolated from foods, food raw materials, water, soil and the environments of houses, hospitals and food production lines (Huertas *et al.*, 2015). Its presence in powdered infant formula is the greatest concern as it has been implicated in causing meningitis, septicaemia and enterocolitis in neonates. Alarmingly, the mortality rate of infants suffering from meningitis has been reported to be 33-80% (Lai, 2001). Since *C. sakazakii* does not survive HTST pasteurisation (with a D_{72} of 1.3 s, pasteurisation at 72 °C for 15 s would give an 11-D kill), contamination of powdered infant formula must occur post-pasteurisation, either during drying or packaging (Nazaro-White & Farber, 1997). This was confirmed by Fei *et al.* (2015) who used molecular typing techniques to show that the *C. sakazakii* strains in infant formulae originated from the spray drying, fluidized-bed drying and packing areas. The organism has been found in warm and dry environments in the vicinity of drying equipment (Shaker *et al.*, 2007). Furthermore it is able to form biofilms which increases its survival in the processing environment and its risk of contaminating milk powder (Iversen *et al.*, 2004) (see Section 6.2.2.8).

The main concern with this organism is in infant formulae after reconstitution. Because it is one of the most thermostable members of the Enterobacteriaceae, the

Food and Agriculture Organization (FAO) and World Health Organization (WHO) have recommended reconstitution of powdered infant formulae with water at ≥70 °C. This was also a recommendation of Osaili *et al.* (2009) after assessing its stability at various temperatures. Huertas *et al.* (2015) found that reconstituting powder at 70 °C destroyed a stress-resistant strain of *C. sakazaki* to undetectable levels but a small number of survivors were able to grow to dangerous levels when the reconstituted product was held at room temperature for a long time. Hence under normal usage conditions, the FAO/WHO recommendation would appear reasonable.

4.4.2 ESL Milk

Extended-shelf-life (ESL) milk is produced by a heat treatment which is more intense than pasteurisation but less intense that UHT. Conditions used commercially vary but are generally in the range of 123-145 °C for <1 to 5 s (see Section 3.3).

The microbiological issues associated with ESL milk can be divided into those related to the bactericidal nature of the heating conditions and those associated with post-processing contamination. The relative importance of these for the shelf-life of ESL milk depends on whether the milk is packaged aseptically or under very clean, but not aseptic, conditions. The nature of the heating conditions is the only factor relevant to aseptically packaged milk while both the heating conditions and post-processing contamination are important for milks not packaged aseptically. ESL milk is usually not packaged under aseptic conditions but in an ultra-clean environment. The bacteria in such ESL milk include spore-forming bacteria which are not destroyed by the heating process, together with spore-forming and non-spore-forming bacteria entering the milk after processing.

4.4.2.1 Microbiological Issues Related to the Heating Process

Bacterial growth in ESL milk is by psychrotrophic organisms only as ESL milk is stored under refrigeration. Since ESL heating at ≥120 °C destroys all non-sporeforming bacteria but does not destroy all spores, psychrotrophic spore-formers are a major concern. They are the only bacteria which should cause spoilage when post-processing contamination is eliminated, as in aseptically packaged milk. The ESL heat treatment can also activate spores to germinate and thus allow the vegetative cells to grow during storage.

There have been a small number of investigations on the spore population of ESL milks collected or packaged under aseptic conditions. Mayr *et al.* (2004b) aseptically collected milk samples from a commercial direct processing plant operated at 127 °C for 5 s. They found that *B. licheniformis* (the dominant organism at 73% of isolates), followed by *B. subtilis, B. cereus, Brevibacillus brevis,* and *B. pumilus,* were the most common spore-formers. When the ESL milks were held at 10 °C for 23 weeks, no growth occurred, indicating absence of psychrotrophic sporeformers. However, when the milks were incubated at 30 °C, growth occurred. Bacteria isolated from the 30 °C incubated milks were found to be able to grow at 8 or 10 °C, suggesting some spores were sub-lethally injured during the heat treatment and were revived at 30 °C. *B. cereus* were the most psychrotrophic of the sporeformers isolated. Blake *et al.* (1995) found *B. licheniformis, B. insolitus, B. coagulans* and *B. cereus/thuringensis* in poor quality milk directly heated at 120-132 °C for 4 s and packaged aseptically. No organisms grew in milk processed at temperatures ≥134 °C. In a trial with good quality milk, Blake *et al.* (1995) observed no psychrotrophic growth in milks processed at ≥128 °C. While not dealing with ESL milk as it is known today, the study of Cromie *et al.* (1989) is instructive. They found that

B. circulans was the dominant spoilage organism in milks heated at 72 to 88 °C for 15 s, aseptically packaged and stored at 3 or 7 °C for up to 7 weeks.

Fredsted *et al.* (1996) introduced the APV Pure-Lac™ system which involves heating milk at 130-145 °C for ≤1 s by steam infusion. Those conditions resulted in 8-log reduction of psychrotrophic aerobic spores. It was concluded that a minimum 6-log reduction of psychrotrophic aerobic spores was required to ensure a milk shelf-life of 10- >40 days at 10 °C. This related to ESL milk which was packed under very clean but not aseptic conditions; however, when Pure-Lac™ milk is packaged aseptically, a shelf-life of ≥45 days can be achieved (Rysstad and Kolstad, 2006).

An extreme high-temperature treatment was reported by Huijs *et al.* (2004) and van Asselt *et al.* (2008) using an Innovative Steam Injection (ISI) system in which milk was heated at 150-180 °C for ≤0.2 s. This produced a sterile product that had a refrigerated shelf-life of >60 days. It could not be stored at room temperature as it became unacceptably bitter in <30 days due to plasmin action (Huijs *et al.*, 2004). ISI-treated milk stored at 7 °C (ESL milk) showed no bitterness during storage for 28 days, as plasmin has very low activity at low temperatures.

4.4.2.2 Optimum Processing Conditions for High Microbiological Quality and Safety of ESL Milk

The important question is what heating conditions will assure microbiological stability and safety of ESL milks if packaged aseptically. From the above discussion it is apparent that different studies have arrived at different minimal treatment temperatures to produce milks with a long shelf-life under refrigeration. The differences can be attributed to different initial bacterial counts and different bacterial species composition. One reason for the different species being present is the ambient climatic conditions. Blake *et al.* (1995) commented on their discovery of *B. insolitus* in their ESL milk which has very low optimal growth temperatures of 0-10 °C. They attributed this to acclimatisation of the organism to the low ambient temperatures in that region (Logan, Utah); average daily winter temperature is -2.2 °C. Commercial treatments at 127 °C have been successful, even with packaging under non-sterile conditions. However, as reported by Blake *et al.* (1995), treatment at ≤132 °C can in some circumstances result in some viable psychrotrophic sporeformers remaining in the milk which could cause spoilage. Of particular importance is the treatment which will eliminate spores of *B. cereus*, a potential pathogen. As discussed in Section 4.2.2.2.1, the reported heat resistances (D-values) of this organism vary widely and hence it is virtually impossible to design a process based on these values. Interestingly, Bergere and Cerf (1992), who reviewed the *B. cereus* literature, indicated that *B. cereus* spores should not be present in milk heated at ≥134 °C. This temperature matches the temperature concluded by Blake *et al.* (1995) as the minimal temperature and appears to represent a safe target temperature.

The optimum processing conditions must also take into account the chemical effect, particularly the effect on the flavour compounds in the ESL milk. As discussed in Section 3.3.1, it is recommended that the denaturation of β-lactoglobulin should be ≤50%. Other temperature–time combinations with an equal bactericidal effect, as shown in Figure 3.1, and below the 50% β-lactoglobulin denaturation line would also be ideal.

4.4.2.3 Microbiological Issues Associated with Post-Process Contamination

Non-spore-forming bacteria have been found to be common spoilage organisms in some commercial ESL milks (Mayr *et al.*, 2004a). They include the Gram-positive *Rhodococcus*,

Anquinibacter, Arthrobacter, Microbacterium, Enterococcus, Staphylococcus, Micrococcus and coryneforms. These bacteria appeared to enter the ESL milk from the air, the equipment and/or the packaging material. Mugadza and Buys (2014) also isolated Gram-positive bacteria from ESL milk, namely *Arthrobacter, Annerococcus* and *Mycobacterium*. Mugadza and Buys (2014, 2015) also found the spore-formers *B. pumilus* (the dominant spore-former), *B. subtilis, Paenibacillus* and *B. cereus* in ESL milk. *Paenibacillus* and *B. cereus,* as well as the non-sporeformer *Micrococcus luteus,* were isolated from filler nozzles which the authors concluded was a source of contamination of the milk.

The contaminating bacteria in ESL milk have been identified to be processor-specific (Mayr *et al.,* 2004a) which is consistent with the above discussion where *B. circulans, B. licheniformis* or *B. pumilus* was found to be dominant in each of three different reports. It is similar to findings for contaminants in pasteurised milk (Griffiths & Phillips, 1996; Craven & Macauley, 1992a,b). These findings suggest the need for processor-specific remedial action to reduce bacterial contamination in market milk. In this regard, the reports by Mugadza and Buys (2014, 2015) that some of the bacteria found in ESL milk could also be isolated from the filler nozzles is very pertinent.

It is worth noting that virtually all of the organisms isolated from thermally produced ESL milk are Gram-positive. This contrasts starkly with the situation in pasteurised milks and ESL milk produced by microfiltration where the dominant spoilage organisms are Gram-negative psychrotrophs such as *Pseudomonas* species. The reason for this is not known.

The common occurrence of *B. cereus* in ESL milks is an issue of concern because of its potential to be pathogenic and to cause spoilage. Mugadza and Buys (2015) identified several *B. cereus* isolates from ESL milk and associated equipment. From discriminatory PCR, all isolates were shown to have the csp*A* gene indicating that they were psychrotrophs. In addition, these isolates produced proteases and hence have spoilage potential. There was a close relationship between *B. cereus* isolates from filler nozzles and those in milk indicating the equipment as a source of contamination of the ESL milk (see also Section 4.2.2.2.1).

The temperature of storage of the ESL milk has a major effect on its shelf-life. This was demonstrated by Rysstad and Kolstad (2006) who stored pasteurised milk, which had been aseptically packaged, at 6, 8 and 10 °C. The times taken to reach a total count of 10^6 cfu/mL were 40, 15, and 7 days respectively. Accordingly, they recommended that ESL milk should be stored at ≤6 °C.

4.4.3 UHT Milk

4.4.3.1 Spores in UHT Milk Produced From Fresh Milk

UHT processing aims to produce a product which does not contain micro-organisms capable of growing under the normal conditions of storage, that is, to be "commercially sterile". This means that UHT milk could contain thermophilic spore-forming bacteria which will normally not grow at room temperature. However, there have been several reports of UHT milk being subjected to high ambient temperatures during storage and consequently being affected by growth of the thermophile *G. stearothermophilus.* This has occurred when UHT milk has been transported in container ships across the equator, during storage in very hot climates and during storage in uninsulated, non-temperature-controlled warehouses in which temperatures can reach ≥50 °C. Growth of

G. stearothermophilus is the major cause of the "'flat sour" defect in low-acid foods in which acid but no gas is produced (Pflug & Gould, 1999). This defect has been observed in temperature-abused UHT milk and also in canned evaporated milk (Kalogridou-Vassiliadou, 1992). In addition to *G. stearothermophilus, B. licheniformis, B. coagulans, B. macerans* and *B. subtilis* were isolated from flat-sour canned evaporated milk (Kalogridou-Vassiliadou, 1992). All of these spore-formers produced acid in milk at temperatures from 20 to 55 °C and hence are capable of causing the flat-sour defect if present in UHT milk.

Another thermophilic spore-former which can cause spoilage of UHT milk is *B. thermoamylovorans.* It was first isolated from palm wine (Combetblanc *et al.*, 1995), but has since been isolated from animal feed and milking equipment (Scheldeman *et al.*, 2005, Coorevits *et al.*, 2011), and UHT milk (Flint, 2016, Pers Com). It has an optimum growth temperature of ~50 °C and a maximum growth temperature of ~58 °C (Combetblanc *et al.*, 1995). It has been shown to have a D_{120} of 1.9 min and a z-value of 12.2 °C (Berendsen *et al.*, 2015). In UHT milk it can produce gas and cause blown packs (Flint, 2016, Pers Com). It is interesting that *B. thermoamylovorans* has also been isolated from cocoa and hence UHT chocolate-flavoured products could be at risk from this organism (Witthuhn *et al.*, 2011). Coorevits *et al.* (2011) identified four isolates from dairy farm environments as *B. thermoamylovorans* through 16S rRNA gene sequence similarity with a type strain; however, they also identified a further 18 thermophilic isolates from raw milk which showed 99.3% similarity between them but only 93.9% similarity with the *B. thermoamylovorans* type strain. The authors proposed the name *B. thermolactis* for this new species. It is possible that this species is a variant of *B. thermoamylovorans* which has adapted to the dairy environment (S. Flint, 2016, Pers Com).

Bacterial contaminants of UHT milk result from either survival of heat-resistant spores or post-process contamination. While it is mostly due to the latter, Cerf and Davey (2001) determined statistically that a very small percentage of spores could pass through the holding tube too fast to be destroyed and cause spoilage of UHT packs at a calculated rate of 1-4 per 100 000. Contamination by this means is most likely when there is a high number of spores in the raw milk. In such situations, the very small percentage which survive the UHT treatment, may be sufficient to cause spoilage of the milk (see Section 4.3). It is of interest to note that reported common spoilage rates for UHT milk are 1-2 per 10 000 (von Bockelmann & von Bockelmann, 1998; Muir, 1990) with ≤1 in 10,000 considered to be a reasonable target (Robertson, 2003; Pujol & Membré, 2014).

Several spore-formers including *B. sporothermodurans, Paenibacillus* spp, *G. stearothermophilus, B. licheniformis, B. coagulans, B. circulans, B. badius, B. subtilis, B. cereus, B. polymyxa* and *B. sphericus* have been isolated from UHT milks (Lück *et al.*, 1978; Forschino *et al.*, 1990; Skladal *et al.*, 1993; Meier *et al.*, 1995; von Bockelmann & von Bockelmann, 1998; Coelho *et al.*, 2001). Some of these may have survived the heating process while others would have been due to post-sterilisation contamination. The spores capable of surviving the UHT process are mainly *G. stearothermophilus, B. subtilis* (Muir, 1990), *B. megaterium* (Hassan *et al.*, 1993), *B. sporothermodurans* (Pettersson *et al.*, 1996) and *Paenibacillus lactis* (Scheldeman *et al.*, 2004). Spores produced by *G. stearothermophilus* are very heat-resistant with some being shown to withstand heating at 156 °C for 6 s (Intaraphan, 2001); however, this organism is unable to grow below 40 °C. On the other hand, *B. sporothermodurans* and some *Paenibacillus* species produce highly heat-resistant spores and can grow at or below ~30 °C (Pettersson *et al.*, 1996) after activation and germination.

The spores of *B. sporothermodurans* have been reported to have D-values at 140 °C of 5.0 and 4.7 s and z-values from 13.1 and 14 °C (Huemer *et al.*, 1998; Scheldeman *et al.*, 2006). Hammer *et al.* (1996) found that UHT holding conditions of 148 °C for 10 s or 150 °C for 6 s were required to inactivate the spores to a reasonable level. *B. sporothermodurans* is usually present at very low levels in raw milk and hence should normally not be a problem in UHT milk. Furthermore, it does not cause spoilage although it can cause a pink discolouration of the milk. However, when out-of-date UHT milk containing the organism is reprocessed, elevated levels may result in the reprocessed product and the processing equipment can become contaminated (Pearce, 2004; Tabit & Buys, 2011); such reprocessing should not occur. Once *B. sporothermodurans* contaminates UHT equipment, it is difficult to eliminate and has resulted in closure of some UHT processing lines (IDF, 2000).

While some spores may survive UHT conditions through their innate heat resistance or through the Cerf and Davies statistical route, some arise from post-sterilisation contamination. This can occur through contamination of equipment in the post-sterilisation section of the plant. For example, spores can be trapped under seals in downstream homogenisers and contaminate sterilised product (Kessler, 1994). Frequently changing the seals is an effective way of preventing contamination from this source. Furthermore, some spores acquire significantly enhanced heat resistance when attached to stainless steel and this may explain the appearance in UHT of spores which do not usually show high heat resistance (Simmonds *et al.*, 2003).

Spoilage of UHT milk by heat-resistant spore-forming organisms first requires activation and germination of the spores and growth of the vegetative cells. It has been suggested that this may occur in some cases where the UHT heat treatment does not destroy the spores but activates them to germinate and grow (Hersom & Hulland, 1980).

Contamination of UHT milk products causing product recalls is rare but does occur. Recent examples are UHT chocolate milk contaminated with *B. subtilis*, which could be pathogenic, and UHT skim milk contaminated with *B. circulans* which developed an objectionable flavour and odour.

4.4.3.2 Spores in Milk Powders Used for UHT Reconstituted Milk

Considerable importance is placed on the numbers of bacterial spores in milk powders destined for reconstituted UHT milk manufacture. For example, specifications of <500 to <2,000 cfu/g aerobic thermophilic spores in skim milk powder and whole milk powder for this purpose have been set by some international powder buyers (Watterson *et al.*, 2014).

Sterile milk products reconstituted from powders involves somewhat different microbiological considerations from those produced from fresh liquid milk. Hill and Smythe (2012) reported that the major thermophilic spore-formers in powders (*Geobacillus* species, *A. flavithermus* and *B. licheniformis*) differed from those in liquid milk where *B. licheniformis* is commonly the dominant organism. This finding was supported by an extensive survey of milk powders produced in New Zealand (Ronimus *et al.*, 2003) and another of powders produced in 18 different countries (Rueckert *et al.*, 2004) which found that strains of *A. flavithermus*, *G. stearothermophilus*, and *B. licheniformis* accounted for the majority of thermophilic spores (defined as resistant to heating at 80 °C for 10 min), with *A. flavithermus* being the dominant organism. *A. flavithermus* (or *A. flavothermus*, formerly *Bacillus flavothermus* [Heinen *et al.*, 1982]) is a thermophilic spore-former which has attracted considerable attention in the dairy industry in recent years.

Hill and Smythe (2012) concluded that the spores of thermophilic bacteria in milk powders were mostly derived from the powder processing operation and not from the raw milk; however, Pearce (2004) indicated that they can be present in raw milk, and germinate, grow and re-sporulate during milk powder production. Parts of the process where these organisms can accumulate are cream separators, parts of evaporators and heat regeneration sections of heat exchangers where the operating temperatures are in the growth temperature range for thermophiles, 45–65 °C. This is consistent with the reports of Murphy *et al.* (1999) who isolated *G. stearothermophilus* and *B. licheniformis* and of Scott *et al.* (2007) who isolated *A. flavithermus* and *Geobacillus* species from the pre-heater sections and the evaporators in milk powder processing plants.

A major factor responsible for the entry of the spores into the final powder is formation of fouling deposits during processing, particularly during long production runs. The fouling can be either biofouling involving biofilms of these organisms directly, or chemical fouling of surfaces and under seals which prevent access to the spores by cleaning agents. Sloughing of spores from the fouling deposits releases them into the product (Watterson *et al.*, 2014). Of the major thermophiles in milk powders, *Geobacillus* spp. are the most likely to survive UHT and in-container sterilisation (Hill, 2004). A further factor is the propensity of spores to attach to stainless steel and to fouling deposits (Flint *et al.*, 1997, 2001).

Thermophilic spores have traditionally been enumerated by heating at 100 °C for 30 min; however, Hill (2005) showed that the spores that survived this heating included both *A. flavithermus* and *Geobacillus* species but only *Geobacillus* could withstand UHT or in-container sterilisation processing. Therefore the traditional method overestimated the number of thermophilic spores. On this basis, Hill (2005) proposed a more discriminating heating process of 108 °C for 30 min to inactivate *A. flavithermus* but not *Geobacillus* species. A similar heating process, 106 °C for 30 min, followed by incubation at 55 °C for 48 h, was adopted in an IDF/ISO Technical Specification for enumerating "specially thermoresistant spores of thermophilic bacteria" (IDF/ISO, 2009). Because all mesophilic and *A. flavithermus* spores are destroyed by sterilisation processes, Hill (2005) questioned the necessity of routine spore testing of milk powders destined for manufacturing sterilised recombined milk where the products are unlikely to be consistently exposed to temperatures of >37 °C. These are the conditions which favour growth of *Geobacillus*.

4.4.3.3 Spores in Non-Milk Ingredients Used in UHT Milk Products

Ingredients used in UHT- and in-container-sterilised milk-based products may also contain spores as they are seldom pre-sterilized. Of particular significance is cocoa powder used in preparing chocolate-flavoured milk as it contains spores which can affect the quality of the final product. Some spores isolated from cocoa and chocolate-flavoured milk products are given in Table 4.5. Inactivation of spores in cocoa powder present quite a challenge to the milk processor as the spores can be intimately associated with particles of powder and thereby be protected from heat during processing. Furthermore, the fibrous particles are difficult to hydrate; hydration is necessary as the spores can only be heat-inactivated in a moist environment. Therefore, before being added to a product,

Table 4.5 Spores isolated from cocoa and chocolate products.

Spore	Source	Reference
B. subtilis	ESL Chocolate milk	Witthuhn *et al.*, 2011
B. circulans	Chocolate dessert	
B. thermoamylovorans	Chocolate milk (before heat treatment)	
Geobacillus pallidus	Cocoa powder	
B. thermoamylovorans	Cocoa powder	
B. licheniformis group (dominant)	Cocoa	Lima *et al.*, 2012
B. subtilis group (dominant)		
B. cereus group		
B. barbaricus		
B. coagulans		
B. drentensis		
B. oleroneus		
Brevibacillus thermofuber		
Geobacillus spp		
Thermoactinomyces		
B. subtilis group (dominant, 66%)	Cocoa powder	Lima *et al.*, 2011
B. licheniformis group (next most dominant)		
Cl. thermosaccharolyticum	Sterilised chocolate milk	Langevel, 1967

Source: Fweja, 2007. Reproduced with permission of Cambridge University Press.

cocoa powder should be hydrated and then heat-treated to activate and germinate the spores to render them much more susceptible to a subsequent heat treatment. It is important that a time period is allowed for after the heat treatment for the spores to germinate. A tyndallisation process in which this treatment is repeated two or three times is sometimes used (see Section 3.3.6). Other approaches such as incubation with hydrogen peroxide have also been proposed but this may not be permitted in some jurisdictions. Another consideration is the type of alkalisation with which the cocoa has been treated. Alkalisation, particularly nib alkalisation, which involves adding water and alkali and heating at 85-115 °C (ICMSF, 2011), greatly reduces the number of spores in cocoa (Dyer, 2003). Other modern technologies which combine roasting with a steam treatment also reduce spore numbers (ICMSF, 2011). Judicious cocoa selection in terms of specifications for spore content and particle size is important to minimize spore-related problems. Several manufacturers now produce cocoa powders specifically for dairy applications.

4.4.3.4 Other Microbial Contamination

Non-spore-forming bacteria, both Gram-positive and Gram-negative, have also been isolated from UHT milk in several studies and can only originate from post-sterilisation sources. The isolated bacteria include *Micrococcus, Corynebacterium*-like bacteria, *Str. lactis, St. aureus, Enterococcus faecalis* and *Enterobacter sakazakii* (now *C. sakazakii,* see Section 4.4.1) (Skladal *et al.,* 1993; von Bockelmann & von Bockelmann, 1998; Coelho *et al.,* 2001).

A filamentous fungus, *Fusarium oxysporum,* is another organism which has caused post-sterilisation contamination in UHT milk from several plants. It produces a cheesy flavour and gas, resulting in swollen or 'blown' packages. It typically enters UHT milk packages through the air supplied to filling machines. It can be avoided by maintaining positive (filtered) air pressure to the filling machine. Unfortunately, *Fusarium* is difficult to remove from contaminated equipment.

4.4.4 In-Container Sterilised Milk

As the bactericidal conditions of in-container sterilisation are equivalent to those in UHT processing, the issue of highly heat-resistant spore-formers that survive the heat treatment is the same for both processes. Where the two processes differ is in the mode of post-sterilisation contamination; for UHT products this occurs in the down-stream processing and packaging sections of the UHT plant while for in-container sterilised products this mainly occurs through "leakage" of bacteria into the container. The leakage can occur at different stages but the most common is during the cooling process where cooling water can enter the container through faulty seals (Put *et al.*, 1972). Leakage of bacteria is facilitated by the high vacuum generated in the containers during processing (Jackson & Shinn, 1979).

As previously mentioned, destruction of *Cl. botulinum* is the primary aim of in-container sterilisation and is the reason for the development of the 12D botulinum-cook process. A count of 10^{-9}/mL is considered reasonable and this can be achieved in a 12D process if the initial level of *C. botulinum* in the raw material is 10^3 cfu/mL (Jackson & Shinn, 1979). As indicated in Chapter 2, such a level would be a very rare occurrence in raw milk.

Surveys of sterilised milks have shown *B. subtilis* to be the major mesophilic spore-former followed by *B. licheniformis* and *B. circulans.* Other mesophiles isolated included *B. megaterium, B. cereus, B. coagulans, B. macerans, B. circulans* and *B. brevis.* The major thermophilic spore-formers isolated are *B. stearothermophilus* and *B. calcidolactis* (both now classified as G. *stearothermophilus*) and the anaerobic spore-former *Cl. thermosaccharolyticum* (Candy & Nichols, 1956; Ridgeway, 1958; Langevel, 1967). In contrast to G. *stearothermophilus* which produces acid but no gas (only flat-sour defect). *Cl. thermosaccharolyticum* produces gas, which can explode containers, if incubated at 45 or 55 °C. Neither G. *stearothermophilus* nor *Cl. thermosaccharolyticum* grow at lower temperatures (Langevel, 1967).

Like UHT packages, sterilised packs have a failure rate which has been reported to be 1 in 1000 units, or closer to 1 in 5 000 if flat-souring is taken into account (von Bockelmann & von Bockelmann, 1998). Higher failure rates have been attributed to high counts of highly heat-resistant spore-formers in the raw material (Ashton & Romney, 1981).

4.5 Sterilisation of Equipment and Packaging to Prevent Microbial Contamination of UHT Products

Apart from using an appropriate temperature–time profile in UHT plants for inactivation of spores which could affect the commercial sterility of the product, the equipment downstream of the sterilisation section must be properly sterilised. Sterilisation of equipment is performed with saturated steam at ≥121 °C for 30 min (David & Carlson, 2012), with treatment at 130 °C being common. In addition, packaging material is also sterilised before product filling. This can be achieved in several ways but commonly hydrogen peroxide at ≥30% combined with a thermal treatment, is used (see Section 5.5 for more information).

Verification of sterilisation of the processing and packaging equipment, including the aseptic tank (or A tank), is essential when a new plant is commissioned or the plant has been modified. Guidelines for verification of UHT plants using a product such as skim milk have been established by equipment manufacturers (e.g., Tetra Pak, undated). They commonly involve three separate runs and a statistically based sampling plan. For example, for the commonly used defect level of 1 in 1000 and a confidence level of 95%, Tetra Pak recommends a total number of samples of 7,720 to be taken over the three runs. If the number of non-sterile packs is not greater than three, plant sterility is verified. Samples are tested for sterility after incubation at 30 °C for 7 days (see Chapter 8 for more information on sampling).

References

Agata, N., Ohta, M. & Yokoyama, K. (2002) Production of *Bacillus cereus* emetic toxin (cereulide) in various foods. *International Journal of Food Microbiology* **73**, 23–27.

Ahmed, A.A.H., Moustafa, M.K. & Marth, E.H. (1983) Incidence of *Bacillus cereus* in milk and some milk products. *Journal of Food Protection* **46**, 126–128.

Anderson Borge, G.I., Skeie, M., Sorhaug, T., Langsrud, T. & Granum, P.E. (2001) Growth and toxin profiles of *Bacillus cereus* isolated from different food sources. *International Journal of Food Microbiology* **69**, 237–246.

Ashton, T.R. & Romney A.D.T (1981) In-container sterilization. In: *Factors Affecting the Keeping Quality of Heat Treated Milk*. IDF Document 130, pp. 55–70. International Dairy Federation, Brussels.

Bartoszewiez, M., Hansen, B.M. & Swiecicka, I. (2008) The members of the *Bacillus cereus* group are commonly present contaminants of fresh and heat-treated milk. *Food Microbiology* **25**(4), 588–596.

Berendsen, E.M., Krawczyk, A.O., Klaus, V., de Jong, A., Boekhorst, J., Eijlander, R.T., Kuipers, O.P. & Wells-Bennik, M.H.J. (2015) *Bacillus thermoamylovorans* spores with very-high-level heat resistance germinate poorly in rich medium despite the presence of ger clusters but efficiently upon exposure to calcium-dipicolinic acid. *Applied and Environmental Microbiology* **81**(22), 7791–7801.

Becker, H., Schaller, G., von Wiese, W. & Terplan, G. (1994) *Bacillus cereus* in infant foods and dried milk products. *International Journal of Food Microbiology* **23**(1), 1–15.

Bergere, J.-L. & Cerf, O. (1992) Heat resistance of *Bacillus cereus* spores. In: *Bacillus cereus in milk and milk products*. IDF Bulletin 275, pp 23–25. International Dairy Federation, Brussels.

Bergere, J.-L. (1992) Spore formation and germination of *Bacillus cereus*: the spore cycle. In: *Bacillus cereus in milk and milk products*. IDF Bulletin 275, pp. 9–14. International Dairy Federation, Brussels.

Blake, M.R., Weimer, B.C., McMahon, D.J. & Savello, P.A. (1995) Sensory and microbial quality of milk processed for extended shelf-life by direct steam injection. *Journal of Food Protection* **58**, 1007–1013.

Bosshard, P., Zbinden, R. & Altwegg, M. (2002) *Paenibacillus turicensis* sp. *nov*., a novel bacterium harbouring heterogeneities between 16S rRNA genes. *International Journal of Systematic and Evolutionary Microbiology* **52**, 2241–2249.

Bradshaw, J.G., Peeler, J.T. & Twedt, R.M. (1975) Heat-resistance of ileal loop reactive *Bacillus cereus* strains isolated from commercially canned food. *Applied Microbiology* **30**, 943–945.

Bradshaw, J.G., Peeler, J.T. & Twedt R.M (1991) Thermal resistance of *Listeria* spp. in milk. *Journal of Food Protection* **54**, 12–14.

Bramley, A.J. & McKinnon, C.H. (1990) The microbiology of raw milk. In: *Dairy Microbiology*, Vol 1. (ed. R.K. Robinson), pp. 163–208. Elsevier Applied Science, New York.

Burgess, S.A., Flint, S.H. & Lindsay, D. (2014) Characterization of thermophilic bacilli from a milk powder processing plant. *Journal of Applied Microbiology* **116**(2), 350–359.

Burton, H. (1988) *Ultra-high Temperature Processing of Milk and Milk Products*. Elsevier Applied Science, London.

Button, P.D., Roginski, H., Deeth, H.C. & Craven H.M (2011) Improved shelf-life estimation of UHT milk by prediction of proteolysis. *Journal of Food Quality* **34**, 229–235.

Candy, M. R. & Nichols, A. A. (1956) Some bacteriological aspects of commercially sterilized milk II. Type of spore-forming bacteria isolated. *Journal of Dairy Research* **23**(3), 329–335.

Cerf, O. & Davey, K.R. (2001) An explanation of non-sterile (leaky) milk packs in well-operated UHT plants. *Transactions of the Institute of Chemical Engineers* **79** (Part C), 197–210.

Champagne, C.P., Laing, R.R., Roy, D. & Mafu, A.A. (1994) Psychrotrophs in dairy products: Their effect and their control. *Critical Reviews in Food Science and Nutrition* **34**, 1–30.

Coelho, P.S., Silva, N., Brescia, M.V. & Siqueira, A.P. (2001) [Microbiological aspects of UHT whole milk commercialized in Belo Horizonte, Brazil]. *Arquivo Brasileiro de Medicina Veterinária e Zootecnia* **53**(2), 256–262.

Coghill, D. & Juffs, H.S. (1979) Incidence of psychrotrophic spore-forming bacteria in pasteurized milk and cream products and effect of temperature on their growth. *Australian Journal of Dairy Technology* **34**,150–153.

Combetblanc, Y., Ollivier, B., Streicher, C., Patel, B.K.C., Dwivedi, P.P., Pot, B., Prensier, G. & Garcia, J.L. (1995) *Bacillus thermoamylovorans* sp *nov*, a moderately thermophilic and amylolytic bacterium. *International Journal of Systematic Bacteriology* **45**(1), 9–16.

Cook, G.M. & Sandeman, R.M. (2000) Sources and characterisation of spore-forming bacteria in raw milk. *Australian Journal of Dairy Technology* **55**, 119–126.

Coorevits, A., Logan, N.A., Dinsdale, A.E., Halket, G., Scheldeman, P., Heyndrickx, M., Schumann, P., Van Landschoot, A. & De Vos, P. (2011) *Bacillus thermolactis* sp nov., isolated from dairy farms, and emended description of *Bacillus thermoamylovorans*. *International Journal of Systematic and Evolutionary Microbiology* **61**, 1954–1961.

Cousin, M.A. (1980) Proteolytic activity of psychrotrophic micro-organisms on milk and dairy products. In: *Psychrotrophic Micro-organisms in Spoilage and Pathogenicity*. (eds T.A. Roberts, G. Hobbs, J.H.B Christian & N. Skovgaard), pp. 63–72. Academic Press, London.

Craven, H.M. & Macauley, B.J. (1992a) Micro-organisms in pasteurised milk after refrigerated storage. 1. Identification of types. *Australian Journal of Dairy Technology* **47**, 38–45.

Craven, H.M. & Macauley, B.J. (1992b) Micro-organisms in pasteurised milk after refrigerated storage. 3. Effect of milk processors. *Australian Journal of Dairy Technology* **47**, 50–55.

Craven, H.M., Forsyth, S.R., Drew, P.G. & Macauley, B.J. (1994) A new technique for early detection of Gram-negative bacteria in milk. *Australian Journal of Dairy Technology* **49**, 54–56.

Cromie, S.J., Dommett, T.W. & Schmidt, D. (1989) Changes in the microflora of milk with different pasteurisation and storage conditions and aseptic packaging. *Australian Journal of Dairy Technology* **44**, 74–77.

Cumming, M., Kludt, P., Matyas, B., DeMaria, A., Stiles, T., Han, L., Gilchrist, M., Neves, P., Fitzgibbons, E. & Condon, S. (2008) Outbreak of *Listeria monocytogenes* infections associated with pasteurized milk from a local dairy - Massachusetts, 2007. *Morbidity and Mortality Weekly Report* **57**, 1097–1100.

David, J.R.D. & Carlson, V.R. (2012) Validation and establishment of aseptic processing and packaging operations. In: *Handbook of Aseptic Processing and Packaging.* (eds J.R.D David, R.H. Graves & T. Szemplenski), 2nd edn, pp. 139–159. CRC Press, Boca Raton.

Deeth, H.C. (1997) The role of phospholipids in the stability of the milk fat globule membrane. *Australian Journal of Dairy Technology* **52**, 44–46.

Deeth, H.C., Khusniati, T., Datta, N. and Wallace, R.B. (2002) Spoilage patterns of skim and whole milks. *Journal of Dairy Research* **68**, 227–241.

Dufeu, J. & Leesment, H. (1974) Growth and resistance characteristics of some psychrotrophic spore-formers isolated from raw milk. *XIX International Dairy Congress*, 535–536.

Dufrenne, J., Bijwaard, M., Te Giffel, M., Beumer, R. & Notermans, S. (1995) Characteristics of some psychrotrophic *Bacillus cereus* isolates. *International Journal of Food Microbiology* **27**(2-3), 175–183.

Dyer, B. (2003) Alkalized cocoa powders. 57th PMCA Production Conference, pp.128–135. (http://www.blommer.com/_documents/alkalized-cocoa-powders-article.pdf accessed January 2015).

EFSA (2005) *Bacillus cereus* and other *Bacillus* spp. in foodstuffs. Opinion of the Scientific Panel on Biological Hazards on *Bacillus cereus* and other *Bacillus* spp in foodstuffs. *The EFSA Journal* **175**, 1–48.

FAO (2005) Benefits and potential risks of the lactoperoxidase system of raw milk preservation. (ftp:/ftp.fao.org/docrep/fao/009/a0729e/a0729e00.pdf).

Fei, P., Man, C., Lou, B., Forsythe, S. J., Chai, Y., Li, R., Niu, J. & Jiang, Y. (2015) Genotyping and source tracking of *Cronobacter sakazakii* and *C. malonaticus* isolates from powdered infant formula and an infant formula production factory in China. *Applied and Environmental Microbiology* **81**(16), 5430–5454.

Flint, S. H., Palmer, J., Bloemen, K., Brooks J. & Crawford, R. (2001) The growth of *Bacillus stearothermophilus* on stainless steel. *Journal of Applied Microbiology* **90**, 151–157.

Flint, S.H., Bremer, P.J. & Brooks J.D. (1997) Biofilms in dairy manufacturing plants—Description, current concerns and methods of control. *Biofouling* **11**, 81–97.

Foddai, A., Elliott, C.T. & Grant, I.R. (2010) Rapid assessment of the viability of *Mycobacterium avium* subsp. *paratuberculosis* cells after heat treatment, using an optimized phage amplification assay. *Applied and Environmental Microbiology*, **76**, 1777–1782.

Food Standards Agency. (2002) http://acmsf.food.gov.uk/sites/default/files/mnt/drupal_data/sources/files/multimedia/pdfs/ACM485.PDF

Food Standards Agency (2002) *Draft strategy for the control of* Mycobacterium avium *subsp.* paratuberculosis *(MAP) in cows' milk.* FSA, London, UK.

Forschino, R., Gaill, A. & Ottogalli, G. (1990) Research on the microflora of UHT milk. *Annali di Microbiologia ed Enzymologia*, **40**, 47–59.

Frank, J.F. & Hassan, A.N. (2002) Micro-organisms associated with milk. In: *Encyclopedia of Dairy Sciences.* (eds H. Roginski, J.W. Fuquay & P.F. Fox), pp. 1786–1796. Academic Press, London.

Frank, J.F. (1997) Milk and dairy products. In: *Food microbiology: Fundamentals and Frontiers.* (ed. M.P. Doyle, L.R. Beuchat & T.J. Montville), ASM Press, Washington, DC.

Franklin, J.G. (1969) Some bacteriological problems in the market milk industry in the UK. *Journal of the Society of Dairy Technology* **22**(2), 100–112.

Franklin, J.G. (1970) Spores in milk - problems associated with UHT processing. *Journal of Applied Bacteriology* **33**(1), 180–191.

Fredsted L.B., Rysstad G. & Eie T. (1996) Pure-Lac(TM): The new milk with protected freshness and extended shelf-life. In: *Heat Treatments and Alternative Methods.* IDF Document 9602, pp. 104–125. International Dairy Federation, Brussels.

Fromm, H.I. & Boor, K. (2004) Characterization of pasteurized fluid milk shelf-life attributes. *Journal of Food Science* **69**, 207–214.

Fweja, L.W.T., Lewis, M.J. & Grandison, A.S. (2007) Alternative strategies for activation of the natural lactoperoxidase system in cows' milk: trials in Tanzania. *Journal of Dairy Research* **74**(4), 381–386.

Grant, I.R., Rowe, M.T., Dundee, L. & Hitchings, E. (2001) *Mycobacterium avium* subsp. *paratuberculosis*: its incidence, heat resistance and detection in raw milk and dairy products. *International Journal of Dairy Technology* **54** (1), 2–13.

Granum, P.E. & Lund, T. (1997) *Bacillus cereus* and its food poisoning toxins. *FEMS Microbiology Letters* **157**(2), 223–228.

Greenstein, R.L. (2003) Is Crohn's disease caused by a mycobacterium? Comparisons with leprosy, tuberculosis, and Johne's disease. *Lancet Infective Diseases* **3**(8), 507–514.

Griffiths, M.W. & Phillips, J.D. (1986) The application of the preincubation test in commercial dairies. *Australian Journal of Dairy Technology* **41**(2-3), 71–79.

Griffiths, M.W. & Phillips, J.D. (1988) Modeling the relation between bacterial growth and storage temperature in pasteurized milks of varying hygienic quality. *Journal of the Society of Dairy Technology*, **41**(4), 96–102.

Griffiths, M.W. & Phillips, J.D. (1990) Strategies to control the outgrowth of spores of psychrotrophic *Bacillus* spp in dairy-products. 1. Use of naturally-occurring materials. *Milchwissenschaft* **45**(10), 621–626.

Griffiths, M.W. (2006) *Mycobacterium paratuberculosis.* In: *Emerging Foodborne Pathogens.* (eds. Y. Motarjemi & M. Adams). pp. 522–556. Woodhead Publishing, Cambridge, U.K.

Guirguis, A.H., Griffiths, M.W. & Muir, D.D. (1983) Spore-forming bacteria in milk .1. Optimization of heat-treatment for activation of spores of *Bacillus* species. *Milchwissenschaft* **38**(11), 641–644.

Haddadin, M.S., Ibrahim, S.A. & Robinson, R.K. (1996) Preservation of raw milk by activation of the natural lactoperoxidase system. *Food Control* 7(3), 149–152.

Hammer, P., Knappstein, K. & Hahn, G. (1998) Significance of *Mycobacterium paratuberculosis* in milk. IDF Bulletin 330, pp. 12–16. International Dairy Federation, Brussels.

Hammer, P., Lembke, F., Suhren, G. & Heesschen, W. (1996) Characterization of a heat-resistant mesophilic *Bacillus* species affecting the quality of UHT milk. In: *Heat Treatments and Alternative Methods*. IDF Document 9602, pp. 9–16. International Dairy Federation, Brussels.

Haryani, S., Datta, N, Elliott, A.J. & Deeth, H.C. (2003) Production of proteinases by psychrotrophic bacteria in raw milk stored at low temperature. *Australian Journal of Dairy Technology* **58**, 16–20.

Hassan, A.N., Zahnran, A.S., Metwalli, N.H. & Shalabi, S.I. (1993) Aerobic spore-forming bacteria isolated from UHT milk produced in Egypt. *Egyptian Journal of Dairy Science* **21**, 109–121.

Heinen, W., Lauwers, A.M. & Mulders, J.W.M. (1982) *Bacillus flavothermus,* a newly isolated facultative thermophile. *Antonie van Leeuwenhoek* **48**, 265–272.

Hersom, A.C. & Hulland, E.D. (1980) *Canned Foods: Thermal Processing and Microbiology*. Churchill Livingstone, Edinburgh, New York.

Hickey, M. (2009) Current legislation of market milks. In: *Milk Processing and Quality Management* (ed A.Y. Tamine). pp. 101–138. SDT, Wiley-Blackwell, Chichester, UK.

Hill, B.M. & Smythe, B.W. (2012) Endospores of thermophilic bacteria in ingredient milk powders and their significance to the manufacture of sterilized milk products: An industrial perspective. *Food Reviews International* **28**(3), 299–312.

Hill, B.M. (2005) Thermophilic spores: UHT and retort processing of recombined milks. In: *Proceedings of 4th International Symposium on Recombined Milk and Milk Products,* Cancun, Mexico, May 9–12, 2004. (ed P. Tong). pp. 184–193. US Dairy Export Council, Arlington, VA.

Huemer, I., Klijn, N., Vogelsang, H. & Langeveld, L. (1998) Thermal death kinetics of spores of *Bacillus sporothermodurans* isolated from UHT milk. *International Dairy Journal* **8**, 851–855.

Huertas, J.-P., Alvarez-Ordonez, A., Morrissey, R., Ros-Chumillas, M., Esteban, M.-D., Mate, J., Palop, A., & Hill, C. (2015) Heat resistance of *Cronobacter sakazakii* DPC 6529 and its behavior in reconstituted powdered infant formula. *Food Research International* **69**, 401–409.

Huijs, G., Van Asselt, A. J., Verdurmen, R. E. M. & De Jong, P. (2004) High speed milk. *Dairy Industries International* **69**, 30–32.

ICMSF (International Commission on Microbiological Specifications for Foods) (2011) Cocoa, chocolate and confectionery. In: *Microorganisms in Foods 8. Use of Data for Assessing Process Control and Product Acceptance*. pp. 241–246. Springer, New York.

IDF (2000) *Bacillus sporothermodurans* – a *Bacillus* forming highly heat-resistant spores. IDF Bulletin 357, pp. 1–27, International Dairy Federation, Brussels.

IDF (2001) *Mycobacterium paratuberculosis*. IDF Bulletin 362. International Dairy Federation, Brussels.

IDF (1998) *Significance of Mycobacterium paratuberculosis in milk*. IDF Bulletin 330. International Dairy Federation, Brussels.

IDF/ISO (2009) Dried milk – Enumeration of the specially thermoresistant spores of thermophilic bacteria. Technical Specification. 1st edn, ISO/TS 27265:2009(E), IDF/RM 228:2009(E).

Intaraphan, T. (2001) *Heat-resistant Spore-formers in UHT Milk.* PhD thesis, The University of Queensland, Brisbane, Australia.

Iversen, C., Lane, M. & Forstyhe, S.J. (2004) The growth profile, thermotolerance and biofilm formation of *Enterobacter sakazakii* grown in infant formula milk. *Letters in Applied Microbiology* **38**, 378–382.

Iversen, C., Mullane, N., McCardell, B.D., Tall, B., Lehner, A., Fanning, S., Stephan, R. & Joosten, H. (2008) *Cronobacter* gen. nov., a new genus to accommodate the biogroups of *Enterobacter sakazakii,* and proposal of *Cronobacter sakazakii* gen. nov., comb. nov., *Cronobacter malonaticus* sp. nov., *Cronobacter turicensis* sp. nov., *Cronobacter muytjensii* sp. nov., *Cronobacter dublinensis* sp. nov., *Cronobacter genomospecies* 1, and of three subspecies, *Cronobacter dublinensis* subsp. *Dublinensis* subsp. nov., *Cronobacter dublinensis* subsp. *lausannensis* subsp. nov. and *Cronobacter dublinensis* subsp. *lactaridi* subsp. nov. *International Journal of Systematic and Evolutionary Microbiology* **58**, 1442–1447.

Ivy, R.A., Ranieri, M.L., Martin, N.H., den Bakker, H.C., Xavier, B.M., Wiedmann, M. & Boor, K.J. (2012) Identification and characterization of psychrotolerant spore-formers associated with fluid milk production and processing. *Applied and Environmental Microbiology* **78**(6), 1853–1864.

Jackson, J.M. & Shinn, B.M. (1979) *Fundamentals of Food Canning Technology.* AVI Publishing Company, Westport, Connecticut.

Janštová, B. & Lukášová, J. (2001) Heat resistance of *Bacillus* spp. spores isolated from cow's milk and farm environment. *Acta Veterinaria Brno* **70**, 179–184.

Jay, J.M. (1996) *Modern Food Microbiology.* 5th edn, Chapman and Hall, New York.

Jooste, P.J. & Anelich, L.E.C.M. (2008) Safety and quality of dairy products. In: *Advanced Dairy Science and Technology.* (eds T.J. Britz & R.K. Robinson), pp. 153–182. Blackwell Publishing Professional, Oxford.

Juffs, H. & Deeth, H.C. (2007) *Scientific Evaluation of Pasteurisation for Pathogen Reduction in Milk and Milk Products.* FSANZ, Canberra (ISBN 978-0-642-34555-4).

Kalogridou-Vassiliadou, D. (1992) Biochemical activities of *Bacillus* species isolated from flat sour evaporated milk. *Journal of Dairy Science* **75**, 2681–2686.

Kelly, A., Datta, N. & Deeth, H.C. (2012) Thermal processing of dairy products. In: *Thermal Food Processing: New Technologies and Quality Issues* 2nd Edn (ed. D-W. Sun), pp. 273–307. CRC Press, Taylor & Francis Group, Boca Raton, FL.

Kessler H.G. (1994) Heat resistance of micro-organisms under seals. In: *Proceedings of the 24th International Dairy Congress, Melbourne.* International Dairy Federation, Brussels.

Labots, H. & Hup, G. (1964) *Bacillus cereus* in raw and pasteurized milk. II. The occurrence of slow and fast germinating spores in milk and their significance in the enumeration of *B. cereus* spores. *Netherlands Milk and Dairy Journal* **18**, 167–176.

Lai, K.K. (2001) *Enterobacter sakazakii* infections among neonates, infants, children and adults. *Medicine* **80**, 113–122.

Langevel, L. P. (1967) *Clostridium thermosaccharolyticum* in sterilized chocolate milk. *Netherlands Milk and Dairy Journal* **21**(3-4), 245–247.

Larsen, H.D. & Jørgensen, K. (1999) Growth of *Bacillus cereus* in pasteurized milk products. *International Journal of Food Microbiology* **46**(2), 173–176.

Larsen, H.D., & Jørgensen, K. (1997) The occurrence of *Bacillus cereus* in Danish pasteurized milk. *International Journal of Food Microbiology* **34**(2), 179–186.

Lewis, M.J. (1999) Microbiological issues associated with heat-treated milks. *International Journal of Dairy Technology* **52**, 121–125.
Lewis, M.J. & Deeth, H.C. (2009) Heat treatment of milk. In: *Market Milks – Processing and Quality Management* (ed. A.Y. Tamime), pp. 168–204. Blackwell Publishing Ltd, Oxford.
Lewis, M.J. & Heppell, N. (2000) *Continuous Thermal Processing of Foods: Pasteurization and UHT Sterilization*. Aspen Publishers, Gaithersburg, MD.
Lima, L., Kamphuis, H., Nout, M. & Zwietering, M. (2011) Microbiota of cocoa powder with particular reference to aerobic thermoresistant spore-formers. *Food Microbiology* **28**, 573–582.
Lima, L., van der Velpen, V., Wolkers-Rooijackers, J., Kamphuis, H., Zwietering, M. & Nout, M. (2012) Microbiota dynamics and diversity at different stages of industrial processing of cocoa beans into cocoa powder. *Applied and Environmental Microbiology* **78**(8), 2904–2913.
Lindström, M., Myllykoski, J., Sivelä, S. & Korkeala, H. (2010) *Clostridium botulinum* in cattle and dairy products. *Critical Reviews in Food Science & Nutrition* **50**, 281–304.
Lück, H., Mostert, J.F. & Husmann, R.A. (1978) Bacteriological evaluation of UHT milk, *South African Journal of Dairy Technology* **10**, 83–85.
Lund, B.M. (1990) The prevention of foodborne listeriosis. *British Food Journal* **92**, 13–22.
Markland, S.M., Farkas, D.F., Kniel, K.E. & Hoover, D.G. (2013) Pathogenic psychrotolerant spore-formers: An emerging challenge for low-temperature storage of minimally processed foods. *Foodborne Pathogens and Disease* **10**(5), 413–419.
Marks, N. E., Grandison, A.S. & Lewis, M.J. (2001) Challenge testing of the lactoperoxidase system in pasteurized milk. *Journal of Applied Microbiology* **91**, 735–741.
Marshall, R.T. (1996) Critical point to control contamination, growth and enzymatic activity. In: *Symposium on Bacteriological Quality of Raw Milk*. Wolfpassing, Austria (13-15 March), International Dairy Federation, Brussels.
Martin, N.H., Ranieri, M.L., Murphy, S.C., Ralyea, R.D., Wiedmann, M. & Boor K.J. (2011) Results from raw milk microbiological tests do not predict shelf-life performance of commercially pasteurized fluid milk. *Journal of Dairy Science* **94**, 1211–1222.
Mayr, R., Gutser, K., Busse, M. & Seiler, H. (2004a) Gram-positive non-spore-forming bacteria are frequent spoilage organisms of German retail ESL (Extended Shelf-life) milk. *Milchwissenschaft* **59**, 262–266.
Mayr, R., Gutser, K., Busse, M. & Seiler, H. (2004b) Indigenous aerobic spore-formers in high heat treated (127 °C, 5 s) German ESL (Extended Shelf-life) milk. *Milchwissenschaft* **59**, 143–146.
McGuiggan, J. T. M., McCleerey, D.R., Hannan, A. & Gilmour, A. (2002) Aerobic spore-forming bacteria in bulk raw milk: factors influencing the numbers of psychrotropic, mesophilic and thermophilic Bacillus spores. *International Journal of Dairy Research* **55**, 100–107.
McKellar, R.C. (1989) Regulation and control of synthesis. In: *Enzymes of Psychrotrophs in Raw Food*. (ed. R.C. McKellar), CRC Press, Boca Raton, Fl.
Meer, R.R., Baker, J., Bodyfelt, F.W. & Griffiths, M.W. (1991) Psychrotrophic *Bacillus* spp in fluid milk-products - a review. *Journal of Food Protection* **54**(12), 969–979.
Meer, R.R., Wodburn, M.J. & Bodyfelt, F.W. (1993) Identification and characterization of heat-resistant psychrotrophic bacteria in Oregon Grade A raw milk. *Dairy, Food and Environmental Sanitation* **13**(11), 631–637.

Meier, J., Rademacher, B., Walenta, W. & Kessier, H.G. (1995) Heat resistant spores under UHT treatment. In: *Heat Treatments and Alternative Methods*. IDF Document 9602, pp. 17–25. International Dairy Federation, Brussels.

Mikolajcik, E.M. (1970) Thermodestruction of *Bacillus* spores in milk. *Journal of Milk and Food Technology* **33**, 77–82.

Mitchell, G.E. & Ewings, K.N. (1985) Quantification of bacterial proteolysis causing gelation in UHT-treated milk. *New Zealand Journal of Dairy Science and Technology* **20**(1), 65–76.

Moats, W.A. (1971) Kinetics of thermal death of bacteria. *Journal of Bacteriology* **105**(1), 165–171.

Mugadza D. & Buys E. (2014) Diversity of spore former and non-spore former bacteria in Extended Shelf-life (ESL) milk. South African Society of Dairy Technology 2014 Dairy Symposium. (http://www.sasdt.co.za/wp-content/uploads/2015/04/Session-8-Speaker-26-Mugadza1.pdf)

Mugadza D. & Buys E. (2015) Characterization of *B. cereus* in Extended Shelf-life (ESL) milk. International Association for Food Protection Conference Portland, Oregon PP2-138 (Abs). (https://iafp.confex.com/iafp/2015/webprogram/Paper8975.html)

Muir, D.D. (1990) The microbiology of heat treated fluid milk products. In: *Dairy Microbiology*, Vol 1. (ed. R.K. Robinson), pp 209–243. Elsevier Applied Science, New York.

Muir, D.D. (1996a) The shelf-life of dairy products: I. Factors influencing raw milk and fresh products. *Journal of the Society of Dairy Technology* **49**(1), 24–32.

Muir, D.D. (1996b) The shelf-life of dairy products: II. Raw milk and fresh products. *Journal of the Society of Dairy Technology* **49**(1), 44–48.

Muir, D.D. (1996c) The shelf-life of dairy products: III. Factors influencing intermediate and long life dairy products. *Journal of the Society of Dairy Technology* **49**(1), 67–72.

Murphy, P.M., Lynch D. & Kelly P.M. (1999) Growth of thermophilic spore forming bacilli in milk during the manufacture of low-heat powders. *Journal of the Society of Dairy Technology* **52**, 45–50.

Nazaro-White, M. & Farber, J. (1997) Thermal resistance of *Enterobacter sakazakii* in reconstituted dried infant formula. *Letters in Applied Microbiology* **24**, 9–13.

Nazaro-White, M. & Farber, J. (1997) Thermal resistance of *Enterobacter sakazakii* in reconstituted dried infant formula. *Letters in Applied Microbiology* **24**, 9–13.

Notermans, S., Dufrenne, J., Teunis, P., Beumer, R., Te Giffel, M. & Peeters Weem, P. (1997) A risk assessment study of *Bacillus cereus* present in pasteurized milk. *Food Microbiology* **14**(2), 143–151.

Odumeru, J.A., Toner, A.K., Muckle, C.A., Griffiths, M.W. & Lynch, J.A. (1997) Detection of *Bacillus cereus* diarrheal enterotoxin in raw and pasteurized milk. *Journal of Food Protection* **60**, 1391–1393.

Osaili, T. M., Shaker, R. R., Al-Haddaq, M. S., Al-Nabulsi, A. A. & Holley, R. A. (2009) Heat resistance of *Cronobacter* species (*Enterobacter sakazakii*) in milk and special feeding formula. *Journal of Applied Microbiology* **107**(3), 928–935.

Overcast, W.W. & Atmaram, K. (1974) The role of *Bacillus cereus* in sweet curdling of fluid milk. *Journal of Milk and Food Technology* **37**(5), 233–236.

Pearce L E. (2004) Survey of data on heat resistance of dairy pathogens. IDF Bulletin 392, pp. 37–41. International Dairy Federation, Brussels.

Pettersson, B., Lembke, F., Hammer, P., Stackebrant, E. & Priest, G. (1996) *Bacillus sporothermodurans*, a new species producing highly heat-resistant endospores. *International Journal of Systematic Bacteriology* **46**, 759–764.

Pflug, I.J. & Gould, G.W. (1999) Heat treatment. In: *Microbiological Safety and Quality of Food*. (eds B. Lund, T.C. Baird-Parker & G.W. Gould), pp. 36–64. Aspen, Gaithersburg, Maryland, USA.

Prentice, G.A. (1994) Listeria monocytogenes. In: *Monograph on the Significance of Pathogenic Microorganisms in Raw Milk*. pp 101–112. IDF Document S.I 9405. International Dairy Federation, Brussels.

Pujol, L. & Membré, J.-M. (2014) Advanced microbial modelling techniques and risk-based management applied to aseptic-UHT process. *New Food Magazine* **17**(1), 13–16.

Pujol, L., Albert, I., Magras, C., Johnson, N.B. & Membre, J.M. (2015) Probabilistic exposure assessment model to estimate aseptic-UHT product failure rate. *International Journal of Food Microbiology* **192**, 124–141.

Put, H.M., Van Doren, H., Warner, W.R. & Kruiswijk, J.Th. (1972) The mechanism of microbiological leaker spoilage of canned foods: a review. *Journal of Applied Bacteriology* **35**, 7–27.

Ranieri, M.L. & Boor, K.J. (2009) Short communication: bacterial ecology of high-temperature, short-time pasteurized milk processed in the United States. *Journal of Dairy Science* **92**, 4833–4840.

Ridgway, J.D. (1958) The incidence and thermal resistance of mesophilic spores found in milk and related environments. *Journal of Applied Bacteriology* **21**(1), 118–127.

Robertson, G. L. (2003) Ultra-high temperature treatment (UHT) – aseptic packaging. In: *Encyclopedia of Dairy Sciences* (eds H. Roginski, J.W. Fuquay & P.F. Fox), pp. 2637–2642. Academic Press, London.

Ronimus, R.S., Parker, L.E., Turner, N., Poudel, S., Rueckert, A. & Morgan, H.W. (2003) A RAPD-based comparison of thermophilic bacilli from milk powders. *International Journal of Food Microbiology* **2644**, 1–17.

Rowan, N.J. & Anderson, J.G. (1998) Diarrhoeal enterotoxin production by psychrotrophic *Bacillus cereus* present in reconstituted milk-based infant formulae (MIF). *Letters in Applied Microbiology* **26**(2), 161–165.

Rueckert, A., Ronimus, R.S. & Morgan, H.W. (2004) A RAPD-based survey of thermophilic bacilli in milk powders from different countries. *International Journal of Food Microbiology* **96**, 263–272.

Rueckert, A., Ronimus, R.S. & Morgan, H.W. (2005) Rapid differentiation and enumeration of the total, viable vegetative cell and spore content of thermophilic bacilli in milk powders with reference to *Anoxybacillus flavithermus*. *Journal of Applied Microbiology* **99**, 1246–1255.

Rysstad G. & Kolstad J. (2006) Extended shelf-life milk - advances in technology. *International Journal of Dairy Technology* **59**, 85–96.

Salustiano, V.C., Andrade, N.J., Soares, N.F.F., Lima, J.C., Bernardes, P.C., Luiz, L. M.P. & Fernandes, P.E. (2009) Contamination of milk with *Bacillus cereus* by post-pasteurization surface exposure as evaluated by automated ribotyping. *Food Control* **20**(4), 439–442.

Scheldeman, P., Goossens, K., Rodriguez-Diaz, M., Pil, A., Goris, J., Herman, L., De Vos, P., Logan, N.A. & Heyndrickx, M. (2004) *Paenibacillus lactis* sp. *nov*. isolated from raw and heat-treated milk. *International Journal of Systematic & Evolutionary Microbiology* **54**, 885–891.

Scheldeman, P., Herman, L., Foster, S. & Heyndrickx, M. (2006) *Bacillus sporothermodurans* and other highly heat-resistant spore formers in milk. *Journal of Applied Microbiology* **101**, 542–555.

Scheldeman, P., Pil, A., Herman, L., De Vos, P. & Heyndrickx, M. (2005) Incidence and diversity of potentially highly heat-resistant spores isolated at dairy farms. *Applied and Environmental Microbiology* **71**(3), 1480–1494.

Scott, S.A., Brooks, J.D., Rakonjac, J., Walker, K.M.R. & Flint, S.H. (2007) The formation of thermophilic spores during the manufacture of whole milk powder. *International Journal of Dairy Technology* **60**, 109–117.

Shaker, R., Osaili, T., Al-Omary, W., Jaradat, Z. & Al-Zuby, M. (2007) Isolation of *Enterobacter sakazakii* and other *Enterobacter* sp. from food and food production environments. *Food Control* **18**, 1241–1245.

Shehata, A.E., Magdoub, M.N.I., Sultan, N.E. & EI-Samragy, Y.A. (1983) Aerobic mesophilic and psychrotrophic spore-forming bacteria in buffalo milk. *Journal of Dairy Science* **66**, 1228–1231.

Shelley, A.W., Deeth, H.C. & MacRae, I.C. (1987) A numerical taxonomic study of psychrotrophic bacteria associated with lipolytic spoilage of raw milk. *Journal of Applied Bacteriology* **62**, 197–207.

Simmonds, P, Mossel B.L., Intaraphan T. & Deeth H.C. (2003) The heat resistance of *Bacillus* spores when adhered to stainless steel and its relationship to spore hydrophobicity. *Journal of Food Protection* **66**, 2070–2075.

Skladal, P., Mascini, M., Salvadori, C. & Zannoni, G., (1993) Detection of bacterial contamination in sterile UHT milk using an L-lactate biosensor. *Enzyme & Microbial Technology* **15**, 508–512.

Slaghuis, B.A., Te Giffel, M., Beumer, R.R. & Andre, G. (1997) Effect of pasturing on the incidence of *Bacillus cereus* spores in raw milk. *International Dairy Journal* **7**, 201–205.

Sorhaug, T. & Stepaniak, L. (1997) Psychrotrophs and their enzymes in milk and dairy products: quality aspects. *Trends in Food Science and Technology* **8**, 35–41.

Stadhouders, J. & Hup, G. (1980) Effect of milk storage before pasteurization on the formation of flocs (bitty cream) by *Bacillus cereus* in pasteurized milk. *Zuivelzicht* **72**(8), 178–180.

Stadhouders, J., Hup, G. & Langeveld, L.P.M. (1980) Some observations on the germination, heat resistance and outgrowth of fast-germinating and slow-germinating spores of *Bacillus cereus* in pasteurized milk. *Netherlands Milk and Dairy Journal* **34**(4), 215–228.

Stewart, D.B. (1975) Factors influencing the incidence of *B. cereus* spores in milk. *Journal of the Society of Dairy Technology* **28**(2), 80–85.

Stone, J.M. & Rowlands, A. (1952) Broken or bitty cream in raw and pasteurised milk. *Journal of Dairy Research* **19**, 52–62.

Suhren, G. (1989) Producer micro-organisms. In: *Enzymes of Psychrotrophs in Raw Food*, (ed. R.C. McKellar), pp. 3–34. CRC Press, Boca Raton, Fl.

Sung, N. & Collins, M. T. (1998) Thermal tolerance of *Mycobacterium paratuberculosis*. *Applied and Environmental Microbiology* **64**, 999–1005.

Tabit, F.T. & Buys, E.M. (2011) Incidence and survival of *Bacillus sporothermodurans* during processing of UHT milk. *British Food Journal* **113**(4-5), 505–518.

Te Giffel, M., Beumer, R.R., Bonestroo, M.H. & Rombouts, F.M. (1996) Incidence and characterization of *Bacillus cereus* in two dairy processing plants. *Netherlands Milk and Dairy Journal* **50**(4), 479–492.

Te Giffel, M., Beumer, R.R., Slaghuis, B.A. & Rombouts, F.M. (1995) Occurrence and characterization of (psychrotrophic) *Bacillus cereus* on farms in the Netherlands. *Netherlands Milk and Dairy Journal* **49**(2/3), 125–138.

Te Giffel, M.C., Beumer, R.R., Granum, P.E. & Rombouts, F.M. (1997) Isolation and characterisation of *Bacillus cereus* from pasteurised milk in household refrigerators in the Netherlands. *International Journal of Food Microbiology* **34**(3), 307–318.

Teh, K. H., Lindsay, D., Palmer, J., Andrewes, P., Bremer, P. & Flint, S. (2014) Proteolysis in ultra-heat-treated skim milk after exposure to multispecies biofilms under conditions modelling a milk tanker. *International Journal of Dairy Technology* **67**(2), 176–181.

Tetra Pak (undated) *Procedures & Guidelines Guideline for Microbiological Evaluation of Commercially Sterile Products.* Tetra Pak QAM – 588002-0107.

Tompkin. R.B. (2002) Control of *Listeria monocytogenes* in the food-processing environment. *Journal of Food Protection* **65**, 709–725.

Touch, V. & Deeth, H.C. (2009) Microbiology of raw and market milks. In: *Market Milks – Processing and Quality Management* (ed. A.Y. Tamime), pp. 48–71. Blackwell Publishing Ltd, Oxford.

van Asselt, E. D. & Zwietering, M. H. (2006) A systematic approach to determine global thermal inactivation parameters for various food pathogens. *International Journal of Food Microbiology* **107**(1), 73–82.

Van Netten, P., van de Moosdijk, A., van Hoensel, P., Mossel, D.A.A. & Perales, I. (1990) Psychrotrophic strains of *Bacillus cereus* producing enterotoxin. *Journal of Applied Bacteriology* **69**(1), 73–79.

Vasavada, P.C. & Cousin, M.A. (1993) Dairy microbiology and safety. In: *Dairy Science and Technology Handbook*, Vol 2 *Product Manufacturing*. (ed. Y.H. Hui), pp. 301–426. VCH Publishers, USA.

von Bockelmann, B. & von Bockelmann, I.A. (1998) *Long-life Products: Heat-treated, Aseptically Packed: A Guide to Quality*. Dr Bernard von Bockelmann, Åkarp, Sweden.

Vyletelova, M., Svec, P., Pacova, Z., Sedlacek, I. & Roubal, P. (2002) Occurrence of *Bacillus cereus* and *Bacillus licheniformis* strains in the course of UHT milk production. *Czech Journal of Animal Science* **47**(5), 200–205.

Watterson, M.J., Kent, D.J., Boor, K.J., Wiedmann, M. & Martin, N.H. (2014) Evaluation of dairy powder products implicates thermophilic spore-formers as the primary organisms of interest. *Journal of Dairy Science* **97**, 2487–2497.

Wescott, G.G., Fairchild, T.M. & Foegeding, P.M. (1995) *Bacillus cereus* and *Bacillus stearothermophilus* spore inactivation in batch and continuous flow systems. *Journal of Food Science* **60**, 446–450.

Wilkinson, G. & Davies, F.L. (1973) Germination of spores of *Bacillus cereus* in milk and milk dialysates: effect of heat treatment. *Journal of Applied Bacteriology* **36**(3), 485–496.

Witthuhn, M., Lucking, G., Atamer, Z., Ehling-Schulz, M. & Hinrichs, J. (2011) Thermal resistance of aerobic spore formers isolated from food products. *International Journal of Dairy Technology* **64**(4), 486–493.

5

UHT Processing and Equipment

5.1 The UHT Process

The UHT process is an integrated series of unit processes which transforms a non-sterile product such as raw milk into a packaged commercially sterile product which is stable at room temperature for several months. From the inlet pump to the final package, the system is closed to the external environment with the sections following the sterilisation section being maintained is an aseptic condition (Chandarana *et al.*, 2010). As sterilisation occurs at around 140 °C, the system is pressurized at 0.4-0.5 MPa to ensure water in the product does not boil. In some applications, the raw material is deaerated before being heated to improve thermal efficiency of the heat exchangers and reduce oxidation of the product (Dogan & Kokini, 2007).

The major steps in a UHT process are as follows:

- Raw material input
- Pre-heating, with or without a holding time
- Deaeration (optional)
- Homogenisation (for indirect systems)
- Heating to sterilisation temperature
- Holding at sterilisation temperature
- Initial cooling (vacuum flash-down in direct systems)
- Homogenisation (for direct or indirect systems)
- Final cooling
- Aseptic holding
- Aseptic packaging
- Ambient storage of UHT product

The pre-heating stage takes the temperature of the product from ~5°C to 75-95°C using the hot product after it emerges from the sterilisation step to heat the incoming cold product. In some UHT plants, over 90% of the heat is regenerated in this way which is important for the energy efficiency of the plant. After pre-heating, the milk may enter a holding tube which maintains the temperature for a set period of time, from 30 to 120 s. The major reasons for this pre-heat holding step are to "stabilize" the proteins in order to reduce the amount of fouling, or deposit formation, in subsequent heat exchangers, and to inactivate plasmin, a natural milk protease which can cause defects in the product during storage (see more detail on pre-heating in Section 5.2.1.3 and Chapter 7).

High Temperature Processing of Milk and Milk Products, First Edition. Hilton C. Deeth and Michael J. Lewis.

The final heating step to the required sterilisation temperature is achieved by one of two major types of heating, *direct* steam heating or *indirect* heating using heat exchanger systems.

When the sterilisation temperature is reached, the milk enters a holding tube. The temperature of the milk and the time it takes to pass through this holding tube are the nominal sterilisation conditions which are usually cited for a UHT process, for example, 140 °C for 5 s.

The initial cooling of the product in direct systems is achieved by passing it through a vacuum chamber which removes the water condensed into the product during the steam heating and reduces the temperature of the product to the temperature from which it was heated, usually around 75 °C. Initial cooling in indirect systems occurs through heat exchange with the cold incoming product as discussed above. In the final cooling steps, in both direct and indirect systems, cooling is provided by indirect heat exchange with chilled water (Glaser & Kreutz, 2010).

After cooling, the product enters the aseptic packaging section. In most commercial plants, the product is stored in an aseptic tank (A tank) before being aseptically packaged. The A tank provides a buffer for the product to accommodate the different rates of product sterilisation and packaging and also any stoppages of the aseptic packaging equipment. Various packaging options are available with the most common being paperboard and multilayered plastic bottles and pouches. The packages used are either pre-sterilised or sterilised with a sterilant such as hot hydrogen peroxide before being filled (see Section 5.5).

Fat-containing products such as whole milk are homogenised during the UHT process. Homogenisation is carried out at 60-75 °C either before or after the sterilisation step. In direct systems, homogenisation is always carried out downstream of the sterilisation section while in indirect systems, the homogeniser can be placed upstream or downstream (more detail on homogenisation is in Section 5.3).

5.2 Heating

There are two major energy sources used for UHT heating systems, steam/hot water and electrical. The steam/hot-water systems are by far the most common type used commercially (JBT, undated).

5.2.1 Steam-/Hot-Water-Based Heating Systems

These systems use either steam directly or hot water, normally heated by stream. They are divided into two types, direct and indirect. In direct systems, heating occurs through direct contact between steam and the product and in indirect systems the heat is transferred to the product from the steam or hot water through a stainless steel barrier in the heat exchanger.

5.2.1.1 Direct Heating

In direct heating UHT plants, milk which has been pre-heated indirectly to 70-80 °C is heated to sterilisation temperatures by direct contact with culinary superheated steam. The heating occurs through transfer of the latent heat of evaporation of the steam to the product. The heating is very rapid with a rise in temperature of 60-70 °C occurring in around 0.5 s. At the same time, water is condensed from the steam and dilutes the product. The amount of water condensed depends on the initial and final temperatures but is of the order of 11%. The added water is removed from the product in a vacuum vessel (item 4 in Figure 5.1) placed immediately after the sterilisation holding tube which also performs the initial cooling of the product.

Direct systems can be either an *injection* type, in which steam is injected into the product (steam into product), or an *infusion* type, in which product is infused into a chamber of steam (product into steam).

5.2.1.1.1 Steam Injection

In steam injection, steam is mixed with a stream of product via a steam injector nozzle (Figure 5.2a). At the point of mixing in the steam injection system, steam vapour bubbles condense and the associated implosions cause cavitation and extensive turbulence which causes some homogenisation of the product (van Boekel & Folkerts, 1991). Some fouling deposit can also form at the point of mixing which can limit the run time of the plant.

5.2.1.1.2 Steam Infusion

In steam infusion, the product is passed in the form of streams into an infusion chamber (Figure 5.2b and item 2 in Figure 5.1) filled with steam at a temperature a little higher than the target sterilisation temperature. The infusion method of mixing product with steam is gentler than that of steam injection and no homogenisation occurs.

The product is introduced through a distribution device in circularly distributed strings or streams. Some distribution devices incorporate a deaeration mechanism to remove air from the product. Free air in the product has a major detrimental effect on heat transfer. It has been reported that the heat transfer rate into product containing air can be only 90 °C/s while a rate of 500 °C/s can be achieved with a properly deaerated product (Fredsted *et al.*, 1996) (deaeration is discussed in more detail in Section 5.4).

Little fouling occurs during steam infusion processes. Some deposit may form at the base of the infusion chamber where the hot product makes contact with the chamber wall. However, this is minimized on some plants by reducing the temperature of this part of the chamber to 110-130 °C. This is indicated by the "Cooling water" jacket in Figure 5.1. This allows the formation of a thin film of condensate at the base of the chamber which prevents fouling and also foaming (Fredsted *et al.*, 1996). Another point at which fouling can occur is in the product distribution plate at the entry to the infusion chamber. However, this has been shown to be avoided if there is turbulent flow at the outlet (Reynolds number >10,000; de Jong *et al.*, 1996) (see Section 5.2.1.8.3.3).

The set up of a direct heating system using steam infusion is shown in Figure 5.1. A system using steam injection is similar to that shown in Figure 5.1, except that it has a steam injector nozzle rather than the stream infusion chamber (item 2). As shown in Figure 5.1, direct heating is only used for the high-heat section; other heating sections utilize indirect heating. The indirect heating sections in Figure 5.1 are shown as plate heat exchangers but they are commonly tubular.

As indicated above, in direct UHT plants, water is condensed into the product and must be removed in the vacuum flash-down chamber (item 4 in Figure 5.1). This requires close attention from the operator as removal of insufficient water leads to adulteration of the

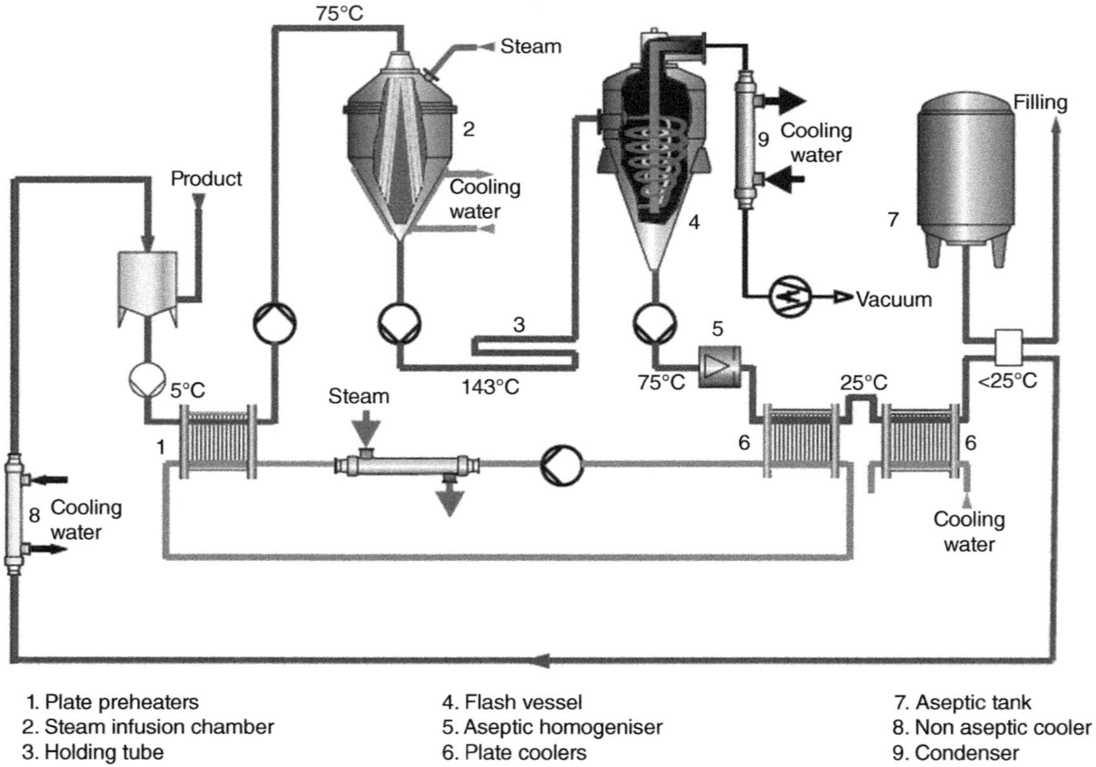

Figure 5.1 Set up of a direct (infusion) UHT system with plate heat exchangers in the preheat and final cooling stages. (Source: Reproduced with permission of SPX Flow Inc.)

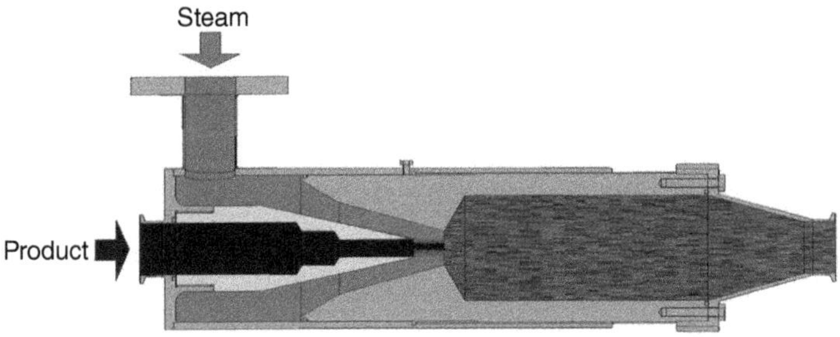

Figure 5.2a Steam injector nozzle. (Source: Reproduced with permission of SPX Flow Inc.)

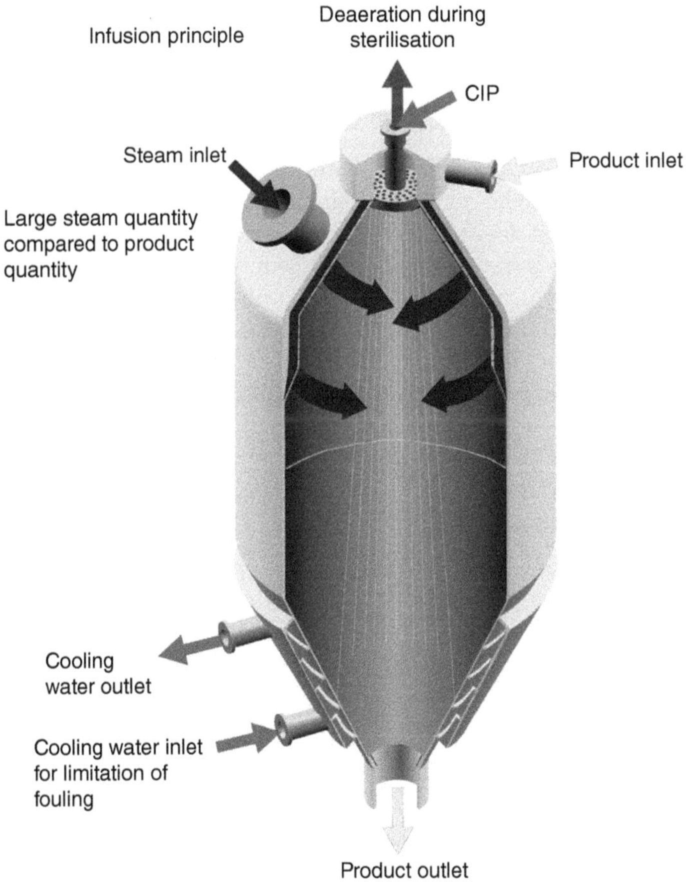

Figure 5.2b Steam infusion chamber. (Source: Reproduced with permission of SPX Flow Inc.)

final product and removal of too much water results in concentration of the product which reduces its volume and hence financial returns. The water content of the UHT product can be conveniently checked during processing by total solids determination using a rapid drying method or by freezing point determination using a cryoscope (see Section 11.2.8).

An important practical consideration with direct UHT plants is the requirement for high-quality, culinary steam. Several operators have experienced severe flavour problems in directly processed UHT milk when the steam was contaminated. The contamination may originate in the water used or from chemical treatment of water in the boiler.

A temperature–time profile of a commercial direct UHT plant is shown in Figure 5.3; however, the profiles of commercial plants can vary considerably. The major features are the slow indirect pre-heating followed by the very rapid heating to sterilisation temperature, rapid initial cooling due to the vacuum flash-down and the slower indirect final cooling. Some plants may also include a pre-heat holding section for "proten stabilisation" and plasmin inactivation (see Section 5.2.1.3).

A further feature in Figure 5.3 is the slight rise in temperature after the initial cooling step which is due to heat gained in the downstream homogeniser. As indicated above, the homogeniser is usually placed downstream in a direct UHT plant. The reason for this is to break up protein aggregates which form during direct steam injection or infusion heating of milk that can impart a chalky/astringent mouth-feel to the final product. Furthermore, direct processing tends to lead to more sediment in the final product than indirect processing; this is also reduced by downstream homogenisation. The sediment in directly processed milk is believed to be caused by a similar mechanism to fouling in plate or tubular heat exchangers but in direct heating it forms in the bulk of the product and is less likely to attach to the stainless steel surfaces. While it is generally recognised that UHT products which are homogenised after the sterilisation step have better physical stability during storage, excessive homogenisation may decrease product stability due in part to the increased density of very small fat globules caused by attachment of protein causing them to sediment in the container. A major disadvantage of downstream homogenisation is that the homogeniser must be aseptic so that it does not contaminate the sterilised product. This requires special attention to be paid to the maintenance of the homogeniser, particularly to the seals which can harbour bacteria and create a microenvironment favourable to their growth (Kessler, 1994).

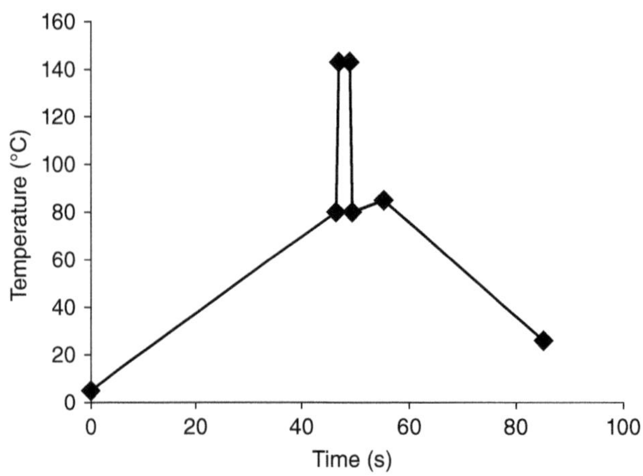

Figure 5.3 Temperature–time profile of a commercial direct UHT plant.

5.2.1.2 Indirect Heating

Indirect UHT plants are the most common type used in the dairy industry. In these plants, heat is transferred to the product by conduction and convection from steam or hot water. When hot water is used, the water flows in the reverse direction to product. This counter-currrent flow minimises the temperature differential between the product and the heating medium which in turn minimises fouling or burn-on deposit on the heat exchanger surfaces. Hot water is a significantly better heating medium than steam with respect to burn-on and flavour of the product as it enables a smaller temperature differential between the milk and the heating medium (Dentener, 1984).

The heat exchangers used in indirect heating can be either plate or tubular, although tubular are more common. Tubular heat exchangers can have different configurations such as tube-in-shell, in which several small tubes are contained in parallel inside a larger tubular shell, and concentric tubes. The layout of an indirect tubular UHT plant is shown in Figure 5.4. Salient features include the upstream homogeniser (2) between two pre-heating sections (1) followed by a pre-heat holding section (shown at 95 °C) (3), the high temperature heating section (4) which heats the product to sterilisation temperature, the sterilisation holding tube (shown at 140 °C), and initial and final cooling sections (5 and 6). It also shows the heat regeneration loop from the initial cooling section to the initial pre-heat section.

Plate heat exchangers consist of a series of parallel, channelled plates clamped together with gasket spacers. Product and heating medium flow in opposite directions through the alternate spaces between plates. Plate packs are depicted in Figure 5.1 in the initial heating and final cooling sections. While they have the advantage of relatively fast heating, the inter-plate gaskets require constant attention. Furthermore, the gaskets limit the temperatures and pressures at which the plant can be operated. The pressure limit

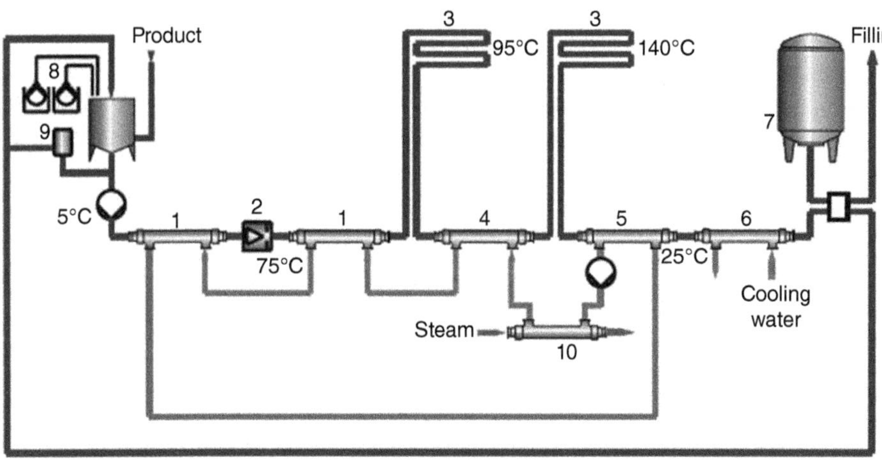

1. Tubular regenerative preheaters
2. Homogeniser
3. Holding tubes
4. Tubular final heater
5. Tubular regenerative cooler
6. Final cooler
7. Sterile tank
8. CIP unit
9. Sterilising loop
10. Water heater

Figure 5.4 Set up of a indirect (tubular) UHT system. (Source: Reproduced with permission of SPX Flow Inc.)

is partially related to another limitation, the viscosity of the product. Plate heat exchangers are much less suitable for viscous products that tubular heat exchangers.

The heat transfer rate in tubular heat exchangers is slower than in plate heat exchangers as the distance between the walls of the tube and product at the centre of the tube is large compared with the space between plates in a plate heat exchanger. Furthermore, the configuration of the plates creates a high level of turbulence in the product which also aids heat transfer to the product. These differences lead to shorter processing times in plate systems than in tubular systems. Thus the rate of heating in plate heat exchangers is between the relatively slow heating of tubular heat exchangers and the very rapid heating of direct steam heating.

A practical issue for heat exchangers is the development of pinholes which allow the heating medium to contaminate the product. This is of major concern in the downstream sections of the plant where the product is sterile. It is a greater issue for plate than tubular heat exchangers. Regular testing for pinholes is therefore required, especially for plate systems. Some tubular heat exchangers are set up with a double jacket or closed loop containing sterile water so that the heating/cooling medium heats/cools sterile water in the jacket which then heats/cools the product. Any pinholes in the wall of the heat exchanger in contact with the product will then only allow contamination of the product with sterile water.

Fouling is a major issue for indirect UHT plants and limits their run times. A comparison of the typical run times of direct and indirect UHT plants is as follows:

Direct steam infusion	>20 h
Direct steam injection	>18 h
Indirect plate	8–12 h
Indirect tubular	12–16 h

(Source: APV)

Note that run times can be further extended by introducing a preheating section for protein stabilisation. (See 5.2.1.3)

The temperature–time profile of a commercial indirect UHT plant is shown in Figure 5.5. It must be stressed however, that the shape of the profiles can vary considerably as illustrated by Lewis and Heppell (2000) and Tran *et al.* (2008) for several commercial

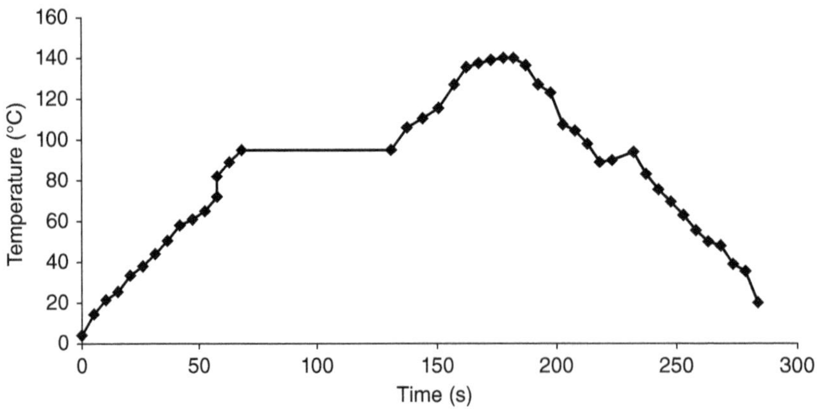

Figure 5.5 Temperature–time profile of a commercial indirect UHT plant.

plants. The major differences between this profile and that of the direct plant shown in Figure 5.3 are the pre-heat holding tube of about 20 s, and the slower heat-up and cool-down sections of either side of the sterilisation holding tube. As illustrated in Chapter 6, considerable information can be derived from such profiles.

Some indirect UHT plants are capable of operating at between 50 and 100% of the nominal capacity, but with the same bactericidal efficiency at all flow rates. This enables the plant to reduce its output if problems are experienced with a packaging line. It is achieved using a *split heater* design whereby the heating section is split into sub-sections. When a 50% reduction in flow is required, a valve is activated which diverts some of the heating medium to flow outside of a part of the heating section. This effectively increases the pre-heat holding time. The product is then heated to the sterilisation section in which the temperature is adjusted to allow for the increased holding time at the lower flow rate, but to provide the same heat load to the product (Tetra Pak, 2015).

5.2.1.3 Pre-Heating

Many UHT plants are fitted with a pre-heat holding section which is designed to hold the milk for periods of time from 30 s to 120 s at a temperature between 75 and 95 °C. It is claimed that this increases the run time and makes the milk less prone to forming deposits in the high temperature section. It may also reduce the amount of sediment in the product. There is no precise time and temperature conditions stated for optimum performance, but as noted below some recommendations have been published, particularly in relation to inactivation of plasmin.

To date, pre-heat holding has been mostly used on indirect UHT plants. This is presumably because fouling is a greater issue with indirect than direct plants (see Section 6.2.2). However, inactivation of plasmin is important for the shelf-life of UHT milk (see Sections 7.1.3.4, 7.1.4 & 7.2.2.1) and hence a pre-heat holding section is very beneficial for direct plants also.

Table 5.1 shows the contributions of various temperature–time pre-heat holding combinations to some process and product parameters (see Sections 3.2.2 & 3.2.3). It shows that there is no contribution to the bactericidal values (B* and F_0) and

Table 5.1 Contribution of various pre-heat temperature–time (°C/s) combinations to process and product parameters in a UHT plant.

	80/60	80/300	85/120	90/30	90/60	95/30	95/60
B*	0	0	0	0	0	0	0
F_0	0	0	0	0	0	0	0
C*	0.03	0.18	0.1	0.04	0.07	0.05	0.1
Lactulose (mg/100 g)	0.1	0.4	0.3	0.13	0.3	0.24	0.5
Browning (s at 121 °C)	1.6	8	5	2	4	3	6
Denaturaton of β-Lg (%)[1]	14	42	51	55	71	61	76
Denaturaton of α-La (%)[2]	5	24	14	5	10	7	13

[1] based on kinetics data of Lyster (1970); higher percentage denaturation is predicted by the kinetics data of Dannenberg & Kessler (1988). Cumulative data – includes heat-up of 120 s.
[2] based on kinetics data of Lyster (1970)

contributions to some of the chemical parameters are small, but not negligible. On an indirect UHT plant with a sterilisation holding of 138 °C for 4 s, a pre-heat holding section at 95 °C for 120 s would increase the parameters as follows. The first and second figures give values with and without the pre-heat holding sections: browning (s at 121 °C) 141, 128; F_0 12.9, 12.9; B* 3.1, 3.1; C* 1.56, 1.33; lactulose (mg/100 mL) 38.8, 34.5; HMF (μmol/L) 38.5, 36; and β-Lg denaturation (%) 95.9, 94.6.

It is thought that protein stabilisation is achieved by a combination of whey protein denaturation and a reduction in ionic calcium, although this has not been verified experimentally. A further suggestion is that some precipitation of calcium phosphate onto the surface of the micelle occurs in this section, which makes it less available to precipitate onto the heat exchange surface in the high-temperature section.

The first significant study of the effect of pre-heat holding was undertaken by Mottar and Moermans (1988). They concluded that an optimum temperature–time for the pre-heating process was within the range of 76-80 °C/40-70 s in which adverse effects were minimal and positive effects were maximized. The beneficial effects included minimal deposit formation in the high-temperature sections, minimal protein destabilisation (sediment formation) during storage, minimum cooked flavour development and reasonable inactivation of indigenous plasmin. Sediment was measured after 3 months but not initially. Reduction in proteolytic activity was more influenced by time than by temperature.

It is interesting that the conditions used by many processors are more severe than this. However, this is not surprising as the optimum conditions would be also influenced by natural variations in milk composition, which were not taken into account. The authors also do not comment whether they consider fouling to be excessive, moderate, or low. More recently, Prakash *et al.* (2015) investigated the effect of pre-heat holding temperatures of 65-95 °C and two holding times (5 and 25 s) on fouling of reconstituted skim milk at sterilisation temperatures of 135-150 °C. A bench-top pilot plant was used and fouling was monitored by observing the variation in overall heat transfer coefficient (OHTC) with time. A significantly shorter run-time was obtained with pre-heating at 65 °C than with pre-heating at 75, 85 and 95 °C. A longer holding time (25 s) in the pre-heating section significantly increased the run-time of the plant compared with 5 s holding. The sterilisation temperature also affected the OHTC and run-time; similar run-times were observed at 135, 140, 142 and 145 °C but at 150 °C there was no induction phase and fouling occurred rapidly. Overall, pre-heating at 85 °C or 95 °C for 25 s, combined with a high-heat temperature of 145 °C, and pre-heating at 95 °C for 25 s combined with a high-heat temperature of 142 °C produced the longest run-times.

The interaction with β-lactoglobulin during heating was used as the basis of a method of inactivating plasmin during the pre-heating stages of UHT processing by Van Asselt *et al.* (2008). They devised a pre-treatment of 80 °C for 300 s which inactivated plasmin and prevented proteolysis in milk treated by Innovative Steam Injection (150-180 °C for 0.2 s). Previously this very-high-temperature treatment had been shown to effectively inactivate highly heat-resistant bacterial spores but led to bitter flavours in stored UHT milk due to plasmin action (Huijs *et al.*, 2004). An alternative pre-heat treatment of 90-95 °C for 30-60 s was found to be required to inactivate milk plasmin and prevent proteolysis of directly processed UHT reconstituted skim milk during storage (Newstead *et al.*, 2006).

The effects of the contributions of pre-heating conditions close to those suggested by Mottar and Moermans (1988) (80 °C/60 s), Newstead *et al.* (2006) (90 °C/60 s; 95 °C/30 s; 95 °C/60 s), Van Asselt *et al.* (2008) (80 °C/300 s) and Prakash *et al.* (2015) (95 °C/30 s) to the overall product parameters are shown in Table 5.1. It is interesting to note that the conditions suggested by Van Asselt *et al.* (2008) (80 °C for 300 s) cause the most denaturation of α-lactalbumin but not of β-lactoglobulin. Processing UHT milk with the most severe of these pre-heat holding conditions would contribute significantly to the milk having a negative Aschaffenberg turbidity test, illustrating a limitation of that test (see Sections 3.4.3.1 & 11.2.28.2).

5.2.1.4 Comparison of Indirect and Direct UHT Plants

A summary comparison of direct and indirect UHT plants is shown in Tables 5.2 and 5.3. Of particular note is the higher propensity of indirect plants to foul because of the large area of hot surfaces. In the first stages of the plant where the temperature reached is <100 °C, the main reaction is denaturation of whey proteins, mainly β-lactoglobulin, with subsequent deposition of the denatured protein. In the higher-temperature (>100 °C) sections of the plant, the major reaction is deposition of calcium phosphate, which has reduced solubility at high temperatures. Consequently, the deposit in the high-temperature sections is predominantly mineral, whereas the deposit in the lower-temperature section is predominantly protein. The build up of fouling deposit causes a reduction in heat transfer and an increase in pressure. In commercial plants, the product temperature is maintained by increasing the temperature of the heating medium. This exacerbates the fouling because of the increased temperature differential between the heating medium and the product and, eventually, the plant has to be shut down for cleaning (Ansari *et al.*, 2006).

The degree of heat energy recovery is another difference between the two systems. In direct heating systems, less regeneration of heat is possible since the steam flashed-off in the vacuum chamber is condensed, and the useful heat is lost from the system. In indirect systems, almost all the heat in the hot sterile product at the sterilisation temperature (135-150 °C) can be used in the regeneration section. Thus, heat regeneration in indirect systems is usually >90%, whereas it is only about 50% in direct systems (Lewis & Heppell, 2000).

A major difference between direct and indirect systems is the rate at which the milk is heated; direct systems heat milk from pre-heat temperature to sterilisation temperature in less than one second (Biziak *et al.*, 1985) whereas indirect systems can take several seconds to minutes (Tran *et al.*, 2008). The major consequence of this difference is that, for the same bactericidal effect, the direct systems produce much less chemical change in the milk constituents than the indirect systems. This is reflected in the characteristics of UHT milk produced on the two type of plants. These characteristics are summarized in Table 5.3. The various characteristics are discussed in detail in Chapters 6 and 7.

5.2.1.5 Combination Direct–Indirect Systems

UHT systems have been developed which incorporate both direct and indirect heating (and cooling) in the high-heat section. They were introduced to take advantage of the benefits of each system, in particular, the superior flavour characteristics and greater bacterial spore destruction capability of direct heating (product can be rapidly heated

Table 5.2 Comparison of the processing characteristics of direct and indirect UHT heating systems.

Processing characteristic	Direct systems	Indirect systems
Sterilising temperature for equal bactericidal effect	3-4 °C higher than in indirect systems	
Heating rate from pre-heat to high heat and initial cooling rate	Fast (<0.5 s)	Slow (~30-120 s); rate for plate heat exchangers is greater than for tubular systems
Heat regeneration	~50%	≥90%
Pre-heat hold (at ~90°C) 'protein stabilisation' step	Uncommon	Common
Homogeniser placement	Generally downstream of high-heat section (requires aseptic homogeniser)	Upstream or downstream of high-heat section
Ability to process viscous product	Reasonable, especially with infusion	Good capability with tubular heat exchangers but little with plate heat exchangers
Fouling/burn-on	Usually minimal. Some may occur around the steam injector and the steam infuser distribution plate. Some also occurs in indirect heat exchanger sections and holding sections	A major problem. Less fouling occurs in tubular heat exchangers than in plate heat exchangers
Run time	Longer than for indirect; 18-20 h	Shorter than for direct; 8-16 h; tubular longer than plate type
Steam quality requirement	Very high	No specific requirement
Energy requirement	Higher than indirect	
Ability to reach very high temperature (i.e.>145 °C)	Capable	Limited
Ability to destroy heat-resistant sporeformers without excessive chemical damage	Capable	Limited
Process control issues	Careful control of water removal after sterilization holding tube is required to prevent concentration or dilution	Need to control pressure increase and temperature differential between product and heating tube or plate as fouling layer builds up
Possibility of contamination from heating medium through pinholes	Nil for sterilising section Possible in regeneration and other indirect heating and cooling sections	Significant especially with plate heat exchanger; minimized in closed-loop systems
Water requirement	Greater (~1500 L water per 1000 L product) than for indirect system	
Other process features	Steam injection causes some homogenisation	Tubular is most common UHT heating system. Corrugated tubes are used to increase turbulence

(adapted from Datta *et al.*, 2001)

Table 5.3 Comparison of direct and indirect UHT heating systems in relation to the characteristics of the final products.

Product characteristic	Direct systems	Indirect systems
Flavour (assuming same bactericidal effect)	Mild cooked flavour; chalky if homogenised before high heat section	Strong cooked/heated flavour
Oxygen level. (assuming, no headspace in package, no use of aseptic tank, package not permeable to O_2)	Low (~1 ppm)	High (7-9 ppm)
Sediment formation	Higher than for indirect	
Susceptibility to age gelation	Higher than for indirect	
Plasmin and plasminogen level	Neither completely inactivated	Plasmin generally inactivated but some residual plasminogen may remain
Fat separation	Low, especially for steam injection in which some homogenization occurs	More than for direct
Heat indices – HMF, lactulose, furosine	Lower than for indirect	

(adapted from Datta *et al.*, 2001)

to and cooled down from higher temperatures without excessive chemical damage), and the higher heat regeneration capacity of indirect systems which makes such systems more operationally economic.

In each of two variations, the temperature rise to sterilisation temperature produced by direct heating is 20-30 °C (compared with 60-70 °C for normal direct heating systems); this increases the amount of heat regeneration possible. However, the sequence of heating and cooling steps is different for the two variations. In one (APV's High Heat Infusion system), the product enters the vacuum chamber after the pre-heating step and before the high-heat section, while in the other (Tetra Pak's Tetra Therm® Aseptic Plus 2 system), it enters the vacuum chamber after an initial cooling step following the high-heat (steam injection) section. The amount of heat regenerated (up to 75%) and the extent of chemical change to the milk constituents in these direct-indirect combination systems fall between those of the conventional direct and indirect systems.

The development of these combination systems was partly prompted by the discovery of very heat-resistant bacterial spores in UHT milk in Europe and other countries. The mesophilic *Bacillus sporothermodurans* was the major organism of concern although *Paenibacillus lactis* had also been isolated with it (Scheldeman *et al.*, 2004). The high temperatures (up to 150 °C) which can be achieved in the combination systems are more likely to destroy these highly heat-resistant spores than are the highest temperatures which can be realistically used with indirect heating alone.

5.2.1.6 Scraped-Surface Heat Exchanger Systems

Continuous heat processing of viscous and particulate products provides some special challenges. For viscous products, a tubular heat exchanger can be used, provided there

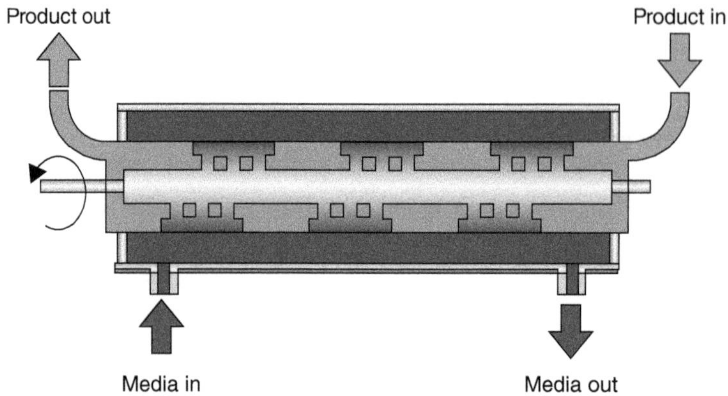

Figure 5.6 Scraped-surface heat exchanger. (Source: Reproduced with permission of SPX Flow Inc.)

are no pressure drop restrictions. However, for very-high-viscosity fluids, a scraped-surface heat exchanger (SSHE) is more suitable. Figure 5.6 shows a typical SSHE, incorporating a scraper blade which continually removes product away from the heat transfer surface. These heat exchangers are mechanically more complex than tubular heat exchangers due to the moving parts, and seals are needed at the inlet and outlet ends of the scraper blade shaft. The overall heat transfer coefficients tend to be low and the flow patterns complex, and there may be regions of streamline flow and turbulent flow, despite the induced agitation. The fluid being processed is likely to be non-Newtonian (see Section 5.2.1.8.2.3). The heating and cooling rates are quite low and the process is generally more expensive to run because of the higher capital and maintenance costs. There is also limited scope for energy savings by regeneration. SSHEs may be used for pasteurisation or sterilisation processes. Two other more specialised uses for SSHEs are ice-cream freezing and manufacture of margarine and low-fat spreads. In most heating applications, increasing the agitation speed improves the heat transfer efficiency by increasing the overall heat transfer coefficient. However, in cooling applications, the products may become very viscous. In some circumstances, where product crystallization may also occur, the use of higher agitation speeds may create additional frictional heat, warming the product rather than cooling it. This is known as viscous dissipation. As well as handling very viscous products, scraped-surface heat exchangers can also handle systems with particulates up to 25 mm diameter.

Problems arise with particulate systems because the solid phase conducts heat more slowly than the liquid phase so it takes longer to sterilise the particles compared to the liquid. Also, it is not straightforward to determine the heat film coefficient (see Section 5.2.1.8.3.3) due to the uncertainty in the relative velocity of the solid with respect to the liquid. There are also problems determining the residence time distribution (see Section 5.2.1.8.4) of the solid particles. The user would need to be clear whether the entire particle needs to be sterilised or just the surface of the particle.

A second approach is to have a selective holding tube system, whereby larger particles are held up in the holding tube for a longer period of time. A third approach involves heating the solid and liquid phases separately and recombining them. One such system is the Jupiter heating system. These systems have been discussed in more detail by Lewis and Heppell (2000). An alternative approach is ohmic heating, where particulate material is pumped through a non-conducting tube in which electrodes are placed. This is discussed in more detail in Section 5.2.2.2.

A variation of the scraped-surface heat exchanger is the RotaTherm® process (Gold Peg International, undated) in which the column containing the scraper blade is heated by multiple direct steam injectors. This reduces burn-on due to the formation of a film of condensate on the column wall and causes even heating of the product. Because of the direct use of steam, the product passes into a vacuum flash vessel after being heated to remove the entrained water and to cool the product; this is the same as for direct steam injection or steam infusion UHT processing (see Section 5.2.1.1).

5.2.1.7 Pilot-Scale Equipment

Plot-scale UHT plants are widely used for experimental purposes when it is not convenient or economic to use commercial high-capacity plants. The small capacities of the pilot plants (10-200 L/h) allow several runs to be made in a reasonably short time and also allows replicates to be processed, thereby increasing the validity of the data obtained. It also allows the same bulk material, often raw milk, to be used for all samples and obviates the day-to-day variation which can compromise product and process development trials. Some of the many uses for UHT pilot plants are as follows:

- product development
- process development
- product improvement. This could be in terms of improving keeping quality, reducing the number of additives, improving heat stability, economising on formulation costs, and changing the sensory characteristics. It may also involve investigating different processing conditions.
- fouling studies. The narrow tubes and small capacities of pilot plants, particularly bench-top units, make them very suitable for use in fouling studies which often involve several hours of running (Kastanas *et al.*, 1995; Pakash *et al.*, 2007; Prakash, 2015). Fouling is commonly monitored by pressure increase using pressure transducers installed over the high temperature section, by a fall in product outlet temperature or an increased demand for hot water or steam in circumstances where the plant is set up to maintain a constant sterilisation temperature.
- as a teaching aid. UHT processing integrates the principles of fluid dynamics and heat transfer and allows calculations be performed to determine residence times, pressure drops, F_0, B^* and C^* values, and energy balances. It also demonstrates the effect of UHT processing on product characteristics.

UHT pilot plants can be categorised into two throughput ranges, 10-20 L/h and 100-500 L/h. The former are bench-top units and can be located in a normal laboratory environment, whilst the latter need to be located in a pilot plant or factory environment, as they have more significant requirements for electricity, steam and refrigeration.

Several commercial UHT pilot plants are available. They include SPX, Tetra Pak, GEA, Yu Hui, Microthermics, Armfield, Powerpoint, Omve and Cooktube. Heating is usually by hot water or steam, with some having their own boiler to generate either steam or pressurised hot water. Some, for example, Cooktube, uses two oil baths to achieve pre-heat and sterilisation temperatures; at the designed flow rate, there is a 10 to 15 °C temperature differential between the hot oil and the final UHT temperature.

Some have microwave heating as an alternative heating method. Most do not have regeneration sections. Homogenisers are usually incorporated separately.

All UHT pilot plants require some form of aseptic packaging. Laminar-flow cabinets operating with air outflow (the opposite direction to the flow in microbiological safety laminar-flow cabinets) are commonly used with manual filling of sterile containers. The cabinets are usually sterilised with alcohol and UV (Note that UV is used before filling commences and not during filling due to the risk of eye damage to the person filling the containers). In the authors' experience, this arrangement can, with care, be used successfully with a low non-sterility failure rate[1]. Other alternatives include aseptic laminated bag fillers such as the Intersept™ system which has the advantage of being plumbed in to the UHT pilot plant and hence the risk of contamination during manual filling is eliminated.

The authors have had considerable experience with UHT pilot plants, both bench-top and pilot plant scale; some of this is discussed below to provide an indication of the scope and usefulness of these units. Also discussed is how pilot plants can be best used to simulate commercial UHT plants.

In relation to bench-top UHT plants, Kastanas *et al.* (1995) and Prakash *et al.* (2015) described two different miniature tubular units constructed at the authors' laboratories at the Universities of Reading and Queensland, respectively. These plants, which operate at about 10 L/h, have been used for fouling and flavour studies, and product development trials. Diagrams of the plants are shown in Figures 5.7 and 5.8. Both plants are based on narrow-bore stainless steel tubing but differ in their mode of heating; the Reading unit

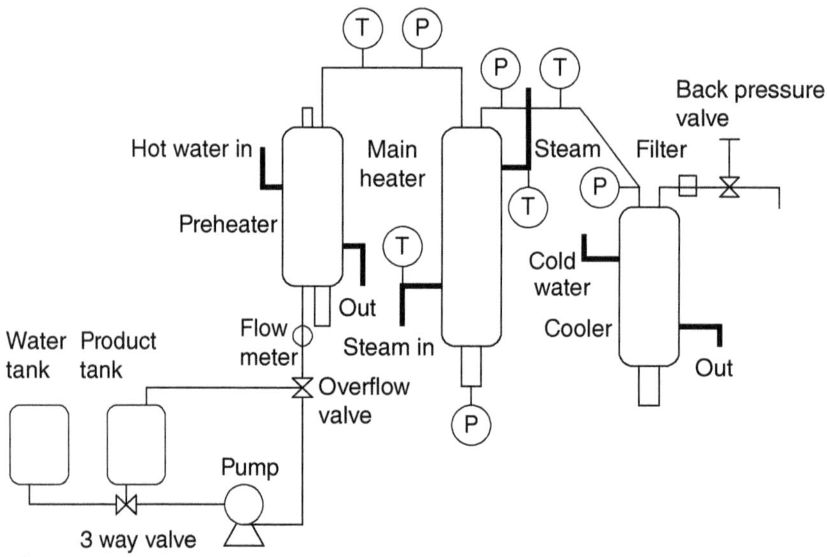

Figure 5.7 Flow diagram of the bench-top UHT plant constructed at the University of Reading. (Source: Kastanas, 1996. Reproduced with permission.)

1 One instance in which samples filled into screw-top containers in a laminar-flow cabinet can become contaminated is when the samples are subsequently air-freighted in a non-pressured part of a plane. An explanation is that the low pressure at altitude creates a partial vacuum in the containers which facilitates access of air, and bacteria, when the containers are later exposed to atmospheric pressure.

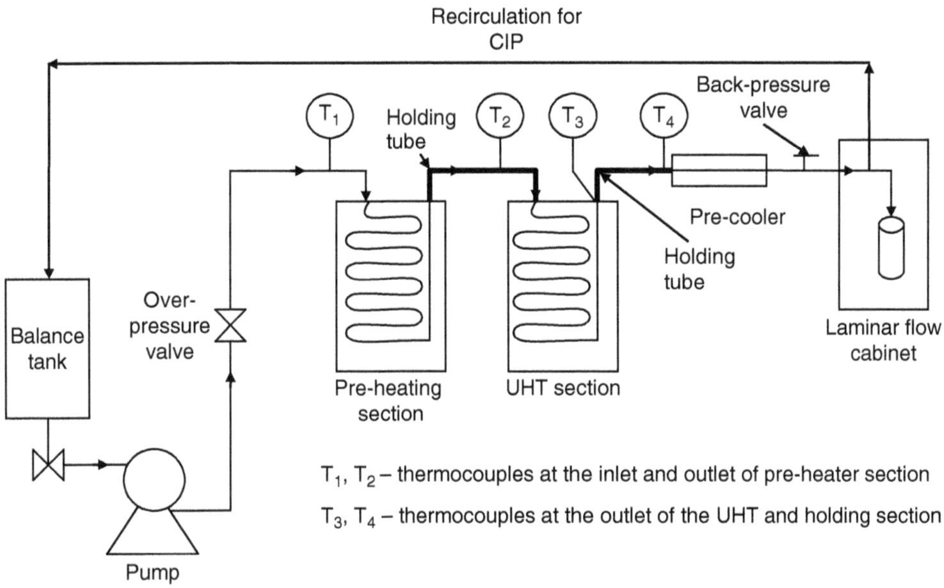

Figure 5.8 Flow diagram of the bench-top UHT plant constructed at the University of Queensland. (Source: Prakash *et al.*, 2015. Reproduced with permission of Elsevier.)

uses hot water and steam jackets while the Queensland unit uses a water bath and an oil bath for pre-heating and sterilisation, respectively.

In relation to larger-capacity pilot plants, considerable use has been made of an APV Junior UHT plant at the University of Reading. It has a capacity of 100–200 L/h, and capabilities of indirect (plate) and direct infusion heating, where the milk is injected through a small orifice into steam. It allows batches of as little as 10 L to be processed on the indirect system and about 25 L on the direct system. It was used extensively for flavour studies (Gaafar, 1987; Prasad, 1989) and more recently for investigating ultrafiltration (UF) and fouling of milk at high temperatures (up to 140 °C) (Wang, 2012; On-Nom, 2012). For the latter, a small UF module was incorporated into the holding tube of the UHT plant. The Reading pilot plant has been used to assist many companies with product and process development. The UHT trials have included development of flavoured milk products, UHT-stable flavours, chocolate and confectionery-type products, dairy ice-cream mixes, non-dairy milk products, and the use of additives for improving the heat stability of goat's milk. Despite the success with many products using this plant, one product, a UHT-stable batter mix for use in pancakes or Yorkshire puddings, has so far eluded successful development.

This pilot plant was also used by Browning *et al.* (2001) in relation to development of a spreadsheet for calculating B* and C* values (see Sections 3.2.2 & 3.2.3), and some product parameters including lactulose (see Section 6.3). Lactulose levels were calculated according to two different reported activation energies, 122 and 151 kJ/mol, and compared with actual analytical results on UHT milks produced on the pilot plant at various temperature–time combinations as shown in Table 5.4. The experimentally derived data clearly support the use of an activation energy of 122 kJ/mol. The plant was also used to show that no difference in browning could be detected by eye ($\Delta E = 0.3$) in

Table 5.4 Determination of appropriate kinetic data for lactulose formation for use in Excel predictive model. (Source: Browning *et al.*, 2001. Reproduced with permission of John Wiley & Sons.)

	Lactulose (mg/L)		
Processing conditions (temperature, °C/time s)	**Predicted (Ea = 151 kJ/mol)**	**Predicted (Ea = 122 kJ/mol)**	**Actual concentration**
135/10	258	74	76
130/27	300	136	103
130/58	492	212	201
140/58	1248	453	480
150/58	3317	930	1103

milks subjected to heat treatments associated with C* values of 0.5 and 1.5, despite the significantly different predicted browning values of 54 and 143 s (equivalent times at 121 °C). This finding was consistent with report of Fink and Kessler (1988) of a threshold browning value of 400 s (at 121 °C). Similarly, it was shown from sensory analysis that these milks (C* = 0.5 and 1.5) were not perceived to have different flavours.

A multipurpose APV pilot plant (100 L/h) with direct (injection, infusion) and indirect (plate, tubular) heating capabilities was used at the University of Queensland (SPX 2013) (see Figure 5.9). This is a very versatile unit which has been used for a range of product and process development trials including ESL processing, investigation of chemical heat indices (Elliott *et al.*, 2003) and flavour studies (Perkins *et al.*, 2007).

Based on the methodology of Browning *et al.* (2001), Tran *et al.* (2008) calculated B*, C* and product parameters for three different UHT pilot plants for milk heated to different temperatures. The pilot plants were a 100 L/h multipurpose APV pilot plant operating in indirect (tubular) and direct (infusion) modes, and a 10 L/h bench-top indirect (tubular) unit. The effects of increasing the holding temperature from 135 to 144 °C are shown in Table 5.5. A subtle difference between pilot plants and production units which was taken into account by Tran *et al.* (2008) is the difference in Reynolds number; the flow may be streamline on a pilot plant (Re <2000), as the flow rates are much lower than in commercial plants where the flow is usually turbulent (Re >10,000). The authors compared the parameters determined for the pilot plants with those determined for several commercial plants.

This methodology can also be used to determine how a pilot plant can be set up to mimic B* and C* values of a commercial plant. Figure 5.10b shows the temperature–time profile of an indirect pilot plant which could be used to achieve almost identical B* and C* values to a commercial indirect plant with the profile shown in Figure 5.10a. Note that a very different temperature–time profile had to be devised for the pilot plant to match the commercial plant because of the different rates of heating and cooling in the two plants. This demonstrates the importance of not using the same nominal holding tube temperature–time combination of a pilot plant to simulate a commercial operation.

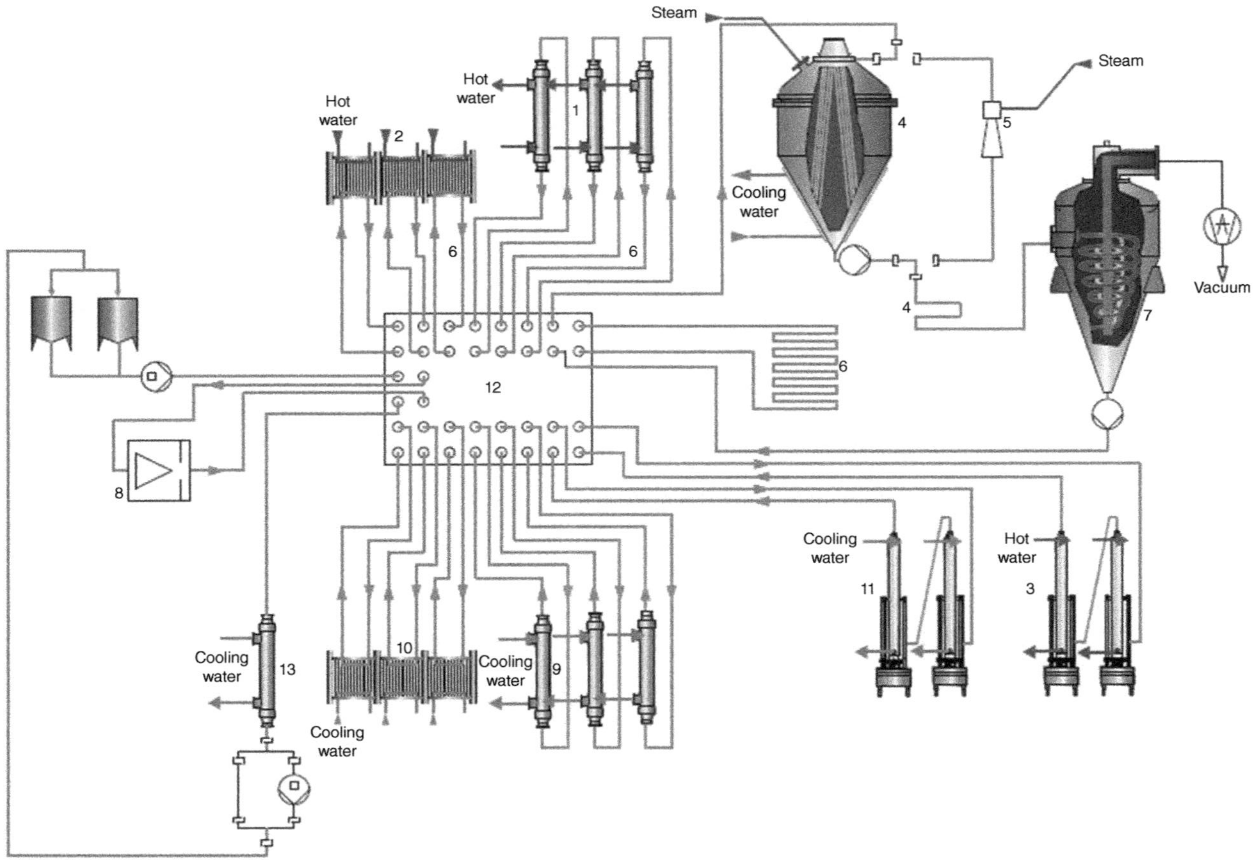

1. Tubular heaters
2. Plate heaters
3. Scraped surface heaters
4. Steam infusion chamber
5. Steam injector
6. Holding tubes
7. Flash vessel
8. Aseptic homogeniser
9. Tubular coolers
10. Plate coolers
11. Scraped surface coolers
12. Multipurpose flow plate
13. Non-aseptic cooler

Figure 5.9 Flow diagram of APV multipurpose UHT pilot plant. (Source: Reproduced with permission of SPX Flow Inc.)

Table 5.5 Effects of increasing temperature from 135 °C to 144 °C on three pilot plants (Tran *et al.*, 2008).

Parameter	Pilot plant 1- indirect 100 L/h; holding time 4.2 s	Pilot plant 2- direct 100 L/h; holding time 4.2 s	Pilot plant 3- indirect 10 L/h; holding time 5.4 s
B* adjusted[1]	1.2–7.4	0.5–3.7	1.0–6.8
C*	1.1–1.8	0.3–0.5	0.7–1.2
Browning (s at 121 °C)	97–184	28–56	63–122
Lactulose (mg/L)	150–310	40–100	100–210
Furosine (mg/100 g protein)	37–57	12–20	23–35
α Lactalbumin (% denaturation)	60–72	28 – 56	42–53

[1] B* adjusted for fastest particle according to Reynolds number, = B*/1.2.

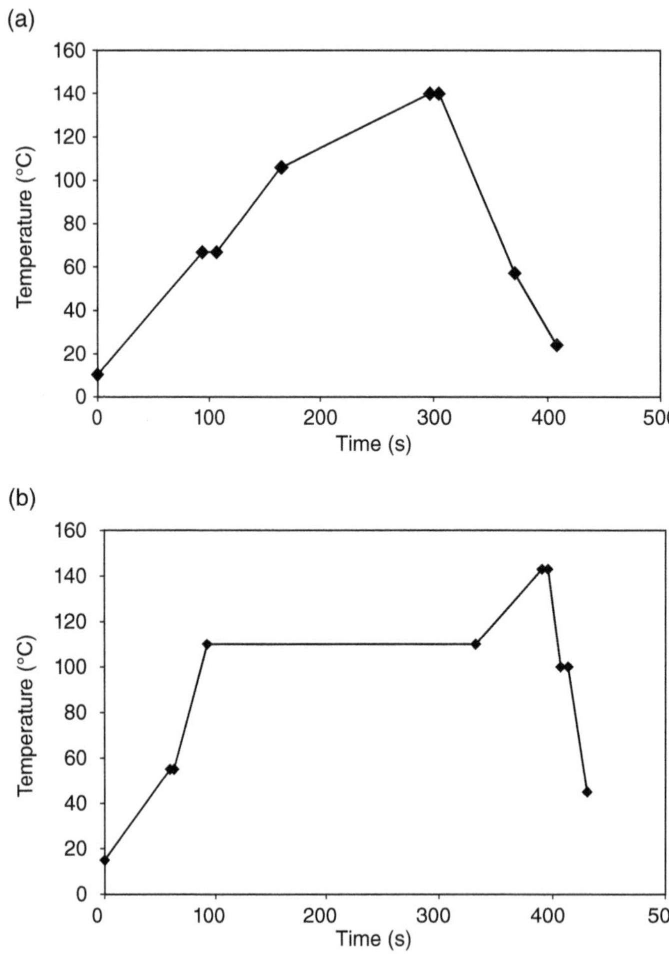

Figure 5.10 Temperature–time profiles of indirect commercial (a) and pilot-scale (b) plants with matching B* (8.4) and C* (3.2) values.

5.2.1.8 Engineering Aspects

This section reviews some properties of foods that influence how they behave when subjected to thermal processes. It also discusses residence time distributions and steam requirements for UHT plants.

5.2.1.8.1 Food Composition

Food composition tables provide detailed information on the major components (water, protein, fat, carbohydrate) as well as minerals, vitamins and trace nutrients in foods. This information is useful for products which are to be heat treated as it illustrates their chemical complexity and thus the numerous heat-induced reactions which can take place. It is crucial to appreciate that a major limitation of these tables is that they report average values which do not account for the biological variability that is found for all food products. As an example, data in Chen *et al.* (2014) shows the variations in composition of bulk milk from one farm over a one year period (Table 1.6).

This natural variation in composition leads to challenges in thermal processing, especially with respect to heat stability (see Section 6.2.1). However, the variations are of less significance for estimating the physical properties of foods, as discussed later. It would be misleading (incorrect) to use values in these tables as the starting value for estimating nutritional changes during processing. A much better solution for assessing the severity of a heating process is to make use of an "indicator" component which is not present in the raw material but which is produced on after heat treatment, for example, lactulose in milk (see Section 6.1.7).

Food composition tables do not provide information on the enzymes that are found in foods or pathogenic and spoilage organisms which may be present. This can be found in more specialised literature on specific commodities. They also provide no information about the presence of anti-nutritional compounds (e.g., trypsin inhibitor in soy beans) and environmental pollutants which may need to be inactivated or removed as part of the process.

5.2.1.8.2 Physical Properties of Foods

It is important to be able to heat and cool foods as quickly as possible. In this context, the heat transfer rates are influenced by the physical properties of a food. In particular, its moisture content has a pronounced effect, as will be shown later. Many models for predicting physical properties are based on food composition. The simplest models consider the food to be a two-component system (water and solids) and these are sufficient in many cases. More complex models break down the solid fraction into protein, carbohydrate, fat and minerals. In this regard, food composition tables provide a useful starting point. Also, whether the food is solid or liquid is very important, as this determines whether heat transfer is by conduction or convection. It is quite difficult to establish whether a food is solid or liquid from inspection of its composition. For example. canned foods may be classified as either conduction or convection packs (see Section 5.2.1.8.3.3). Heat transfer by convection is inherently faster than by conduction.

5.2.1.8.2.1 Size, Shape, Surface Area Both heat and mass transfer rates increase as surface area increases. A food product heats and cools more quickly as its surface area to volume (SAV) ratio is increased. This is one potential advantage offered by a plate heat exchanger over a tubular heat exchanger, which leads to more rapid heating in the plate system, making plate heat exchangers still a popular choice for some products.

For in-container sterilised foods, the SAV ratio decreases as the size of the can increases. New packaging materials and configurations such as plastic trays or flexible pouches also increase the SAV ratio, allowing more rapid heating and cooling, which in turn may give rise to an improvement in product quality. One downside to a larger SAV ratio is for gas permeation (oxygen and water vapour) through plastic films or containers, which might accelerate oxidation and other reactions during storage.

Homogenisation increases the SAV ratio of the fat phase. Conversely, clumping of bacteria leads to a decrease in the SAV ratio, as does the presence of clumps of poorly dispersed or dissolved solid matter. Heat transfer is slower into bacterial clumps or into poorly dispersed powders and such situations may lead to survival of heat-resistant spores (see Chapter 8).

5.2.1.8.2.2 Density An awareness of product density is important and to overlook it may lead to some serious defects in product quality. Also, density differences between components in a formulated food may lead to separation, for example, cream plug in milk and cream, sediment formation in UHT milk and separation of the solid fraction, such as rice or tapioca, in diary desserts. Also, bacteria are more dense than water and bacterial spores are even more dense than vegetative cells. This fact is used in removing bacteria from milk by centrifugation (bactofugation) (see Section 10.6)

Density (ρ) is defined as mass/volume, with SI units of kg m^{-3} . Water has a density of 1000 kg m^{-3}. It can also be expressed as specific gravity (SG) by reference to water (ρ_w); this is dimensionless:

$$SG = \rho/\rho_w \qquad (5.1)$$

The densities of other major components in foods are (kg m^{-3}): oils and fats (900-950); sucrose (1590); starch (1500); cellulose (1270-1610); protein (1400); salt (2160) and citric acid (1540). It can be seen that most are substantially different to water and this may lead to separation if they are not fully dissolved. The separation velocity can be estimated from Stokes equation:

$$v = \frac{d^2 g(\rho_s - \rho_f)}{18\mu} \qquad (5.2)\ (also\ 7.1)$$

where v = terminal velocity (m s^{-1}); D= particle diameter (m) and ρ_s and ρ_f are the solid and fluid densities respectively (kg m^{-3}); μ = fluid viscosity (Pa s)

This shows how the separation velocity is affected by density differences between the phases, particle diameter and fluid viscosity. Its applications to gravitational and centrifugal separation methods are discussed in Chapter 11 and are significant for defects such as fat separation in UHT milk (see Section 7.2.4).

Solids and liquids are often assumed to be incompressible, that is, their density is not affected by moderate changes in temperature and pressure. However, in practice, the density of liquids decreases as temperature increases and a heated liquid rises; these small density changes provide the driving force for natural convection.

The presence of dissolved air (ρ = 1.27 kg m^{-3}) substantially decreases the density of a fluid or solid. Hence, whipping and foaming lead to a decrease in density of a liquid. Many products which are to be heat treated are mixed beforehand; it is important to avoid too much air incorporation by excessive agitation as this decreases the heat transfer rate in the product. This is relevant to the rate of heating of milk in UHT plants (see Section 5.2.1.1.2).

In contrast to liquids and solids, gases and vapours are readily compressible. This is best illustrated by reference to steam tables, where compressibility is presented in terms of specific volumes (see Table 5.6). For example, at atmospheric pressure, steam has a specific volume of 1.67 $m^3\ kg^{-1}$, whereas at 70 °C (under vacuum) it increases to 5.05 $m^3\ kg^{-1}$. These conditions approximate the conditions in the flash cooling vessel in a direct UHT plant.

For processes involving direct contact with steam, it is important to check that the product has not been diluted (or concentrated). This is particularly relevant to direct UHT heating processes (see Section 5.2.1.1). Freezing point depression or accurate density measurement is useful for this purpose (see Section 11.2.8).

Table 5.6 Properties of saturated steam. (Source: Lewis, 1990. Reproduced with permission of Elsevier.)

Temp °C	Pressure * Absolute bar	Specific Volume (m^3kg^{-1})	Enthalpy (kJ/kg) h_f	h_{fg}	h_g	Entropy (kJ/kg K) s_f	s_{fg}	s_g
0	0.006	206.30	0.0	2502	2503	0.00	9.16	9.16
2	0.007	179.92	8.4	2497	2505	0.03	9.07	9.10
10	0.012	106.43	42.0	2478	2520	0.15	8.75	8.90
20	0.023	57.84	83.9	2454	2538	0.30	8.37	8.67
30	0.042	32.93	125.7	2431	2556	0.44	8.02	8.45
40	0.075	19.55	167.5	2407	2574	0.57	7.69	8.26
50	0.123	12.05	209.3	2383	2592	0.70	7.37	8.08
60	0.199	7.68	251.1	2359	2610	0.83	7.10	7.93
70	0.311	5.05	293.0	2334	2627	0.95	6.80	7.75
80	0.474	3.41	334.9	2309	2644	1.08	6.54	7.62
90	0.701	2.36	376.9	2283	2660	1.19	6.29	7.40
100	1.013	1.67	419.1	2257	2676	1.31	6.05	7.36
110	1.433	1.21	461.3	2230	2691	1.42	5.82	7.24
120	1.985	0.89	503.7	2202	2706	1.53	5.60	7.13
130	2.701	0.67	546.3	2176	2720	1.63	5.39	7.03
140	3.610	0.51	589.1	2144	2733	1.74	5.19	6.93
150	4.760	0.39	632.1	2113	2745	1.84	4.99	6.83
175	8.924	0.22	741.1	2031	2772	2.09	4.53	6.62
200	15.549	0.13	852.4	1939	2791	2.33	4.10	6.43
225	25.501	0.08	966.9	1834	2801	2.56	3.68	6.25
250	39.776	0.05	1086	1715	2800	2.79	3.28	6.07
300	85.927	0.02	1345	1406	2751	3.26	2.45	5.71

* absolute pressure = gauge pressure+atmospheric pressure

For foods which contain little entrapped air, the density of a food can be estimated from a knowledge of its composition (mass fractions of the main components (m_i) and their densities (ρ_i) (see values after equation 5.1). This is useful for most liquid formulations.

$$\rho = \frac{1}{\sum \frac{mi}{\rho i}} \tag{5.3}$$

There are some unusual density scales that still may be encountered, for example, Brix, Baume, and Twaddell. These are discussed in Lewis (1990) and Hayes (1987).

Density or specific gravity can be measured simply and accurately by means of a density bottle or density hydrometer. This may be useful for checking whether a product mixing procedure has incorporated too much air into the mix (see Section 11.2.3).

One important function of blanching and exhausting (for canned foods) is to remove air from the food and the container headspace. This minimises strain during the subsequent sterilisation process as well as increases the density of the food, ensuring that it packs better. For liquids, the level of dissolved oxygen may also be important. Note that the solubility of dissolved oxygen decreases as temperature increases. Therefore, air will come out of solution during heating and may cause air bubbles or air locks. Dissolved oxygen is an interesting parameter to measure in UHT milk as it has a major role in development of stale and oxidised flavours and destruction of some vitamins such as ascorbic acid and folic acid (see Sections 7.1.2 and 11.2.4).

5.2.1.8.2.3 Viscosity The main resistance to heat transfer by convection is by the viscosity (μ) of the fluid. Viscosity is a measure of the internal friction within a fluid. It can be thought of as how easily the fluid will flow when it is subject to shear force, such as that due to gravity.

Dynamic viscosity is defined as the ratio of shear stress (τ) (Pa) to shear rate (dv/dy) (s^{-1}). It has SI units of Pa s (Nsm^{-2}). The cgs unit, centipoise (cP) is still widely used; the conversion factor being: 1 mPa s = 1 cP. Kinematic viscosity is equal to dynamic viscosity/density.

Water is a low-viscosity fluid with a viscosity of 1.0 mPa s at 20°C. Milk has a value of about twice that of water. The most accurate viscometer for measuring low viscosity milk products is the capillary flow viscometer, which measures kinematic viscosity (cSt or m^2s^{-1}) (Lewis, 1990). Viscosity values for some fluids are given in Table 5.7.

Viscosity decreases as temperature increases. Therefore, milk will have a much lower viscosity in the holding tube of a UHT plant compared to that in the feed tank. The viscosity of milk at 140 °C is rarely reported. At 100 °C it is about 0.5 mPa s, whereas at 20 °C it is 2 mPa s. Figure 5.11 shows the viscosities of milk and some dairy liquids over the temperature range of 0 to 100 °C.

One problem that might occur for products with thickening agents, such as custard, is the development of high viscosities in the cooling section of a heat exchanger. This may also occur for products containing substantial amounts of fat which may undergo crystallisation. Not only will the viscosity be increasing, but the material will be giving out its latent heat of crystallisation. More information about the temperature dependence of viscosity for dairy products is given by Lewis (1993).

Many fluids are non-Newtonian, where the viscosity may change with both the shear rate and the time of application of the shear. Non-Newtonian behaviour can be

Table 5.7 Viscosities of some fluids. (Source: Lewis, 1990. Reproduced with permission of Elsevier.)

Fluid	Viscosity (mPa s)	Temperature (°C)	Fluid	Viscosity (mPa s)	Temperature (°C)
Carbon dioxide (gas at 3 MPa)	0.018	27	Sucrose (20% w/w)	2	20
Carbon dioxide (liquid at 10 MPa)	0.072	27	Sucrose (40% w/w)	6.2	20
Water	1	20	Sucrose (60% w/w)	57	20
Water	0.28	100	Peanut oil	60	25
Milk (homogenised)	2	20	Safflower oil	52	25
Milk (homogenised)	1.5	40	Corn oil	57	25
Milk (homogenised)	0.8	60	Glycerol	1490	20
Diethyl ether	0.23	20	Honey	6000	25

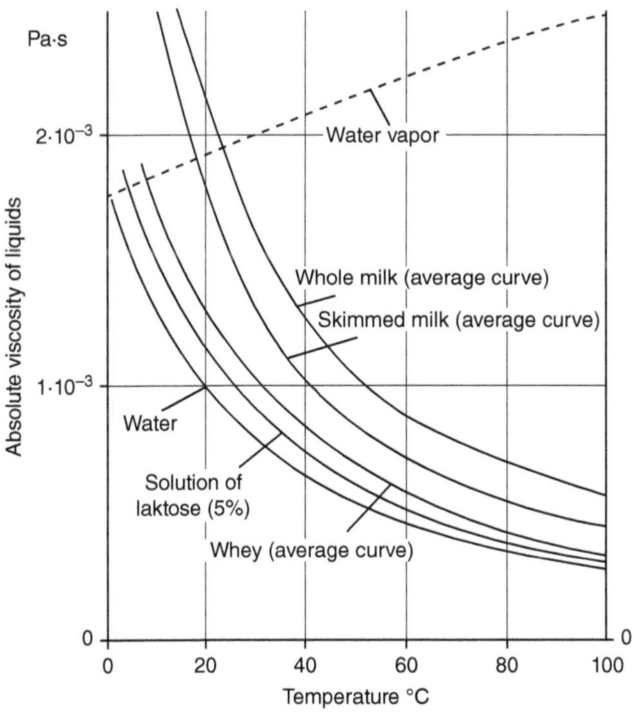

Figure 5.11 Change in viscosity of water and dairy liquids with temperature (from Kessler, 1981).

classified as time-independent or time-dependent (see Figure 5.12). The two types of time-independent behaviour are pseudoplastic (shear thinning) and dilatant (shear thickening). Pseudoplastic or shear-thinning behaviour is the most common type of behaviour encountered. These fluids decrease in viscosity with increasing shear rate. Many dilute solutions of macromolecules show this type of behaviour.

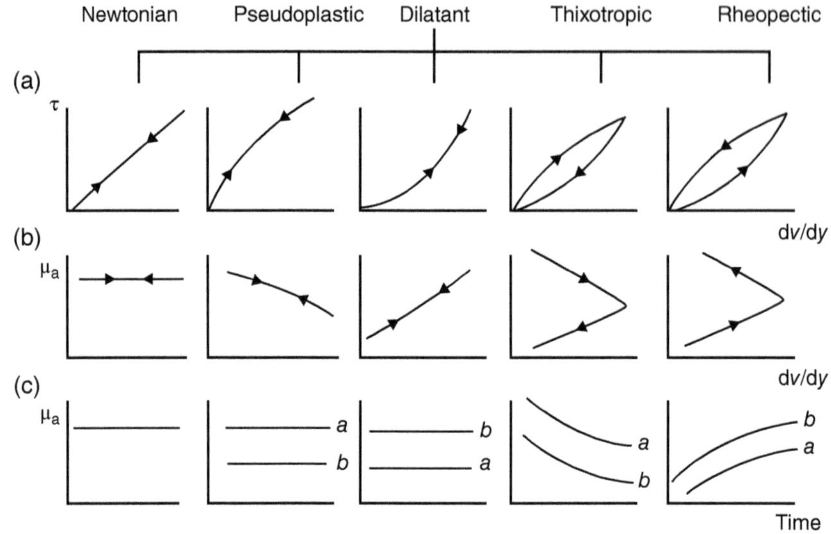

Figure 5.12 Rheograms for non-Newtonian fluids: (a) Shear stress (τ) against shear rate (dv/dy); (b) Apparent viscosity (μa) against shear rate; (c) Apparent viscosity against time for two different shear rates, a and b, where b>a. (Source: Lewis, 1993. Reproduced with permission of Elsevier.)

Table 5.8 Some values for k and n in the Power Law equation [$\tau = k\,(dv/dy)^n$] for various foods.

Product	Temperature (°C)	k (Ns^nm^{-2})	n
Custard	80	7.24	0.36
Tomato juice (12%TS)	32	2.0	0.43
Tomato Juice (25% TS)	32	12.9	0.40
Gravy	80	2.88	0.39
Orange juice (65 Brix)	19.9	0.72	1.6
Mango puree (9.3 Brix)	24.2	20.6	0.334
Peach puree (11.7 % TS)	30.0	7.2	0.28

Data from various sources

Dilatant or shear-thickening behaviour in foods is comparatively rare. This can occur with cream products which are rapidly cooled. It is fortunate that it is rarely encountered, as it hinders heat transfer in situations where there are high shear rates. Examples of models which have been used to characterise time-independent non-Newtonian liquids are given in Lewis and Heppell (2000). The simplest of these is the power law equation, where

$$\tau = k\left(dv/dy\right)^n \tag{5.4}$$

Some values for k and n for the power law equation (Equation 5.4) for various foods are given in Table 5.8. A further complication is time-dependent behaviour. The most commonly observed is shear-thinning time dependency, known as thixotropy. When viscosity is measured with a rotational viscometer at a constant speed, such fluids will show a decreasing viscosity value with time. These fluids may eventually reach an equilibrium viscosity value. They may also recover their viscosity when rested. They show hysteresis when subject to increasing shear stress followed by decreasing shear stress.

Some fluids may also exhibit some elastic characteristics, superimposed on their viscous nature; such fluids are termed viscoelastic and this property makes them more difficult to pump in some circumstances. However, these situations are not commonly found with products subjected to UHT treatment.

The viscosity of a fluid affects the selection of heat exchanger, pumps and ancillary equipment. As a fluid becomes more viscous, more energy is required to pump it and the pressure drops become much larger. During UHT processing, temperatures change rapidly and different shear conditions prevail, making it difficult to predict what the viscosity of the product might be at any specific location in the plant. The viscosity strongly influences the Reynolds number and determines whether the flow regime is streamline or turbulent. This in turn influences the rate of heat transfer and the distribution of residence times. This is discussed in more detail in Section 5.2.1.8.4. Methods for measuring viscosity are discussed in Section 11.2.26.

5.2.1.8.3 Thermal Properties of Foods

All heating and cooling processes involve the transfer of energy, so the thermal properties of foods are very important. These influence both the total amount of energy that is required as well as the rate of energy transfer that is required.

Specific heat is defined as the amount of energy required to raise unit mass by unit temperature rise. It has SI units of $kJ\ kg^{-1}\ K^{-1}$. Water has a high value ($4.18\ kJ\ kg^{-1}\ K^{-1}$) compared to most other components of food. This fact, combined with its ubiquitous nature, makes water a very effective cooling medium and to a lesser extent as a heating medium. When used as a coolant, close attention should be paid to its microbial quality.

A number of models are available to predict the specific heat of a food from its composition. The simplest model treats the food as a two-component system: water and solid as follows:

$$c = m_w c_w + m_s c_w \tag{5.5}$$

where c = specific heat and m = mass fraction.

For this purpose, the specific heat of solids is taken as $1.46\ kJ\ kg^{-1}\ K^{-1}$.

For multi-component systems, a more accurate form of the equation becomes: $c = \Sigma\ m\ c$, that is, the sum of the products of the mass fraction and the specific heats for water, fat, carbohydrate, protein and minerals. The specific heats ($kJ\ kg^{-1}\ K^{-1}$) for these components are 4.18 (water); 1.7 (fat); 1.4 (carbohydrate); 1.6 (protein) and 0.8 (minerals). The specific heats of all other food components are much lower than that of water. Ice has a specific heat of about half that for liquid water, as also has water vapour. Some specific heat values for various foods are given in Table 5.6. Where specific heat data are not readily available, the food composition tables, together with the models above, are useful for predicting the values for specific foods.

Specific heat is a measure of sensible heat change. For substances which may be subject to some crystallisation, for example, fat in milk or cream, an apparent specific heat may be used (Table 5.9). This takes into account changes brought about by crystallisation (latent heat) and sensible heat changes.

Table 5.9 Specific heat values for various materials and foods. (Source: Lewis, 1990. Reproduced with permission of Elsevier.)

Material	Temperature	Specific heat ($kJ\ kg^{-1}\ K^{-1}$)	Specific heat ($kcal\ kg^{-1}\ K^{-1}$) (BTU/1b °F)
Water	59°F	4.18	1.000
Ice	32°F	2.04	0.487
Water vapour	212°F	2.05	0.490
Air	−10°F–80°F	1.00	0.240
Copper	20°C	0.38	0.092
Aluminum	20°C	0.89	0.214
Stainless steel	20°C	0.46	0.110
Ethylene glycol	40°C	2.21	0.528
Alcohol	0°C	2.24	0.535
Oils	20°C	1.73	0.414
Corn	0°C	1.86	0.446
Sunflower	20°C	1.93	0.460
Apples (84.1% m.c.)	above F.pt.	3.59	0.860
	below F.pt.	1.88	0.45
Potatoes (77.8% m.c.)	above F.pt.	3.43	0.82
	below F.pt.	1.80	0.43
	dried (10.9% m.c.)	1.85	0.443
Lamb (58.0% m.c.)	above F.pt.	2.80	0.67
	below F.pt.	1.25	0.30
Cod	above F.pt.	3.76	0.90
	below F.pt.	2.05	0.49
Milk (87.5% m.c.)	above F.pt.	3.89	0.930
	below F.pt.	2.05	0.490
Soya beans, 8.7% m.c.		1.85	0.442
Wheat, 10.0% m.c.		1.46–1.80	0.35–0.43

5.2.1.8.3.1 Latent Heat and Steam Tables Latent heat changes assume importance when evaporating or condensing liquids or during crystallisation or melting. **Water has very high latent heat of vaporisation** and as a result steam is widely used as a heating medium. The properties of steam are summarised in the saturated steam tables (Table 5.6). These give the relationship between steam pressure and temperature, as well as information on the specific volumes, enthalpy and entropy values for the liquid (f) and vapour (g) states (specific volumes of liquids are often not recorded). Special attention must be paid to the quality of steam for direct heating purposes.

It is useful to keep a copy of the steam tables for reference when UHT processing. They can be used to cross-check the accuracy of steam gauges and temperature probes. For example, if a UHT steam gauge reads 3.76 bar (g, guage), this is 4.76 bar absolute and so the steam temperature should be 150 °C. If the two readings do not coincide, one or other is incorrect. In direct UHT processes, steam tables can be used to set the required temperature in the flash cooling chamber. If a temperature of 70 °C is required, the pressure should be set at 0.311 bar (a, absolute). It also illustrates the large volume of water vapour that needs to be removed in the flash cooling process. For example, at 70 °C, the specific volume of water vapour is 5.05 m^3 kg. Thus there will be a massive expansion when 1 kg of water (occupying about 1 L) results in 5050 L vapour when it is flashed off. Also, mass balances can be calculated on the theoretical amounts of steam to be used.

5.2.1.8.3.2 Enthalpy and Specific Enthalpy Enthalpy (H) is a thermodynamic function which is defined as: H = U+PV, where U = internal energy, P = pressure and V = volume. For any chemical or physical reaction, the enthalpy change (ΔH) is of real interest. **Specific enthalpy changes can be equated to heat changes, when processes take place at constant pressure.**

$$\Delta H = q = \int c_p d\theta \tag{5.6}$$

This is useful, since most reactions take place at constant pressure. The heat given out when steam condenses at a constant pressure is equal to (h_g – h_f). The same applies for refrigerants when they evaporate or condense. As mentioned, water has a very high latent heat of vaporisation (see h_{fg} column in Table 5.6) compared to most other substances. Similarly amongst the refrigerants, ammonia has a high latent heat of vaporisation.

The unit of enthalpy is (J). In most compilations of thermodynamic properties, specific enthalpy is used, which is the enthalpy per unit mass (J kg^{-1}).

Enthalpy data are particularly useful when crystallisation takes place during cooling; for example, fat crystallisation during cooling of emulsions and water crystallisation during freezing. Most enthalpy data also provide information on the percentage crystalline solids (α) present at any temperature. An interesting system is ice-cream mix, which is pasteurised or sterilised, homogenised, cooled, aged, cooled and frozen. During the cooling and freezing, fat crystallisation takes place from about 40 °C downwards and water ice formation starts at about −2 °C. Enthalpy changes can be determined by differential scanning calorimetry (DSC) (Biliaderis, 1983, Kaletunc, 2009).

Enthalpy data for milk are shown in Figure 5.13. They show how enthalpy is affected by both changes in temperature and solids content. Thus they can be used to calculate the amount of energy required or to be removed when heating and cooling milk products of different solids content. More information on performing these calculations was given by Lewis (1990). This diagram is extremely useful for showing how much water is frozen at different temperatures, which could happen to UHT milk under adverse storage conditions (see Chapter 7). Similar information for fruit juices of different solids contents is available (ASHRAE, 1993).

Riedel (1955) presented detailed enthalpy data for over 20 different oils and fats. These data are particularly useful during cooling of products containing fat, when crystallisation also takes place. Table 5.10 shows enthalpy data for various fats and oils, as well as their percentage crystallisation at different temperatures.

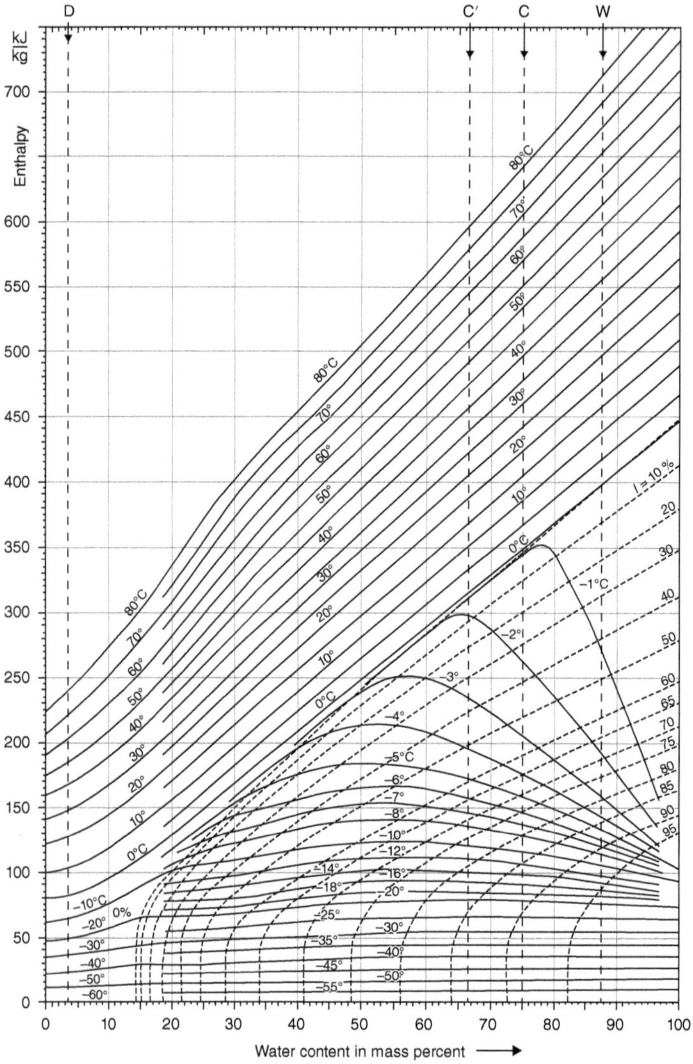

Figure 5.13 Change in enthalpy in milk with water content (reproduced from Riedel, 1976, with permission).

5.2.1.8.3.3 Heat Transfer The three mechanisms of heat transfer are conduction, convection and radiation. Conduction is the predominant mechanism in solids and convection in liquids. Most processes involve both mechanisms. Radiation involves heat transfer by electromagnetic radiation which can cover the whole electromagnetic spectrum: radiofrequency, microwaves, infrared, ultraviolet, X-rays and gamma rays.

- Conduction and thermal conductivity

In solids, the main mechanism of heat transfer is by conduction. However, the thermal conductivity of foods is low compared to that of metals and, in general, solid foods

Table 5.10 Enthalpy values for some fats and oils. (Source: Lewis, 1990. Reproduced with permission of Elsevier.)

	Butter fat		Sunflower oil		Coconut oil		Lard	
Temperature (°C)	Enthalpy H ($kcal\ kg^{-1}$)	Amount α of crystalline solids (%)	Enthalpy H ($kcal\ kg^{-1}$)	Amount α of crystalline solids (%)	Enthalpy H ($kcal\ kg^{-1}$)	Amount α of crystalline solids (%)	Enthalpy H ($kcal\ kg^{-1}$)	Amount α of crystalline solids (%)
−40	3	100	4	100	4	100	3	100
−20	11	98	11	94	10	100	10	100
−10	17	90	25	18	15	98	15	94
0	24	75	37	0	20	87	22	82
10	32	56	42	0	25	69	33	59
20	45	20	47	0	45	35	39	50
30	54	10	52	0	59	0	50	33
40	60	0	57	0	65	0	60	10
50	65	0			70	0	67	2

heat very slowly. The thermal conductivity of a material is defined as the rate of heat transfer passing through 1 m^2 cross-sectional area of the material at steady state, when a temperature gradient of 1 K (1 °C) is maintained over a distance of 1 m. The SI units are $Wm^{-1} K^{-1}$. The values for a number of materials are given in Table 5.11. The highest values are for silver, copper and aluminium. It is notable that stainless steel has a much lower value than all these metals. In comparison, it can be seen that foods are poor conductors of heat compared to metals, with typical values between 0.4 to 0.6 $Wm^{-1} K^{-1}$. As moisture content is reduced, the thermal conductivity decreases. Air is a very poor conductor of heat, so this means that porous foods and foams are very poor conductors of heat.

There are several models for predicting thermal conductivity from food composition. These have been reviewed by Rahman (1995) and Singh (2007). Thermal conductivity increases slightly as temperature increases. Models for a wide range of foods are listed by Rahman (1995). It is noteworthy that values are also cited for liquids. This is done by ensuring that temperature distribution by convection is eliminated.

Table 5.11 Thermal conductivity values for various materials and foods. (Source: Lewis, 1990. Reproduced with permission of Elsevier.)

Material	Temperature (°C)	Thermal conductivity ($Wm^{-1} K^{-1}$)
Silver	0	428
Copper	0	403
Copper	100	395
Aluminum	20	218
Stainless steel	0	8–16
Glass	0	0.1–1.0
Ice	0	2.3
Water	0	0.573
Corn oil	0	0.17
Glycerol	30	0.135
Ethyl alcohol	20	0.24
Air	0	2.42×10^{-2}
Cellular polystyrene	0	3.5×10^{-2}
Freeze-dried peach (1 atm)	0	4.18×10^{-2}
Freeze-dried peach (10^{-2} Torr)	0	1.35×10^{-2}
Whole soya beans	0	0.097–0.133
Starch (compact powders)	0	0.15
Beef, parallel to fibers	0	0.491
Frozen beef	−10	1.37
Fish		$0.0324+0.3294m_w$
Sorghum		$0.564+0.0858m_w$

The resistance to heat transfer due to conduction can be characterised by L/k, where L = the half thickness (or radius of the food). Therefore the resistance to heat transfer is increased by increasing the dimension of the food and by decreasing the thermal conductivity. Changing the material of construction, from tin plate to glass, usually increases the resistance significantly as the glass has a lower thermal conductivity and a greater wall thickness.

Air has a very low thermal conductivity and is effectively a good insulator. The presence of air in steam or in foods reduces the efficiency of heat transfer processes.

- Thermal diffusivity

Thermal diffusivity (α) is a more complex property. It is defined as k/c ρ, with SI units of $m^2\ s^{-1}$. It is an unsteady-state heat-transfer property which is a measure of how quickly temperature changes with time, when a substance is heated or cooled. It can be measured directly by using a number of methods which involve unsteady-state heat-transfer principles (Mohsenin, 1980; Rahman, 1995; Sweat, 1995).

Copper has a very high thermal diffusivity of $1.1 \times 10^{-4}\ m^2\ s^{-1}$ whereas most foods have a much lower thermal diffusivity; typical values for fresh meat are $1.80 \times 10^{-7}\ m^2\ s^{-1}$ and for frozen meat $1.2 \times 10^{-6}\ m^2 s^{-1}$. It is noteworthy that frozen food has a much higher thermal diffusivity than fresh food. More detailed compilations are provided by Mohsenin (1980), Rahman (1995), and Holdsworth (1997).

- Convection and heat film coefficient

Convection is inherently much faster than conduction as a heat transfer mechanism. In convection, heat is distributed from the hotter to the colder parts by the bulk movement of the molecules. In natural convection, the driving force is density difference, whereas in forced convection, energy is supplied, for example, by pumping or by agitation.

The effectiveness of heat transfer by convection is given by the heat film coefficient; this depends on a number of factors. The resistance to heat transfer is considered to take place in a film of the fluid in contact with the solid. All the temperature gradient occurs across this film. The heat film coefficient (h) describes the rate of heat transfer from the bulk of the fluid (θ_b) to the solid surface (θ_s) in terms of the surface area (A) and temperature driving force. The SI units are $W\ m^{-2}\ K^{-1}$. The rate of heat transfer (Q) is given by the equation:

$$Q = h\,A\left(\theta_b - \theta_s\right) \quad (5.7)$$

The heat film coefficient depends mainly upon the type of fluid and the degree of turbulence. In general, gases are poor heat transfer fluids, liquids have intermediate values and boiling liquids and condensing vapours have very high heat film coefficient values. Some typical values are given in Table 5.12.

As heat transfer fluids, steam is more effective than hot water which in turn is more effective than hot air. **The presence of air in steam is detrimental for two reasons**: firstly, it reduces the heat film coefficient and secondly it lowers the temperature of the steam at a fixed pressure. Both situations may lead to a reduced rate of heat transfer and the possibility of under-processing. One might suspect the presence of air if steam pressures and temperatures do not coincide (according to steam tables, see Table 5.6). This has been discussed in more detail by Holdsworth (1997).

Table 5.12 Heat film coefficients for some different flow situations.

Flow situation	Heat film coefficient ($Wm^{-2}K^{-1}$)
Gases – free convection	6-23
Liquids –free convection	114-681
Gases – forced convection	11-114
Viscous liquids- forced convection	57-568
Water – forced convection	568-11,360
Boiling water	1136-23,000
Condensing vapours	1136-113,600
Still air	6
Moving air	30
Liquid through pipes	
Low viscosity	1200-1600
High viscosity	120-1200
Condensing steam	12,000
Condensing steam+3% air	3500

Taken from data in Rahman (1995) and Fellows (2009).
Correlations for heat film coefficients for different flow situations are given in Heldman and Lund (2007)

There are many empirical relationships for predicting heat film coefficient values for different flow situations. Some of these are as follows:

- Dimensionless groups

In fluid flow and heat transfer problems, use is made of a technique known as dimensionless analysis. Some important dimensionless groups which result from this work are as follows:

Reynolds number (Re), which is defined in equation 2.17 is an important dimensionless number, which indicates whether the flow is streamline or turbulent. Reynolds numbers influence heat transfer rates and residence time distributions. More detail is provided in Section 5.2.1.8.4.

Two other important dimensionless groups are the Nusselt number (Nu) and the Prandtl number (Pr).

The **Nusselt number** is defined as the ratio of heat transfer by convection to heat transfer by conduction or hD/k where h is the heat film coefficient, D is the tube diameter and k is the thermal conductivity of the fluid. The Nusselt number is used mainly for estimating the heat film coefficient.

The **Prandtl number** is the ratio of momentum diffusivity to thermal diffusivity and is defined as cμ/k where c is specific heat of the fluid, μ is fluid viscosity and k is thermal conductivity. It takes into account the factors affecting the heat transfer into a fluid:

One widely used correlation for estimating the heat film coefficient for a fluid passing along a tube is as follows:

$$\frac{hD}{k} = 0.023\left(\frac{vDp}{\mu}\right)^{0.8}\left(\frac{c\mu}{k}\right)^{0.4} \qquad (5.8)$$

It can be alternatively written as:

$$\mathrm{Nu} = 0.023\,\mathrm{Re}^{0.8}\,\mathrm{Pr}^{0.4} \tag{5.9}$$

It is a very useful correlation which can be used for any fluid, provided its physical properties, flow rates and tube dimensions are known. A large number of other common flow situations are encountered; for example, the flow of fluids through tubes, between parallel plates and heat transfer from fluids to suspended particles. Many of these are also based on dimensionless analysis. A more thorough coverage of these correlations has been given by Kessler (1981), Rahman (1995), and Singh (2007).

- Multiple resistances and overall heat transfer coefficient

In most heat transfer processes, there is more than one resistance to heat transfer. In a simple heat exchanger, where heating is indirect, there are three resistances, two due to convection and one to conduction. These can be combined in an overall heat transfer coefficient (OHTC)(U).

For thin-walled tubes, U can be calculated from the equation:

$$\frac{1}{U} = \frac{1}{h_1} + \frac{1}{h_2} + \frac{L}{k} \tag{5.10}$$

Each term, for example, $1/h_1$, in this expression is termed a resistance. In principle, the rate of heat transfer can be increased by reducing any of the resistances. This approach also helps to identify **the limiting resistance**, which is the largest of the resistance terms. This is one which controls the overall transfer of heat and would be the one to focus upon in order to gain the largest increase in heat transfer efficiency. Where fouling occurs, the additional resistance associated with the fouled deposit may become the limiting resistance and control the overall process (see Section 6.2.2).

The OHTC also provides a simple expression for the rate of heat transfer in terms of the temperature in the bulk of the hot fluid and cold fluid, which eliminates surface temperatures which are difficult to measure. The following equation is the basic design equation which is used to estimate the surface area of a heat exchanger (A), in terms of the duty (Q′), the OHTC and the log mean temperature driving force ($\Delta\theta_m$).

$$Q' = U\,A\,\Delta\theta_m \tag{5.11}$$

Note that the use of counter-current flow, whereby the two fluids flow in opposite directions, is usually employed as it gives a larger log mean temperature difference and allows a closer approach temperature.

The duty Q′ (or rate of heat transfer – J s^{-1}) is calculated from the equation:

$$Q' = m\,c\,\Delta\theta \tag{5.12}$$

where m = mass flow rate (kg s^{-1}), c = specific heat (J kg^{-1} K^{-1}), $\Delta\theta$ = temperature change (K).

A special case is direct contact heating, where only two resistances are involved, the rate of heat transfer can be expressed in terms of the **Biot number** (Bi) which is the ratio of the heat transfer resistances inside of and at the surface of a body. It is defined as hL/k where L is a characteristic length dimension (e.g., half thickness for a slab or radius for a sphere), h is the heat film coefficient and k the thermal conductivity.

If the Biot number is below 0.2, convection is the limiting resistance and most of the temperature gradient takes place over the boundary layer. If the Biot number is above 0.2, the limiting resistance is conduction and most of the temperature gradient takes place between the surface and centre of the particle. One direct application of this is in the heat treatment of suspended particles or particulate systems.

It is also not quite so straightforward to predict the relative velocity between the solid and the liquid phase in a continuous flow situation, that is, the residence time distribution for the particles may not be the same as that for the liquid phase. This introduces extra uncertainty into the estimate of the heat film coefficient. Hence it is much more complicated to heat-treat particulate systems; some interesting methods have been developed to overcome the problems (see Section 5.2.1.6).

5.2.1.8.3.4 Water Activity (a_w) Water activity is a measure of the availability of water for microbial, chemical and enzymatic reactions. It is usually defined as p/p_s, where p = water vapour pressure exerted by the food and p_s is the saturated water vapour pressure. One method of measuring water activity involves equilibrating a sample of the food in a sealed container with a relatively small free volume and measuring the equilibrium relative humidity (RH). The water activity is given by the equation:

$$a_w = RH/100 \tag{5.13}$$

Water activity may have an influence on thermal processing in three ways:

- it will affect the heat resistance of microorganisms and hence the extent of their inactivation during processing;
- it will affect the growth rate of any organisms surviving the heat treatment (more of importance for short-shelf-life products); and
- it will affect the rate of chemical reactions during both processing and subsequent storage

Most foods which pass through a heat exchanger have a relatively high water activity, that is, greater than 0.95. Some water activity values are given in Table 5.13. Water activity is reduced by the presence of low-molecular-weight solutes. Humectants lower the

Table 5.13 Water activity values of some foods. (Source: Lewis, 1990. Reproduced with permission of Elsevier.)

a_w	Food
0.98–1.00	Fresh vegetables, fruit, meat, fish, poultry, milk, cottage cheese
0.93–0.96	Cured meats, most cheese varieties
0.86–0.93	Salami, some dry cheeses
0.8–0.87	Flour, cakes, rice, beans, cereals, sweetened condensed milk
0.72–0.88	Intermediate-moisture foods, jams, old salami
0.6–0.66	Dried fruits
0.6	Dehydrated foods

water activity when added to food systems. Substances which reduce water activity also increase freezing point depression (FPD) (see Section 11.2.8).

5.2.1.8.3.5 Electrical Properties The electrical properties of foods are of interest in continuous heating as they influence direct electrical heating (ohmic heating) processes and microwave heating processes. Both of these technologies can be used for single-phase materials but their main benefits arise from the opportunity of generating energy directly within the solid phase in a particulate system and thus accelerating the heating of the solid phase. Applications for ohmic heating and microwave heating are discussed in Sections 5.2.2.2 and 5.2.2.3, respectively.

- Electrical conductance/resistance

Electrical resistance and conductance provide a measure of the ability of a material to transport an electric current; resistance is usually preferred for solids and conductance for liquids. The specific conductance (K) or electrical conductivity has units of S m^{-1}, where S is Siemen which is the same as reciprocal ohm (ohm^{-1}). Specific conductance is measured by resistance techniques, the cell usually being calibrated with a liquid of known specific conductance.

Electrical conductivity increases as temperature increases, according to the following equation:

$$K_\theta = K_{25}\left(1 + M\left(\theta - 25\right)\right) \tag{5.14}$$

where K_θ = electrical conductivity at θ °C (S m^{-1}), M = proportionality constant, usually about 0.02 (°C and θ = temperature (°C). Some electrical conductivity values for different foods are given in Table 5.14.

Conductivity measurement can be useful in other applications; for example, for checking water purity in deionisation and for monitoring the presence of detergents in cleaning processes.

- Dielectric constant and loss factor

Table 5.14 Electrical conductivity of some foods (from Kim *et al.*, 1996. Reproduced with permission of Institute of Food Technologists.)

Material	Electrical conductivity (S/m)
Potato	0.037
Carrot	0.041
Pea	0.17
Beef	0.42
5.5% starch solution	
With 0.2% salt	0.34
With 0.55% salt	1.3
With 2% salt	4.3

The dielectric properties of foods are of relevance in the context of microwave and dielectric heating. Microwave systems are available for continuous thermal processing. Permitted frequencies are 915 and 2450 MHz. The two properties of interest are the dielectric constant (ε') and the dielectric loss factor (ε'') (see Section 5.2.2.3).

The dielectric constant is a measure of the amount of energy that can be stored when the material is subjected to an alternating electric field. It is the ratio of the capacitance of the material being studied to that of a vacuum or air under the same conditions. In an AC circuit containing a capacitor, the current leads the voltage by 90°. When a dielectric is introduced, this angle may be reduced. The loss angle (θ) is a measure of this reduction and is usually recorded as the loss tangent ($\tan\theta$). Energy dissipation within the dielectric increases as the loss tangent increases. A second property, known as the dielectric loss factor (ε'') is a measure of the energy dissipated within the sample. The relationship between these properties is given by the equation:

$$\varepsilon'' = \varepsilon' \tan\theta \tag{5.15}$$

Values for the dielectric constant and the dielectric loss factor for a wide variety of foods were reported by Mohsenin (1984) and Mudgett (1982). Both properties are affected by the temperature and moisture content of the sample, as well as the frequency of the electric field. The level of salts has a pronounced effect. Dielectric properties of materials are measured over a wide range of frequencies, using a variety of instrumental methods.

During microwave and dielectric heating, the power dissipated (P_o) within the sample is given by the equation:

$$P_o = 55.61 \times 10^{-14} f E^2 \varepsilon'' \tag{5.16}$$

where P_o = power dissipated (W cm^{-3}), f = frequency (Hz) and E = electric field strength (V cm^{-1})

Materials with a high dielectric loss factor absorb microwave energy well and are sometimes termed "lossy" materials. It should be noted that the rate of heating also depends on the specific heat capacity. In general, foods with high moisture content also have higher specific heats. Special fibre optic probes have been developed to measure temperatures during microwaving processes.

5.2.1.8.3.6 Diffusion Properties Although mass transfer operations do not play a major role in heat processing, they may become significant during storage. There may also be some leaching of plasticisers or metals from packaging materials to food components. For example, there was recently some concern (largely unfounded) about the amount of phthalate plasticiser in some milk powders.

During storage, there may be oxygen transfer from dissolved air into the bulk solution, followed by utilisation of this dissolved oxygen for microbial and chemical reactions. There may also be further gas transmission (especially oxygen) through the packaging material.

5.2.1.8.3.7 Surface Properties Surface tension is concerned with the forces acting within the fluid. Molecules at the surface of a fluid are subject to an imbalance of molecular forces and are attracted into the bulk of a fluid. Consequently the surface is said to be under a state of tension. Surface tension can be regarded in two ways: as the force per unit length acting on a given length of surface, or as the work done in increasing its surface area under

Table 5.15 Surface tension values of some fluids. (Source: Lewis, 1990. Reproduced with permission of Elsevier.)

Liquid	Surface tension (mN m^{-1})	Liquid	Surface tension (mN m^{-1})
Water*	72.75	Milk*	42.3–52.1
Ethyl alcohol*	22.75	Skim milk 0.04% fat)†	51.0
Methyl alcohol*	22.65	Whole milk (2.4% fat)†	46.7
Chloroform*	27.14	Cream (34% fat)t	44.8
Carbon tetrachloride*	26.95	Cottonseed oil‡	35.4
Glycerol*	63.4	Coconut oil‡	33.4
Mercury*	435.5	Olive oil‡	33.0
		Oleic acid‡	32.5

* From the data of Weast (1982).
† From the data of Jenness et al. (1974).
‡ From the data of Powrie and Tung (1976).

isothermal conditions. The SI units are Nm^{-1} or Jm^{-2}. It is the surface tension forces which cause most finely dispersed liquids to form spherical droplets, the shape which has the minimum surface area to volume ratio.

The surface tension of a range of fluids is given in Table 5.15. Water has a high value, although it readily becomes contaminated with surface active agents. Milk and other materials have much lower values. In general, surface tension decreases slightly as temperature increases.

Surface active agents lower surface tension which is the basis of emulsification and detergency. Foaming can be a problem with some substances, especially in pumping and filling operations. Interfacial tension deals with the forces acting at an interface, usually between two immiscible liquids: it assumes importance in emulsions.

Other examples of surface reactions concern the attachment of food components to the walls of heat exchangers (fouling) and their subsequent removal (cleaning). Biofilms may assume importance; for example, algae and other materials in cooling water. There is an increasing interest in biofilms. They are relevant to thermal processing operations, as improperly cleaned deposits may form breeding grounds for microbial growth, which may compromise product safety; there are also implications for corrosion. There have been a number of recent articles devoted to the removal of biofilms and to the inactivation of different microroganisms in biofilms. In thermal processing, the effective removal of biofilms (if they are present) would seem to be the most important priority. Biofilm formation and its elimination from food processing equipment have been reviewed by Gibson *et al.* (1995), Sjoeberg *et al.* (1995), and Teh *et al.* (2015). Biofilms are further discussed in Section 6.2.2.8.

5.2.1.8.3.8 Optical Properties and Sensory Characteristics Of immediate impact is the appearance of food. For example, for milk, this is the presence or absence of visible defects such as fat separation or coagulation, and its colour. Colour, like texture, is not strictly a physical property, but a sensory or pyschological characteristic. As such it should be evaluated by human assessors (see Section 11.2.23). This can be extremely time consuming, so a number of instruments have been designed to measure both texture and colour. Those for measuring colour are based on the spectral signal which results from light transmitted through, or reflected by, the sample. Other optical properties include refractive index and light scattering.

Refractive index for a transparent material is defined as the ratio of the velocity of light in air to that in the material. It is normally measured using light at a constant wavelength (589.3 nm – the sodium D line) to four decimal places; values are usually quoted at a constant temperature of 20 °C. It is affected by both temperature and wavelength.

The refractive index of some materials are: water (1.333); milk (1.338) and milk fat (1.462). The presence of dissolved or suspended solids (below 0.1 μm) increases refractive index above that of water. Larger particles such as fat globules, air bubbles and sugar crystals have no effect.

Contrary to this, light scattering is influenced (caused) by larger particles whose refractive indices are different from that of the surrounding medium. Such scattering is a random process, but it is possible to measure the intensity of the light at some angle to the incident beam, as well as the intensity of the light transmitted through the sample. The same principles apply to other forms for infrared and ultra violet radiation. Both transmitted and scattered light (radiation) form the basis of a range of techniques for analysis of milk and dairy products. Laser scattering techniques form the basis for particle size measurements, fat globule distributions in emulsions (see Section 11.2.5.2) or particle size determinations for powders.

The sensory characteristics of appearance, colour, flavour and texture are influenced by many of the physical properties described above. Changes in sensory characteristics resulting from heat treatment and storage are discussed in more detail in Chapters 6, 7 and 11.

5.2.1.8.4 Residence Time Distribution

In UHT processing, the flow of the product being pumped through the heat exchanger will either be streamline or turbulent. It is extremely useful to know whether it is streamline or turbulent as this will influence both heat transfer rates and the distribution of residence times. Turbulent flow is highly desirable but not always possible to achieve.

The nature of the flow, whether streamline or turbulent, can be established in the holding tube, for example, by evaluating the Reynolds number (Re) (see Section 5.2.1.8.3.3), using either the average velocity, or the volumetric flow rate, according to equation:

$$\mathrm{Re} = \frac{vDp}{\mu} = \frac{4Q\rho}{\Pi\mu D} \tag{5.17}$$

where v = average fluid velocity (m s^{-1}), ρ = fluid density (kg m^{-3}), D = tube diameter (m), μ = fluid viscosity (Pa s) and Q = volumetric flow rate (m^3 s^{-1}). The most difficult parameter to estimate is viscosity, because viscosity is temperature-dependent (see Section 5.2.1.8.2.3). It also varies throughout the UHT plant.

The average residence time in the holding tube (t_{av}), which is based on the average velocity or plug flow velocity, can be determined from the equation:

$$t_{av} = \frac{V}{Q} \tag{5.18}$$

where V= the volume of the holding tube (m^3) and Q = the volumetric flow rate (m^{3}s^{-1})

However, in real flow situations, there is a velocity distribution across the tube which gives rise to a distribution of residence times, both in the holding tube, and in the heating and cooling sections.

For viscous fluids, the flow is likely to be streamline, that is, its Reynolds Number (Re) would be less than 2000 and there will be a wide distribution of residence times. For Newtonian fluids the minimum residence time will be half the average residence time. Turbulent flow (Re >4100) results in a narrower distribution of residence times, with a minimum residence time of 0.83 times the average residence time.

Figure 5.14 illustrates residence time distributions for three situations, plug flow, streamline flow and turbulent flow. Plug flow is the ideal situation, as there is no distribution of

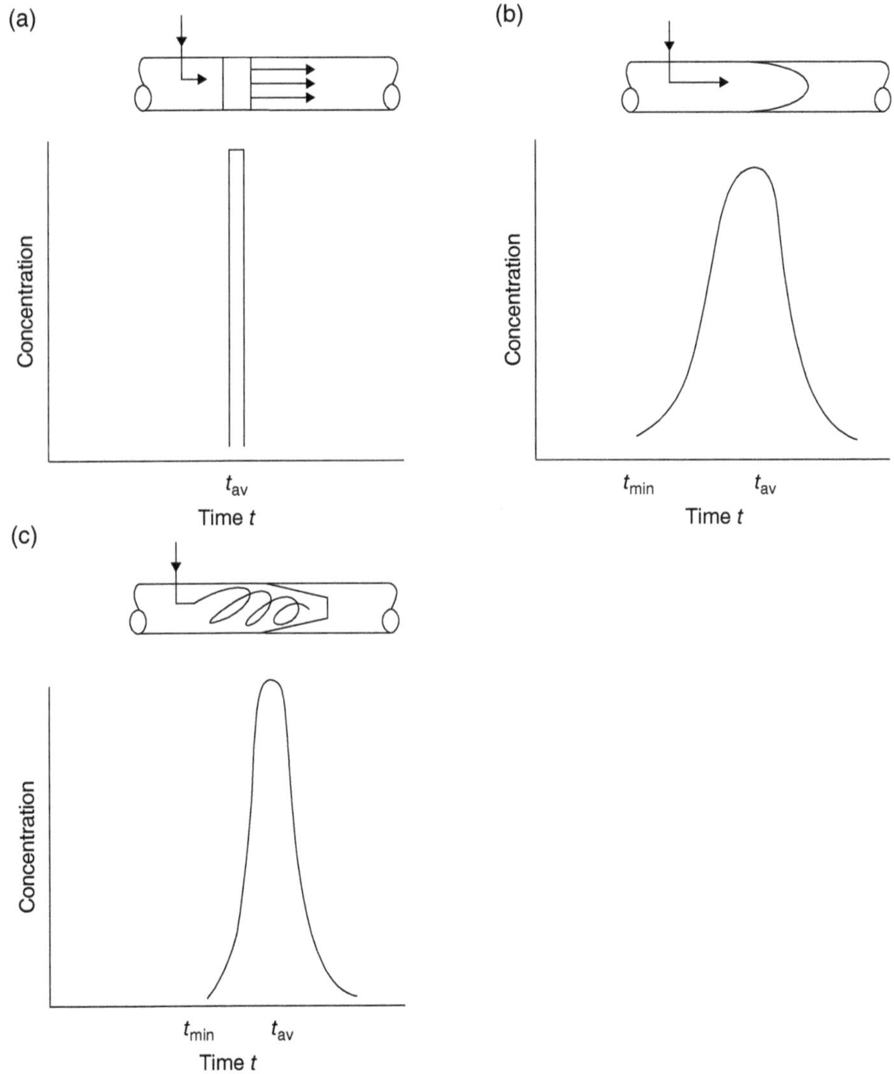

Figure 5.14 Residence time distributions for three types of flow: (a) plug flow; (b) streamline flow; (c) turbulent flow. (Source: Lewis, 1990. Reproduced with permission of Elsevier.)

residence times. However, for both streamline and turbulent flow, the minimum residence time should be greater than the stipulated residence time, to ensure that all elements of the fluid are held at the desired temperature for the desired time. This will ensure that the fluid is not under-processed, which is fundamental for ensuring food safety. Residence time distributions and their implications for UHT processing are discussed in more detail by Lewis and Heppell (2000) and Burton (1988).

5.2.1.8.5 Steam Requirements

Steam utilisation for heating a UHT product can be estimated from the energy balance: heat lost by steam = heat gained by product

$$m_s h_{fg} = m_p c\, \Delta\theta \tag{5.19}$$

where m_s = mass of steam; h_{fg} = latent heat of vaporisation, m_p = mass of product; c = specific heat of product and $\Delta\theta$ = temperature rise

As an example of a calculation, the amount of steam (m_s) required to heat 1000 kg milk from 20 to 140 °C is given by:

$$m\,s \times 2144 = 1000 \times 3.9 \times 120$$

$$ms = 218 \text{ kg steam } (\text{assuming no heat losses}).$$

This is independent of whether it is done by direct or indirect heating. However, if it is done by direct heating, there would be 1218 kg of diluted milk and if the original solids had been 12.5 % then the final solids would be just under 10.3%. This added water would need to be removed by flash cooling.

Of course, energy is saved by regeneration. If, for example, the regeneration efficiency was 90%, then the amount of steam used would be only 10%, that is, in this example, 21.8 kg. A high regeneration efficiency also contributes to reduced refrigeration costs. Regeneration efficiencies are not as high for direct plants as for indirect plants. The approximate energy cost for producing 1 tonne of steam can be as low as £20, but may be considerably higher.

The quality of the steam used for direct steam injection and infusion processes is a major consideration. It is also important to ensure that the steam is free of water droplets and any other chemical contaminants.

5.2.1.8.6 Pressure Measurement

There are many processes where the performance is influenced by the operating pressure, so pressure is a very important processing variable. Examples of such processes are membrane separation processes, homogenisation and high-pressure processing. Also, many different units of pressure might be encountered, which can cause some confusion.

Absolute pressure is defined as force/area. The SI units of absolute pressure are N m^{-2} or Pascals (Pa). This is a very small unit of pressure and thus for convenience the units of bar are often used, although it is not an SI unit:

$$1\,\text{bar} = 10^5\,\text{Pa} = 0.1\,\text{MPa} = 100\,\text{kPa}$$

Absolute pressure can also be expressed as a height of head of fluid

$$P = \rho g h \tag{5.20}$$

where P = absolute pressure (Pa); h = pressure head (m); ρ = fluid density (kg m^{-3}) and g = acceleration due to gravity = 9.81 m s^{-2}

Thus a pressure of 1 bar (10^5 Pa) would support a head of water of about 10.3 m, or 76 cm (29.92 inches) of mercury. What makes the bar a popular pressure measurement is that 1 bar is approximately equal to 1 atmosphere. Homogenisation conditions are often recorded as bar or MPa (where 1 bar = 0.1 MPa).

Many food processes are performed at atmospheric pressure. One standard atmosphere (in different units) is given as:

- 760 mm Hg
- 101.3 kPa
- 1.013 bar
- 1.033 kg (f)/cm^2
- 14.69 psi

Since most pressure gauges measure pressure above (or below) atmosphereic pressure, the reading is often given as (g), for example, steam pressure of 9 bar (g) is 9 bar above atmospheric pressure. The relationship between **gauge pressure** and absolute pressure is:

$$\text{Absolute pressure} = \text{gauge pressure} + \text{atmospheric pressure.}$$

The reason this distinction is made is in situations where the pressure may be measured as a gauge pressure, but where the absolute pressure is required in some equations, for example, in the ideal gas equation (PV = RT), or when interpreting data from steam tables (Table 5.6). As discussed in Section 5.2.1.8.3.1, the steam tables measure absolute pressure, which is pressure above an absolute vacuum. It is analogous to the situation for temperatures, which are measured in degree Celsius, but need occasionally converting to an absolute temperature (K).

We also encounter situations where the pressure is reduced below atmospheric pressure, for example, the flash cooling vessel in direct steam injection and infusion. Again, it is important to distinguish whether pressure is measured as a gauge pressure or an absolute pressure. The units of pressure used in the steam tables are usually absolute pressure, which is pressure above an absolute vacuum.

For vacuum measurement units of inches of mercury ("Hg) may still be encountered. It is measuring the pressure below atmospheric pressure. In UHT processing, it is essential to keep the pressure high to prevent the fluid boiling at the UHT temperatures (see steam tables). This is also important to prevent air coming out of solution (Burton, 1988). Should the fluid pressure fall for whatever reason below that of the saturated vapour pressure at the desired UHT processing temperature, then the temperature will also fall. Pressure control is also used to establish the temperature in the flash cooling vessel and also the amount of water removed (see also Section 5.2.1.8.3.1).

Diaphragm gauges should be used on UHT product lines, as these are hygienic gauges, whereas Bourdon gauges may be used for measuring pressures of steam, processing and cooling water, compressed air and refrigerants. These must not be used on UHT product lines as they are not hygienic, as the processing fluid will enter the Bourdon tube and this will be a dead end and impossible to clean.

5.2.2 Electrically Based Heating Systems

5.2.2.1 Electrical Tube Heating

An electrical tubular heating system which is used for UHT processing of milk uses electrical energy to heat a stainless steel tube carrying the product. The stainless steel tube acts as an electrical resistor and is heated by an electrical current by the Joule effect. Heat transfer to the product in the tube occurs by conduction and mixed convection (Lefebvre and Leuliet, 1997). While the generic name given to the technology is Current Passage Tube (Tube à Passage de Courant [TPC] in French), it is also referred to as Actijoule®, the brand name of one of the companies manufacturing the equipment. It was developed in France in the late 1980s and commercialised in the early 1990s (Deeth, 1999).

Notable characteristics reported for this technology are the linear rise in temperature of the product, the small and constant temperature differential between the tube and the product, a variable heating flux rate for different products enabling fast heating for low-viscosity products such as milk, and absence of thermal inertia in the system for start-up and shut-down; when the power is switched off, heating ceases immediately (Montaron *et al.*, 1991). A UHT system used for production of UHT milk includes a pre-heat/regeneration section heated by hot milk and an electrically heated section raising the temperature from 100 to 140°C in 4 seconds. The energy efficiency of the plant is ≥95% (Montaron *et al.*, 1991). The equipment is capable of achieving temperatures of up to 170 °C. Sterilisation of the plant before use is achieved by using the electrical heating capacity to heat a weak solution of an inorganic salt which is a strong electrical conductor.

An attractive feature of the system is the lack of need for steam generation and reticulation, and the associated specialist personnel. A disadvantage, however, is the cost of running the plant. In most countries, electrical energy is more expensive that fossil fuels for heating and hence unless the plant has access to a supply of cheap electricity, it may be uneconomic to operate.

5.2.2.2 Ohmic Heating

Ohmic heating of food, which also goes by the names of Joule heating, or electrical resistance heating (Herrick *et al.*, 2000), refers to a process whereby food materials are heated by passage of an electric current. The technology itself dates back to the 1800s and was revived in 1980s. Since the mid-1990s, commercial installations have been introduced (Sastry, 2008). Ohmic heating is an attractive food processing technology because of its technical simplicity, uniform, rapid heating patterns, high energy efficiency, and its applicability to liquids or particulate mixtures. However, the products still need to be cooled and there are no rapid means of doing this.

The electrical energy is almost entirely dissipated within the heated materials during ohmic heating, hence yielding a very high energy transfer efficiency close to 100% (Jun & Sastry, 2005); in comparison, microwave heating is ≤65% efficient (Salengke, 2000). Ohmic heating can be used for shear-sensitive foods or food mixtures containing a high percentage of particulates (Marcotte, 1999) and heat-sensitive nutrients. Volumetric heat generation resulting in uniform and rapid heating can produce foods which retain better quality attributes and nutrients than those produced by conventional thermal processes. Ohmic heating has been utilized for several food processing applications such as blanching, sterilisation, pasteurisation, thawing, dehydration and extraction, and commercial products such as diced, sliced fruits, pasteurised liquid egg processed by ohmic heating have appeared in the market.

5.2.2.2.1 Principles of Ohmic Heating

Heat generation by the passage of an electric current during ohmic heating is governed by Ohm's law:

$$Q = VI = IR^2 \tag{5.21}$$

If the voltage gradient is known, the heat can be calculated from voltage gradient and specific or electrical conductivity:

$$Q = |\Delta V|^2 K \tag{5.22}$$

Knowledge of electrical conductivity of food materials is critical for ensuring uniform and efficient ohmic heating. The electrical conductivity of foods, σ (*S/m*), can be determined by:

$$K = \frac{I}{V} \cdot \frac{L}{A} \tag{5.23}$$

where V is applied voltage, *I* is electric current (*A*), *L* is distance between two electrodes (*m*) and *A* is the area of electrode (m^2). The electrical conductivities of most food materials are dependent on their temperature, as given by Palaniappan and Sastry (1991):

$$K(T) = K_{ref}(1 + mT) \tag{5.24}$$

where K_{ref} is the electrical conductivity at a reference temperature and *m* is a temperature coefficient. Foods containing a high amount of ionic materials such as salts have high electrical conductivity. On the contrary, foods composed of non-ionized constituents and having high viscosity such as honey and syrups will thermally lag other components in the system.

5.2.2.2.2 Electrochemical Reaction on Electrodes

The application of alternating current (AC) at low frequencies during ohmic heating induces electrochemical reactions at electrode/solution interfaces including electrode corrosion, migration of metal ions, and generation of O_2, H_2 and free radicals (Ghnimi *et al.*, 2008). Electrochemical reactions at the electrodes may lead to metal ions migrating into the foods and are considered as contaminants and, possibly, toxic substances. These metal ions can also form various complexes with food constituents and catalyze undesired chemical reactions such as lipid oxidation (Samaranayake, 2003). Oxygen and free radicals adversely affect food constituents such as lipids and vitamins through oxidation reactions. Electrochemical reactions occurring at the interfaces between electrodes and solution are greatly influenced by the physical and chemical properties of the electrodes. Other critical factors include pH of solution, frequency, pulse width and pulse delay time. Studies on various electrode materials such as titanium, stainless steel, platinized-titanium, and graphite at different pH values demonstrated that platinized-titanium was relatively inert to electrochemical corrosion (Samaranayake & Sastry, 2005). It is also known that the migration of metal ions such as Fe, Cr, Ni, Mn, and Mo from stainless steel electrode into foods in a reusable pouch is minimized when a high frequency pulsed waveform is used (Jun *et al.*, 2007). Heating patterns and the modelling of ohmic heating have been described by Lewis and Jun (2012).

In conclusion, ohmic heating is an attractive technology for pasteurising and sterilising foods. Despite some challenges which still exist, such as non-uniform heating, ion migration into the foods and the identification of cold spots, the technology has become better understood and remarkably improved due to continued research efforts in food properties, process modeling, electrochemical reactions and field characteristics. Future research is required to investigate the combination of ohmic heating with other technologies and to develop a further understanding in modeling and food property changes under various process conditions. One can forsee some potential application for UHT products, for example to satisfy the demand for innovative particulate foods.

5.2.2.3 Microwave Heating

5.2.2.3.1 Principles of Microwave Heating

Microwaves are electromagnetic waves with a frequency band of 300 MHz to 30 GHz, which is located between the radio frequency and infrared regions (Schubert & Regier, 2005). Heating with microwaves involves primarily two mechanisms – dielectric and ionic. Dielectric properties, dielectric constant and dielectric loss factor, are the relevant material properties for explaining interactions with electric fields (Sumnu & Sahin, 2012) (see Section 5.2.1.8.3.5). Most molecules are electric dipoles, meaning that they have a positive charge and a negative charge, and therefore vibrate as they try to align themselves with the alternating electric field induced by a microwave beam. Water in food is often the primary component responsible for dielectric heating. Due to their dipolar nature, water molecules try to follow the electric field associated with microwave radiation as it oscillates at high frequencies. Such oscillations of the water molecules produce heat. The dielectric properties of the food also depend on the frequency of the applied alternating electric field, the temperature of the material, and the density, composition and structure of the material (Datta & Anatheswaran, 2001). The second major mechanism of heating with microwaves is through the oscillatory migration of ions in the food that generates heat under the influence of the oscillating electric field. Ionic polarization differs from the electronic mechanism in that it occurs due to the relative motion of the atoms instead of a shift of the electron clouds surrounding atoms.

The rate of heat generation per unit volume (Q) at a particular point in the food generated by microwave heating is given by the equation:

$$Q = 2\pi\varepsilon''\varepsilon_0 fE^2 \tag{5.25}$$

where ε'' is the dielectric loss factor, ε_0 is the dielectric constant of free space, f is the frequency (Hz) and E is the root-mean-squared value of the electric field intensity. Therefore, the higher the loss factor, the greater the microwave absorption. The loss factor is made up of two components, the dipolar loss factor and the ionic loss factor, in line with the two mechanisms of microwave heating discussed above. In foods with a low ionic strength, the dipolar loss factor dominates whereas in foods with a high ionic strength, such as high-salt foods, the ionic loss factor dominates.

To prevent interferences with radio frequencies used for broadcasting and telecommunications, special frequency bands are reserved for industrial, scientific and medical applications. Typically, microwave food processing uses two frequencies of 2450 and 915 MHz. Of these two, the 2450 MHz frequency is used for domestic microwave ovens, and both frequencies are used in industrial heating. The 915 MHz frequency has considerable advantages for industrial applications (Schubert & Regier, 2005).

Microwave heating has been investigated for pasteurisation and sterilisation heating because it is rapid and therefore requires less time to reach the desired process temperature than when conventional indirect heat exchangers are used. This is particularly true for solid and semi-solid foods that depend on the slow thermal diffusion process in conventional heating. Microwave heating is volumetric, that is, the microwave energy can penetrate food to generate heat internally as well as at the surface of the treated material. Conventional indirect heating methods apply heat to the surface only and hence product near the surface may become overheated and may foul or burn on to the surface of the heat exchanger.

A problem encountered with microwave heating is the non-uniformity of the heating, which in microwave ovens can result in cold spots. This has also presented difficulties with sterilisation with regard to meeting regulatory requirements which require evidence that the entire contents are held at the sterilisation temperature for the required period of time (Datta & Anatheswaran, 2001). The development of cylindrical microwave heating systems has resulted in more uniform heating in continuous flow systems. It is claimed that these systems rapidly heat food products without contact with a hot heating surface, and allows for ultra-rapid pasteurisation or sterilisation with minimal product degradation (Stone, 2009).

Another issue is that microwaves have limited penetration. The penetration depth depends on the type of food but also the temperature. For example, the penetration depth, defined as the depth at which the power density reduces to 37% (1/e) of that at the surface, in water at 45 and 95 °C is 1.4 and 5.7 cm, respectively. The penetration decreases as the frequency increases. Hence, commercial systems operating at 915 MHz have greater penetrating power than the domestic appliances operating at 2450 MHz. A related electromagnetic radiation, radiofrequency, has much greater penetration than microwave waves.

5.2.2.3.2 *Applications of Microwave Heating*

Applications of microwave heating in the food industry include drying, baking, blanching, thawing and tempering, cooking, pasteurisation and sterilisation (Summu & Sahin, 2012). Only the last applications are discussed here. There has been a limited number of reports on the use of microwave technology for heat treatment of milk. Villamiel *et al.* (1996) reported the continuous-flow microwave pasteurisation of milk and concluded that microwave heating caused less chemical damage than conventional heating. Lin and Ramaswamy (2011) also compared continuous-flow microwave with conventional pasteurisation of milk and found that the D-values for alkaline phosphatase inactivation were an order of magnitude lower under microwave heating than under conventional heating. This could have ramifications for the use of alkaline phosphatase inactivation as an index of effective pasteurisation of microwave-treated milk. However,

Villamiel *et al.* (1996) found that pasteurisation by both microwave treatment and conventional heating inactivated alkaline phosphatase. These authors did however show that microwave heating caused less denaturation of β-Lg. They attributed the differences in effects of the microwave and conventional pasteurisation treatments to differences in heat distribution.

The first paper to report the effects of microwave processing on the sensory, microbiological, rheological, and biochemical parameters of UHT milk during long-term storage was by Clare *et al.* (2005). They compared microwave and conventional indirect heating to produce UHT skim milk. For the microwave heating, a 60-kW continuous flow microwave-heating unit operating at 915 MHz was used. Microwaves were focused to a cylindrical applicator and delivered to the product by a waveguide of rectangular cross-section. The rates of heating in the two systems were almost the same (3.51 and 3.26 °C/s) and the calculated F_0 values were the same for both heating modes (8 for white milk; sterilised at 137.8 °C for 20 s). Plasmin was inactivated by both treatments and the thiol contents and viscosity values were similar for both. Differences were, however, observed in the sensory characteristics; the conventionally heated milks were darker and more astringent, and developed more caramelised and stale/fatty flavours during storage.

A commercial microwave system for sterilising packaged food, known as Microwave-Assisted Thermal Sterilisation (MATS), has been introduced. It operates at 915 MHz and treats the packaged food in pouches in pressurized hot water. The process was approved by the US Food and Drug Administration (FDA) for production of shelf-stable food in 2010 (Harrington, 2010).

5.3 Homogenisation

Homogenisation is a critical stage in the processing of foods which contain substantial amounts of fat. It can improve product stability, shelf-life, flavour release, digestiblity, taste and viscosity, and control particle size. In some fomulations, it also permits lower levels of additives to be used. In UHT processing, the major aim of homogenisation is to prevent fat separation during storage but is also useful for dispersing other particulate material in the product.

Homogenisation is a process which reduces the fat globule size, usually by mechanical disruption. This leads to a large increase in the interfacial area. Fat globules in raw milk range between 1 and 10 μm, with an average size of about 3.5 μm. Such milk is subject to creaming if it is left to stand even for a few hours and this was why a visible cream layer is observed when unhomogenised pasteurised milk is sold in clear glass or plastic bottles. After homogenisation, the average fat globule size is reduced to around 1.0 μm and preferentially to less than this. Below this value further reductions in size are small compared to the increased energy costs involved in obtaining higher pressures and that would just end up generating more heat. New valve designs are claimed to be more effective at lower pressures, thereby reducing energy costs.

Importantly, there should not be an appreciable number of droplets >1.0 μm in order to avoid fat separation during subsequent storage. As an illustration, if particle size analysis shows that there are some fat globules which are larger than 1 μm, these are the ones that are likely to separate. This would suggest that the homogenisation process was defective.

Hooi *et al.* (2004) recommended that the d(0.9) value for homogenised milk should be less than 1.7 μm, whereas that for raw milk is 5 to 6 μm. They claim that a properly functioning homogeniser should give a value below 1.3 μm. (A d(0.9) value of 1.3 μm indicates that 90 % of the fat is in globules below that size, or 10% of the fat globules (by volume) have diameters greater than that). This recommendation might be of use in interpreting the particle size measurement shown in Figure 11.1a. As the d(0.9) was 1.98 micron, it would suggest that homogenisation was not ideal.

Homogenisation should take place above 50 °C to ensure that the milk fat is in the liquid state, as this facilitates disruption of the fat globules. In UHT processing, homogenisation also helps to reduce sediment formation. The nature of the dispersion is affected by the pressure, temperature, type of homogeniser and the nature of the product. Two-stage homogenisation is usually used, especially for products with higher fat contents and also to disrupt clumps that might form in the first stage. Two-stage homogenisation is expressed as two pressures, e.g., 200 bar (20 MPa) and 50 bar (5 MPa), the total pressure being 250 bar (25 MPa). For two-stage homogenisation there is the option of using the first stage upstream and the second stage downstream, which might be a good option for some products. Walstra and Jenness (1984) discuss these effects in more detail.

Most homogenisers are positive displacement pumps and as such their presence ensures that flow rate, and hence the residence time, remain constant (see Equation 5.18). They function by subjecting the product to high pressure and forcing it through a narrow restriction (the homogenisation valve) at high velocity, which produces intense turbulence. It is estimated that the energy density is 10^{10} - 10^{11} W/m^3 and this determines the size and velocity of the eddies which disrupt the fat globules. Homogenisation also increases product temperature, by about 1 °C per 580 psi (40 bar or 4 MPa). Under normal UHT processing, a temperature increase of 4-5 °C could be expected (see rise in temperature–time profile in Figure 5.3).

During homogenisation, the newly formed fat globules are covered and stabilised by new membrane material, which in milk is predominantly casein. It has been estimated that this process takes place within a short time period of less than 10 μs. Large micelles are adsorbed in preference to small ones. This may cause the homogenised fat globules to behave more like large casein micelles under some circumstances. For example, conditions which will cause aggregation of casein micelles may also cause homogenised fat globules to aggregate. This may lead to incorporation of fat into deposit or sediment during UHT treatment. For example Boumpa *et al.* (2009) reported that the sediment formed during UHT treatment of goat's milk contained a substantial amount of fat. It may also be a factor in fat separation which can accompany age gelation (see Section 7.2.2). In some food products with higher fat contents (e.g., ice cream mix) additional emulsifiers and stabilisers may be required to ensure that the formed emulsion has good stability. Another important reason to homogenise UHT products is to disperse protein aggregates which may form during heat treatment. In this way, homogenisation helps to reduce sediment formation in milk and is best performed after heat treatment (Zadow, 1975). However, care should be taken as excessive homogenisation might result in more sediment formation.

Therefore, an important decision which has to be made in UHT processing is whether to place the homogeniser upstream or downstream of the sterilisaton section. If it is positioned upstream, it is not in the sterile part of the plant and does not have to perform under aseptic conditions. Where downstream homogenisation is selected, it is important for the homogeniser to be of aseptic design, for example, having a sterile block and pistons moving through sterile water or an atmosphere of steam. This downstream positioning, however, increases the risk of microbial contamination, but it may result in a product with better stability. In terms of reducing potential spoilage, it is best to position the homogeniser upstream and remove the requirement to incorporate a sterile block.

Typically in UHT processing, the product is heated to 60 to 80 °C before homogenisation and final heating to the sterilisation temperature. However, the drawback of homogenisation in this position is that the higher temperatures used later in the process, or the high shear rates found in the heat exchangers (especially plates) or the induced turbulence in steam injection may cause some destabilisation of the emulsion, which might affect its stability during storage. Heating to high temperature is reported to destabilise emulsions. However, many UHT plants operate with the homogeniser in the upstream position and this is the preferred position for situations where stability and sedimentation are not considered to be major problems.

For tubular UHT plants there is another option, which is to place the pumping part of the homogeniser upstream and the homogeniser valve downstream. Thus the entire section between the homogeniser and its valve is at high pressure, but it saves the complication of needing to keep the homogeniser sterile. The option of separating the homogeniser itself from the homogenisation valve is not an option for plate heat exchangers, as they will not withstand the high pressures used in homogenisation.

With direct UHT processing, downstream homogenisation is essential (Burton, 1988). It is thought that the injection or infusion process itself may damage or destabilise the emulsion produced by homogenisation. It may cause reassociation of fat globules and formation of casein aggregates which have a chalky, astringent character; these aggregates are broken down by downstream homogenisation and the chalky sensation disappears (IDF, 1972). The aggregates can also lead to excessive sediment formation if the milk is not homogenised downstream. Perkin (1978) compared injection and infusion methods for sediment formation and found no differences in the amounts, but considerable difference in their appearance. Other situations where downstream homogenisation is preferable are when there is excessive destabilisation during the high-temperature heating steps or when excessive sediment may be formed in the product because of circumstances other than direct processing mentioned above. Furthermore, double homogenisation with one homogeniser upstream of the sterilisation section and one downstream can be used for increasing the stability of some UHT products such as coffee cream and evaporated concentrated milk (Tetra Pak, 2015).

An interesting observation is that steam injection achieves some homogenisation and reduces fat globule size. Zadow (1969) calculated that the input from a steam injector was equivalent to 1100-1600 psi (7.5-11 MPa). As a consequence, he concluded that direct injection processes require less homogenisation pressure to achieve the same degree of fat stabilization compared with indirect processes.

In a UHT process using plate heat exchangers, it has been shown that the fat globule size in whole milk is influenced by the position of the homogeniser, with downstream homogenisation giving the better homogenisation effect. The difference was much more marked for direct steam injection, where upstream homogenisation was not effective. Lewis and Heppell (2000) reported that at equal homogenisation pressures, the fat globule size was slightly smaller for indirect compared to direct processes. In contrast, Hillbrick *et al.* (1999) showed that the homogenised UHT milks processed by three different UHT treatments (indirect heating with upstream homogenisation, indirect heating with downstream homogenisation and direct heating with downstream homogenisation) had similar microstructures when examined with transmission electron microscopy and immunogold labelling, and formed cream layers of similar thicknesses on storage.

Kaw *et al.* (1996) observed that for whole milk, UHT sterilisation alone had no effect on particle size distribution (PSD). Homogenisation prior to UHT processing, resulted in considerable aggregation of fat globules; these aggregates were not easily broken down by a second homogenisation process. Homogenisation after heating was most effective, reducing the fat globule size considerably, as well as the tendency toward aggregation. Addition of a protein dissociating agent reduced these aggregates, indicating the role of casein in their formation. The suggestion was that the aggregates were intact fat globules in a protein matrix.

Downstream homogenisation was also most effective for UHT treatment of reblended concentrated whole milk, made by mixing skim milk and cream (no composition was given) and for RO concentrates. The reblended milk behaved differently to whole milk, in that UHT treatment alone caused aggregation. Also the aggregates did not incorporate fat globules. The reverse osmosis process alone was also found to have a homogenisation effect, caused by the passage through the back-pressure valve. However, UHT processing of RO concentrates produced an increase in particle size in the PSD indicating considerable aggregation. The large aggregates were not easily broken down by subsequent homogenisation (Kaw *et al.*, 1996).

Homogenisation has been reported to reduce the heat stability of milk concentrates, so it is preferable to homogenise after UHT processing. This option is commonly used in the manufacture of heat-evaporated concentrates (Muir, 1984).

One might expect homogenisation to increase a product's susceptibility to oxidation reactions. Walstra and Jenness (1984) reported the opposite, that homogenisation reduces its proneness to autoxidation, whether induced by copper or light. This effect must result from the change in surface coat of the fat globules, although the exact mechanism is unknown. Homogenisation makes milk whiter. It also improves the mouthfeel and makes it less watery and more creamy.

Problems can arise when the homogeniser does not function correctly and this leads to fat deposits and fat separation in the package. Depending on the storage conditions, the fat may form a thin layer on the surface or, in more extreme cases, may even form a solid cream plug. These undesirable situations are more likely to arise if the fat content of the product is high and the storage period is long. However fat separation is hindered as product viscosity is increased, as can be seen from Stokes equation (see Equation 5.2).

Homogenisers are discussed in more detail by Kessler (1981) and Wilbey (2011). High pressure homogenisers are discussed by Huppertz (2011) and in an NIRD publication (1985). Methods for measuring homogenisation efficiency are discussed in Section 11.2.5. Small homogenisers operating at pressures up to 4000 bar (400 MPa) are now available and hold considerable promise if large-scale units can be produced. High-pressure homogenisation is described in Section 10.5.

5.4 Deaeration

A deaeration unit is incorporated into some UHT processes to remove dissolved oxygen (see Section 7.1.2). Two claimed advantages are an added efficiency of heating and improvement in flavour. As noted in Section 5.2.1.1.2, during steam infusion, the heat transfer rate into product containing air can be several times lower than into properly deaerated product (Fredsted *et al.*, 1996). Renner (1978) showed that the organoleptic quality and ascorbic acid retention in indirect UHT milk was improved by including a deaeration step in the UHT process.

Carlson (1996) listed four reasons for deaeration: to minimise chemical reactions (especially oxidation) whose rate increases with increased processing temperature; to maintain constant filling conditions and prevent foaming; to maintain specific volume of product - if considerable amount of air is present, the holding time may be too short; and to reduce fouling. The last point was investigated by Prakash (2007) who found that that the air content of milk had no effect on fouling. Since UHT plants operate at about 4 bar (0.4 MPa) pressure, air is not released as bubbles on the heat exchanger surface and hence fouling is not initiated.

One of the earliest commercial indirect UHT systems was the Ahlborn process which incorporated two stages of deaeration, one upstream immediately after homogenisation and one downstream after an initial cooling step. The first stage would increase the efficiency of the plant and the second stage was included to remove volatile sulfur compounds produced during the heating steps (Hsu, 1970). Ikezumi *et al.* (2006) patented a downsteam deaeration step in UHT processing to suppress "deterioration of flavour of cow's milk and milk beverage" in glass and polyethylene terephthalate (PET) bottles during storage. According to the patent, deaeration is performed at a temperature below the boiling point and reduces the dissolved oxygen concentration in milk to ≤3 mg/kg.

Raw milk has been estimated to contain, on average, 6% dissolved gases by volume. Some milk tankers are equipped with deaerators to remove air to obtain a more accurate measure of volume; however, much of the dissolved gases in milk is removed by procedures in the milking parlour and in the reception area on arrival at the dairy (Tetra Pak, 2003). Deaerators in heat treatment equipment work by subjecting the milk to a vacuum flash cooling process, typically from 68 °C down to 60 °C. This is effective in removing dissolved gases but also removes a small amount of water, as vapour. This can be condensed and added back to the milk (Tetra Pak, 2003). In the APV Parasol Deaerator, the product is sprayed as a thin film in a form of a parasol into a vessel under vacuum. The thin film of product provides a large surface area and maximises deaeration efficiency (SPX, 2008). It is claimed to reduce oxygen levels to 0.5 mg/kg.

The vacuum flash cooling vessel of a direct UHT plant works on a similar principle, but its primary aim is to remove the water that has condensed from the steam during the steam infusion or injection step. Steam tables (see Table 5.6) are useful for determining the appropriate flash cooling pressures. The vacuum flash vessel also removes dissolved oxygen and some heat-induced volatile components (see Section 7.1.2). Renner (1978) maintained that the superior flavour of direct UHT milk compared with indirect UHT milk is due to oxygen removal in the direct UHT processes.

Despite the apparent advantages of deaeration, Carlson (1996) warned of its downside. A deaerator is another item in the processing system that has to be balanced with respect to flow in and flow out and, if it is not be sealed adequately, air leaks can occur which reduce the device's effectiveness and may cause foaming. It was suggested that if the value of a deaerator was questionable in a particular situation, it should not be used.

5.5 Aseptic Packaging

The aseptic packaging system is an integral part of a UHT plant. It has a major role in quality of the product as the spoilage rate of UHT products is largely determined by the performance of this system (von Bockelmann & von Bockelmann, 1998). It includes the filling machine and associated equipment and in most cases it also includes an aseptic tank (or surge tank) often referred to as the "A-tank". The A-tank provides a buffer to overcome the problem of the filling machine running at a different speed and at a different time to the UHT plant. Every aspect of the aseptic packaging system has to maintain sterility at all times. Any contamination can have disastrous effects on the quality of the packaged product. This is achieved by sterilisation of all contact surfaces from the sterilisation holding tube to the final product package, the packaging material or preformed packages, and the aseptic zone in the filler. Sterility is maintained in the filler with sterile air and in the A-tank headspace with sterile air or inert gas. Thus the sterile product is filled into a sterile container in a sterile environment and the container is sealed hermetically to ensure continued sterility during handling, distribution and storage.

While aseptic packaging is correctly associated with UHT products, it is recognized that it is highly desirable, though not essential, for ESL products (Brody, 2000, 2006) (see Section 3.3.1.1). Aseptic packaging eliminates post-heating contamination and hence the shelf-life of the product will be determined solely by the microorganisms which survive the heat treatment and can grow at low temperature, that is, psychrotrophic spore-formers such as *B. circulans* (Cromie *et al.*, 1989) and *B. cereus*. In this way, a refrigerated shelf-life of 90 days is possible (Brody, 2000) whereas with ultraclean rather than aseptic filling, a shelf-life of 30-40 days can be expected. Brody (2000) cites defect rates for ESL product packaged in ultra-clean and aseptic fillers as 1 in 1000 and 1 in 10,000, respectively.

5.5.1 Types of Packaging

Several different aseptic packaging systems are used commercially for UHT products. An overview of the different types is provided by Floros *et al.* (2010), Robertson (2011, 2013) and David *et al.* (2013). The four main types of containers are: paperboard cartons, plastic bottles, sachets/pouches/bags and cups. Each of these can be either pre-formed or be of the form-fill-seal type (von Bockelmann & von Bockelmann, 1998). A fifth type of aseptic packaging is bulk packaging used for transport of large volumes of sterile product (Nelson, 2014)

5.5.1.1 Paperboard Cartons

This is a very common type of aseptic packaging which is used throughout the world. The paperboard is coated on both sides with polyethylene, to make it impermeable to liquids and to facilitate sealing, and incorporates a thin aluminum foil layer (~6 μm) which makes it impermeable to light and oxygen (Robertson, 2011).

5.5.1.1.1 Form-Fill-Seal Cartons

Tetra Pak's Tetrabrik® was introduced in 1969 and is the best known commercial brand of this type of paperboard carton (Robertson, 2002). The carton is made on the aseptic filling machine from a roll of paperboard which has been creased to facilitate folding when made into the carton on the machine. After being sterilised (see below), the paperboard is formed into a tube and heat-sealed by a plastic strip attached to the inside of one edge of the material on the roll forming the longitudinal seal. Product is then fed into the tube and transverse heat-induced seals are made through the product to form the bottom and top of the carton. This ensures a very small volume of headspace (7-8 mL per 1-L carton; Perkins *et al.*, 2005) which limits the oxygen available to the product and has implications for the keeping quality of the product (see Sections 7.1.2 & 7.1.3.3). The filled package is then formed into a rectangular shape in a mold and the flaps on the bottom and top are folded down and heat-sealed.

5.5.1.1.2 Preformed Cartons

Preformed cartons are also widely used, Combibloc being a well-known brand. They are usually made off-site and transported to the filling machine in the form of lay-flat blanks with the longitudinal seals already in place. On the filling machine, they are formed into the rectangular carton shape and sealed at one end. The carton blank is then sterilised (see below), filled with UHT product, the top sealed and the flaps folded down and sealed. These cartons have much more headspace than the form-fill-seal cartons, 21-40 mL per 1-L carton (Perkins *et al.*, 2005).

5.5.1.2 Plastic Bottles

Plastic bottles have become increasingly popular in recent times. One reason for this is the increased production of UHT flavoured milk which can be readily consumed from plastic bottles. The bottles are made of either high density polyethylene (HDPE), polypropylene (PP) or polyethylene terephthalate (PET). One issue with these plastics is that they are clear and hence allow passage of light which affects the milk constituents, particularly vitamins, and causes light-induced off-flavours. Consequently, measures are taken to exclude light. These include incorporation of pigments, use of multilayer plastics with a light barrier, and use of opaque sleeves which also contain all the labelling information and branding features. A second issue is the relatively large headspace estimated by Perkins *et al.* (2005) to be 58 mL in a 1-L bottle. As for cartons, plastic bottles can be formed in situ and immediately filled or preformed, transported to the filling machine and filled.

5.5.1.2.1 Preformed Bottles

Preformed non-sterile bottles are conveyed to the sterile environment of the aseptic filler, inverted and sterilised (see below) inside and outside, filled and immediately capped with a sterile closure which is heat-sealed to the top of the bottle.

Some preformed bottles are sterile inside and sealed when conveyed to the filling machine. In the filling machine, the sealed bottles are sterilised on the outside, the seal at the top of the bottle is cut away, the bottle is filled and the filled bottle is capped as above.

5.5.1.2.2 Bottles Blown on Site

In some filling machines, plastic bottles are formed in situ by heating and blowing plastic parisons (tube-shaped injection preforms), filled and sealed in sequence. In this case, there is no need for a chemical sterilisation step as the temperature reached (165-235 °C) during the blow moulding is sufficient to sterilise the bottle (Robertson, 2013). In some cases however, the performs have the bottle thread already formed and that part does not reach high temperatures during the blow moulding. These performs are either sterilised before being blow moulded or the bottle is sterilised after being formed.

5.5.1.3 Pouches

Plastic pouches are increasing in use and popularity and it has been predicted that they may become the major aseptic packaging form. In some countries this is already the case (Astley, 2015). Drivers for the adoption of pouches include their low cost, light weight and their ecological benefits. Some published comparisons with other packaging are as follows: the cost of a UHT pouch is <20% of the cost of a paperboard carton (Elecster, undated); the weight of a 1-L pouch (Ecolean Air Aseptic®) is 16 g compared with a 1-L HDPE bottle at 32 g and a 1-L PET bottle at 36 g (Reynolds, 2012); and the "ecological balance sheet" compared with paperboard cartons according to Elecster (undated) is: energy requirement, 22.8%; air pollution, 19%; and water pollution, 1.9%. Furthermore, the equipment required for forming, filling and sealing the pouches is much simpler than those used with other aseptic packaging (Floros *et al.*, 2010).

Plastic pouches are available in different formats such as "pillow pouches" with and without spouts or closures or filling fitments, and stand-up packs; some have a handle such as Ecolean's Air Aseptic® which has an air-filled spine for ease of handling. The smaller pouches are designed as bag-alone packages but because of the risk of damage during handling and transport, larger bags are usually protected by an outer more-durable material forming the so-called bag-in-box systems. Such "boxes" may be steel or paper drums, paperboard or wooden boxes, or plastic bins. These packs vary in size from 1 to 300 gallons (4 to 1100 L) (Floros *et al.*, 2010).

The plastic film used is usually multi-layered based on polyethylene or polypropylene with oxygen and light barriers made of material such as EVOH. Some have calcium carbonate incorporated to provide opacity and stiffness.

The packaging material for pouches is fed into the filling machine in one of two ways: as a continuous roll or as a reel of thousands of preformed pouches. The film on the continuous roll is not sterile and is usually sterilised by passing through a hydrogen peroxide bath. The film is formed into pouches with two, three or four sides sealed by either heat or ultrasonication. The inside surface of the preformed pouches is sterile, having been treated with either electron beam or gamma radiation. The outside surfaces, however, have to be sterilised, usually with hydrogen peroxide vapour, with or without UV light. After sterilisation, the top of the preformed pouch is cut with rotary knives before the pouch is filled and resealed.

In some pouches, fitments are attached which provide a method of filling the sterile product into a pre-sterilised pouch or bag and in some cases a method of removing the contents by insertion of a specially designed tap. One such fitment contains a membrane which is penetrated to fill the bag and after the bag is filled is resealed with either a foil cap or a plastic membrane heat sealed onto the base of the fitment. Before and after the filling of the bag, the fitment is sterilised by steam.

5.5.1.4 Bulk Aseptic Packaging

Bulk aseptic packaging was introduced to facilitate transport of large quantities of sterile food products. It was originally developed for tomato products but has now been extended to other food products (Nelson, 2014). The bulk containers were initially metal tanks but more recently multilayer flexible plastic bag-in-box systems with a capacity of up to 3000 gallons (~11,000 L) have been used. The capacity of the aseptic packages can vary considerably; ships with bulk aseptic storage capacity of up to 8 million gallons (~30 million L) have been built (Szemplenski, 2012; Nelson, 2014).

5.5.2 Sterilisation of Packaging

Sterilisation of the packaging material is achieved by chemical treatment, irradiation, heat treatment or a combination of these. Hydrogen peroxide with heat is a very effective chemical sterilant and is widely used for sterilising the surface of paperboard packaging material. The main method of application is the immersion system in which the packaging material is fed from the roll through a bath of 35% H_2O_2 at 70 °C for 6 s. Residual H_2O_2 is removed with hot air at ~125 °C. Recently, Tetra Pak released filling machines in which the paperboard is sterilised by electron beam irradiation which is claimed to be more economical than the hydrogen peroxide system it replaced.

A chemical alternative to H_2O_2 is oxonia, a mixture of peracetic acid with H_2O_2 and acetic acid to stabilise it. It can be used at 40-60 °C for sterilising filling machines as well as PET and HDPE bottles (Fox, 2013). It is particularly effective against bacterial spores and is useful for sterilising surfaces which are difficult for sterilants to access. It produces a low level of harmless residuals although these can be removed with a sterile water rinse (Robertson, 2013).

Electron beam and gamma radiation are used to sterilise some packaging materials such as preformed plastic bottles and pouches, and bags used in "bag-in-box" systems. Aluminium foil used in closures on aseptic packages can be sterilised by gamma radiation followed by UV irradiation (Brody, 2000). UV can also be used in conjunction with hydrogen peroxide to increase its effectiveness.

5.5.3 Establishing and Maintaining a Sterile Environment

A sterile environment is of paramount importance to ensure the sterility of the product is maintained from the processing line to the filling station and in the aseptic zone during packaging. The aseptic zone is the area within the filling machine in which the packaging material or preformed package is sterilised, and the package is filled and sealed. Before production, the air in this zone and the numerous surfaces, including those of moving machine parts, have to be sterilised. This is achieved by two techniques: heat using hot air and steam or heat in combination with hydrogen peroxide vapour. With heat, air is heated by incinceration at 330-360 °C and the aseptic zone is sterilised by the hot air at ~240 °C for at least 30 min. Steam at ~130 °C is used to sterilise the product valve. Complex filling systems which cannot be sterilised by heat are sterilised with heat and hydrogen peroxide. The peroxide is introduced into the filler as a fog or a gas, which condenses on the surfaces, with the aid of heated sterile air. The hot air is also used to remove the peroxide from the system.

The aseptic zone must be maintained in a sterile condition throughout the process. This is achieved by an overpressure of sterile air. The air can be sterilised by incineration as above or by filtration using a HEPA (High Efficiency Purified Air) filter. These filters should remove 99.99% of all particles >0.3 μm which include most bacteria and dust particles. The filtered air is passed through the aseptic zone at low velocity (0.5 m/s) and low pressure in laminar flow (von Bockelmann & von Bockelmann, 1998). Laminar flow across the filling chamber prevents microorganisms from a contaminated package or packaging area causing further contamination. By controlling the direction of flow of the air, the microbial contaminants are prevented from contaminating the sterile product. A positive pressure of about 0.5 bar (0.05 MPa) should be is maintained in the aseptic zone.

5.5.4 Aseptic Package Integrity

Package integrity is paramount for an aseptic system. Consequently, routine testing of the integrity of the packages is essential. This includes constant visual inspection but also destructive testing of a sample of the packages. While this applies to all types of package, paperboard cartons receive the most attention. In these, the longitudinal seal and the two transverse seals are regularly tested in tear-down tests to ensure the seal is stronger than the packaging material when subjected to physical tearing apart of the seal, and there are no unsealed parts which can be detected by a dye test. Leaks in the carton can also be checked with a conductivity test where the carton is filled with water, placed in a brine bath and the conductivity checked to determine if any salt has entered the carton. Non-destructive tests which can be performed on-line have attracted considerable interest but their commercial adoption to date has been limited (Robertson, 2013).

5.5.5 Validation of Aseptic Packaging Operations

UHT processing and aseptic packaging operations are complex and require close attention to detail to ensure they operate as intended and produce a safe product of the quality specified. Validation of the process is therefore essential and is some countries is mandated by Government regulations (David & Carlson, 2013). The whole validation process starts from the design of the plant and continues through commissioning to verification of the commercial sterility of the final product. It requires close cooperation between the processor and the equipment supplier. Generally the equipment supplier will have the necessary knowledge, systems and protocols available to facilitate the validation.

Following equipment selection, a process schematic is prepared which describes how the product is be processed and packaged, how the heating, holding and cooling steps are to be designed and how the system is to be cleaned and sterilised. This is followed by the preparation of the process and instrument diagrams, and construction and installation of the equipment. Following installation, all aspects of the equipment are reviewed and tested to ensure correct construction and operation; this includes checking piping welds and any parts which would be difficult to clean, instrumentation and control systems. The equipment is then tested, usually on water, to check all aspects of the operation perform as designed and required. This includes checks on the integrity of the aseptic packages At this point any identified deficiencies are corrected before the final validation step, the verification of commercial sterility.

An overview of the verification procedure is given here but for more detail the reader is referred to equipment manufacturers' manuals such as Tetra Pak's *Procedures and*

Guidelines Guideline for Microbiological Evaluation of Commercially Sterile Products and other published information (e.g., von Bockelmann & von Bockelmann, 1998; David & Carlson, 2013).

Sterility verification is based on data from trials of the plant run on product such as skim milk. The data are usually microbiological results on product processed and packaged during the trials. Three trials are commonly carried out and a prescribed number of samples are taken for testing according to the target criterion of unsterile packages; this is commonly 1 per 1,000 packages. Tetra Pak (undated) recommends trial 1 be carried out on day 1 and trials 2 and 3 be carried out on day 2 with an intermediate clean. It is further recommended that events such as stops, as would occur under normal operating conditions, be included in the trial. These should include use of the A tank for at least one of the trials in which product should be stored overnight before aseptic packaging on the following day.

The numbers of samples to be taken to achieve the target criterion of 1 per 1,000 packages with 95% confidence are given in Table 5.16. Accordingly, if 3,000 samples are tested, then the number of allowable unsterile packs is zero. However, the number of samples recommended by Tetra Pak is 7,720 from which a maximum of three unsterile packs is allowed for the sterility verification. The numbers of samples must be accumulated from the three different trials. It should be noted that if the target criterion is changed from 1 in 1,000 to 1 in 10,000, the number of samples to be tested increases 10-fold (Tetra Pak, undated).

A verification method which does not involve a "real' product involves sterilisation and packaging of a sensitive microbiological medium such as Linden Grain and monitoring packages for microbial growth. If the equipment and packages have been adequately sterilised there should be no growth in the Linden Grain medium (Chaven & Sedarati, 2012).

In addition, some legislative directives require evidence of testing of the equipment's capability of eliminating pathogens. This testing is usually performed in challenge testing using a non-pathogenic spore-former which has a similar or greater heat resistance to *Cl. botulinum*, the target pathogen in sterilised food products (Rha, 1975). The common organism used is *Cl. sporogenes* which conveniently produces gas and a putrid odour which facilitate detection of non-sterile packs. It is used in what is known as the inoculated pack test where a medium which will support growth of the organism, which

Table 5.16 Sampling plan for verification of commercial sterility trials (assuming a 95% detection probability with a defect rate of 1 per 1,000).

Total number of samples to be taken in 3 separate trials	Maximum number of non-sterile packs allowed
3000	0
4730	1
6270	2
7720	**3**
9151	4

From: Tetra Pak (undated)

may be skim milk, is inoculated with the organism and processed and packaged aseptically. The packages are incubated at 35 °C for 3-4 weeks and then checked for sterility (David & Carlson, 2012).

5.6 Plant Cleaning and Sanitisation

5.6.1 Introduction

At the end of the heat treatment run, all surfaces in contact with product, and especially the heat transfer surfaces need to be cleaned and disinfected, as preparation for the next production run. Cleaning involves the removal of fouling deposits from the surface and may take place in one or more stages.

Heat exchangers should be designed with cleanability in mind, as this is fundamental for producing safe products. The hygienic design of food processing equipment has been discussed in more detail by Jowitt (1980) and Romney (1990). Particular care and attention to avoid deadspaces in pipelines, valves and other fittings is essential.

The normal cleaning procedure is to first rinse or flush out the product with water. This is followed by pumping detergent solutions (commonly an alkali such as caustic soda and an acid such as nitric acid) through the equipment, with water rinses between the different detergents. These rinsing processes are necessary to avoid any cross-contamination, but are unproductive and are regarded as downtime. Cleaning uses valuable resources, for example water for rinsing, detergents for cleaning, and energy. It also produces effluent which needs to be treated. These are now considered to be important environmental issues. Following cleaning, water is used to flush detergent from the plant and the surfaces may then either be disinfected, for example, for pasteurisation processes or sterilised for UHT processes, prior to shut-down. This may not always be necessary at the end of the process, but is essential prior to the next processing run. Disinfecting involves reducing the microbial count on the surface to predetermined low levels, whereas sterilisation involves their complete removal.

There is scope for optimising the entire production process. For example, the simple view might be to ensure that the processing run is as long as possible. However, this may create fouling deposit which is more difficult to remove, thereby increasing the time for cleaning and the costs for detergents and effluent treatment.

5.6.2 Rinsing

Rinsing may be done either at the end of production, or when changing from one product to another. It can be very wasteful of water and produce significant effluent. It is particularly important to rinse correctly, particularly between batches of different products to avoid cross-contamination. Rinsing has been studied less than fouling and cleaning.

Changeover processes of any kind, for example, between product and water or detergent and water, can be monitored by conductivity, turbidity or fluoresecence measurements; conductivity is most widely used. For pipeline flow, the transition from product to water can be envisaged as a step change in input function (provided the flow rate of the two fluids remains the same). If the concentration is measured against time at the outlet, the curve shown in Figure 5.15 results (Loncin, 1979). The ideal situation for maximum

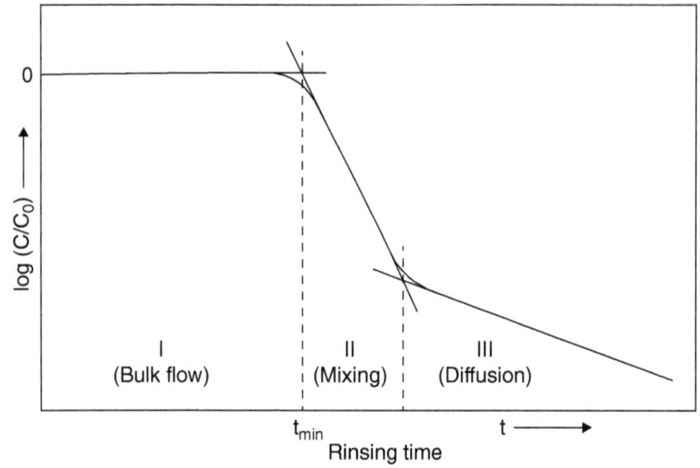

Figure 5.15 Stages during rinsing of plant. (Source: Loncin, 1979. Reproduced with permission of Elsevier.)

washout of the material is plug flow. Any item of equipment which extends the distribution of residence time will make rinsing more difficult. One way to analyse the rinsing process is to plot log C/C_o) against time at the outlet, following a step change from C_o to C at the inlet (C is the concentration of solids in the rinse water). There are three distinct phases when log (C/C_o) is plotted against time, which have been modelled by Loncin (1979) (see Figure 5.15):

- Phase 1 - no change in composition. This ends at a time corresponding to the minimum residence time through the equipment;
- Phase 2 - removal of the bulk of the fluid by displacement of water. There is mixing of the water and the material; and
- Phase 3 - a slower removal of material from the boundary layer, by molecular diffusion. This is the rate limiting process.

5.6.3 Water–Product Changeover

This can be considered to be the reverse of the rinsing process. After the plant has been sterilised with water at the start of the process, and the appropriate temperature and flow conditions have been established, there is a changeover from water to product. It may take some time for steady state to be re-established. From water to milk, there is little difference in terms of flow characteristics, but a changeover from water to a viscous product may involve some large differences in product viscosity. If the product is collected too early, it will be diluted. If it is not collected early enough, there may be substantial waste. The problems are almost the same as those encountered in rinsing, the slow depletion of material from the boundary layers.

5.6.4 Cleaning

Cleaning is the removal of deposits from the fouled surface of a heat exchanger. This section will focus on cleaning-in-place methods (CIP), whereby deposits are removed by contacting them with hot detergent solutions. This has now almost completely replaced cleaning-out-of-place, where pipe unions are taken apart by hand and lengths of pipeline are manually cleaned.

Cleaning combines three energy sources: thermal energy, chemical energy and mechanical energy. The steps involved are transferring the cleaning agent from the bulk of solution to the surface of the deposit, its diffusion into the deposit, and its reaction with and breaking down of the deposit. The final step is transfer of the dispersed or dissolved deposit back into the bulk solution. According to Loncin and Merson (1979), cleaning is achieved by a progressive soaking and dispersion of the soil layer and not by simple dissolution. The objectives are to achieve a surface which is both clean and hygienic. Thus the surface must be free of soil, pathogenic and spoilage bacteria, as well as detergents and disinfecting agents (Plett, 1985). The types of deposit have been summarised by Plett (1985) and are presented in Table 5.17.

Simple deposits may be cleaned with hot water, although this is rarely the case with milk products where detergents are used. Cheow and Jackson (1982a) studied water - cleaning of a heat exchanger fouled with tomato juice at 20, 70 and 90 °C. The most effective temperature was 70 °C because some protein denaturation occurred at 90 °C. For most operations, detergents are required. In fact when the same deposits were cleaned with 2% caustic soda, the rate of cleaning was faster at 90 °C than at 70 °C and no protein denaturation was observed (Cheow and Jackson, 1982b). It was also found that when pressure was increased, the rate of cleaning decreased.

A detergent is defined as any substance that either alone, or in a mixture, reduces the work requirement of a cleaning process. A wide range of detergents are available. Simple ones are based on either caustic soda or nitric acid. These form the basis of the widely accepted two-stage, alkali-acid procedure which involves at least five steps: pre-rinse, alkali circulation, intermediate rinse, acid circulation and a final rinse. It is designed to remove the types of deposit listed in Table 5.17. Typical detergent concentrations are about 2.0% for caustic soda and 1.0% for nitric acid. Other alkaline (e.g., sodium

Table 5.17 Types of deposit, ease of removal and alteration by heat. (Source: Plett 1985. Reproduced with permission of University of Wisconsin.)

Component	Solubility	Removal	Heat alterations
Sugar	Water: soluble	Easy	Caramelization
Fat	Water: insoluble	Difficult (good with surfactants)	Polymerization
	Alkali: good		
	Acid: poor		
Protein	Water: poor	Difficult	Denaturation
	Alkali: good	Good	
	Acid: medium	Difficult	
Mineral salts, monovalent	Water: soluble	Easy	
	Acid: soluble		
Mineral salts, polyvalent	Water: insoluble	Difficult	Precipitation
	Acid: soluble	Good	

orthosilicate, sodium carbonate) or acidic (phosphoric, hydroxyacetic) agents may be incorporated. Caustic soda and nitric acid are also effective bactericides (Tamime, 2008).

Fouling deposits from milk found on heated surfaces consist mainly of protein, minerals and some fats. Alkali (caustic)-based detergents are effective in dissolving and removing fat (by hydrolysis/saponification) and protein, whereas acid-based detergents are effective in removing minerals. In some circumstances, a caustic plus acid process alone may not remove all the soil. Other additives may be used to improve the performance of the detergents. These include surface active agents, to improve wetting, emulsification and dispersion, and sequestering agents to remove water scale and to give good rinsing properties. Table 5.18 lists the properties of some of the main components of detergents. For specialised applications proteolytic enzymes may be used although they are less widely used on heat exchangers, compared to membrane processes. Particular care needs to be taken when using enzyme-containing detergents as trace residues can have deleterious effects on products (Tran *et al.*, 2003).

The two-stage cleaning procedure was considered to be the most effective in removing milk deposits by Kane and Middlemiss (1985). However, a single-stage procedure has been in use in the UK for many years, reducing the number of steps to three. Typically, a single-stage detergent consists of sodium or potassium hydroxide together with sequestering and surfactant agents and is used at 1-3% concentration at 70-82 °C. It is claimed to reduce the need for acid cleaning in cold milk areas and allow for occasional acid cleaning of heat exchangers (https://cydan.com.au/en/cip-cleaners/275-snglex-cip-sngle-stage high-caustc.html). However, there are many proprietary single-stage commercial cleaners, usually detergent-based, which are more complex and contain many of the additives listed in Table 5.18 to improve their performance. These should be more efficient and save time, energy and produce less effluent. The complexity of such proprietary detergents is illustrated by the large number of formulations which are available from detergent suppliers.

Timperley and Smeulders (1987) have compared the single- and two-stage processes on a bench-scale tubular heat exchanger and pilot-scale plate heat exchanger. The use of a single-stage detergent was shown to produce physically clean surfaces in half the time taken by the two-stage alkali plus acid procedure, which did not remove completely calcium deposits. Similar findings were reported by Timperley e*t al.* (1994) on pilot-scale and production-scale UHT sterilisers, leading to significant reductions in cleaning time (20-30%) in addition to lower water and energy consumption.

5.6.5 Methods of Measuring Cleaning Effectiveness

In principle, the effectiveness of a cleaning programme can be evaluated by measuring the amount of soil remaining after cleaning. This can be done directly by weighing the material removed or by use of some property which is dependent upon the amount of soil still present. Examples are visual inspection, microscopic observations and optical, microbiolgical, radiological or chemical analysis. However, such methods can only be used if the surface is accessible; they are more suited to laboratory testing procedures. It would be extremely useful if production-scale plant could be made accessible, as it would be invaluable to be able to inspect internal heat exchangers and holding tube sections at regular intervals. However, in practice, this has not yet been achieved. Without being able to inspect these surfaces, it is difficult to establish how effective the cleaning process has been. Testing procedures and equipment for this purpose have been reviewed by Romney (1990) and Shapton and Shapton (1998).

Table 5.18 Cleaning detergents and their properties. (Source: Romney, 1990. Reproduced with permission of Society of Dairy Technology.)

Detergent Raw Material	Chemical Formulation	pH 1% Solution	Chemical Reactivity	Wetting	Soil-Lifting Power	Dispersion	Emulsification	Wafer Softening	Threshold Effect	Foam
Caustic soda	NaOH	13.3	5	1	1	1	1	2	—	—
Sodium orthosilicate, monohydrate	$2Na_2O{:}SiO_2{:}H_2O$	12.8	4	2	3	3	2	2	—	—
Sodium metasilicate, pentahydrate	$Na_2SiO_3{:}5H_2O$	12.3	3	2	3	3	2	2	—	—
Trisodium orthophosphate, crystalline	$Na_3PO_4{:}12H_2O$	11.95	3	2	3	3	2	2	—	—
Soda ash										
Sodium hexametaphosphate (Calgon)	Na_2CO_3	11.5	1	1	1	1	1	2	—	—
	$Na(PO_3)_6$	7.6	1	1	4	4	2	5	4	—
Sodium tripolyphosphate	$Na_5P_3O_{10}$	9.6	1	1	4	4	2	5	4	—
Sodium gluconate		7.1	1	1	4	4	2	5	—	—
Tetra sodium salt of EDTA*		11.2	1	1	4	4	2	5	—	—
Sodium dodecyl benzene sulphonate		6.9	1	5	3	4	4	1	—	5
Nonyl phenol/9EO*										
Disodium salt of acetodiphosphonic acid		7.1	1	5	3	4	5	1	—	2
		8.5	1	1	1	5	2	3	5	—

5.6.6 Kinetics of Cleaning

The dynamics of the cleaning process and how quickly the soil is removed are important topics. This can be investigated by monitoring properties which are dependent upon the amount of soil which is solubilised, such as optical density, or the concentrations of protein, fat or minerals in the bulk circulating cleaning fluid.

Loncin and Merson (1979) have reviewed the kinetics of cleaning. Within certain limits, soil removal can be described by a first order reaction. Thus if m represents the mass of soil per unit area, then the rate of removal is directly proportional to the mass as per the equation:

$$\mathrm{dm} / \mathrm{dt} = -\mathrm{km} \tag{5.26}$$

or

$$ln(m_i / m) = kt \tag{5.27}$$

where m_i = initial mass, m = mass after time (t) and k is a rate constant.

The rate constant depends upon several factors which are:

- the type of soil and its state. There are numerous differences between soils from different foods and their ease of removal (see Table 5.17).
- the nature of the support materials. These will affect the forces of adhesion between the deposit and the surface, for example, highly polished surfaces may be assumed to clean easier although they may also be more susceptible to corrosion. The use of coatings such as teflon may also ease cleaning.
- the type and concentration of the detergents used. For simple caustic soda solutions, k was found to vary in a linear fashion with concentration. However, this finding is not likely to apply to surface active components. Note also that in some cases a decline in the cleaning rate was observed when the concentration was increased. This was attributed to transport inhibition, perhaps caused by a very rapid swelling of the outer soil layers at these higher concentrations.
- temperature of cleaning. Note that the rate constant follows Arrhenius kinetics (straight line relationship between ln k and 1/T). However (as for concentration), there may be an optimal cleaning temperature beyond which cleaning rates decline due to heat-induced physical alterations of some soils.
- hydrodynamic factors. All researchers have found an improvement in cleaning rates with mechanical action; thus increasing the flow rate (hence the flow velocity and Reynolds number) will improve cleaning action but will also increase the pumping and energy costs. There has been some discussion on how to characterize this effect better, that is, by wall shear stress or by Reynolds number. Note that the wall shear stress (τ_W) is given by D ΔP/4 L, where D and L are the pipe diameter and length and ΔP is the pressure drop.

Plett (1985) reviewed some of the mathematical models for predicting the effect of mechanical action on cleaning rates. He concluded that the preference is toward correlations involving wall shear stress which can be related much better to mean velocity than to Reynolds numbers.

One of the first reported practical suggestions was to obtain a high shear stress at the wall by ensuring that the fluid velocity is high, a minimum of 1.5 ms^{-1} being recommended for effective cleaning. Jackson and Low (1982) found a wall shear stress threshold of 0.8 Pa, above which there was a remarkable effect on cleaning for tomato juice. High shear stresses can be induced through turbulent flow conditions, where the shear stress is proportional to the mean velocity. On the other hand, Jennings *et al.* (1957) postulated a minimum Reynolds number of 25,000, above which the mechanical effect is effective. It is noteworthy that at a constant volumetric throughput, both the Reynolds number and the wall shear stress will decrease as pipe diameter increases. Therefore any expansions or enlargements of the pipe system may not be so easy to clean. These matters should be considered at the design stage.

Improvements in cleaning rates have also been observed when the direction of flow is reversed every 5 or 15 seconds (Grasshoff, 1997). There are some claimed advantages of dispersing air bubbles into the cleaning solution (two phase flow). Under appropriate conditions, annular flow takes place. In such cases the rate of momentum transfer in the annular liquid layer becomes very intense and a strong cleaning action can be expected, with limited pumping costs. Plett (1985) suggested that the amplitude and frequency of the shear stress peaks (arising from turbulent bursts) may play a more important role and be a more significant parameter than the average wall shear stress. On some equipment, it may be necessary to install an additional pump to obtain these cleaning velocities or to bypass the homogeniser in situations where it is limiting the flow.

Plett (1985) in his review of kinetic models found that deviations from first order reaction kinetics occur, which is not surprising considering the overall complexity of the process. The ability of first order kinetics to model the overall dynamic behaviour has been attributed to the fact that the rate limiting process which occurs at the end of the process is the diffusion of cleaning reaction products. It has often been noted that soil removal is more rapid than predicted at the beginning of the process and that it decreases toward the end. One explanation for this is that at the beginning of the process the soil adheres to another layer of soil, whereas at the end of the operation the forces of adhesion between the soil and the support must be broken and that these are greater.

Thus the cleaning process is the reverse process to fouling (see Section 6.2.2). It results in a restoration of the overall heat transfer coefficient and a reduction in the pressure drop over the once fouled section. Thus, cleaning can be monitored indirectly by measuring the restoration of OHTC, or ensuring that a minimum pressure drop has been attained. Grasshoff (1997) observed that pressure drop initially increases during cleaning due to significant swelling of the deposit. It was also observed that the rate of cleaning was initially low, reaching a maximum after a short time, followed by a steady decline.

5.6.7 Disinfecting and Sterilising

Disinfecting and sterilising the equipment (as appropriate) should always be done immediately before the process commences and, if possible after cleaning. There are two major ways of doing this: with heat or chemicals.

5.6.7.1 Use of Heat

Wherever possible, disinfecting and sterilising by heat is the simplest and most effective procedure and this is also the most commonly used for continuous processing. For this purpose, steam or superheated water up to 140 °C is used. One of the main advantages over chemical cleaners is that there is no residue to wash out and no risk of contaminating the product with chemical cleansers and disinfectants.

For pasteurised and ESL products, post-pasteurisation contamination is the most important determinant of keeping quality (see Sections 2.3.7, 4.4.1 & & 4.4.2); this is reduced by ensuring all surfaces which come into contact with the product downstream of the holding tube reach a temperature of 95 °C for 30 min. For UHT plants, it is crucial to sterilise the equipment downstream of the holding tube. To achieve this requires 130 °C for 30 min. It may also take some time to actually achieve this temperature. This is a very severe process, as can be determined by calculating the F_0 value, which would be about 230 (cf 3-18 for low-acid foods). It may be difficult to apply such temperatures to the cooling sections with direct refrigeration sections. This requires that the relevant sections can be pressurised to obtain these temperatures (see steam tables – Table 5.6). The traditional view for UHT processing is that all sections of the plant downstream of the holding tube should reach a temperature of 130 °C for 30 min prior to every production run. On some UHT plants it is now possible to sterilise all sections of the plant at these temperatures. This might be effective in eliminating biofilms.

As a general design principle, the aim is to simplify the pipework downstream of the holding tube: minimise pipework to ensure that it can be sterilised in a practicable time. Positioning the homogeniser downstream increases the time for plant sterilisation. Instruments based on thermal imaging for measuring surface temperatures are useful for checking whether all pipework is adequately sterilised. Temperature-sensitive stick-on indicators provide an inexpensive useful alternative for this purpose.

5.6.7.2 Use of Chemicals

There are situations where chemical disinfecting or sterilising agents are required, such as where materials may be sensitive to heat. For example, some refrigeration systems may be difficult to sterilise by heat due to expansion of the refrigerant. One issue is being able to remove the chemical while retaining a sterile environment. One possibility is sterile water, but this then requires a supply of sterile water.

Note that caustic soda and nitric acid are themselves bactericides. Some data on their effectiveness is presented by Tamime (2008). Active chlorine is also used, most usually in the form of sodium hypochlorite, in the active chlorine concentration range of 10-200 ppm. It should be used cold as it is corrosive at high temperature. Note that a concentration of 1000 ppm of chlorine has no sporicidal effect if the pH is high. It should also be noted that the concentration may diminish with time due to chemical reaction, for example, chlorine reacts with oxidisable material.

Hydrogen peroxide is used to sterilise surfaces and for aseptic packaging (see Section 5.5.2). Others used are peracetic acid, peroxyacetic acid and iodophores. Information on their chemistry and uses can be found in Romney (1990) and Shapton and Shapton (1998).

References

Ansari, I.A., Sahoo, P.K. & Datta, A.K. (2006) Fouling of heat exchangers in the dairy industry: an overview. *Indian Dairyman* **56**, 41–47.

ASHRAE (1993) Handbook, *Fundamentals - Thermal Properties of Foods*. pp. 1–26, Atlanta, Georgia.

Astley, M (2015) *Dairy Reporter* March 2015. http://www.dairyreporter.com/Processing-Packaging/In-some-countries-it-already-is-Aseptic-pouch-could-become-the-standard-for-milk-Ecolean.

Biliaderis, C.G. (1983) Differenetial scanning calorimetry in food research – a review. *Food Chemistry* **10**, 239–266.

Biziak, R.B., Swartzel, K.R. & Jones, V.A. (1985) Energy use for continuous thermal processing of milk. *Journal of Food Science* **50**, 1607-1610, 1614.

Boumpa, T., Tsioulpas, A., Grandison, A.S. & Lewis, M.J. (2008) Effects of phosphate and citrates on sediment formation in UHT goat's milk. *Journal of Dairy Research* **75**, 160–166.

Brody, A.L. (2000) The when and why of aseptic packaging. *Food Technology* **54**(9), 101–102.

Brody, A.L. (2006) Aseptic and extended-shelf-life packaging. *Food Technology* **60**(2), 66–68.

Browning, E., Lewis, M. & MacDougall, D. (2001) Predicting safety and quality parameters for UHT-processed milks. *International Journal of Dairy Technology* **54**, 111–120.

Burton, H. (1988) *Ultra High Temperature Processing of Milk and Milk Products.* Elsevier Applied Science, London.

Carlson, V.R. (1996) Food plant organization – Present operations. In: *Aseptic Processing and Packaging of Food: A Food Industry Perspective.* (eds J.R.D. David, R.H. Graves & V.R. Carlson), pp. 51–94. CRC Press, Boca Raton.

Chandarana, P.I., Unverferth, J.A., Knap, R.P., Deniston, M.F., Weise, K.L. & Shafer, B. (2010), Establishing the aseptic processing and packaging operation. In: *Principles of Aseptic Processing and Packaging.* (ed P.E. Nelson), 3rd edn, pp.135–156. GMA Science and Education Foundation, US.

Chaven, S. & Sedarati, A. (2012) Food safety systems for low-acid aseptic beverages. *Food Safety Magazine* **18**(3), 68–73.

Chen, B.Y., Grandison, A.S. & Lewis, M.J. (2015) Effect of seasonal variation on some physical properties and heat stability of milk subjected to ultra-high temperature and in-container sterilisation. *Food Chemistry* **181**, 227–234.

Cheow, C.S. & Jackson, A.T. (1982a) Cleaning of plate heat exchanger fouled by tomato juice, 1) cleaning with water. *Journal Food Technology* **17**, 417–430.

Cheow, C.S. & Jackson, A.T. (1982b) Cleaning of plate heat exchanger fouled by tomato juice, 2) cleaning with caustic soda solution. *Journal Food Technology* **17**, 431–440.

Clare, D.A., Bang, W.S., Cartwright, G., Drake, M.A., Coronel, P. & Simunovic, J. (2005) Comparison of sensory, microbiological, and biochemical parameters of microwave versus indirect UHT fluid skim milk during storage. *Journal of Dairy Science.* **88**, 4172–4182.

Cromie, S.J., Dommett, T.W. & Schmidt, D. (1989) Changes in the microflora of milk with different pasteurisation and storage conditions and aseptic packaging. *Australian Journal of Dairy Technology* **44**, 74–77.

Dannenberg, F. & Kessler, H. (1988) Reaction kinetics of the denaturation of whey proteins in milk. *Journal of Food Science* **53**(1), 258–263.

Datta, A.K. & Anatheswaran, R.C. (2001) *Handbook of Microwave Technology for Food Applications.* Marcel Dekker, Inc., New York.

Datta, N., Elliott, A.J., Perkins M.L. & Deeth, H.C. (2002) Ultra-High Temperature (UHT) treatment of milk: Comparison of direct and indirect modes of heating. *Australian Journal of Dairy Technology*, **57**, 211–227.

David, J.R.D. & Carlson, V.R. (2013) Validation and establishment of aseptic processing and packaging operations. In: *Handbook of Aseptic Processing and Packaging.* (eds J.R.D David, R.H. Graves & T. Szemplenski), 2nd edn, pp. 130–159. CRC Press, Boca Raton.

David, J.R.D., Graves, R.H. & Szemplenski, T. (2013) *Handbook of Aseptic Processing and Packaging.* 2nd edn, CRC Press, Boca Raton.

de Jong, P., Waalewijn, R. & van der Linden, H.J.L.J. (1996) Occurrence and behaviour of volatile sulfur-containing compounds in milk by heat treatment. In: *Heat Treatments and Alternative Methods.* Doc. 9602, pp. 51–72. International Dairy Federation, Brussels.

Deeth H.C. (1999) Electrical heating: Current passage tube technology. *Australian Journal of Dairy Technology* **54**, 66–68.

Dentener, A. (1984). Influence of milk protein on quality aspects of UHT milk products. *Food Technology in New Zealand* **19**, 33–35.

Dogan, H. & Kokini, J.L. (2007) Rheological properties of foods. In: *Handbook of Food Engineering,* 2nd edn. (eds D.R. Heldman & D.B. Lund), pp.1–124, CRC Press, Boca Raton.

Elecster (undated) Modern aseptic processing and packaging technology. http://www.elecster.fi/Yleisesite_eng.pdf (accessed June 2015).

Fellows, P.J. (2009), *Food Processing Technology.* 3rd edn, Woodhead Publishers, Oxford, UK.

Elliott, A.J., Dhakal, A., Datta, N. & Deeth, H.C. (2003) Heat-induced changes in UHT milks – Part 1. *Australian Journal of Dairy Technology* **58**, 3–10.

Fink, R. & Kessler, H.G. (1988) Comparison of methods for distinguishing between UHT treatment and sterilisation of milk. *Milchwissenschaft* **43**, 275–279.

Floros, J.D, Weiss, I & Mauer, L.J. (2010) Aseptic packaging technology. In: *Principles of Aseptic Processing and Packaging.* (ed P.E. Nelson), 3rd edn, pp. 101–133.

Fox, R. (2013) Aseptic packaging materials and sterilants. In: (2013) *Handbook of Aseptic Processing and Packaging.* (eds J.R.D David, R.H. Graves & T. Szemplenski), 2nd edn, pp. 103–118. CRC Press, Boca Raton.

Fredsted, L.B., Rysstad, G. & Eie, T. (1996) Pure-Lac: the new milk with protected freshness and extended shelf-life. *Heat Treatment and Alternative Methods.* Document 9602, pp. 104–125, International Dairy Federation, Brussels.

Gaafar, A.M.M. (1987). *Investigation into the Cooked Flavour in Heat-Treated Milk.* PhD thesis, University of Reading, Reading, UK.

Ghnimi, S., Flach-Malaspina, N., Dresch, M., Delaplace, G. & Maingonnat J.F. (2008) Design and performance evaluation of an ohmic heating unit for thermal processing of highly viscous liquids. *Chemical Engineering Research and Design* **86**, 626–632.

Gibson, H., Taylor, J.H., Hall, K.H. & Holah, J.T. (1995) Removal of bacterial biofilms. Report, No 2, Campden and Chorleywood Food Research Association, Chipping Campden, UK.

Glaser, W. & Kreutz, M. (2010) UHT milk in PET. *European Dairy Magazine Issue* **4**, 30–33.

Gold Peg International (undated) RotaTherm® continuous cooker. http://www.goldpeg.com/rotatherm/rotatherm.htm (accessed September 2015).

Grasshoff, A. (1997) Cleaning of heat treatment equipment. In: IDF Bulletin 328, pp. 32–34, International Dairy Federation, Brussels.

Harrington, R. (2010) Microwave sterilisation system may revolutionise food processing. http://www.foodproductiondaily.com/Processing/Microwave-sterilisation-system-may-revolutionise-food-processing.

Hayes, G.D. (1987) *Food Engineering Handbook*. Longman Scientific and Technical, Harlow, UK.
Heldman, D.R. & Lund, D.B. (eds) (2007) *Handbook of Food Engineering*. 2nd edn, CRC Press, Boca Raton, Florida, USA
Herrick, J.P., Sastry, S.K., Clyde, G.F. & Wedral, E.R. (2000) *On-demand direct electrical resistance heating system and method thereof*. US Patent **6**,130,990.
Hillbrick, G.C., McMahon, D.J. & McManus, W.R. (1999) Microstructure of indirectly and directly heated ultra-high-temperature (UHT) processed milk examined using transmission electron microscopy and immunogold labelling. *Lebensmittel-Wissenschaft & Technologie* **32**, 486–494
Holdsworth, S.D. (1997) *Thermal Processing of Packaged Foods*. Blackie Academic and Professional, Chapman & Hall, London, UK.
Hooi, R., Barbano, D.M., Bradley, R.L., Budde, D., Bulthaus, M., Chettiar, M., Lynch, J. & Reddy, R. (2004) Chemical and physical methods. In: *Standard Methods for the Examination of Dairy Products*. 17th edn. (eds H.M. Mehr & J.F. Frank), Amercian Public Health Association, Washington, USA.
Hsu, D.S. (1970) *Ultra-high-temperature (U.H.T) Processing and Aseptic Packaging (A.P.) of Dairy Products*. Damana Tech Inc, New York.
Huijs, G., van Asselt, A., Verdurmen, R. & de Jong, P. (2004) High speed milk. *Dairy Industries International* **69**(11), 30–32.
Huppertz, T. (2011) High-pressure homogenisers. In: *Encyclopedia of Dairy Sciences* (eds J.W. Fuquay, P.F. Fox & P.L.H. McSweeney) 2nd edn, Volume **3**, pp. 755–764. Academic Press, San Diego.
IDF (1972) IDF Annual Bulletin, part V *Monograph on UHT milk*, International Dairy Federation, Brussels.
Ikezumi, M., Nakagawa, T., Matsura, H. & Mochiji, K. (2006) Cow's milk flavor deterioration suppression method in store and supermarket, involves performing deaeration processing at specific temperature to reduce dissolved oxygen concentration in cow's milk. Patent number JP2006042814-A. Assignee: Nippon Milk Community KK.
Jackson, A.T. & Low, M.W. (1982) Circulating cleaning of a plate heat exchanger fouled by tomato juice III The effect of fluid flow rate on cleaning efficiency. *Journal of Food Technology* **17**, 745–752
JBT. (undated) SPACE aseptic filler. http://www.jbtfoodtech.com/en/Solutions/Equipment/SPACE aseptic filler\UHT milk in PET.
Jenness, R., Shipe,W.F. Jr., & Sherbon, J.W. (1974) Physical properties of milk. In: *Fundamentals of Dairy Chemistry* (eds B.H. Webb, A.H. Johnson & J.A. Alford), pp. 409–460, AVI. Westport, Conneticut
Jennings, W.G., McKillop, A.A. & Luick, J.R. (1957) Circulating cleaning. *Journal of Dairy Science* **40**, 1471–1479.
Jowitt, R. (ed.) (1980) *Hygienic Design and Operation of Food Plant*. Ellis Horwood, Chichester.
Jun, S. & Sastry, S.K. (2005) Modeling and optimizing of pulsed ohmic heating of foods inside the flexible package. *Journal of Food Process Engineering* **28**, 417–436.
Jun, S., Sastry, S.K. & Samaranayake, C. (2007) Migration of electrode components during ohmic heating of foods in retort pouches. *Innovative Food Science & Emerging Technologies* **8**(2), 237–243.

Kane, D.R. & Middlemiss, N.E. (1985) Cleaning chemicals- State of knowledge in 1985, In: *Fouling and Cleaning in Food Processing* (eds. D. Lund, A Plett & C. Sandu). Pp. 312–335. University of Wisconsin Extension Duplicating, Madison WI.

Kastanas, P. (1996) *Studies on the Fouling of Heat Exchangers during the Sterilisation of Cow's and Goat's Milks.* PhD thesis, University of Reading, UK.

Kastanas, P., Lewis, M.J. & Grandison, A.S. (1995) Design and development of a miniature UHT plant for fouling studies and its evaluation using milk adjusted to different pH values, goats' milk and milk permeate. *Transactions of the Institute of Chemical Engineers* **73, Part C**, 83–92.

Kaletunc, G. (2009) *Calorimetry in Food Processing: Analysis and Design of Food Systems.* Wiley-Blackwell.

Kaw, A.K., Singh, H. & McCarthy, O.J. (1996) The effects of UHT sterilisation on milks concentrated by reverse osmosis. In: *Heat Treatments and Alternative Methods.* IDF Doc S.I. 9602. pp. 299–314, International Dairy Federation, Brussels

Kessler, H.G. (1981) *Food Engineering and Dairy Technology.* Verlag A Kessler, Freising, Germany

Kessler, H.G. (1994) Heat resistance of microorganisms under seals. *Proceedings of the 24th International Dairy Congress (Melbourne),* International Dairy Federation, Brussels.

Kim, H.-J., Choi, Y.-M., Yang, T.C.S, Taub, I.A., Tempest, P., Skudder, P., Tucker, G. & Parrott, D.L, (1996) Validation of ohmic heating for quality enhancement of food products. *Food Technology* **50**(5), 253–261.

Lefebvre, S. & Leuliet, J.C. (1997) Laminar mixed convection in constant wall heat flux tubular heat exchangers. In *Engineering and Food at the Seventh International Congress of Engineering and Food (ICEF 7),* Part 1. (ed R. Lowitt). Pp. C97–C100. Academic Press, Sheffield.

Lewis, M.J. & Heppell, N. (2000) *Continuous Thermal Processing of Foods: Pasteurisation and UHT Sterilisation.* Aspen Publishers, Gaithersburg, MD.

Lewis, M.J. (1990) *Physcial Properties of Foods and Food Processing Systems.* Woodhead Publishers, Cambridge, UK.

Lewis, M.J. (1993) Physical properties of dairy products. In: *Modern Dairy Technology,* Volume 2 (ed R.K. Robinson), pp. 331–380, Elsevier Applied Science, Amsterdam.

Lewis, M.J. & Jun, S. (2012) Thermal processing. In: *Food Processing Handbook.* 2nd edn, Volume **1**, (eds J.G. Brennan &A.S. Grandison), pp. 31–75, Wiley-VCH, Weinheim, Germany.

Lin, M.G. & Ramaswamy, H.S. (2011) Evaluation of phosphatase inactivation kinetics in milk under continuous flow microwave and conventional heating conditions. *International Journal of Food Properties* **14**, 110–123.

Loncin, M. & Merson, R.L. (1979) *Food Engineering,* Academic Press, London, UK.

Lyster, R.L.J. (1970) The denaturation of α-lactalbumin and β-lactoglobulin in heated milk. *Journal of Dairy Research* **37**, 233–243.

Marcotte, M. (1999) *Ohmic Heating of Viscous Liquid Foods.* PhD thesis. McGill University, Canada.

Mohsenin, N.N. (1980) *Thermal Properties of Foods and Agricultural Materials.* Gordon & Breach, London

Mohsenin, N.N. (1984) *Electromagnetic Radiation Properties of Foods and Agricultural Materials.* Gordon and Breach, London

Montaron M P, De Stoutz J, Cosson J L & Gervais P (1991) Le traitement thermique des fluides par effect Joule. *Industries Alimentaires et Agricoles* **108**, 833–837.

Mottar, J. & Moermans, R. (1988) Optimization of the forewarming process with respect to deposit formation in indirect ultra high temperature plants and the quality of milk. *Journal of Dairy Research* **55**, 563–568.

Mudget, R.E. (1982) Electrical properties of foods in microwave processing. *Food Technology* **36**(2), 109–115.

Muir, D.D. (1984) UHT sterilised milk concentrate: are view of practical methods of production. *Journal Society Dairy Technology* **37**, 135–141.

Nelson, P. E. (2014). From tomato king to world food prize laureate. *Annual Review of Food Science and Technology* **5**, 1–20.

Newstead, D.F., Paterson, G., Anema, S.G., Coker, C.J. & Wewala, A.R. (2006) Plasmin activity in direct-steam-injection UHT-processed reconstituted milk: effects of pre-heat treatment. *International Dairy Journal* **16**, 573–579.

National Institute of Urban Development. (1985) Available at: http://www.nird.org.in/(accessed November 7, 2016).

On-Nom, N. (2012) *Partitioning of Milk at High Temperature.* PhD thesis, University of Reading, Reading, UK.

Palaniappan, S. & Sastry, S.K. (1991) Electrical conductivity of selected solid foods during ohmic heating. *Journal of Food Process Engineering* **14**, 221–236.

Perkin, A.G. (1978) Comparison of milks homogenised before and after UHT sterilisation, *20th International Dairy Congress*, 709. International dairy Federation, Brussels.

Perkins, M.L., Elliott, A.J., D'Arcy, B.R. & Deeth, H.C. (2005) Stale flavour volatiles in Australian commercial UHT milk during storage. *Australian Journal of Dairy Technology* **60**, 231–237.

Perkins M.L., Zerdin K., Rooney, M.L., D'Arcy, B.R. & Deeth H.C. (2007) Active packaging of UHT milk to prevent the development of stale flavour during storage. *Packaging Technology & Science* **20**(2), 137–146.

Plett, E.A. (1985) Cleaning of fouled surfaces. In: *Fouling and Cleaning in Food Processing* (eds D. Lund, A Plett & C. Sandu) University of Wisconsin Extension Duplicating, Madison WI.

Powrie, W.D. & Tung, M.A. (1976) Food dispersions. In: *Principles of Food Science*, Part 1, *Food Chemistry* (ed. O.R. Fennema), pp. 539–575. Marcel Dekker, New York.

Prakash, S, Kravchuk, O. & Deeth, H.C. (2015) Influence of pre-heat temperature, pre-heat holding time and high-heat temperature on fouling of reconstituted skim milk during UHT processing. *Journal of Food Engineering* **153**, 45–52.

Prakash, S. (2007) *Burn-on in Ultra-High-Temperature Processing of Milk.* PhD thesis, University of Queensland, Brisbane, Australia.

Prasad, K.S. (1989) *The Sensory Characteristics of Heat-treated Milks, with Special Reference to UHT Processing.* PhD Thesis, University of Reading, Reading, UK.

Rahman, S. (1995) *Food Properties Handbook.* CRC Press, Boca Raton.

Renner, E. (1978) [Effect of differences in technology on the organoleptic and nutritive quality of UHT milk]. *Rivista della Societa Italiana di Scienza dell'Alimentazione* **7**, 499–504.

Reynolds, P. (2012) Chinese dairy introduces a new kind of aseptic pouch. http://www.packworld.com/package-design/structural/chinese-dairy-introduces-new-kind-aseptic-pouch

Rha, C.K. (1975) *Theory, Determination and Control of Physical Properties of Food Materials.* D Reidel, Dordrecht.

Riedel, L. (1955) Calorimetric investigations of the melting of fats and oils. *Fette Seifen Anstrichmittel* **57**, 771–782.

Riedel, L. (1976) [Calorimetric measurements of milk and milk products]. *Chemie Mikrobiologie Technologie der Lebensmittel* **4**, 177–186.

Robertson, G. (2002) The paper beverage carton: Past and future. *Food Technology* **56**(7), 46–48, 50–52.

Robertson, G.L. (2011) Heat treatment of milk: Ultra-High Temperature treatment (UHT): Aseptic packaging. In: *Encyclopedia of Dairy Sciences* (eds J.W. Fuquay, P.F. Fox & P.L.H. McSweeney) 2nd edn, Volume 2, pp. 708–713. Academic Press, San Diego.

Robertson, G.L. (2013) Aseptic packaging of foods. In: *Food Packaging: Principles and Practice*. 3rd edn, pp. 367–382. CRC Press, Boca Raton.

Romney, A.J.D. (ed.) (1990) *CIP: Cleaning In Place*. 2nd edn, Society of Dairy Technology, Huntingdon UK.

Salengke, S. (2000) *Electrothermal Effects of Ohmic Heating on Biomaterials: Temperature Monitoring, Heating of Solid–Liquid Mixtures, and Pre-treatment Effects on Drying Rate and Oil Uptake*. PhD thesis, Ohio State University, Ohio, USA.

Samaranayake, C.D. (2003). *Electrochemical Reactions during Ohmic Heating*. PhD thesis, Ohio State University, Ohio, USA.

Samaranayake, C.P. & Sastry, S.K. (2005) Electrode and pH effects on electrochemical reactions during ohmic heating. *Journal of Electroanalytical Chemistry* **577**, 125–135.

Sastry, S. (2008) Ohmic heating and moderate electric field processing. *Food Technology International* **14**(5), 419–422.

Scheldeman, P., Goossens, K., Rodriguez-Diaz, M., Pil, A., Goris, J., Herman, L., De Vos, P., Logan, N.A. & Heyndrickx, M. (2004) *Paenibacillus lactis* sp. *nov.* isolated from raw and heat-treated milk. *International Journal of Systematic & Evolutionary Microbiology* **54**, 885–891.

Schubert, H. & Regier, M. (2005) *The Microwave Processing of Foods*. Woodhead Publishing Limited, England.

Shapton, D.A. & Shapton, N. F. (1998) *Principles and Practices for the Safe Processing of Foods*, Butterworth Heinemann, Woodhead Publishing, Cambridge Ltd

Singh, P.R. (2007) Heating and cooling processes for food. In *Handbook of Food Engineering*, 2nd edn. (eds D.R.Heldman & D.B.Lund), pp. 397–426. CRC Press, Boca Raton

Sjoeberg, A.M., Wirtanen, G. & Matilla-Sandholm, T. (1995) Biofilm and residue investigations of detergents on surfaces of food processing equipment. *Food and Bioproducts Processing*, **73**(C1), 17–21.

SPX (2008) *APV Technolology update. Long Life Dairy, Food and Beverage Products*. SPX Corporation.Getzville, NY. http://www.apv-tapflo.ro/userfiles/file/Technology_Update_6750_01_08_2008_US.pdf.

SPX (2013) *APV Multipurpose Pilot Plant*. SPX Corporaton, Silkeborg, Denmark. http://www.spx.com/en/assets/pdf/APV_Multipurpose_UHT_Pilot_Plant_SPP_6752_06_04_2013_GB_tcm11-7192.pdf (accessed September 2015).

Stone, M. (2009) Microwave processing improves quality of low-acid. foodshttp://www.foodproductiondaily.com/Processing/Microwave-processing-improves-quality-of-low-acid-foods

Sumnu, S.G. & Sahin, S. (2012) Microwave heating. In: *Thermal Food Processing: New Technologies and Quality Issues* 2nd edn (ed. D.-W. Sun) pp. 555–582. CRC Press, Taylor & Francis Group, Boca Raton, FL.

Sweat, V.E. (1995) Thermal properties of foods. In: *Engineering Properties of Foods*, (eds M.A.Rao & S.S.H. Rizvi). Marcel Dekker, New York.

Szemplenski, T. (2112) Aseptic bulk packaging. In: *Handbook of Aseptic Processing and Packaging*. (eds J.R.D David, R.H. Graves & T. Szemplenski), 2nd edn, pp. 1190–127. CRC Press, Boca Raton.

Tamime, A.Y. (ed) (2008) *Cleaning-in-Place: Dairy, Food and Beverage Operations*. Blackwell Publshing Professional, Oxford, UK.

Teh, K.H., Flint, S., Brooks, J & Knight, G. (eds) (2015) *Biofilms in the Dairy Industry*, Society of Dairy Technology Series, Wiley, Oxford, UK.

Timperley, D.A., Hasting, A.P.M. & Degoederen, G. (1994) Developments in the cleaning of dairy sterilization plant. *Journal of the Society of Dairy Technology* **47**(2), 44–50.

Timperley, D.A. & Smeulders, C.N.M. (1987) Cleaning of dairy HTST plate heat-exchangers - Comparison of single-stage and 2-stage procedures. *Journal of the Society of Dairy Technology* **40**(1), 4–7.

Tetra Pak (2003) *Dairy Processing Handbook*. Tetra Pak Processing Systems, Lund, Sweden.

Tetra Pak (2015) Long-life milk. Chapter 9 in: *Dairy Processing Handbook*,Tetra Pak processing Systems AB, Lund, Sweden (On-line at http://www.tetrapak.com/au/about/newsarchive/dairy-processing-handbook-free-online)

Tetra Pak (undated) *Procedures & Guidelines.Guideline for Microbiological Evaluation of Commercially Sterile Products*. Document number QAM – 588002-0107.

Tran, H., Datta, N., Lewis, M.J. & Deeth, H.C. (2008) Predictions of some product parameters based on the processing conditions of ultra-high-temperature milk plants. *International Dairy Journal* **18**, 939–944.

Tran, T.T. H., Datta, N & Deeth, H.C. (2003) Amylases in enzyme detergents and their effects on a simulated dairy dessert. *Milchwissenschaft* **58**, 527–531.

van Asselt, A.J., Sweere, A.P J., Rollema, H.S. & de Jong, P. (2008) Extreme high-temperature treatment of milk with respect to plasmin inactivation. *International Dairy Journal* **18**, 531–538.

van Boekel, M.A.J.S. & Folkerts, T. (1991). Effect of heat-treatment on the stability of milk-fat globules. *Milchwissenschaft* **46**(12), 758–765.

Villamiel, M., LopezFandino, R., Corzo, N. & Olano, A. (1996) Continuous flow microwave pasteurisation of milk - Comparison between microwave and conventional heating. In: *Heat Treatments & Alternative Methods*, IDF Doc S.I. 9602 .pp. 260–265. International Dairy Federation, Brussels.

von Bockelmann, B. & von Bockelmann, I. (1998) Aseptic packaging. In: *Long-Life Products: Heat-treated, Aseptically Packed: A Guide to Quality*, pp. 9–13. Bernhard von Bockelmann, Åkarp, Sweden.

Walstra, P. & Jenness, R. (1984) *Dairy Chemistry and Physics*. John Wiley, New York.

Wang, D.K. (2012) *Ultrafiltration of Milk at High Temperature*. M.Phil. thesis, University of Reading, UK.

Weast, R.C. (ed) (1982) *Handbook of Physics and Chemistry*, 63rd edition, CRC Press, Cleveland, Ohio.

Wilbey (2011) Homogenisation of milk. In: *Encyclopedia of Dairy Sciences* (eds J.W. Fuquay, P.F. Fox & P.L.H. McSweeney) 2nd edn, Volume 3, pp. 750–754. Academic Press, San Diego.

Zadow, J.G. (1969) Studies on the ultra-heat treatment of milk. I. Comparison of direct and indirect heating of whole milk. *Australian Journal of Dairy Technology* **24**(2), 44–49.

Zadow, J.G. (1975) Ultra-heat treatment of dairy products. *CSIRO Food Research Quarterly* **35**(2), 41–47.

6

Changes During Heat Treatment of Milk

Unlike lower-temperature heat treatments such as pasteurisation, UHT processing causes several changes in the chemical components and physical attributes of milk and milk products. These changes are dependent on the both the mode and severity of the heat treatment. Some changes are relatively small but may initiate further changes during subsequent storage at ambient temperature. Other changes, such as whey protein denaturation, can be marked and change little during storage. This chapter discusses the heat-induced changes under two broad categories: chemical and physical.

6.1 Chemical

6.1.1 pH and Ionic Calcium

It has been known for a long time that heat stability is influenced by milk pH. However, it is now better understood that ionic calcium is also an important determinant, and that their interrelationship is also important. Heat stability is discussed further in Section 6.2.1.

There are considerable variations in both pH and Ca^{2+} in milk from individual cows (White & Davies, 1958a; Lin, 2002; Tsioulpas *et al.*, 2007, Nian *et al.*, 2012). Bulking of milk reduces these variations but does not eliminate them (Chavez *et al.*, 2004, Tsioulpas *et al.*, 2007, Grimley *et al.*, 2009). Much of the earlier literature on UHT processing focused on the influence of pH on heat stability and paid much less attention to that of Ca^{2+}. Both calcium and hydrogen ions are positively charged and tend to neutralise the negative charge on the casein micelle. Potassium and sodium are also positively charged and are present at much higher concentrations, but are nowhere near as influential. Pyne and McHenry (1955) pointed out that heat coagulation was slow in milk which was low in calcium ion concentration and colloidal phosphate (Fink & Kessler, 1986a).

Tsioulpas *et al.* (2007) found that the Ca^{2+} concentration for 234 milk samples from individual cows ranged between 1.05 and 5.29 mM: the average was 1.88 mM. Chen (2013) found for 25 bulk milk samples collected over a one-year period from the same farm had an average Ca^{2+} concentration of 2.05 mM and a range of 1.68 to 2.55 mM. White and Davies (1958a) reported an inverse relationship between pH and ionic calcium. Others have observed the same, although none has reported a strong correlation. Thus, pH and ionic calcium are independent parameters; both individual and bulk

High Temperature Processing of Milk and Milk Products, First Edition. Hilton C. Deeth and Michael J. Lewis.

milk samples with a particular pH can have a wide range of Ca^{2+} concentrations and vice versa (Lin, 2002; Nian *et al.*, 2012), as shown in Figure 6.1 for cow's milk samples (Nian *et al.*, 2012). Figure 6.2 shows a comparison of the relationship for cow's milk taken at 20 °C and at 115 °C. Values at 115 °C were measured on dialysates, as described in Secton 11.2.25.

When the pH of any milk sample is reduced, by whatever means, its Ca^{2+} concentration is increased. This was investigated by Zadow *et al.* (1983) for goat's milk and for cow's milk by Geerts *et al.* (1983) and Tsioulpas *et al.* (2007). The last authors found a linear relationship between pH and log (Ca^{2+}), with a slope of -0.62 (Figure 6.3). Ionic

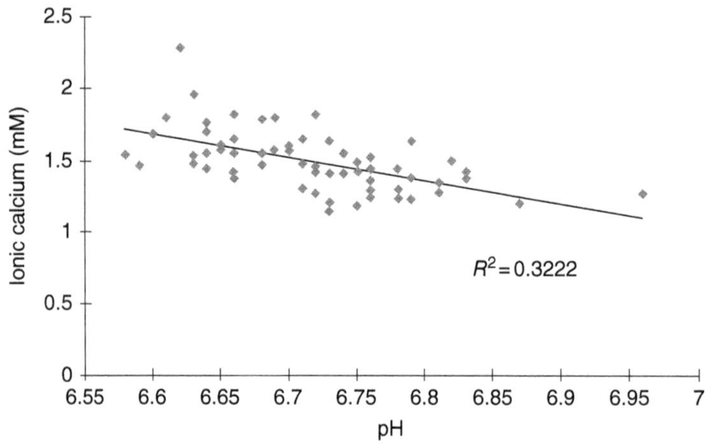

Figure 6.1 Variations in pH and ionic calcium in milk from individual cows. (Source: Nian *et al.*, 2012. Reproduced with permission of John Wiley & Sons.)

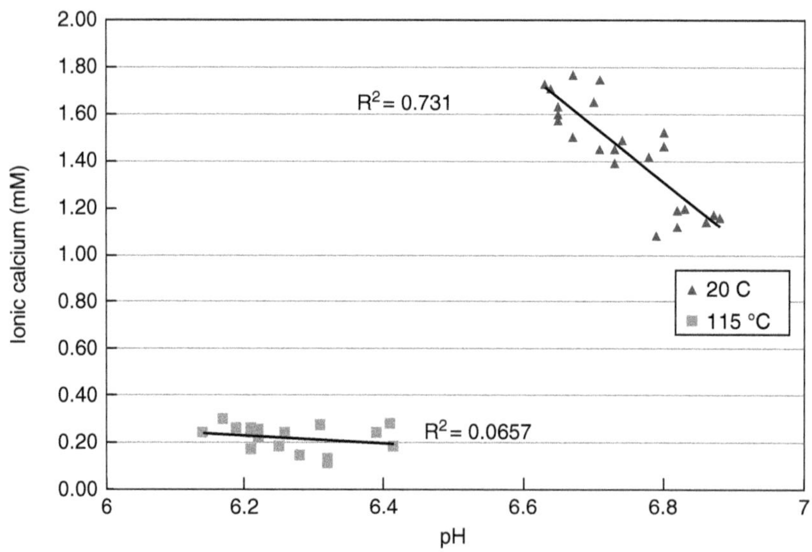

Figure 6.2 Some variations in ionic calcium in milk from some individual cows, determined at 20 °C and 115 °C. (Source: Nian *et al.*, 2012. Reproduced with permission of John Wiley & Sons.)

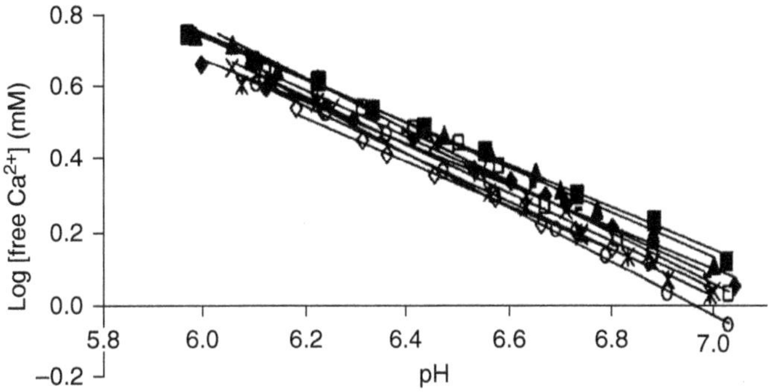

Figure 6.3 Effect of pH on log Ca^{2+} concentration in 9 individual cow's milk samples. (Source: Tsioulpas *et al.*, 2007. Reproduced with permission of Cambridge University Press.)

calcium in milk from other species has been reviewed by Lewis (2010) and its measurement is discussed in more detail in Section 11.2.17.1.

Holt (2004) showed how Ca^{2+} concentration could be predicted to change in milk over the pH range 5.0 to 7.0. Ca^{2+} concentration is influenced by the amount and type of casein, by phosphates and citrates, and by the pH of the milk. Therefore, milk will have a Ca^{2+} concentration and pH which are determined by its own unique composition and which can be predicted by models developed by Holt *et al.* (1981) and Holt (2004). One practical drawback of these models is that they are applicable to milk samples at 25 °C. At this point such models are unable to account for changes in temperature.

Two further important considerations are that both pH and Ca^{2+} concentration of milk decrease as its temperature increases. This is less well appreciated but it is an important consideration as the milk reaches 140 °C in the holding tube of a UHT plant. We would speculate that it is the milk pH and Ca^{2+} concentration at 140 °C that will influence whether that milk is likely to be heat stable at that temperature. When pH was measured directly in the holding tube of a heat exchanger, it decreased from 6.57 at 40 °C to 6.26 at 80 °C (Ma & Barbano, 2003). Walstra & Jenness (1984) reported that milk pH decreases to just below 6.0 when milk is heated to 100 °C. More recent data suggest that it may be as low as 5.6 at 140 °C (On-Nom, 2012). Thus, increasing the temperature of milk induces two changes which have competing effects: reduction in both pH and calcium phosphate solubility. One might expect that the reduction in pH would increase calcium phosphate solubility but calcium phosphate becomes less soluble at high temperature. This latter factor dominates and soluble calcium decreases as temperature increases (Pouliot *et al.*, 1989). Rose (1962) proposed that Ca^{2+} concentration would increase with increasing temperature, due mainly to the fall in pH, despite some contradictory evidence reported by Tessier and Rose (1959). Only recently have measurements of Ca^{2+} at high temperatures been reported; directly using electrodes up to 60 °C (Chandrapala *et al.*, 2010), and on UF permeates up to 80 °C and on dialysates up to 120 °C (On-Nom *et al.*, 2010). Both pH and Ca^{2+} decreased as temperature increased. Furthermore, ultrafiltration (UF) has been performed on milk up to 140 °C by placing the UF module in the holding tube of a UHT plant (On-Nom, 2012), as described in Section 11.2.25. One interesting observation is that dialysates and UF permeates collected at high temperature show hardly any change in their pH and Ca^{2+} whilst they cooled to room temperature.

Figure 6.4 and Table 6.1 show how pH and ionic calcium change after various heat treatments and other processes. Rose (1963) concluded that a definite correlation between calcium ion concentration at ambient temperature and heat stability at high temperature could not be established. This opinion was also expressed by McKinnon *et al.* (2009) who monitored pH when milk was heated to 90 °C. Thus future research should focus on how pH and ionic calcium concentration reached at high temperatures influences heat stability.

UHT processing has little effect on the pH of white milk. This fact is seldom reported in the literature. In fact, the main interest in milk pH in earlier publications on UHT processing was its effect on heat stability. Gaucher *et al.* (2008a) reported raw milk used for heat treatment had a pH range 6.69 to 6.76. After pasteurisation the range was 6.70 to 6.78 and after UHT treatment it was 6.66 to 6.76.

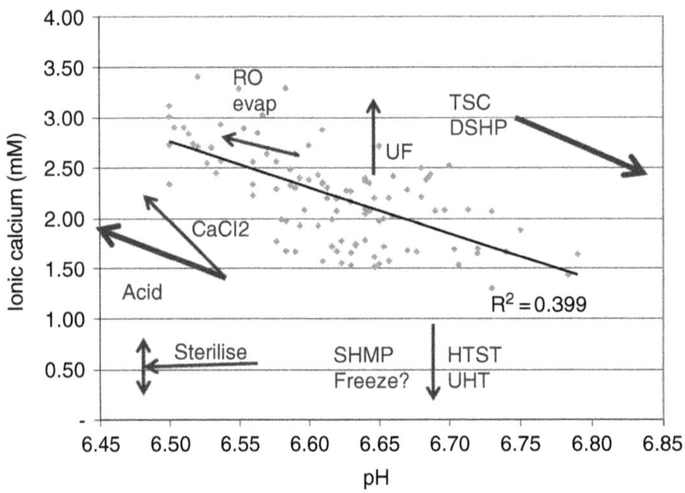

Figure 6.4 Indicative changes in pH and ionic calcium following heat treatment and some other processes.

Table 6.1 Effects of some processes on the pH and ionic calcium of milk.

Process	Change in pH	Change in Ca^{2+}
HTST	little change	decrease
UHT	little change	decrease but recovers during storage
In-container sterilisation	decreases	may increase or decrease
Evaporation and RO	decreases	slight increase
UF	no change	slight increase
Freezing	no change	Decrease
pH reduction and recovery	no change	Increase
High pressure	increase	no change

The fact that UHT processing does not change the pH of milk substantially is in marked contrast to the fall in pH of 0.2 to 0.3 units when milk is subjected to in-container sterilisation. This has been observed in over 200 batches of raw milk processed at Reading University.

von Bockelmann and von Bockelmann (1988) stated that UHT-treated milk showed a slight increase in titratable acidty due to formation of sulfhydryl groups. However, Pyne and McHenry (1955) estimated that half the acidity developed on heating in the temperature range of 100 to 130 °C was due to lactose decomposition, a third due to liberation of phosphate from the casein micelle and a sixth due to displacement of the calcium phosphate equilibrium.

A small reduction in pH may occur in flavoured milk and milk subject to severe UHT processing (high C* value). Allowing UHT milk to remain above 30 °C for some time immediately after processing may also lead to a reduction in pH, caused by the Maillard reaction. In contrast, soy milk increases in pH after UHT treatment (see Section 9.17.1). The reason for this increase is not known.

Heating milk usually leads to a reduction in Ca^{2+} concentration. Bringing milk to the boil and then immediately cooling it was found to decrease Ca^{2+} from 2.78 mM to 2.09 mM (Demott, 1968). UHT treatment produces a larger reduction in Ca^{2+} concentration than does pasteurisation (Geerts *et al.,* 1983; Ranjith, 1995). However, during subsequent storage, the Ca^{2+} concentration recovers, but not to its original value. A linear relationship was found between the logarithm of time and the recovery of the calcium ion activity following heat treatment (Geerts *et al.,* 1983). In-container sterilisation may result in either an increase or decrease in Ca^{2+} concentration, but there is always a considerable decrease in milk pH (Tsioulpas *et al.,* 2010). Nieuwenhuijse *et al.* (1988) reported that forewarming slightly reduced Ca^{2+} but had little effect on pH.

6.1.1.1 Effects of Addition of Phosphates, Citrate and EDTA

Salts of phosphoric acid and citric acid (both tri-basic) are frequently added to milk products to improve their heat stability. This has been practised for a long time to prevent coagulation during the production of evaporated milk. In fact, in the Codex regulations, these salts are described as either stabilisers or acidity regulators. More recently they have been added to normal strength milk products to improve their stability to UHT treatment.

Christiansson *et al.* (1954) drew attention to the role of trisodium citrate (TSC) in reducing Ca^{2+} and Tessier and Rose (1958) reported that addition of phosphate and citrate both decreased Ca^{2+}. However phosphate addition precipitated calcium, whereas citrate dissolved colloidal calcium. Singh and Fox (1987) and Udabage *et al.* (2000) reported that the amount of soluble casein increased with the addition of TSC, whereas the addition of calcium or mixtures of disodium hydrogen phosphate (DSHP) and dihydrogen sodium phosphate (DHSP) had the opposite effect.

These salts are rarely required for UHT processing or in-container sterilisation of cow's milk but they may be required for formulated or fortified milk drinks and are definitely required for goat's milk. Tsioulpas *et al.* (2010) reported that addition of TSC and DSHP to milk reduced Ca^{2+}, increased pH and increased ethanol stability (see Section 6.2.1.4) in a concentration-dependent fashion. Sodium hexametaphosphate (SHMP) also reduced Ca^{2+} concentration considerably, but its effect on pH was less noticeable. In contrast, DHSP reduced pH but had little effect on Ca^{2+}. In-container

Table 6.2 Effects of addition of some salts on pH and ionic calcium in milk.

Salt	Molecular weight (of anhydrous salt)	0.1 % addition (mM)	Effect on pH	Effect on Ca^{2+}
DHSP	118	8.47	decrease	little change
DSHP	142	7.04	increase	decrease
SHMP	612	1.63	no change	decrease
TSC	258	3.88	increase	decrease
EDTA disodium	336	2.98	reduce	little change
EDTA tetrasodium	380	2.63	increase	decrease
$CaCl_2$	111	9.01	decrease	increase

sterilisation had variable effects on Ca^{2+}. For DSHP and DHSP, Ca^{2+} decreased after sterilisation, but for SHMP it remained little changed or increased. Milk containing 3.2 mM SHMP or more than 4.5 mM $CaCl_2$ coagulated on sterilisation. It is interesting that the latest EU regulations permit the addition of some phosphates (but not citrates) to UHT milk (Hickey, 2009).

EDTA (ethylenediamine tetraacetic acid) is a strong calcium-chelating agent, which has been much investigated, although rarely added to milk. Most often it is the disodium salt which is added, which decreases pH, whereas the tetrasodium salt increases pH. Thus, addition of both these salts will reduce Ca^{2+} concentration provided that the pH is adjusted to its original value. It could be argued that it is immaterial which form is added, provided that the pH is readjusted, although this has not been established experimentally. For both salts, excessive additions (>20 mM) cause destabilisation of the casein micelle and the milk becomes whey-like in appearance. Udabage *et al.* (2000) investigated EDTA, mixtures of DSHP and DHSP, and TSC. They found significant reductions in Ca^{2+}activity in milk adjusted to pH 6.65.

Gaucher *et al.* (2008a) showed that UHT treatment improved heat stability, ethanol stability and also stability to phosphate (potassium dihydrogen phosphate) addition (Ramsdell test). These improvements most probably arose from a reduction in ionic calcium brought about by the UHT process, although this was not measured.

In summary, it is possible to manipulate the pH and Ca^{2+} by addition of different salts. These effects are summarised in Table 6.2. As discussed, for white milk, addition of phosphates are now permitted in UHT milk in some countries, including the EU, and for milk-based drinks a wider variety of additives can be used. However, it is prudent to establish how any additive will affect the pH and Ca^{2+} of the formulation and how this will influence its heat stability.

6.1.2 Mineral Salts

6.1.2.1 Mineral Partitioning and Associated Changes

Gaucher *et al.* (2008a) also reported that UHT milk contained slightly less soluble calcium but the soluble phosphate hardly changed during UHT treatment. No reports have been found on the effects of UHT processing on freezing point depression (FPD) which could provide information on changes in the concentration of soluble salts.

However, from our experience with pilot plant runs, FPD values for UHT milk are very close to those of raw milk. This would confirm that changes in mineral partitioning and lactose are slight during UHT treatment. In fact, even though a small proportion of the lactose is converted to lactulose, one would not expect this to affect FPD, as they are isomers. For milk heated at the lower temperatures of 70 °C, 80 °C and 90 °C, FPD increased by 2.3 m°C, 3.6 m°C and 5.4 m°C, respectively. These changes are very small and one explanation for them is movement of a small amount of soluble calcium or phosphate to the colloidal phase.

Another change which occurs in precipitation of calcium phosphate whose solubility decreases with temperature. If this occurs on the walls of the heat exchanger it can act as an initiator in the formation of a fouling deposit (see Section 6.2.2). However, some also associates with casein which results in some subtle changes to the surface of the micelle.

Methods for partitioning the mineral fraction of milk, such as dialysis and ultrafiltration, are discussed in greater detail in Chapter 11.2.24.

6.1.2.2 Addition of Mineral Salts

6.1.2.2.1 Calcium Salts

Calcium supplementation of milk is challenging, because milk is already saturated with calcium. A wide range of calcium salts could be added (see Table 9.2). These have a wide range of solubility values and can be broadly classified as salts of strong acids, salts of weak acids and salts which have very limited solubility (Deeth & Lewis, 2015).

Calcium chloride is a salt of a strong acid and has good solubility. When calcium chloride is added to milk, hydrogen ions are generated, which reduce pH and increase Ca^{2+} concentration (Phillipe *et al.*, 2003; Tsioulpas *et al.*, 2010; On-nom *et al.*, 2012). In contrast, calcium hydroxide addition results in a pH increase and a reduction in Ca^{2+}. Another approach is to add calcium salts of weaker acids such as calcium gluconate and calcium lactate. In a study where six calcium salts were compared, calcium chloride showed the largest destabilizing effect, followed by calcium lactate and calcium gluconate. Milk became unstable to UHT processing at lower calcium additions compared to in-container sterilisation (Omoarukhe *et al.*, 2010). However, addition of calcium carbonate, calcium phosphate and calcium citrate (30 mM), which are considered to be insoluble, did not change any of the properties of milk influencing heat stability, such as pH and Ca^{2+}. There were no major signs of instability associated with coagulation, sediment formation or fouling when the milk was subjected to UHT and in-container sterilisation. The buffering capacity was also unaltered. On-Nom *et al.* (2012) studied the combined effects of calcium chloride addition (up to 25 mM) and heat treatment (60 to 120 °C) on heat stability and found that the amount of calcium chloride needed to induce poor heat stability decreased as temperature increased. Ramasubramanian *et al.* (2012, 2014) found that addition of calcium chloride to milk which had been heated to ≥70 °C formed a gel at low calcium addition levels (<20 mM) but formed a coagulum from which whey separated when >20 mM calcium was added. The properties of the gel and coagulum depended on the heat treatment history of the milks with UHT milk producing the weakest products.

Potassium calcium citrate (Gadocal K®) has the unusual combination of properties of being soluble but does not increase Ca^{2+} (Ramasubramanian *et al.*, 2008). It is therefore a suitable candidate for producing heat-stable, calcium-fortified milk and milk products.

Calcium removal from milk can be achieved by ion exchange resins (Ranjith *et al.*, 1999). Changes in Ca^{2+} and pH depend upon the exchanging counter ions: effects are different for Na^+ or K^+, compared to H^+. Although it may seem counterintuitive from a nutritional viewpoint to remove calcium, it has been found to improve heat stability, presumably by reduction in Ca^{2+}.

This has been reported by Ranjith *et al.* (1999), Jeurnick and DeKrief (1995), Prakash *et al.* (2007) and Grimley *et al.* (2010). Demott (1968) reported that treatment with an anionic exchanger increased Ca^{2+} concentration, because of removal of phosphates. An alternative procedure, which may be less disruptive to the micelle, is to separate milk by UF and to then remove calcium from the permeate, rather than from the milk (Ranjith, 1995). This de-calcified permeate can then be blended with the UF concentrate. Deeth and Lewis (2015) have recently reviewed calcium supplementation and its partial removal from milk.

6.1.2.2.2 Other Mineral Salts

Abdulghani *et al.* (2014) produced UHT milk fortified with iron, magnesium and zinc to give total concentrations (natural plus added) of approximately 25, 50, 75 and 100% of the recommended daily intakes (RDI) per litre of the UHT milk. The highest concentrations (100% RDI) were 8, 320 and 16 mg/L for iron, magnesium and zinc, respectively. Zinc and iron preferentially partitioned into the casein micelles, whereas most of the magnesium was found in the serum phase (see also Section 9.5.2).

6.1.3 Proteins

Under the most commonly used conditions, UHT processing causes denaturation of the whey proteins and interaction of the whey proteins with casein via disulfide bonding, but only minor changes to the casein molecules. Lactosylation, the first step in the Maillard reaction, occurs to a minor extent in both whey proteins and caseins, largely at the ε-amino group of lysine residues (see Section 6.1.4.1).

6.1.3.1 Whey Protein Denaturation

It is common practice to discuss the effect of UHT treatment on whey proteins in terms of "denaturation" and this practice is followed here. However, it must be remembered that true denaturation, which is the unfolding of the protein molecule, is only the first change in the whey proteins caused by heat. In most cases, of greater significance are the subsequent changes involving protein–protein interactions which occur mostly through sulfhydryl–disulfide interactions with formation of intermolecular disulfide bonds but also through hydrophobic interactions. The most significant interaction is between whey proteins and caseins, chiefly between β-lactoglobulin (β-Lg) and κ-casein, but interactions between whey proteins also occur, yielding whey protein aggregates.

Irreversible denaturation of whey proteins commences at ~70 °C and the extent of denaturation increases with the severity of processing in terms of temperature and time. This means, in general, that more denaturation occurs in indirectly processed UHT milk than in directly processed UHT milk. Within each of these heating modes, the extent of denaturation also depends on the temperature–time profile of the heating process. As noted below in the discussion of fouling in indirect plants, much of the denaturation occurs, or at least should occur, in the pre-heat section of the plant (see Section 6.2.2). Because of this, the level of denaturation is not necessarily a good indicator of the severity of the overall heat treatment. Hence, care needs to be exercised when using the Aschaffenberg turbidity test which gives an estimate of undenatured whey protein, as an indicator of the total heat treatment that a UHT milk has received (see Section 11.2.28).

The order of the susceptibility of the major whey proteins to denaturation, from the most to the least heat-sensitive, is immunoglobulins > bovine serum albumin > β-Lg > α-lactalbumin (α-La). Proteose peptones are not affected (Donovan & Mulvihill, 1987). Immunoglobulins and bovine serum albumin are completely denatured during most UHT treatments but as these proteins are present at much lower concentrations than β-Lg and α-La, the effects of their denaturation and interaction with other proteins is minimal.

Tran *et al.* (2008) estimated the extent of denaturation of β-Lg and α-La in milk processed on several commercial UHT plants. Their calculations were based on the temperature–time profiles of the plants and used the kinetic data of Dannenberg and Kessler (1988) and Lyster (1970). β-Lg was assessed to be almost completely denatured in the indirect UHT plants but to only 74-92% in the direct plants. The estimated percentages of denaturation of α-La were 25-90% and 27-58% respectively for indirect and direct plants.

The calculated percentage denaturation values of β-Lg and α-La in milk processed on two different UHT plants, one direct and one indirect, having the same B* value of 2.6, are shown in Table 6.3. The temperature–time profiles for these plants are shown in Figure 6.5. Also shown are the values at the end of the pre-heat sections and at the end of sterilisation holding tube. These data illustrate that most of the whey protein denaturation occurs in the pre-heat section in the indirect plant but in the sterilisation section of the direct plant, where a pre-heat section is not usually included. Furthermore, they show the lower denaturation of α-La than of β-Lg.

Denaturation of whey proteins is affected by the concentration of milk components. According to McKenna and O'Sullivan (1971), β-Lg was denatured at a lower rate but α-La was denatured at a similar rate in concentrated milk compared with single-strength

Table 6.3 Denaturation (%)[1] of β-lactoglobulin and α-lactalbumin during processing on indirect and direct UHT plants run with a B* of 2.6[2].

	β-Lactoglobulin		α-Lactalbumin	
Position in UHT plant	**Indirect**	**Direct**	**Indirect**	**Direct**
After the preheating steps	78/95	1/<1	16	<1
After the sterilisation holding tube	89/99	74/58	37	10
After final cooling (Total for plant)	92/99	76/59	51	11

[1] The first values for β-Lg and all values for α-La are based on the kinetics data of Lyster (1970); the second values shown for β-Lg are based on the kinetics data of Dannenberg and Kessler (1988).
[2] The temperature–time profiles are given in Figure 6.5.

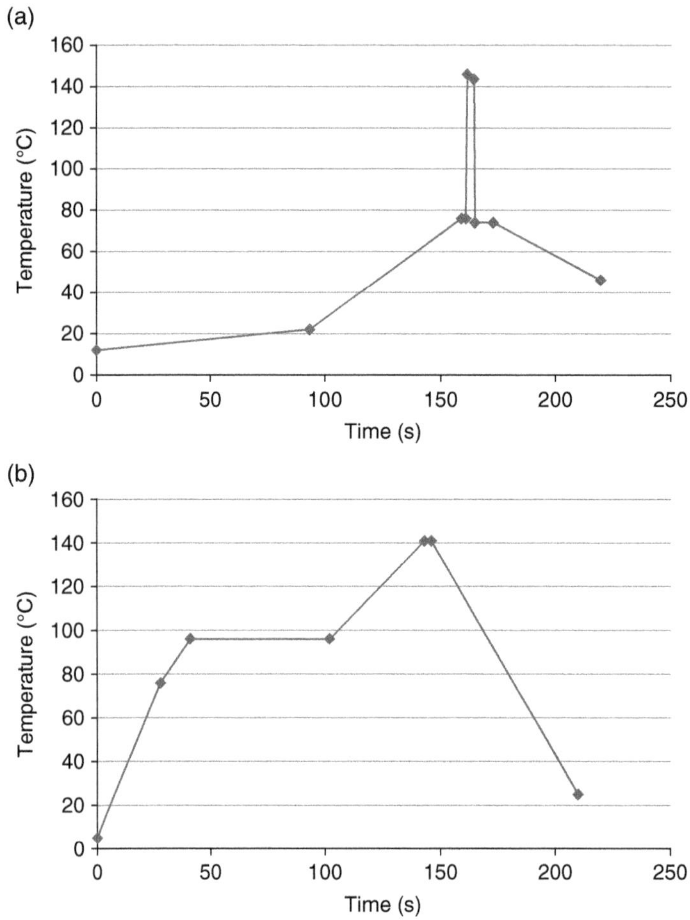

Figure 6.5 Temperature–time profiles of (a) a direct UHT plant and (b) an indirect UHT plant, each with a B* of 2.6.

milk. This was explained by Anema (2009) in terms of the effects of proteins and non-protein components. The denaturation rate of both β-Lg and α-La decreases with increased non-protein solids level but increases with increasing protein content; for α-La, these effects cancel each other out but for β-Lg, the effect of the non-protein components is greater than that of the proteins.

During heating, the native dimeric β-Lg unfolds and dissociates into monomers, each with an exposed –SH group and exposed hydrophobic residues. Molecules of the reactive β-Lg monomer react with other reactive β-Lg molecules, with κ-casein and with α-La through -SH–S-S interactions. The reaction between the whey proteins may occur before or after β-Lg reacts with κ-casein. Thus, in UHT milk produced without pH modification, several or all of the casein micelles have whey proteins attached. This fact is utilised in determining the amount of undenatured whey protein in UHT milk as acidification precipitates the casein with the attached denatured whey proteins leaving the undenatured whey proteins in solution for analysis. Since β-Lg and κ-casein have similar molecular weights (~18,000 and 19,000), are present in milk in similar concentrations and react with each other with a stoichiometry of 1:1 (β-Lg has one –SH group and κ-casein has one disulfide bond (one cystine) per molecule, almost all κ-casein molecules in UHT milk will have one β-Lg attached if all β-Lg is denatured during processing. The effects of heating on whey proteins and approaches to reducing the extent of denaturation and protein–protein interaction were reviewed by Wijayanti *et al.* (2014).

Heating releases some of the κ-casein with attached whey proteins from the micelle into the milk serum. Two major factors, pH and the heating method, determine how much of the denatured whey proteins remains bound to the casein micelle and how much becomes associated with κ-casein in the serum phase (Corredig & Dalgleish, 1996a,b). When milk is heated at pH ~6.5, the denatured whey proteins preferentially attach to the casein micelle while at higher pH, >6.8, most of the whey proteins are found in the serum attached to κ-casein (Kudo, 1980). Whether the whey proteins link with the κ-casein before or after it is released from the casein micelle has not been definitively determined (Oldfield *et al.*, 1998; Corredig & Dalgleish, 1999; Anema, 2009).

Singh and Latham (1993) studied the aggregation and dissociation of protein in milk heated at 140 °C and found that initial heating gave rise to the formation of high-molecular-weight complexes of whey proteins and κ-casein. With continued heating, the quantities of these complexes remained more or less constant but the amounts of intermediate-sized protein material cross-linked through covalent (non-disulfide) bonds increased gradually. Increasing the pH at heating resulted in increased quantities of whey protein–κ-casein complexes and monomeric protein in the ultra-centrifuged supernatant. More discussion on dissociation of caseins from the micelle is given in Section 6.1.3.4.

Relatively slow heating as in indirect UHT processing causes a high proportion of denatured whey proteins to become associated with casein micelles while rapid heating such as in direct UHT systems results in around 50% of the whey proteins remaining in the milk serum and 50% attaching to casein micelles. The nature of the heating also affects the ratio of β-Lg to α-La attached to the micelle, with low heat intensities, such as in direct UHT heating, resulting in a high β-Lg:α-La ratio while higher intensities, such as in indirect UHT treatments, cause a lower ratio, that is, more α-La, less β-Lg, to become attached (Mottar *et al.*, 1989). This may have significant practical implications for products made from UHT milk and may explain why yogurt made from UHT milk has weaker body than yogurt made with traditional heating of 90-95 °C for 5-10 min (Krasaekoopt *et al.*, 2003). It may also explain why calcium-induced gels and coagulums made from UHT milk are weaker than those made from milk heated at 90 °C for 10 min (Ramasubramanian *et al.*, 2012, 2014).

The denaturation of whey proteins during UHT processing is problematic for some products due to fouling of heat exchangers since denatured β-Lg is a major component of fouling deposits in the first stages of heating. In fact, models of fouling are based on the kinetics of denaturation of β-Lg (see also Section 6.2.2). This is a major consideration when products with high whey contents are UHT processed. Such products have become very popular because of the nutritive value of the whey proteins. Consequently, research has been conducted into producing heat-stable whey protein powders for incorporation into high-whey-protein UHT products. Several approaches have been used, some of which are used commercially; these have been reviewed by Wijayanti *et al.* (2014). These include modification by microparticualation of whey proteins by shear and heat into small spherical particles, enzymatic hydrolysis, partial denaturation with formation of soluble whey protein aggregates, ultrasonication of partially denatured whey proteins, microencapsulation of denatured proteins into nanoparticles and cross-linking with transglutaminase. In addition, methods of increasing the heat stability of whey proteins in products during UHT processing include addition of carbohydrates, chelation of minerals and addition of chaperone proteins. The last involves the addition of proteins, chiefly caseins and caseinates, to products such as high-whey-protein beverages to "protect" the whey proteins from (harmful) heat during UHT processing. This approach fits well with normal dairy processing and has considerable potential for such products (see Section 9.7).

6.1.3.2 Coagulation of Caseins

In contrast to whey proteins, the caseins in bovine milk are quite stable to heat. According to the classical heat-stability test, single-strength cow's milk should be stable to heating at 140 °C for at least 20 min before coagulation occurs. However, this does not apply to goat's milk which readily coagulates because of a high level of ionic calcium and requires chelation or removal of ionic calcium before it can be successfully UHT processed (see Section 6.1.1).

Coagulation and/or insolubilisation of the casein can also occur during UHT processing if the milk pH is too low (< ~6.5), which can occur in milk with a high bacterial count, and in milk preparations with an unfavourable mineral balance or inappropriate stabilisation. Dispersions of milk protein concentrate and micellar casein tend to coagulate during UHT processing and this is believed to be due to high ionic calcium levels (Beliciu *et al.*, 2012, Deeth & Lewis, 2015).

6.1.3.3 Protein Cross-linking

In addition to the cross-linking of proteins through disulfide bonds discussed above, cross-linking can also occur during UHT heating through formation of dehydroalanine or Maillard reaction products. As they are not broken by reducing agents such as dithiothreitol or mercaptoethanol which reduce disulfide bonds, these cross-links can be readily detected by amino acid analysis, GC/MS, HPLC and other methods after digestion with acid (Friedman, 1999) or enzymatic proteolysis.

Hydrolysis of proteins cross-linked via dehydroalanine produces isodipeptides such as lysinoalanine (LAL), histidinoalanine (HAL) and lanthionine (Friedman, 1999). LAL is the most commonly determined of these compounds (Lauber *et al.*, 2001). The dehydroalanine involved in cross-linking with protein-bound lysine, histidine or cystine can be produced by elimination of phosphate from *O*-phosphorylserine, a sugar moiety from *O*-glycosylserine or hydrogen sulfide from cysteine. It is well known that cross-linking via dehydroalanine proceeds faster at pH values higher than that 6.7, the pH of normal milk, and with high-intensity heating. Little such cross-linking occurs during normal UHT processing of milk.

Protein cross-linking also occurs via advanced Maillard reaction products. Of the many chemical products produced in the Maillard reaction, some such as formaldehyde, glyoxal and methyl glyoxal are effective cross-linking agents. While the involvement of the Maillard reaction in cross-linking during UHT processing was proposed some time ago (Andrews & Cheeseman, 1971; Andrews, 1975), proof of such cross-links in freshly processed UHT milk has been difficult to obtain. Rather, the evidence that cross-links can be formed in this way has been obtained indirectly. For example, it has been concluded from heating milk which contained lactose and milk from which lactose had been removed, at 95-100 °C for several hours, that cross-linking via Maillard reaction products occurred because of the much greater degree of cross-linking in the milk containing lactose than in milk without lactose (Andrews & Cheeseman, 1972, Al-Saadi *et al.*, 2013). However, some cross-linking via dehydroalanine also occurs in milk with and without lactose indicating that both types of cross-linking occur (Al-Saadi *et al.*, 2013).

Reports vary on the content of non-disulfide covalently cross-linked proteins in UHT milk. Some, but not all, authors have reported cross-linked proteins in freshly processed milk but all authors agree that cross-linking increases during storage (see Section 7.1.5). Using electrophoretic methods under reducing conditions, negligible amounts of these cross-linked proteins are observed in fresh UHT milk (Al-Saadi & Deeth, 2008; Holland *et al.*, 2011). However, Zin El-Din *et al.* (1991) using HPGPC found that 5% of the casein in UHT milk was cross-linked during indirect UHT processing with holding tube conditions of 138 °C for 2 min. This is much less than the levels of 14% determined by Andrews (1975) and Lauber *et al.* (2001) in freshly processed UHT milk using size-exclusion chromatography under denaturing reducing conditions. Of the percentage polymerised casein in the freshly prepared UHT milk, Zin El-Din *et al.* (1991) reported that raw milk contributed 2.5% while Andrews reported the contribution to be 6%. The differences may be attributable to the different methodologies used and whether oligomers (e.g., dimers and trimers) were included in the polymerised protein figure.

The extent of polymerisation increases with the intensity of the heat treatment. For example, in a comparison of the levels in differently processed milk, directly processed UHT milk had negligible amounts of LAL, indirectly processed milk had up to 50 mg LAL/kg protein while in-container sterilised milk had 110 - 710 mg LAL/kg protein (Fritsch *et al.*, 1983).

Non-disulfide protein cross-linking in milk has some practical consequences. Nutritionally, it reduces the digestibility of the proteins and decreases the availability of some essential amino acids such as lysine (Friedman *et al.*, 1981). However, it may increase the viscosity of milk which may be significant for the manufacture of yogurt (Lauber *et al.*, 2001).

6.1.3.4 Dissociation of Caseins from the Casein Micelle

Heat treatment of milk causes both protein association and disaggregation of caseins from the casein micelle. The extent and nature of these effects depends on the temperature and time of heating, pH and milk solids concentration. Several papers have been published on the effects in concentrated milk products which tend to be greater than those in single-strength milk. There is little available information on dissociation of caseins from unconcentrared milk processed under normal UHT conditions.

In unheated milk, some caseins exist in non-micellar form, the proportions varying from 3 to 15% for the individual caseins (de la Fuente *et al.*, 1998). These proportions increase during heating. More dissociation occurs in in-container sterilised unconcentrated milk and concentrated milk products than in UHT-processed unconcentrated milk (Aoki and Imamura, 1974a, b). An important dissociation is that of κ-casein, before or after it associates with whey proteins (Anema, 2008), as discussed in Section 6.1.3.1, as it makes the surface of the casein micelle more susceptible to aggregation induced by calcium ions (Anema, 2009). Kappa-casein has been shown to dissociate from the micelle even in whey protein-free concentrated milk. In this case, large casein micelles with low κ-casein contents were formed by aggregation of smaller micelles (Aoki *et al.*, 1977).

6.1.3.5 Effects on Enzymes

6.1.3.5.1 Proteases

Of the native enzymes in milk, one which can survive UHT heating conditions is alkaline protease or plasmin. This is very significant for the storage stability of UHT milk as residual plasmin may cause proteolysis producing peptides which impart a bitter flavour and may also destabilise the protein and cause age gelation (see Section 7.2.2). In addition to plasmin, the native proteases elastin and cathespins G, B and D have been implicated in producing peptides in stored UHT milk (Gaucher *et al.*, 2008b). The other proteases which survive UHT treatment are bacterial proteases, particularly those produced by psychrotophic bacteria such as *Pseudomonas* species in raw milk during storage before heat processing. These, like plasmin, also cause bitterness and age gelation (see Sections 7.1.3.4 & 7.2.2.1).

Plasmin in raw milk is a component of a complex protein system which consists of the inactive precursor, plasminogen, plus proteolytic enzymes, and protease activators and inhibitors as shown in the following scheme:

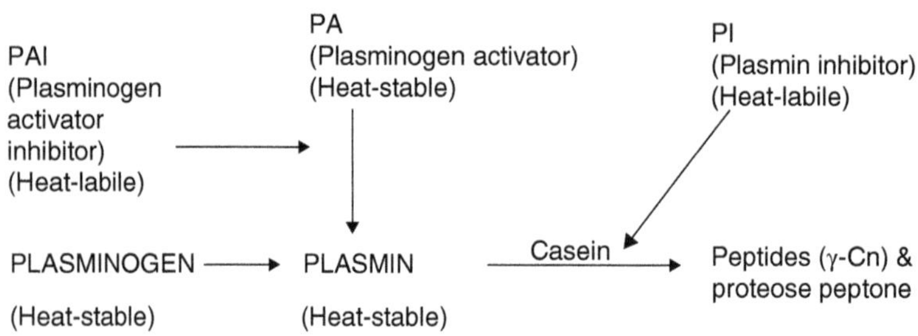

Thus the activity of plasmin in raw milk is determined by the interplay of these components. It is also influenced by cow factors such as stage of lactation and udder health status, particularly mastitis as indicated by the somatic cell count of the milk. In milk as it emerges from the cow, most of the plasmin exists as the precursor, plasminogen, which is converted to the active plasmin by plasminogen activators. Reported ratios of plasminogen to plasmin vary from 9:1 to 2:1 (Richardson, 1983; Auldist *et al.*, 1996). The ratio changes during lactation and with somatic cell count; milk from cows in late lactation and milk with a high somatic cell count have the lowest ratios (more plasminogen is converted to plasmin) and highest plasmin activities (Auldist *et al.*, 1996). Somatic cells contain a protease, plasminogen activator, which enhances the conversion of inactive plasminogen to plasmin (Kelly & Foley, 1997).

Plasmin activity in milk is affected by the heat treatment applied; the more severe the treatment, the greater inactivation of plasmin. Topçu *et al.* (2006) reported that when milk with high plasmin levels was processed commercially at 150 °C on an indirect UHT plant, the resulting milk showed less proteolysis during storage than when processed indirectly at 140 °C. Furthermore, less active plasmin remains after indirect UHT treatment than after direct heat treatment. This was demonstrated by Manji *et al.* (1986) who found no plasmin activity in indirectly processed UHT milk but 19% (of the original activity) in directly processed milks; the corresponding figures for plasminogen were 19 and 37% respectively.

During heat treatment, plasmin inactivation and complexation with β-Lg through sulfhydryl–disulfide interactions occur simultaneously (Kelly & Foley, 1997). Therefore plasmin can be inactivated under conditions which denature β-Lg such as in the pre-heat section of UHT plants. Effective pre-heat conditions for this inactivation of plasmin include 90 °C for 30 or 60 s (Newstead *et al.*, 2006), 80 °C for 300 s (van Asselt *et al.*, 2008) and 95 °C for 180 s (Rauh *et al.*, 2014). The heating conditions in the high-heat section can also be used to control plasmin activity; heating at 150 °C has been found to inactivate more plasmin than heating at 140 °C (Topçu *et al.*, 2006). Therefore, knowledge of the effects of the different heating conditions is important for devising UHT heating conditions to minimise plasmin-induced problems during storage of UHT milk.

Addition of potassium iodate to raw milk has been shown to increase plasmin activity in the UHT milk (Grufferty & Fox, 1986; Kelly & Foley, 1997). Iodate prevents sulfydryl–disulfide interchange between denatured β-lactoglobulin and plasmin by oxidising the free sulfydryl group of β-lactoglobulin and thus reduces inactivation of plasmin (Grufferty & Fox, 1986). This is also relevant to fouling in UHT plants (see Section 6.2.2).

Plasmin action on β-casein is inhibited by lactosylation of this protein. This is due to the modification of the enzyme's substrate making it unrecognisable to the substrate-binding pocket of plasmin (Bhatt *et al.*, 2014). This may partly explain why UHT milks which receive a severe (indirect) heat treatment are less prone to plasmin hydrolysis than milk which receives a milder (direct) heat treatment.

Some bacterial proteases are extremely resistant to UHT heating conditions. For example, decimal reduction times at 140 °C (D_{140}) of 2-300 s were reported for *Pseudomonas* proteases by Mitchell and Ewings (1985) while Griffiths *et al.* (1981) reported that after heating at 140 °C for 5 s, 20–40% of the activity of *Pseudomonas* proteases remained.

The proteases are produced in raw milk at the end of the log phase of growth of psychrotrophic bacteria. The presence of such proteases in milk can be avoided if the psychrotrophic bacterial count of the raw milk is not allowed to exceed ~10^6 cfu/mL. This is a "rule-of-thumb" figure as counts as low as 10^5 cfu/mL have been shown to result in measurable protease activity while milk with a high count, up to 10^7 cfu/mL, may contain no protease (Haryani *et al.*, 2003; Baglinière *et al.*, 2012). Like plasmin, more active bacterial protease is retained in directly processed than in indirectly processed UHT milk.

As discussed below (Section 6.2.2.8), proteases may also enter milk from biofilms in milk tankers and other equipment. Teh *et al.* (2014) demonstrated that such enzymes could cause proteolysis in UHT milk.

A novel heat treatment known as low temperature inactivation (LTI) is effective in inactivating some bacterial proteases. The conditions of the heat treatment, which can be performed before or after UHT processing, are ~55 °C for 30–60 min (Barach *et al.*, 1976). However, Kocak & Zadow (1985) found that the effect of LTI varied considerably and concluded that different proteases may require different conditions. A similar conclusion was reached by Fitz-Gerald *et al.* (1982) for bacterial lipases. LTI is effective when applied one day after UHT processing although the logistics of doing this preclude its use commercially. While the mechanism of LTI is unclear, it may be due to autodigestion of the enzyme or changes in the conformation of the enzyme through interaction with milk proteins (Schokker & van Boekel, 1999). West *et al.* (1978) found no gelation in LTI-treated UHT milk and the time required for a bitter flavour to develop was greatly increased. There was no adverse effect on the quality of the UHT milk and there was no evidence of reactivation of the protease in milk stored for 300 days.

6.1.3.5.2 Lipases

The natural milk lipase, a lipoprotein lipase, is not stable to high temperature treatments. It is almost completely inactivated by commercial HTST pasteurisation conditions of ~72 °C for 15 s (Deeth & Fitz-Gerald, 2006). However, bacterial lipases, like their corresponding proteases, are much more heat-stable and retain some of their activity after UHT treatments. Reported percentage retentions after UHT processing vary from 7 to 35% (Mottar, 1981; Panfil-Kuncewicz *et al.*, 2005; Button, 2007).

Several authors have reported a LTI treatment to be effective in increasing inactivation for bacterial lipases as well as proteases when combined with UHT treatment. Some low-temperature treatments found to be effective were 55 °C for 1 h (Griffiths *et al.*, 1981; Reddy *et al.*, 1991), 60 °C for 5 min (Bucky *et al.*, 1987) and 55 °C for 15 min (Zhang *et al.*, 2007). However, bacterial lipases vary considerably in their heat stabilities and hence a low-temperature heating process may not be appropriate for all of them. Fitz-Gerald *et al.* (1982) found that of 20 lipases which were inactivated to < 25% by heating at 100 °C for 30 s, three were more stable at 55 °C for 1 h while the other 17 were less stable to heating at 55 °C for 1 h than at 100 °C for 30 s.

6.1.3.5.3 Amylases

Thinning can occur in some starch-based UHT and ESL desserts during storage due to amylase-catalysed degradation of the added starch (see Section 7.2.3 for more details). The amylase obviously withstands the high-temperature treatments. There are a few possible sources of the amylase but bacteria, especially spore-forming bacteria, are the most likely as bacterial amylases are very heat stable and can remain active after UHT-like heat treatments. Some examples of the high heat stability of bacterial amylases are as follows: an α-amylase from *B. licheniformis* retained 40-50% of its activity after heating at 110 °C for 60 min (Medda & Chandra, 1980); amylases of *B. licheniformis* retained 98% of their activity after heating at 85 °C for 60 min (Morgan & Priest, 1981); α-amylases produced by *B. licheniformis* had D_{95} of 6-157 min (De Cordt *et al.*, 1992); and an α-amylase from *B. amyloliquefaciens* had a D_{140} of 60-70 min (Saraiva *et al.*, 1996).

These amylases have enhanced heat stability in desserts. For example, when *G. stearothermophilus* α-amylase was added to pudding, 26% survived a heat treatment of 143 °C for 22.2 s, which is in excess of commercial UHT treatments (Anderson *et al.*, 1983). α-Amylases produced by *B. licheniformis* and *B. coagulans* also had enhanced activity in the presence of starch.

Milk also contains a natural α-amylase and a β-amylase, with α-amylase being the dominant enzyme (Farkye, 2003). α-Amylase is less heat-stable than some other milk enzymes such as xanthine oxidase and is inactivated after heating at 75 °C for 30 min. It is much less heat stable than the microbial α-amylases (Griffiths, 1986) and would be unlikely to be a problem in long-shelf-life products. However, there is one report of it withstanding in-container sterilisation conditions of 115 °C for 15 min (Wüthrich *et al.*, 1964).

6.1.3.6 Effect on Rennet Coagulation of Casein

It has been well established that heat treatment of milk impairs its rennet coagulation properties. As the temperature of the heat treatment is increased, the coagulation time increases and the gel strength of the coagulum decreases. Milk heated at 140 °C forms either a very weak coagulum or none at all. The effect has been largely attributed to denaturation of the whey proteins and their attachment to the casein micelle, inhibiting aggregation of casein micelles and subsequent coagulation. Waungana *et al.* (1996) found that there was no impairment of rennet coagulation of milk with up to 60% denaturation but at higher denaturation levels, gelation times increased. It has also been shown that heat treatment causes a redistribution of calcium which can have a significant effect on coagulation (Schreiber, 2001). The holding times used in this investigation, up to 600 s, were much longer that those used in normal UHT processing and the trials were carried out on a 3% casein suspension which contained only 0.02% whey proteins. The calcium distribution effect may be less important in UHT milk heat-treated at the same temperatures but for holding times of <10 s.

Samel *et al.* (1971) reported that the rennet coagulation time increased from 7 to 50 min after UHT treatment. However, this change was reversed after storage of the UHT milk at 20 °C for 6 months when the coagulation time decreased to 4 min.

The effect of whey proteins on rennet coagulation has been studied on milks from which whey proteins have been partially or fully removed by membrane filtration For example, Bulca *et al.* (2004) used microfiltration/diafiltration on skim milk to concentrate the casein and remove up to 96% of the whey proteins. They found that the maximum temperature to which the retentate could be heated with minimal impairment of coagulation time and gel strength, increased with the level of concentration and extent of removal of whey proteins. With 82% removal of whey proteins, the retentate could be heated to 120 °C while with 96% removal of whey proteins it could be heated at 140 °C.

It is generally considered that the effect of UHT heating is on the second stage of rennet coagulation, not on the first stage, the proteolysis step. Vasbinder *et al.* (2003) reported that denaturation of whey proteins caused only a slight reduction in enzymic degradation of κ-casein. Interestingly, the caseinomacopeptide (CMP) released from UHT-treated milk differs from that released from raw milk; it contains 40% less of the glycosylated forms (Ferron Baumy *et al.*, 1992). It was concluded that after UHT treatment, the denatured β-Lg influenced release of only the glycosylated form of CMP. Another factor implicated in the impaired rennetability of UHT milk is precipitation of calcium phosphate on the casein micelle which is known to occur during UHT treatment due to the decreased solubility of this salt at high temperature. Vasbinder *et al.* (2003) showed that heat-induced calcium phosphate precipitation does not affect the enzymic cleavage step but has a clear effect on the clotting process.

Concentration of the protein partially offsets the effect of heat on rennet coagulation properties of milk. For example, McMahon *et al.* (1991, 1993) found that UHT-processed single-strength milk did not coagulate but UHT-processed 3X concentrated milk formed a weak gel and gelation proceeded more slowly. The microstructure of the gels formed from UHT-processed concentrated milk was different from that of gels made from unheated milk. Much of the casein was present as a homogeneous mass around the fat droplets and there were areas without a protein network, which presumably weakened the UHT milk gels.

Other approaches investigated to improve the rennetability of UHT milk include addition of calcium ions and pH adjustment (Ohashi *et al.*, 1989; Waungana *et al.*, 1998). Addition of calcium ions alone has little effect but has more effect when combined with reduction of pH to around 6.2. A further improvement is obtained with the addition of unheated casein micelles (Ohashi *et al.* 1989).

A major reason for heating milks intended for cheesemaking at high temperature is to inactivate *Clostridium tyrobutyricum* which causes the late-blowing defect in cheese. So the challenge is to heat-treat milk with properties and at conditions which give sufficient inactivation of the spores of this organism but enable sufficient rennet coagulation. Schreiber (2001) used kinetics data to predict a heating temperature–time range in which a "milk" (3% casein with 0.02% whey proteins) could be heated to retain high rennetability (gel strength at least equal to that of a pasteurised skim milk gel) but cause a 4-log inactivation of *Cl. tyrobutyricum.* This is shown in Figure 6.6. Using pilot-scale

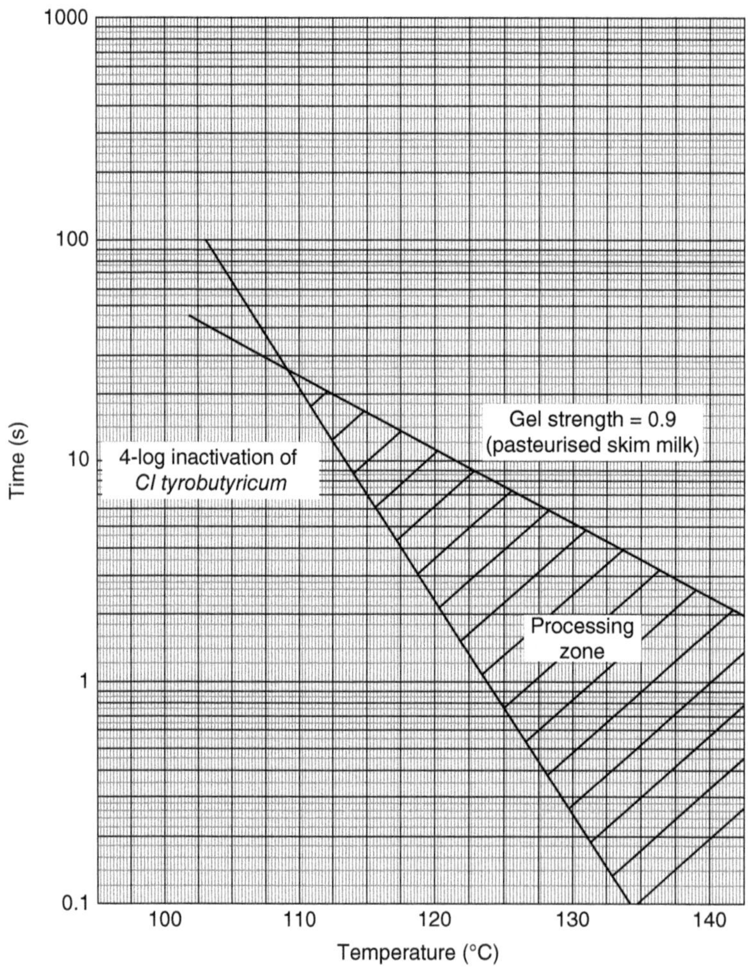

Figure 6.6 Predicted processing zone for heating a whey-protein-free, 3% casein suspension to inactivate *Cl. tyrobutyricum* but cause minimal impairment of rennet coagulation properties. (Source: Schreiber, 2001. Reproduced with permission of Elsevier.)

equipment Schreiber confirmed the suitability of "cheese milk" UHT-heated within this range shown for cheese manufacturing. This is a demonstration of the practical value of process modelling based on reaction kinetics (see also Section 6.3).

6.1.4 Lactose

6.1.4.1 Lactosylation and the Maillard Reaction

Lactose reacts with proteins during heat processing of milk to produce lactosylated proteins in the first step of what is known as the Maillard reaction. Presumably, it also reacts with free lysine (0.02-0.05 μmoles/mL [Walstra & Jenness, 1984]) and peptides (~0.82 μmoles/mL [McKellar, 1981) but information is lacking on this. The Maillard reaction is very significant during the processing and storage of UHT milk. Several reviews on the topic are available (e.g., O'Brien, 1995; van Boekel, 1998). A recent review discussed lysine blockage caused by the lactosylation (Mehta & Deeth, 2015).

The lactosylation reaction occurs mainly through the ε-amino group of lysine and to a lesser extent with arginine, methionine, tryptophan and histidine. The first major product formed with lysine is the protein-bound Schiffs base lactosyl-lysine which undergoes an Amadori rearrangement to produce the more stable protein-bound product lactulosyl-lysine or ε-N-deoxylactulosyl-L-lysine. This step is a highly significant one as the lysine in the Amadori product is biologically unavailable because the lactosylated lysine hinders proteolysis of the protein by digestive proteases. Hence the more lactosylation the greater the negative effect on the nutritional quality of the milk. The Maillard reaction is highly temperature-dependent; the more severe the milk heat treatment, the greater the extent of lactosylation. Table 6.4 shows the amount of blocked lysine in different heat treated milk products.

Because of the nutritional implications of lactosylation, the extent of formation of the Amadori product, lactulosyl-lysine, is sometimes cited in terms of "lysine blockage" (Mehta and Deeth, 2015). There are various ways of determining the extent of lysine blockage but one of the most common is through measurement of **furosine**,

Table 6.4 Blocked lysine in milk and dairy products (data from several reports, based on furosine measurement (adapted from Mehta and Deeth (2015).

Product	% Blockage
Raw milk	0
Pasteurised milk	0-2
UHT milk	0-10
In-container sterilised	10-15
Evaporated milk	15-20
Condensed milk	14-36
Spray-dried powder	0-7
Roller-dried powder	10-50
Infant formulae	5-34

ε-N-(furoylmethyl)-L-lysine, an amino acid not present in milk but formed by the acid hydrolysis of lactulosyl-lysine (Erbersdobler & Somoza, 2007). It has been widely, though not universally, accepted that acid hydrolysis (~6 M HCl at 105 °C for 20 hours) of the blocked lysines in milk proteins yields about 32% furosine and 40% lysine (plus also ~12% pyridosine). Therefore, when the furosine analysis is combined with amino acid analysis, the amount of blocked lysine can be calculated from the furosine value (using a factor of 100/32 or 3.1) and the total lysine can be calculated from the amount determined by amino acid analysis after acid hydrolysis plus the blocked lysine determined from the furosine value. A percentage figure for blocked lysine can then be calculated. The analytical procedures are described in more detail in Section 11.2.16.

An alternative method of estimating blocked lysine was proposed by Mehta and Deeth (2015). It is based on a knowledge of the lysine content of the protein in the product. This will vary according to the ratio of casein to whey proteins, since the lysine contents of casein and whey proteins differ, around 8.2% and 12.2%, respectively. The method assumes that the furosine yield from acid hydrolysis of the protein is 32%. The formula is as follows (see also Section 11.2.16.1):

$$\text{Blocked lysine}\,(\%) = \frac{3.1\ \text{furosine}}{\text{Theoretical lysine in the milk proteins}} \times 100$$

While furosine is still used as a convenient measure of lactosylation, there are several other methods available such as the direct measurement of lactulosyl-lysine in protein hydrolysates (Henle *et al.*, 1991). However, in recent times, sophisticated methods based on mass spectrometry have been used. These are outlined in an excellent review by Siciliano *et al.* (2013). They have been used to observe the various lactosylated forms of whey proteins and caseins as each lactose adds 324 Da to the molecular weight which can be observed in the mass spectrographs. Furthermore, by analysing tryptic digests of the lactosylated proteins, it is possible to locate the positions of the lactosylated lysines. The mass spectrometry-based methods have been largely qualitative but methods are now available for using them quantitatively (Section 11.3.3). These methods include measuring the relative intensities of mass spectrometry peaks (Scaloni *et al.*, 2002) and multiple reaction monitoring in which specific lactosylated peptides in tryptic digests of the milk proteins are measured (Le *et al.*, 2013). However, as pointed out by Le *et al.* (2013) and Siciliano *et al.* (2013), absolute quantification requires the use of isotopically labelled lactosylated peptides as internal standards.

From mass spectrometric analysis, it has also been shown that more molecules of lactose attach to the milk proteins with increased severity of heat treatment. Fogliano *et al.* (1998) reported mono-lactosylated forms of β-Lg in commercial UHT milk (which accounted for about 3% of total β-Lg) and di- as well as mono-lactosylated forms in in-container sterilised milks. They also showed that Lys100 was preferentially lactosylated in these milks. Scaloni *et al.* (2002) found mono- and di-lactosylated forms of α_{s1}- and β-caseins in UHT and in-container sterilised milks. In contrast to the report of Fogliano *et al.* (1998) on β-Lg, lactosylation of the caseins in these milks seemed to be non-specific, with seven of 14 lysines in α_{s1}-casein and five of 11 lysines in β-casein being lactosylated. The authors found that in less-severely heated, pasteurised, milk, where only mono-lactosylated forms were detected, the lactosylation was specific with Lys34 and Lys107 being preferentially targeted in α_{s1}- and β-casein, respectively.

Alternative methods of estimating the Amadori product in milk are measurement of hyroxymethyl furfural (HMF) and carboxymethyllysine (CML). They are members of a group of compounds collectively known as Advanced Glycation Endproducts or AGEs. CML is also known as a glycoxidation product as it is formed by oxidation of sugar moieties. Other AGEs which have been found in heated milk include lysylpyrraline, maltosine, maltol, β-pyranone, 3-furanone, cyclopentanone, galactosyl-isomaltol, acetylpyrrole, and pentosidine (van Boekel, 1998). AGEs have attracted considerable attention in recent times in medical research because of their purported links with several diseases, such as inflammatory diseases and diabetes (Davis *et al.*, 2014), as well as to aging (Nagai *et al.*, 2012). However, AGEs are formed endogenously as well as being ingested from heated foods. Hence the focus of much of the medical research has been on the endogenously formed AGEs rather than on AGEs originating from the diet. Although the term AGE is now widely used it should be noted in the context of heated milk that these compounds are also referred to as Advanced Maillard Reaction products as they emerge in the *advanced* stage of the Maillard reaction, that is, after the *early* stage in which the Amadori product is formed and before the *final* stage in which brown pigments (melanoidins) are formed and protein cross-linking occurs (van Boekel, 1998).

HMF has been commonly used as a chemical index of the heat treatment of milk. There are two measures of HMF, "free" and "total". Free HMF is formed in low concentrations by degradation of sugars and the Amadori product and is not bound to any other compound, whereas total HMF includes free HMF plus potential HMF which can be produced during acid digestion, usually by boiling the milk with oxalic acid. The digestion process generates HMF mostly from the Amadori compound but also from other Maillard browning intermediates and lactose. The levels of free HMF are much lower than those of total HMF. For example, Elliott *et al.* (2003) found the levels of free HMF in raw, pasteurised, direct UHT and indirect UHT milks to be 0.5, 0.5, 0.6 and 1.3 μmol/L, respectively, while the corresponding total HMF levels were 2.4, 3.8, 6.5 and 15.2 μmol/L, respectively. The pasteurised and UHT milks were all produced from the same raw milk. From these data it is apparent that the free HMF level is not very sensitive to heat treatment while the total HMF levels increase with the intensity of heat treatment. However, total HMF is, in effect, a surrogate measure of lactosylation and an indirect measure of the Amadori product (Morales & Jiménez-Pérez, 1998).

Carboxymethyllysine (CML) is formed by oxidation of the Amadori product with periodic acid followed by acid hydrolysis of the protein for the purpose of analysis (Büser & Erbersdobler, 1986; Badoud *et al.*, 1991). In addition, a small amount is produced during heating of milk. For example, Hewedy *et al.* (1994) and Drusch *et al.* (2008) found that the CML in experimentally produced UHT milk was below the detection limit although severe heating, at 148 °C for 128 s, caused a CML level of 5.66 mg/L (161 mg/kg protein assuming 3.5% protein; Hewedy *et al.*, 1994). However, Fenaille *et al.* (2006) reported levels of 29-46 mg/kg protein in three commercial UHT milk samples. Higher levels of CML (164-1015 mg/kg protein) have been reported in sterilised, evaporated and condensed milks (Drusch *et al.*, 2008; Nguyen *et al.*, 2014). A summary of reported CML data is presented in Mehta and Deeth (2015).

The percentages of blocked lysine increase with intensity of heat treatment (see Table 6.4 for a summary of data based on furosine measurement). The percentages vary but are of the order of 0-10% for UHT and 10-15% for in-container sterilised milk. Powders show a wide range with the highest being roller dried powder and infant formulae. The levels are not alarming as milk proteins are a rich source of lysine and hence the available lysine level is still high. However, the higher level of blocked lysine in infant formulae (15-20% according to Henle *et al.*, 1991) has attracted considerable attention because of the child's reliance on the product for its essential amino acids and also because of the composition of infant formulae, with high levels of whey proteins and lactose or other reducing sugars, and the conditions of drying and heating used in their manufacture. The amount of lactosylation is generally higher in powdered formulae than in liquid formulae, presumably because the Maillard reaction proceeds more efficiently at low water activities (French *et al.*, 2002). It should be borne in mind that lactosylation continues during storage (see Section 7.1.7) and hence the reported blocked lysine percentages include contributions from both processing and storage.

Lactose-hydrolysed milks in which the lactose is hydrolysed before heat treatment are more susceptible to lysine blockage than lactose-containing products because the products of hydrolysis, glucose and galactose, are more reactive than lactose (Messia *et al.*, 2007). In this case, hexose adducts (molecular weight increase of 162 Da) of whey proteins are observed by mass spectrometry (Carulli *et al.*, 2011). However, where the β-galactosidase is added aseptically after the heat treatment, the level of lysine blockage in freshly manufactured UHT milk is similar to that of milk containing lactose (Messia *et al.*, 2007).

A major consideration for the level of blocked lysine is that during the ambient storage of milk products such as UHT, in-container sterilised milk and milk powders, including infant formulae, the Maillard reaction continues and the level of blocked lysine increases (this is discussed further in Chapter 7).

The kinetics of lactosylation have been determined and hence it is possible to predict the levels of furosine (Browning *et al.*, 2001) associated with freshly prepared UHT milk if the temperature–time profile is known. Tran *et al.* (2008) predicted the furosine level for milk produced on 17 indirect and 5 direct commercial UHT plants to be 32-120 and 14-50 mg/100 g protein respectively.

Dissolved oxygen appears to decrease the Maillard reaction in milk. Rada-Mendoza *et al.* (2002) showed that less lactosylation, as measured by furosine, occurred in oxygenated milk than deaerated milk when batch-heated at 110 and 120 °C. They attributed this difference to oxidation of a precursor of lactulosylysine in the oxygenated milk. Similarly, Katsuno *et al.* (2013) found that deaeration increased the production of aminoreductone, a Maillard reaction product in heated milk. These authors found that this occurred in milk batch-heated at 130 °C and also in milk heated in an indirect UHT pilot plant with pre-heating at 80 °C for 3 min and high-temperature holding conditions at 130 °C for 2 s.

6.1.4.2 Lactulose Formation

During heat treatment of milk, some of the lactose (galactose-glucose) is epimerised to lactulose (4-O-β-D-galactopyranosyl-β-D-fructofuranose), a disaccharide of galactose and fructose. The lactulose content in milk increases with the severity of heating. Lactulose is not present in raw (unheated) milk and changes little during storage of heated milks. Hence it is a useful indicator of heat treatment of processed milk (see Section 6.1.7). The concentration of lactulose in freshly manufactured UHT milk shows a reasonable correlation with the intensity of cooked flavour (Burton, 1988) and also with lactosylation, as measured by furosine (Cattaneo *et al.*, 2008).

In general, indirectly heated UHT milk has a higher lactulose content than directly heated milk. In pilot plant trials using the same milk with the steriliser holding tube conditions for the indirect and direct heating being 138 °C/6 s and 143 °C/6 s respectively, the resulting lactulose concentrations were <40 mg/L and 239 mg/L (Elliott *et al.*, 2003). The heating conditions were chosen to provide approximately the same bactericidal effect (B*). Typical data for commercial UHT milks produced on direct and indirect plants are, respectively, in mg/L, 90–250 and 310–570 (Calvo *et al.*, 1987), 99–175 and 195–669 (Andrews, 1989), and 125 ± 20 and 466 ± 217 (Elliott *et al.*, 2005). In-container sterilised milks contain much higher levels with typical values being 460-1880 (Calvo *et al.*, 1987) and 1121 ± 56 (Morales *et al.*, 2000).

The concentration of lactulose has been proposed by the International Dairy Federation and other bodies to be used for assessing the heat treatment applied to milk, particularly during UHT or in-container sterilisation. Levels of 100-600 mg/L have been proposed for UHT milk and >600 mg/L for in-container sterilised milk. However, 400 mg/L has also been suggested as the upper limit for UHT milk. Milk with a level in excess of the proposed upper limit is considered over-processed and would have strong cooked flavours.

Surveys of commercial UHT milks have shown several to be over-processed by either suggested upper limit, 400 or 600 mg/L, for lactulose. In three surveys involving analysis of a total of 87 UHT milk samples, 38-58% were found to have >400 mg/L and 20-33% had >600 mg/L of lactulose (Biewendt, 1994; Elliott *et al.*, 2005; Cattaneo *et al.*, 2008). From these figures it is apparent that over-processing does occur. This can have an adverse effect on the quality of the product, especially on flavour and sediment formation, but this may be balanced by the positive effect of increasing the inactivation of heat-resistant proteases, thereby decreasing the risk of bitterness and age gelation (see Sections 7.1.3.4 & 7.2.2).

As for furosine, the kinetics of the epimerisation of lactose to lactulose have been reported by several authors. The activation energies and rate constants vary but the differences are probably insignificant when the errors in measurement of processing temperatures are taken into account. There appears to be general agreement that the reaction is zero order. Tran *et al.* (2008) used the kinetics reported by Rombaut *et al.* (2002) to estimate the lactulose levels in milks produced on 22 commercial UHT plants after establishing their temperature–time profiles. The values ranged from 60 to 550 mg/L. From the calculations, it appeared that almost all of the lactulose is formed in sections of the UHT plant at temperatures >90 °C.

The amount of lactulose formed during heat processing varies with the composition of the product so care needs to be exercised in interpreting lactulose data on non-standard milk products. Adding citrate or phosphate increases the formation of lactulose even when the pH is kept constant, with citrate having the greatest effect. For example, for a given heat process, a 10% increase in the concentration of citrate or a 20-40% increase in the phosphate approximately doubles the rate of lactulose formation (Andrews & Prasad, 1987). This may be significant for UHT milk in which citrate or phosphates (monophosphates such as DSHP or polyphosphates such as SHMP) are regularly added to improve protein stability. The lactulose-enhancing effect can be explained in terms of base-catalysis of the epimerisation reaction.

The effect on lactulose formation of addition of calcium salts depends on the nature of the salts added. Addition of the soluble ionic salt, calcium chloride, reduces the formation of lactulose, presumably through reduction of phosphate and/or citrate, or through lowering the pH (Martinez-Castro *et al.*, 1986; Olano *et al.* 1987). By contrast, addition of the insoluble alkaline calcium carbonate enhances formation of lactulose, presumably through base-catalysis (Paseephol *et al.*, 2008). Calcium carbonate is sometimes added to fortify UHT milk with calcium (Deeth & Lewis, 2015).

Another factor which may influence lactulose formation is dissolved oxygen. It has also been reported to decrease the level of lactulose produced during heat treatment at 110 and 120 °C for up to 20 min using glass tubes in a laboratory setting (Rada-Mendoza *et al.*, 2002). This was attributed to oxidation of precursors of the enediol precursor of lactulose.

6.1.5 Vitamins

A common belief is that high-temperature processing destroys the vitamins in milk. While some destruction occurs, the levels of destruction are modest except for the most intense heat treatments. A few general points are worth noting: the more intense the heat treatment, the greater the destruction which increases in the order pasteurisation < direct UHT < indirect UHT < in-container sterilisation; only the water-soluble vitamins are labile to heat as the fat-soluble vitamins are quite stable; an important consideration is the presence of oxygen as vitamins such as ascorbic acid and folic acid are relatively stable to heat in the absence of oxygen but quite unstable in the presence of oxygen (Viberg *et al.*, 1997); and more destruction occurs during storage of shelf-stable milk than during heat processing.

Table 6.5 shows the range of levels of destruction reported by several authors. It should be noted that some of the data were reported before modern methods of analysis were available and this may account for the large range for some vitamins.

Table 6.5 Losses (%) of water-soluble vitamins caused by pasteurisation, UHT treatment and in-container sterilisation.

Vitamin	Pasteurisation[1]	UHT[1]	Direct UHT[2]	Indirect UHT[3]	In-container sterilisation[1]
B1 Thiamine	<10	10	12	6 ± 5.2	30
B2 Riboflavin	NS	NS	2.5	NS	NS
B3 Nicotinic acid	NS	NS	4	NS	NS
B5 Pantothenic acid	NS	NS	3.5	NS	NS
B6 Pyridoxine	<10	10	5 ± 2.8	7	20
B8 Biotin	NS	NS	NS	NS	NS
B9 Folic acid	<10	15	16 ± 5	22 ± 15	50
B12 Cobalamin	<10	10	17 ± 0.8	13 ± 18	<90
C Ascorbic acid	20	25	25 ± 10	24 ± 5	90

NS: Not Significant

[1] Burton, 1988;

[2] Ford *et al.*, 1969; Görner F. & Uherova R., 1981; Mottar & Naudts, 1979; Oamen *et al.*, 1989;

[3] Andersson & Öste, 1992a; Ford *et al.*, 1969; Haddad & Loewenstein, 1983.

The destruction of thiamine is of particular interest as the chemical index C*, introduced by Kessler (1981), is based on this reaction and used as a measure of the chemical effect of a particular heat treatment on milk (see Section 3.2.3). A C* value of 1 is equivalent to 3% destruction of thiamine. Thiamine appears to have the most stable degradation kinetics of the heat labile vitamins in milk. It has been found to generally follow first-order kinetics (Lewis & Heppell, 2000), although second-order kinetics have been reported (Horak & Kessler, 1981). Horak and Kessler (1981) found that the loss in thiamine followed second order kinetics but when it was < 10%, it could be expressed to a good degree of approximation by a first-order reaction. In long-shelf-life milk, most of the loss of thiamine occurs during the heat treatment; storage causes no further loss if the milk is not exposed to light.

In the survey by Tran *et al.* (2008) of 22 commercial UHT plants, the calculated C* values varied from 0.4 to 3.0 indicating the predicted thiamine loss ranged from <1% to ~10%. However, some reported values for thiamine loss during UHT processing exceed these (see Table 6.5). Sierra *et al.* (2000) reported that thiamine losses in milk are normally between 3 and 10 % for pasteurised milk, between 10 and 20 % for UHT milk and about 20 % for sterilised milk.

Despite the significance attached to the destruction of thiamine in UHT milk, its concentration is seldom measured. The reported concentrations for pasteurised, UHT and in-container-sterilised milk are 370, 330 and 240 µg/100 L (Nohr *et al.*, 2011). Thiamine exists in multiple forms but mostly as free thiamine (50-70%) and thiamine monophosphate (18-45%) with 5-17% bound to protein. Therefore, to measure the concentration of total thiamine, pre-treatment with enzymes and/or acid is required before analysis, typically by HPLC (Lynch & Young, 2000). In addition, an oxidation step is usually included to convert the thiamine to the fluorescent thiochrome which is readily detected fluorometrically. The thiochrome can also be determined fluorometrically without HPLC separation but this is less accurate because of interfering fluorescent compounds.

6.1.6 Flavour

The most notable heat-induced change during high-temperature processing of milk from a consumer viewpoint is in flavour. Many consumers refer to the typical flavour as "heated" or "cooked" although food scientists have classified the flavour into several categories. The heated flavour which originates fully or in part from the heating process includes three types: 1. cooked/sulfurous/cabbagey; 2. rich/heated; 3. scorched/sterilised. Type 1 is attributable to volatile sulfur compounds (discussed below) and type 2 to mostly carbonyl compounds (discussed below). Type 3 flavour, scorched, occurs when there is a large amount of burn-on or fouling deposit in the heat exchangers.

In addition to "heated" flavours, other flavour notes develop in UHT milk during storage. These include stale, bitter and rancid flavours which result from oxidation of fats, proteolysis and lipolysis respectively. In addition, Maillard reactions continue and enhance the rich/heated note. These are discussed in Section 7.1.3.

The literature varies considerably on the nomenclature of the different flavours encountered in heated milk produucts. Table 6.6 summarises the different flavours which are discussed in this book. Some authors have included flavours other than those in the table, for example, caramelised/sterilised (Andersson & Öste, 1995) which is

Table 6.6 Types of UHT milk flavour.

Flavour/taste	Chemical basis	Causative agent	Time when flavour is present	Comments
Cooked/ sulfurous,	Volatile sulphur compounds, e.g., H_2S, CH_3SH	Degradation of S-amino acids of whey proteins	0-12 days	Disappears due to oxidation by dissolved oxygen; can be reduced with oxidising agents such as iodate, hydrogen peroxide and the enzyme sulfhydryl oxidase
Heated/rich/ sterilised	Diacetyl, methyl ketones, δ-lactones, benzaldehyde, furfural, phenylacetaldehyde, vanillin, oct-1-en-3-ol, n-heptanol, 2-butoxyethanol, maltol, acetophenone, benzonitrile, benzothiazole	Heat-induced Maillard reaction and lipid oxidation	From manufacture to end of shelf-life but most noticeable after 1-3 weeks storage when the sulphurous note decreases	Possibly the most important but least understood.
Scorched	Unknown	Occurs from excessive burn-on in heat exchangers	From manufacture to end of shelf-life	Can be prevented by minimising fouling deposits and frequent cleaning of heat exchangers
Stale	Short-chain aldehydes and methyl ketones	Lipid oxidation	From 4-6 weeks to end of shelf-life	Reduced if oxygen is reduced
Bitter	Hydrophobic peptides	Proteolysis of caseins by bacterial proteases or natural milk plasmin	After a few days if the protease activity level is high or after months if the levels of protease is low.	Bacterial proteases are only present if the raw milk prior to heat treatment contains $> \sim 10^6$ cfu/mL psychrotrophs and the proteases are not completely inactivated by the heat treatment. Plasmin may be present if the heat treatment is mild

Rancid	Free fatty acids	Lipolysis of the milk fat by bacterial lipases	After a few days if the lipase activity level is high or after months if the levels of lipase is low	Bacterial lipases are only present if the raw milk prior to heat treatment contains $> \sim 10^6$ cfu/mL psychrotrophs and the lipases are not completely inactivated by the heat treatment.
Chalky, astringent	Calcium phosphate, casein particles	Heat-induced insolubilisation of calcium phosphate and aggregation of casein micelles	Immediately after manufacture, and may persist during storage	Reduced by downstream homogenisation; more common with direct heating (steam infusion of injection)

included here under "rich/heated' and oxidised (Adhikari & Singhal, 1992) which is included here under "stale". Flavours due to bacterial contamination are not considered here.

Another sensory characteristic, although not strictly a flavour, which can occur in freshly processed UHT milk, particularly directly processed milk, is a chalky/astringent/powdery/dry puckery oral sensation in the mouth (Shipe *et al.*, 1978, Adhikari & Singhal, 1992). Richards *et al.* (2014) likened it to the inner surface of banana peel. It is attributable to large protein and/or calcium phosphate particles. The homogenizer is always placed downstream in direct UHT processing (Tetra Pak, 2015), although no reasons are given for this. One suggestion is that it disperses these particles and reduces the chalkiness.

6.1.6.1 Volatile Sulfur Compounds

When UHT milk is freshly processed, it has a strong sulfurous flavour, sometimes described as cooked or cabbagey. It is largely due to volatile sulfur compounds produced from milk proteins which contain the sulfur amino acids cystine, cysteine and methionine. In addition, the structures of some proteins such as β-Lg are modified by the heating process exposing some previously buried sulfhydryl groups. These sulfhydryls are attached to large molecules and are unlikely to contribute to the heated flavour of UHT milk. However, the total sulfhydryl content correlates well with the intensity of heated flavour (Andersson & Öste, 1992b).

The volatile sulfur compounds in UHT milk include hydrogen sulfide, dimethyl disulfide, dimethyl sulfide, methanethiol and, to a lesser extent, carbon disulfide, carbonyl sulfide, dimethyl trisulfide, dimethyl sulfoxide and dimethyl sulfone (Vazquez-Landaverde *et al.*, 2006; Al-Attabi *et al.*, 2014). The formation and levels of these compounds in UHT milk have been reviewed by Al-Attabi *et al.* (2009). All of these compounds, with the exception of the odourless carbonyl sulfide and dimethyl sulfone, are capable of contributing to the flavour of heat-treated milks. Individually, however, many will be below their flavour threshold (determined independently in water) but it is possible that mixtures of compounds in low concentrations could make a contribution to flavour. Immediately after UHT processing, the concentrations of hydrogen sulfide, methanthiol, dimethyl sulfide and dimethyl trisulfide are higher than their respective flavour thresholds and hence contribute to the flavour of the milk; however, they decrease during storage to levels below their thresholds concentrations (Al-Attabi *et al.*, 2014).

The reported concentrations of volatile sulfur compounds in UHT milk vary considerably (summarised by Al-Attabi *et al.*, 2009). This could be due to a variety of reasons, not least of which are the sampling and analytical methodologies used. Modern sampling methods such as Solid Phase Microextraction (SPME) and Solid Phase Dynamic Extraction (SPDE) combined with gas chromatographic methods using sensitive detection equipment such as pulsed flame photometric detectors have greatly facilitated analysis of low concentrations of these compounds. A second reason for the different reported concentrations is that the levels are largely determined by the mode and intensity of the UHT thermal treatment. Even where a temperature–time combination for the sterilisation holding tube is given, the chemical effect on production of volatile sulfur compounds can differ considerably between direct and indirect plants. As an illustration, for direct and indirect plants nominally operating at 138 °C for 1 s, the chemical effects, C^* values, are approximately 0.2 and 2.0, respectively. A third reason for the differences is the age of the samples analysed as most volatile sulfur compounds decrease during storage, presumably due to oxidation and interaction with other milk constituents. Hence analyses of commercial samples of unknown age can lead to a wide range of concentrations. A fourth reason is that the practical difficulties of handling the highly volatile and reactive volatile sulfur compounds limits their accurate quantification in milk (Vazquez-Landaverde *et al.*, 2006).

β-Lactoglobulin (β-Lg) is the main source of volatile sulfydryl (-SH) compounds in UHT milk, with the concentration of these compounds being related to the extent of denaturation of β-Lg and hence to the intensity of heat treatment. For this reason, UHT milk produced by the indirect process generally has a more cooked flavour than directly processed UHT milk.

Another source of volatile sulfur compounds is the milk fat globule membrane proteins. Hydrogen sulfide, methanethiol, dimethyl sulfide, carbon disulfide, carbonyl sulfide, dimethyl sulfoxide and dimethyl disulfide, but not dimethyl sulfone, are produced in higher concentrations in UHT whole than UHT skim milk (Al-Attabi *et al.*, 2014). This is illustrated by the concentrations of hydrogen sulfide produced in UHT whole and skim milk: ~50 and 12 µg/L respectively, and the corresponding figures for methanethiol, ~10 and 4 µg/L. Furthermore, higher concentrations of hydrogen sulfide, methanethiol and dimethyl trisulfide have been reported to be produced in UHT milk with higher fat content, for example, the dimethyl trisulfide contents of 1% and 3.25% fat UHT milk were 32.9 ng/kg and 47.3 ng/kg respectively (Gaafar, 1987; Vazquez-Landaverde *et al.*, 2005). This is consistent with the reported good correlation between fat content, hydrogen sulfide concentration and cooked flavour intensity in UHT milk (Thomas *et al.*, 1976).

A third source of volatile sulfur compounds in heated milk is the Maillard reaction (Calvo & de la Hoz, 1992). Dimethyl disulfide, dimethyl sulfide, methanethiol and dimethyl trisulfide are formed from methionine while hydrogen sulfide is formed from cysteine (Nursten, 1981).

6.1.6.1.1 Hydrogen Sulfide

Hydrogen sulfide (H_2S) is formed via a Strecker degradation of cysteine with a diketone. The diketones present in heated milk include diacetyl and 2,3-pentadione which also contribute to the flavour of UHT milk. It can also be formed through degradation of thiamine and methionine (Samuelsson, 1962; Dwivedi & Arnold, 1973).

The concentration of H_2S increases in UHT milk with the severity of heat treatment, with indirectly processed milk having higher levels than directly processed milk. However, Gaafar (1987) found that it increased with the level β-lactoglobulin denaturation up to 100% denaturation and then decreased. This is consistent with the finding that it is absent or present in very low concentration (e.g., 0.3 µg/L) in sterilised milk (Badings *et al.*, 1981; Badings & de Jong 1984). H_2S may be produced initially during sterilisation but it subsequently reacts with milk components or is oxidised by the dissolved oxygen in the product.

H_2S has been generally regarded as a major contributor to the cooked or cabbagey flavour of fresh UHT milk. This is consistent with the fact that H_2S and cooked flavour decrease during storage. However, H_2S may not be the only contributor to the cooked flavour which persists after ambient storage when H_2S is undetectable by gas chromatography/flame photometric detection (Slinkard, 1976).

6.1.6.1.2 Methanethiol

Strecker degradation of methionine is responsible for formation of methanethiol (or methyl mercaptan, CH_3SH) via the intermediate methional, together with some dimethyl sulfide and dimethyl disulfide (Ballance, 1961; Scarpellino & Soukup, 1993). Methanethiol has been described as having a putrid, cooked and cabbage-like flavour and hence it is reasonable to assume it could contribute to the cabbagey flavour of freshly processed UHT milk (Samuelsson, 1962). This was supported by the findings of Jaddou *et al.* (1978) that there was a correlation between the cabbagey flavour intensity and the concentration of methanethiol and that its concentration in UHT milk (heated at 140 °C for 3 s) (~12 µg/L) was higher than the reported sensory threshold value in water of 0.02 – 2.1 µg/L. Vazquez-Landaverde *et al.* (2006) detected a higher concentration (23.9 µg/kg) in 3.25% fat UHT milk than in 1% fat milk (16.1 µg/kg). Methanethiol may also contribute to the flavour of in-container sterilised milk as its concentration has been reported to be 5 µg/kg, considerably higher than its flavour threshold value (Badings & de Jong, 1984).

6.1.6.1.3 Dimethyl Sulfide and Dimethyl Disulfide

Dimethyl sulfide (DMS) and dimethyl disulfide, (DMDS) are formed as a result of the Strecker degradation of methionine during UHT treatment. It has also been suggested that dimethyl disulfide in heated milk is due to oxidation of methanethiol (Calvo & De la Hoz, 1992). DMS appears to be formed from β-lactoglobulin since its concentration in UHT milk correlates with the level of β-lactoglobulin denaturation (Gaafar, 1987).

Most of the reported levels of DMS are below the flavour threshold value of 20 µg/L; the ones that are higher are for freshly processed UHT milk (~327 µg/L in whole milk and ~71 µg/L in skim milk; Al-Attabi *et al.*, 2014), milk processed at high intensity, e.g., 142 °C for 30 s (>50 µg/L Badings *et al.*, 1985) and commercial indirectly processed UHT milk (conditions not reported) (120 µg/kg; Bosset *et al.*, 1996). High DMS concentrations (98 and 178 µg/kg) were also detected in retort-sterilised milk (150 °C for 1 h and 0.5 h respectively) (Bosset *et al.*, 1996). Based on sensory data, Bosset *et al.* (1996) concluded that DMS did not make a significant contribution to cooked flavour in UHT milk, despite the levels detected in commercial indirect UHT milk exceeding the reported flavour threshold concentration.

The concentrations of DMDS are the lowest of the volatile sulfur compounds detected in UHT milk. However, unlike the other compounds, DMDS was higher in skim milk than in whole milk (Al-Attabi *et al.*, 2014). Recently it has been detected at < 0.5 µg/kg (Vazquez-Landaverde *et al.*, 2006; Al-Attabi *et al.*, 2014) which is lower than those reported earlier for commercial direct and indirect UHT milks (3 and 6 µg/kg respectively). A much higher level (18 µg/kg) was recorded for milk retort-sterilised at 150 °C for 1 h (Bosset *et al.*, 1996). Since the reported flavour threshold concentration in water is 0.2-50 µg/kg, it is unclear if DMDS makes a significant contribution to the cooked flavour of UHT milk.

6.1.6.1.4 Dimethyl Sulfone, Dimethyl Sulfoxide and Dimethyl Trisulfide

Dimethyl sulfone and dimethyl sulfoxide (DMSO) are formed by oxidation of dimethyl sulfide. Dimethyl sulfone is flavourless and hence is not considered further here. DMSO has been detected at 1.46 mg/kg in whole UHT milk (Vazquez-Landaverde *et al.*, 2006). Higher levels were found in freshly processed commercial UHT whole and skim milk

(203 and 36 mg/kg respectively) (Al-Attabi *et al.*, 2014). While these concentrations are by far the highest of all known volatile sulfur compounds in UHT milk, the contribution of DMSO to the flavour of the milk is unknown as its flavour threshold concentrations have not been reported.

Dimethyl trisulfide has been detected in 3.25% fat UHT milk (age unknown) at 47.3 ng/kg (Vazquez-Landaverde *et al.*, 2006) and in freshly prepared UHT whole milk at 1.9 μg/L by Al-Attabi *et al.* (2014). As these concentrations are higher than its reported sensory threshold value in water of 8 ng/kg, it appears that dimethyl trisulfide makes a significant contribution to the sulfurous flavour of UHT whole milk. Interestingly, Al-Attabi *et al.* (2014) did not detect it in UHT skim milk.

6.1.6.1.5 *Carbonyl Sulfide and Carbon Disulfide*

Carbonyl sulfide is flavourless and is assumed not to contribute to the flavour of UHT milk. Carbon disulfide has been detected at 58.9 ng/kg and 3-6 μg/L in UHT milk (Vazquez-Landaverde *et al.*, 2006; Al-Attabi *et al.*, 2014). These values are well below the reported threshold value in milk of 100-1000 μg/L and hence would not be expected to contribute to the sulfurous flavour of freshly prepared UHT milk.

6.1.6.1.6 *Minimising Sulfurous Flavour in UHT Milk*

In the belief that the sulfurous note is the main component of flavour which influences consumer acceptance of UHT milk, several methods of reducing it have been proposed. With the exception of minimising the severity of heat treatment, these all involve additives, some of which impart unwanted flavours or cause other quality issues and some are not legal additives in some jurisdictions. Furthermore, most consumers are opposed to the use of additives.

The most common approach has been the addition of sulfhydryl blocking and oxidising agents such as of sodium or potassium iodate, sodium or potassium bromate (at 5-25 mg/kg) before or after heating (Samuelsson & Borgstrom, 1973), and organic thiosulfonates and thiosulfates at 30-500 mg/kg before heating (Ferretti, 1973). Another sulfhydryl-specific agent which has been found effective and may be acceptable to consumers is sulfydryl oxidase, a natural enzyme in milk, which catalyses the conversion of sulfydryl groups to disulfides. However, this enzyme is unstable at high temperatures and hence must be added aseptically after UHT treatment (Swaisgood, 1977). Another additive which may be readily accepted is the polyphenol epicatechin which has been shown to reduce cooked flavour and not impart a bitter flavour at an addition rate of 100-1000 mg/kg. A bitter flavour was detected on addition of 2000 mg/kg (Colahan-Sederstrom & Peterson, 2005).

An effective additive which is legal in several foods in several countries is hydrogen peroxide. Low concentrations (10 or 50 mg/kg) significantly reduce the level of sulfur volatiles in the UHT milk when added either before or after processing. For example, UHT milk with 50 mg/kg hydrogen peroxide added before or after processing contained no detectable hydrogen sulfide or methanethiol by the second day after manufacture (Al-Attabi, 2009).

6.1.6.2 Monocarbonyl Compounds

Short-chain aldehydes and methyl ketones are important volatile components in UHT milk. They have attracted most attention because of their association with the stale flavour that develops during storage of UHT milk. However, they are present in freshly prepared UHT milk, with the methyl ketones, which are generated during the heat treatment, being in higher concentration than the aldehydes.

Methyl ketones are produced by the decarboxylation of β-keto fatty acids. Triacylglycerols containing one β-keto acid and two fatty acid moieties account for 0.045% of the lipid fraction of milk (Crossley *et al.,* 1962). Methyl ketones may also originate via thermal β-oxidation of liberated saturated fatty acids, followed by decarboxylation (Contarini *et al.,* 2002).

Seven methyl ketones, 2-pentanone, 2-hexananone, 2-heptanone, 2-octanone, 2-nonanone, 2-decanone and 2-dodecanone, are present in UHT milk. Of these, 2-heptanone and 2-nonanone are in the highest concentration (Perkins *et al.,* 2005). It has been suggested that these two compounds are the main contributors to the overall flavour of UHT milk (Badings *et al.,* 1981; Moio *et al.,* 1994).

The concentration of methyl ketones increases with the intensity of heat treatment and has therefore been suggested as a chemical heat index (Langler & Day, 1964; Contarini & Povolo, 2002). The suggestion was based on the fact that the concentration of one methyl ketone, 2-heptanone, could be used to distinguish between pasteurised, UHT and in-bottle sterilised milk. Furthermore, indirectly processed UHT milk has much higher concentrations of methyl ketones than directly processed milk (Badings *et al.,* 1981). Perkins *et al.* (2005) found that both individual and total odd-carbon-numbered (C5, C7, C9 and C11) methyl ketones correlated well with other heat indices: a strong positive correlation with furosine and lactulose and a strong negative correlation with undenatured whey proteins.

Another carbonyl compound which contributes to the rich, heated flavour of UHT milk is diacetyl which is a major contributor to the flavour of butter. Other carbonyl compounds which have been associated with this flavour are lactones, acetophenone, 4-hydroxy-3-methoxybenzaldehyde (vanillin), maltol and benzaldehyde (Anderson & Öste, 1995).

6.1.7 Chemical Heat Indices

The preceding sections outlined several changes that occur as a result of the heat treatment of milk, particularly under sterilisation conditions. It has therefore been of interest to many researchers and dairy industry personnel to have convenient methods of estimating the intensity of the heat treatment in terms of the effects on chemical compounds (Pellegrino *et al.,* 1995b). This should not be confused with measures of heat treatment in terms of its effect on bacteria as the kinetics of destruction of bacteria differ considerably from those of chemical changes.

The chemical changes of interest can be divided into those involving destruction of milk constituents and those involving production of new compounds (Pellegrino *et al.,* 1995b). Examples of the former are destruction of thiamine, the basis of the parameter C^*, and denaturation of whey proteins, whereas examples of the latter include formation of lactulose, lactulosyl lysine (often determined as furosine which is not naturally in milk), 5-hydroxymethylfurfural (HMF), brown pigments, volatile sulfur compounds, methyl ketones, galactose and advanced glycation end products such as carboxymethyl lysine (CML). To be useful as a practical index of the severity of heat treatment, a compound representing a chemical change needs to fulfil criteria such as the following: the kinetics of the change should be such that differences in high-temperature treatment can be readily determined, it should be easily and accurately measured and it should not change during storage of the product. Its presence in unheated milk is also a consideration.

Table 6.7 summarises the characteristics of various possible chemical heat indices. From the table it is apparent that the most useful index is **lactulose** (see Sections 6.1.4.2 and 11.2.12) as most of the others suffer from changing during storage and/or difficulty of measurement. Of the others, **furosine, HMF** and **whey protein denaturation** have been used most but as the table indicates, care must be exercised in interpreting data on these indices.

Furosine is a widely used chemical heat index. When used on freshly processed UHT milk it gives a reliable measure of lactulosyl lysine, the first stable product of the Maillard reaction, and hence of the heat load applied. However, as indicated in Section 6.1.4.1, it increases on storage, the increase being considerable at high storage temperatures. For this reason it is unsuitable for all but freshly processed milk.

HMF has often been measured as an index of heat treatment in UHT milk. Initially it was popular because it could be readily estimated spectrophotometrically due its strong absorbance at 280 nm or by a colourimetric assay with TBA. However, these methods lack accuracy and have been mostly replaced by HPLC methods (Morales *et al.*, 1992) (see Section 11.2.11). As discussed in Section 6.1.4.1, there are two measures of HMF, "free" and "total". The free HMF level is not very sensitive to heat treatment while the total HMF levels increase with the intensity of heat treatment. During storage, HMF is subject to both formation and degradation and so it is not a reliable index of heat treatment in stored products. When determining HMF in UHT milk it must be borne in mind that raw and pasteurised milk yield a substantial amount of total HMF on acid treatment, and this must be taken into account when estimating the total HMF attributable to the heat treatment (Fink & Kessler, 1986b).

The amount of **denaturation of whey proteins,** indirectly assessed by measuring the residual undenatured whey proteins, is another widely used index of heat treatment which gives processors a useful guide to the effect of their heat treatement. However, care needs to be taken in interpreting the denaturation data as it is possible for a UHT milk to have a very high level of denaturation but to be inadequately heated with respect to its bactericidal effect. This can happen if the pre-heating conditions (which have little effect on bacterial spores but considerable effect on whey proteins) are severe but the sterilisation conditions are mild.

Oliveira *et al.* (2015) assessed the use of the whey protein nitrogen index (WPNI), which is an index of undenatured whey protein, as a heat treatment indicator for UHT milk in Brazil. This test was developed by the American Dairy Products Institute (2009) and is generally used for grading milk powders into low-, medium- and high-heat powders. They found that the average WPNI of commercial samples of UHT milk analysed ($n = 60$) was 2.39 ± 1.41 mg WPN/mL, with 90% being classified as medium-heat (1.51-5.99 mg WPN/mL) and 10% as high-heat (<1.5 mg WPN/mL). The authors commented that the variability in the results indicated a lack of standardization of commercial UHT heat treatment conditions, and suggested that some of the products may develop undesirable characteristics during storage as a result of their heat treatments. This would apply particularly to the products with the highest WPNI which could

Table 6.7 Assessment of various chemical heat indices for UHT milk.

Compounds measured	Chemical change	Ease of measurement[1]	Kinetics suitability[2]	Presence in unheated milk	Change during storage[3]	Comments	Reference
Undenatured β-Lg & α-La	Denaturation of whey proteins	4 4	2 4	Not usually known if assessing a commercial milk. Measuring % reduction usually not possible.	2	β-Lg is commonly used but can be misleading if intense pre-heating is used, resulting in complete denaturation in that step. α-La is a better index as it reaches 100% only under extreme heating conditions	Lyster (1970); Dannenberg & Kessler (1988) Tran *et al.* (2008)
Trypophan	Denaturation of whey proteins	5	2	As above	2	Measures trypophan in undenatured whey proteins by fluorescence of pH 4.6 filtrate	Birlouez Aragon *et al.* (1998); Elliott *et al.* (2003)
Thiamine	Destruction of thiamine	2 Colorimetric method easy but lacks accuracy)	5	Not usually known if assessing commercial milk. Measuring % reduction usually not possible	Thiamine decreases with increased storage temperature; little loss at 20 °C	Calculated from known kinetics to determine C*. Seldom measured	Fink & Kessler (1985)
Lactulose	Epimerisation of lactose	4	5	Negligible	1	Possibly the best chemical heat index May be combined with furosine	Elliott *et al.* (2005); Pellegrino *et al.* (1995a,b)
Furosine, lactulosyl lysine	Lactosylation/ Maillard reaction	3	5	Low concentration	5	A good index for freshly produced product. Good for estimating 'lysine blockage'	Erbersdobler & Somoza (2007); Henle *et al.* (1991)
Available lysine	Lactosylation/ Maillard reaction	3	5	Can be determined or estimated from proteins present	5	Several methods available, including dye bindng and via furosine	Mehta & Deeth (2015); Fink & Kessler (1986a)

Total HMF	Maillard reaction	4	4	Negligible	2	A reasonable index for freshly produced product. Both formed and destroyed during storage.	Morales *et al.* (2000); Elliott *et al.* (2003)
Brown pigment	Maillard reaction	5	4	Negligible	5	Suitable only for severely heated product immediately after production	Morales *et al.* (1997)
CML	Maillard reaction	2	3	Negligible	5	Most applicable to severely heated products products. Can be increased by ingredients such as phosphates, citrate, sucrose, ascorbic acid & cocoa	Hewedy *et al.* (1994); Erbersdobler & Somoza (2007); Hegele *et al.* (2008a,b); Drusch *et al.* (2006)
Methyl ketones	Thermal degradation of fatty acids	3	4	Low concentration	5	Has been suggested for freshly processed milk but not used in practice	Contarini & Povolo (2002); Perkins *et al.* (2003)
Volatile sulphur compounds	Thermal degradation of sulphur amino acids	1	4	Negligible	5	Suitable as heat index immediately after processing only	Andersson. & Öste. (1992b); Vazquez-Landaverde *et al.* (2006)
Monosaccharides (galactose)	Thermal degradation of lactose	4	3	Low concentration	2	Suitable as heat index immediately after processing only	Olano *et al.* (1989); Troyano *et al.* (1996)

[1] 1 = very difficult, 5 = easy;
[2] 1 = not suitable, 5 = very suitable;
[3] 1 = little or no change, 5 = considerable change.

indicate considerable residual plasmin activity and hence a high risk of bitterness and/or age gelation developing during storage (see Sections 7.1.3.3 and 7.22).

UHT processing, particularly indirect processing, of milk results in almost complete denaturation of the main whey protein, β-Lg, and hence the amount of undenatured β-Lg is not very discriminatory with regard to the effect of the heat treatment. However, the effect of heat on α-La and the amount of undenatured α-La has been suggested as a much better indicator of heat treatment than the total amount of whey proteins (Tran *et al.*, 2008).

Some authors have suggested using combinations of indices such as furosine and lactulose (Corzo *et al.*, 1994; Pellegrino *et al.*, 1995a,b). These two correlate well in freshly manufactured milks; Pellegrino *et al.* (1995a) reported a correlation coefficient of 0.99 and a regression equation of $y = 3.02\,x - 41.51$ where lactulose (y) was expressed as mg/L and furosine (x) as mg/100 g protein. Furosine is more affected by drying, storage and intense heating than lactulose and consequently the use of both indices has been suggested for detecting overheating and the presence of reconstituted milk powder in sterilised milk. Erbersdobler and Somoza (2007) suggested that for assessing the intensity of heat treatment from a nutritional perspective, furosine should be linked with other compounds such as CML, pyrraline and pentosidine as these are better markers of severe heat treatment.

6.2 Physical Changes

6.2.1 Heat Stability

Heat stability can be defined as the ability of milk to withstand high processing temperatures without visible coagulation or gelation (Singh, 2004). Arguably it is the most important property of milk destined to be sterilised. Inherently, cow's milk shows good heat stability, and it can take more than 20 min heating at 140 °C to initiate coagulation, although heat stability may deteriorate quickly if the milk is not stored correctly. This high heat stability is due to the loose, ill-defined three-dimensional structure of its principal proteins, the caseins. This is despite the fact that the whey protein fraction is heat labile and starts to irreversibly denature at about 70 °C. The interactions that take place between denatured whey protein and casein during the heating process are also of great importance (see Section 6.1.3.1).

6.2.1.1 Measurement of Heat Stability

There are many methods available for measuring heat stability. The most common method involves the measurement of the heat coagulation time (HCT). HCT is defined as the time that elapses between placing a sample of milk in an oil bath at a definite temperature and the onset of visible coagulation. The heating temperature is usually 140 °C for unconcentrated milk or 120 °C for concentrated milk (O'Connell & Fox, 2003). The usual procedure is to measure HCT for the sample after it has been adjusted to a number of different pH values. Typical relationships between HCT and adjusted pH are shown in Figure 6.7.

Early research led to the distinction between Type A and Type B milk samples. Rose (1961) performed experiments where the pH of milk samples was adjusted to a range of

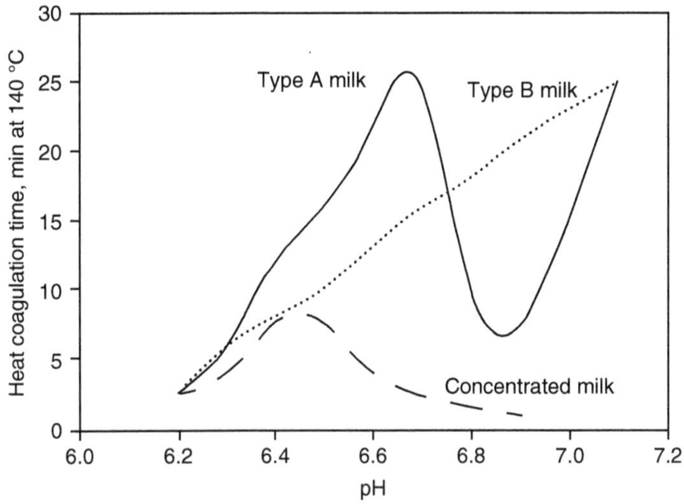

Figure 6.7 Relationship between HCT and adjusted pH of milk. (Source: Fox & McSweeney, 2003. Reproduced with permission of John Wiley & Sons.)

pH values (typically between 6.4 and 7.0), prior to determination of their HCT. This was to observe the relevance of starting pH to heat stability. It was found that for the majority of milk samples (Type A), the heat stability passes through a maximum and a minimum as the pH is increased from 6.4 to 7.0; for a minority of samples (Type B), there is a steady increase in heat stability as pH increases. Type A milks predominate in most countries (with the exception of Japan), accounting for 70% of Scottish, 80% of Canadian and ~100% of Irish and Australian milks of individual cows (Horne & Muir, 1990). At a pH well below the maximum, milk coagulates rapidly. This is attributed to the fact that the low pH (and high calcium ion activity) decreases electrostatic repulsions between the proteins. In addition, relatively high amounts of whey proteins associated with the micelles at low pH may promote aggregation of casein particles through cross-linking of whey proteins bound to different micelles. The occurrence of a maximum (at ~pH 6.7) in the HCT–pH profile of normal milk is considered to be due to the greater stability of whey protein-coated micelles, as the formation of a complex between κ-casein and β-lactoglobulin on the surface of the casein micelles alters the steric and electrostatic interactions and prevents the dissociation of micellar κ-casein. At pH values above 6.7, the stability decreases due to the dissociation of micellar κ-casein, thus reducing the stabilizing effect of this protein. The κ-casein-depleted micelles are sensitive to calcium ions. Therefore, the minimum in the HCT–pH profile is a result of coagulation of κ-casein-depleted micelles, in the presence of calcium ions. At higher pH, although dissociation of micellar κ-casein increases, the HCT increases due to the increase in protein charge and low calcium ion activity. It is also possible that the dissociated κ-casein may reassociate during extended heating because of a heat-induced decrease in pH (Singh, 2004). Some more observations relating to casein dissociation at higher pH values are discussed in Sections 6.1.3.1 and 6.1.3.4.

Many studies have investigated how the HCT–pH coagulation profile is influenced by factors such as milk composition (proteins, lactose, salts, urea), pre-heating conditions, concentration and other parameters; these have been reviewed by Singh (2004). Good earlier reviews are those of Rose (1962), Fox and Morrissey (1977) and Horne and Muir (1990). Sievanen *et al.* (2008) compared such profiles for normal milk and concentrated milk with added calcium chloride and found that it reduced the heat coagulation time.

From an analytical and practical perspective, there is an element of subjectivity in determining the point at which coagulation first takes place. Consequently, the HCT test is not straightforward to perform for determining the suitability of milk for thermal processing. However, it is claimed by regular users that it is possible to obtain reproducible results with practice. von Bockelmann and von Bockelmann (1998) reported that milk should have a HCT value in excess of 9 min (at 140 °C) for it to be suitable for UHT processing. This appears to be the only such guideline given for using this procedure, and later evidence would suggest that this procedure for establishing the suitability of milk for UHT processing should be treated with caution.

A variant on the HCT procedure is determination of the heat coagulation temperature, which is the temperature required to coagulate milk in a given time period (Rose, 1962; Horne & Muir, 1990). Other methods for determining heat stability include the ethanol test, a whitening test, sediment formation and viscosity determination (Singh, 2004). These tests are discussed in more detail in Section 11.2.25.

However, despite all the research work, problems related to poor heat stability of milk are still encountered in commercial processes. In many countries, the vast majority of good quality raw milk used for producing UHT white milk shows good heat stability and withstands commercial sterilisation procedures without causing excessive fouling of heat exchangers or producing too much sediment which would be noticed by the consumer.

6.2.1.2 Is HCT a Good Predictor of Heat Stability in UHT Treatment?

Singh (2004) stated that the heat coagulation time (heat stability) often correlates very poorly with the stability of milk to commercial sterilisation. He also pointed out that from an industry point of view, the use of a pilot- scale or a laboratory-scale steriliser which simulates sterilisation conditions used in practice provides more reliable results and prediction of behaviour of milk in commercial plants, compared with laboratory-based subjective HCT tests. Some reasons for this poor relationship are discussed below.

Milk can show very good heat stability and can take over 20 min to coagulate at 140 °C in HCT studies. Even poor quality milk may take up to 5 min to coagulate. However, during UHT processing, milk only reaches a temperature of ~140 °C for a few seconds. In the case of direct processes, milk reaches 140 °C very quickly and even in most indirect processes, it takes less than 3 min to attain that temperature. There are differences in the heating and cooling rates for different UHT plants, as discussed by Tran *et al.* (2008). Thus the heating profiles found in the HCT test are totally different to those found for UHT processes. Some UHT plants also incorporate a "stabilisation section" (see Section 5.2.1.3). Thus any HCT studies on milk should take into account pre-heating conditions in the range of 75 to 95 °C for up to 120 s.

During in-container sterilisation, temperatures rarely exceed 120 °C and relatively few HCT studies have been carried out on unconcentrated milk at this temperature. In fact, one might speculate that HCT times would be extremely long at 120 °C for unconcentrated milk.

Most investigators interpret the mechanisms of heat stability from measurements of the properties of milk at ambient temperature. Thus, important determinants such as pH, ionic calcium and conditions on the surface of the micelle at sterilisation temperatures are little considered.

Although it was recognised some time ago that pH decreases as temperature increases, this factor has been hardly considered in terms of explaining the mechanisms. For example, Pyne and McHenry (1955) found that at the point of coagulation, the pH of milk (heated at 130 °C) decreased by approximately 0.7 units. Fox (1981) reported that pH decrease is the most important single factor which leads to the coagulation of milk during heating. Indeed, if the pH of milk is readjusted periodically to its original value, it may be heated at 140 °C for at least 3 hours without coagulating (Fox, 1981). It was also concluded that different factors control the heat coagulation process at different initial pH values.

Therefore, in our opinion, heat stability will be better understood when important variables can be measured at high temperatures and a better understanding is gained about how pH and ionic calcium change with temperature and how this affects conditions on the surface of the micelle, and the dissociation of soluble caseins from the micelle. It can be argued that factors which minimize the pH changes caused when the temperature of milk increases will have a beneficial effect (a temperature buffering effect) on heat stability. Salaün *et al.* (2005) wrote an excellent review on the buffering capacity of dairy products and mentioned that addition of citrate and phosphate salts to milk induces quantitative changes in the buffering capacity. This is important, as milk with a higher buffering capacity would be subject to a smaller decrease in pH as its temperature is increased, which may be contributory factor to its heat stability at high temperature. Further investigation of factors affecting the buffering capacity over the pH range 6.8 to 5.6 is worthy of further investigation, as this is the change that UHT milk undergoes when it is heated from 20 to 140 °C. They also reported heat treatment (120 °C for 10 min) increases the buffering capacity and shifts the maximum peak from pH 5.0–5.2 for untreated milk to pH 4.3–4.5 for heated milk. It is not clear how this is helpful in terms of explaning heat stability, but it may be useful to follow up for yogurt and other fermented dairy products as these changes fall within their pH range.

Another compounding influence is the biological variability of milk, so those contributory factors influencing heat stability may be different for every batch of milk which is processed. To illustrate this, Chen *et al.* (2015) measured the heat stability of 25 samples of bulk milk, subjected to UHT processing and in-container sterilisation, collected from one farm throughout the year. The variation in milk composition is reported in Table 1.5 and results for the physical properties of these milk samples is shown in Table 6.8. Variations in heat stability of milk taken at different times of the year have been reported by Grandison (1988), Grimley *et al.* (2009) and Chen *et al.* (2015), but no single compositional factor has been identified as being responsible.

In an ideal world, it would be desirable to be able to determine whether milk is likely to be stable to any particular thermal process from a knowledge of its composition. Some reports which discuss differences in heat stability of milk samples from individual cows and bulk milk samples are those of White and Davies (1958b), Burton (1988), Chavez *et al.* (2004), Grandison (1988) and Grimley *et al.* (2009). White and Davies (1958b) reported that HCT for bulk milks ranged from 17.2 to 59 min at 130 °C, whereas from individual cows it ranged from 0.6 to 86.2 min. They also pointed out that, in

Table 6.8 Some physical properties of bovine milk. (Source: Chen *et al.*, 2015. Reproduced with permission of Elsevier.)

Physical properties	Mean + SD	Range	Seasonal variation
Density (g/cm^3)	1.028 ± 0.01	1.026–1.031	NS
Casein micelle size (d.nm)	163 ± 16	132–202	NS
Viscosity (cp)	1.93 ± 0.21	1.52–2.36	NS
FPD (m°C)	523 ± 3	514–530	W > SP and SM
Foaming ability (s)	88 ± 47	24–205	SP > SM, A and W
RCT (min)	19 ± 3.1	12–24	SP > A
Ethanol stability %	93 ± 5	84–100	SP > A
Sediment (%)	0.09 ± 0.02	0.03–0.13	SM and A > W
Buffering capacity	0.84 ± 002	0.78–0.88	NS

general, HCT was not related to the concentration of ionic calcium, except when the calcium phosphate content of the caseinate complex was very low, in which case it was inversely related. It has been suggested that the pH maximum often corresponds to the natural pH of the milk (Horne and Muir, 1990), although this has not been specifically investigated.

In UHT treatment, milk having a poor heat stability will give rise to heat exchanger fouling and sediment formation and should not be used. Some suggestions for avoiding milk which may be unsuitable for sterilisation are given in Section 6.2.1.3. Overall, good quality raw milk will have a high heat stability, but this may not be the case when it is used in formulated milk drinks, arsing from the use of a wide range of components, which will modify the pH and Ca^{2+}. The vast majority of the literature covering heat stability of milk deals with measuring HCT. Only a small fraction deals specifically with UHT processes, making it a challenge to suggest practical solutions to improve milk drink formulations. We hope that the information discussed here and in the next section will provide some insights into how to achieve this.

6.2.1.3 Stability to UHT Processing and Some Comparisons with In-Container Sterilisation

Usually good quality cow's milk shows few problems related to poor heat stability when UHT processed. Chen *et al.* (2015) measured sediment formation as an indicator of heat stability for 25 bulk milk samples collected from one supplier over a period of 1 year. Some of their physical properties are given in Table 6.8. Their ethanol stability ranged from 84 to 100% and there were considerable variations in rennet coagulation time (RCT). In all cases, the heat stability was good as measured sediment in UHT milk ranged from 0.10 to 0.29% (dry weight basis).

Lewis *et al.* (2011) determined sediment in UHT cow's milk to be 0.29%, which indicates that heat stability is good. Adding 1.8 mM calcium chloride had little effect on sediment but adding 2.7 mM brought about a drastic increase. Addition of 4.5 mM calcium chloride resulted in a sediment level of 4.8%. Sediment was reduced almost back to its original value when the ionic calcium concentration was restored to its original value (by addition of NaOH). Changing cows from indoor feeding to grazing outdoors was found to reduce the amount of sediment in milk subjected to in-container sterilisation (Grimley *at al.,* 2009). These studies suggest that sediment arising from UHT treatment of good quality cow's milk is below 0.5% (on-Nom *et al.,* 2012), whereas in goat's milk it is much higher, at levels found in cow's milk when adding 4.5 mM of calcium chloride. Boumpa *et al.,* (2008) found that sediment ranged from 3.5 to 6.5% (dwb) for four batches of UHT goat's milk, but that it was possible to reduce this to below 0.5% by addition of DSHP or TSC. This sediment was usually found to contain less than 5% minerals on a dry weight basis and a fat/protein ratio in the range 1.43:1 to 1.67:1. Fat is incorporated into the sediment as a result of homogenisation, which coats the fat globules with casein. On-Nom *et al.* (2012) analysed sediment from in-container sterilised cow's milk with added calcium chloride and found both casein and whey protein fractions to be present.

Heat stability problems related to in-container sterilisation of normal (unconcentrated) milk are rarely reported, but for milk concentrates this is a major concern, which is further considered in Section 11.2.25. Tsioulpas *et al.* (2010) sterilised normal cow's milk which contained TSC, DSHP, SHMP, DHSP and $CaCl_2$ at 121 °C for 15 min and results are described in Section 6.1.1.1. Milk containing 3.2 mM SHMP or more than 4.5 mM $CaCl_2$ coagulated on sterilisation. It was interesting that DHSP reduced pH slightly but had no adverse effect on sediment and resulted in less browning during storage, because it resulted in a lower pH.

Grimley *et al.* (2009) examined changes in heat stability of milk from five farms over the Spring flush period in the UK, when cows move from indoor to pasture feeding. Sediment formation ranged from 0.41% to 0.77% when milk was sterilised at 120 °C for 15 min; heat stability was from 21 to 26 min when measured by the HCT method at 140 °C. No correlation between the methods was found. Although there were clear variations in milk composition over this period, there was no evidence that milk became unstable to in-container sterilisation as a result of these changes. Even though the levels of sediment showed some variation, the overall heat stability of all samples was acceptable. On-Nom *et al.* (2012) heated reconstituted skim milk powder (9% total solids) containing up to 25 mM $CaCl_2$ from 60 to 120 °C for different periods of time. At 60 °C, milk could accommodate 16.2 mM $CaCl_2$ without coagulating whereas at 120 °C, milk containing only 3.6 mM of calcium chloride coagulated. One interesting observation was that at all temperatures, poor heat stability was manifested first by sediment formation, which was then followed by coagulation at higher calcium chloride additions. Milk samples were also dialysed during their respective heat treatments to recover the soluble phase and to measure pH and ionic calcium. No coagulation was observed if the measured Ca^{2+} concentrations were <0.5 mM and pH was >6.3, measured at their respective coagulation temperatures.

Milk was found to become unstable to UHT processing at lower calcium additions compared to in-container sterilisation (Omoarukhe *et al.,* 2010). The range of addition, as calcium gluconate and calcium lactate, used was 0-10 mM. Nian *et al.* (2012) evaluated the heat stability of milk from individual cows and found that milk with the highest Ca^{2+} produced the least sediment on in-container sterilisation ($R^2 = 0.41$). Ezeh and Lewis (2011) compared the heat stability of milk which had been reduced to different pH values and then restored. Sediment ranged from 0.1% up to 0.19% for UHT treatment and from 0.1% to 0.23% for in-container sterilisation.

In summary, milk could be adjusted to a pH as low as 5.95 and then restored without adversely affecting its heat stability. During this pH reduction and restoration, considerable amounts of calcium and phosphorus move out of and back into the micelle without adversely affecting heat stability. In contrast, rennet coagulation time was considerably reduced by such readjustments and ionic calcium was higher.

There has been little reported research on comparing the heat stability of the same milk subjected to UHT and in-container sterilisation. Two recent studies highlight differences in heat stability when the same milk samples were subjected to UHT and in-container sterilisation. Chen *et al.* (2012) showed that goat's milk produced significantly more sediment following UHT processing compared to in-container sterilisation. When ionic calcium was increased slightly by adding 2 mM $CaCl_2$, sediment increased drastically after UHT treatment, but by much less following in-container sterilisation. Addition of up to 12.8 mM DSHP or TSC resulted in a progressive increase in sediment for in-container sterilisation. In contrast, for UHT milk, adding 6.4 mM of these salts reduced sediment formation but sediment increased on addition of 12.8 mM of these salts. Thus it is also possible to add too much DSHP and TSC, as sediment formation went through a minimum with increasing additions of these compounds. Results for casein micelle size showed that an increased amount of sediment was accompanied by an increased casein micelle size. Some sediment levels are shown in Figure 6.8.

Similar observations were found for cow's milk (Chen *et al.*, 2014). In a comparison study, 25 batches of bulk cow's milk showed good heat stability when subjected to UHT treatment and in-container sterilisation. The average sediment level was 0.19% for UHT treatment (range 0.1 to 0.29%) and 0.24% for in-container sterilisation (range 0.02 to 0.56%). Adding 10 mM DSHP or TSC increased sediment much more for in-container sterilisation than for UHT treatment. Thus, trends were similar to those found for goat's milk, although sediment levels were lower in the control cow's milk. Adding only 2 mM calcium chloride increased sediment significantly for UHT treatment but not for in-container sterilisation. However, although variations were found in amounts of sediment, the overall heat stability of these milk samples was consistently high and none of

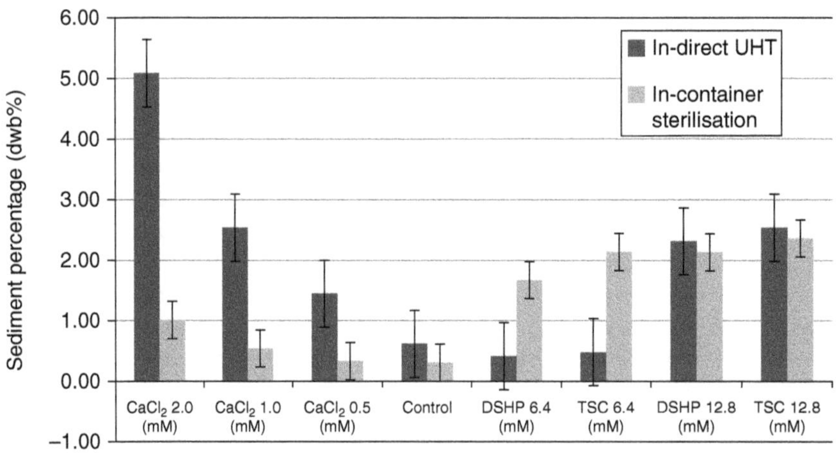

Figure 6.8 Sediment levels in goat's milk after sterilisation and UHT treatments. (Source: Chen *et al.*, 2012. Reproduced with permission of Elsevier.)

them produced sediment levels that would be detectable to consumers. Some sediment values are shown in Figure 6.9. Figure 6.10 shows the corresponding casein micelle size measurements for these samples and supports the premise that sediment results from aggregation of the casein micelles. It is of further interest that there was not a good

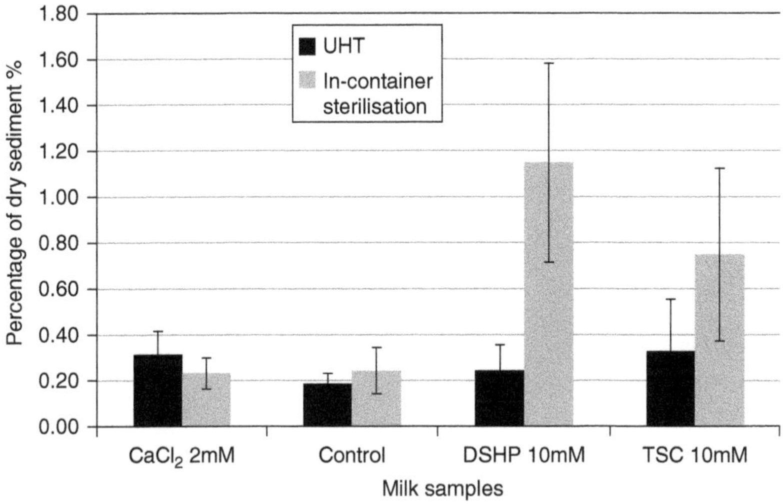

Figure 6.9 Sediment levels in cow's milk after sterilisation and UHT treatments. (Source: Chen *et al.*, 2015. Reproduced with permission of Elsevier.)

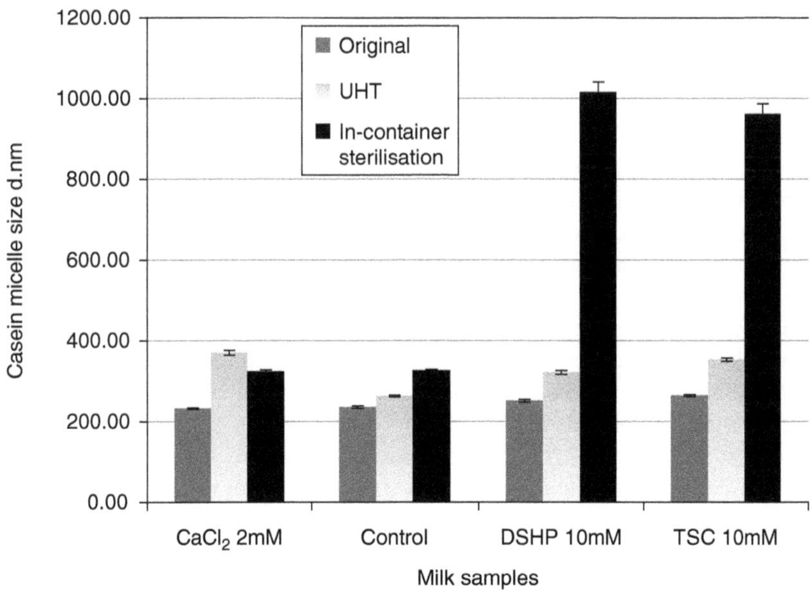

Figure 6.10 Casein micelle size in milk with added calcium chloride, disodium hydrogen phosphate and trisodium citrate after UHT and in-container sterilisation. (Source: Chen *et al.*, 2015. Reproduced with permission of Elsevier.)

correlation when sediment found after in-container sterilisation was plotted against sediment found after UHT treatment (see Figure 6.11).

In conclusion, it has been shown that when pH is adjusted in a more acidic direction, the product is more likely to be more unstable to UHT processing, but when adjusted in an alkaline direction it becomes more unstable to in-container sterilisation. Why this is happening is not yet clear, but is worthy of further investigation. There was also a poor correlation between that amounts of sediment produced by the two sterilisation procedures. It is not clear why there are these differences and why relatively small reductions in ionic calcium make milk more susceptible to sediment formation during in-container sterilisation compared to UHT treatment.

Another difference is that in-container sterilisation is essentially a static process (although for more viscous products the containers may be agitated), which will encourage coagulation and the formation of a gel. In contrast, during UHT processing, milk is subjected to more shear, which might disrupt gelation and result in sediment formation. Some of these findings for in-container sterilisation are in agreement with HCT–pH relationships for Type A milk samples in that narrow pH range between their minimum and maximum values. The speculated pH and ionic calcium profiles for these two sterilisation procedures are shown in Figure 6.12.

6.2.1.4 Is Ethanol Stability a Good Predictor of Heat Stability in UHT Treatment?

There is a sound argument in favour of avoiding using milk for sterilisation processes which has poor heat stability, but for the reasons discussed, assessing heat stability is not straightforward in practice and the results may then not be reliable. Since measuring HCT is not simple and it may not be a reliable guide (see Section 6.2.1.2), one suggestion is to monitor the ethanol stability of the milk. Horne and co-workers have investigated the factors affecting ethanol stability, starting in the 1980s and summarised by Horne (2003). Several workers have observed that ethanol stability is influenced by ionic calcium, as reviewed by Lewis (2010). The use of ethanol stability for this purpose was discussed by Shew (1981) who recommended that milk should be stable in 74% ethanol to be suitable for UHT processing. As far as the authors are aware, there are no similar recommendations for in-container sterilisation.

This suggests that ethanol stability is a reliable indicator of stability to UHT processing conditions, so a key question becomes is this the case? Opinion is divided about this; some consider ethanol stability to be a surrogate for pH. Some further thoughts and observations follow.

Results from pilot plant and laboratory experiments on cow's milk and goat's milk show that reducing ionic calcium is beneficial in terms of reducing fouling of heat exchangers and sediment formation (Boumpa *et al.*, 2008; Prakash *et al.*, 2007). Also, reducing ionic calcium increases ethanol stability and there is a good correlations between ethanol stability and Ca^{2+}. Therefore, in situations where sediment formation or fouling is a problem, the following proposals have been offered (Lewis & Deeth, 2009). pH, ethanol stability and ionic calcium should be routinely monitored in raw milk to establish their effects on sediment and fouling-related problems in UHT milk. Over time, this should provide data to assess, understand and eventually reduce problems arising from poor heat stability. Chavez *et al.* (2004) grouped bulk milk samples into those having alcohol stability values above and below 72%. Over 30% of bulk milk samples had ethanol stabilities less than 72%; these had Ca^{2+} values between 1.84 and

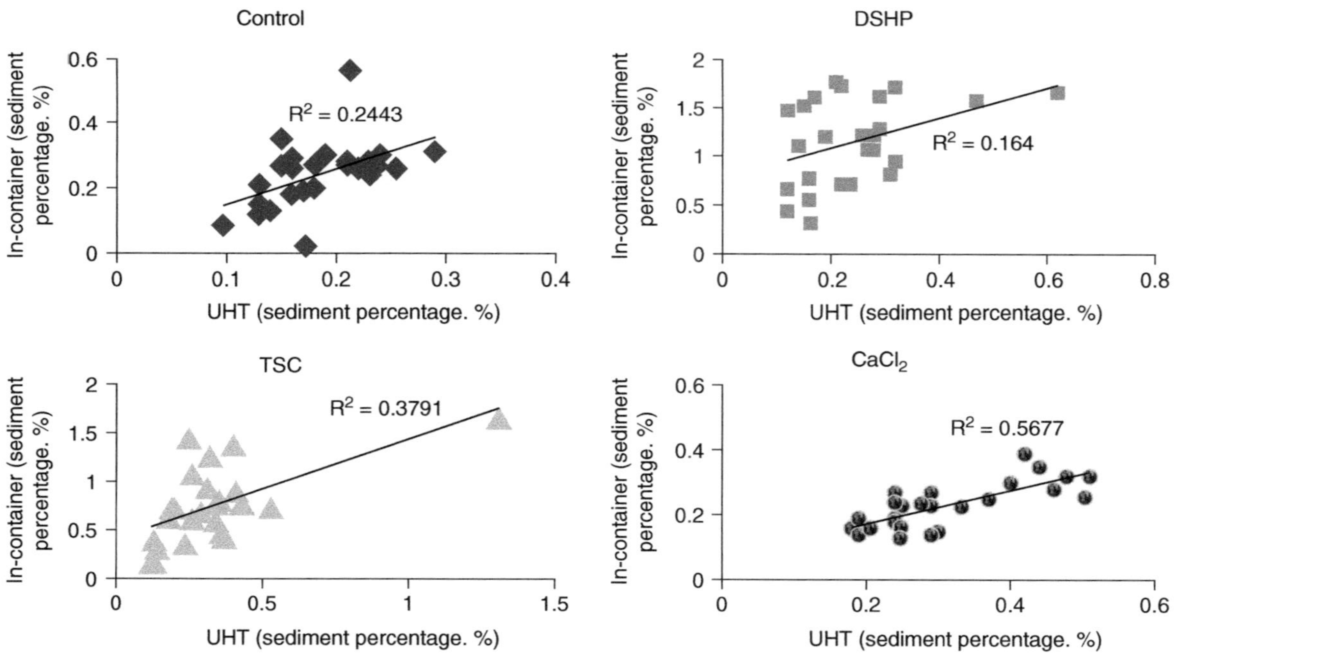

Figure 6.11 Comparison of sediment levels produced by UHT and in-container sterilisation of the same milk samples. (Source: Chen, 2013. Reproduced with permission.)

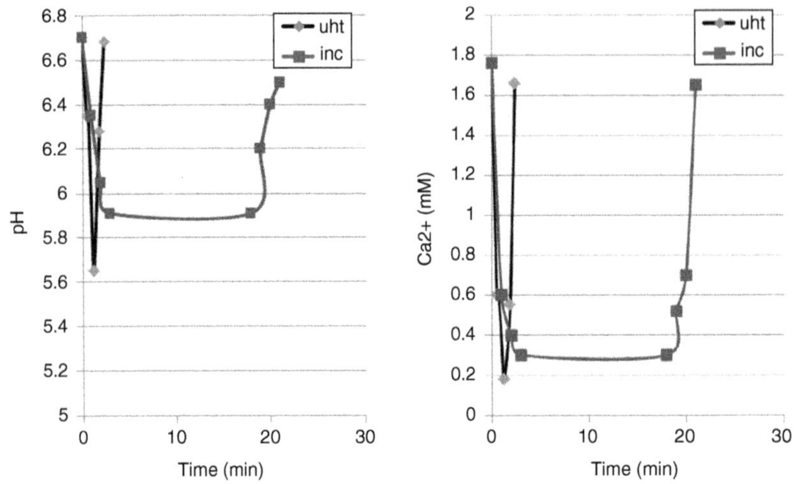

Figure 6.12 Speculated pH and ionc calcium profiles for UHT and in-container sterilisation (inc).

2.59 mM and pH values in the narrow range of 6.67 to 6.69. Those showing ethanol stability greater than 72% had Ca^{2+} from 1.66 mM to 2.04 mM and pH from 6.70 to 6.72. The mean heat coagulation times were 23.8 min and 19.9 min for samples with ethanol stabilities above and below 72%, respectively. Unfortunately, none of these milks were subjected to UHT treatment

There are two main reasons why milk may have a low ethanol stability (<74%). The first is due to poor microbial quality which is accompanied by a fall in pH and the second is a salt imbalance (Horne, 2003). The former situation is likely to arise with milk of poor hygienic quality or milk stored with poor refrigeration. As raw milk bacteriological quality deteriorates, its pH is usually reduced, which in turn increases ionic calcium and reduces ethanol stability. It is important to establish whether the microbial count is high and if so to solve the problem by improving raw milk quality. However, as Horne (2003) cautioned, milk having a low ethanol stability may just have a salt imbalance. For example, any factors which reduce the negative charge, such as H^+, Ca^{2+}, Mg^{2+}, Na^+ and K^+ ions, as well as changes in the proportions of different casein fractions in the micelle, may reduce ethanol stability.

For formulated products, such as flavoured milk, reconstituted milk or ice cream mix, there is also the opportunity to manipulate composition to improve heat stability. Measuring alcohol stability might be useful for these products. Also, with reconstituted products, water quality may also influence heat stability, especially with regard to the amount of calcium and magnesium which may be present and this could also be investigated by measuring alcohol stability.This is discussed further in Section 11.2.25.1.

6.2.2 Fouling

6.2.2.1 Introduction

When milk is heated, aggregation reactions may take place which give rise to deposit formation on heat exchanger surfaces, which is termed fouling. Fouling reduces run times of UHT plants, particularly indirect heating plants and this has significant economic effects. If the aggregates or deposits do not attach or become dislodged, they may cause problems downstream of the holding tube or in the final product and be perceived as sediment. Swartzel (1983) linked sediment formation in aseptic packages to the severity of heat treatment and the extent of fouling. Burton (1968) supposed that sediment is similar to the matter deposited on the hot surfaces of heat exchangers. Thus situations which give rise to fouling of heat exchangers often also result in more sediment in the product. Sediment measurement has been shown to be a reliable indicator of milk heat stability (see Section 11.2.22).

Fouling may take time to manifest itself, whereas sediment may be noticeable in a product which has poor heat stability, even if it is only processed for short period of time. This might lead to a situation where excessive sediment is found in the product, even though no evidence of fouling was seen from the recorded UHT plant data.

The problem of fouling was discussed over 70 years ago, that is, before UHT processing was a commercial realisation. The fouling deposit is also referred to as milk scale or milk stone. Where fouling occurs to any extent, the deposit must be subsequently removed by the cleaning process. Surfaces must finally be disinfected or sterilised. Cleaning and disinfecting are two separate but very important procedures (see Section 5.6). In terms of optimising the UHT process for reducing water, energy and detergent use, and minimising waste, the full cycle of events should be considered from plant sterilisation to product processing to plant rinsing to cleaning and disinfecting.

Fouling occurs in most continuous heat exchangers, but particularly in evaporation and UHT sterilisation. It may result in a fall in the overall heat transfer coefficient (OHTC) and a drop in the product outlet temperature. This can have a marked influence on product safety and quality, especially on the microbiological flora surviving. Also the flow channels become narrower and in extreme cases they may become completely blocked. For tubes that are inaccessible to manual cleaning this can be disastrous. If the product flow rate is kept constant, fouling may result in an increase in pressure drop over the fouled sections. Fouling has several economic implications:

- increased capital outlay - the reduction in heat transfer efficiency means that the heat exchanger capacity is reduced. If fouling is not controlled properly an oversized heat exchanger is required for a particular heating duty;
- increased energy costs to overcome the thermal resistance of the fouling layer;
- increased cleaning costs in terms of the amounts of detergents used. This leads to an escalation of detergent costs. Effluent disposal costs are also increased. These factors will contribute to a negative environmental score; and
- reduced processing times and increased down-time.

If fouling leads to a compromise in product safety, for example, due to under-processing, the costs involved in recalling product and the adverse publicity can be high. There are also safety implications with regard to process plant operators. For plate heat exchangers the additional pressures may lead to gasket leakage or even failure, with release of hot fluids under considerable pressure. Quality implications may result from overheating of some elements of fluid (those in the thermal boundary layer) which may adversely influence colour and flavour.

If not properly removed, the fouling layer may act as a source of nutrients or "breeding ground" for micro-organisms. Although most fouling deposits associated with milk and milk products are chemical in nature, the term biofilm is becoming more widespread, to describe a film which comprises primarily microorganisms and has microbial and enzymic activity associated with it (see Section 6.2.2.8).

In general, the severity of fouling increases as the processing temperature increases, or with increasing total solids in the feed. Also, fouling is product-dependent and it is also influenced by naturally occurring variations in food composition. For example as raw milk is stored for longer, its titratable acidity increases, pH falls and Ca^{2+} increases. Direct UHT plants are capable of longer run times than indirect plants because of low fouling potential and are suited to circumstances where it is not as easy to control raw milk quality. However, longer processing times can be achieved on indirect plants by manipulating the pre-heating conditions (see Section 5.2.1.3). Some foods may foul so excessively at normal UHT temperatures that they cannot be processed; examples are cheese whey, eggs and starch-containing products such as desserts, soups and batters.

Thus the extent of fouling is influenced by processing conditions and the choice of heat exchanger, together with factors such as flow rate and product viscosity which influence the extent of turbulence. It may also be reduced by incorporation of a suitable pre-heat treatment, which is discussed below and in Section 5.2.1.3.

It would be very useful to be able to predict whether fouling is likely to be a problem for each product prior to processing. If this were possible, unsuitable raw material could be diverted to other products or processes where fouling is not likely to be encountered. For formulated milk products, such as flavoured milk drinks, or ice cream mix, there is also the opportunity to manipulate composition to reduce fouling. Also, with reconstituted products, water quality may play an important role, especially with regard to the amount of calcium and magnesium which may be present in hard water. Fouling can also be a problem when hard water has to be used for part of the process, such as in production of steam or for plant sterilisation. In such situations, boiler water needs to be treated to remove these salts, as does water which is to be used for sterilising the plant. Note that UHT plants are sterilised by circulating hot water at ~130 °C.

There has been some success in controlling water scale, using magnetic (electronic) water descaling systems, although the mechanisms are far from clear. These methods have been evaluated with milk and other products (Yoon & Lund, 1994).

Some fouling will always occur with some products. However, it is important to understand the processes taking place in order to reduce the problems caused by fouling. Overviews of fouling have been published by Grandison (1996), Bansal and Chen (2006) and Sadeghinezhad *et al.* (2013).

6.2.2.2 Terms Used in Fouling

Fouling is an accumulation of deposits on the surface of a heat exchanger. The deposits interfere with flow and they have a low thermal conductivity and there may be a considerable temperature gradient set up over the deposit, with the wall temperature being considerably higher than the bulk product temperature. In general, a number of terms have been used to describe and explain the fouling process.

Since many fouling experiments have shown two distinct periods, reference is often made to an induction period where the measured fouling parameter does not change with time and a fouling period, where the measured property does change with time. In a tubular system, the end of the induction period coincides with a Biot number of 0.05, where the Biot number is defined as the ratio of the resistance due to fouling (R_f) and the overall heat transfer coefficient (U) (see Section 6.2.2.3.3). Fryer (1989) reported that sometimes these periods are obvious and sometimes not. For example, experiments in tubes generally give obvious induction periods, whereas experiments with plate configurations gave no obvious induction periods, although fouling increases with time. Kastanas (1996) and Prakash *et al.* (2015) also found that the induction period was very short, or even absent in some fouling situations.

Fouling can be viewed as a dynamic process and the rate of fouling is a balance between deposition and erosion. Fouling may be **mass-transfer controlled,** where the rate of movement of the foulant to the wall is rate controlling (probably less temperature-dependent) more shear-dependent or **reaction rate controlled,** where fouling is more influenced by either the bulk or the wall temperature. Surface reaction is concerned with what is happening at the surface and is influenced by the condition at the surface, especially the effect of temperature on the substances responsible for fouling. If the controlling reaction for fouling takes place in the fluid bulk itself, this could be in the thermal boundary layer or the bulk of the fluid (turbulent core). Thus fouling is likely to be temperature-, concentration- and shear-dependent. In experiments with a small tubular heat exchanger, Kastanas (1996) showed that fouling manifests itself in a number of ways. The extremes are as follows:

- no change in OHTC but an increase in pressure, which in the end limits the extent of the process. This is more likely to occur in smaller tubular or plate heat exchangers; and
- a decrease in OHTC, with no accompanying substantial increase in pressure; in such cases, the rate of fall of OHTC is temperature-dependent.

Often both situations occur, for example, a period where OHTC remains constant, followed by a period where the OHTC decreases and the pressure may or may not increase.

6.2.2.3 Measurement of Fouling

When fouling occurs, the two main effects are a reduction in the cross sectional area of the flow passage and a decrease in the overall rate of heat transfer. A number of methods have been described in the literature to monitor fouling. These have been reviewed by Lewis and Heppell (2000) and Prakash *et al.* (2005). They include use of production-scale plant, pilot plant, heat flux meters and purpose-designed experimental rigs for measuring the formation of deposits on heated surfaces. Some of the methods for measuring fouling will now be reviewed.

6.2.2.3.1 Amount of Fouling Deposit Formed

To measure the amount of fouling deposit formed during the process, it is necessary to have access to the heat exchanger surface, or, where this is not possible, to be able to solubilise the deposit and determine the amount of dissolved material. One method is to weigh the total deposit produced, preferably on a dry weight basis. It is also possible to measure the amount of deposit and the deposit composition at different locations within the heat exchanger and determine how the composition of the deposit changes through the thickness of the deposit. It is interesting to note that the amount of any component in the deposit is usually a minute fraction of the total amount of that component which has passed through the heat exchanger during the processing period. Thus, observing changes in milk composition during processing is not a reliable method for measuring what may have deposited onto the surface. The thickness of the deposit can also be measured directly, using a microscope or other optical equipment - or indirectly by selecting some physical characteristic of the deposit that is different to that of the metal wall, for example, electrical conductivity or thermal conductivity. It is noteworthy that the thermal conductivity of a fouling deposit might be in the order of 0.5 $Wm^{-1}K^{-1}$, which is about 30-40 times lower than that of stainless steel.

Correlations between the thickness of the deposit (e, m) and the fouling resistance (R_f, W^{-1} m^2 K) have been established by Corrieu *et al.* (1981) as follows:

$$R_f = 1.378\,e + 1.603 \qquad (6.1)$$

Fibre optics may be used to inspect deposit at more inaccessible locations and could prove to be very useful in determining whether deposits have been removed effectively by the cleaning procedure.

6.2.2.3.2 Change in Temperature

Ideally a continuous HTST or UHT plant will operate at steady state and the temperature at any particular location will remain constant. If the heating medium conditions (temperature and flow rate) remain constant, the holding tube temperature will fall as fouling proceeds. Eventually the temperature will fall to below that which is considered to be safe, signifying the end of the run and the plant must be cleaned.

In commercial practice, it is more normal to keep the product outlet temperature constant, using a temperature controller. In this case, when fouling occurs, it is accompanied by an increased demand for steam (or hot water) in order to maintain the same rate of heat transfer. Thus the steam control valve may open more or the steam pressure will rise or the hot water set temperature will increase. Changes in these temperatures and pressures can be used to monitor whether fouling is taking place. There will also be a considerable temperature gradient across the fouled layer, and in the heating section the temperature at the wall will be considerably higher than the temperature in the bulk fluid. In theory, measuring this temperature gradient should provide an efficient way of measuring fouling, but accomplishing this presents many practical difficulties. The higher wall temperatures may also affect product quality, as mentioned earlier.

6.2.2.3.3 Change in Pressure

Pressure monitoring is extremely useful as it provides an indirect indication that the cross-sectional area of the heat exchanger flow passage may have diminished. Some typical data are shown in Figure 6.13. As fouling deposit accumulates, the flow passage becomes narrower. If the same flow rate is maintained, then the pressure drop over the fouled section will increase. Thus the extent of fouling can be determined by monitoring both the inlet pressure and the pressure drop over various sections of the heat exchanger. In some cases, processes are terminated when the pressure reaches a preset limit. Pressure build-up is likely in plate heat exchangers and at worst it may cause gaskets to fail. The more voluminous protein deposits make a major contribution to the increase of pressure (see Section 6.2.2.5).

Thus, observations of temperatures and pressures can provide useful information on the development of fouling both in pilot plant and in commercial UHT equipment. In

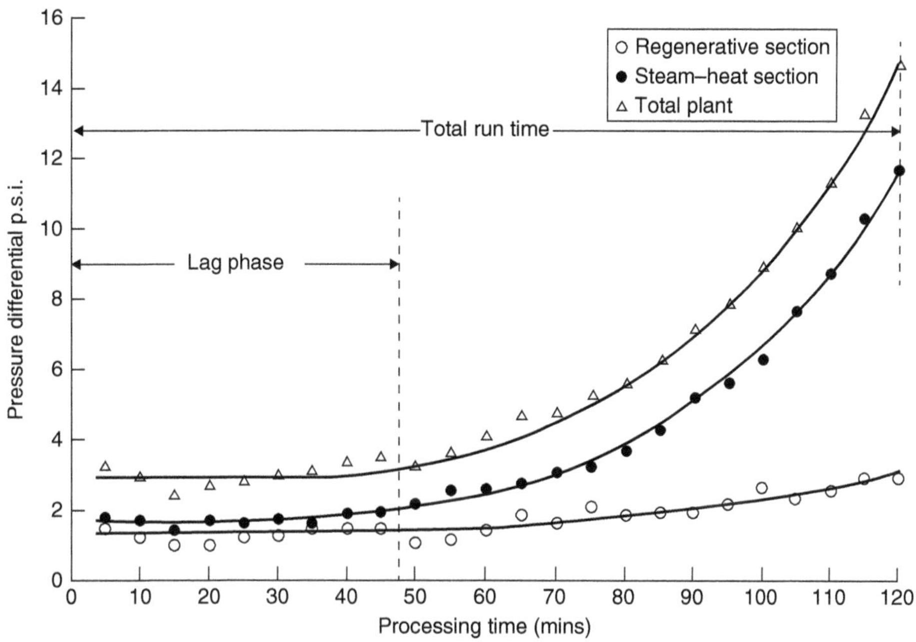

Figure 6.13 Pressure differentials across the regenerative and steam-heated sections during a typical UHT run. (Source: Grandison, 1988. Reproduced with permission of John Wiley & Sons.)

situations where the control system is geared to keep the UHT holding temperature constant, this results in an increase in steam pressure and temperature when fouling occurs.

6.2.2.3.4 Change in the Overall Heat Transfer Coefficient (OHTC)

Fouling can be monitored by a measuring a decrease in the overall heat transfer coefficient. In situations where there is no or minimal fouling, OHTC will remain constant. Some typical data are shown for fouling of goat's milk at different temperatures in Figure 6.14.

Traditionally, heat exchangers have been designed, sized and costed by adding a fouling resistance (R_t) to the initial or clean overall heat transfer coefficient. For a clean heat exchanger: $1/U = 1/h_1 + 1/h_2 + L/k$. For a fouled heat exchanger: $1/U_f = 1/U + R_f$ where R_f is termed the fouling resistance, which is equivalent to L_f/k_f where L_f is the thickness of the boundary layer, k_f is its thermal conductivity and U is the overall heat transfer coefficient.

For modelling purposes a Biot number (Bi) has been proposed, where $Bi = (R_f U)$. See Figure 6.15 (Fryer, 1989).

Thus, when a fouling deposit forms, there is a reduction in the overall heat transfer coefficient. This can be used to indicate when the fouling resistance becomes the limiting resistance. The overall heat transfer coefficient can be measured at a specific location, or over an entire heat exchange section.

6.2.2.3.5 Use of Fouling Rigs

Some methods used to investigate fouling involve test rigs which are connected to the main UHT plant. Such a test rig can be more easily dismantled than the main plant. Some thought needs to be paid to where it should be located (i.e., in the heating section

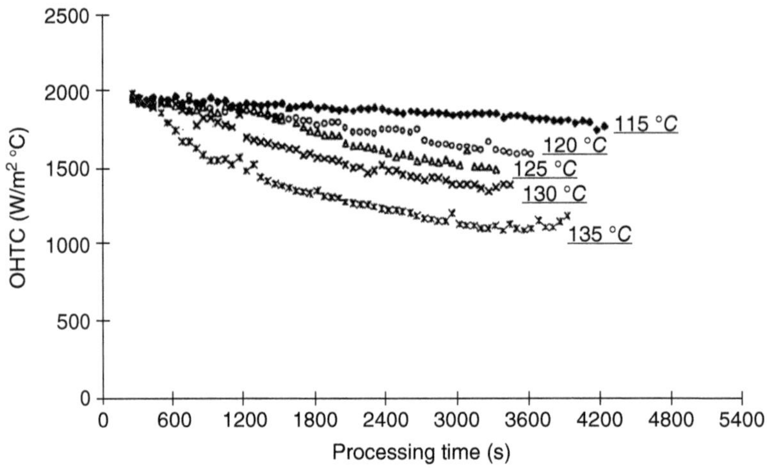

Figure 6.14 Change in OHTC with time of processing of goat's milk at various temperatures. (Source: Reproduced with permission of International Dairy Federation.)

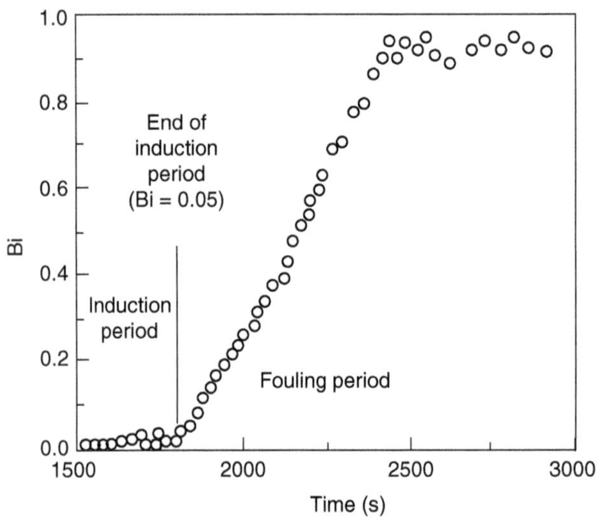

Figure 6.15 Change in Biot number (Bi) with time of processing. (Source: Fryer, 1989. Reproduced with permission of John Wiley & Sons.)

or the holding tube) and the flow regime through the test rig may well be different to that in the main plant. Other test rigs involve the use of laboratory scale or pilot-scale equipment. Some of these are summarised in Table 6.9.

6.2.2.3.6 Fouling Sensors

The concept of an on-line sensor which indicates that fouling has occurred and is impairing the efficiency of the heating process is an attractive prospect. It would be useful to indicate exactly when to start the cleaning procedure. A number of principles have been investigated to accomplish this.

Table 6.9 Some experimental equipment used to investigate fouling (and cleaning). Some information is taken from Lewis & Heppell (2000).

Electrically heated wires	Burton (1968)
Radial flow cell	Fryer *et al.* (1984)
Heated plates	Shilton *et al.* (1992)
Miniature UHT plant	Kastanas *et al.* (1995); Wadsworth and Bassette (1985); Prakash *et al.* (2015)
Pilot scale fouling rigs	Schreier *et al.* (1994); Deplace *et al.* (1994)
Commercial plant	Timperley *et al.* 1994
Sediment formation	Boumpa *et al.* (2008); Chen *et al.* (2012); Chen *et al.* (2015)

Jones *et al.* (1994) developed a fouling monitoring device based on a heat flux sensor which monitored the change of heat flux between a heated copper block and the product. The change in OHTC can be measured from the change in heat flux. Davies *et al.* (1997) used a heat flux sensor to measure the thermal resistance of whey protein deposits in situ. Thermal resistance is defined as the product of density and thermal conductivity. This method gave values of 470 $W\,kg\,m^{-4}\,K^{-1}$ for deposits exposed to wall temperatures of less than 85 °C. Deposits formed at higher wall temperatures showed the effects of ageing and gave larger thermal resistance values.

Withers *et al.* (1994) developed an ultrasonic sensor for detecting fouling during UHT processing. This is a non-invasive technique based on the time required for ultrasound to travel through deposits as they accumulate on a surface. This still needs further development to accurately measure the change in deposit at UHT conditions.

Daufin *et al.* (1985) studied the effect of electrochemical potential on metal surfaces. They concluded that electrochemical potential and current intensity were not useful parameters to monitor fouling, as there was no correlation between these parameters and fouling deposits.

Corrieu *et al.* (1988) described an on-line fouling sensor based on evaluating disturbances in hydrodynamic functioning resulting from fouling. It had been used in the heat regeneration section (80 - 120 °C) and the high temperature section (120 - 140 °C) on a plant of capacity of 10,000 L/h.

6.2.2.3.7 Indirect Indicators

As discussed earlier, another approach is to try to establish whether the product is susceptible to heat-induced fouling prior to heat treatment to either avoid using it or to modify its properties to reduce the fouling. Some examples of properties which could be tested are heat stability, alcohol stability, sedimentation, gelation, pH and titratable acidity (see Section 11.2.27).

6.2.2.4 Factors Affecting Fouling

In general, fouling is influenced by a number of factors, which can be categorised as process- or product-dependent. **Process-dependent** variables include operating conditions, such as flow rate, turbulence, shear rate, temperature and pressure. **Product-dependent** variables are related to the composition of the product and the effects of heat and shear on reaction rates. Some of these product-dependent variables are discussed in more detail in Section 6.2.2.5.

6.2.2.4.1 Plant Construction, Design and Operation

There are a number of process-related parameters which affect fouling. Wherever possible, these should be considered at the design stage. However, once a plant is installed, there may not be much scope for changing parameters to reduce fouling (especially temperatures or flow rates), as they will be pre-determined by safety and quality issues, as well as residence times and pressure drops.

The heat exchanger surface is one of the major participants of the fouling process. Stainless steel is the normal material of construction. It is a common belief that a rough surface is likely to increase fouling by providing more surface to which material may adsorb and by allowing better "keying" of the deposit on the surface. However, the evidence for this is conflicting (Burton, 1988; Kastanas, 1996) and may apply more to mineral fouling than protein fouling. One reason for these conflicts might be that once an initial deposit has formed, its subsequent development will be dependent on the deposit composition and not on the nature of the surface.

Stainless steel has a lower thermal conductivity than copper and aluminium; lower thermal conductivities give rise to higher temperature gradients, which result in increased fouling. Covering the surface with PTFE has been found not to eliminate fouling but to make cleaning easier, as adhesion to it is poorer. The PTFE layer will also decrease the thermal conductivity of the heat transfer barrier. Yoon and Lund (1994) compared the fouling rates on three stainless steel surfaces (titanium plate, standard 304 and electropolished stainless steel) and found very little difference. However, when the stainless steel was coated with polysiloxane and teflon, there was about a 14 and 20% increase in the fouling rates, respectively.

Stainless steel is a hydrophobic surface. The adsorption of whey proteins onto hydrophilic and hydrophobic surfaces has been studied by a number of workers, who found that more adsorbed fouling mass occurs on hydrophobic surfaces than on hydrophilic surfaces. There is evidence that highly polished surfaces are easier to clean, but there is no evidence that it gives an increased resistance to fouling.

6.2.2.4.2 Temperature

Temperature is considered to be an important operating variable in terms of influencing fouling. In addition to the bulk product temperature, the wall surface temperature, the heating medium temperature (steam or hot water) and the temperature difference between the product and the heating medium are also important.

In a heat exchanger, the bulk product temperature changes along its length. However, there is no direct relationship between the bulk product temperature and the amount of deposit formed at that location. Some examples of this are shown in Figures 6.16 and 6.17. The nature and location of deposits may also be influenced by the rate at which the product is heated, which in turn may influence the location where most of the β-Lg is denatured. Lewis and Heppell (2000) showed that when milk is heated over the temperature range 70 to 140 °C, the amount of denatured β-Lg varies at different locations along the heating section as its initial concentration (undenatured) changes. More β-Lg denaturation occurs in the first section for high initial concentrations,

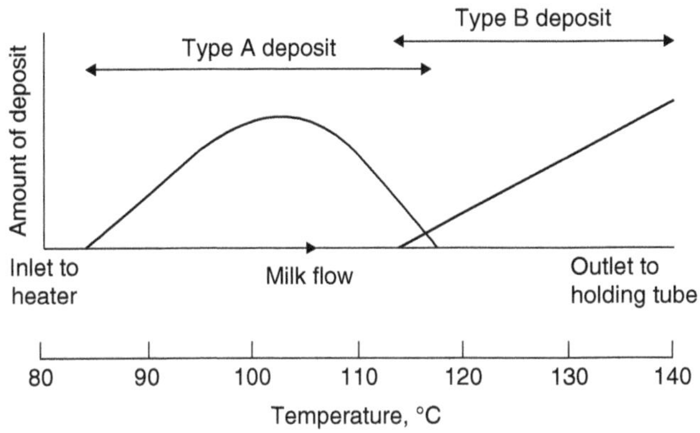

Figure 6.16 Fouling distribution in an indirect heat exchanger operating on raw milk (see Section 6.2.2.5 for an explanation of Type A and Type B deposits). (Source: Burton, 1988. Reproduced with permission of Harold Burton.)

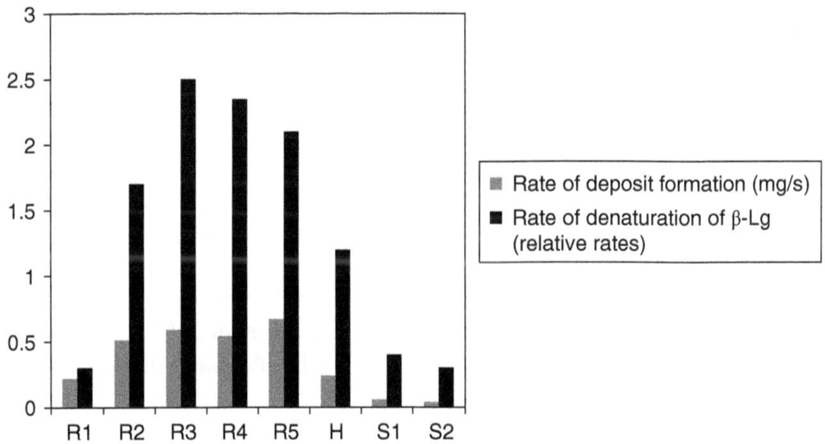

Figure 6.17 Relationship between rate of fouling deposit formation and β-Lg denaturation in different sections of a heat treatment plant. (R1 to R5, regeneration sections, from 85 to 114 °C; H, preholding section, 111°C; and S1 and S2, steam heating sections at 118 °C and 121°C). Data taken from Burton (1988) and Hiddink *et al.*, (1986).

more in the middle section for intermediate concentrations and more in the end (higher temperature) section for lower concentrations. This illustrates the complexity of the situation.

The most common method used to study the effect of product temperature on fouling involves heating the product at UHT conditions, up to a predetermined temperature and studying the deposit formation at the various sections of the plant; this procedure can then be repeated at different temperatures, but it is time consuming. Fouling is no doubt related to the bulk temperature and the severity of fouling increases as the processing temperature increases. Some products are more susceptible to changes in temperature than others (see Section 9.2).

6.2.2.4.3 Flow Rate

Product flow rate was one of the first parameters studied for its effect on fouling, but its effect is still not clear. In general fouling is high at low fluid velocities and Reynolds numbers and is considerably reduced at high flow rates and Reynolds numbers. However this trend has not been quantified successfully and the transition point from extensive fouling to reduced fouling remains unknown. Also, visual inspection of fouled plates prior to cleaning shows that there is more deposit formed where flow is relatively stagnant. However, some reports have shown that fouling rate was influenced only slightly by flow rate, compared to other factors (Lalande & Corrieu, 1981).

6.2.2.4.4 Product Chartacteristics and Other Factors

The use of pressurised hot water rather than steam and the reduction of temperature differentials between the heating medium and the process fluid reduces fouling rates in general. Hiddink *et al.* (1986) suggested that temperature differentials only affect fouling when they exceed 10-15 °C. This is unlikely to be the case where regeneration efficiencies are high and ΔT values are low. This is different for the pilot plants which generally do not have regeneration (see Section 5.2.1.7). However, reduced temperature differentials mean that the heat exchange surface is larger and there is a greater area for the deposit to be spread over, thereby diluting the overall effect.

More fouling has been reported in tubular **heat exchangers** than plate heat exchangers, attributed to the higher turbulence and shear rates found in plate heat exchangers (Bansal & Chen, 2006). However, the total cross sectional area for flow is also much higher in tubular heat exchangers and this may be the overall controlling factor. Direct UHT plants are considered to be capable of dealing with poorer quality raw milk much better than indirect plant, as longer run times can be achieved on a direct plant. Fouling in direct plants is much reduced, due to the reduced heat transfer surface at high temperatures and the very rapid rise in temperature. However this may lead to more sediment in the final product compared to indirect plant (Perkin *et al.*, 1973; Ramsey & Swartzel, 1984). Deposits may still form in the holding tube, the back-pressure valve and the beginning of the cooling section. In steam injection plants, they also occur around the stream injection point. Higher proportions of fat (0-35%) are found in deposits from direct steam injection plants, attributed to fat destabilisation during the injection process; post-sterilisation homogenisation at high temperature also produces deposits with a higher proportion of fat.

The role of **pH** in deposit formation was found not to be straightforward in earlier studies, most probably because it was considered in isolation and not together with ionic calcium. Fresh milks with the same pH showed considerable variations in their susceptibility to fouling (Burton, 1988; Grandison, 1988). However, for any individual batch of milk, in general, a reduction in pH of about 0.15 units during storage, for example, from 6.67 to 6.52 significantly increases fouling. Reducing pH was also found to increase the amount of fat within the deposit. The addition of sodium hydroxide to increase the pH by about 0.1 unit prior to processing had little effect, whereas addition of TSC (4 mM) increased the pH by 0.7 and increased both processing times and product quality, and slightly increased product viscosity. However, the effect of TSC is partly due to its ionic calcium-binding effect. Burton (1968) showed that for the same milk sample, the amount of deposit formed increased as the pH was reduced, using hydrochloric acid, with significant changes taking place below pH 6.5. Kastanas *et al.* (1995)

showed that fouling increased as pH was reduced in milk heated at 140 °C. This was measured by a decrease in overall heat transfer coefficient (OHTC) and an increase in pressure drop in a miniature tubular UHT plant. Skudder *et al.* (1986) observed that it was extremely useful to measure pH, as a slight increase in pH improves processing times. Note that this was in contradiction to what Grandison (1988) reported above and emphasises the need to take into account both pH and ionic calcium. Similar results were obtained for reconstituted milks (Zadow & Hardham, 1981). Other additives have been used to reduce fouling, but they may not be legally acceptable in all countries. Sodium and potassium pyrophosphates, added at 100 ppm, doubled the running time in a plate heat exchanger (Burdett, 1974). Skudder *et al.* (1981) found that the addition of iodate extended running time considerably by interfering with the formation of type A deposit (see Section 6.2.2.5), which is largely polymerised β-Lg. Addition of only 10 ppm iodate doubled the running time before requiring cleaning but 20 ppm caused bitterness due to plasmin-induced proteolysis during subsequent storage (Grufferty & Fox, 1986).

Grandison (1988) suggested that pH was not a reliable indicator by itself but that there were other factors in milk which make a significant contribution to fouling. Experiments designed to investigate seasonal variations in milk composition showed that there were significant differences in fouling behaviour throughout a complete year (see Figure 6.18). However it was not possible to correlate fouling with any physical or chemical parameters, so the reasons for this observation were not clearly established. Milks with high levels of κ-casein were more prone to deposit formation. Note that ionic calcium was not measured by Grandison (1988). Jeurinck and DeKruif (1995) produced low- and high-calcium milk and found that both had reduced HCT. In terms of deposit formation, the high-calcium milk produced significantly more than the control whereas the

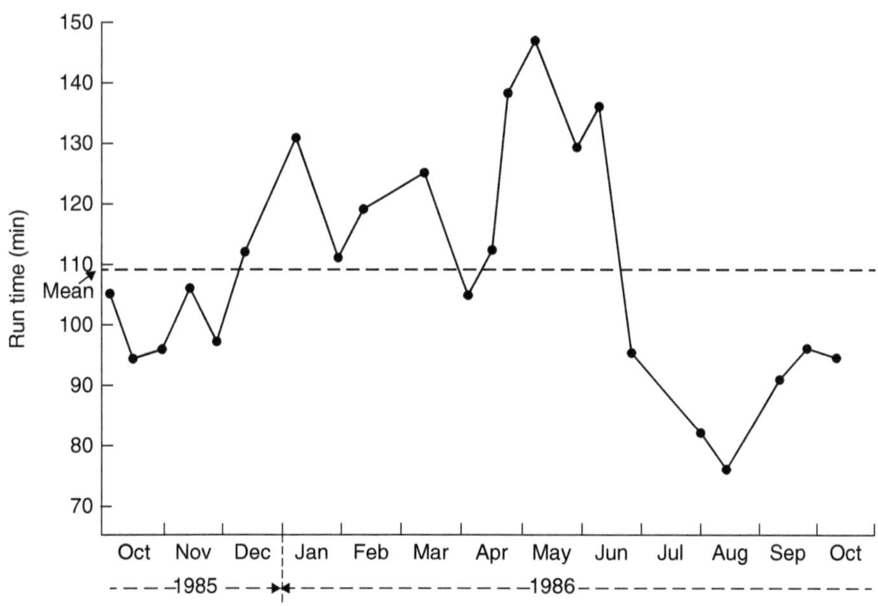

Figure 6.18 Variation in run time of a UHT processing plant over a 1-year period. (Source: Grandison, 1988. Reproduced with permission of John Wiley & Sons.)

low-calcium milk produced only slightly more. Ionic calcium was also measured but only three samples were investigated. It is recommended that both pH and ionic calcium should be monitored to gain a better understanding of the fouling process.

Lalande and Corrieu (1981) showed that during pasteurisation of milk there was a strong positive correlation between the rate of deposit formation and the concentration of **ammonia**; an increase in ammonia concentration from 3.7 to 7.2 ppm was associated with a doubling of the fouling rate constant. One suggestion is that ammonia concentration is related to the urea content of milk; natural urease may break down the urea producing ammonia, thereby increasing its susceptibility to fouling, even though the pH should increase. Added urea has been found to increase the heat stability of cows' milk heated at 120 °C (Muir & Sweetsur, 1976). Prakash (2007) investigated the effect of added ammonia and urease (added to produce ammonia) on fouling of reconstituted skim milk on a bench-top UHT plant. The levels of added ammonia were 4.56, 6.76, 8.98 and, from added urease, 20.3 mg/L. No significant effect was observed during a 5-h UHT run. This suggests that the effects of ammonia noted during pasteurisation and in heat stability tests are not applicable to UHT processing.

Aging raw milk at 4 °C for 12 - 24 h, without change of pH, was found to reduce its susceptibility to fouling. This was thought to be due to lipolysis and the production of fatty acids (Burton & Burdett, 1974). Addition of capric acid reduced fouling whereas addition of stearic acid increased fouling (Al-Roubaie & Burton, 1979). Capric acid was thought to associate with the casein micelle and prevent interactions which would lead to the build-up of the deposit. de Jong *et al.* (1993) also reported that fouling increased strongly with aging of the milk. In contrast, Kastanas *et al.* (1996) found that (good quality) raw milk could be stored chilled for a considerable time without change in its susceptibility to fouling. In fact some raw goat's milk was kept for almost 19 days before it became unsuitable for heat treatment. Similarly, Prakash (2007) stored raw cow's milk for 8 days and found its fouling potential was not changed.

Gynnig *et al.* (1958) showed that removal of **air** from milk reduced the total amount of deposit produced by between 50 and 75% in a laboratory pasteuriser operating at 85 °C. Fouling is believed to result from the presence of bubbles on the heat exchanger surface as the air becomes less soluble at increased temperature and provides nucleation sites for deposit formation. On commercial UHT plants air can be removed by using a deaerator (see Section 5.4); however, there is no reported evidence that this will reduce fouling. In fact, Prakash (2007) altered the dissolved oxygen content of milk from 1.15 to 7.15 ppm and found no difference in fouling in a bench-top UHT pilot plant. Burton (1968, 1988) suggested that the pressure under which UHT processing is performed (~0.5 MPa) prevents air bubbles escaping from the product and facilitating fouling. He recommended operating pressures in UHT plants of at least 0.1 MPa higher than that corresponding to the highest product temperature in the plant to minimise fouling (Burton, 1988).

Goat's milk is difficult to UHT process because of its high **ionic calcium** content. It is very susceptible to both fouling and sediment formation. Its alcohol stability is well below that of cow's milk, at between 40 and 60%. Zadow *et al.* (1983) studied the problem of UHT processing of goat's milk and concluded that either pH adjustment to well above 7.0 or addition of 0.2% disodium hydrogen phosphate before processing was necessary to reduce sediment formation. The higher levels of ionic calcium, in part due to its lower levels of citrate, were thought to be responsible for its instability.

Kastanas *et al.* (1996) found that goat's milk fouling caused a rapid decrease in OHTC, with no initiation period. Reconstituted goat's milk, made from freeze-dried powder, which had a lower ionic calcium level was less susceptible to fouling than fresh goat's milk. Kastanas *et al.* (1996) showed that a number of treatments, such as citrate addition, forewarming and pH adjustment, all reduced fouling in goat's milk. Prakash *et al.* (2007) showed that SHMP, TSC and calcium reduction by ion-exchange resin treatment were also effective. The findings of Boumpa *et al.* (2008) on sediment in UHT goat's milk are in agreement with these findings. On the other hand, buffalo milk is widely available in some countries (e.g., Egypt and India), some of which is UHT processed, often mixed with cow's milk. There are no reports that this causes a severe fouling problem, despite its higher total solids of 16-18%. In fact the alcohol stability and heat stability of buffalo milk compare favourably with those of cow's milk (Laxminarayana and Dastur, 1968).

6.2.2.5 Fouling Mechanism

Fouling of heat exchangers has been widely studied but the mechanisms are still not fully understood. Burton's review articles (Burton, 1968, 1988) provide a good summary of the state of knowledge at those times, and an update was provided by Bansal and Chen (2006). Early investigations on fouling of UHT milk led to the recognition of two distinct types of deposit, Type A and Type B deposit (Lyster, 1965; Burton, 1968). The major area where deposits occur is the pre-heating section. Maximum deposit formation is in the temperature range 95-110 °C (Burton, 1988) and at 110 °C according to Skudder *et al.* (1986). The deposit forming between 80 °C and 105 °C is a white voluminous deposit, which has a high protein content (50-70%) and a significant mineral content (30-40%) and tends to block the flow passages. This is known as Type A deposit (see Figure 6.16). At the lower end of the temperature spectrum, the protein is predominantly denatured β-Lg but toward the top end of the range it is predominantly casein. Tissier *et al.* (1984) also found two major temperature zones for deposit formation; the first at ~90 °C (predominantly protein (50%) and the second at ~130 °C, which was predominantly mineral (75%) (Type B deposit, see Figure 6.16). The major protein contributing to the lower temperature peak was β-Lg (62%) while in the second peak, β-casein (50%) and α_{s1}-casein (27%) were predominant. Work by Lalande *et al.* (1984) was in general agreement with this. Type B deposits, which form at higher temperatures, are finer, more granular and predominantly mineral in origin (70-80%), with only small amounts of protein (10-20%). The mineral component is probably β-tricalcium phosphate whose solubility decreases with increasing temperature.

However, this situation may be totally changed if a pre-heat holding or protein stabilisation section is included. This is discussed in more detail in Section 5.2.1.2.1. Clearly such an inclusion can considerably change the nature of the deposit downstream of that section. The conditions in this section would also cause considerable whey protein denaturation and produce some deposit in this section.

It is interesting that fat does not feature significantly in any of the deposits, usually less than 5%, despite it being present in equal concentrations to protein and in greater concentration than minerals in whole milk. However, it has been found in greater proportions in the sediment from full cream milk. Newstead *et al.* (1998) reported that fouling deposits from recombined milk had higher levels of fat (up to 60%) compared to fresh milk (10% or less). Fouling in creams, where the fat might represent up to 75% of the dry matter is reported to be predominantly protein and mineral in character, with the fat playing only a minor role.

Fouling in milk occurs under conditions where the fouling precursors, principally the denatured, molten-state form of β-Lg, formed by the action of heat in the bulk solution, are unable to associate with casein micelles or other whey proteins before they move toward, and deposit on, the wall of the heat exchanger. For this reason it has been proposed that, to minimise fouling, the pre-heat conditions in a UHT plant should be such as to minimise the time spent by the denatured, non-aggregated, "sticky" form of β-Lg in the plant. This in accord with the reports discussed above of reduction of fouling when relatively severe pre-heating conditions were used.

There is still uncertainty whether protein denaturation or protein aggregation is the key reaction in fouling and whether it is denatured whey protein or minerals that deposit first on the heat exchange surface. This is discussed in detail in the review by Bansal and Chen (2006). They cite several reports which conclude that deposition of whey protein, particularly in the denatured non-aggregated form, initiates fouling. Lewis and Deeth (2009) also revewed the evidence that fouling is initiated by denatured whey protein adsorbing onto the heated surface. However, there are also reports which agree with the early suggestion of Burton (1968) that deposit formation is initiated by the nucleation and growth of crystals of mineral salts such as tricalcium phosphate from their supersaturated solutions. Any effect which impedes nucleation and growth might be expected to lessen the amount of deposit formed.

The following anecdote supports the role of precipitation of calcium salts on fouling. All sections of the UHT plant must be sterilised downstream of the holding tube prior to processing milk. This is most often done by circulating hot water through the cooling section at 130 °C for 30 min. In hard water areas, it is important that this water is softened. In pilot plant practical classes, over 200 trials were conducted with good quality cow's milk and no fouling was observed over a period of 20-30 min, except for one occasion, when excessive fouling was observed. It was subsequently discovered that hard water had been used for plant sterilisation due to a breakdown of the water softener. This resulted in calcium being deposited during the plant sterilisation procedure, which then accelerated accumulation of proteins at the surface. This is an argument for mineral deposits initiating the fouling process although in other circumstances it may be initiated by protein deposition.

6.2.2.6 Methods to Reduce Fouling

Although many of the following observations are specific to milk, the general principles and approaches may be relevant to, and help provide explanations for, the fouling of other products. The first approach is to avoid the use of raw materials which may be susceptible to fouling and to divert them to other products or processes where fouling is not a major issue. For this, suitable test methods are required. One such test is the ethanol stability test, which is discussed in more detail in Sections 6.2.1.4 and 11.2.25.1.

For formulated products, such as flavoured milk, reconstituted milk or ice cream mix, there is also the opportunity to manipulate composition to improve heat stability. Also, with reconstituted products, water quality may also influence heat stability, especially with regard to the amount of calcium and magnesium which may be present.

As whey proteins are responsible for much of the type A fouling deposits, a reduction in undenatured whey proteins reduces fouling. When milk is forewarmed for 4-6 min at 85 °C, sufficient to denature most of the whey proteins, the nature of the fouling deposit changes to that of a similar composition to the Type B deposit described earlier, throughout the heat exchanger. Such conditions have been found to extend the run times for UHT plants (see Section 5.2.1.2.1). Similarly, reconstituted high-heat skim milk powder fouls much less readily than the corresponding low-heat powder due to the denaturation and aggregation of the whey proteins prior to UHT processing (Prakash, 2007).

One approach to reducing fouling in UHT plants is to carry out a heat treatment prior to high-temperature sterilisation. This has been termed forewarming, pre-holding and pre-heating and is discussed in more detail in Section 5.2.1.3. While some authors have used these terms interchangeably, forewarming is a term used in early literature to refer to a procedure performed on milk prior to evaporation to improve the heat stability of evaporated milk. In UHT processing such a heat treatment, often called pre-heating, has been found to decrease fouling and increase run times (Burton, 1968; Patil & Reuter, 1986; Mottar & Moemans, 1988; Foster & Green, 1990; Prakash *et al.*, 2015), with higher temperatures being more effective than lower temperatures. For example, Patil and Reuter (1986) investigated the effect of pre-heat temperatures of 70 to 110 °C and holding times of 30 to 120 s during treatment of raw cow's milk in an indirectly heated UHT plant. They found the longest run times before cleaning occurred when the pre-heat conditions were 90 or 100 °C for 90 or 120 s. Mottar and Moemans (1988), using response surface methodology, found that the optimum pre-heating/forewarming conditions to reduce fouling were 70-90 °C for 40-80 s. However, because the more intense treatments can lead to off-flavour production, they concluded that the optimum pre-heating conditions are 70-80 °C for 40-70 s. In contrast to the above, Srichantra *et al.* (2006) reported the opposite effect of pre-heating, that is, the more severe the pre-heating the greater the rate of fouling. They suggested the reason for their different results could be related to the different processing procedures, including the stage of homogenisation; they pre-heated milk which had been previously homogenised (and pasteurised). Prakash *et al.* (2015) investigated the effect of pre-heat holding temperatures of 65-95 °C and two holding times (5 and 25 s) in combination with sterilisation temperatures of 135–150 °C on fouling of reconstituted skim milk using a bench-top UHT pilot plant. They found that pre-heating at 85 °C or 95 °C for 25 s, combined with a sterilisation temperature of 145 °C, or pre-heating at 95 °C for 25 s combined with sterilisation at 142 °C produced the longest run-times (see also Section 5.2.1.3).

A further approach is to prevent the whey proteins aggregating by intercepting the denatured, "sticky" form of β-Lg when formed by the heating process. This form has an exposed sulfydryl group which participates in interactions with disulfide and sulfydryl groups on caseins and other whey proteins as well as heat exchanger surfaces. As mentioned above, addition of potassium iodate reduces fouling and this is due to oxidation of the sulfydryl group of the denatured β-Lg.

Other approaches to inhibiting whey protein aggregation and interactions during heat treatment which may be applicable were recently reviewed by Wijayanti *et al.* (2014) (see also Section 6.1.3.1). One interesting approach to reduce fouling is to use chaperone proteins such as caseins which associate with the whey proteins (Mounsey & O'Kennedy, 2010). The authors have found that this can be particularly effective for UHT processing of high whey protein beverages using caseinate as the chaperone protein.

6.2.2.7 Fouling in Other Products

Fouling may also be an issue in products other than cow's mik. This will largely depend on the heat stability of the product which will affect the duration of the run, which in many cases will be much shorter than for liquid milk. Although fouling may not always be an issue, it is likely that products showing poor heat stability will contain more sediment which might be noticed by the consumer. Goat's milk and camel milk are special cases and are discussed in Section 9.16.

In general, fouling during processing of UHT cream is not reported to be a major problem, again suggesting that the fat phase does not play an important part in fouling processes. Cream products are not produced in such abundance as milk products, so processing runs are generally shorter and this may be a significant factor to explain the shortage of literature on fouling during cream production. Pre-heat holding was found to be effective in reducing fouling in cream, the recommended conditions being 95 °C for a few minutes (Hiddink *et al.*, 1986). Also fouling deposits from whipping cream contained little fat and were predominantly protein and minerals. Therefore, cream fouling may be influenced by similar factors to those for milk fouling. However, with cream being more viscous, the thermal boundary layer will be thicker and cream may thus be more susceptible to fouling than milk, at similar flow rates.

6.2.2.8 Biofilms

Biofilms are surface-associated multicellular microbial communities whose structural integrity depends on an extracellular matrix produced by the constituent microorganisms. While the composition of the matrix, sometimes referred to as slime, varies considerably, polysaccharides and proteins are the major components (Branda *et al.*, 2005). In the context of this book, biofilms are important as they may colonise a heat exchanger and associated equipment and compromise the safety and quality of heat-treated products.

Stoodley *et al.* (2002) described the development of biofilms in five stages: reversible attachment; irreversible attachment; early development of biofilm structure; maturation of biofilm structure; and detachment of single cells from the biofilm. Of particular relevance here are attachment of spores to a surface, usually stainless steel but can also be other materials such as those with which seals are made, and the release of bacterial cells, spores or vegetative cells, from the biofilm into the product. The spores of interest are *Bacillus* and *Geobacillus.* Some of these are highly hydrophobic and this allows them to adhere firmly to surfaces such as stainless steel (Doyle *et al.*, 1984; Flint *et al.*, 2001; Parkar *et al.*, 2001). However, hydrophobicity is not the only factor involved in the attachment and is not a good predictor of attachment of bacteria to a surface (Seale *et al.*, 2008). A complicating factor is that there are several methods for assessing hydrophobicity and these show poor correlations between them (Simmonds *et al.*, 2003). Once attached, spores acquire increased resistance to heat (Simmonds *et al.*, 2003) and to cleaning and sanitation procedures (Wirtanen *et al.*, 1996; Te Giffel *et al.*, 1997; Lindsay & von Holy, 1999). Pfeifer and Kessler (1995) reported increased heat resistance of *B. cereus* spores trapped between a silicone-rubber seal and a stainless-steel surface.

Biofilms can form on any surfaces but the type of bacteria involved varies. For example *Pseudomonas* are more frequently associated with biofilms on surfaces such as raw milk tankers and silos while *Bacillus* and *Geobacillus* form biofilms in heated sections of processing equipment as well as cold surfaces of tankers and silos. Flint *et al.* (2001)

reported that *Bacillus* biofilms can form on clean stainless steel surfaces and release bacteria, predominantly vegetative cells but also spores, into milk. Fouled surfaces attracted up to 100 times more bacteria than cleaned surfaces. Conditions for growth of *G. stearothermophilus* are favourable in biofilms.

In a review of biofilms in dairy processing, Marchand *et al.* (2012) reported that biofilms are one of the main recontamination sources of milk. This applies particularly to *Pseudomonas* and contamination of pasteurised and ESL milk which are not packaged aseptically. It has been established that for each planktonic bacterium detected, there might be close to 1000 organisms present in biofilms. In the dairy industry, mono- as well as multi-species biofilms can occur. Pathogenic bacteria can coexist within a biofilm with other environmental organisms; an example of this is *L. monocytogenes* surviving in *Pseudomonas* biofilms.

The major products affected by the release of bacterial cells from biofilms are pasteurised milk and cheese made from it, ultrafiltered milk, evaporated milk and powders made from it, ESL milk and, indirectly, UHT milk. A major biofilm problem in pasteurised milk arises from long run times where thermoduric organisms build up in the regeneration section and contaminate the pasteurised product (Lehmann *et al.*, 1992). Major organisms responsible are streptococci (Bouman *et al.*, 1982) and *Bacillus* species (e.g., *B. lichiniformis* [Lehmann, 1996]). Introduction of an intermediate hot caustic wash was found to control the build up of a *B. licheniformis* biofilm in UF units and pasteurisers (Lehmann, 1996) and periodic temperature step changes were found to be effective in controlling the development of a *Str. thermophilus* biofilm in pasteurisers (Knight *et al.*, 2004). The temperature step changes effectively prevent the bacteria from entering the exponential phase of growth.

Biofilm development in evaporators used in the production of powders is responsible for the major contaminants in powders, namely, *G. stearothermophilus, B. licheniformis, B. subtilis* and *Anoxybacillus flavithermus* (Ruckert *et al.*, 2004). As discussed in Section 4.4.3.2, these are also the major contaminants of UHT reconstituted milk. Burgess *et al.* (2009) reported that at 55 and 60 °C, evaporator operating temperatures, both biofilms and spores of *A. flavithermus* can form very rapidly and simultaneously.

ESL milk packaged under non-sterile conditions can become contaminated with bacteria released from biofilms built up in sections of the plant. Mugadza & Buys (2014, 2015) reported that *Paenebacillus, B. pumilus* and *B. cereus* were isolated from both filler nozzles and ESL milk; in the case of *B.* cereus there was a close relationship between isolates from the ESL milk and those from the filler nozzles.

Teh *et al.* (2014) reported that raw milk exposed to biofilms in milk tankers can be contaminated with heat-stable enzymes produced by the biofilm bacteria. They showed that UHT milk produced from raw milk exposed to the tanker biofilms exhibited more proteolysis during storage than did UHT milk produced from raw milk which had not been exposed to the biofilm.

Teh *et al.* (2013) also demonstrated considerable lipase production within biofilms of milk tanker surfaces. This indicates that extracellular enzymes can enter the milk from such biofilms without substantial planktonic growth. In such cases, bacterial counts would not give a true indication of the possible presence of these enzymes.

Apart from the situation of UHT milk reconstituted from powder contaminated with biofilm-derived spore-formers and the issue of bofilm-derived heat-resistant enzymes reported by Teh *et al.* (2013, 2014), there appears to be no other evidence that biofilms are an issue in UHT processing. However, a possible source of spores in UHT milk are biofilms forming in parts of the UHT plant which operate at temperatures conducive to their formation as occurs in pasteurisers, UF plants and evaporators. The most likely locations are the section prior to the protein stabilisation/pre-heat section and any holding tube in that section.

Biofilms are difficult to remove from milk processing environments due to the constituent exoploysaccharides and the difficulties associated with cleaning complex processing equipment and processing environments. In CIP studies, Gibson *et al.* (1995) found that bacteria remained on the surface after prolonged cleaning times and high flow velocities and a small proportion of cells remained viable even after treatment at high temperature and alkaline pH. Bremer *et al.* (2006) also found that a standard CIP procedure did not remove biofilms of *Bacillus* and thermoresistant streptococci; however, a modified CIP procedure in which chelating and sequestering agents, and surface-active wetting agents were added to the caustic agent, and surfactants were added to the nitric acid, was much more effective.

Johansen *et al.* (1997) investigated the use of enzymes for removal of biofilms and inactivation of the constituent bacteria. They found that a complex mixture of polysaccharide-hydrolysing enzymes (which removed the biofilms) and oxidoreductases (which inactivated the bacteria) was effective. Ultrasound has also been investigated for removal of biofilms. Oulahal *et al.* (2004) found that ultrasound alone was not completely successful but the effectiveness was enhanced when combined with the use of a chelating agent. Similarly, Bauman *et al.* (2009) combined ultrasound with ozone treatment to improve its effectiveness. A range of options for removing and sanitising dairy biofilms was reviewed by Marchand *et al.* (2012); they suggested that improved cleaners and sanitizers, including the use of EPS- and protein-degrading enzymes, should be a focus of future research.

In summary, "biofilm control relies in the end on the design of storage and processing equipment, effective cleaning and sanitising procedures, and the correct implementation and application. The management of these factors is important to ensure safe and good-quality milk and dairy products" (Marchand *et al.*, 2012).

6.3 Kinetics and Computer Modelling

There are programmes available for calculating B*, C* and F_0 values (see Section 4.1). Data reported by Tran *et al.* (2008) were based on an Excel spreadsheet which was first developed by Browning *et al.* (2001). This programme allows the evaluation of different heating and cooling periods and different combinations of holding times and temperatures. Once the information on heating conditions (initial temperature, final temperature and heating time), holding time and temperature and cooling conditions (as for heating) have been entered, it calculates the heating and cooling profiles and performs a graphical integration. It is assumed that the sections of the heating profile are linear. The programme was developed using the formulae for B*, C* and F_0 values. In addition to these parameters, information on the effects of the heating regime on a range of chemical components can also be obtained. These include β-Lg and α-La denaturation, lactulose, furosine, HMF and browning. The programme uses published kinetics data for these calculations.

Mullan (2011) has presented a programme which takes the time, up to 5 time periods, at up to 5-temperatures and sums the thermal indicators to give a total value. The time intervals can be selected in either in minutes or seconds. This is a free application.

NIZO have produced a modular software platform, *Premia®*, that contains modeling applications on membrane transport, spray drying, falling film evaporation, heat treatment, cooling, cheese making and more. It started 20 years ago as an internal software tool and has now grown into a simulation platform that is widely used in food companies all over the world. They claim that reductions in production costs from 20% to as high as 50% have been realised through its use. The specific application to UHT processing predicts not only B* and C* values but also changes occurring during the heat treatment including microbial inactivation and the extent of fouling. Thus there are opportunities to optimize the process to maximise microbial inactivation and minimize the amount of fouling. This will also reduce cleaning costs.

A number of models have been proposed for the fouling process, related to changes in some of the measured characteristics described earlier. Modelling fouling processes is complex because the fouling rate at any point is a function of temperature, but temperatures change as a result of fouling. The models describing fouling dynamics can be classified into three groups, according to the parameter they predict: the amount of

Table 6.10 Simulated impact of direct and indirect UHT sterilisation on product quality and fouling. (Source: Hotrum, 2010. Reproduced with permission of Elsevier.)

	Direct steam injection	Indirect heating
Processing parameter		
Holding time (s)	6.4	51.9
Holding temperature (°C)	141.0	132.3
F_0 (min)	12	12
Component		
Geobacillus stearothermophilus[a] (decimal reduction)	6.4	6.3
Thiamine[b] (% remaining)	95	77
Free lysine content[c] (% remaining)	94	74
Methanethiol[d] (%, normalised to indirect heating)	20	100
Hydrogen sulphide[d] (%, normalised to indirect heating)	27	100
5-Hydroxymethylfurfural[e] (μmol/L)	7.6	24.6
Native β-lactoglobulin[f] (% remaining)	20	1
Deposit due to fouling[g] after 12 h run time (kg), applying a product flow of 125 L/h	0.08	0.27

[a] Inactivatation kinetics of Peri *et al.* (1985)
[b] Destruction kinetics of Kessler and Fink (1986)
[c] Destruction kinetics of Bayoumi and Reuter (1985)
[d] Formation kinetics of de Wit and Nieuwenhuijse (2008)
[e] Formation kinetics of de Jong (1996)
[f] Denaturation kinetics of Dannenberg and Kessler (1988)
[g] Fouling kinetics of de Jong *et al.* (2002)

deposit; the thickness of the deposit; or changes in the overall heat transfer coefficient. Models of fouling have been reviewed by Grijspeerdt *et al.* (2004).

The NIZO and other models of fouling are mainly based on denaturation of β-Lg. The kinetic parameters for the fouling reaction scheme as given by Grijspeerdt *et al.* (2004) are, for the temperature range 85-115 °C, $E_a = 45{,}100 \pm 4{,}500$ J/mol and $\ln(k_o) = -0.82 \pm 1.45$. De Jong *et al.* (1993) evaluated such a model experimentally using temperatures from 70 to 130 °C and found it was valid up to 115 °C. Others have reported it to be valid up to 90 °C (Grijspeerdt *et al.*, 2004). It is not surprising that is not valid at high temperatures as it does not take account of mineral deposition which largely occurs at these temperatures. However, the weight of deposit formation in the regenerative section of a UHT plant operating at 80-110 °C has been found to show a significant correlation (r = 0.792) with run time while the deposit in the high-heat section was not correlated with run time (Grandison, 1988). This provides some justification for the fouling models based on denaturation of β-Lg.

Hotrum *et al.* (2011) presented a case study of this model to UHT sterilisation and compared a direct heating process and indirect heating process, which had the same F_0 value. The results are summarised in Table 6.10. One of limitations of this fouling model is that it does not take into account the roles and variations of pH and ionic calcium in raw milk (see Section 6.2.2.4.4).

References

Abdulghani, A.H., Ali, M.Y., Prakash, S. & Deeth, H.C. (2015) Sensory evaluation and storage stability of UHT milk fortified with iron, magnesium and zinc. *Dairy Science and Technology* **95**(1), 33–46.

Adhikari, A.K. & Singhal, O.P. (1992) Effect of dissolved oxygen content on the flavour profile of UHT milk during storage. *Australian Journal of Dairy Technology* **47**(1), 1–6.

Al-Attabi, Z., D'Arcy, B.R. & Deeth, H.C. (2009) Volatile sulphur compounds in UHT milk. *Critical Reviews in Food Science and Nutrition* **49**, 28–47.

Al-Attabi, Z., D'Arcy, B.R. & Deeth, H.C. (2014) Volatile sulfur compounds in pasteurised and UHT milk during storage. *Dairy Science and Technology* **94**(3), 241–253.

Al-Attabi, Z.H.R. (2009) *Volatile Sulphur Compounds in UHT milk.* PhD thesis, University of Queensland, Australia.

Al-Roubaie, S.M. & Burton, H. (1979) Effect of free fatty acids on the amount of deposit formed from milk on heated surfaces. *Journal of Dairy Research* **46**, 463–471.

Al-Saadi, J.S. & Deeth, H.C. (2008) Cross-linking of proteins and other changes in UHT milk during storage at different temperatures. *Australian Journal of Dairy Technology* **63**(3), 93–99.

Al-Saadi, J.S., Easa, A.M. & Deeth, H.C. (2013) Effect of lactose on cross-linking of milk proteins during heat treatment. *International Journal of Dairy Technology* **66**, 1–6.

Aoki, T., Hatanaka, C., & Imamura, T. (1977) Large casein micelles and soluble casein in heated concentrated whey protein-free milk. *Agricultural and Biological Chemistry* **41**, 2349–2355.

American Dairy Products Institute (ADPI) (2009) *Standards for Grades of Dry Milks.* 3rd edn. Bulletin 916. ADPI, New York.

Anderson, L.E., Adams, D.M. & Walter, W.M. (1983) Conditions under which bacterial amylases survive ultrahigh temperature sterilisation. *Journal of Food Science* **48**, 1622-1625, 1631.

Andersson I. & Öste R. (1992a) Loss of ascorbic acid, folacin and vitamin B12, and changes in oxygen content of UHT milk. II. Results and discussion. *Milchwissenschaft* **45**, 299–302.

Andersson, I. & Öste, R. (1992b) Sensory changes and free SH-groups in UHT milk. *Milchwissenschaft* **47**(7), 438–441.

Andersson, I. & Öste, R. (1995) Sensory quality of UHT milk. In: *Heat-Induced Changes in Milk* IDF Document 9602, (ed. P.F. Fox), pp. 318–330. International Dairy Federation, Brussels.

Andrews, A.T. (1975) Properties of aseptically packed ultra-high-temperature milk .3. Formation of polymerized protein during storage at various temperatures. *Journal of Dairy Research* **42**, 89–99.

Andrews, A.T. & Cheeseman, G.C. (1971) Properties of aseptically packed UHT milk: casein modification during storage and studies with model systems. *Journal of Dairy Research* **38**, 193–207.

Andrews, A.T. & Cheeseman, G.C. (1972) Properties of aseptically packed ultra-high-temperature milk. 2. Molecular weight changes of casein components during storage. *Journal of Dairy Research* **39**, 395–408.

Andrews, G.R. & Prasad, S.K. (1987) Effect of the protein, citrate and phosphate content of milk on formation of lactulose during heat treatment. *Journal of Dairy Research* **54** 207–218.

Andrews, G.R. (1989) Lactulose in heated milks. *International Dairy Federation Bulletin* **238**, 45–52.

Anema, S. G. (2008) On heating milk, the dissociation of kappa-casein from the casein micelles can precede interactions with the denatured whey proteins. *Journal of Dairy Research* **75**(4), 415–421.

Anema, S.G. (2009) The whey proteins in milk: thermal denaturation, physical interactions and effects on the functional properties of milk. In: *Milk proteins: From Expression to Food.* (eds A. Thompson, M. Boland & H. Singh), pp. 230–281. Academic Press/Elsevier, Amsterdam, Boston.

Aoki, T. & Imamura, T. (1974) Changes of casein complex in sterilized concentrated skim milk during storage. *Agricultural and Biological Chemistry* **38**(2), 309–314.

Aoki, T. & Imamura, T. (1974) Changes of casein complex during storage of sterilized skim milk. *Agricultural and Biological Chemistry* **38**(10), 1929–1934.

Auldist, M.J., Coats, S.J., Sutherland, B.J., Hardham, J.F., McDowell, G.H. & Rogers, G.L. (1996) Effect of somatic cell count and stage of lactation on the quality and storage life of ultra high temperature milk. *Journal of Dairy Research* **63**, 377–386.

Badings, H.T. & de Jong, C. (1984) Headspace analysis for the study of aroma compounds in milk and dairy products. In: *Analysis of Volatiles: Methods and Applications.* pp. 401–417. Walter de Gruyter, Berlin.

Badings, H.T., de Jong, C., Dooper, R.P.M. & De Nijs, R.C.M. (1985) Rapid analysis of volatile compounds in food products by purge-and-cold trapping/capillary gas chromatography. *Developments in Food Science* **10** (Issue: Progress in Flavour Research), 523–532.

Badings, H.T., van der Pol, J.J.G. & Neeter, R. (1981) Aroma compounds which contribute to the difference in flavour between pasteurized milk and UHT milk. In: *Flavour '81: 3rd Weurman Symposium: Proceedings of the International Conference.* pp. 683–692. Walter de Gruyter, Munich.

Badoud, R., Fay, L., Richli, U. & Husek, P. (1991) Gas chromatographic determination of N-carboxymethyl amino acids, the periodate oxidation products of Amadori compounds. *Journal of Chromatography* **552**, 345–351.

Bagliniere, F., Tanguy, G., Jardin, J., Mateos, A., Briard, V., Rousseau, F., Robert, B., Beaucher, E., Humbert, G., Dary, A., Gaillard, J.L., Amiel, C. & Gaucheron, F. (2012) Quantitative and qualitative variability of the caseinolytic potential of different strains of *Pseudomonas fluorescens*: Implications for the stability of casein micelles of UHT milks during their storage. *Food Chemistry* **135**(4), 2593–2603.

Ballance, P.E. (1961) Production of volatile compounds related to the flavour of foods from the strecker degredation of $_{DL}$-methionine. *Journal of the Science of Food and Agriculture* **12**, 532–536.

Bansal, B. & Chen, X.D. (2006) A critical review of milk fouling of heat exchangers. *Comprehensive Reviews of Food Science and Food Safety* **5**, 27–33.

Barach, J.T., Adams, D.M. & Speck, M.L. (1976) Low temperature inactivation in milk of heat-resistant proteases from psychrotrophic bacteria. *Journal of Dairy Science* **59**, 391–395.

Baumann, A.R., Martin, S.E. & Hao, F. (2009) Removal of *Listeria monocytogenes* biofilms from stainless steel by use of ultrasound and ozone. *Journal of Food Protection* **72**(6), 1306–1309.

Bayoumi, E.S. & Reuter, H. (1985) [Thiamin losses during UHT treatment of whole milk.] *Milchwissenschaft* **40**(12), 713–716.

Beliciu, C.M., Sauer, A. & Moraru, C.I. (2012) The effect of commercial sterilization regimens on micellar casein concentrates. *Journal of Dairy Science* **95**(10), 5510–5526.

Bhatt, H., Cucheval, A., Coker, C., Patel, H., Carr, A. & Bennett, R. (2014) Effect of lactosylation on plasmin-induced hydrolysis of β-casein. *International Dairy Journal* **38**, 213–218.

Biewendt, H.G. (1994) [Problems associated with limiting values for milk heating processes.] *Deutsche Milchwirtschaft* **45**(17), 786–790.

Birlouez Aragon, I., Nicolas, M., Metais, A., Marchond, N., Grenier, J. & Calvo, D. (1998) A rapid fluorimetric method to estimate the heat treatment of liquid milk. *International Dairy Journal* **8**(9), 771–777.

Bosset, J.O., Eberhard, P., Gallmann, P., Gauch, R., Rattray, W. & Sieber, R. (1996) Occurrence and behaviour of volatile sulfur-containing compounds in milk by heat treatment. In: *Heat Treatments and Alternative Methods*. Doc. 9602, pp. 409–421. International Dairy Federation, Brussels.

Bouman, S., Lund, D.B., Driessen, F.M. & Schmidt, D.G. (1982) Growth of thermoresistant streptococci and deposition of milk constituents on plates of heat exchangers during long operating times. *Journal of Food Protection* **54**, 806–812.

Boumpa, T., Tsioulpas, A., Grandison, A.S. & Lewis, M.J. (2008) Effects of phosphate and citrates on sediment formation in UHT goat's milk. *Journal of Dairy Research* **75**, 160–166.

Branda, S., Vik, A., Friedman, L. & Ko, R. (2005) Biofilms: the matrix revisited. *Trends in Microbiology* **13**(1), 20–26.

Bremer, P.J., Fillery, S. & McQuillan, A.J. (2006) Laboratory-scale clean-in-place (CIP) studies on the effectiveness of different caustic and acid wash steps on the removal of dairy biofilms. *International Journal of Food Microbiology* **106**(3), 254–262.

Browning, E., Lewis, M. & MacDougall, D. (2001) Predicting safety and quality parameters for UHT-processed milks. *International Journal of Dairy Technology* **54**(3), 111–120.

Bucky, A.R., Hayes, P.R. & Robinson, D.S. (1987) A modified ultra high temperature treatment for reducing microbial lipolysis in stored milk. *Journal of Dairy Research* **54**(2), 275–282.
Bulca, S., Leder, J. & Kulozik, U. (2004) Impact of UHT or high heat treatment on the rennet gel formation of skim milk with various whey protein contents. *Milchwissenschaft* **59**, 590–593.
Burdett, M. (1974) The effect of phosphates in lowering the amount of deposit formation during the heat treatment of milk. *Journal of Dairy Research* **41**, 123–129.
Burgess, S. A., Brooks, J. D., Rakonjac, J., Walker, K. M. & Flint, S. H. (2009) The formation of spores in biofilms of *Anoxybacillus flavithermus*. *Journal of Applied Microbiology* **107**(3), 1012–1018.
Burton, H. & Burdett, M. (1974) The effect of ageing of milk on the amount of deposit formation on heated surfaces. *XIX International Dairy Congress* **1E**, 167–168.
Burton, H. (1968) Reviews of progress of dairy science, Section G, Deposits from whole milk in heat treatment plant - a review and discussion. *Journal of Dairy Research* **35**, 317–330.
Burton, H. (1988) *Ultra High Temperature Processing of Milk and Milk Products*, Elsevier Applied Science, London.
Büser, W. & Erbersdobler, H. F. (1986) Carboxymethyllysine, a new compound of heat damage in milk products. *Milchwissenschaft* **41**, 780–785.
Button, P.D. (2007) *UHT milk spoilage owing to lipase & protease from psychrotrophic Pseudomonas species: Characteristics of enzymatic breakdown, improved assays for enzyme detection and sources of the bacteria*. PhD thesis, University of Melbourne, Australia.
Calvo, M.M. & de la Hoz, L. (1992) Flavour of heated milks. A review. *International Dairy Journal* **2**, 69–81.
Calvo, M.M., Klett, J.M., Santos, M.P. & Olano, A. (1987) [Comparative study of lactulose determination in milk by means of gas chromatography and liquid chromatography]. *Revista Espanola de Lecheria* **16**, 11–13.
Carulli, S., Calvano, C. D., Palmisano, F., & Pischetsrieder, M. (2011) MALDI-TOF-MS characterization of glycation products of whey proteins in a glucose/galactose model system and lactose-free milk. *Journal of Agricultural and Food Chemistry* **59**, 1793–1803.
Cattaneo, S., Masotti, F. & Pellegrino, L. (2008) Effects of overprocessing on heat damage of UHT milk. *European Food Research and Technology* **226**(5), 1099–1106.
Chandrapala, J., McKinnon I, Augustin, M.-A. & Udabage P. (2010) The influence of milk composition on pH and calcium activity measured *in situ* during heat treatment of reconstituted skim milk. *Journal of Dairy Research* **77**, 257–264.
Chavez, M.S., Livia, M.N., Taverna, M.A. & Cuatrin, A. (2004) Bovine milk composition parameters affecting the ethanol stability. *Journal of Dairy Research.* **71**, 201–206.
Chen B.Y. (2013) *Best Use for Milk and Some Associated Issues*. PhD thesis, University of Reading, UK.
Chen, B., Lewis, M.J. & Grandison, A.S. (2014) Effect of seasonal variation on the composition and properties of raw milk destined for processing in the UK. *Food Chemistry* **158**, 216–223.
Chen, B.Y., Grandison, A.S. & Lewis, M.J. (2012) Comparison of heat stability of goat's milk subjected to UHT and in-container sterilisation. *Journal of Dairy Science* **95**, 1057–1063.

Chen, B.Y., Grandison, A.S. & Lewis, M.J. (2015) Effect of seasonal variation on some physical properties and heat stability of milk subjected to ultra-high temperature and in-container sterilisation. *Food Chemistry* **181**, 227–234

Christiannson, G., Jenness, R. & Coulter, S.C. (1954) Determination of ionic calcium and magnesium in milk. *Analytical Chemistry* **26**, 1923–1927.

Colahan-Sederstrom, P.M. & Peterson, D.G. (2005) Inhibition of key aroma compound generated during ultrahigh-temperature processing of bovine milk via epicatechin addition. *Journal of Agricultural and Food Chemistry* **53**(2), 398–402.

Contarini, G. & Povolo, M. (2002) Volatile fraction of milk: Comparison between purge and trap and solid phase microextraction techniques. *Journal of Agricultural and Food Chemistry* **50**, 7350–7355.

Corredig, M. & Dalgleish, D.G. (1996a) The binding of α-lactalbumin and β-lactoglobulin to casein micelles in milk treated by different heating systems. *Milchwissenschaft* **51**, 123–127.

Corredig, M. & Dalgleish, D.G. (1996b) Effect of temperature and pH on the interactions of whey proteins with casein micelles in skim milk. *Food Research International* **29**, 49–55.

Corredig, M. & Dalgleish, D.G. (1999) The mechanisms of the heat-induced interaction of whey proteins with casein micelles in milk. *International Dairy Journal* **9**, 233–236.

Corrieu, G., Lalande, M. & Ferret, R. (1981) New monitoring equipment for the control and automation of milk pasteurization plants. In: *Food Process Engineering*, Volume 1, (eds P. Linko, Y., Malkki & J. Larinkari), pp. 165–171, Applied Science Publishers, London.

Corrieu, G., Lalande, M. & Ferret, F. (1988) On-line measurement of fouling and cleaning of industrial U.H.T. exchangers. In: *Fouling Science & Technology*, pp 575–587, Springer, Netherlands.

Corzo, N., Delgado, T., Troyano, E. & Olano, A. (1994) Ratio of lactulose to furosine as indicator of quality of commercial milks. *Journal of Food Protection* **57**, 737–739.

Crossley, A., Heyes, T.D. & Hudson, J.F. (1962) The effect of heat on pure triglycerides. *Journal of the American Oil Chemists Society* **39**, 9–14.

Dannenberg, F. & Kessler, H. (1988) Reaction kinetics of the denaturation of whey proteins in milk. *Journal of Food Science* **53**(1), 258–263.

Daufin, G., Kerherve, F.L. & Quemerais, A. (1985) Electrochemical potential influence upon fouling of stainless steel and titanium heat surfaces by sweet whey. In: *Fouling and Cleaning in Food Processing*. (eds D.B. Lund, E. Plett & C. Sandu., pp. 263–275. Madison University Press, Wisconsin, USA.

Davies, T.J, Hensstridge, C., Gillham, C.R. & Wilson, D.I., (1997) - Investigation of Whey protein deposit properties using heat flux sensors. *Trans IchemEng*, Vol 75, Part C, 106–110.

Davis, K., Prasad, C., Vijayagopal, P., Juma, S. & Imrhan, V. (2014) Advanced glycation end products, inflammation, and chronic metabolic diseases: Links in a chain? *Critical Reviews in Food Science and Nutrition* DOI: 10.1080/10408398.2012.744738.

De Cordt, S. Vanhoof, K., Hu, J., Maesmans, G., Hendrickx, M. & Tobback, P. (1992) Thermostability of soluble and immobilised á-amylase from *Bacillus licheniformis. Biotechnology and Bioengineering* **40**, 396–402.

de Jong, P. (1996) Modelling and optimization of thermal treatments in the dairy industry. Thesis, Technical University, Delft, The Netherlands.

de Jong, P., te Giffel, M.C., Straatsma, H. & Vissers, M.M.M. (2002) Reduction of fouling and contamination by predictive kinetic models *International Dairy Journal* **12**(2-3), 285–292.

de Jong, P., Waalewijn, R. & Van Der Linden, H. (1993) Validity of a kinetic fouling model for heat-treatment of whole milk. *Lait* **73**, 293–302.

de la Fuente, M.A. (1998) Changes in the mineral balance of milk submitted to technological treatments. *Trends in Food Science and Technology* **9**, 281–288.

de Wit, R. & Nieuwenhuijse, H. (2008) Kinetic modelling of the formation of sulphur-containing flavour components during heat-treatment of milk. *International Dairy Journal* **18**(5), 539–547.

Deeth, H.C. & Lewis, M.J. (2015) Protein stability in sterilised milk and milk products. In: *Advanced Dairy Chemistry*. Volume 2. *Proteins.* (eds P.L.H. McSweeney & J.A. O'Mahony) pp. 247–286, Springer, New York.

Deeth, H.C. & Fitz-Gerald, C.H. (2006) Lipolytic enzymes and hydrolytic rancidity. In: *Advanced Dairy Chemistry 2 Lipids.* (eds P.F. Fox and P.L.H. McSweeney), pp 481–556. Springer, New York.

Deeth, H.C. & Lewis, M.J. (2015) Practical consequences of calcium addition to and removal from milk and milk products. *International Journal of Dairy Technology* **65**(1), 1–10.

Delplace F, Leuliet J C & Tissier J P (1994) Fouling experiments of a plate heat exchanger by whey protein solutions, *Transactions of the Institute of Chemical Engineers* **72**, Part C, 163–169.

Demott, B.J. (1968) Ionic calcium in milk and whey. *Journal of Dairy Science* **51**, 1008–1012.

Donovan, M. & Mulvihill, D.M. (1987) Thermal-denaturation and aggregation of whey proteins *Irish Journal of Food Science and Technology* **11**(1), 87–100.

Doyle, R. J., Nedjat-Haiem, F. & Singh, J. (1984) Hydrophobic characteristics of *Bacillus* spores. *Current Microbiology* **10**, 329–332.

Drusch, S., Faist, V. & Erbersdobler, H.F. (1999) Determination of N-epsilon-carboxymethyllysine in milk products by a modified reversed-phase HPLC method. *Food Chemistry* **65**, 547–553.

Dwivedi, B.K. & Arnold, R.G. (1973) Chemistry of thiamine degradation in food products and model systems: A review. *Journal of Agricultural and Food Chemistry* **21**(1), 54–60.

Elliott, A. J., Datta, N. & Deeth, H. C. (2005) Heat-induced and other chemical changes in commercial UHT milks. *Journal of Dairy Research* **72**, 442–446.

Elliott, A.J., Dhakal, A., Datta, N. & Deeth, H.C. (2003) Heat-induced changes in UHT milks – Part 1. *Australian Journal of Dairy Technology* **58**, 3–10.

Erbersdobler, H. F. & Somoza, V. (2007) Forty years of furosine - Forty years of using Maillard reaction products as indicators of the nutritional quality of foods. *Molecular Nutrition & Food Research* **51**, 423–430.

Ezeh, V.N. & Lewis, M.J. (2011) Milk reversibility following reduction and restoration of pH. *International Journal of Dairy Technology* **64**, 179–187.

Farkye, N.Y. (2003) Other enzymes. In: *Advanced Dairy Chemistry, 1A Proteins*, (ed. P.F. Fox & P. McSweeney), pp. 571–603. Kluwer Academic/Plenum Publishers, New York.

Fenaille, F., Parisod, V., Visani, P., Populaire, S., Tabet, J.C. & Guy, P. A. (2006) Modifications of milk constituents during processing: A preliminary benchmarking study. *International Dairy Journal* **16**, 728–739.

Ferretti, A. (1973) Inhibition of cooked flavour in heated milk by use additives. *Journal of Agricultural and Food Chemistry* **21**(6), 939–942.

Ferron Baumy, C., Molle, D., Garric, G. & Maubois, J. L. (1992) Characterization of caseinomacropeptides released from renneted raw and UHT treated milks. *Lait* **72**(2), 165–173.

Fink, R. & Kessler, H. G. (1985) Reaction kinetics study of thiamine losses in stored UHT milk. **40**(12), 709–712.

Fink, R. & Kessler, H.G. (1986a) [Determination and mathematical analysis of lysine losses of stored heated milk.] *Ernaehrungs Umschau* **33**(4), 121–124

Fink, R. & Kessler, H.G. (1986b) HMF values in heat treated and stored milk. *Milchwissenschaft* **41**(10), 638–641.

Fitz-Gerald, C.H., Deeth, H.C., & Coghill, D.M. (1982) Low temperature inactivation of lipases from psychrotrophic bacteria. *Australian Journal of Dairy Technology* **37**(2), 51–54.

Flint, S., Palmer, J. Bloemen, K. Brooks, J. & Crawford, R. (2001) The growth of *Bacillus stearothermophilus* on stainless steel. *Journal of Applied Microbiology* **90**, 151–157.

Fogliano, V., Monti, S.M., Visconti, A., Randazzo, G., Facchiano, A.M., Colonna, G. & Ritieni, A. (1998) Identification of a β-lactoglobulin lactosylation site. *Biochimica et Biophysica Acta* **1388**, 295–304.

Ford J.E., Porter J.W.G., Thompson S.Y., Toothill J. & Edwards-Webb J. (1969) Effects of ultra-high-temperature (UHT) processing and of subsequent storage on the vitamin content of milk. *Journal of Dairy Research* **36**, 447–454.

Foster, C.L. & Green, M.L. (1990) A model heat exchange apparatus for the investigation of fouling of stainless steel surfaces by milk. II. Deposition of fouling material at 140 °C, its adhesion and depth profiling. *Journal of Dairy Research* **57**, 339–348.

Fox, P.F. & Morrissey, P.A. (1977) Reviews of the progress of dairy science: the heat stability of milk. *Journal of Dairy Research* **44**, 627–646.

Fox, P.F. (1981) Heat stability of milk: significance of heat-induced acid formation in coagulation. *Irish Journal of Food Science and Technology* **5**, 1–12.

Fox, P.F. & McSweeney, P.L.H (2003) *Advanced Dairy Chemistry*. Volume 1, 3rd edn, Kluwer Acadamic/Plenum Publishers, New York.

French, S.J., Harper, W.J., Kleinholz, N.M., Jones, R.B. & Green-Church, K.B. (2002) Maillard reaction induced lactose attachment to bovine beta-lactoglobulin, electrospray ionization and matrix-assisted laser desorption/ionization examination. *Journal of Agricultural and Food Chemistry* **50**, 820–823.

Friedman, M. (1999). Chemistry, biochemistry, nutrition, and microbiology of lysinoalanine, lanthionine and histidinoalanine in food and others proteins. *Journal of Agricultural and Food Chemistry* **47**, 1295–1349.

Friedman, M., Zahnley, J. & Masters, P. (1981) Relationship between *in vitro* digestibility of casein and its content of lysinoalanine and D-amino acids. *Journal of Food Science* **46**, 127–131.

Fritsch, R.J., Hoffmann, H., & Klostermeyer, H. (1983) Formation of lysinoalanine during heat treatment of milk. *Zeitschrift für Lebensmittel-Untersuchung und Forschung* **176**, 341–345.

Fryer P.J., Pritchard, A.M., Slater, M.K. H. & Laws, J.F. (1984) A new device for studying the the effects of surface shear stress on fouling in the presence of heat transfer. In *Fouling in*

heat exchanger equipment. Symposium of American Society of Mechanical Engineers, pp. 27–32.
Fryer, P. (1989) The uses of fouling models in the design of food process plant, *Journal of the Society of Dairy Technology.* **42**, 23–28.
Gaafar, A.M.M. (1987) *Investigation into the Cooked Flavour in Heat-treated Milk.* PhD thesis, University of Reading, UK.
Gaucher, I., Boubellouta, T., Beaucher, E., Piot, M., Gaucheron, F. & Dufour, E. (2008a) Investigation of the effects of season, milking region, sterilisation process and storage conditions on milk and UHT milk physico-chemical characteristics: a multidimensional statistical approach. *Dairy Science & Technology* **88**(3), 291–312.
Gaucher, I., Molle, D., Gagnaire, V. & Gaucheron, F. (2008b) Effects of storage temperature on physico-chemical characteristics of semi-skimmed UHT milk. *Food Hydrocolloids* **22**, 130–143.
Geerts, J.P., Bekhof, J. J. & Scherjon, J.W. (1983) Determination of calcium ion activities in milk with an ion selective electrode. *Netherlands Milk and Dairy Journal* **37**, 197–211.
Gibson, H., Taylor, J.H., Hall, K.H. & Holah, J.T. (1995) Removal of bacterial biofilms. *R&D Report, Campden & Chorleywood Food Research Association* **2**(34).
Görner F. & Uherova R. (1980) Vitamin changes in ultra high temperature sterilized milk during storage. *Nahrung-Food* **24**(4/5), 373–379.
Grandison, A.S. (1988) UHT processing of milk: seasonal variation in deposit formation in heat exchanger. *Journal of the Society of Dairy Technology* **41**, 43–49.
Griffiths, M.W. (1986) Use of enzymes as indices of heat treatment. *Journal of Food Protection* **449**, 696–705.
Griffiths, M.W., Phillips, J.D. & Muir, D.D. (1981) Thermostability of proteases and lipases from a number of species of psychrotrophic bacteria of dairy origin. *Journal of Applied Bacteriology* **50**(2), 289–303.
Grijspeerdt, K., Mortier, L., De Block, J & Van Renterghem, R. (2004) Applications of modelling to optimise ultra high temperature milk heat exchangers with respect to fouling. *Food Control* **15**(2), 117–130.
Grimley, H., Grandison, A.S. & Lewis, M.J. (2009) Changes in milk composition and processing properties during the spring flush period. *Dairy Science & Technology* **89**, 405–416.
Grimley, H.J., Grandison, A.S. & Lewis, M.J. (2010) The effect of calcium removal from milk on casein micelle stability and structure. *Milchwissenschaft* **65**, 151–154.
Grufferty, M. & Fox, P. (1986) Potassium iodate-induced proteolysis in ultra-heat-treated milk during storage: the role of beta-lactoglobulin and plasmin. *Journal of Dairy Research* **53**, 601–613.
Gynnig, K., Thorne, K.E. & Sammuelsson (1958) The burning of milk in plate pasteurizers. *Milchwissenschaft* **13**, 62–70.
Haddad G.S. & Loewenstein M. (1983) Effect of several heat treatments and frozen storage on thiamine, riboflavin, and ascorbic acid content of milk. *Journal of Dairy Science* **66**, 1601–1606.
Haryani, S., Datta, N, Elliott, A.J. & Deeth, H.C. (2003) Production of proteases by psychrotrophic bacteria in raw milk stored at low temperature. *Australian Journal of Dairy Technology* **58**, 16–20.
Hegele, J., Buetler, T. & Delatour, T. (2008) Comparative LC-MS/MS profiling of free and protein-bound early and advanced glycation-induced lysine modifications in dairy products. *Analytica Chimica Acta* **617**(1-2), 85–96.

Hegele, J., Parisod, V., Richoz, J., Forster, A., Maurer, S., Krause, R., Henle, T., Butler, T. & Delatour, T. (2008) Evaluating the extent of protein damage in dairy products: simultaneous determination of early and advanced glycation-induced lysine modifications. *Annals of the New York Academy of Sciences* **1126**(1), 300–306.

Henle, T., Walter, H. & Klostermeyer, H. (1991) Evaluation of the extent of the early Maillard reaction in milk products by direct measurement of the Amadori product lactulosyllysine. *Zeitschrift für Lebensmittel-Untersuchung und Forschung* **193**, 119–122.

Hewedy, M.M., Kiesner, C., Meissner, K., Hartkopf, J. & Erbersdobler, H.F. (1994) Effects of UHT heating of milk in an experimental plant on several indicators of heat treatment. *Journal of Dairy Research* **61**(2), 305–309.

Hickey, M. (2009) Current legislation of market milks. In: *Milk Processing and Quality Management.* (ed. A.Y. Tamime) pp. 101–138, SDT, Wiley Blackwell, Chichester, UK

Hiddink, J., Lalande, M., Maas, A.J.R. & Streuper, A. (1986) Heat treatment of whipping cream. I. Fouling of the pasteurisation equipment. *Milchwissenschaft* **41**(9), 542–546.

Holland, J, Gupta, R., Deeth, H.C. & Alewood, P.F. (2011) Proteomic analysis of temperature-dependent changes in UHT milk. *Journal of Agricultural and Food Chemistry* **59**, 1837–1846.

Holt, C. (2004) An equilibrium thermodynamic model of the sequestration of calcium phosphate by casein micelles and its application to the calculation of the partition of salts in milk. *European Biophysics Journal* **33**, 421–434.

Holt, C., Dalgleish, D.G. & Jenness, R. (1981) Calculation of the ion equilibria in milk diffusate and comparison with experiment. *Analytical Biochemistry* **113**, 154–163.

Horak, F. P. & Kessler, H. G. (1981) Thermal thiamine destruction - A second order reaction. *Zeitschrift fur Lebensmittel Untersuchung und Forschung* **173**(1), 1–6.

Horne, D.S (2003) Ethanol stability. In: *Advanced Dairy Chemistry, Volume 1 Proteins,* (ed. P.F. Fox), pp. 975–1000. Elsevier Applied Science Publishers, London.

Horne, D.S. & Muir, D.D. (1990) Alcohol and heat stability of milk protein. *Journal of Dairy Science* **73**, 3613–3626.

Hotrum, N., Fox, M., van Lieverloo, H., Smit, E. & de Jong, P. (2010) Modelling heat processing of dairy products. In: *Improving the safety and quality of milk,* Volume 1: *Milk production and processing* (ed. M.W. Grifiths), pp. 330–348, Woodhead Publishing, CRC, Oxford, UK.

Jaddou, H.A., Pavey, J.A. & Manning, D.J. (1978) Chemical analysis of flavour volatiles in heat-treated milks. *Journal of Dairy Research* **45**, 391–403.

Jeurnink, T.J.M. & DeKruif, K.G. (1995) Calcium concentration in milk in relation to heat stability and fouling. *Netherlands Milk and Dairy Journal* **49**, 151–165.

Johansen, C., Falholt, P & Gram, L. (1997) Enzymatic removal and disinfection of bacterial biofilms. *Applied and Environmental Microbiology* **63**(9), 3724–3728.

Jones, A.D., Ward, N.J. & Fryer, P.J. (1994) The use of a heat flux sensor to monitor milk fluid fouling. In: *Fouling and Cleaning in Food Processing.* pp. 199–205. Jesus College, Cambridge, UK.

Kastanas, P. (1996) *Studies on the Fouling of Heat Exchangers during the Sterilisation of Cow's and Goat's Milks.* PhD thesis, University of Reading, UK.

Kastanas, P., Lewis, M.J. & Grandison, A.S. (1995) Design and development of a miniature UHT plant for fouling studies and its evaluation using milk adjusted to different pH values, goat's milk and milk permeate. *Transactions of the Institute of Chemical Engineers* **73**, Part C, 83–92.

Kastanas, P., Lewis, M.J. & Grandison, A.S., (1996) Comparison of heat exchanger performance for goat and cow milk. In: *Production and Utilization of Ewe and Goat Milk*. Doc 9603, pp. 221–230. International Dairy Federation, Brussels.

Katsuno, S., Shimamura, T., Kashiwagi, T., Izawa, N. & Ukeda, H. (2013) Effects of dissolved oxygen on the Maillard reaction during heat treatment of milk. *International Dairy Journal* **33**, 34–37.

Kelly, A.L. &d Foley, J. (1997) Proteolysis and storage stability of UHT milk as influenced by milk plasmin activity, plasmin/beta-lactoglobulin complexation, plasminogen activation and somatic cell count. *International Dairy Journal* **7**, 411–420.

Kessler, H.G. (1981) *Food Engineering and Dairy Technology*. Verlag A. Kessler, Freising, Federal Republic of Germany.

Kessler, H.G. & Fink, R. (1986) Changes in heated and stored milk with an interpretation by reaction kinetics. *Journal of Food science* **51**(5), 1105–1111.

Knight, G.C., Nicol, R.S. & McMeekin, T.A. (2004) Temperature step changes: a novel approach to control biofilms of *Streptococcus thermophilus* in a pilot plant-scale cheese-milk pasteurisation plant. *International Journal of Food Microbiology* **93**, 305–318.

Kocak, H.R. & Zadow, J.G. (1985) The effect of low-temperature-inactivation treatment on age gelation of UHT whole milk. *Australian Journal of Dairy Technology* **40**, 53–58.

Krasaekoopt, W., Bhandari, B. & Deeth, H.C. (2003) Yogurt from UHT milk: A review *Australian Journal of Dairy Technology* **58**, 26–29.

Kudo S. (1980) The heat stability of milk: formation of soluble proteins and protein-depleted micelles at elevated temperatures. *New Zealand Journal of Dairy Science and Technology* **15**, 255–263.

Lalande, M. & Corrieu, G. (1981) Fouling of plate heat exchanger by milk. In: *Proceedings of Conference Fundamentals and Application of Surface Phenomena Associated with Fouling and Cleaning in Food Processing*., 6-9 April 1981, Tlosand, Sweden. pp. 279–288.

Lalande, M., Tissier, J.P. & Corrieu, G. (1984) Fouling of plate heat exchanger used in ultra-high-temperature sterilisation of milk. *Journal of Dairy Research* **51**, 557–568.

Langler, J.E., & Day, E.A. (1964) Development and flavour properties of methyl ketones in milk fat. *Journal of Dairy Science* **47**, 1291–1296.

Lauber, S., Klostermeyer, H. & Henle, T. (2001) On the influence of non-enzymatic crosslinking of caseins on the gel strength of yogurt. *Nahrung/Food* **45**, 215–217.

Laxminarayana, H. & Dastur, N.N., (1968) Buffaloes' milk and milk products. *Dairy Science Abstracts* **30**(5), 231–241.

Le, T.T., Deeth, H.C., Bhandari, B., Alewood, P.F. & Holland, J.W. (2013) Quantification of lactosylation of whey proteins in stored milk powder using multiple reaction monitoring. *Food Chemistry* **141**, 1203–1210.

Lehmann, F.C. (1996) Recontamination of industrial ultrafiltration units and pasteurisers by thermoduric-thermophilic bacteria. In: *Heat Treatments & Alternative Methods*, pp. 126–128. Doc 9602, International Dairy Federation, Brussels.

Lehmann, F.C., Russel, P.S., Solomon, L.S. &Murphy, K.D. (1992) Bacterial growth during continuous milk pasteurisation. *Australian Journal of Dairy Technology* **47**, 28–32.

Lewis, M.J. & Deeth, H.C. (2009) Heat treatment of milk. In: *Milk Processing and Quality Management*, (ed. A.Y. Tamine), pp 168–204. Society of Dairy Technology, Wiley-Blackwell, Chichester, UK.

Lewis, M.J. & Heppell, N. (2000) *Continuous Thermal Processing of Foods: Pasteurization and UHT Sterilization*, Aspen Publishers, Gaithersburg, MD,

Lewis, M.J. (2010) The measurement and significance of ionic calcium in milk - A review. *International Journal of Dairy Technology* **64**, 1–13.

Lewis, M.J., Grandison, A.S. Lin, M.-J. & Tsioulpas, A. (2011) Ionic calcium and pH as predictors of stability of milk to UHT processing. *Milchwissenschaft* **66**, 197–200.

Lin, M.-J. (2002) *Measurement of Ionic Calcium and its Role in Milk Stability*. PhD thesis, University of Reading, UK.

Lindsay, D. & von Holy, A. (1999) Different responses of planktonic and attached *Bacillus subtilis* and *Pseudomonas fluorescens* to sanitizer treatment. *Journal of Food Protection* **62**, 368–379.

Lynch, P.L.M., & Young, I.S. (2000) Determination of thiamine by high-performance liquid chromatography. *Journal of Chromatography A* **881**(1-2), 267–284.

Lyster, R.L.J. (1965) The composition of milk deposits in an ultra-high-temperature plant. *Journal of Dairy Research* **32**, 203–210.

Lyster, R.L.J. (1970) The denaturation of α-lactalbumin and β-lactoglobulin in heated milk. *Journal of Dairy Research* **37**, 233–243.

Ma, Y. & Barbano, D.M. (2003) Milk pH as a function of CO_2 concentration, temperature, and pressure in a heat exchanger. *Journal of Dairy Science* **86**, 3822–3830.

Manji, B., Kakuda, Y. & Arnott, D.R. (1986) Effect of storage temperature on age gelation of ultra-high temperature milk processed by direct and indirect heating systems. *Journal of Dairy Science* **69**, 2994–3001.

Marchand, S., De Block, J, De Jonghe, V., Coorevits, A., Heyndrickx, M. & Herman, L. (2012) Biofilm formation in milk production and processing environments; influence on milk quality and safety. *Comprehensive Reviews of Food Science and Food Safety* **11**, 133–147.

Martinez-Castro, I., Olano, A. & Corzo, N. (1986) Modifications and interactions of lactose with mineral components of milk during heating processes. *Food Chemistry* **21**, 211–221.

McKellar, R.C. (1981) Development of off-flavors in ultra-high temperature and pasteurized milk as a function of proteolysis. *Journal of Dairy Science* **64**, 2138–2145.

McKenna, B.M. & O'Sullivan, A.C. (1971) Whey protein denaturation in concentrated skimmilks. *Journal of Dairy Science* **54**(7), 1075–1077.

McKinnon, I.R., Yap, S.E., Augustin, M.-A. & Hemar, Y. (2009) Diffusing-wave spectroscopy investigation for heated reconstituted skim milks containing calcium chloride. *Food Hydrocolloids* **23**, 127–133.

McMahon, D. J., Yousif, B. H. & Kalab, M. (1993) Effect of whey protein denaturation on structure of casein micelles and their rennetability after ultra-high temperature processing of milk with or without ultrafiltration. *International Dairy Journal* **3**, 239–256.

McMahon, D., Savello, P.A., Brown, R.J. & Kalab, M. (1991) Effects of phosphate and citrate on the gelation properties of casein micelles in renneted ultra-high-temperature (UHT) sterilized concentrated milk. *Food Structure* **10**(1), 27–36.

Medda, S. & Chandra, A.K. (1980) New strains of *Bacillus lichiniformis* and *Bacillus coagulans* producing thermostable α- amylase at alkaline pH. *Journal of Applied Bacteriology* **48**, 47–58.

Mehta B.M. & Deeth H.C. (2015) Blocked lysine in dairy products – formation, occurrence, analysis and nutritional implications. *Comprehensive Reviews in Food Science and Food Safety* **15**, 206–218.

Messia, M. C., Candigliota, T. & Marconi, E. (2007) Assessment of quality and technological characterization of lactose-hydrolyzed milk. *Food Chemistry* **104**(3), 910–917.

Mitchell, G.E. & Ewings, K.N. (1985) Quantification of bacterial proteolysis causing gelation in UHT-treated milk. *New Zealand Journal of Dairy Science and Technology* **20**, 65–76.

Moio, L., Etievant, P., Langlois, D., Dekimpe, J. & Addeo, F. (1994) Detection of powerful odorants in heated milk by use of extract dilution sniffing analysis. *Journal of Dairy Research* **61**, 385–394.

Morales, F.J., Romero, C. & Jimenez Perez, S. (1997) Application of the clarifying reagent index as a heat-induced parameter in commercial milk. *Journal of Food Protection* **60**(10), 1241–1246.

Morales, F.J., Romero, C., & Jiménez-Pérez, S. (2000) Characterization of industrial processed milk by analysis of heat-induced changes. *International Journal of Food Science and Technology* **35**(2) 193–200.

Morales, F.J. & Jiménez-Pérez, S. (1998) Study of hydroxymethylfurfural formation from acid degradation of the Amadori product in milk-resembling systems. *Journal of Agricultural & Food Chemistry* **46**(10), 3885–3890.

Morales, F.J., Romero, C. & Jiménez-Pérez, S. (1992) An enhanced liquid-chromatographic method for 5- hydroxymethylfurfural determination in UHT milk. *Chromatographia* **33**(1-2), 45–48.

Morales, F.J., Romero, C. & Jiménez-Pérez, S. (2000) Characterization of industrial processed milk by analysis of heat-induced changes. *International Journal of Food Science & Technology* **35**(2), 193–200.

Morgan, F.J. & Priest, F.G. (1981) characterization of a thermostable alpha-amylase from *Bacillus licheniformis* NCIB-6346. *Journal of Applied Bacteriology* **50**(1), 107–114.

Mottar J. & Naudts M. (1979) La qualité du lait chauffé à ultra-haute temperature comparée à celle du lait pasteurisé et stérilisé dans la bouteille. *Le lait.* **588**, 476–488.

Mottar, J. & Moermans, R. (1988) Optimization of the forewarming process with respect to deposit formation in indirect ultra high temperature plants and the quality of milk. *Journal of Dairy Research* **55**, 563–568.

Mottar, J. (1981) Heat resistant enzymes in UHT milk and their influence on sensoric changes during uncooled storage. *Milchwissenschaft* **36**(2), 87–91.

Mottar, J., Bassier, A., Joniau, M. & Baert, J. (1989) Effect of heat-induced association of whey proteins and casein micelles on yogurt texture. *Journal of Dairy Science* **72**, 2247–2256.

Mounsey, J.S. & O'Kennedy, B.T. (2010) Heat-stabilisation of β-lactoglobulin through interaction with sodium caseinate. *Milchwissenschaft* **65**(1), 79–83.

Mugadza D. & Buys E. (2014) Diversity of spore former and non-spore former bacteria in Extended Shelf-life (ESL) milk. South African Society of Dairy Technology 2014 Dairy Symposium. http://www.sasdt.co.za/wp-content/uploads/2015/04/Session-8-Speaker-26-Mugadza1.pdf

Mugadza D., Buys E. (2015) Characterization of *B. cereus* in Extended Shelf-life (ESL) milk, International Association for Food Protection Conference Portland, Oregon PP2-138 (Abs) https://iafp.confex.com/iafp/2015/webprogram/Paper8975.html

Muir, D.D. & Sweetsur, A.W.M. (1976) Influence of naturally occurring levels of urea on heat stability of bulk milk. *Journal of Dairy Research* **43**, 495–499.

Mullan, W.M.A. (2011) Calculating the Fo, B* and C*values following ultra high temperature (UHT) processing of milk. [On-line] UK: Available: https://www.dairyscience.info/uht1.asp.

Nagai, R., Jinno, M., Ichihashi, M., Koyama, H., Yamamoto, Y. & Yonei, Y. (2012) Advanced glycation end products and their receptors as risk factors for aging. *Anti-Aging Medicine* **9**(4), 108–113.

Newstead, D.F., Groube, G.F., Smith, A.F. & Eiger, R.N. (1998) Fouling of UHT plants by recombined and fresh milk: some effects of preheat treatment. In: *Fouling and Cleaning in Food Processing.* (eds. D.I. Wilson, P.J. Fryer & A.P.M. Hasting), pp. 17–24. Cambridge, UK.

Newstead, D.F., Paterson, G., Anema, S.G., Coker, C.J. & Wewala, A.R. (2006) Plasmin activity in direct-steam-injection UHT-processed reconstituted milk: effects of preheat treatment. *International Dairy Journal* **16**, 573–579.

Nguyen, H.T., van der Fels-Klerx, H.J., & van Boekel, M.A.J.S. (2014) N^{ε}-(carboxymethyl) lysine: A review on analytical methods, formation, and occurrence in processed food, and health impact. *Food Reviews International* **30**(1), 36–52.

Nian, Y., Chen, B., Aikman, P., Grandison., A & Lewis, M.J. (2012) Naturally occurring variations in milk pH and ionic calcium and their effects on some properties and processing characteristics of milk. *International Journal of Dairy Technology* **65**, 490–497.

Nieuwenhuijse, J.A., Timmermans, W. & Walstra, P. (1988) Calcium and phosphate partitions during the manufacture of sterilized milk and their relations to the heat stability. *Netherlands Milk and Dairy Journal* **42**, 387–421.

Nohr, D., Biesalski, H.K. & Back, E.I. (2011) Thiamine. In: *Encyclopedia of Dairy Sciences* Vol. 4 (eds J.W. Fuquay, P.F. Fox and P.L.H. McSweeney), 2nd edn, pp. 701–703. Academic Press San Diego.

Nursten, H.E. (1981) Recent developments in studies of the Maillard reaction. *Food Chemistry* **6**, 263–277.

O'Brien, J. (1995) Heat-induced changes in lactose: Isomerisation, degradation, Maillard browning. In: *Heat-induced Changes in Milk.* (ed P.F. Fox) Doc. 9501, pp. 134–170. International Dairy Federation, Brussels.

Oamen E.E., Hansen A.P. and Swartzel K.R. (1989) Ultra-high-temperature steam injection processing and aseptic storage on labile water-soluble vitamins in milk. *Journal of Dairy Science* **72**, 614–619.

O'Connell, J.E. & Fox, P.F. (2003) Heat-induced coagulation of milk. In: *Advanced Dairy Chemistry. I. Proteins.* (eds P.F. Fox & P.L.H. McSweeney), 3rd edn, pp. 879–930, Kluwer Academic/Plenum Press, New York.

Ohashi, T., Yamauchi, K., Haga, S., Matsuki, K., Aoki, I., Ogawa, T. & Fujino, H. (1989) [Effects of pH, calcium and unheated casein micelles on rennet coagulation of UHT-heated milk.]. *Japanese Journal of Dairy and Food Science [Rakuno Kagaku Shokuhin no Kenkyu]* **38**(3), A103–A109.

Olano, A., Corzo, N., Paez, M.I. & Martinez-Castro, I. (1987) Isomerization of lactose during heat treatment of liquid and freeze-dried simulated milk ultrafiltrates. Effect of pH and calcium. *Milchwissenschaft* **42**, 628–630.

Oldfield, D.J., Singh, H. & Taylor, M.W. (1998) Association of β-lactoglobulin and α-lactalbumin with the casein micelles in skim milk heated in an ultra-high temperature plant. *International Dairy Journal* **8**, 765–770.

Oliveira, L.N., Marinho, V.T., Zamagno, M.V., Lauro, M.A., Barboza, J.A.N. & da Silva, P.H.F. (2015) Assessment of whey protein nitrogen index as an indicator of heat treatment for UHT milk and milk powder. *Journal of Candido Tostes Dairy Institute* **70**(2), 78–84.

Omoarukhe, E.D., On-Nom, N., Grandison, A.S. & Lewis, M.J. (2010) Effects of different calcium salts on properties of milk related to heat stability. *International Journal of Dairy Technology* **63**, 504–511.

On-Nom, N. (2012) *Partitioning of Milk at High Temperature*, PhD thesis, University of Reading, UK.

On-Nom, N., Grandison, A.S., & Lewis, M.J. (2010) Measurement of ionic calcium, pH and soluble divalent cations in milk at high temperature. *Journal of Dairy Science* **93**, 515–523.

On-Nom, N., Grandison, A.S. & Lewis, M.J. (2012) Heat stability of milk supplemented with calcium chloride. *Journal of Dairy Science* **95**, 1623–1631.

Oulahal, N., Martial-Gros, A., Bonneau, M. & Blum, L.J. (2004) Combined effect of chelating agents and ultrasound on biofilm removal from stainless steel surfaces. Application to "*Escherichia coli* milk" and "*Staphylococcus aureus* milk" biofilms. *Biofilms* **1**, 165–173.

Panfil-Kuncewicz, H., Kuncewicz, A. & Juskiewicz, M. (2005) The influence of the sterilisation method on the changes in UHT milk fat fraction. *Milchwissenschaft* **60**(1), 33–36.

Parkar, S. G., Flint, S. H. Palmer J. S, and Brooks. J. D. (2001) Factors influencing attachment of thermophilic bacilli to stainless steel. *Journal of Applied Microbiology* **90**, 901–908.

Paseephol, T., Small, D.M. & Sherkat, F. (2008) Lactulose production from milk concentration permeate using calcium carbonate-based catalysts. *Food Chemistry* **111**(2), 283–290.

Patil, G. R. & Reuter, H. (1986) Deposit formation in UHT plants. I. Effect of forewarming in indirectly heated plants. *Milchwissenschaft* **41**(6), 337–339.

Pellegrino, L., De Noni, I. & Resmini, P. (1995a) Coupling of lactulose and furosine indices for quality evaluation of sterilized milk. *International Dairy Journal* **5**, 647–659.

Pellegrino, L., Resmini, P. & Luf, W. (1995b) Assessment (indices) of heat treatment of milk. In: *Heat-induced Changes in Milk*. Docl. 9501 (ed P.F. Fox). pp. 409–453. International Dairy Federation, Brussels.

Peri, C., Pagliarini, E. & Pierucci, S. (1988) A study of optimising heat treatment of milk. I. Pasteurisation. *Milchwissenschaft* **43**, 636–639.

Perkin, A.G., Henschel, M.J. & Burton, H. (1973) Comparison of milks processed by the direct and indirect methods of ultra-high-temperature sterilization. VI. Effects on sediment formation and clotting with enzymes. *Journal of Dairy Research* **40**, 215–220.

Perkins, M.L., Elliott, A.J., D'Arcy, B.R. & Deeth, H.C. (2005) Stale flavour volatiles in Australian commercial UHT milk during storage. *Australian Journal of Dairy Technology* **60**, 231–237.

Pfeifer, J. & Kessler, H. (1995) Heat resistance of *Bacillus cereus* spores located between seals and seal surfaces. *Journal of Food Protection* **58**(11), 1206–1210.

Philippe, M., Gaucheron, F, Le Graet, Y, Michel, F. & Garem, A. (2003) Physicochemical characterization of calcium supplemented skim milk. *Lait* **83**, 45–59.

Pouliot,Y., Boulet, M. & Paquin, P. (1989) Observations on the heat induced salt balance changes in milk 1. Effect of heating time between 4 and 90 °C. *Journal of Dairy Research* **56**, 185–192.

Prakash, S, Kravchuk, O. & Deeth, H.C. (2015) Influence of pre-heat temperature, pre-heat holding time and high-heat temperature on fouling of reconstituted skim milk during UHT processing. *Journal of Food Engineering* **153**, 45–52.

Prakash, S. (2007) *Burn-on in Ultra-high-temperature Processing of Milk*. PhD thesis, University of Queensland, Australia.

Prakash, S., Datta, N. & Deeth, H.C. (2005) Methods of detecting fouling caused by heating of milk. *Food Reviews International* **21**, 267–293.

Prakash, S., Datta, N., Lewis, M. & Deeth, H. (2007) Reducing fouling during UHT treatment of goat's milk. *Milchwissenschaft* **62**, 16–19.

Pyne, G.T. & McHenry, K.A. (1955) The heat coagulation of milk. *Journal of Dairy Research* **22**, 60–68.

Rada-Mendoza, M., Villamiel, M. & Olano, A. (2002) Dissolved air effects on lactose isomerisation and furosine formation during heat treatment of milk. *Lait* **82**, 629–634.

Ramasubramanian, L., D'Arcy, B.R & Deeth, H.C. (2012) Heat-induced coagulation of whole milk by high levels of calcium chloride. *International Journal of Dairy Technology* **65**(2), 184–190.

Ramasubramanian, L., Restuccia, C. & Deeth H.C. (2008) Effect of calcium on the physical properties of stirred probiotic yogurt. *Journal of Dairy Science* **91**, 4164–4175.

Ramasubramanian, R., D'Arcy, B.R., Deeth, H.C & Oh, H.E. (2014) The rheological properties of calcium-induced milk gels. *Journal of Food Engineering* **130**, 45–51.

Ramsey, J. A. & Swartzel, K.R. (1984) Effect of ultra high temperature processing and storage conditions on rates of sedimentation and fat separation of aseptically packaged milk. *Journal of Food Science* **49**, 257–262.

Ranjith, H.M.P. (1995) *Assessment of some Properties of Calcium-reduced Milk and Milk Products from Heat Treatment and other Processes*. PhD thesis, University of Reading, UK.

Ranjith, H.M.P., Lewis, M.J. & Maw, D. (1999) Production of calcium-reduced milks using ion exchange. *Journal of Dairy Research* **66**, 139–144.

Rauh, V. M., Sundgren, A., Bakman, M., Ipsen, R., Paulsson, M., Larsen, L. B. & Hammershoj, M. (2014) Plasmin activity as a possible cause for age gelation in UHT milk produced by direct steam infusion. *International Dairy Journal* **38**(2), 199–207.

Reddy, K.K., Nguyen, M.H., & Kailasapathy, K. (1991) The effects of some treatments and storage temperatures on UHT whole milk. *Australian Journal of Dairy Technology* **46**(2), 57–63.

Richards, M., De Kock, H. L. & Buys, E. M. (2014) Multivariate accelerated shelf-life test of low fat UHT milk. *International Dairy Journal* **36**(1), 38–45.

Richardson, B.C. (1983) The proteases of bovine milk and the effect of pasteurisation on their activity. *New Zealand Journal of Dairy Science and Technology* **18**, 233–245.

Rombaut, R., Dewettinck, K., De Mangelaere, G. & Huyghebaert, A. (2002) Inactivation of heat resistant spores in bovine milk and lactulose formation. *Milchwissenschaft* **57**, 432–436.

Rose, D. (1961) Factors affecting the pH-sensitivity of the heat stability of milk from individual cows. *Journal of Dairy Science* **44**, 1405–1413.

Rose, D. (1962) Factors affecting the heat stability of milk. *Journal of Dairy Science* **45**, 1305–1311.

Rose, D. (1963) Heat stability of bovine milk: a review. *Dairy Science Abstracts* **25**, 45–52.
Ruckert, A., Ronimus, R.S. & Morgan, H.W. (2004) A RAPD-based survey of thermophilic bacilli in milk powders from different countries. *International Journal of Food Microbiology* **96**, 263–272.
Sadeghinezhad, E., Kazi, S.N., Badarudin, A., Zubair, M.N.M., Dehkordi, B.L. & Oon, C.S. (2013) A review of milk fouling on heat exchanger surfaces. *Reviews in Chemical Engineering* **29**(3), 169–188.
Salaün, F., Mietton, B. & Gaucheron, F. (2005) Buffering capacity of dairy products. *International Dairy Journal* **15**, 95–109.
Samel, R., Weaver, R.W.V. & Gammack, D.B. (1971) Changes on storage in milk processed by ultra-high-temperature sterilization. *Journal of Dairy Research* **38**, 323–332.
Samuelsson, E. G. & Borgstrom, S. (1973) Method for protecting milk against flavour deterioration when heat treated. *Milchwissenschaft* **28**, 25–26.
Samuelsson, E. G. (1962) Experiments on sunlight flavour in milk - S^{35}-labelled milk. *Volkswirtschaftlicher verlag* **17**, 401–405.
Saraiva, J., Oliveira, J.C., Hendrickx, M., Oliveira, F.A.R. & Tobback, P. (1996) Analysis of the inactivation kinetics of freeze-dried alpha-amylase from *Bacillus amyloliquefaciens* at different moisture contents. *Lebensmittel-Wissenschaft & Technologie* **29**(3), 260–266.
Scaloni, A., Perillo, V., Franco, P., Fedele, E., Froio, R., Ferrara, L. & Bergamo, P. (2002) Characterization of heat-induced lactosylation products in caseins by immunoenzymatic and mass spectrometric methodologies. *Biochimica et Biophysica Acta* **1598**, 30–39.
Scarpellino, R. & Soukup, R. J. (1993) Key flavors from heat reactions of food ingredients, In: *Flavor Science: Sensible Principles and Techniques*. (eds T.E. Acree & R. Teranishi), pp. 309–335. American Chemical Society, Washington, DC.
Schokker, E.P. & van Boekel, M. (1999) Influence of pH on low-temperature inactivation of the extracellular proteinase from *Pseudomonas fluorescens* 22F. *Milchwissenschaft* **54**(7), 377–379.
Schreiber, R. (2001) Heat-induced modifications in casein dispersions affecting their rennetability. *International Dairy Journal* **11**, 553–558.
Schreier, P.J.R., Green, C.H., Hasting A.P.M., Pritchard, A.M. & Fryer, P.J. (1994) Development of a fouling and cleaning test bed for milk fluids - Continuous flow processing, DTI Food Processing Sciences. Available at: http://www.dti.gov.ph/dti/index.php/news/810-dti-unveils-food-innovation-hub (accessed 7 November 2016).
Seale, R.B., Flint, S.H., McQuillan, A.J. & Bremer, P.J. (2008) Recovery of spores from thermophilic dairy bacilli and effects of their surface characteristics on attachment to different surfaces. *Applied and Environmental Microbiology* **74**, 731–737.
Shew, D.I. (1981) *New monograph in UHT milk*, Document No. 133, pp 115–121, International Dairy Federation, Brussels.
Shilton, N.C, Johnson, A. & Lewis, M.J. (1992) An investigation of a possible relationship between the ethanol stability of milk and the fouling of milk in an ultra high temperature process. *Journal Society Dairy Technology* **45**, 9–10.
Shipe, W.F., Bassette, R., Deane, D.D., Dunkley, W.L., Hammond, E.G., Harper, W.J., Kleyn, D.H., Morgan, M.E., Nelson, J.H. & Scanlan, R.A. (1978) Off flavors of milk: nomenclature, standards, and bibliography. *Journal of Dairy Science* **61**, 855–869.
Siciliano, R.A., Mazzeo, M.F., Arena, S., Renzone, G., & Scaloni, A. (2013) Mass spectrometry for the analysis of protein lactosylation in milk products. *Food Research International* **54**(1), 988–1000.

Sierra, I., Vidal-Valverde, C. & Lopez-Fandino, R. (2000) Effect of high pressure on the vitamin B1 and B6 content of milk. *Milchwissenschaft* **55**(7), 365–367.

Sievanen, K., Huppertz, T., Kelly, A.L. & Fox, P. F. (2008) Influence of added calcium chloride on the heat stability of unconcentrated and concentrated bovine milk. *International Journal of Dairy Technology* **61**, 151–155.

Simmonds, P, Mossel B.L., Intaraphan T. & Deeth H.C. (2003) The heat resistance of *Bacillus* spores when adhered to stainless steel and its relationship to spore hydrophobicity. *Journal of Food Protection* **66**, 2070–2075.

Singh, H & Fox, P.F. (1987) Heat stability of milk – Role of β-lactoglobulin in the pH-dependent dissociation of micellar κ-casein. *Journal of Dairy Research* **54**, 509–521.

Singh, H. & Latham, J.M. (1993) Heat stability of milk: Aggregation and dissociation of protein at ultra-high temperatures. *International Dairy Journal* **3**, 225–237.

Singh, H. (2004) Heat stability of milk. *International Journal of Dairy Technology* **57**, 111–119.

Skudder, P.J, Brooker, B.E, Bonsey, A.D. & Alvarez-Guerrero, R. (1986) Effect of pH on the formation of deposit from milk on heated surfaces during ultra high temperature processing. *Journal of Dairy Research* **53**, 75–87.

Skudder, P.J., Thomas, E.L., Pavey, J.A. & Perkin, A.G. (1981) Effects of adding potassium iodate to milk before UHT treatment. I. Reduction in the amount of deposit on the heated surfaces. *Journal of Dairy Research* **48**, 99–113.

Slinkard, M.S. (1976) *Changes in the Headspace Concentration of Volatile Sulfur Compounds during Aseptic Storage of Ultra-High-Temperature Processed Milk*, MS Thesis, University of Minnesota.

Srichantra, A., Newstead, D.F., McCarthy, O.J. & Paterson, A.H.J. (2006) Effect of preheating on fouling of a pilot scale UHT sterilizing plant by recombined, reconstituted and fresh whole milks. *Food Bioproducts Processing* **84**, 279–285.

Stoodley, P., Sauer, K., Davies, D.G. & Costerton, J.W. (2002) Biofilms as complex differentiated communities. *Annual Review of Microbiology* **56**, 187–209.

Swaisgood, H.E. (1977) Process of removing the cooked flavour from milk. United States Patent Application Number 4053644.

Swartzel, K.R. (1983) The role of heat exchanger fouling in the formation of sediment in aseptically processed and packaged milk. *Journal of Food Processing &. Preservation* **7**, 247–257.

Te Giffel, M.C., Beumer, R.R. Langeveld, L.P.M. & Rombouts. F.M. (1997) The role of heat exchangers in the contamination of milk with *Bacillus cereus* in dairy processing plants. *International Journal of Dairy Technology* **50**, 43–47.

Teh, K.H., Lindsay, D., Palmer, J., Andrewes, P., Bremer, P. & Flint, S. (2013) Lipolysis within single culture and co-culture biofilms of dairy origin. *International Journal of Food Microbiology* **163**(2-3), 129–135.

Teh, K.H., Lindsay, D., Palmer, J., Andrewes, P., Bremer, P. & Flint, S. (2014) Proteolysis in ultra-heat-treated skim milk after exposure to multispecies biofilms under conditions modelling a milk tanker. *International Journal of Dairy Technology* **67**(2), 176–181.

Tessier, H & Rose D (1958) Calcium ion concentration in milk. *Journal of Dairy Science* **41** 351–359.

Tetra Pak (2015) *Dairy Processing Handbook*, Tetra Pak processing Systems AB, Lund, Sweden.

Thomas, E.L., Reineccius, G.A., DeWaard, G.J. & Slinkard, M.S. (1976) Determination of hydrogen sulfide in heated milks by gas-liquid chromatographic headspace analysis. *Journal of Dairy Science* **59**(11), 1865–1869.

Timperley, D.A., Hasting, A.P.M. & De Goederen, G. (1994) Developments in the cleaning of dairy sterilization plant. *Journal of the Society of Dairy Technology* **47**, 2, 44–49.

Tissier, J.P., Lalande, M. & Corrieu, G. (1984) A study of milk deposit on heat exchange surface during UHT treatment. In: *Engineering and Food*, Volume 1, (ed. B.M. McKenna) pp. 49–58. Elsevier Applied Science Publishers, London.

Topçu, A., Numanoglu, E., & Saldamli, I. (2006) Proteolysis and storage stability of UHT milk produced in Turkey. *International Dairy Journal* **16**(6), 633–638.

Tran, H., Datta, N., Lewis, M.J. & Deeth, H.C. (2008) Predictions of some product parameters based on the processing conditions of ultra-high-temperature milk plants. *International Dairy Journal* **18**, 939–944.

Tsioulpas, A., Koliandris, A., Grandison, A.S. & Lewis, M.J. (2010) Effects of stabiliser addition and in-container sterilisation on selected properties of milk related to casein micelle stability. *Food Chemistry* **22**, 1027–1034.

Tsioulpas, A., Lewis, M.J. & Grandison, A.S. (2007) Effect of minerals on casein micelle stability of cow's milk. *Journal of Dairy Research* **74** 167–173.

Udabage, P., McKinnon, I. & Augustin, M.-A. (2000) Mineral and casein equilibria in milk: effects of added salts and calcium chelating agents. *Journal of Dairy Research* **67**, 361–370.

van Asselt, A.J., Sweere, A.P J., Rollema, H.S. & de Jong, P. (2008) Extreme high-temperature treatment of milk with respect to plasmin inactivation. *International Dairy Journal* **18**, 531–538.

van Boekel, M.A.J.S. (1998). Effect of heating on Maillard reactions in milk. *Food Chemistry* **62**(4), 403–414.

Vasbinder, A.J., Rollema, H.S. & de Kruif, C.G. (2003) Impaired rennetability of heated milk: Study of enzymatic hydrolysis and gelation kinetics. *Journal of Dairy Science* **86**, 1548–1555.

Vazquez-Landaverde, P., Torres, J.A. & Qian, M.C. (2006) Quantification of trace volatile sulfur compounds in milk by solid-phase microextraction and gas chromatography-pulsed flame photometric detection. *Journal of Dairy Science* **89**, 2919–2927.

Vazquez-Landaverde, P.A., Velazquez, G., Torres J.A. & Qian M.C. (2005) Quantitative determination of thermally derived off-flavor compounds in milk using solid-phase microextraction and gas chromatography. *Journal of Dairy Science* **88**, 3764–3772.

Viberg, U., Jagerstad, M., Osta, R. & Sjoholm, I. (1997) Thermal processing of 5-methyltetrahydrofolic acid in the UHT region in the presence of oxygen. *Food Chemistry* **59**(3), 381–386.

von Bockelmann, B. & von Bockelmann, I. (1998*) Long-life Products: Heat Treated: Aseptically Packed: A Guide to Quality*. von Bockelmann, Linnevagan, Sweden.

Wadsworth, K.D. & Bassette, R. (1985) Laboratory scale system to process ultra-high-temperature milk. *Journal of Food Protection* **48**, 530–531.

Walstra, P. & Jenness, R. (1984) *Dairy Chemistry and Physics*, John Wiley and Sons, New York.

Waungana, A., Singh, H. & Bennett, R. J. (1996) Influence of denaturation and aggregation of beta-lactoglobulin on rennet coagulation properties of skim milk and ultrafiltered milk. *Food Research International* **29**, 715–721.

Waungana, A., Singh, H. & Bennett, R.J. (1998) Rennet coagulation properties of skim milk concentrated by ultrafiltration: effects of heat treatment and pH adjustment. *Food Research International* **31**, 645–651.

West, F.B., Adams, D.M. & Speck, M.L. (1978) Inactivation of heat-resistant proteinases in normal ultra-high temperature sterilized skim milk by a low temperature treatment. *Journal of Dairy Science.* **61**, 1078–1084.

White, J.C.D. & Davies, D.T. (1958a) The relation between the chemical composition of milk and the stability of the caseinate complex. 1. General introduction, description of samples, methods and chemical composition of samples. *Journal of Dairy Research* **25**, 236–255.

White, J.C.D. & Davies, D.T. (1958b) The relation between the chemical composition of milk and the stability of the caseinate complex, IV. Coagulation by heat. *Journal of Dairy Research* **25**, 281–296.

Wijayanti, H.B., Bansal. N.& Deeth, H.C. (2014) Stability of whey proteins during thermal processing: a review. *Comprehensive Reviews in Food Science & Food Safety* **13**, 1235–1251.

Wirtanen, G., Husmark, U. & Mattila Sandholm, T. (1996) Microbial evaluation of the biotransfer potential from surfaces with *Bacillus* biofilms after rinsing and cleaning procedures in closed food-processing systems. *Journal of Food Protection* **59**(7), 727–733.

Withers P.M, Taylor J.H, Richardson P.S. & Holah J.T. (1994) Ultrasonic detection and monitoring of fouling during UHT processing. In: *Fouling and Cleaning in Food Processing.* Jesus College, Cambridge, 23–25 March 1994, pp 181-190.

Wüthrich, S., Richterich, R., & Hostettler, H. (1964) Studies of milk enzymes. II.Enzymes in centrifuged & heat-treated cows' milk. *Zeitschrift fur Lebensmittel-untersuchung und -forschung* **124**(5), 345–348.

Yoon, J. & Lund, D.B. (1994) Magnetic treatment of milk and surface treatment of plate heat exchangers: effects on milk fouling. *Journal of Food Science* **59**(5), 964–969, 980.

Zadow, J.G. & Hardham, J.F. (1981) Studies on the stability of concentrated and reconstituted concentrated skim milks toward UHT processing. *Australian Journal of Dairy Technology* **36**, 30–33.

Zadow, J.G., Hardham. J., Kocak, H. & Mayes, J. (1983) The stability of goat's milk to UHT processing. *Australian Journal of Dairy Technology* **38**, 20–23.

Zhang, S., Jin, Y., & Yun, Z. (2007) Effect of low temperature heat treatment on the activity of lipase VIII from *Pseudomonas* species in UHT milk. *Transactions of the Chinese Society of Agricultural Engineering* **23**(8), 231–235.

Zin El-Din, M., Aoki, T. & Kako, Y. (1991) Polymerization and degradation of casein in UHT milk during storage. *Milchwissenschaft* **46**, 284–287.

7

Changes During Storage of UHT Milk

The UHT process itself, including the heating, holding and cooling periods, lasts a relatively short period of time, from under 2 min and usually no longer than 6 min (Tran *et al.*, 2008). During this period, the product spends only a few seconds at the high temperature of approximately 140 °C. After cooling, there may be some intermediate storage in an aseptic storage tank before it is aseptically filled, or it may be filled directly into retail or bulk packaging.

In contrast to this, these "commercially sterile" products may then be stored at ambient temperature for up to one year. The intended shelf-life of the product is indicated on the package by a "best before" date. Thus the quality of the product which is finally consumed (and no doubt evaluated) will depend upon the culmination of the reactions which have taken place during raw material storage, during thermal processing and packaging, and during storage. The product may be consumed at any time up to the "best before" date, or even later. Although an acceptable "best before" time has been generally accepted as six months, there is increasing pressure to extend this to up to 12 months and even longer. This should be done with caution, as it takes many UHT products out of their comfort zone and causes unnecessary problems for the UHT processor, particularly after 6 months storage when problems become most apparent. However, UHT milk which has been stored for more than one year may still be acceptable to drink, although this may not be the case for all products of this age or even all samples from the same production batch.

Ambient temperature may range from below 0 °C to over 50 °C; in some locations, some products may occasionally freeze. Many countries have several different climatic zones. For example, in a large country such as America, the temperatures may range from below -20 °C to ≥30 °C at a certain time of the year. Many countries have very hot summers and very cold winters, so ambient storage conditions impose some tough conditions on UHT products. In addition, UHT products are often transported by ship and not refrigerated and may spend longer than a month to reach their destination. Temperature loggers have recorded values approaching 60 °C on some of these journeys (Kessler & Fink, 1986).

Therefore, in terms of the effect on UHT product quality, storage is by far the most variable and therefore the most indeterminable factor. It is possible for the food manufacturer to select good quality raw materials and control the processing conditions and packaging by putting in place quality assurance systems (Chapter 8), thereby ensuring that a top quality product leaves the premises; however, the manufacturer has much less

High Temperature Processing of Milk and Milk Products, First Edition. Hilton C. Deeth and Michael J. Lewis.

influence over its storage conditions. Therefore, it is desirable to establish the likely condition of the product at the end of its shelf-life, even under adverse storage conditions.

Currently, smart packaging or time-temperature indicators are not yet routinely incorporated into UHT packaging (see Section 8.12). These would advise both the consumer and the manufacturer if the storage conditions had been much less than ideal for some of the storage period, although certain product characteristics, for example, brown colour of white milk products, may strongly suggest that this had been the case (see Section 7.2.5).

The aim of this chapter is to explore the changes that take place in UHT products during storage, especially those that influence product safety and quality. The use of product incubation to accelerate these changes is discussed in Section 7.4 and Chapter 8, as this can provide useful information about the stability of a product in a relatively short time period. For UHT products where there has been a major failure of heat treatment or gross post-processing contamination, changes in the product can occur rapidly and are accentuated by incubation at elevated temperatures. For example, such faulty UHT milk products may sour or produce gas within 24 to 72 h at 30 °C. This is easily detected by a fall in pH or dissolved oxygen and possibly made more noticeable by a foul smell, coagulation or swelling of the container. Where such a gross failure does occur, it should be detected before the product is released. The vast majority of UHT products are commercially sterile, but they are still subject to changes brought about by chemical reactions, residual enzyme activity and physical changes. These may well reduce the nutritional value of the food as well as change its sensory characteristics. *It is important to emphasise that all chemical reactions are temperature-dependent and any initiated during processing will continue to take place during storage.* Thus a sequence of reactions, for example, those involved in both flavour and colour changes in UHT milk, take place much more quickly at elevated temperatures. Consequently, foods generally become unacceptable more quickly as the storage temperature increases. This is particularly noticeable above 30 °C.

Several changes occur during storage of milk which is commercially sterile, that is, not affected by growth of microorganisms. If it is contaminated, further changes occur which usually limit its shelf-life. The major changes which affect the sensory characteristics and/or nutritional value are:

- colour changes, with browning being the predominant reaction
- flavour changes: cooked flavour and changes in sulfhydryl group activity and the development of stale and oxidised flavours (These changes result from chemical reactions involving proteins and fats and may lead to the production of volatile components with unpleasant odours and flavours); oxygen- and light-induced changes; changes induced by enzymes producing off-flavour defects such as bitter and rancid flavours
- texture changes such as sediment formation, thickening, thinning or gelation
- changes in nutritional value arising from loss of vitamins and changes in proteins, lactose and fats

These changes are discussed in more detail in the following sections. The main body of research has been on changes in milk and milk-based products, but many of the changes described also affect other foods and so the principles involved are also applicable to them.

The rate of these changes is largely dependent on the storage temperature but several other factors such as the UHT heating mode and conditions, the dissolved oxygen content, the milk composition, particularly fat content, and concentration of heat-resistant enzymes are also important. An indication of the importance of storage temperature on sensory quality is given by Richards *et al.* (2014) who monitored the flavour deterioration of commercial, UHT low-fat milk packaged in HDPE bottles at 25, 35 and 45 °C. They concluded that from the rates of sensory deterioration that the shelf-lives at three temperatures for the low-fat milks were 211, 73 and 27 days, respectively.

There is no doubt that refrigerated storage will maintain UHT products acceptable for a long time period, as chemical reactions giving rise to undesirable changes in the sensory characteristics of the foods will be slowed down. However, one of the main advantages of UHT products is that they do not require refrigerated storage. Such refrigerated storage would introduce additional energy costs and have a negative impact on environmental issues.

Questions often asked about UHT milk are "how long will UHT milk keep for?" and "what happens when UHT milk is kept for a long time?". The scientist's answer to the first question is that it depends on several factors such as those alluded to above. However, commercially, companies have to make a decision on the "use-by" or "best before" date to place on the package. Some companies assume shelf-lives of 9 and even 12 months while other more conservative companies base their date stamps on a 3-month shelf-life. Valero *et al.* (2001) stated that 90 days was the legal shelf-life of UHT milk in Spain. When setting "best before" dates, companies need to be aware of the keeping quality of their milk. If the quality of the raw milk is good, that is, low bacterial and somatic cell counts, and they use processing conditions which favour a high degree of enzyme destruction, that is, a UHT system with high C* with suitable pre-heat conditions, the product should still be acceptable after 9-12 months.

Consumption of UHT milk as a percentage of total drinking milk varies widely between countries from less than 10% in the UK, USA, Australia and New Zealand to over 90% in some European countries. As a consequence, the significance of changes during storage also varies. Provided the raw milk quality is reasonable and the UHT processing conditions are adequate, the changes impacting on consumer acceptance of the UHT milk during the first few months of ambient storage will be minimal and hence in countries with a high proportion of UHT milk, changes during storage will have far less relevance than in situations where shelf-lives of 9 to 12 months are required.

When samples of UHT milk are kept at room temperature for long periods of time, >12 months, most show two types of deterioration, coagulation/separation/sedimentation and brown discoloration. The first can be attributed to protein instability, such as age gelation, sedimentation and fat emulsion destabilisation, and the second to Maillard reactions which continue during storage. If they do not have bitter or rancid flavours, aged milks have a bland but not unpleasant taste. Discussed below are changes which occur in UHT milk, and the conditions under which they occur, during normal storage periods.

7.1 Chemical Changes

7.1.1 pH

It is well established that the pH of UHT milk decreases during storage (e.g., AlKanhal *et al.*, 1994; Gaucher *et al.*, 2008b). This has been largely attributed to formation of formic and acetic acids, intermediates in the Maillard reaction (van Boekel, 1998; van Boekel & Brands, 1998). Examples of this decrease are given in Section 7.5.

Proteolysis may be another cause of a reduction in pH of UHT milk during storage. While this has not been observed by all researchers, some have shown a small (≤1 pH unit) reduction during storage for ≥6 months due to plasmin-induced proteolysis (Kohlman *et al.*, 1991; Newstead *et al.*, 2006).

7.1.2 Dissolved Oxygen Content

Dissolved oxygen is very significant in UHT milk. In fact, Kessler (1989) stated that storage changes in heat-treated foods fall into two categories: those which are influenced by dissolved oxygen and those which are not. Flavour changes during storage and the rate at which some vitamins are lost are examples of those affected by the level of dissolved oxygen, whereas browning and gelation are apparently not influenced by dissolved oxygen concentration (Andrews, 1986).

The oxygen content in good quality raw milk is normally close to saturation, at about 9-11 mg/L. [Some authors have reported lower levels of 6-7 mg/L in saturated milk (Rada-Mendoza *et al.*, 2002; Perkins *et al.*, 2005)]. Pumping and gentle agitation will ensure that it remains close to saturation. Dissolved oxygen is less soluble at higher temperatures and may come out of solution at UHT temperatures; however, it will largely re-dissolve on cooling as it cannot escape due to the enclosed environment, except in flash cooling. According to Burton (1988), oxygen is prevented from coming out of solution during UHT processing by the application of about 1 bar (0.1 MPa) in excess of the saturated vapour pressure at the corresponding processing temperature (see Table 5.6).

During UHT processing, if deaeration is not used, the oxygen content of indirectly processed milk remains close to saturation level, but may be considerably lower in directly processed milk due to the vacuum treatment step after steam injection or steam infusion to remove water condensed during the heating stage. The level may be less than 1 mg/kg (Burton, 1988) but it may be higher if the milk is allowed to absorb air when stored for some time in an aseptic tank before being packaged, or the package has a large air headspace.

Provided the packaging material is impermeable to oxygen, the oxygen content in UHT milk during storage is determined by two factors operating in opposite directions. The first is the amount of oxygen in the headspace of the package and the second is the use of oxygen in oxidation reactions of oxidisable compounds in the milk. The amount of oxygen in the headspace is determined mostly by the volume, which in turn is determined by the type of package. For example, a Tetra Pak Tetra Brik® paperboard carton which is made in situ from a roll of paperboard immediately before filling has very little headspace, 7-8 mL in a 1-L carton, a 1-L Combibloc preformed carton has a headspace of 21-40 mL and a 1-L plastic bottle has 55-63 mL of headspace (Perkins *et al.*, 2005). Another factor which affects the amount of oxygen in the headspace is flushing with nitrogen before closure of the package. This is sometimes done to reduce oxidation in the product during storage, particularly if the product contains added oxygen-sensitive ingredients such as ascorbic acid.

If the headspace volume is small, as for Tetra Brik cartons, the concentration of dissolved oxygen in UHT milk decreases during storage (Adhikari & Singhal, 1992; Perkins *et al.*, 2005; Al-Attabi *et al.*, 2014). This is due to its use in oxidation of sulfhydryl-containing compounds and oxidisable vitamins such as ascorbic acid and folic acid

(Andersson & Öste, 1992a). The sulfhydryl compounds formed during heating include both volatile compounds discussed in Section 6.1.6.1 and the proteins with exposed sulfhydryl groups, for example β-Lg. If the headspace volume is small, the oxygen in the headspace is insufficient to keep the milk saturated with oxygen; conversely if the headspace volume is large, as in plastic bottles, the dissolved oxygen lost through oxidation reactions can be replenished with oxygen from the headspace. This was illustrated by Perkins *et al.* (2005) where directly processed commercial UHT milks packed in plastic bottles with a large headspace (average of 58 mL) did not show a decrease in dissolved oxygen during 4-months' storage whereas indirectly processed milks in Tetra Brik cartons showed a decrease to ~1 mg/L over that time period.

If the packaging material is oxygen-permeable, for example polyethylene, there will be an almost infinite supply of oxygen. However, most UHT containers are not permeable to oxygen as shown by Rysstad *et al.* (1998). They measured the oxygen permeability of three types of paperboard carton: polyethylene-coated, polyethylene coated with a non-foil paperboard light barrier and polyethylene coated with an aluminium foil light barrier. The oxygen transmission figures were >1000, 10-20 and 0 mg $O_2/m^2/24\,h$ respectively. The polyethylene coated container with an aluminium foil light barrier is typical of the most common paperboard cartons used for UHT milk.

Because of the importance of dissolved oxygen in oxidation and development of oxidised/stale flavour (see Section 7.1.3.3) ways of decreasing it have been sought. As indicated above, direct UHT processing with its vacuum treatment is one way of decreasing dissolved oxygen. However, deaeration/degassing before high heat treatment has been the most used technique, reducing the dissolved oxygen to 2-3 mg/kg. This has been done by vacuum treatment (Ikezumi *et al.*, 2006, Katsuno *et al.*, 2013), bubbling with nitrogen (Rada-Mendoza *et al.*, 2003), and a combination of nitrogen sweep and sonication (Wadsworth & Bassette, 1985) (see also Section 5.4).

Deaeration has been shown to reduce the development of stale flavour (Wadsworth & Bassette, 1985), and loss of ascorbic acid and folacin (Andersson & Öste, 1992b). Deaeration may be particularly useful in situations where considerable aeration takes place in the preparation stage; for example, in the mixing of dried ingredients, such as ice-cream mix, fortified milk drinks or drinks made by powder reconstitution.

The effects of deaeration may not all be positive. Some dissolved oxygen is necessary for oxidising the volatile sulfur compounds responsible for the intense cooked flavour of UHT milk over the first few days of storage, so a low dissolved oxygen level will delay or prevent this oxidation, causing the cooked flavour to remain for a longer time. Zadow and Birtwistle (1973) concluded that for optimum flavour in the milk, the headspace should have a partial pressure of oxygen of 60-100 mm Hg. There is some evidence that deaeration increases the production of Maillard reaction products during heating of milk (Rada-Mendoza *et al.*, 2002; Katsuno *et al.*, 2013) (see Section 6.1.4.1); however, the effect of the level of dissolved oxygen on the Maillard reaction during storage of sterilised products is unclear.

The above discussion assumes that the UHT milk is commercially sterile, that is, no growth of bacteria occurs. It has been reported that early evidence of bacterial contamination in such a product is a low dissolved oxygen content (Zadow & Birtwistle, 1973). This is important for trouble shooting defective product (see Chapter 8).

The effect of dissolved oxygen in heat-treated milk also depends on the antioxidant – prooxidant balance in milk which may be affected by heat treatment and storage of UHT milk. It is apparent that while heat and storage are known to increase some antioxidant properties due to formation of sulfhydryl compounds (Taylor &Richardson, 1980) and Maillard reaction products (Vandewalle & Huyghebaert, 1980; McGookin & Augustin, 1997), natural antioxidants may also be destroyed and some prooxidants may be formed. Caligaris *et al.* (2004) reported that short heat treatments may deplete the antioxidant properties of milk but severe heat treatments involving formation of brown pigments, melanoidins, may increase antioxidant activity. This is consistent with the findings of Aloglu (2013) that the antioxidant capacities of raw, pasteurised and sterilised milks were not significantly different.

7.1.3 Flavour

In addition to the flavours produced during the heat treatment (Section 6.1.6), flavours are produced or modified during storage of UHT milk which affect its acceptability during the storage period. It is generally accepted that the flavour of UHT milk changes during storage but is most acceptable after about 2-3 weeks. It is then affected by other flavours such as stale and oxidised. A graphical representation of UHT flavours during storage is shown in Figure 7.1. A good account of the flavour changes in milk on heating and during storage was given by Ashton (1965) who recognised two phases, each with a number of distinct stages. These phases, which are summarised below, indicate the dynamic nature of the process:

Primary phase

a) Initial heated flavour, accompanied by a strong sulfhydryl or cabbagey smell.
b) Weaker sulfhydryl or cabbage odour with residual cooked flavour
c) Residual cooked flavour with normal, acceptable agreeable flavour

Secondary phase

d) Normal, acceptable to agreeable, flat flavour
e) Flat, acceptable to mild oxidised flavour
f) Incipient oxidised flavour (or rancidity) to pronounced rancidity.

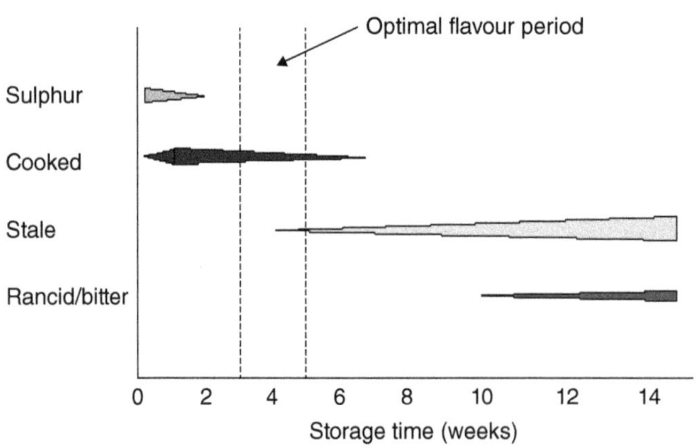

Figure 7.1 Profile of organoleptic changes in UHT milk during storage at 25 °C. (Source: Blanc, 1981. Reproduced with permission of International Dairy Federation.)

The flavour changes which take place are considerable. Ashton considered that UHT milk is best consumed while it is in stages (c) and (d). However, this is difficult to assure because the rate at which any sample of milk progresses through this sequence of events will depend largely on the storage temperature and the level of dissolved oxygen. Because the rate of change of flavour is so dependent on temperature, deterioration of flavour is much reduced at refrigeration temperatures. However, one of the major benefits of UHT processing is eliminating the need for refrigeration.

In an investigation of the effect of storage temperature on flavour of UHT milk, Prasad (1989) collected UHT milk from a commercial dairy on its day of production and stored it at 4, 22 and 30 °C. The milk samples were evaluated by a trained panel (20 assessors) over a period of 6 months. The results from the hedonic testing are given in Table 7.1. At 4 °C and 22 °C, the milks were most preferred when 2 weeks old, whereas at 30 °C the milk was most preferred when 4 days old. At all three storage temperatures, the general preference was for milks which were days or weeks old rather than months. At 22 °C and 30 °C (typical ambient temperatures) milks which were two months old or older were preferred less and milks that were six months old were least preferred. However, for milks stored at 4 °C, those which were one week old or less also scored poorly, indicating that the initial objectionable cooked flavour disappeared slowly at low temperature.

The chemical bases of the flavour changes during storage are described below. The major flavours identified in stored UHT milk are the sulfurous flavour remaining from the heat-treatment, the cooked/heated/sterilised flavour, the stale/oxidised flavour, and, if the product contains sufficient residual active proteases or lipases, bitter or rancid flavours respectively. Unfortunately, the descriptors used for the flavour of UHT milk vary considerably with different terms being used for the same flavour (see Table 6.6).

Table 7.1 Hedonic results for UHT milk samples stored at three different temperatures for 6 months. (Source: Prasad, 1989. Reproduced with permission.)

	Storage temperature (°C)		
	4	22	30
Hedonic order	**Sample age**		
1.	2 weeks	2 weeks	4 days
2.	1 month	4 days	2 weeks
3.	2 months	1 week	1 month
4.	3 months	1 day	1 day
5.	4 days	1 month	1 week
6.	1 week	3 months	2 months
7.	4 months	4 months	4 months
8.	5 months	2 months	5 months
9.	6 months	5 months	3 months
10.	1 day	6 months	6 months

7.1.3.1 Sulfurous Flavour

As indicated in Chapter 6, the intensity of the sulfurous flavour decreases markedly during the first week of storage. This is due to the decrease in the volatile sulfur compounds, produced during heat processing, through oxidation and reactions with other milk components. The exact nature of the reactions and the resultant compounds formed which then remain in the milk and may contribute to the flavour of UHT milk during storage have not been elucidated.

During the first 1-2 weeks, many of the flavoursome volatile sulfur compounds (hydrogen sulfide, methanethiol, dimethyl sulfide and dimethyl sulfoxide) decrease rapidly while others (carbon disulfide, dimethyl disulfide) decrease more slowly (Al-Attabi *et al.*, 2014). In particular, hydrogen sulfide which is a major contributor to the sulfurous flavour immediately after processing disappears rapidly to be negligible after 12 days. After this initial period of reduction in sulfurous flavour there is a period of optimum flavour of UHT milk which according to some estimates lasts around 3 weeks, before the onset of other flavours, principally stale flavour. This period of optimum flavour is dependent on the storage temperature; the higher the temperature, the shorter the optimum flavour period and the earlier its onset (Adhikari & Singhal, 1992). When UHT milk is stored at low temperature, the initial improvement in flavour due to reduction in volatile sulfur compounds does not occur and hence it has been suggested that there is no benefit in storing the milk at low temperatures rather than at moderate room temperature (Andersson & Öste, 1995). However this is contradicted by Blanc *et al.* (1980) who found that UHT milk stored at 5 °C has better flavour than milk stored at 25 °C.

Al-Attabi *et al.* (2014) found that the concentrations of all volatile sulfur compounds, with the exception of dimethyl trisulfide (and possibly dimethyl sulfoxide whose flavour threshold is not available), decreased to levels below their flavour thresholds in 2 months. However, a sulfurous note appears to be present after this time which may be due to a combination of sulfur compounds at low concentrations plus dimethyl trisulfide (and possibly dimethyl sulfoxide) or to some other compounds not detected to date.

The sulfur-containing compounds are formed from proteins during heating. The major source in skim milk is β-lactoglobulin and in whole milk it is β-lactoglobulin and milk fat globule membrane proteins. This is reflected in the higher level of the sulfur-containing compounds such as hydrogen sulfide and methanethiol in UHT whole milk than in UHT skim milk soon after processing (Al-Attabi *et al.*, 2014). The mechanisms for the formation of some of these sulfur volatiles have been discussed by Walstra and Jenness (1984) and Al-Attabi *et al.* (2009).

Gaafar (1987) made a detailed investigation into sulfur-containing compounds in relation to the onset of cooked flavour. He concluded that the threshold of cooked flavour corresponds to a β-lactoglobulin denaturation of about 60%, a hydrogen sulfide concentration of 3.4 μg/L and a reactive sulfhydryl concentration of 0.037 mmol/L.

7.1.3.2 Cooked/Heated/Sterilised Flavour

This appears to be the flavour which most consumers associate with UHT milk and refer to as "cooked". However, the vocabulary used for describing this flavour varies, with terms such as cooked, boiled, cabbagey, sulfury, heated, sterilised and caramelised being used. This flavour is unpopular with many consumers and is a major reason why

UHT milk sales remain low in some countries. The flavour is easily detected in liquid milk and by many consumers in hot drinks such as tea and coffee, despite relatively small amounts of milk being used; it is less easy to identify in other formulated milk drinks, particularly those with strong characteristic flavours such as chocolate or coffee. This cooked flavour which appears to be present to some extent in all UHT milks is somewhat ill-defined chemically but appears to be largely due to Maillard reaction products plus, to some extent, lipid oxidation and some residual sulfur-containing compounds. Maillard reaction products are formed in only low concentrations during UHT processing as the major Maillard reaction at that stage mostly involves lactosylation of proteins forming, with Amadori rearrangement, protein-bound lactulosyl lysine which does not contribute to flavour. Subsequent Maillard reactions which continue during storage produce a range of compounds, some of which are flavoursome.

Some of the compounds which have been identified with this flavour are diacetyl, methyl ketones, δ-lactones, benzaldehyde, furfural, phenylacetaldehyde, vanillin, oct-1-en-3-ol, n-heptanol, 2-butoxy ethanol, maltol, acetophenone, benzonitrile, benzothiazole. However, the relative contributions and the origins of each are still unclear. Valero *et al.* (2001) analysed the volatile components of commercial UHT skim and whole milk samples during storage. A large number of compounds were identified with some, for example, furanic compounds, being attributed to Maillard reactions.

7.1.3.3 Stale/Oxidised Flavour

At some time in the first two months of ambient storage, the flavour quality of UHT milk begins to deteriorate due to an increase in the intensity of "stale" flavour, also described as oxidized, tallowy, coconut-like and cardboardy (Wadsworth & Bassette, 1985). Staleness is a major cause of criticism of UHT milk flavour (Rerkrai *et al.*, 1987). It becomes superimposed on the "cooked" flavour described above but many consumers identify it as a dominant flavour in aged UHT milk.

The stale flavour in UHT milk is generally attributed to volatile lipid oxidation products, particularly the $C_{3\text{-}7,9\text{-}11,13}$ methyl ketones and the $C_{2\text{-}7,9}$ saturated aldehydes (Mehta, 1980). In total, 18 stale flavour volatiles have been identified in UHT milk; these include seven methyl ketones (i.e., 2-pentanone, 2-hexanone, 2-heptanone, 2-octanone, 2-nonanone, 2-decanone, 2-dodecanone) and six aldehydes (i.e., pentanal, hexanal, heptanal, octanal, nonanal, decanal). All have been shown to increase in commercial UHT milks during ambient storage (Perkins *et al.*, 2005).

While the methyl ketones are produced by decarboxylation of β-keto fatty acids, the saturated aldehydes are formed via oxidation of saturated or unsaturated fatty acids to hydroperoxides. The saturated fatty acid hydroperoxide decomposes to an aldehyde and a short chain carboxylic acid radical, which in turn can oxidise, dehydrate and decarboxylate to a produce a further aldehyde molecule (Grosch, 1982). Aldehydes can also be formed through autoxidation of unsaturated fatty acids which forms hydroperoxides which decompose to straight-chain aldehydes (Forss, 1979). In this way, oleic acid (C18:1) yields nonanal, octanal, decanal and heptanal while linoleic acid (C18:2) predominantly yields hexanal and, to a lesser extent, pentanal and heptanal (Badings, 1970).

Methyl ketones have been shown to be more abundant in stored UHT milk than aldehydes; however, aldehydes contribute more to the stale flavour than the methyl ketones because of their lower flavour thresholds (Badings *et al.*, 1981; Gaafar & El-Sayed, 1991,

Perkins *et al.*, 2005). Accordingly, the aldehydes have been shown to correlate more with stale flavour than the methyl ketones (Jeon *et al.*, 1978; Rerkrai *et al.*, 1987).

However, according to Moio *et al.* (1994) who analysed commercial UHT milk by gas chromatography–olfactometry and used the CharmAnalysis™ technique, the methyl ketones 2-heptanone and 2-nonanone were the main contributors to the flavour of UHT milk. Using odour activity values, Vazquez-Landnverde *et al.* (2005) concluded that 2-heptanone and 2-nonanone are very important contributors to UHT milk flavour but they also showed that aldehydes, especially nonanal, decanal, octanal, hexanal, 2-methylbutanal, 3-methylbutanal and 2-methylpropanal, also contribute to the flavour of UHT milk. Interestingly, methyl ketones and aldehydes have been shown to exhibit an additive flavour interaction in UHT milk whereby a mixture of a class of compounds has a lower flavour threshold concentration than the sum of the threshold concentrations of each compound in the mixture (Day *et al.*, 1963). Therefore, a mixture of aldehydes may have an effect on the flavour of milk even when the concentrations of each of the individual compounds are lower than their respective threshold concentrations.

7.1.3.4 Bitterness

As discussed in Chapter 6, the native protease, plasmin, and some bacterial proteases have high heat stabilities and can survive UHT heat treatment and remain active in the product during storage. Given that UHT products are stored at room temperature over several months, even trace amounts of residual active enzyme can cause proteolysis of milk proteins. The main effects of proteolysis are flavour change with production of bitter flavours, due to hydrophobic peptides, and physical change due to gelation during storage (see Section 7.2.2.1).

In general, in stored UHT milk, bitterness is highly correlated with proteolysis, which in turn is highly correlated with the count of psychrotrophic bacteria in the raw milk before UHT treatment (Collins *et al.*, 1993). Furthermore there is a strong (inverse) relationship between the level of proteolysis and the intensity of heat treatment. This is evident in the fact that more proteolysis occurs in directly heated UHT milk than in indirectly heated milk, due to the greater inactivation of the proteases by the higher heat intensity of the indirect treatment (Corradini & Pecchini, 1981; Driessen, 1981; McKellar *et al.*, 1984). According to experiments carried out by Driessen (1981), bitterness occurred after 6 weeks in UHT milk directly processed at 137 °C but after 12 weeks when indirectly heated at 143 °C. Proteolysis of UHT milk was more affected by the duration of heating (duration varied 0-16 s at 140 °C) than by the temperature of heating (temperature varied 134-146 °C for 4.8 s). Based on a D-value for the inactivation of protease of 7 s at 142 °C, a holding time of >16 s would be necessary to prevent proteolysis by milk plasmin in UHT milk during storage. This is clearly excessive; a more effective strategy is to inactivate the plasmin in the pre-heat section (see Section 5.2.1.3).

The temperature of storage has a major effect on the level of proteolysis. Langsrud and Hadland (1977) studied UHT goat's milk and found more proteolysis at 37 °C than at 30 °C. During storage at 37 °C, bitter flavour developed after 21 days, after 71 days at 30 °C and after 105 d at ambient temperature; no bitterness developed in milk after 2 years when stored at 4 °C. These authors observed bitterness after the content of proteose peptone, their measure of proteolysis, doubled. Similarly, McKellar (1981) detected off flavours in UHT milk to which three different proteases had been added when the increases in proteolysis were 0.287, 0.355 and 0.554 μmoles/mL against a background of 0.82 μmoles/mL.

Another factor affecting proteolysis and development of bitterness in UHT milk is the fat content. More proteolysis occurs in UHT skim milks than in UHT whole milks processed under the same UHT conditions. The same effect occurred with proteolysis by plasmin and bacterial proteases (López Fandiño *et al.*, 1993a). Therefore UHT heating conditions should be more severe for skim milk than for whole milk, a practice followed by some processors.

Bitterness can arise in lactose-hydrolysed products from a different source. Several authors have reported this to be due to the presence of protease contamination in the β-galactosidase preparation. A good example is the report by Jansson *et al.* (2014).

7.1.3.5 Hydrolytic Rancidity (Lipolysis)

Hydrolytic rancidity due to free fatty acids can develop in UHT milk during storage. Lipolysis and hydrolytic rancidity in milk and milk products was comprehensively reviewed by Deeth and Fitz-Gerald (2006). The free fatty acids are the result of lipase-catalysed degradation of triacylglycerols. The free fatty acids vary in length from C4 (butyric) to C18 (stearic, oleic and linoleic) but only the short- and medium-chain-length acids are very flavoursome. Scanlan *et al.* (1965) found that C4 (butyric) to C12 (lauric) acids accounted for most of the rancid flavour and that the higher fatty acids contributed very little. It has been suggested that the flavour of the shorter fatty acids is detected on the front of the palate and the longer fatty acids exert a sensation at the back of the palate which remains for some time after a lipolysed milk is consumed.

Consumers vary widely in their perception and detection of the flavour resulting from lipolysis as it is quite a subtle flavour. Free fatty acids in non-lipolysed milk, 0.5-1.0 mmol/L, contribute to the normal flavour of milk but when present at much higher levels they impart an off-flavour. Experienced milk grading personnel may detect the flavour at around 1.5 mmol/L free fatty acids but most consumers will only detect it as unpleasant and unacceptable when the free fatty acid level reaches ≥2.0 mmol/L. Curiously, high levels of free fatty acids are essential for the typical flavour of some foods such as parmesan and feta cheese, and some chocolates.

Most of the research performed on lipolysis has been on raw and pasteurised milk and cream, butter and cheese where lipolysis causes off-flavours which are described as rancid, butyric, unclean, soapy and astringent (Deeth & Fitz-Gerald, 2006). The flavour has also been described as bitter but that term should be reserved for the off-flavour caused by proteolysis. The lipolysis problems encountered with raw and pasteurised milk and cream, and, to some extent, butter and cheese are due to lipolysis by the natural milk lipoprotein lipase, although microbial lipases can cause problems in butter and cheese which are stored for longer periods of time than raw and pasteurised milk and cream. Most of the milk lipoprotein lipase-related issues originate on-farm through activation of the milk lipase system, but can be exacerbated by inappropriate raw milk handling practices in factories. While the flavour of UHT milk will be affected if lipolysed raw milk is used, this flavour will not change during storage and hence is considered to be a raw milk quality issue (see Section 8.5.1).

The native milk lipoprotein lipase is not involved in lipolysis during storage of UHT milk as it does not survive UHT heat treatment; in fact it is almost completely inactivated by commercial pasteurisation. The enzymes involved in lipolysis in UHT milk are heat-resistant lipases produced by psychrotrophic bacteria during growth in raw milk (see Chapter 6).

Several authors have shown that free fatty acids and rancid off-flavours result when a bacterial lipase is deliberately added to UHT milk (e.g., Driessen & Stadhouders, 1974; Anderssson *et al.,* 1981; Christen *et al.,* 1986; Button, 2007) or there is residual heat-resistant lipase in the milk (Blake *et al.,* 1995; Celestino *et al.,* 1997). Furthermore, significant correlations have been found between psychrotroph counts in raw milk and lipolysis in UHT milk (e.g., Mottar, 1981). However, instances of rancid off-flavours in UHT milk are rare. A possible explanation for this is that the psychrotrophic bacteria which produce the heat-resistant lipases also produce heat-resistant proteases which cause the development of bitter flavour during storage; this may dominate or mask any rancid flavour. This could be the reason why Collins *et al.* (1993) were unable to find a significant correlation between lipolysis and rancid off-flavour in commercially processed UHT skim milk but found significant correlations between proteolysis and bitter flavour. These authors suggested that proteases were more important than lipases in determining the taste and acceptability of UHT skim milk. However, Celestino *et al.* (1997) reported that the flavour of UHT reconstituted milk made from powder which had been produced from stored raw milk was more affected by lipolysis than by proteolysis. Interestingly, they implicated reactivated bacterial lipase as the cause of the lipolysis.

The extent of lipolysis is influenced by the amount of active lipase present and the temperature and duration of storage. The amount of active lipase remaining after UHT heat treatment is influenced by the effectiveness of the heat treatment in inactivating the enzyme. Less active enzyme remains after indirect processing than after direct processing (Panfil-Kuncewicz *et al.,* 2005) and, as indicated in Section 6.1.3.5, less may remain if a low-temperature inactivation step at 50-60 °C is included. More lipolysis occurs at higher temperatures of storage and longer periods of storage.

Choi *et al.* (1993) detected residual lipase activity in commercial UHT milks. The highest activity was found in the cream fraction, but activity was also found in the aqueous supernatant and the casein fraction. During storage, lipolytic activity was much higher at 35 °C, than at 23 °C. A similar acceleration of lipolysis during higher-temperature storage was reported by Singh *et al.* (2004).

7.1.3.6 Flavour Improvement Approaches

There have been a number of attempts to improve the flavour of UHT milks. One approach is to add substances to milk before heat treatment to reduce the intensity of the "cooked" flavour. Badings (1977) reported that cooked flavour could be reduced by addition of L-cystine to milk prior to heating. For indirect heating, addition of 30 and 70 mg cystine/kg of milk reduced the hydrogen sulfide concentration in the heated milk from 82.5 to 9.5 and 1.7 μg/g. Again, the hydrogen sulfide was found to disappear quickly during the first 24 h of storage. There was also no tendency to oxidation or other flavour defects; it was proposed that hydrogen sulfide was removed by L-cystine with L-cysteine as the reaction product. On the other hand, addition of L-cysteine resulted in a massive increase in a most unacceptable cabbagey or sulfury flavour.

Skudder *et al.* (1981) found that potassium iodate (10 -20 ppm) also reduced cooked flavour, by causing the oxidation of exposed sulfhydryl groups. Unfortunately, bitter components were formed about 14 days after processing at these levels of iodate addition. This was attributed to either increased stability of proteases within the milk or suppression of natural protease inhibitors.

Another additive which may be readily accepted is the polyphenol epicatechin which has been shown to reduce cooked flavour and not impart a bitter flavour at an addition rate of 100-1000 mg/kg. A bitter flavour was detected on addition of 2000 mg/kg (Colahan-Sederstrom & Peterson, 2005).

An effective additive which is legal in several foods in several countries is hydrogen peroxide. Low concentrations (10 or 50 mg/kg) significantly reduce the level of sulfur volatiles in the UHT milk when added either before or after processing. For example, UHT milk with 50 mg/kg hydrogen peroxide added before or after processing contained no detectable hydrogen sulfide or methanethiol by the second day after manufacture (Al-Attabi, 2009).

Swaisgood (1977) patented a process to remove cooked flavour from milk by immobilised sulfhydryl oxidase attached to glass beads. Sulfhydryl oxidase is a natural enzyme in milk. Since it is unstable to sterilisation conditions, Swaisgood (1977) suggested placing a sulfhydryl oxidase enzyme reactor downstream of the UHT holding tube. This process does not appear to have been adopted commercially.

In a novel approach Perkins *et al.* (2007) used an oxygen-scavenging film incorporated into aseptic pouches to reduce dissolved oxygen in UHT milk. This reduced the level of stale flavour volatiles (methyl ketones and aldehydes) in indirectly processed UHT milk.

7.1.4 Proteolysis

This section discusses proteolysis caused by the two different proteases relevant to UHT processing, plasmin and bacterial proteases. Proteolysis is also discussed in this chapter in relation to development of bitterness (Section 7.1.3.4) and age gelation (Section 7.2.2.1).

Plasmin and bacterial proteases have different specificities for individual caseins. The preferred substrate for plasmin is β-casein; however, α_{S1}- and α_{S2}-casein are also hydrolysed (de Rham & Andrews, 1982; Richardson, 1983) but κ-casein is generally considered to be resistant (Eigel, 1979). Plasmin does not hydrolyse whey proteins; they may actually have an inhibitory effect (Politis *et al.*, 1993). γ-Casein and proteose peptones are the peptides formed by plasmin.

Bacterial proteases generally have a preference for κ-casein, which is similar to that of chymosin in rennet, but they can also attack other caseins (Law *et al.*, 1977). A common product of proteolysis of milk proteins by bacterial proteases and chymosin is para-κ-casein, fragment 1-105 by cleavage of the Phe 105-Met 106 bond of κ-casein. This peptide is easily detected by PAGE (Figure 7.2). Other peptides released from κ-casein by *Ps. fluorescens* protease are fragments 1-103, 1-104, 1-106 and 1-107 (Miralles *et al.*, 2003). It should be noted however, that the reported specificities of bacterial proteases vary. This may be due to different bacterial species and strains tested or to different experimental conditions used (Nicodème *et al.*, 2005). Some authors (Carini & Todesco, 1977; Deepa & Mathur, 1994; Baglinière *et al.*, 2012) have found that β-casein was most preferred. Baglinière *et al.* (2012) reported the specificity of proteases from five strains of *Ps. fluorescens* was in the order β- > α_{S1}- > κ- > α_{S2}-casein.

Based on the difference in the peptides released by plasmin and most bacterial proteases, analytical methods for distinguishing between the two types have been developed. Plasmin produces peptides which are larger and more hydrophobic than

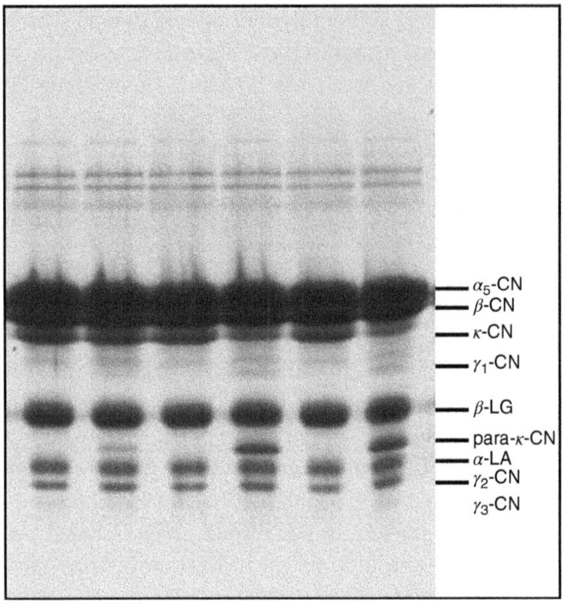

Figure 7.2 Protein degradation of UHT milk samples before & during storage at 25 °C, analysed with SDS-PAGE. a) Lane 1: Day 0 - control, Lane 2: Day 0 - treatment, Lane 3: Day 3 - control; Lane 4: Day 3 - treatment, Lane 5: Day 6 - control, Lane 6: Day 6 - treatment. (Source: Button, 2007. Reproduced with permission.)

those released by bacterial proteases. Plasmin peptides are soluble at pH 4.6 but not in 4% TCA while peptides from bacterial proteases are soluble at pH 4.6 and in 4% TCA. Because of their difference in hydrophobicity, the peptides produced by bacterial proteases elute earlier than those produced by plasmin on reversed-phase HPLC (López Fandiño *et al.*, 1993b; Datta & Deeth, 2003). The peptides in pH 4.6 and 4% TCA filtrates of milk can also be quantified by methods such as fluorescamine which detect free amino groups in peptides. These methods are outlined in Section 11.2.21.1.

A method of estimating low levels of protease activity in UHT milk which produce proteolysis during long storage periods has been sought by the dairy industry for some time in order to be able to detect potentially defective milk before storage. Several methods have been investigated (Cliffe & Law, 1982a; Kwan *et al.*, 1983; Christen & Senica, 1987; Fairbairn, 1989; Chen *et al.*, 2003) but none has been universally adopted. A suitable method would be very sensitive in order to detect trace levels of activity and correlate well with the protease's action in milk. Ideally, it would use the natural substrate, casein or casein derivative, rather than a non-natural substrate. One such method which uses a fluorogenic casein derivative, FITC-casein, and long assay incubation times has been shown to be able to detect very low levels of added bacterial protease in UHT milk (Button *et al.*, 2011). It was able to detect 0.0003% of a *Ps. fluorescens* cell-free culture supernatant added to UHT milk using a 10-day incubation period; a 0.003% addition could be detected in an 8-hour incubation. Addition of 0.0003% was the lowest level to produce proteolysis in UHT milk, which was measurable by the TNBS method, during room temperature storage. Another sensitive method which could be used in conjunction with the incubation conditions used by Button *et al.* (2011) is RP-HPLC.

In this method, the protease activity is quantified by integration of the areas of the peptide peaks in the chromatograms and relating the total to an internal standard dipeptide such as Tyr-Leu (Le *et al.,* 2006) (see also Section 11.2.19).

Proteolysis in UHT milk is easier to measure than protease activity. Methods of measuring proteolysis do not have to be extremely sensitive. Several methods have been developed and found useful. These include measurement of non-protein nitrogen (NPN) (12% TCA-soluble N) and non-casein nitrogen (NCN) (pH 4.6-soluble N) using a nitrogen method such as Kjeldahl, measurement of free amine groups using fluorescamine, trinitrobenzenesulfonic acid (TNBS) and ortho-phthahalaldehyde (OPA), and peptide measurement by HPLC and electrophoretic methods (Recio *et al.,* 1996; AlKanhal, 2000; Datta & Deeth, 2003; Le *et al.,* 2006) (see Section 11.2.21).

In summary, proteolysis is one of the major concerns for UHT processors because of its involvement in development of bitterness and age gelation. However, it is also one change during storage which processors are able to control. If it is caused by bacterial proteases, the raw milk quality needs to be improved and if it is caused by plasmin the process conditions, particularly the pre-heat conditions, need to be changed.

7.1.5 Protein Cross-Linking

Cross-linking of proteins during sterilisation heat treatments was discussed in Section 6.1.3.3. In general, only low levels of cross-linking occur during UHT sterilisation but higher levels are produced by in-container sterilisation (Fritsch *et al.,* 1983). During storage, however, considerable cross-linking occurs and the percentage of cross-linked casein can reach a high proportion of the total protein, particularly in milk stored at higher temperatures. Andrews (1975) reported that after storage of UHT milk at 37 °C for nine months, over 50% of the casein was polymerized or cross-linked. This represented an increase over that in raw milk of some six-fold. By contrast, storage at 4 °C for nine months resulted in 23% being cross-linked, an increase of 9%; this increase is similar to the level in raw milk (6%). Zin El-Din *et al.* (1991) reported similar but smaller increases during storage for 150 days: from 7.5% in freshly processed UHT milk to 11% and 23% during storage at 5 °C and 30 °C, respectively. Using densitometry of 1-D PAGE gels, Al-Saadi and Deeth (2008) detected increases in high-molecular weight proteins during storage of UHT milk for 12 weeks at 5, 20, 37 and 45 °C with the greatest increase of 36% at 45 °C; little increase occurred at 5 and 20 °C. These results were reflected in the lysinoalanine (LAL) results which showed little development at 5 and 20 °C but substantial increases at 37 and 45 °C, from 60 mg/kg to 130 and 170 mg/kg, respectively. The effect of storage temperature was also shown by Holland *et al.* (2011) using 2-D PAGE. An example of the gels obtained and the appearance of bands of high-molecular-weight proteins which increase in intensity during storage is shown in Figure 7.3 for a UHT milk stored at 28 °C for two months.

As discussed in Section 6.1.3.3, the cross-linking of proteins in UHT milk can occur via isodipeptide linkages such as LAL and by the reaction of the proteins with reactive cross-linking compounds such as glyoxal and methyl glyoxal formed during the advanced stages of the Maillard reaction. It has been shown that LAL increases with storage and there is extensive evidence of the Maillard reaction continuing during storage; therefore it is clear that both cross-linking mechanisms operate during storage of UHT milk and that both proceed faster at higher temperatures of storage. That Maillard

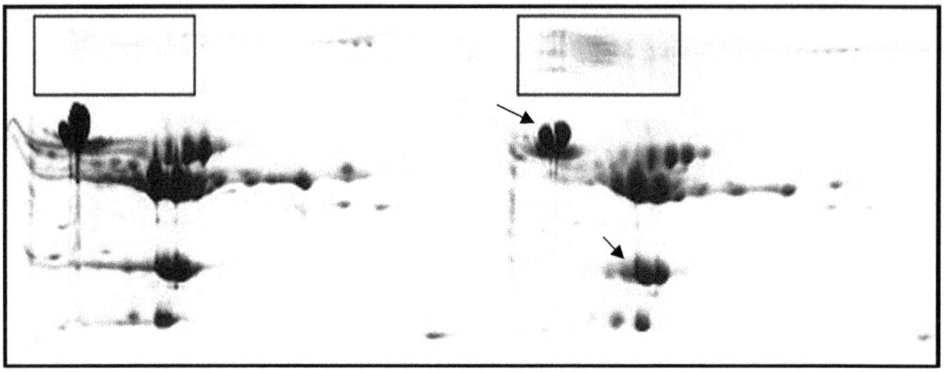

Figure 7.3 2-D PAGE showing non-disulfide covalent cross-linking in stored UHT milk. Horizontal dimension by charge, pH 4-7, L to R; vertical dimension by size. Control milk on the left and milk stored at 28 °C for 2 months on the right. Boxes show the region containing covalently non-disulfide cross-linked proteins. The arrows indicate deamidated proteins [Gupta, R., Holland, J., Deeth, H. and Alewood, P. unpublished; similar electrophoretograms appear in Holland *et al.* (2011)].

reactions are involved in the cross-linking was demonstrated by Le *et al.* (2013) by incubating milk proteins with methylglyoxal and producing similar polymerised proteins to those which form in UHT milk during storage.

Andrews (1975) found that the percentage of proteins in UHT milk cross-linked via disulfide bonds decreased during storage. He attributed this to disulfide-linked proteins becoming involved in polymers formed by non-disulfide cross-links and then not being susceptible to disruption by the disulfide bond reducing agent, 2-mercaptoethanol.

RP-HPLC chromatograms of the proteins in UHT milk change considerably during storage such that the chromatograms after several months' storage at room temperature or after a shorter time at higher temperatures are unrecognizable as those of milk proteins. This is clearly demonstrated in the chromatograms published by Gaucher *et al.* (2008b) and Al-Saadi and Deeth (2008). These dramatic changes are due to various reactions on the proteins but cross-linking is a major factor. Other changes include lactosylation, deamidation and proteolysis. Interestingly, cross-linking has not been associated with changes in the functional properties of UHT milk, with the possible exception of age gelation where McMahon (1996) suggested cross-linking may retard age gelation of UHT milk when stored at >30 °C. However, Venkatachalam *et al.* (1993) concluded that the Maillard reaction which causes protein cross-linking (Le *et al.* 2013) had little or no effect on age gelation.

Andrews (1975) suggested that α_{s1}-casein is the casein most involved in the cross-linking reactions and β-casein is involved to a lesser extent. This was confirmed by Holland *et al.* (2011) who used a proteomic approach to show that the high-molecular-weight bands on 2D-PAGE gels were dominated by α_{s1}-casein and contained lesser amounts of cross-linked forms of β-casein and α_{S2}-casein.

7.1.6 Deamidation

Non-enzymatic deamidation involves hydrolysis of the amide groups in asparagine and glutamine residues of proteins to aspartic acid and glutamic acid, respectively, and ammonia. This can be caused by very intense heating of milk such as at 140 °C for several minutes but to only a small degree during normal sterilisation processes, either UHT (Holland *et al.*, 2011) or in-container sterilisation (van Boekel, 1999). However, significant deamidation occurs during storage of UHT milk (Holland *et al.*, 2011). Asparagine residues adjacent to glycines show the highest rates of deamidation but asparagines adjacent to serines and histidines are also deamidated significantly.

Because ammonia is released during deamidation, the reaction can be monitored by increases in NPN (Metwalli & van Boekel, 1998). Alternatively, the new protein species formed, which has a higher negative charge than the parent protein, can be detected using isoelectic focussing as used in the first dimension of 2-D PAGE (Holland *et al.*, 2011, 2012). New spots for deamidated α_{S1}-casein and β-lactoglobulin can be clearly seen on 2-D gels of stored UHT milk (shown by arrows in Figure 7.3). Deamidation increases with temperature and time of storage.

There are some practical implications of deamidation of milk proteins which have not been fully explored for UHT milk processing. It has been shown to increase their heat stability, reduce the extent of fouling deposit formation, maintain the nutritional value of proteins through reduced blockage of lysines and improve some physical functional properties (Miwa *et al.*, 2010; Timmer-Keetels *et al.*, 2011). It is now possible to deliberately deamidate proteins using specific enzymes such as protein-glutaminase to improve the solubility, viscosity and emulsification capacity of skim milk solutions (Miwa *et al.*, 2010). The extent to which the deamidation, which occurs during storage of UHT milk, affects its functional properties is not known.

7.1.7 Lactosylation

Protein lactosylation or glycation with lactose, the first step in the Maillard reaction, is initiated during processing (see Section 6.1.4.1) and continues during storage. This is often indicated by furosine analysis and by the mass spectrometric methods discussed in Section 6.1.4.1. In all UHT milks, the level of lactosylated, and hence blocked, lysine residues increases with time of storage with more lactosylation occurring at higher storage temperatures. Several authors have reported this but typical of the data are those of Corzo *et al.* (1994) where storage of four batches of UHT milk for 90 days at 20 °C yielded an increase in furosine of 2.7-15.8 mg/L (~8-49 mg/100 g protein), at 30 °C the increase was 25.2-51.1 mg/L (~76-153 mg/100 g protein) and at 40 °C it was 54.1-101.4 mg/L (~162-304 mg/100 g protein). Much less lactosylation occurs at refrigeration temperatures. The lactosylated lysines, in the form of the Amadori product lactulosyl lysine, are also degraded during storage to other Maillard reaction products which include the brown pigments, melanoidins (see Section 7.2.5). Hence the level of lactosylated lysines measured in UHT milk after a period of storage is the resultant of both formation and degradation.

In addition to the total amount of blocked lysine, the number of lactose residues attaching to the proteins also increases. While one or two attach to the whey proteins during UHT processing as discussed in Section 6.1.4.1, at least five attach during storage. Furthermore these seem to attach in a non-specific manner. It has been shown that 9 of the 15 lysine residues in β-Lg and 8 of the 12 lysines in α-La were lactosylated (Holland *et al.*, 2012).

In lactose-hydrolysed milks, lysines are glycated by glucose and galactose and during storage this glycation occurs at a faster rate than lactosylation in lactose-containing milks. Messia *et al.* (2007) reported much higher levels of glycated lysine, as determined by furosine analysis, in lactose-hydrolysed UHT milks after 4-months' storage at 20 °C than at 4 °C and recommended refrigerated rather than ambient storage for UHT lactose-hydrolysed products.

7.1.8 Formation of Monosaccharides

During UHT and in-container sterilisation, the concentration of glucose remains the same as in raw milk (3-7 mg/100 mL) but the galactose level increases; reported levels of galactose in raw, UHT and sterilised milk were 7.1, 12.5 and 21.2 mg/100 mL (Olano *et al.*, 1989). Tagatose and 3-deoxypentulose have also been detected in sterilised milk but not UHT milk (Troyano, 1996). During storage of UHT milk for 3 months, Recio *et al.* (1998) found that the monosaccharides galactose, N-acetylglucosamine and N-acetylgalactosamine in milk with high residual protease activity increased but no changes occurred in these compounds in milk with negligible protease activity. Glucose showed no change in either milk type.

Belloque *et al.* (2001) attributed increases in the concentrations of galactose and N-acetylglucosamine increase during storage due to dephosporylation of their respective phosphates by heat-resistant microbial phosphatases. However, galactose levels in UHT milk stored at >20 °C increased faster than could be accounted for by dephosphorylation of galactose phosphate and the authors suggested that glycosidases may also be involved. This is consistent with the report of Romero *et al.* (2001) that at 40 and 50 °C, galactose increased from ~10 mg/100 mL to ~90 mg/100 mL over 12 weeks. The evidence of the action of a glycosidase of unknown origin in UHT milk was also reported by Bushnell (1980). It would therefore appear that the monosaccharide levels in good quality milks are unlikely to increase during storage but galactose and N-acetyl hexosamines will increase in UHT milk made from poor quality raw milk containing residual heat-resistant bacterial enzymes.

7.1.9 Reactivation of Alkaline Phosphatase

Inactivation of alkaline phosphatase (AP) is universally used as an indicator of the effectiveness of HTST pasteurisation based on the notion that the inactivation signifies destruction of non-sporeforming pathogenic bacteria. Because AP is inactivated by pasteurisation, it is reasonable to assume that it would be inactivated by more severe heat treatments such as UHT. While this is true for freshly processed UHT milk, it is not true for stored UHT milk as the enzyme is reactivated (Burton, 1988). The extent of reactivation depends on the storage temperature with virtually no reactivation at 6-10 °C but an increasing amount of reactivation occurs at temperatures from 17 to 34 °C. Over 20 days' storage at 32-34 °C, AP activity increased about 12 fold (Burton, 1988; Lorenzen *et al.*, 2011).

A factor which inhibits reactivation is the presence of air. Hence reactivation occurs more readily in directly processed than indirectly processed UHT milk (although it still does occur in indirectly processed milk [Gandon *et al.*, 1974]) and in milk packages with a reasonable headspace (Gallusser & Bergner, 1981). The explanation given was that –SH groups were required for the reactivation and these are oxidized in milk with a high dissolved oxygen content. This accords with the fact that the extent of AP reactivation has been reported to be seasonal and to closely correlate with the level of free –SH groups in UHT milk measured 4-6 h after manufacture (Burton, 1988).

A further factor which affects the extent of reactivation is the fat content; less reactivation occurs in high-fat than in low-fat (<0.3%) milk (Gallusser & Bergner, 1981).

Reactivated AP can be distinguished from native AP by an official AOAC method (AOAC, 1984) involving the addition of magnesium ions; these increase the activity of the reactivated enzyme ~15-fold but the native enzyme only 2-fold (Fox and McSweeney, 1998).

While reactivation of AP has been reported by several researchers, no practical significance has been placed on it except that the activity of AP in stored UHT milk cannot be used as in indicator of destruction of pathogens. AP seems to have little or no effect on caseins in milk (Lorient and Linden, 1977) and hence its reactivation is unlikely to cause significant dephosphorylation in UHT milk during storage.

7.1.10 Vitamins

Some losses in vitamins occur during storage of UHT milk. Most of these losses are accelerated by light and the presence of oxygen. Since most modern UHT packaging materials incorporate light and oxygen barriers, the major factor influencing vitamin losses during storage is the level of dissolved oxygen (see Section 7.1.2). Ideally, slight improvements in nutritional value may result if products are deaerated prior to UHT processing; however, Burton (1988) concluded that the effects on flavour may be of much greater commercial significance than changes in nutritional characteristics.

The fat-soluble vitamins, A, D and E, are stable in the absence of light for at least 3 months at ambient temperature. Vitamin A has been found to be stable in the dark for 8 months at 25 °C but, in diffuse daylight it suffered 40% loss after 14 days.

However, the water-soluble vitamins are less stable. Burton (1988) summarised the published data on decreases in these vitamins. Thiamine and riboflavin have been found to decrease by ≤10% over 3 months, biotin ≤20%, pantothenic acid ≤30%, vitamin B12 15-49% and vitamin B6 (35-50%). Marked losses of folic acid and ascorbic acid occur during storage of UHT products; these are interlinked and depend upon the availability of oxygen (see Section 7.1.2) and hence also to the method of processing. Ford *et al.* (1969) showed that the loss of both ascorbic acid and folic acid was dependent upon dissolved oxygen concentration and therefore took place at a much higher rate in indirect processed milk, or milk without any deaeration.

Where UHT milk contained a high level of dissolved oxygen (about 9 mg/L), there was a rapid destruction of ascorbic acid and folic acid. Ascorbic acid and folic acid disappear within about 14 days; ascorbic acid disappears slightly more quickly than folic acid. If the container is oxygen-impermeable, the dissolved oxygen level will decrease as these vitamins are oxidised. Thomas *et al.* (1975) showed that as the initial dissolved oxygen level fell, the rate of loss of ascorbic acid also fell. There was a sharp fall when the dissolved oxygen level fell below 2 mg/L and at 1 mg/L there was virtually no loss. Similar results have been obtained for loss of folic acid. Note that whilst ascorbic acid oxidation reactions are taking place, fat oxidation reactions are prevented. Thus ascorbic acid acts as an anti-oxidant and helps to reduce off-flavours produced by oxidation reactions.

Ascorbic acid is oxidised to dehydroascorbic acid, which still retains its vitamin activity. It is much more heat labile than ascorbic acid and its vitamin activity may be lost if the milk is subsequently heat treated, for example, if the milk is used in heated drinks. However milk is not considered to be an important source of ascorbic acid or folic acid.

It was also reported that vitamin B_{12} losses increased slightly at high dissolved-oxygen levels. A reduction in loss from 20% to 10% was observed over a 9-week storage period when dissolved oxygen was reduced from 8-9 mg/L to below 1 mg/L. Andersson and Öste (1992b) reported that vitamin B_{12} losses were dependent upon the concentration after processing. Losses ranged from no loss for milk with ~1.5 µg/L to almost 70% loss for milk with 2.7 µg/L after 21 weeks. These variable results were thought to be due to the different proportions of the different forms of the vitamin in milk.

The oxygen permeability of the container will influence the course of these reactions, although most UHT containers are impermeable to oxygen. Materials which are highly permeable to oxygen are polyethylene, polyethylene-paper laminates and polypropylene. If milk which is low in oxygen is packaged into such containers, the oxygen level will increase and may approach saturation after a few days. Therefore, they offer no protection against oxidation reactions, as oxygen replacement is more rapid than its depletion. The headspace in the container may also be a significant source of oxygen, especially in oxygen-impervious containers (see Section 7.1.2). It has been estimated that if the headspace volume is 15% and milk is saturated with oxygen at 10 °C, then the ratio of headspace oxygen to dissolved oxygen is about 5:1.

The stability of vitamins in stored UHT skim milk was studied by enriching it in the following vitamins: A, D_3, B_1, B_2, B_6, B_{12}, C, α-tocopherol, nicotinic acid, calcium pantothenate and folic acid (Dolfini *et al.*, 1991). Each vitamin was added at a concentration of at least twice the recommended daily allowance (RDA) per litre of milk and then subjected to a direct UHT process. The vitamin contents were in general still higher than the RDA values per litre after 3 months storage at 20 °C. Contents of vitamins A and C were found to decrease considerably during storage. Curiously, no vitamin C (initial added concentration. 300 mg/L) could be detected in samples of direct UHT-treated milk after 3 months but its concentration in indirect UHT-treated milk was 180 mg/L. Addition of vitamin C (as a means of protecting other vitamins) to indirect UHT milk generally had little protective effect on the contents of other vitamins during storage, with the exceptions of vitamin A and folic acid.

In summary, some water-soluble vitamins lose some activity during storage. The loss is generally enhanced by the presence of dissolved oxygen. Deaeration can reduce but not prevent losses of vitamins and hence nutritive value.

7.1.11 Light-Induced Changes

While most UHT packages contain light barriers, this is not the case for all pasteurised and ESL milk packages. Major reasons why milk may be packaged in light-pervious containers are the added cost of incorporating a light barrier and the desire of some consumers to be able to see the contents of the package. It is therefore instructive to consider the ramifications of allowing milk access to light.

Light-induced off-flavours in milk, sometimes described as medicinal, result from oxidation and degradation of proteins and lipids (Wishner, 1964; Dimick, 1982). The effect on proteins is catalysed by riboflavin which is the only component in milk which has appreciable absorbance in the visible region (maximum at ~450 nm). A major point of attack is methionine which is oxidised to methional which has an unpleasant, broth-like flavour itself but it also degrades into volatile sulfur-containing compounds such as mercaptans, sulfides and disulfides which have strong flavours (see Section 6.1.6.1).

Light also catalyses the oxidation of unsaturated lipids via a free radical mechanism forming hydroperoxides which decompose into carbonyl compounds responsible for the oxidized, stale flavour in milk. The chief carbonyl compounds are unsaturated and saturated aldehydes such as pentanal and hexanal which are used as lipid oxidation markers (see Section 6.1.6.2).

Light-induced flavours in milk can develop quite rapidly under fluorescent light which has a wavelength range of 300-750 nm with a peak at ~440 nm which is close to the maximum for riboflavin. Chapman *et al.* (2002) reported that flavour defects were detected in 2% fat pasteurised milk in clear HDPE containers at 6°C under fluorescent light in 15-30 min by trained tasters and in 54 min to 2 h by untrained consumers. These authors indicated that the average time pasteurised milk remains in a display cabinet was about 8 h. This of course can be much longer for longer shelf-life products such as ESL and UHT milk.

To overcome the risk of light-induced flavours, most milk is now packaged in opaque or coloured plastic containers, plastic containers with a light-impervious over-sleeve or paperboard cartons. Paperboard containers with an aluminium light barrier are the most effective. However, manufacturers would like to allow consumers to able to see the product in the container and for that reason some milk is packaged in clear plastic or glass bottles. Glass bottles have been found to be suitable for some products such as sterilised flavoured milk; this may be due to the masking of any light-induced flavour by the added flavours or it may relate to a protective effect of the glass. In fact, Andrews (1996) found that after storing sterilised milk in polyethylene bottles for 119 days in an illuminated cabinet, it was bleached by the light while milk in glass bottles was unaffected. However, brown glass bottles have been shown to be preferable to clear glass bottles for UHT milk (Biewendt & Prokopek, 1992).

Rysstad *et al.* (1998) assessed the development of light-induced flavour of UHT milk packaged aseptically in three types of paperboard carton: polyethylene-coated (PE), polyethylene coated with a non-foil paperboard light barrier (X-board) and polyethylene coated with an aluminium foil light barrier (foil). The light transmission through the cartons at 400 nm were 0.2, 0 and 0% respectively, and at 500 nm were 6, 0.2 and 0% respectively. These were all below the IDF recommendations of 2 and 8% at 400 and 500 nm respectively. However, when the cartons were exposed to strong light at 6°C, light induced flavours were detected after 2 and 8 weeks for the PE and X-boards cartons but no change in flavour could be detected in the milk in the foil cartons. These results show clearly the benefit of the aluminium foil light barrier which is a common feature of UHT paperboard-based packaging.

Petrus *et al.* (2009) studied the stability of UHT packaged in HDPE pigmented with titanium dioxide. They concluded that despite the limited light barrier, the milk exhibited good stability with shelf-life of 4-11 weeks. Many would not consider such light protection to be acceptable for UHT milk but it may be suitable for ESL milk.

van Aardt *et al.* (2005) investigated the effects of addition of antioxidants on light-induced odors in ESL milk by measuring pentanal, hexanal, heptanal and 1-octen-3-ol contents. No one antioxidant was effective in reducing formation of all of these compounds. They concluded that weekly addition of butylated hydroxyanisole (BHA) and butylated hydroxytoluene (BHT) was more effective than initial single additions of BHA and BHT or α-tocopherol and ascorbyl palmitate in reducing light-induced odours. They suggested controlled migration of antioxidants from packaging material as a possible strategy for minimising light-induced oxidation in ESL milk.

Light also has a detrimental effect on some vitamins in milk, in particular, riboflavin and vitamin A, but also vitamin B_{12} and vitamin B_6 (Burton, 1988). Curiously, riboflavin is destroyed at the same wavelength as it catalyses light-induced oxidation. Vitamin A is destroyed at wavelengths of less than 415 nm but is not affected at >455 nm. It has been reported that 40% of vitamin A in milk was destroyed in 14 days at room temperature when exposed to diffuse sunlight but no loss occurred when it was stored in the dark (Burton, 1988).

Light induced changes involve both the protein and fat phases so it would be expected that the effect will be greater in fat-containing milk than in skim milk. This has been shown by the greater production of pentanal and hexanal in 2% fat milk than in skim milk (Chung *et al.*, 2004).

7.2 Physical

7.2.1 Sedimentation

Almost all UHT products show some sediment if left undisturbed during storage. This was partially explained by Dalgeish (1992) who calculated theoretically the rate at which casein micelles would sediment. He calculated that the largest natural casein micelles would sediment at a rate of 1.3 mm/week. While this rate is small, a significant number of these large micelles and any casein aggregates would be expected to sediment during storage for several months. Smaller micelles would sediment at a much slower rate with some not sedimenting at all because their diffusion rate is higher than their sedimentation rate.

UHT products made from milk showing poor heat stability result in considerable sediment as a result of casein micelle aggregation. Blanc and Odet (1981) reported that the longer milk is stored, the greater is the amount of sediment. Sediment has also been found to increase as storage temperature increased (Ramsey & Swartzel, 1984; Hawran *et al.*, 1985; Vesconsi *et al.*, 2012). However, for products which show good heat stability, sedimentation may not be affected by storage temperature. Milk used in the two storage trials reported in Sections 7.5.1 and 7.5.2 showed no indication that sediment formation was influenced by storage temperature.

Hawran *et al.* (1985) investigated sediment in indirectly processed UHT milk. They found that initial sediment (at 1 week) increased as the processing time increased (samples were taken at 30-45-min intervals during the run) but the holding time (0.75, 7.32 or 18.6 s) at the sterilisation temperature (143 °C) had no significant effect on sediment formation. Sediment increased with storage time and the rate of increase was higher at higher storage temperatures. Sediment formation was not substantially affected by the amount of fat, when comparing levels of 0.5%, 1.5% and 3.2%. They concluded that since the sediment increased linearly with processing time, and fouling of heater 1 (a tubular heat exchanger heating milk from 74 to 114 °C) also increased with processing time, the sediment was largely due to erosion of deposit from that heat exchanger. Furthermore, they confirmed an earlier report by Ramsey and Swartzel (1984) that there are two rates of sedimentation during storage, a fast rate which results in sediment before 6 days and a slower rate which results in sediment forming after 6 days. They suggested that the eroded fouling deposit sediments rapidly as the initial sediment, presumably because of larger particle size; this would be consistent with the predictions of Dalgleish (1992).

Ramsey and Swartzel (1984) reported that the initial sedimentation was higher for direct processed than indirect processed UHT milk while Perkin *et al.* (1973) reported twice as much sediment in direct processed than indirect processed milk (processed to the same B*) after 100 days' storage.

Sediment formation in buffalo milk was studied by Sharma and Prasad (1990). They showed that UHT treatment increased the viscosity of milk. Over the processing temperature range of 120-150 °C for 2 s, there was little effect on sediment formation. Sediment formation increased with increased storage time and the amount of sediment was slightly higher after storage at 15 °C compared to 30 °C. This contrasts with the reports on cow's milk where sedimentation rate increases with temperature of storage.

Sediment formation is a major problem when goat's milk is UHT processed; this is discussed in more detail in Chapters 6 & 9. In this case, the sediment usually appears almost immediately after processing.

7.2.2 Age Gelation

Age gelation, characterised by an increase in viscosity during storage followed by formation of a gel limits the shelf-life of UHT sterilised milks. It has been a major problem for UHT milk and has consequently attracted a considerable amount of research. It has also been extensively reviewed (Harwalker, 1992; Datta & Deeth, 2001; Nieuwenhuijse & van Boekel, 2003; Chavan *et al.*, 2011).

Almost all UHT milks if left long enough at room temperature will undergo gelation. However, the aim of all manufacturers is for this not to occur within the intended shelf-life of the milk. This is quite a challenge when the expected shelf-life is 9-12 months but of much less concern when a much shorter shelf-life such as 12 weeks (McKellar *et al.*, 1984) is required. Because of the risk of age gelation, Erasmo *et al.* (1981) suggested that the storage time for UHT milk should be limited to two months. In many situations this would be impractical and would defeat the purpose of producing so-called long-life milk.

During storage, UHT milk may initially decrease in viscosity, a thinning effect, before its viscosity increases (de Koning *et al.*, 1985). The increase is typically gradual at first but at some point becomes more rapid until the viscosity reaches ~10 mPa.s at which gelation occurs (Kocak & Zadow, 1985a). After the gel forms, the viscosity drops because the gel breaks up into protein curds and serum. The gelation cannot be reversed and marks the end of the milk's shelf-life.

The nature of the gel has been described in different ways. It may be like a set yogurt or it may be a protein coagulum such as occurs in rennet coagulation (Snoeren *et al.*, 1979). It has also been described as a curd or custard-like formation in the lower portion of the containers (Harwalker, 1992). Hardham (1998) observed a creamy layer on the surface of the samples which thickened into a curd-like layer 3-5 mm deep. Furthermore, it is sometimes difficult to determine whether what is referred to as sedimentation in sterilised milks is the same as or different from gelation. In general, however, the particle size of sediments is much smaller than that of coagulated or gelled proteins.

The variations in description and nomenclature make it difficult to elucidate the mechanism (or mechanisms) of so-called gelation. Nieuwenhuijse and van Boekel (2003) identified four types of gel in sterilised milks: particle gel, fine-stranded gel, gel by bridging flocculation and gelled sediment and assigned the causes to the four types as follows: bacterial proteolysis, plasmin proteolysis, low-temperature storage and high-temperature storage, respectively. These together with several other factors which have been reported to promote gelation, have been identified and mechanisms of the gelation process have been proposed. In addition, some factors which delay or inhibit gelation have also been identified. Despite this, the exact biochemical and physical changes involved in age gelation have not been fully elucidated. A further complication is that gelation of concentrated milk appears to proceed by a different mechanism to that of unconcentrated milk.

The major factors which have been identified to affect gelation are proteolysis, milk production factors, severity of the sterilisation heating and temperature of storage, while addition of some chemical additives delay or accelerate gelation.

7.2.2.1 Proteolysis

Proteolysis is the factor which has been most widely implicated in gelation. Several, but not all, researchers have shown a good correlation between proteolysis and gelation in unconcentrated milks. By contrast, gelation in concentrated milk appears to be seldom related to proteolysis although McKenna and Singh (1991) concluded that proteolysis may cause gelation of UHT reconstituted concentrated skim milk.

As discussed in Section 7.1.3.4, proteolysis causes bitterness due to released hydrophobic peptides. Age gelation in unconcentrated milk is often, but not always, associated with development of bitter flavour (McKellar *et al.*, 1984; Mitchell and Ewings, 1985). Bitterness may occur in milk which shows no obvious gelation. Conversely, gelation may occur without bitterness.

Proteolysis by plasmin has been identified as a cause of gelation in milks known to be free of bacterial enzymes. This has been demonstrated by addition of plasmin to UHT milk (Kohlmann *et al.*, 1991). As discussed in Chapter 6, directly processed UHT milk generally contains more plasminogen/plasmin than indirectly UHT processed milk. Consequently, the residual plasmin causes age gelation to occur earlier in directly processed UHT milk than in indirectly processed UHT milk (McKellar *et al.*, 1984). However, the time to gelation does not always correlate with the extent of proteolysis (Manji *et al.*, 1986).

As for plasmin, bacterial proteases added in the form of cell-free culture supernatants can induce gelation in UHT milk (Law *et al.*, 1977; Button *et al.*, 2011). The addition can be either to the raw milk or aseptically to the sterilised product. Gelation can also be induced by growth of proteolytic psychrotrophic bacteria in the raw milk before sterilisation.

Very low concentrations of protease can cause sufficient proteolysis to cause gelation in UHT milk during storage at room temperature. Levels of ≤1-2 ng/mL and ≤0.3 ng/mL have been reported to limit shelf-lives to around 4 and 3 months respectively (Richardson & Newstead, 1979; Mitchell & Ewings, 1985). These levels provide a useful insight into how little enzyme can lead to gelation but they have little practical use as the concentrations of enzymes are normally not measured. Furthermore, the threshold enzyme concentrations will vary because the enzymes differ in specific activity and also in substrate specificity which dictates the peptides produced and hence the effects they have in the UHT milk.

Measuring the very low levels of protease activity is a major challenge and is seldom attempted in the industry because of the lack of a reliable method. If such a method was available and standardised, it may be possible to predict the shelf-life of a UHT milk from its protease activity. As discussed in Section 7.1.3.4, Button *et al.* (2011) developed a sensitive method based on the fluorogenic substrate FITC-casein. Using a long incubation time (12 h), they were able to detect protease activity which led to proteolysis in UHT milk sufficient to cause gelation (about 12 mM free amino groups). Another approach to predicting the shelf-life of UHT milk is measuring the rate of increase in proteolysis during the early stages of storage. McKellar (1981) found that the TNBS method of determining free amino groups was suitable for this purpose but methods based on NPN determination and the Hull test lacked sufficient sensitivity.

It has been reported that the gels produced in UHT whole milks as a result of proteolysis by plasmin and bacterial protease differ in appearance. Bacterial proteases lead to formation of a custard-like gel (Harwalkar, 1992) whereas plasmin proteolysis causes a creamy surface layer which thickens into a curd-like layer (Hardham 1998). According to Nieuwenhuijse and van Boekel (2003), bacterial proteolysis leads to particle gels while plasmin proteolysis leads to fine-stranded gels. Gels caused by bacterial proteases have a tighter protein network with thicker strands and contain more intact casein micelles and micelle aggregates than plasmin-initiated gels which have a thread-like structure containing casein micelles which were almost completely disintegrated (de Koning *et al.*, 1985). Concentrated milk gels showed a network of aggregated, partly-deformed casein micelles which had not been degraded by proteolysis.

7.2.2.2 Milk Production Factors

There have been several reports of the influence of milk production factors on age gelation in UHT milk. The effect of some of these factors such as stage of lactation and level of mastitic infection have been correlated with plasmin and plasminogen levels but in other cases, no correlation with protease level was found (Auldist *et al.*, 1996). Some authors have reported a seasonality effect. For example, Hardham and Auldist (1996) reported more age gelation (and sedimentation and burn-on) in spring and late-autumn milk. This was attributed to stage of lactation as spring corresponded to early lactation for the majority of cows. These authors did not find a relationship between age gelation onset and levels of protease or proteolysis. They suggested changes in mineral balance or mineral/protein ratios which occur during lactation as possible causes. It is always difficult to unravel the effects of season and those of stage of lactation, particularly in regions of seasonal calving.

Auldist *et al.* (1996) investigated the relative effects of stage of lactation and somatic cell count (a measure of mastitic infection) on age gelation in UHT milk. They concluded that stage of lactation was the dominant factor but at a particular stage of lactation, higher somatic cell counts hastened the onset of age gelation. This is attributable to the higher plasmin activity in mastitic milk than in normal milk (Bastian & Brown, 1996).

7.2.2.3 Severity of Sterilisation Heating

For unconcentrated milk, the risk of age gelation decreases as the severity of the heat treatment increases; that is, the higher the C* (Tran *et al.*, 2008) the longer is the time required for gelation to occur. As indicated above, milks subjected to direct UHT treatment are more susceptible to age gelation than milks subjected to an indirect process with the same bactericidal effect (B*) because the latter will have a higher C*.

Furthermore, whether direct or indirect, more severe heating reduces the incidence of age gelation (Topçu *et al.*, 2006). This is thought to be due to the greater inactivation of proteases at the higher heat intensity but it may also relate to stabilisation of the casein against proteolysis through a greater degree of lactosylation (Bhatt *et al.*, 2014) or greater protein cross-linking (McMahon, 1996). Other as-yet unidentified physico-chemical factors may also be involved.

The effect of the intensity of heating is most clearly seen with in-container sterilised milk in which gelation rarely occurs (Samel *et al.*, 1971). However, in-container sterilised evaporated milk has been reported to sometimes gel during storage but, interestingly, the gelation did not occur if the concentrate was sterilised without prior cold storage (Harwalker *et al.*, 1983).

7.2.2.4 Temperature of Storage

Storage temperature has a major effect on age gelation of UHT milk but unfortunately it is a variable which in most cases cannot be controlled. Low temperature, that is, ~4 °C, greatly retards gelation but keeping UHT milk refrigerated is expensive and negates the benefit of UHT processing which enables the product to be stored at room temperature, unless of course ambient conditions fall in the low-temperature range. Somewhat counterintuitively, high-temperature (> ~30 °C) storage also retards gelation compared with the intermediate temperatures of ~25-28 °C where the rate of gelation is maximised (Samel *et al.*, 1971; Kocak and Zadow, 1985a; Manji and Kakuda, 1986; Gaafar and El Sayed, 1991). Therefore, the temperatures which favour gelation are close to room-temperature in many regions of the world.

Two suggestions have been advanced for the mechanism of retardation of gelation at the higher temperatures. These are that at these temperatures proteolysis proceeds too rapidly for the gel to develop (Manji *et al.*, 1986), and that protein cross-linking occurs at these temperatures which stabilises the casein micelles and prevents release of the β-lactoglobulin–κ-casein complex into the milk serum which would otherwise form into a gel (McMahon, 1996). As discussed in Section 7.1.5, considerable cross-linking occurs during storage via either dehydroalanine-forming linkages between lysine and alanine or histidine on adjacent molecules, or cross-linking by advanced Maillard reaction products such as glyoxal and methylglyoxal (Le *et al.*, 2013). Other reactions which occur during storage and which are accelerated by higher temperature storage are lactosylation of lysine residues (Section 7.1.6) and deamidation of asparagine and glutamine residues of proteins (Section 7.1.5). The involvement of cross-linking and lactosylation are consistent with the suggestion that gelation may be retarded if regions of proteins that would otherwise be involved in protein-protein interactions were blocked by interactions with lysine (Samel *et al.*, 1971; Bhatt *et al.*, 2014). Furthermore, deamidation alters the charge on the proteins and is known to cause changes in the physical properties of proteins.

7.2.2.5 Additives

Because of the major impact of age gelation on the UHT milk industry, "magic bullets" in terms of additives have been sought. The most common additive used commercially is sodium hexametaphosphate (SHMP). It is added routinely by some UHT milk manufacturers. SHMP is sometimes referred to as "polyphosphate" but this is a non-specific term which can also apply to several other phosphates. Addition of about 0.1% (w/w) of

SHMP to milk before UHT treatment considerably extends the time to gelation (Kocak and Zadow, 1985b). This effect of SHMP differed from that reported by Snoeren *et al.* (1979) that SHMP delayed gelation in UHT milk made from milk with a low psychrotroph count (2.7 x 10^3 cfu/mL) but not in UHT milk made from raw milk with a high psychrotroph count (2.06 x 10^6 cfu/mL). The reason for this different result is not known. However, Kocak and Zadow (1985c) warned that SHMP varies in its effectiveness depending on the chemical's manufacturer and hence this could account for different results obtained by different researchers.

SHMP does not influence proteolysis and hence its mechanism of action does not involve inhibition of protease action. In fact, milk to which SHMP has been added may undergo considerable proteolysis without gelling. Kocak and Zadow (1985b) suggested that gelation induced by protease occurs in two stages: the first is proteolysis of caseins in the micelle which releases peptides into the serum and the second involves aggregation of destabilised casein micelles. So if SHMP does not inhibit proteolysis but inhibits gelation, it must act on the second stage by altering the charge on the micelle or by interacting with the casein micelles and preventing aggregation. SHMP is known to stabilise casein because UHT milk containing SHMP shows high ethanol stability throughout storage while milk which gels shows decreased ethanol stability. SHMP also decreases ionic calcium (Mittal *et al.*, 1990, Tsioulpas *et al.*, 2010) which may be related to the high ethanol stability, often used as an indicator of heat stability (see Section 6.2.1.4). Curiously, Mittal *et al.* (1990) found that the heat stability of UHT recombined milk containing SHMP declined more during storage than control milk without SHMP.

Because of the effect of SHMP in reducing ionic calcium, there has been interest in the role of calcium, particularly ionic calcium, in age gelation. Unfortunately there have been conflicting reports on this role. Kocak and Zadow (1985b) found that chelation of calcium with 0.1 and 0.3% trisodium citrate or EDTA (Na_2) accelerated age gelation which is the opposite of the effect of SHMP. They also reported that addition of 0.05% calcium chloride delayed gelation but Samuelson and Holm (1966) reported the opposite effect. Added calcium has also been reported to accelerate age gelation in UHT concentrated milk (Ellerton & Pearce, 1964).

SHMP has also been used to stabilise calcium-fortified UHT milk from age gelation (Sher *et al.*, 2002). It was much more effective than sodium citrate which is consistent with the finding of Kocak and Zadow (1985b) that citrate accelerated age gelation but SHMP delayed it.

In contrast to "polyphosphate", monophosphates accelerate age gelation (Kocak & Zadow, 1985b). Harwalkar and Vreeman (1978) demonstrated the difference between the effects of SHMP and orthophosphate in stored UHT milk by electron microscopy. The structure of the casein micelles in UHT milk containing orthophosphate appeared similar to that of micelles in milk without phosphate; they were slightly distorted, had thread-like tails on their perimeters and aggregated into interconnected chains at gelation. By contrast, the micelles in UHT milk containing SHMP remained unchanged and separate from each other during storage.

The protein-destabilising effect of the monophosphate, potassium dihydrogen phosphate (KH_2PO_4) is the basis of the Ramsdell test for stability of UHT milk. Increasing amounts of 0.5 M KH_2PO_4 are added to the milk with heating (100 °C for 10 min) and the amount added without coagulating the milk is used as a measure of the milk's stability (Gaucher *et al.*, 2008b). Small amounts of added phosphate cause instability in unstable milks, that is, they have low phosphate test values. The mechanism of the destabilisation is unclear but may be due to the reduction in pH caused by the acidic salt, rather than a specific effect of the phosphate.

7.2.2.6 Mechanism

Proposed mechanisms of age gelation mostly involve proteolysis combined with physico-chemical changes which result in gel formation. (Manji & Kokuda, 1988; Nieuwenhuijse & van Boekel, 2003). A variation of the two-stage process (mentioned in Section 7.2.2.5) in which a proteolysis stage is followed by aggregation of destabilised casein micelles (Kocak and Zadow, 1985b) was proposed by McMahon (1996). In his model, the proteolysis stage releases the β-lactoglobulin–κ-casein complex from the casein micelle through degradation of κ-casein anchor points in the micelle. When the released complex in the serum phase reaches a certain critical concentration in the serum, a three-dimensional protein gel network, in which casein micelles are embedded, forms. McMahon (1996) suggested the reason why gelation was retarded at higher storage temperatures was that intramicellar covalent cross-linking occurred which prevented release of the β-lactoglobulin–κ-casein complex.

7.2.2.7 Practical Issues with Gelation

All UHT processors would like to be able to predict whether a batch of UHT milk will form a gel before the "best before" date, ideally on the day of production, but certainly before the product is released. To achieve this still remains one of the major challenges.

It is likely that more than one mechanism is involved in gelation of UHT milk. Whatever the mechanism, two factors need to be present for gelation to occur; the UHT milk needs to be susceptible to gelation and the causative agent of gelation must be present. In fact, cheesemakers know that UHT milk is much more difficult to gel than pasteurised milk and raw milk and the normal amounts of chymosin used for cheesemaking will not cause it to gel and usually at best the milk will not change in viscosity and at worse form a paste-like consistency (see Section 9.15.2).

One approach to establishing whether UHT milk is likely to form a gel within 12 months is to subject it to reagents that are known to cause gelation/coagulation of milk (e.g., $CaCl_2$, alcohol, chymosin (rennet), trypsin, ginger, pawpaw) and observe whether gelation occurs. To accelerate the test, much higher concentrations of these reagents than would be normally used could be added. If none of these inducements bring about gelation, one might conclude that the product is simply not susceptible to gelation and therefore it is not likely to occur. On the other hand, if gelation can be induced, then one should be alerted to the fact that this might occur during storage. The applicability of such tests requires investigation. As noted in Section 7.4.3, Samel *et al.* (1971) reported that UHT milk stored at 20 °C showed marked changes in susceptibility to added $CaCl_2$, rennet coagulation time and alcohol stability before gelation. In contrast, milk stored at 37 °C did not gel and showed an increase in rennet coagulation time.

There is a massive body of literature on gelation of UHT milk. Many authors have reported bitterness in association with gelation but this may not always be the case. In our experience, some UHT milk samples gel with no accompanying change in pH and detectable off-flavours. In one instance, a sample was acceptable after about 7 months but at 8 months it gelled, after a period when it increased in viscosity and had a pastey mouthfeel but no off-flavour. There was also no noticeable production of peptides. While gelation in such a case is not related to proteolysis, the mechanism of the gelation is not clear. It is often just attributed to a physicochemical phenomenon (Harwalkar, 1992).

Age gelation appears to be less likely to occur now because of the more widespread recognition of the relationship between poor quality milk and development of gelation and other defects in UHT milk during storage. Thus processors tend to use better quality raw milk for UHT processing (low in pyschrotrophic microorganisms and somatic cell count) and, in some cases, select milk from particular producers or tanker runs with a history of high quality. This is particularly important for directly processed UHT milk which is more susceptible to gelation than indirectly processed milk. It is more challenging to prevent gelation when the best-before dates are extended to 12 months.

7.2.3 Thinning

Thinning of ESL/UHT custards and desserts during storage is experienced by many dairy companies. The major thickening agent in these products is starch, but other texture modifying agents such as xanthan gum, carrageenan and locust bean gum are also used. The thinning is often accompanied by a decrease in starch granules.

In an American study of UHT-sterilised, starch-based puddings which showed a "dramatic decrease in viscosity within two weeks to six months", the authors found measurable amounts of amylase activity and attributed the decrease in viscosity to the action of heat-stable amylases on the starch. This activity was greatest between pH 5 and 7, with an optimum at 6. Little activity occurred at pH 4 which was consistent with their observations that little or no thinning occurred in starch-based lemon puddings with a pH of 4.2 - 4.5. There was significant activity over the temperature range tested, 15 to 70 °C, with an optimum at 50 °C (Barefoot & Adams, 1980). The amylase appeared to be a bacterial α-amylase as it was metal-dependent (especially Ca^{2+}) and heat-stable. Other amylases such as β-amylases and non-bacterial amylases are not metallo-enzymes (Hsiu *et al.*, 1964) and are not heat-stable (Miller *et al.*, 1953). Furthermore, the defect in the pudding could be reproduced by the addition of *G. stearothermophilus* α-amylase to unspoiled pudding.

The origin of the amylase in desserts is sometimes not clear. Because thermophilic bacteria such as *G. stearothermophilus* and *B. licheniformis* produce heat-stable α-amylases and are common components of the bacterial flora of milk, the milk and milk powders used in the manufacture of the products are likely sources. The growth of psychrotrophs in raw milk before processing has also been implicated as, in one study, up to 15% of psychrotrophic bacteria produced amylase (Barefoot, 1979).

It has also been reported that *Xanthomonas campestris*, the organism used to produce xanthan gum, produces an amylase which has a similar heat stability to the *Bacillus* α-amylases. Some processors have overcome the problem of thinning by eliminating xanthan gum from their formulations, because it contained a contaminating amylase. Amylase-free xanthan gum is now commercially available. However, other manufacturers have experienced thinning when not using xanthan gum, which suggests the milk ingredients as the likely source of the enzyme.

Another possible amylase source is a residue of enzyme-based detergents which contain amylases as well as proteases and lipases. Addition of low levels of these detergents to a simulated dairy dessert caused an initial increase in viscosity followed by thinning; addition of higher levels caused only thinning (Tran *et al.*, 2003).

Amylase is not the only enzyme implicated in the thinning phenomenon. Native milk protease (plasmin) has also been reported to reduce the viscosity of custard during storage (Driessen *et al.*, 1991). A holding time of 18 seconds at 143 °C was required to eliminate the problem. These authors found no amylase present after this treatment. Since many psychrotrophic bacteria are also active producers of proteases which can degrade milk proteins, they may also cause a decrease in viscosity in custard. Poor quality raw milk could therefore contain both heat-stable amylases and proteases which could decrease the viscosity of dairy desserts.

The detection of amylase has proved difficult because of the very low levels that can produce thinning in desserts during extended storage. Traditionally amylases have been detected using starch as the substrate by: 1. reduction in viscosity of a starch paste; 2. decrease in the iodine-staining properties of a starch substrate; and 3. increase in the reducing power of a starch substrate. However, these methods generally lack sensitivity. There are now several methods based on oligosaccharides as substrates which are more sensitive and are purported to be suitable for bacterial amylases. For more details see Section 11.2.1.

7.2.4 Fat Separation

Fat separation in UHT milk during ambient storage is not uncommon and is a major problem in UHT milk. In most cases, the fat layer represents only a small percentage of the total fat and is readily dispersed back into the milk. On occasions however, a substantial layer of fat, which may form a solid layer, appears on the top of the milk after storage. In paperboard packages, the separated fat may go almost unnoticed by consumers unless it does not readily disperse and forms into lumps of fat which are noticeable when the product is poured from the carton. In other packages such as plastic bottles, the fat layer may be quite noticeable when the lid is removed.

Homogenisation is an integral operation in UHT processing of fat-containing milk. Most cases of fat separation can be related back to inadequate design or maintenance of the homogeniser or to the conditions of homogenisation. Hansen (1990) considered that maximising the efficiency of the homogeniser was the key to reducing fat separation. The homogeniser valve is the critical component in the homogeniser and if it is not functioning efficiently, some large fat globules may escape proper homogenisation and cause fat separation. Regular visual inspection, using a microscope if necessary, and maintenance are necessary to ensure efficient operation.

According to Stokes Law (Equation(7.1)), the terminal velocity at which a particle of fat will rise to the surface in a particular product is directly proportional to the square of the diameter of the fat particle. So the main aim of homogenisation is to decrease the diameter of the fat particles to slow the rate of creaming. An efficient homogeniser on a UHT plant should reduce the diameter of the fat globules from an average of about 3 µm to less than 1 µm and with a narrow range of diameters.

$$v = \frac{d^2 g(\rho_s - \rho_f)}{18\mu} \tag{7.1}$$

(also 5.2)

where: d = diameter of the fat particle; ρ_s = density of the milk serum; ρ_f = density of milk fat; μ = viscosity of milk serum; v = terminal velocity.

Table 7.2 The effect of fat globule diameter on volume and number distribution of fat globules in milk as illustrated by two theoretical milks with the same average globule size (adapted from Hillbrick *et al.*, 1998).

Fat globule diameter (μm)	Number (%)	Volume (%)	Average fat globule diameter
0.5	99	79	0.51
1.5	1	21	
0.5	99	22	0.51
3.5	1	78	

The importance of the size range is illustrated by Table 7.2 which shows that, although the average diameter of ~0.5 μm is acceptable, a small percentage of large globules can represent a large proportion of the fat in the product and present a potential fat separation problem. For this reason, some companies use as a rule of thumb that the average diameter of the fat globules should be <1 μm, but the percentage greater than 5 μm should be <2% and the percentage greater than 2 μm should be <5%.

While large fat particles in the final product can result from their passage through a defective homogeniser valve, they can also be formed by coalescence of smaller particles. Homogenisers on UHT plants usually operate in two-stages where the pressure of the first stage is 5-10 times the pressure of the second stage which operates at up to about 3.5 MPa. The second stage homogenisation breaks up any aggregated globules formed in the first stage and thus reduces their tendency to rise in the product. Similarly, downstream homogenisation is practised on direct UHT plants since the direct heat process can cause aggregation of fat globules. This may be a reason why directly processed UHT milks show less fat separation than indirectly processed milks which may be homogenised upstream of the sterilisation section (Ramsey & Swartzel, 1984). Coalescence of fat particles during storage is a major cause of fat separation. Higher storage temperatures increase the rate of fat separation (Ramsey & Swartzel, 1984; Hansen, 1990; AlKanal *et al.*, 1994).

The efficiency of homogenisation also depends on the temperature of the product and the homogenisation pressure. Homogenisation is most efficient when all the fat is in liquid form and so it is usually carried out at >50 °C. It has been shown that fat separation decreases with increased homogenisation pressure up to ~35 MPa (the combined pressure of the two stages in a two-stage homogeniser) (Hillbrick *et al.*, 1998). Most homogenisers on commercial UHT plants operate below this pressure.

Interestingly, Zhang *et al.* (2007) found that low temperature inactivation treatment which inactivates proteases and lipases also reduces fat separation. The mechanism of this is not known but it suggests fat separation could be associated with protease-initiated gelation. However it is consistent with the report by Kohlmann *et al.* (1991) that UHT milk to which plasmin had been added (at 0.15 mg/L) and gelled during storage showed more fat separation than an ungelled control milk with no added plasmin. Furthermore, Hardham (1998) observed the development of a creamy layer which turned into a curd-like gelled layer in stored UHT milk, suggesting the concomitant development of fat separation and age gelation. We have also observed this in gelled commercial UHT milks. It may occur through linking of casein micelles which become associated with the fat during homogenisation; this would cause clumping of the fat and lead to creaming or fat separation.

7.2.5 Maillard Browning

7.2.5.1 Browning of Milk and Milk Products

The subject of browning is of great interest to the food and dairy industries. For UHT milk-based products, the amount of browning which takes place during storage is much more significant than that resulting from the UHT process itself, especially when storage temperatures are above 30 °C. One might normally expect that sterilisation by whatever means would increase browning. For in-container sterilisation this is found to be the case and is a noticeable difference between sterilised and UHT milk. However, UHT milk immediately after production is whiter than raw milk. This was first observed in skim milk by Rhim *et al.* (1988) but our UHT pilot plant experience has consistently shown that UHT whole milk is also noticeably whiter than raw unhomogenised milk. Although some early-stage Maillard reaction products are produced (see Section 6.1.4.1), other changes such as an increase in casein micelle size and a reduction in fat globule size brought about by homogenisation, where applied (see Section 5.3), alter the light scattering properties of the milk and make it whiter in appearance.

The non-enzymatic browning reaction, also known as the Maillard reaction, takes place between reducing sugars and amino groups, which are abundant in many foods. In milk this is the lactose and primary lysine groups of milk proteins (see also the discussion of lactosylation in Section 6.1.4.1). The source of lysine is usually considered to be the casein and whey protein fractions. The involvement of free amino acids and peptides in this reaction are unclear. In general, the role of free amino acids in milk processing is under investigated and there are very few papers on how free amino acids change after milk has been heat-treated. A better understanding of their role will result when more detailed studies, such as by NMR (see Section 11.3.2), are performed on heat-treated milk.

The reaction scheme for the Maillard reaction is very complex, even in model systems involving one amino acid and a reducing sugar. The reaction proceeds through a large number of intermediate products, with the major end products being brown pigments. Labuza (1994) reviewed the problems involved in interpreting the complexity of the kinetics of the Maillard reaction in a food context.

One problem when monitoring this reaction is selecting the most appropriate compounds to measure. For example, there are many intermediate compounds formed before any brown pigments are observed which then undergo further modification as the reaction proceeds. One of the most commonly examined is hydroxymethylfurfural (HMF) which has been used as a chemical index of heat treatment (Section 6.1.7). However, HMF shows little increase during storage, except where storage temperatures are above 35 °C (Fink & Kessler, 1986; Burton, 1988; Elliott *et al.*, 2003). This has been attributed to the fact that HMF is both formed and degraded during storage. This was substantiated by Jimenez-Perez *et al.* (1992) who stored UHT milks for 90 days at five storage temperatures between 6 and 50 °C. Their results indicated that increases in HMF took place in the temperature range 40 to 50 °C. The most useful indicator of the progression of the Maillard reaction during storage is lactulosyllysine, which is usually measured as acid-hydrolysis product, furosine, as discussed in Section 7.1.6). Furosine has also been used as an indicator of heat treatment but, as discussed in Section 6.3, its increase during storage limits its use for this purpose to freshly processed products.

Since the Maillard reaction decreases the nutritive value of products by chemically blocking and decreasing the biological availability of lysine, monitoring compounds such as lactulosyllysine or furosine during storage has become important. A product in which this is particularly significant is infant formula (Mehta & Deeth, 2015).

The Maillard reaction makes an important contribution to both the flavour and colour of heat-treated products. During storage of UHT milk, the milk itself may become noticeably brown. As a generalization, UHT milk remains substantially unchanged in colour during storage if its temperature is maintained below 20 °C; there are slight changes between 20 and 30 °C but above 30 °C changes are substantial. If it is kept for longer than 4 months at 50 °C, it will have a light chocolate brown colour.

Methods for measuring the colour of milk products have been reviewed by Burton (1988) and more generally in foods by Hutchings (1994) and Hunt (1995). One method used to measure the extent of browning is by extracting the pigments and measuring them in the extract (see Section 11.2.2.2). On-Nom (2012) investigated using dialysis at high temperature as a procedure to extract brown compounds in milk and, at the same time, isolate them from the protein and fat. Figure 7.4 shows milk heated at 90 to 120 °C for 60 min and dialysates from these samples, to illustrate this procedure and the extent of the brown colouration. Thus, dialysis provides a simple procedure for recovering Maillard reaction products for analytical procedures to determine their concentration and identity.

It is now very convenient and straightforward to measure the colour directly using a commercial colour meter such as the Minolta Chromameter (see Section 11.2.2.1). Colour meters are based on the now widely used system for colour measurement, the (CIELAB system), which uses three parameters, L*, a* and b*, where L* is a measure of lightness (0 black to 100 white), a* is a measure of red/green [positive 100 (red) to negative 80 (green)] and b* is a measure of yellow/blue [positive 70 (yellow) to negative 80 (blue)]. Within these spaces, equal distances in space represent approximately equal colour differences.

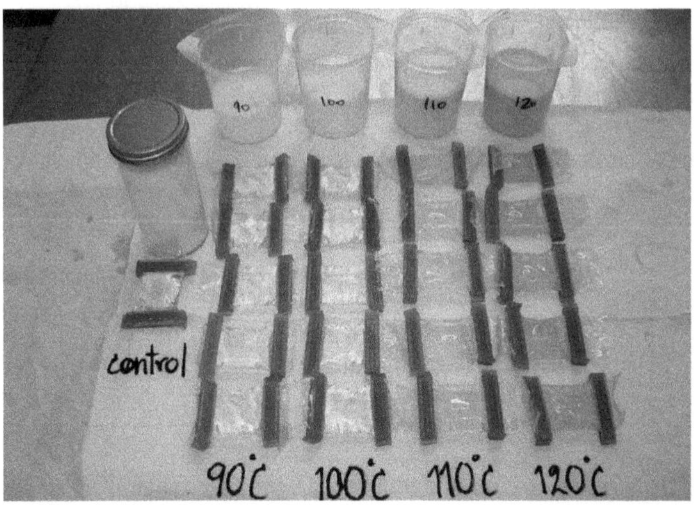

Figure 7.4 Dialysates of milk heated to 90, 100, 110 and 120 °C for 60 min. (Source: On-Nom, 2012. Reproduced with permission.). (*See color plate section for the color representation of this figure.*)

As foods are heated and browning proceeds, the trends are for L* to decrease and for a* and b* to increase. It is often useful to determine whether a significant change has taken place in a food. In the CIELAB colour space, the colour difference between two samples (ΔE) is defined as the difference between two points (1 and 2) representing the colours in the CIELAB space. It is calculated from equation (7.2):

$$\Delta E = \left\{ \left(L_2^* - L_1^*\right)^2 + \left(a_2^* - a_1^*\right)^2 + \left(b_2^* - b_1^*\right)^2 \right\}^{0.5} \tag{7.2}$$

For colour differences to be noticed (perceived), ΔE values should be between 2 and 5. A value of 2.3 has been reported by Sharma (2003) as the JND (Just Noticeable Difference) while Pagliarini *et al.* (1990) showed that a minimum ΔE of 3.8 should be attained before there is a visual perception of milk browning.

Table 7.3 shows colour data for lactose-reduced UHT milk which had been stored at 4, 20, 35 and 50 °C for 4 months. The colour differences (ΔE) were determined by comparing samples stored at 4 °C with those stored at higher temperatures. Figure 7.5 shows the appearance of these samples, with hardly any noticeable difference between samples stored at 20 and 4 °C but more noticeable differences for samples stored at 35 and 50 °C.

Products containing a higher amount of sugar or containing more reducing sugars (for example, with added glucose), lactose-hydrolysed milk, sweetened milk or ice-cream mix or with additional amino groups become browner during storage more rapidly than normal milk. Table 7.4 shows colour difference values (ΔE) for different milk products stored at 35 and 50 °C. In each case the samples were compared with the same product stored at 4 °C. The largest differences were observed for lactose-reduced milk samples. It was interesting that values for skim milk were higher than values for full-cream milk. Also different samples of the same milk type did not show the same extent of browning. The different amounts of browning in a range of milk products which had been stored at 50 °C for 4 months are illustrated in Figures 7.6 and 7.7.

Increasing the pH of a formulation, for example by addition of TSC or DSHP (see Section 6.2.1.3) will also accelerate the Maillard reaction and hence the susceptibility of the product to browning. This is an important fact which is largely ignored in the UHT literature.

Kessler (1989) reported that browning during storage is not affected by the level of dissolved oxygen. However, it is affected by the severity of the heat treatment. For

Table 7.3 Colour values (L*,a*, b*) of lactose-reduced milk stored at 4, 20, 35 and 50 °C for 4 months. For determining the ΔE value of samples stored at 20, 35 and 50 °C, their L*a*b* values were compared those of the same product stored at 4 °C. (Source: DIAL, 2014. Reproduced with permission.)

Temperature (°C)	L*	a*	b*	ΔE
4	100.6	-0.53	1.52	
20	103.0	-0.14	1.5	2.43
35	86.8	7.71	13.9	20.3
50	63.1	8.94	20.1	43.0

Figure 7.5 Lactose-reduced milk samples after storage at 4, 20, 35 and 50 °C for 4 months. (Source: DIAL, 2014. Reproduced with permission.). (*See color plate section for the color representation of this figure.*)

Table 7.4 Colour difference (ΔE) values for different milk products stored at 35 and 50 °C. In each case the samples were compared against the same product stored at 4 °C. (Source: DIAL, 2014. Reproduced with permission.)

Product	Storage temp (35 °C)	Storage temp (50 °C)
Cow's full cream milk	7.22	22.6
Cow's full cream milk	3.45	20.7
Cow's full cream milk	3.77	20.1
Cow's skim milk	7.86	33.7
Cow's skim milk	13.4	33.7
Goat's full cream milk	3.6	20.4
Cow's lactose-reduced milk	20.3	43.0

example, milk with a higher C* value will become browner at a faster rate during storage, even though it is whiter immediately after production. An explanation for this is the increased number of early and intermediate Maillard reaction products formed under UHT conditions with higher C* values, which will be available for further reactions during storage. Thus C* value can be used as an indicator of a product's susceptibility to browning during storage.

Temperature dependence of reactions can be described by z-values or activation energies. z-Values of 21.3 °C (Burton, 1954) and 26.2 °C (Horak & Kessler, 1981) have been reported for the browning reaction. More significantly, according to the latter researchers, browning can be described by one activation energy (107 kJ/mol) over the temperature range 25 to 140 °C. Fink and Kessler (1988) stated that a time of 400 s at

Figure 7.6 Milk samples, skim milk, goat's milk, full cream (cow's) milk (X2) and lactose-reduced milk, after storage at 50 °C for 4 months. (Source: DIAL, 2014. Reproduced with permission.). (*See color plate section for the color representation of this figure.*)

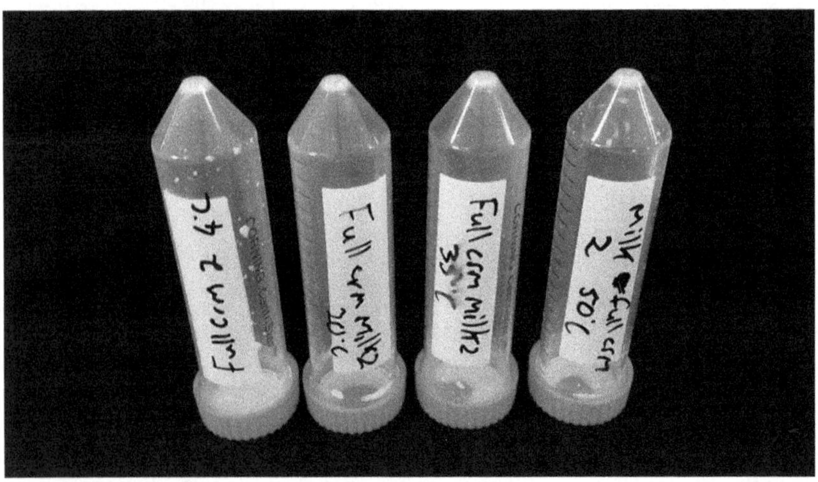

Figure 7.7 Sediment from full cream milk after storage at 4, 20, 35 and 50 °C for 4 months, from approximately 50 mL milk. (Source: DIAL, 2014. Reproduced with permission.). (*See color plate section for the color representation of this figure.*)

121 °C was the threshold for colour detection. If this is the case, some interesting comparisons can be made for different storage conditions and some examples are shown in Table 7.5. In this example, the threshold for detecting a brown colour in milk is taken as 400 s at 121 °C. An activation energy of 107 kJ/mol is used to determine the Arrhenius function $\int e^{-E/RT}.\, dt$, which for the browning threshold is 26.7×10^{-13}. This value can be compared with values for storage temperatures from 10 to 50 °C for 4 months. These calculations predict that browning would be observed in milk stored at 30 °C and above

Table 7.5 Estimated time for brown colour to become noticeable at different storage temperatures (based on Arrhenius factor and two different z-values).

	Based on Arrhenius factor		Based on z = 26.2 °C		Based on z = 21.3 °C	
Temperature (°C)/time	**Arrhenius factor (x 10^{-13})**	**Time to brown (days)**	**Equivalent time at 121 °C (s)**	**Time to brown (days)**	**Equivalent time at 121 °C (s)**	**Time to brown (days)**
121/400 s	26.7		400			
50/4 months	532	6.0	20422	2.4	4860	9.8
40/4 months	149	21.5	8480	5.7	1649	31.5
30/4 months	38.4	83.4	3521	13.6	560	85.7
20/4 months	9.0	355	1462	32.8	190	252
10/4 months	1.9	1677	607	79.1	64.4	745
90/30 min	7.4		63			
140/4 s	1.2		31.5			
150/4 s	2.5		92			

after 4 months, but not in samples stored at 20 °C and 10 °C. It also predicts that browning will be observed after 6 days at 50 °C, after 21.5 days at 40 °C, after 83.4 days at 30 °C and after about 1 year at 20 °C and after about 5 years at 10 °C. These predictions show reasonable agreement with what is observed in storage trials (see Section 7.5.1).

Using a z-value of 26.2 °C, equivalent times for observing browning (at 121 °C for 400 s) can be calculated and are shown in Table 7.5. These predict considerably shorter times to observe browning than found from the Arrhenius equation. In fact a predicted time of 32.8 days at 30 °C is much shorter than found in practice, so this z value is questionable. When the procedure is repeated for a z-value of 21.3 °C, the results are much closer to those observed for the Arrhenius relationship and in fact the predicted times for observing browning at 30 °C are 83.4 and 85.7 days respectively, which are very close. Values predicted by this z-value tend to be slightly longer than those predicted by the Arrhenius equation above 30 °C and slightly shorter below 30 °C.

The Arrhenius function can be used to compare the extent of browning that occurs during both processing and storage. For example storage conditions involving long time periods at relatively low temperatures can be directly compared to processing conditions involving short periods at high temperatures. Table 7.5 shows such data for different storage and processing conditions. These figures predict that the extent of browning caused by heating at 140 °C/4 s would be similar to that resulting from storing milk for one week at 25 °C or about 1 day at 40 °C (not shown in table). Browning caused by sterilisation of milk at 121 °C for 10 min would be about 40 times more than treatment at 140 °C for 4 s (or equivalent to about 120 s at 140 °C).

Such calculations can be applied to practical situations with UHT products. For example, one investigation into why some UHT milks were going excessively brown led to the discovery of a failure in the cooling system, with products being packed in excess of 50 °C and then remaining at that temperature for an extended period of time. These types of calculation show that excessive browning is possible within a few days under adverse circumstance.

A drawback of the above approach to predicting the extent of browning is that it does not take into account several factors which, as previously discussed, influence the rate of browning. These include:

- the pH of formulation
- fat content
- the severity of the heating process (C^*) value
- sugar types and concentrations (including extent of lactose hydrolysis)
- variations in lysine groups

It should also be noted that where product browning is observed, it will be accompanied by a significant fall in pH of 0.1 to 0.3 pH units (see Section 7.5.1). Adhikari and Singhal (1991) reported a change after 24 days storage from 6.82 to 6.52 at 22 °C and to 6.45 at 37 °C. In storage trials, pH changes were significantly greater in products once the storage temperature exceeded 35 °C. This decrease in pH has been attributed to acids, principally formic and acetic, formed as intermediate products in the Maillard reaction (van Boekel, 1998; van Boekel & Brands, 1998).

7.2.5.2 Browning of Fruit Juices

Fruit juices are susceptible to enzymatic browning during their production and steps should be taken to control this. The enzymes responsible are largely inactivated during pasteurisation. However, fruit juices are also susceptible to considerable non-enzymic browning during storage. Compounds involved in the browning reaction include ascorbic acid (particularly in its oxidised form) (Wong *et al.*, 1992, Naim *et al.*, 1993), other sugars and some nitrogenous compounds such as amino acids. It has been suggested that juices should be deaerated prior to heat treatment to reduce vitamin C oxidation. An alternative approach is to add a reducing agent such as cysteine which Naim *et al.* (1993) showed significantly reduced browning in orange juice; it also improved the flavour of the juice and increased retention of ascorbic acid during storage. It is therefore apparent that the Maillard reaction in fruit juices differs from that in milk as milk contains a very low concentration of ascorbic acid (~1 mg/100 mL).

Roig *et al.* (1999) reported that amino acids and possibly other amino groups enhance browning. This was consistent with the finding of Wong and Stanton (1993) that removal of amino acid residues (using adsorbent resins) reduced browning in kiwi fruit during storage. In milk, the Maillard reaction is known to be also enhanced by added amino groups.

Rassis and Saguy (1995) processed aseptically concentrated orange juice at 84, 87 and 90 °C for 72 s and stored the juices for up to 7 weeks at 32 °C and 15 weeks at 22 °C. They found no differences in non-enzymic browning attributable to the different thermal treatments. In this respect, the Maillard reaction in fruit juice also differs from that in milk as the Maillard reaction in UHT milk during storage increases with the severity of the processing conditions.

Kyung *et al.* (1995) measured the browning index in stored canned orange juice and found an activation energy of 45.3 kcal/mol (189.4 kJ/mol), between 40 and 50 °C, which is considerably higher than that found for milk of 107 kJ/mol (Horak & Kessler, 1981). Ibarz *et al.* (2000) measured browning in concentrated apples during storage and found an activation energy of 30.22 kcal/mol (~126.3 kJ/mol). The Q_{10} for browning (the factor by which rate increases for a 10 °C rise in temperature) was estimated to be 2.1 (Kyung *et al.*, 1995). For milk, Burton (1954) reported a Q_{10} of 2.95 (and z = 21.3 °C).

7.3 Changes to Some UHT Products Other than Single-Strength Fresh White Cow's Milk

Most of the work on changes during storage has been reported for cow's milk. This section briefly reviews other products where storage changes have been studied.

- Al-Khanal *et al.* (1994) compared changes in the quality of fresh and **recombined** UHT milks during storage at temperatures of 6, 20 and 35 °C. The UHT recombined milks showed a decrease in sensory quality with increased storage temperature and the reduction was higher than that found for UHT fresh milk. Other changes during storage included reduction in pH and increases in titratable acidity, proteolysis, lipolysis, browning, fat separation and viscosity. These changes were greater at the higher storage temperatures, except for viscosity of the recombined milk which was lower at 35 °C. Sedimentation was generally very low and showed considerable variation between samples.
- Changes in **UF reduced-fat concentrated milk** were measured by Reddy *et al.* (1992) during storage for 10 weeks and compared with those of unconcentrated whole milk. Based on the protein contents, 5.18 and 3.31% respectively, the concentration factor for the UF milk was ~1.6. Fat contents were 1.4 and 3.6% respectively. The decrease in pH was similar for both milk types, age gelation occurred faster in the UF milk, and sedimentation was initially higher in the UF milk but increased in both milks during storage. Sensory scores were similar for both milks initially and after refrigerated storage; after storage at room temperature the whole milk was preferred to the UF milk but both were still acceptable, but after storage at 37 °C, both types of milk were considered to be unsuitable for human consumption.
- Changes in UHT **buffalo** milk during storage for 6 weeks have been reported by Singh and Patil (1989a,b). In 6% fat milk, processed indirectly at 140 °C for 2 s, pH decreased from 6.8 to 6.6, HMF (free/total) increased from 2.0/8.89 mmol/L to 6.52/22.38 mmol/L, ADV (a measure of lipolysis) increased from 3.75 to 5.97 mmoles/L, and sensory scores decreased from 89.2 initially to 87.8 at day 17 to 62.0 at day 41. In general, buffalo milk showed a faster deterioration than cow's milk. The formation of lactones which contribute to the stale and coconut flavour during storage of UHT milk has been studied in buffalo milk by Bansal and Sharma (1994). They found a 2.4- and 5-fold increase in the total lactone fraction during storage at 20 and 37 °C for 6 months, respectively. At 4 months, a coconut-like flavour was detected which correlated with the presence of C10 and C12 delta-lactones.
- Changes in physicochemical and sensory characteristics of UHT **flavoured peanut beverages** during storage have been studied by Rustom *et al.* (1995, 1996a,b). Deterioration in the sensory qualities was highly correlated with an increase in the sedimentation index and a decrease in pH and emulsion stability. The viscosity of the strawberry flavoured beverages was constant whereas chocolate flavoured beverages gelled after 19 weeks at all temperatures.

- UHT **soy beverages** were stored at 5, 25, 35 and 45 °C. pH, reflectance and sensory scores decreased during storage, whereas viscosity, proteolysis, lipolysis, oxidation and browning increased. The rates of these reactions were highly temperature-dependent. Proteolysis was the main reason for the decline in acceptability (Narayanan *et al.*, 1993).

7.4 Accelerated Storage Testing

Samples of heat-sterilised foods are routinely incubated at elevated temperatures (30 to 55 °C) to establish whether they have any bacterial contamination. Typical incubation conditions are 30 °C for 14 days, although other temperatures such as 37, 45 and 55 °C may be used (Shew, 1981). Experience has shown that post-processing contamination or a serious failure of heat treatment usually results in high microbial activity. In our experience, where a sterile product is packaged into a sterile container but not in a sterile environment, it will generally coagulate within 2-3 days at 30 °C. An incubation temperature of 55 °C is also usually stipulated to detect thermophilic sporeformers, although Shew (1981) questioned the necessity for use of this temperature. However, as pointed out in Section 8.3, such an incubation is advisable where the product is likely to reach temperatures of ≥ ~45 °C during storage or transportation.

These incubation procedures can also be used for accelerated storage testing, for giving a quicker indication of how the sensory or other characteristics may be adversely changed during storage; especially with regard to development of off-flavours, sedimentation or unacceptable changes in colour. It can be particularly useful for assessing the action of enzymes such as lipases and proteases which may be in very low concentrations. An example of the effect of temperature on lipolysis was provided by Singh *et al.* (2004). UHT milk processed at 140 °C for 2 s was stored at 9, 15, 25, 35 and 45 °C for 16 weeks. Free fatty acids (FFA) increased from 1.27 meq/L to 1.78, 1.90, 2.30, 3.01 and 3.51 meq/L respectively. After storage at 55 °C for 12 weeks, the FFA level was 5.68 meq/L. Naranganan *et al.* (1993) carried out a similar storage trial on a UHT soy beverage at 5, 25, 35 and 45 °C and demonstrated the value of the use of higher temperatures for assessing the various chemical changes. This enabled them to obtain kinetic data for predictive models for shelf-life prediction.

Richards *et al.* (2014) analysed the sensory data from a trained panel obtained during an accelerated shelf-life trial on low-fat UHT milk stored at 25, 35 and 45 °C for 6.5 months. They found that storage at 35 and 45 °C reduced the shelf-life by a factor of 2.9 and 7.8, respectively, compared with storage at 25 °C. They suggested that changes in sensory attributes of milk stored at accelerated temperatures could be used to predict the shelf-life of a product under market conditions.

As for all accelerated storage testing, it has to be borne in mind that reactions occurring at an elevated temperature may not relate well to what occurs at the normal temperature of storage. For example, age gelation occurs optimally at ~25-28 °C; gelation is inhibited at higher (and lower) temperatures (see Section 7.2.2.4). Hence an accelerated test at 40 °C of a milk for its susceptibility to gelation would not be successful. Also the rate of the Maillard reaction at room temperature (~20 °C) is very low but increases rapidly from ~30 °C. Hence the results of an accelerated test for Maillard browning at ≥40 °C may be quite irrelevant to what would happen at 20 °C.

Another form of accelerated testing of UHT milk is the centrifugation test for sedimentation (see also Section 11.2.22). While we use centrifugation at ~3000 g at room temperature for 20-30 min, other conditions are also effective. For example, Tobin *et al.* (2011) centrifuged UHT milk at 1147 g for 3.5 h, conditions which they reported simulated approximately 6 months' ageing under conditions of normal gravity.

7.5 Chemical and Physical Changes During Storage Trials of UHT Milk

This section reports the results of two controlled (unreported) storage trials performed to demonstrate various changes in commercial milk samples over a range of temperatures, as well as results of two previously reported storage trials. It consolidates many of the storage-related changes discussed in previous sections of this chapter and illustrates the changes that can occur, particularly if products are temperature-abused.

7.5.1 Storage Trial 1 (DIAL, 2014)

Ten commercial UHT milk samples were stored at 4, 20, 35 and 50 °C for 4 months. A summary of the results is presented in Tables 7.6 and 7.7.

pH : Increasing storage temperature resulted in a fall in pH, attributable mainly to acids produced in the Maillard reaction. The change was greatest for the highest temperatures but there was even a small decrease in samples stored at 20 °C (Table 7.6). The largest decrease was observed for lactose-reduced milk which reflects the higher rate of the Maillard reaction involving the hydrolysis products, glucose and galactose, compared with lactose.

Ionic calcium: There were changes in ionic calcium during storage but there were no clear trends with storage temperature. The level in the white cow's milk samples stored at 4 and 50 °C were higher than those stored at 20 and 35 °C (Table 7.6). The reason for this is not known.

Viscosity : There were no signs of any excessive age-thickening occurring at high temperature. In fact, the viscosity of samples stored at 50 °C was lower than that of those stored at lower temperatures, even though they were all measured at room temperature (Table 7.7).

Freezing point depression (FPD) : the FPD of all white milk samples were above 514 m°C. FPD changed considerably at different storage temperatures but not consistently. The FPD of all milk products except the lactose-reduced milk increased at 50 °C (Table 7.7).

Colour differences (ΔE values) : These results are shown in Table 7.4; samples stored at 4 °C were used as the reference. They clearly demonstrate the higher values at 50 °C than at 35 °C. There was some consistency in the results at 50 °C for the full cream milk samples (20.1-20.7) and for the skim milk samples (both 33.7); the values for lactose-reduced milk samples were much higher (for definition of ΔE values see Sections 7.2.5.1 & 11.2.2.1).

Sediment : Formation of sediment seemed to be independent of storage temperature (Figure 7.4) and was low for most samples.

Table 7.6 Results of storage trial 1 of commercial UHT milk products – pH and ionic calcium. (Source: DIAL, 2014. Reproduced with permission.)

	pH						Ca^{2+} (mM)					
	Full cream milk[1]	Skim[2]	Lactose-reduced	Goat	Strawberry	Chocolate	Full cream milk[1]	Skim[2]	Lactose-reduced	Goat	Strawberry	Chocolate
4 °C	6.73	6.72	6.75	6.71	6.72	6.77	3.03	2.31	2.60	3.27	3.67	2.91
20 °C	6.65	6.64	6.68	6.61	6.65	6.72	1.99	1.84	2.31	2.31	3.67	2.46
35 °C	6.58	6.58	6.51	6.51	6.64	6.62	1.99	2.06	2.31	2.60	3.27	2.19
50 °C	6.20	6.18	5.87	6.03	6.12	6.21	3.15	2.76	2.91	2.31	2.91	2.61

[1] Full cream milk results are average of three different milk samples;
[2] skim results are average of two different samples

Table 7.7 Results of storage trial 1 of commercial UHT milk products – viscosity and freezing point depression (FPD). (Source: DIAL (2014). Reproduced with permission.)

	Viscosity (cSt)						Freezing point depression (m°C)					
	Full cream milk[1]	Skim[2]	Lactose-reduced	Goat	Strawberry	Chocolate[3]	Full cream milk[1]	Skim[2]	Lactose-reduced	Goat	Strawberry	Chocolate[3]
4 °C	2.04	1.76	2.04	1.73	2.92	3.71; 3.80	531	514	744	534	765	820; 833
20 °C	2.01	1.85	2.01	1.70	2.68	3.25; 3.65	534	518	758	535	766	818; 835
35 °C	2.01	1.98	2.01	1.64	3.34	3.37; 3.65	539	518	752	542	779	821; 841
50 °C	1.67	1.76	1.64	1.55	3.13	4.99; 3.19	566	535	708	570	805	851; 874

[1] Full cream milk results are average of three different milk samples;

[2] Skim results are average of two different samples;

[3] Results for 2 different chocolate milk samples

Ethanol stability : The samples stored at all temperatures (even those stored at 50 °C in which pH was significantly reduced) were stable in 75% ethanol.

7.5.2 Storage Trial 2 (UCC, 2015)

Samples of UHT full cream (FCM), UHT semi-skim (SS) and UHT skim milk (skim) were stored at -18, 20, 35 and 50 °C for 6 months. Summaries of the results are presented in Tables 7.8 and 7.9.

Sediment: Frozen storage (at -18 °C) had a destabilising influence and produced a large amount of sediment in all samples. More sediment was found in the full cream milk compared to skim milk, but at these levels all samples after being stored frozen would be unacceptable. For storage between 4 and 35 °C, sediment was not affected by storage temperature for milk of all fat contents, although it was slightly higher at 50 °C. In fact, there was some evidence that sediment decreased with storage time, which could be due to slight solubilisation of some of its components (Table 7.8).

Colour: These results show little change in the L* value (see Section 11.2.2.1) for samples over the range at 4 and 20 °C but there was a slight decrease at 35 °C and a larger decrease at 50 °C. This is in line with the progression of Maillard browning previously discussed (see Section 7.2.5) (Table 7.8).

Freezing point depression: The FPD of samples stored between -18 °C and 35 °C showed no changes, but values at 50 °C increased substantially, which was also observed in storage trial 1 (see Section 7.3.1) (Table 7.8).

The FPD results after 2 and 6 months are shown in Table 7.9. Changes after two months compared with "fresh" samples (see Table 7.8) were noticeable, but they became substantial after 6 months. The results for frozen storage are unusual, showing a considerable reduction after two months, but returning to normal after 6 months. It would need to be established whether this is a consistent trend. However, it is clear that it would not be advisable to let UHT products freeze during transportation or storage, as large amounts of sediment may result.

Viscosity: There were no signs of age thickening in any of the samples. The viscosity of samples stored at 50 °C was the lowest, which was also observed in storage trial 1. The viscosity of frozen samples was also higher, more so for full cream milk compared to skim milk (Table 7.8). In follow-up observations, none of the samples had formed a gel after 6 or 12 months.

Table 7.9 shows the particle size, cream layer and sediment data obtained from the Lumisizer for the same samples stored for 2 months and 6 months.

Particle size: Particle sizes were higher after 6 months than after 2 months but for frozen samples the change was dramatic. This can be attributed to destabilization of the casein micelle and subsequent casein aggregation. Particle size did not appear to be influenced by the fat content (Table 7.9).

Cream layer: The cream layer results show no major changes, either at high storage temperatures or during frozen storage (Table 7.9).

Table 7.8 Data for controlled storage trial 2 for UHT full cream milk (FCM), semi-skimmed milk (SS) and skim milk (Skim) after no storage ("fresh") and 6 months' storage. (Source: UCC, 2015. Reproduced with permission.)

	Viscosity (mPa s)			Freezing point depression (m°C)			L* value			Sediment (%)		
	FCM	SS	Skim	FCM	SS	Skim	FCM	SS	Skim	FCM	SS	Skim
Fresh (no storage)	2.56	2.50	2.20	520	516	515	85.9	83.7	79.2	0.36	0.34	0.40
−18 °C	5.00	4.42	3.38	522	516	516	77.8	75.1	71.9	6.84	4.06	1.56
4 °C	3.06	2.69	2.71	523	517	515	84.9	82.3	76.8	0.15	0.16	0.16
20 °C	3.31	2.46	2.40	525	518	518	84.5	82.4	77.7	0.17	0.18	0.22
35 °C	2.65	2.82	2.40	518	522	522	82.0	80.2	74.9	0.17	0.18	0.18
50 °C	2.55	2.42	2.16	577	565	563	70.1	65.5	57.9	0.28	0.18	0.22

Table 7.9 Data for controlled storage trial 2 for UHT full cream milk (FCM), semi-skimmed milk (SS) and skim milk (Skim) after 2 months (first entry) and 6 months (second entry). (Source: UCC, 2015. Reproduced with permission.)

	Particle size (μm)*			Cream layer (mm)*			Sediment height (mm)*			Freezing point depression (m°C)		
	FCM	SS	Skim	FCM	SS	Skim	FCM	SS	Skim	FCM	SS	Skim
−18 °C	0.79	0.56	0.26	0.94	0.15	0.02	1.12	1.32	1.15	464	475	512
	50.1	24.2	39.8	0.43	0.10	0	3.66	3.09	1.96	522	516	516
20 °C	0.44	0.44	0.25	1.06	0.27	0	0.97	1.00	0.60	521	515	514
	0.82	0.82	0.68	1.05	0.21	0	1.26	1.20	1.04	525	518	518
35 °C	0.42	17.1	0.24	1.06	0.19	0	0.78	1.06	0.65	521	515	516
	0.81	0.87	0.67	0.88	0.55	0	1.00	1.45	1.35	518	520	522
50 °C	0.25	0.46	0.25	1.10	0.05	0	0.66	1.00	0.85	532	527	526
	0.78	0.86	0.67	0.98	0.38	0	1.04	0.73	1.10	577	565	563

* Data from Lumisizer

Sediment: The sediment height was slightly greater after 6 months for all samples but, as for particle size, was not influenced by the amount of fat (Table 7.9). The sediment level was much higher for samples stored for six months at -18 °C which is consistent with the results shown in Table 7.8.

7.5.3 Other Storage Trials

The results of the two storage trials above are generally consistent with previous reports on changes during storage. However, additional changes have been reported from other storage trials. Two such trials are mentioned here.

Gaucher *et al.* (2008a) reported that milk pH decreased during storage. Alcohol stability remained high but low stability was observed for samples stored at 40 °C, as determined by both a heat stability test and the Ramsdell phosphate test. They also reported an increase in NCN, while casein micelle size and the electric charge (zeta potential) did not appear to change significantly during storage.

Samel *et al.* (1971) stored commercial indirectly processed milk at 4, 20, 30 and 37 °C for 2 years. Their major findings were: an increase in NCN and NPN, with the increase being greater at the higher storage temperature; gelation in samples stored at 4, 20 and 30 °C after 13 months but no gelation in samples stored at 37 °C; marked decrease in sensitivity to ionic calcium in milk stored at 4 and 20 °C; marked decrease in ethanol stability in milk stored at 20 °C; and decreased rennet coagulation time, from 50 to 4 min in milk stored at 20 °C but increased from 50 to >120 min for milk stored at 37 °C (the RCT for raw milk was 7 min).

References

Adhikari, A.K. & Singhal, O.P. (1991) Changes in the flavour profile of indirectly heated UHT milk during storage: effect of Maillard browning and some factors affecting it. *Indian Journal of Dairy Science* **44**, 442–448.

Adhikari, A.K. & Singhal, O.P. (1992) Effect of dissolved oxygen content on the flavour profile of UHT milk during storage. *Australian Journal of Dairy Technology* **47**, 1–6.

Al-Attabi Z., D'Arcy, B.R. & Deeth, H.C. (2009) Volatile sulfur compounds in UHT milk. *Critical Reviews in Food Science and Nutrition* **49**, 28–47.

Al-Attabi, Z., D'Arcy, B.R. and Deeth, H.C. (2014) Volatile sulfur compounds in pasteurised and UHT milk during storage. *Dairy Science and Technology* **94**(3), 241–253.

AlKanhal, H.A. (2000) Suitability of methods used for the measurement of proteolysis in UHT milk. *Australian Journal of Dairy Technology* **55**, 148–152.

AlKanhal, H.A., Abu Lehia, I.H. & Al Saleh, A.A. (1994) Changes in the quality of fresh and recombined ultra high temperature treated milk during storage. *International Dairy Journal* **4**(3), 221–236.

AlKanhal, H.A., Abu-Lehia,I.H. & Al-Saleh, A.A. (1994) Changes in the quality of fresh and recombined ultra high temperature treated milk during storage. *International Dairy Journal* **4**, 221–236

Aloglu, H.S. (2013) The effect of various heat treatments on the antioxidant capacity of milk before & after simulated gastrointestinal digestion. *International Journal of Dairy Technology* **66**(2), 170–174.

Al-Saadi, J.S., & Deeth, H.C. (2008) Cross-linking of proteins and other changes in UHT milk during storage at different temperatures. *Australian Journal of Dairy Technology* **63**(3), 93–99.

Andersson, R. E., Danielsson, G., Hedlund, C. B. & Svensson, S. G. (1981) Effect of a heat-resistant microbial lipase on flavor of ultra-high-temperature sterilised milk. *Journal of Dairy Science* **64**(3), 375–379

Andersson, I & Öste, R, (1992b), Sensory changes and free SH-groups in UHT milk. *Milchwissenschaft* **47**, 438–441.

Andersson, I. & Öste, R. (1992a) Loss of ascorbic-acid, folacin & vitamin-B12, & changes in oxygen-content of UHT milk. 2. Results and discussion. *Milchwissenschaft* **47**(5), 299–302.

Andersson, I. & Öste, R. (1995) Sensory quality of UHT milk. In: *Heat-Induced Changes in Milk* (ed. P.F. Fox), pp. 318–330. International Dairy Federation, Brussels.

Andrews, A.T. (1975) Properties of aseptically packed ultra-high-temperature milk .3. Formation of polymerized protein during storage at various temperatures. *Journal of Dairy Research* **42**, 89–99.

Andrews, G.R. (1986) Effect of storage, light and homogenisation on the colour of ultra heat treated and sterilised milk. *Journal of Dairy Research*, **53**(4), 615–624.

AOAC (1984) Phosphatase (reactivated and residual) in milk. AOAC 961.08-1984. AOAC International, Rockville, MD, USA.

Ashton, T.R. (1965) Practical experience: the processing and aseptic packaging of sterile milk in the United Kingdom. *Journal of the Society of Dairy Technology* **18**(2), 65–85.

Auldist, M.J., Coats, S.J., Sutherland, B.J., Hardham, J.F., McDowell, G.H. & Rogers, G.L. (1996) Effect of somatic cell count and stage of lactation on the quality and storage life of ultra high temperature milk. *Journal of Dairy Research* **63**, 377–386.

Badings, H.T. (1970) Cold-storage defects in butter & their relation to the autoxidation of unsaturated fatty acids. *Netherlands Milk and Dairy Journal* **24**(3/4), 147–257.

Badings, H.T. (1977) A method of preventing the occurrence of cooked flavour in milk and milk products. *Nordeuropaeisk Mejeri Tidsskrift* **43**(11/12), 379–384.

Badings, H.T., van der Pol, J.J.G. & Neeter, R. (1981) Aroma compounds which contribute to the difference in flavour between pasteurised milk and UHT milk. *'Flavour '81' G* (pp. 683–692).

Bagliniere, F., Tanguy, G., Jardin, J., Mateos, A., Briard, V., Rousseau, F., Robert, B., Beaucher, E., Humbert, G., Dary, A., Gaillard, J.L., Amiel, C. & Gaucheron, F. (2012) Quantitative and qualitative variability of the caseinolytic potential of different strains of *Pseudomonas fluorescens*: Implications for the stability of casein micelles of UHT milks during their storage. *Food Chemistry* **135**(4), 2593–2603.

Bansal, M. & Sharma, U.P. (1995) Formation of lactones during storage of UHT processed buffalo milk. *Journal of Dairy Research* **62**(4), 645–650.

Barefoot, S.F. & Adams, D.M. (1980) Amylase activity in sterile spoiled pudding. *Journal of Food Science* **45**, 1658–1661.

Barefoot, S.F. (1979) Amylolytic activity in spoiled ultrahigh temperature sterilised pudding. M.S. Thesis, North Carolina State University, Raleigh, NC.

Bastian, E.D. & Brown, R.J. (1996) Plasmin in milk and dairy products: An update. *International Dairy Journal* **6**, 435–457.

Belloque, J., Villamiel, M., Lopez Fandino, R. & Olano, A. (2001) Release of galactose and N-acetylglucosamine during the storage of UHT milk. *Food Chemistry* **72**, 407–412.

Bhatt, H., Cucheval, A., Coker, C., Patel, H., Carr, A. & Bennett, R. (2014) Effect of lactosylation on plasmin-induced hydrolysis of β-casein. *International Dairy Journal* **38**, 213–218.

Biewendt, H. G. & Prokopek, D. (1992) [Light-induced changes and the light protective effects of brown glass bottles for directly heated UHT milk.] *Molkerei Zeitung Welt der Milch* **46**(23/24), 676–682.

Blake, M.R., Weimer, B.C., McMahon, D.J. & Savello, P.A. (1995) Sensory and microbial quality of milk processed for extended shelf-life by direct steam injection. *Journal of Food Protection* **58**, 1007–1013.

Blanc, B. & Odet, G. (1981) Appearance, flavour & texture aspects: Recent developments. *New Monograph on UHT Milk*. Document **133**, pp. 25–48, International Dairy Federation, Brussels.

Blanc, B., Flueckiger, E., Rueegg, M. & Steiger, G. (1980) [Changes in biochemical, physical, technological & sensory properties of UHT milk during storage.] *Alimenta* **19**, 24–27.

Burton, H. (1954) Colour changes in heated milks. 1. The browning of milk on heating. *Journal of Dairy Research* **21**, 194–203

Burton, H. (1988) *Ultra-high Temperature Processing of Milk and Milk Products*. London: Elsevier Applied Science.

Bushnell, A.C. (1980) A note on an unusual feature found in UHT milk. *Journal of the Association of Public Analysts* **18**(2), 67–68.

Button, P.D. (2007) *UHT Milk Spoilage owing to Lipase and Protease from Psychrotrophic* Pseudomonas *species: Characteristics of Enzymatic Breakdown, Improved Assays for Enzyme Detection and Sources of the Bacteria*. PhD thesis, University of Melbourne, Melbourne, Australia.

Button, P.D., Roginski, H., Deeth, H.C. & Craven H.M. (2011) Improved shelf-life estimation of UHT milk by prediction of proteolysis. *Journal of Food Quality* **34**, 229–235.

Calligaris, S., Manzocco, L., Anese, M. & Nicoli, M.C. (2004) Effect of heat-treatment on the antioxidant and pro-oxidant activity of milk. *International Dairy Journal*, **14**, 421–427.

Carini, S. & Todesco, R. (1977) [Gelation of UHT milk.]. *Latte* **2**(12), 725–730.

Celestino, E.L., Iyer, M.,& Roginski, H. (1997) Reconstituted UHT-treated milk: effects of raw milk, powder quality and storage conditions of UHT milk on its physico-chemical attributes and flavour. *International Dairy Journal* **7**(2/3), 129–140.

Chapman, K.W., Whited, L.J. & Boor, K.J. (2002) Sensory threshold of light-oxidized flavor defects in milk. *Journal of Food science* **67**(7), 2770–2773.

Chavan, R.S., Chavan, S.R., Khedkar, C.D. & Jana, A.H. (2011) UHT milk processing and effect of plasmin activity on shelf-life: A review. *Comprehensive Reviews in Food Science & Food Safety* **10**, 251–268.

Chen, L., Daniel, R. M. & Coolbear, T. (2003) Detection and impact of protease and lipase activities in milk and milk powders. *International Dairy Journal* **13**(4), 255–275.

Choi, I.W. & Jeon, I.J. (1993) Patterns of fatty acids released from milk fat by residual lipase during storage of ultra-high temperature processed milk. *Journal of Dairy Science* **76**, 78–85.

Christen, G.L. & Senica, M. (1987) Casein fluorescein isothiocyanate as a substrate for measuring protease production by psychrotrophic bacteria. *Journal of Food Protection* **50**, 744–749.

Christen, G.L., Wang, W.C. & Ren, T.J. (1986) Comparison of the heat resistance of bacterial lipases and proteases and the effect on ultra-high temperature milk quality. *Journal of Dairy Science* **69**(11), 2769–2778.

Chung, H.Y., Partridge, J.A. & Harte, B.R. (2004) Quantitative evaluation of light-oxidized off-flavors in reduced fat milk using sensory evaluation and the electronic nose. *Journal of Dairy Science* **87**(Supplement 1), 234–234.

Cliffe, A.J. & Law, B.A. (1982) A new method for the detection of microbial proteolytic enzymes in milk. *Journal of Dairy Research* **49**, 209–219.

Colahan-Sederstrom, P.M. & Peterson, D.G. (2005) Inhibition of key aroma compound generated during ultrahigh-temperature processing of bovine milk via epicatechin addition. *Journal of Agricultural and Food Chemistry* **53**(2), 398–402.

Collins, S.J., Bester, B.H. & McGill, A.E.J. (1993) Influence of psychrotrophic bacterial growth in raw milk on the sensory acceptance of UHT skim milk. *Journal of Food Protection* **56**(5), 418–425.

Corradini, C. & Pecchini, G. (1981) Effect on proteinases of different UHT treatments *Netherlands Milk & Dairy Journal* **35**(3/4), 393–395.

Corzo, N., Lopezfandino, R., Delgado, T., Ramos, M. & Olano, A. (1994) Changes in furosine and proteins of UHT-treated milks stored at high ambient-temperatures. *Zeitschrift fur Lebensmittel-Untersuchung und-Forschung,* **198**(4), 302–306.

Dalgleish, D.G. (1992) Sedimentation of casein micelles during the storage of ultra-high temperature milk products - a calculation. *Journal of Dairy Science* **75**(2), 371–379.

Datta, N. & Deeth, H.C. (2001) Age gelation of UHT milk - A review. *Transactions of the Institute of Chemical Engineers C. Food Bioproducts Processing* **79**, 197–210.

Datta, N. & Deeth, H.C. (2003) Diagnosing the cause of proteolysis in UHT milk. *LWT-Food Science and Technology* **36**(2), 173–182.

Day, E.A., Lillard, D.A. & Montgomery, M.W. (1963) Autoxidation of milk lipids. III. Effect on flavor of the additive interactions of carbonyl compounds at sub-threshold concentrations. *Journal of Dairy Science* **46**, 291–294.

de Koning, P.J., Kaper, J., Rollema, H S. & Driessen, F.M. (1985) Age-thinning and gelation in unconcentrated and concentrated UHT-sterilised skim milk. Effect of native milk proteinase. *Netherlands Milk & Dairy Journal* **39**, 71–87.

de Rham, O. & Andrews, A.T. (1982) The roles of native milk proteinase and its zymogen during proteolysis in normal bovine-milk. *Journal of Dairy Research* **49**(4), 577–585.

Deepa, P. & Mathur, M. P. (1994) Age gelation in ultra heat treated milk: some evidence for a plastein induced mechanism. *Journal of Food Science &Technology, India* **31**(1), 77–79.

Deeth, H.C. & Fitz-Gerald, C.H. (2006) Lipolytic enzymes and hydrolytic rancidity. In: *Advanced Dairy Chemistry, 2 Lipids,* (eds P.F. Fox & P.L.H. McSweeney), 3rd edn, pp. 481–556. Springer, New York.

DIAL (2014) UHT and thermal processing – Results of trial performed in association with a Dairy Innovation Australia Ltd (DIAL) workshop. May 21-22, 2014, Werribee, Australia.

Dimick, P.S. (1982) Photochemical effects on flavor and nutrients of fluid milk. *Canadian Institute of Food Science &Technology Journal* **15**(4), 247–256.

Dolfini, L., Kueni, R., Eberhard, P., Fuchs, D., Gallmann, P.U., Strahm, W. & Sieber, R. (1991) Behaviour of supplemented vitamins during storage of UHT skim milk. *Mitteilungen aus dem Gebiete der Lebensmitteluntersuchung & Hygiene* **82**(2) 187–198.

Driessen, F.M. & Stadhouders, J. (1974) Thermal activation and inactivation of extracellular lipases of some Gram-negative bacteria common in milk. *Netherlands Milk and Dairy Journal* **28**, 10–22

Driessen, F.M. (1981) [Enzymic proteolysis in UHT milk products. II. Relationship between heating intensity, inactivation of native milk protease & keeping quality.]. *Zuivelzicht* **73**(33/34), 688–689.

Driessen, F.M., Huoydonk, A. van, Steuper, A. & Kingmar, F. (1991) Influence of native milk protease on the viscosity of UHT-sterilised custard during storage. *Netherlands Milk & Dairy Journal* **35**(2), 121–31.

Eigel, W.N., Hofman, C.J., Chibber, B.A.K., Tomich, J.M., Keenan, T.W. & Mertz, E.T. (1979) Plasmin-mediated proteolysis of casein in bovine milk. *Proceedings of the National Academy of Sciences of USA* **76**, 2244–2248.

Ellerton, M.E. & Pearce, S.J. (1964) Some observations on the physical-chemical stability of sterile concentrated milks. *Journal of Dairy Science* **47**, 564–569.

Elliott, A.J., Dhakal, A., Datta, N. & Deeth, H.C. (2003) Heat-induced changes in UHT milks – Part 1. *Australian Journal of Dairy Technology* **58**, 3–10.

Erasmo, N., Diego, M. & Carini, S. (1981) [Gelation of UHT milk: soluble nitrogen fraction and casein residue.] *Latte* **6**, 670–675.

Fairbairn, D. (1989) Assay methods for proteinases. In: *Enzymes of Psychrotrophs in Raw Food.* (ed R. C. McKellar), CRC Press, Florida.

Fink, R. & Kessler, H.G. (1988) Comparison of methods for distinguishing between UHT treatment and sterilisation of milk. *Milchwissenschaft* **43**, 275–279.

Ford, J.E., Porter, J.W.G., Thompson, S.Y., Toothill, J. & Edwards-Webb, J. (1969) Effects of ultra-high-temperature (UHT) processing and of subsequent storage on the vitamin content of milk. *Journal of Dairy Research* **36**, 447–454.

Forss, D.A. (1979) Review of the progress of dairy science: mechanisms of formation of aroma compounds in milk & milk products. *Journal of Dairy Research* **46**, 691–706.

Fritsch, R., Hoffmann, H. & Klostermeyer, H. (1983) Formation of lysinoalanine during heat treatment of milk. *Zeitschrift für Lebensmittel-Untersuchung & Forschung* **176**, 341–345.

Fox, P.F. & McSweeney, P.L.H. (1998) *Dairy Chemistry and Biochemistry*. pp. 225–326. Blackie Academic & Professional, London, UK.

Gaafar, A.M. & El Sayed, M.M. (1991) Chemical changes in ultra-heat-treated milk during storage - Properties of the proteins. *Egyptian Journal of Dairy Science* **19**, 257–266.

Gaafar, A.M.M. (1987) *Investigation into the Cooked Flavour in Heat-Treated Milk*, PhD thesis, University of Reading, Reading, UK.

Gallusser, A. & Bergner, K. G. (1981) [Reactivation of alkaline phosphatase in UHT milk, depending on its oxygen content.]. *Deutsche Lebensmittel Rundschau* **77**(12), 441–444.

Gandon, Y., Petit, A. & Dechy, M.F. (1974) [UHT-sterilised milk aseptically packaged in Tetra Brik]. *Lait* **54**(539/540), 645–655.

Gaucher, I., Boubellouta, T., Beaucher, E., Piot, M., Gaucheron, F., & Dufour, E. (2008a) Investigation of the effects of season, milking region, sterilisation process and storage conditions on milk and UHT milk physico-chemical characteristics: a multidimensional statistical approach. *Dairy Science & Technology* **88**, 291–312.

Gaucher, I., Molle, D., Gagnaire, V. & Gaucheron, F. (2008b) Effects of storage temperature on physico-chemical characteristics of semi-skimmed UHT milk. *Food Hydrocolloids* **22**, 130–143.

Grosch, W. (1982) Lipid degradation products and flavour. In: *Food Flavours. Part A. Introduction.* (eds I.D. Morton & A.J. MacLeod). pp. 325–398. Elsevier Scientific Publishing Co., New York, NY.

Hansen, A.P. (1990) Age gelation and fat separation: two major problems in aseptic dairy products. *Journal of Dairy Science* **73**(Supplement 1), 91.

Hardham, J.F. & Auldist, M.J. (1996) The effect of stage of lactation and somatic cell count on the age-gelation of UHT milk. In: *Heat Treatments & Alternative Methods*, IDF Doc SI 9602, International Dairy Federation, Belgium.

Hardham, J.F. (1998) Effect of protein standardisation of milk by addition of UF milk permeate on the composition and storage stability of UHT processed milk. *Australian Journal of Dairy Technology* **53**(1), 22–27.

Hardham, J.F. (1998) Effect of protein standardisation of milk by addition of UF milk permeate on the composition and storage stability of UHT processed milk. *Australian Journal of Dairy Technology* **53**, 22–27.

Harwalkar, V.R. & Vreeman, H.J. (1978) Effect of added phosphates and storage on changes in ultra-high temperature short-time sterilised concentrated skim-milk. I. Viscosity, gelation, alcohol stability, chemical and electrophoretic analysis of proteins. *Netherlands Milk & Dairy Journal* **32**(2), 94–111.

Harwalkar, V.R., Beckett, D.C., McKellar, R.C., Emmons, D.B., & Doyle, G.E. (1983) Age-thickening and gelation of sterilised evaporated milk. *Journal of Dairy Science* **66**(4), 735–742.

Harwalkar, V.R. (1992) Age gelation of sterilised milks. In: *Advanced Dairy Chemistry, Volume 1 Proteins*, 2nd edn, (ed P.F. Fox), pp. 691–734. Elsevier Applied Science Publishers, London.

Hawran, L.J., Jones, V.A. & Swartzel, K.R. (1985) Sediment formation in aseptically processed and packaged milk. *Journal of Food Processing & Preservation* **9**(3), 189–207.)

Hillbrick, G.C, McMahon, D.J. & Deeth, H.C. (1998) Electrical impedance particle size method (Coulter counter) detects the large fat globules in poorly homogenised UHT processed milk. *Australian Journal of Dairy Technology* **53**, 17–21.

Holland, J, Gupta, R., Deeth, H.C. & Alewood, P.F. (2011) Proteomic analysis of temperature-dependent changes in UHT milk. *Journal of Agricultural and Food Chemistry* **59**, 1837–1846.

Holland, J.W., Gupta, R., Deeth, H.C., & Alewood, P.F. (2012) UHT milk contains multiple forms of alpha(S1)-casein that undergo degradative changes during storage. *Food Chemistry* **133**, 689–696.

Horak, F.P. & Kessler, H.G. (1981) [The destruction of meso- and thermophilic spores in ultra-high-temperature UHT-milk and UHT-cream processing]. *Chemie Mikrobiologie Technologie der Lebensmittel* **7**(2), 42–50.

Hsiu, J., Fischer, E.H. & Stein, E.A. (1964) Alpha-amylases as calcium-metallo-enzymes 2. Calcium and the catalytic activity. *Biochemistry* **3**, 61–66.

Hunt, R.W. G. (1995) *Measuring Colour, Second edition*, Ellis Horwood, London.

Hutchings, J.B. (1994) *Food Colour and Appearance*. Chapman Hall, London.

Ibarz, A., Pagan, J. & Garza, S. (2000) Kinetic models of non-enzymatic browning in apple puree. *Journal of the Science of Food and Agriculture* **80**(8), 1162–1168.

IDF (1981) *New Monograph on UHT Milk*. IDF Document 133. International Dairy Federation, Brussels.

Ikezumi, M., Nakagawa, T., Matsura, H. & Mochiji, K. (2006) Cow's milk flavour deterioration suppression method in store and supermarket, involves performing deaeration processing at specific temperature to reduce dissolved oxygen concentration in cow's milk. Japanese Patent 2006-042814.

Jansson, T., Jensen, H. B., Sundekilde, U. K., Clausen, M. R., Eggers, N., Larsen, L. B., Ray, C., Andersen, H. J. & Bertram, H. C. (2014) Chemical and proteolysis-derived changes during long-term storage of lactose-hydrolyzed ultrahigh-temperature (UHT) milk. *Journal of Agricultural and Food Chemistry* **62**(46), 11270–11278.

Jeon, I.J., Thomas, E.L. & Reineccius, G.A. (1978) Production of volatile flavour compounds in ultrahigh-temperature processed milk during aseptic storage. *Journal of Agricultural and Food Chemistry* **26**(5), 1183–1188.

Jimenez-Perez, S., Corzo, N., Morales, F.J, Delgado, T. and Olano, A. (1992) Effect of storage temperature on lactulose and 5-hydroxymethyl-furfural formation in UHT milk. *Journal of Food Protection* **55**, 304–306

Katsuno, S., Shimamura, T., Kashiwagi, T., Izawa, N. & Ukeda, H. (2013) Effects of dissolved oxygen on the Maillard reaction during heat treatment of milk. *International Dairy Journal* **33**, 34–37.

Kessler, H.G. & Fink, R. (1986) Changes in heated and stored milk, with an interpretation by reaction kinetics. *Journal Food Science.* **51**, 1105-1111, 1155.

Kessler, H.G. (1989) Effect of thermal processing on milk. In: *Food preservation -5,* (ed S. Thorne), pp 80–129, Elsevier Applied Science Publisher, London.

Kocak, H.R. & Zadow, J.G. (1985a) Age gelation of UHT whole milk as influenced by storage temperature. *Australian Journal of Dairy Technology* **40**, 14–21.

Kocak, H.R. & Zadow, J.G. (1985b) Controlling age gelation of UHT milk with additives. *Australian Journal of Dairy Technology* **40**, 58–64.

Kocak, H.R. & Zadow, J.G. (1985c) Polyphosphate variability in the control of age gelation in UHT milk. *Australian Journal of Dairy Technology* **40**, 65–68.

Kohlmann, K.L., Nielsen, S.S. & Ladisch, M.R. (1991) Effects of a low concentration of added plasmin on ultra-high temperature processed milk. *Journal of Dairy Science* **74**(4), 1151–1156.

Kwan, K.K.H., Nakai, S. & Skura, B.J. (1983) Comparison of four methods for determining protease activity in milk. *Journal of Food Science* **48**(5), 1418–1421.

Kyung, M., Jae-Young., L, Yoon, J. & Lee, S-R. (1995) Computation of Q10 values and shelf-life for canned and bottled orange juices. *Korean Journal of Food Science and Technology* **27**, 748–752

Labuza, T.P. (1994) Interpreting the complexity of the Maillard reaction. In: *Maillard Reactions in Chemistry, Food and Health.* (eds T.P. Labuza & G.A. Reineccius) pp 176–181, Royal Society of Chemistry, Cambridge.

Langsrud, T. & Hadland, G. (1977) Biochemical and chemical changes in protein fractions in UHT-sterilised goats' milk during storage at different temperatures. *Dairy Science Abstracts* **39**(1), 68–69.

Law, B.A., Andrews, A.T. & Sharpe, M.E. (1977) Gelation of ultra-high-temperature-sterilised milk by proteases from a strain of *Pseudomonas fluorescens* isolated from raw milk. *Journal of Dairy Research* **44**(1), 145–148.

Le T.X., Datta, N. & Deeth, H.C. (2006) A sensitive HPLC method for measuring bacterial proteolysis & protease activity in UHT milk. *Food Research International* **39**, 823–830.

Le, T.T., Holland, J.W., Bhandari, B., Alewood, P.F. & Deeth, H.C. (2013) Direct evidence for the role of Maillard reaction products in protein cross-linking in milk powder during storage. *International Dairy Journal* **31**, 83–91.

López Fandiño, R., Olano, A., Corzo, N. & Ramos, M. (1993a) Proteolysis during storage of UHT milk: Differences between whole and skim milk. *Journal of Dairy Research* **60**, 339–357.

López Fandiño, R., Olano, A., San José, C. & Ramos, M. (1993b) Application of reversed-phase HPLC to the study of proteolysis in UHT milk. *Journal of Dairy Research* **60**, 111–116.

Lorenzen P.C., Clawin-Raedecker I., Einhoff K., Hammer P., Hartmann R., Hoffmann W., Martin D., Molkentin J., Walte H.G. & Devrese M. (2011) A survey of the quality of extended shelf-life (ESL) milk in relation to HTST and UHT milk. *International Journal of Dairy Technology* **64**,166–178.

Lorient, D. & Linden, G. (1976) Dephosphorylation of bovine casein by milk alkaline phosphatase. *Journal of Dairy Research* **43**(1), 19–26.

Manji, B., Kakuda, Y. & Arnott, D.R. (1986) Effect of storage temperature on age gelation of ultra-high temperature milk processed by direct and indirect heating systems. *Journal of Dairy Science* **69**, 2994–3001.

Manji, B. & Kakuda, Y. (1988) The role of protein denaturation, extent of proteolysis, and storage-temperature on the mechanism of age gelation in a model system. *Journal of Dairy Science* **71**(6), 1455–1463.

McGookin, B.J., & Augustin, M.A. (1997) Antioxidant activity of a heated casein-glucose mixture in full cream milk powder. *Australian Journal of Dairy Technology* **52**(1), 15–19.

McKellar, R.C. (1981) Development of off-flavours in ultra-high temperature and pasteurised milk as a function of proteolysis. *Journal of Dairy Science* **64**, 2138–2145.

McKellar, R.C., Froehlich, D.A., Butler, G., Cholette, H. & Campbell, C. (1984) The effect of uncooled storage on proteolysis bitterness & apparent viscosity in ultra-high-temperature milk. *Canadian Institute of Food Science & Technology Journal* **17**(1), 14–17.

McKenna, A.B. & Singh, H. (1991) Age gelation in UHT-processed reconstituted concentrated skim milk. *International Journal of Food Science & Technology* **26**, 27–38.

McMahon, D.J. (1996) Age-gelation of UHT milk: changes that occur during storage, their effect on shelf-life and the mechanism by which age-gelation occurs. In *Heat Treatments and Alternative Methods*, Document 9602, pp. 315–326. International Dairy Federation, Brussels.

Mehta, B.M. & Deeth, H.C. (2016) Blocked lysine in dairy products – formation, occurrence, analysis and nutritional implications. *Comprehensive Reviews in Food Science and Food Safety* **15**, 206–218.

Mehta, R.S. (1980) Milk processed at ultra-high-temperatures - a review. *Journal of Food Protection* **43**(3), 212–225.

Messia, M.C., Candigliota, T. & Marconi, E. (2007) Assessment of quality and technological characterization of lactose-hydrolyzed milk. *Food Chemistry* **104**(3), 910–917.

Metwalli, A.A.M. & van Boekel, M. (1998) On the kinetics of heat-induced deamidation and breakdown of caseinate. *Food Chemistry* **61**, 53–61.

Miller, B.S., Johnson, J.A. & Palmer, D.L. (1953) A comparison of cereal, fungal, & bacterial alpha-amylases as supplements for breadmaking. *Food Technology* 7(1), 38–42.

Miralles, B., Amigo, L., Ramos, M. & Recio, I. (2003) Analysing para-kappa-casein and related peptides as indicators of milk proteolysis. *Milchwissenschaft* **58**(7-8), 412–415.

Mitchell, G.E. & Ewings, K.N. (1985) Quantification of bacterial proteolysis causing gelation in UHT-treated milk. *New Zealand Journal of Dairy Science & Technology* **20**, 65–76.

Mittal, S.B., Hourigan, J.A. & Zadow, J.G. (1990) Effect of added sodium hexametaphosphate on certain technological aspects of UHT recombined milk. *Australian Journal of Dairy Science & Technology* **45**, 1–4

Miwa, N., Yokoyama, K., Wakabayashi, H. & Nio, N. (2010) Effect of deamidation by protein-glutaminase on physicochemical and functional properties of skim milk. *International Dairy Journal* **20**, 393–399.

Moio, L., Etievant, P., Langlois, D., Dekimpe, J. & Addeo, F. (1994) Detection of powerful odorants in heated milk by use of extract dilution sniffing analysis. *Journal of Dairy Research* **61**, 385–394.

Mottar, J. (1981) Heat resistant enzymes in UHT milk and their influence on sensoric changes during uncooled storage. *Milchwissenschaft* **36**(2), 87–91.

Naim,M., Wainish, S., Zehavi, U, Peleg, H. Rouseff, R.L. & Nagy, S, (1993) Inhibition by thiol compounds of off-flavour formation in stored orange juice. I. Effect of L-cysteine and N-acetyl-L-cysteine on 2,5-dimethyl-4-hydroxy-3(2H)-furanone formation. *Journal of Agricultural and Food Chemistry* **41** (9) 1355–1358.

Narayanan, K.R.A., Kumar, A. & Patil, G.R. (1993) Kinetics of various deteriorative changes during storage of UHT soy beverage and development of a shelf-life prediction model. *Lebensmittel Wissenschaft und Technologie* **26**(3), 191–197.

Newstead, D.F., Paterson, G., Anema, S.G., Coker, C.J. & Wewala, A.R. (2006) Plasmin activity in direct-steam-injection UHT-processed reconstituted milk: effects of preheat treatment. *International Dairy Journal* **16**, 573–579.

Nicodeme, M., Grill, J.P., Humbert, G. & Gaillard, J. L. (2005) Extracellular protease activity of different *Pseudomonas* strains: dependence of proteolytic activity on culture conditions. *Journal of Applied Microbiology* **99**(3), 641–648.

Nieuwenhuijse, J.A. & van Boekel, M.A.J.S. (2003) Protein stability in sterilised milk and milk products. In *Advanced Dairy Chemistry, Volume 1 Proteins*, 3rd edn, (eds P.F. Fox and P.L.H. McSweeney), pp. 947–974. Kluwer Academic/Plenum Press, New York.

Olano, A., Calvo, M. M. & Corzo, N. (1989) Changes in the carbohydrate fraction of milk during heating processes. *Food Chemistry* **31**(4), 259–265.

On-Nom, N. (2012) *Partitioning of Milk at High Temperature*, PhD thesis, University of Reading, UK.

Pagliarini, E., Vernille, M. & Peri, C. (1990) Kinetic study on colour changes in milk due to heat. *Journal of Food Science*, **55**, 1766–1767.

Panfil-Kuncewicz, H., Kuncewicz, A. & Juskiewicz, M. (2005) The influence of the sterilisation method on the changes in UHT milk fat fraction. *Milchwissenschaft* **60**(1), 33–36.

Perkin, A.G., Henschel, M.J. & Burton, H. (1973) Comparison of milks processed by the direct and indirect methods of ultra-high-temperature sterilisation. VI. Effects on sediment formation and clotting with enzymes. *Journal of Dairy Research* **40**, 215–220.

Perkins M. L., Zerdin K., Rooney, M.L., D'Arcy, B.R. & Deeth H.C. (2007) Active packaging of UHT milk to prevent the development of stale flavour during storage. *Packaging Technology & Science* **20**(2), 137–146.

Perkins, M.L., Elliott, A.J., D'Arcy, B.R. & Deeth, H.C. (2005) Stale flavour volatiles in Australian commercial UHT milk during storage. *Australian Journal of Dairy Technology* **60**, 231–237.

Petrus, R. R., Walter, E. H. M., Faria, J. A. F. & Abreu, L. F. (2009) Sensory stability of ultra-high temperature milk in polyethylene bottle. *Journal of Food Science* **74**(1), S53–S57.

Politis, I., Zavizion, B., Barbano, D. M. & Gorewit, R. C. (1993) Enzymatic assay for the combined determination of plasmin plus plasminogen in milk: revisited. *Journal of Dairy Science*, **76**(5), 1260–1267.

Prasad, K.S. (1989) *The Sensory Characteristics of Heat-treated Milks, with Special Reference to UHT Processing.* PhD Thesis, University of Reading, Reading, UK.

Rada-Mendoza, M., Villamiel, M. & Olano, A. (2002) Dissolved air effects on lactose isomerisation and furosine formation during heat treatment of milk. *Lait* **82**, 629–634.

Ramsey, J.A. & Swartzel, K.R. (1984) Effect of ultra high temperature processing and storage conditions on rates of sedimentation and fat separation of aseptically packaged milk. *Journal of Food Science* **49**(1), 257–262.

Rassis, D. & Saguy, I.S. (1995) Kinetics of aseptic concentrated orange juice quality changes during commercial processing and storage. *International Journal of Food Science & Technology* **30**, 191–198

Recio, I., Frutos, M., Olano, A. & Ramos, M. (1996) Application of capillary electrophoresis & reversed-phase HPLC to the study of proteolysis in UHT milks. In: *Heat Treatments & Alternative Methods.* IDF Document 9602. pp. 422–428. International Dairy Federation, Brussels.

Recio, M.I., Villamiel, M., Martinez-Castro, I. & Olano, A. (1998) Changes in free monosaccharides during storage of some UHT milks: a preliminary study. *Zeitschrift fur Lebensmittel Untersuchung und Forschung A/Food Research and Technology* **207**(2), 180–181.

Reddy, K.K., Nguyen, M.H., Kailasapathy, K. & Zadow, J.G. (1992) Evaluation of reduced-fat ultrafiltered UHT milk. *ASEAN Food Journal* **7**(3), 152–156.

Rerkrai, S., Jeon, I.J. & Bassette, R. (1987) Effect of various direct ultra-high temperature heat treatments on flavour of commercially prepared milks. *Journal of Dairy Science* **70**, 2046–2054.

Rhim, J.W, Jones, V.A. & Swartzel K.R. (1988) Initial whitening phenomenon of skim milk on heating. *Lebensmittel Wissenschaft und Technologie* **21**, 339–341.

Richards, M., De Kock, H. L. & Buys, E. M. (2014) Multivariate accelerated shelf-life test of low fat UHT milk. *International Dairy Journal* **36**(1), 38–45.

Richardson, B.C. & Newstead, D.F. (1979) Effect of heat-stable proteinases on the storage life of UHT milk. *New Zealand Journal of Dairy Science & Technology* **14**, 273–279.

Richardson, B.C. (1983) The proteases of bovine milk and the effect of pasteurisation on their activity. *New Zealand Journal of Dairy Science and Technology* **18**, 233–245.

Roig, M.G., Bello, J.F., Rivera, Z.S. & Kennedy, J.F. (1999) Studies on the occurrence of non-enzymatic browning during storage of citrus juice. *Food Research International* **32**, 609–619.

Romero, C., Morales, F. J. & Jimenez Perez, S. (2001) Effect of storage temperature on galactose formation in UHT milk. *Food Research International* **34**, 389–392.

Rustom, I. Y. S., Lopezleiva, M. H. & Nair, B. M. (1995) UHT-sterilised peanut beverages - changes in physicochemical properties during storage. *Journal of Food Science* **60**(2), 378–383.

Rustom, I. Y. S., LopezLeiva, M. H. & Nair, B. M. (1996a) Nutritional, sensory and physicochemical properties of peanut beverage sterilised under two different UHT conditions. *Food Chemistry* **56**(1), 45–53.

Rustom, I. Y. S., LopezLeiva, M. M. & Nair, B. M. (1996b) UHT-Sterilised peanut beverages: Kinetics of physicochemical changes during storage and shelf-life prediction modeling. *Journal of Food Science* **61**(1), 198–203, 208.

Rysstad, G., Ebbesen, A. & Eggestad, J. (1998) Sensory and chemical quality of UHT-milk stored in paperboard cartons with different oxygen and light barriers. *Food Additives & Contaminants* **15**(1), 112–122.

Samel, R., Weaver, R.W.V. & Gammack, D.B. (1971) Changes on storage in milk processed by ultra-high-temperature sterilisation. *Journal of Dairy Research* **38**, 323–332.

Samuelson, E.G. & Holm, S (1966) Technological principles for ultra-high-heat treatment of milk. *17th International Dairy Congress* (Munich, Germany), Section B (**1**), 57–65.

Scanlan, R.A., Sather, L.A. & Day, E.A. (1965) Contribution of free fatty acids to the flavour of rancid milk. *Journal of Dairy Science* **48**, 1582–1584.

Sharma, D.K. & Prasad, D.N. (1990) Changes in the physical properties of ultra high temperature processed buffalo milk during storage. *Journal of Dairy Research* **57**(2), 187–196.

Sher, A.A., Mallangi C.R., Jacobson R., Vadehra, D.R. & Wedral, R.E. (2002) Shelf-stable calcium fortified milk and dairy products. Patent No.: US 6,455,082 B1. Assignee: Nestec S A.

Shew, D. (1981) Technical aspects of UHT processing. In: *New Monograph on UHT milk.* IDF Document 133, pp. 115–121. International Dairy Federation, Brussels.

Singh, R.R.B. & Patil, G.R. (1989a) Storage stability of UHT buffalo milk. I. Change in physico-chemical properties. *Indian Journal of Dairy Science* **42**(2), 150–154.

Singh, R.R.B. & Patil, G.R. (1989b) Storage stability of UHT buffalo milk. II. Sensory changes and their relationship with physico-chemical properties. *Indian Journal of Dairy Science* **42**(2), 384–387.

Singh, R.R.B., Patil, G.R. & Balachandran, R. (2004) Kinetics of lipid hydrolysis during storage of UHT milk. *Journal of Food Science and Technology-Mysore* **41**(2), 139–142.

Skudder, P.J., Thomas, E.L., Pavey, J.A. & Perkin, A.G. (1981) Effects of adding potassium iodate to milk before UHT treatment. I. Reduction in the amount of deposit on the heated surfaces. *Journal of Dairy Research* **48**, 99–113.

Snoeren, T.H.M., van Riel, J.A.M. & Both, P. (1979) Proteolysis during storage of UHT-sterilised whole milk. 1. Experiments with milk heated by the direct system for 4 seconds at 142 °C. *Netherlands Milk & Dairy Journal* **33**, 31–39.

Swaisgood, H.E. (1977) Process of removing the cooked flavour from milk. United States Patent Application Number 4053644.

Taylor, M. J. & Richardson, T. (1980) Antioxidant activity of skim milk - effect of heat and resultant sulfhydryl-groups. *Journal of Dairy Science* **63**(11), 1783–1795.

Thomas, E.L., Burton, H., Ford, J.E & Perkin, A.G. (1975) The effect of oxygen content on flavour and chemical changes during aseptic storage of whole milk after ultra-high-temperature processing. *Journal of Dairy Research* **42**, 285–295.

Timmer-Keetels, C.J., Nieuwenhuijse, J.A., Zijtveld-Van Der Wiel, J., Nieuwenhuijse, H.A., Zijtveld-Van Der Wie, J.H., Keetels, C.J.A., Van Der Wiel, J.H. & Zijtveld-Van, D.W.J. (2011) Producing food product having milk protein involves subjecting milk protein to enzymatic deamidation procedure, formulating protein into liquid product, followed by heat sterilisation of food product and spray drying of product into powder. Patents WO2011034418-A2; WO2011034418-A3; NL2003494-C; WO2011034418-A4; EP2477506-A2; CN102595925-A; US2012231117-A1.

Tobin, J.T., Fitzsimons, S.M., Kelly, A.L. & Fenelon, M.A. (2011) The effect of native and modified konjac on the physical attributes of pasteurised and UHT-treated skim milk. *International Dairy Journal* **21** 790–797.

Topçu, A., Numanoglu, E. & Saldamli, I. (2006) Proteolysis and storage stability of UHT milk produced in Turkey. *International Dairy Journal* **16**, 633–638.

Tran, H., Datta, N., Lewis, M.J. & Deeth, H.C. (2008) Predictions of some product parameters based on the processing conditions of ultra-high-temperature milk plants. *International Dairy Journal* **18**, 939–944.

Tran, T.T. H., Datta, N & Deeth, H.C. (2003) Amylases in enzyme detergents and their effects on a simulated dairy dessert. *Milchwissenschaft* **58**, 527–531.

Troyano, E., Villamiel, M., Olano, A., Sanz, J. & Martinez-Castro, I. (1996) Monosaccharides and myo-inositol in commercial milks. *Journal of Agricultural and Food Chemistry* **44**(3), 815–817.

Tsioulpas, A., Koliandris, A., Grandison, A.S. & Lewis, M.J. (2010) Effects of stabiliser addition and in-container sterilisation on selected properties of milk related to casein micelle stability. *Food Chemistry* **22**, 1027–1034.

UCC (2015) Results of trial performed in association with a thermal processing training course, April 15-16, 2015, University College Cork, Ireland.

Valero, E., Villamiel, M., Miralles, B., Sanz, J. & Martinez-Castro, I. (2001) Changes in flavour and volatile components during storage of whole and skimmed UHT milk. *Food Chemistry* **72**(1), 51–58.

van Aardt, M., Duncan, S.E., Marcy, J.E., Long, T.E., O,Keefe, S.F. & Nielsen-Sims, S.R. (2005) Aroma analysis of light-exposed milk stored with and without natural and synthetic antioxidants. *Journal of Dairy Science* **88**, 881–890.

van Boekel, M.A.J.S. & Brands, C.M.J. (1998) Heating of sugar-casein solutions. Isomerisation and the Maillard reaction. In: *The Maillard Reaction in Foods and Medicine* (ed. J. O'Brien) pp. 154–159. Royal Society of Chemistry, Cambridge.

van Boekel, M.A.J.S. (1998) Effect of heating on Maillard reactions in milk. *Food Chemistry* **62**(4), 403–414.

van Boekel, M.A.J.S. (1999) Heat-induced deamidation, dephosphorylation and breakdown of caseinate. *International Dairy Journal* **9**, 237–241.

Vandewalle, L. & Huyghebaert, A. (1980) The antioxidant activity of the non-enzymatic browning reaction in sugar-protein systems. *Mededelingen van de Faculteit Landbouwwetenschappen Rijksuniversiteit Gent* **45**(4), 1277–1286.

Vazquez-Landaverde, P.A., Velazquez, G., Torres J.A. & Qian M.C. (2005) Quantitative determination of thermally derived off-flavour compounds in milk using solid-phase microextraction and gas chromatography. *Journal of Dairy Science* **88**, 3764–3772.

Venkatachalam, N., McMahon, D.J., & Savello, P.A. (1993) Role of protein and lactose interactions in the age gelation of ultra-high temperature processed concentrated skim milk. *Journal of Dairy Science* **76**(7), 1882–1894.

Vesconsi, C.N., Valduga, A.T., & Cichoski, A.J. (2012) Particle sedimentation in semi-skimmed, skimmed and whole milk UHT, during storage. *Ciencia Rural* **42**(4), 730–736.

Wadsworth, K.D. & Bassette, R. (1985) Effect of oxygen on development of off-flavours in ultrahigh-temperature milk. *Journal of Food Protection* **48**(6), 487–493.

Walstra, P. & Jenness, R. (1984) *Dairy Chemistry and Physics.* John Wiley, New York.
Wishner, L.A. (1964) Light-induced oxidations in milk. *Journal of Dairy Science* **47**(2), 216–221.
Wong, M. & Stanton, D.W., (1993) Effect of removal of amino acids and phenolic compounds on non-enzymic browning in stored kiwifruit juice concentrates. *Lebensmittel Wissenschaft &Technologie* **26**, 138–144.
Wong, M., Stanton, D.W. & Burns, D.J.W. (1992) Effect of initial oxidation on ascorbic acid browning in stored kiwifruit juice and model concentrates. *Lebensmittel Wissenschaft & Technologie* **25** 574–578.
Zadow, J.G. & Birtwistle, R. (1973) The effect of dissolved oxygen on the changes occurring in the flavour of ultra-high-temperature milk during storage. *Journal of Dairy Research* **40**, 169–177.
Zhang, S., Jin, Y. & Yun, Z. (2007) Effect of low temperature heat treatment on the activity of lipase VIII from *Pseudomonas* species in UHT milk. *Transactions of the Chinese Society of Agricultural Engineering* **23**(8), 231–235.
Zin El-Din, M., Aoki, T. & Kako, Y. (1991) Polymerization and degradation of casein in UHT milk during storage. *Milchwissenschaft* **46**, 284–287.

8

Quality Control and Assurance

8.1 Introduction

Quality assurance (QA) and quality control (QC) are terms used commonly for ensuring food safety and quality. Harrigan and Park (1991) provide the following definitions, which are internationally agreed. **Quality assurance** is defined as "all those planned and systematic actions necessary to provide adequate confidence that a product or service will satisfy given requirements for quality". **Quality control** is defined as "the operational techniques and activities that are used to fulfil requirements for quality". QA refers to the corporate oversight function to clearly state the corporate product quality objectives and goals as well as to affirm that the QC program is functional and achieving these objectives and goals. QC on the other hand is an on-line or production function that establishes and administers the day-to-day policies, procedures and programmes at plant level (Hubbert *et al.*, 1996). Thus there may be a company-wide QA programme outlining what is expected to ensure safety in all food products, whereas it may be up to each plant manager to ensure that in-plant QC activities in the production of its product are performed satisfactorily to meet the overall corporate QA goal. Quality control is therefore one component of an integrated system of quality assurance.

8.2 Safety and Quality Considerations

The main aim of all heat treatments of foods is to reduce microbial activity to low levels, thereby making them safe for consumption. For sterilisation processes, the more specific aim is to achieve commercial sterility (see Chapters 3 and 4). At the same time, many heat-induced chemical, biochemical and physical changes take place, which tend to reduce food quality and the food's overall acceptability in terms of appearance, flavour and texture (see Chapter 6). Therefore **safety** and **acceptability** are often conflicting requirements in thermal processing. In this chapter quality assurance embraces both these important aspects.

8.2.1 Safety Issues

Safety issues relate primarily to minimising the levels of food poisoning bacteria. In situations where a conscious decision is taken not to apply any heat treatment, this could lead to food poisoning outbreaks. About 50 years ago, over 5% of milk consumed

High Temperature Processing of Milk and Milk Products, First Edition. Hilton C. Deeth and Michael J. Lewis.

Table 8.1 Summary of outbreaks and illnesses associated with raw milk, United States, 1998–2013.

Bacterium	Number of outbreaks	Number of people taken ill
Campylobacter	79	1096
E. coli 0157:H7	21	389
Salmonella	15	363
E.coli (other)	4	29
Brucella	4	14
Listeria	2	24
Cryptosporidium	2	8
Coxiella	1	3
Shigella	1	2

From: http://www.realrawmilkfacts.com/PDFs/Raw-Dairy-Outbreak-Table.pdf

in the UK was raw milk, but this has fallen to about 0.1% currently, and the number of food poisoning cases attributed to raw milk have dropped accordingly. There have been many cases attributed to consumption of raw milk; these were documented by Griffiths (2010). Table 8.1 summarises the outbreaks of illness in America associated with consumption of raw milk between 1998 and 2013. These show 89% of the outbreaks were associated with *Campylobacter, E. coli* 0157:H7 or *Salmonella.*

The range of heat treatments given to milk is extensive (see Table 2.1); pasteurisation is widely practised for inactivating pathogenic microorganisms but it is a mild process and the product has a relatively short shelf-life even at low temperatures. Food poisoning outbreaks can arise from inadequate heat treatment or from post-processing contamination (PPC). The major reported disease outbreaks involving pasteurised milk have been caused by contamination by *Staphylococcus aureus, Campylobacter jejeuni, Salmonella* spp., *Listeria monocytogenes, Yersinia enterocolitica, E. coli* and *Bacillus cereus* (Kelly *et al.*, 2012).

A food spoilage event which involves product being recalled or, worse still, a food poisoning outbreak, is a very expensive and unwelcome occurrence, even if there are no fatalities. The reputation of a product brand can be badly damaged by adverse publicity associated with such an event. One very serious scenario would be a botulism outbreak, but serious botulism cases arising from commercial heat-treated food are very rare. Fortunately, anaerobic spore-formers such as *Clostridium botulinum* are of little importance in the spoilage of milk and milk products but they can pose a risk in cheesemaking (Bintsis *et al.*, 2008; Rowe & Donaghy, 2011). Their significance is clearly indicated in models developed by Pujol *et al.* (2015) which predicted that a botulinum outbreak might occur once in every 367 years in UHT products. As might be expected, the predicted sterility failure rates for *Geobacilus stearothermophilus* and also for *B. cereus* were much higher.

Chemical contaminants are another source of food safety concerns. For example, toxins are naturally present in some foods although fortunately very rarely in milk. One example is trypsin inhibitor in soya beans; fortunately it is largely inactivated by heat treatment. However, inactivation of the inhibitor is not as temperature-dependent as microbial inactivation, so inactivation by UHT processing is less compared to in-container sterilisation processes (see Section 9.17.1).

There have been cases of deliberate chemical contamination of milk. In 2008, melamine was found to be added to milk in China with tragic consequences. Melamine is a toxic chemical that contains a high nitrogen content and when added to milk increases its nitrogen content and hence its apparent protein content. This incident in China was considered by WHO to be one of the largest food safety events it has had to deal with in recent times. MacMahon *et al.* (2012) have identified six potential nitrogen-rich compounds that could be added to milk fraudulently to boost its apparent protein content. These are readily available, economic to add and cause no obvious changes to the sensory characteristics. These are cyromazine, dicyandiamide, urea, biuret, triuret and amidinourea. These can be detected by liquid chromatography-mass spectrometry at addition rates of 1 ppm. Note that levels added to boost protein levels would be much higher than this. Thus, these are compounds that regulatory authorities need to be vigilant about.

Chemical additives are permitted in some UHT products to improve their stability and consumer acceptability; a listing of these on the product label is mandated in most countries. One consequence is that consumers now scrutinise labels more closely and may well avoid products with a long list of additives. Therefore, it is worthwhile for manufacturers to question why each compound is added and whether there is potential for removing it, in order to reduce the number of additives in the product.

8.2.2 Quality Issues

While pasteurisation destroys pathogenic bacteria, thermoduric vegetative bacteria and bacterial spores survive pasteurisation and constitute the total microflora in freshly pasteurised milk (see Section 4.4.1). However, these are seldom the cause of spoilage of pasteurised milk as they tend to grow slowly at refrigeration temperatures. Spoilage is mostly caused by PPC, with pseudomonads being the major bacteria involved. Plate counts are generally in the range of 1-5 x 10^6 cfu/mL when off-flavours are detectable (Hubbell & Collins, 1962). Similar considerations apply to ESL milk (see Section 4.4.2).

Innovative microbiological methods are being developed to identify pasteurised and ESL products which might be susceptible to spoilage before their "use by" date (see Section 8.7.5). In this area, there is a role for rapid methods to provide quick feedback on product microbial quality. On the other hand, monitoring sterilisation processes presents some more interesting and challenging problems, since the aim is to achieve zero defective product items. In this case a defect refers usually to a spoiled product, not an unsafe product. This means that (in theory) large numbers of samples need to be examined to ensure and verify that this is being achieved (see Sections 5.5.5 and 8.4.4).

8.3 Heat Treatment Regulations

Regulations exist for heat treatment of milk. Specific processing conditions such as processing times and temperatures may be specified. They may also include information on permitted additives and on packaging requirements, including sizes of containers and labelling. There may be specified tests for ensuring the efficacy of the heat treatment. To illustrate these points, the current EU heat treatment regulations for milk together with heat treatments developed by Codex (2000, 2004) are given in Table 8.2. The two sets of regulations are very similar.

Table 8.2 EU and Codex heat treatment regulations (Codex, 2000, 2004; EU, 2005).

	EU	Codex
Pasteurisation	• a high temperature for a short time (at least 72 °C for 15 s); • a low temperature for a long time (at least 63 °C for 30 min); or • any other combination of time-temperature conditions to obtain an equivalent effect. The products must show a negative reaction to an alkaline phosphatase test immediately after such treatment.	The minimum pasteurization conditions are those having bactericidal effects equivalent to heating every particle of the milk to 72 °C for 15 seconds (continuous flow pasteurization) or 63 °C for 30 minutes (batch pasteurization). Similar conditions can be obtained by joining the line connecting these points on a log time versus temperature graph. The products subjected to pasteurization should show a negative alkaline phosphatase reaction immediately after the heat treatment
UHT	A continuous flow of heat at a high temperature for a short time (not less than 135 °C in combination with a suitable holding time) such that there are no viable micro-organisms or spores capable of growing in the treated product when kept in an aseptic closed container at ambient temperature. The products should remain microbiologically stable after incubating for 15 days at 30 °C in closed containers or for 7 days at 55 °C in closed containers, or after any other method demonstrating that the appropriate heat treatment has been applied.	The application of heat to a continuously flowing product using such high temperatures for such time that renders the product commercially sterile when combined with aseptic packaging. UHT treatment is normally in the range of 135 to 150 °C in combination with appropriate holding times necessary to achieve commercial sterility. The products must be microbiologically stable at room temperature, either measured after storage until end of shelf life or incubated at 55 °C for 7 days (or at 30 °C for 15 days).

Codex Alimentarius has become the global reference point of national food control agencies, food producers and processors and international food trade. Its influence extends to all continents and has contributed toward to the protection of human health and safety and fair trade practices worldwide (Hickey 2009). Codex has developed a general hygiene code to cover all foodstuffs and a specific code for milk and milk products (FAO/WHO, 2004). The Codex code established performance guidance for pasteurisation designed to achieve five decimal reductions of *Coxiella burnetii.* Conditions between 63 °C for 30 min and 72 °C for 15 s can be used and obtained from a straight line relationship between the log of the time and the temperature (Codex, 2004).

The practicalities of enforcing regulations for heat-treated milk differ from country to country. Control of pasteurisation processes is mediated through regulations which are based upon the long-known requirement that pasteurised milk should be produced under conditions which exceed a certain temperature for a certain time. Effective pasteurisation of milk also inactivates alkaline phosphatase so the requirement that it should be phosphatase-negative is used almost universally. There may also be a requirement to define an upper limit to avoid over-pasteurisation; lactoperoxidase is one

Table 8.3 Standards for the various heat treatments in the European Community in 1995 (EU Directive 92/46). (Source: Wilbey, 1996. Reproduced with permission of John Wiley & Sons.)

Assay	Pasteurised	High-temperature pasteurised	UHT	Sterilised
Alkaline phosphatase	-ve	-ve	-ve	-ve
Lactoperoxidase	+ve	-ve	-ve	-ve
Turbidity	+ve	+ve	+ve or -ve	-ve
Lactulose	Not detected	<50 mg/L	<600 mg/L	>600 mg/L
β-lactoglobulin	>2600 mg/L	>2000 mg/L	>50 mg/L	<50 mg/L

(from Wilbey, 1996)

effective indicator is this respect (see Section 2.3.6.2). In earlier EU regulations there was a requirement for pasteurised milk to be positive for lactoperoxidase activity. If it was not positive, it would have to be described as "high pasteurised" (EU, 1992). More recent regulations (EU, 2005) have seen the removal of these requirements along with the need to use the term "high pasteurised". The reasons why this change was made are not clear. Therefore, there is now no distinction between high-temperature pasteurisation and normal pasteurisation in the new EU hygiene package (Hickey, 2009). Also in 2015, there are no specific regulations in the EU governing heating conditions for ESL milk (see Section 3.3).

As shown in Table 8.2, the regulations related to UHT milk produced within the EU specify a temperature of not less than 135 °C for a suitable holding time such that there are no viable spores capable of growing in the treated product when kept in an aseptic container at ambient temperature (EU, 2005). The product must remain microbiologically stable after incubating at 15 days at 30 °C in closed containers, or 7 days at 55 °C in closed containers or after any other method that demonstrates that appropriate heat treatment has been applied. Another change is that the most current regulations no longer state what level of microbial activity would constitute microbial sterility after these incubation periods. Lewis and Heppell (2000) noted that microbiological specifications at that time required that UHT milk should contain less than 100 cfu/mL after 14 days incubation at 30 °C. This appears to be a prudent limit and is still advocated by some bodies (Food Safety Info, undated). However, if microbial counts are higher than this, the product may still be acceptable for consumption and not hazardous to health, but would most probably not be released for sale. Indeed the fact that samples of UHT milk are incubated during testing, prior to release, under conditions which are very favourable to growth is comforting. In cases where viable bacteria are found, there are now several techniques available for identifying the surviving flora. (see Section 8.7.6). UHT milk with low levels of viable bacteria provides a challenge, as it is uncertain whether they may proliferate in the next 9 to 12 months and be responsible for a product recall. Companies may wish to think about other uses for such UHT milk, for example, for manufacturing drinking yogurt or soft cheese products, where the viable bacteria would be unlikely to lead to any safety issues. To illustrate some subtle changes in legislation, Table 8.3 shows standards for the various heat treatments in the European Community in 1992 (EU Directive 92/46), together with tentative proposals for discussion, involving the measurements of both β-lactoglobulin and lactulose in a range of heat-treated milks (Wilbey, 1996). These stipulate that lactulose level in UHT milk should not exceed 600 mg/L and in sterilised milks it should exceed 600 mg/L, thus distinguishing between UHT and sterilised milk. However, these two chemical tests are not listed in the later EU hygiene directive (EU, 2005).

Another change in the regulations in some countries, for example, Australia and New Zealand, is to remove the stipulated minimum temperatures and times for UHT processing. The only requirement is that the product is to be commercially sterile. For example, the New Zealand regulations state "UHT treatment of liquid dairy material is (A) the application of heat to continuously flowing liquid dairy material using such temperatures for such time that renders the dairy material commercially sterile at the time of processing; then (B) aseptic packaging resulting in commercially sterile product" (NZMAF, 2011). USA also does not specify minimum UHT processing conditions. However, there is a US specification for "ultra-pasteurised" milk of 138 °C for 2 s which are sub-UHT or ESL conditions, equivalent to an F_0 value of about 1.7.

Regulations for milk imported into EU countries (EU, 2010) from countries not recognized as free of foot-mouth-disease state that they should have undergone either: (a) a sterilisation process, to achieve an F_0 value equal to or greater than three, or (b) an ultra-high-temperature (UHT) treatment at not less than 135 °C in combination with a suitable holding time. Interestingly, Canada has set the UHT treatment conditions for milk imported from such countries at a minimum of 140 °C for at least 5 s (CFIA, 2016).

The minimum UHT condition set by China is "132 °C for a short time" (Chinese Ministry of Health, 2010b) although one Chinese standard recommends a minimum of "135 °C for several seconds" (Chinese Ministry of Agriculture, 2005). A heat treatment at 135 °C is preferable to one at 132 °C as the former requires a holding time of about 7.5 s to achieve an F_0 of 3 (which is widely recognised as the minimum level to achieve a 12-log reduction of *Cl. botulinum*) while the latter requires a holding time of about 15 s.

The use of additives in the EU is subject to horizontal legislation, primarily under the additive framework direction. According to Hickey (2009), the only additives permitted in market milks are those specifically permitted by the EU (1995) in sterilised and UHT milk, namely, phosphoric acid (E338), sodium phosphates (E339), potassium phosphates (E340), calcium phosphates (E341), magnesium phosphates (E342), diphosphates (E450), triphosphates (E452) and polyphosphates (E453). Addition of citrates is permitted in goat's milk but not cow's milk. No additives are permitted in pasteurised whole milk, semi-skimmed or skimmed milk.

In the EU and other countries, the term "fresh" is rarely used on milk package labels and there is no official designation for its use. Instead, milk is described by its heat treatment and its fat content. If it is not heat-treated, it is classed as raw milk. However, in parts of the world where both fresh milk and reconstituted milk are available, and where "fresh" milk commands a premium price over reconstituted milk, the term fresh may be allowed on the label as a descriptor of that milk. As mentioned, it is used mainly to distinguish milk originating directly from the cow from milk which is reconstituted from powder or may have some powder added to it. It may be used with all designations of heat treatment. However, in some locations milk which has been heated at conditions which are classed as sterilisation, UHT or in-container, may not be called fresh. Local regulations dictate when the term "fresh" can be used.

Australia New Zealand Dairy Authorities Committee has published food safety guidelines for the validation and verification of dairy heat treatment equipment and processes (ANZDAC, 2007). It was designed to ensure that heat treatment equipment such as pasteurisers are designed, installed, operated and verified in a manner that results in no untreated or partially treated milk or dairy products. Factories would be expected to incorporate systems for the validation and verification of heat treatment as an integral part of their food safety program, where critical heat treatment processes are conducted. The guidelines cover competence requirements for heat treatment evaluators and risk management programme verifiers. They also include corrections to pasteurisation times to account for different fat percentages and particles in the range of less than 200 microns up to 1,000 microns. A selection of the data is given in Table 2.4. They also suggest that for turbulent flow, the maximum velocity may by assumed to be 1.33 times the average velocity for Reynolds number = 4,000 and 1.25 times the average velocity when the Reynolds number exceeds 20,000. These correspond to t_{min}/t_{av} ratios of 0.75 and 0.80 respectively. Further discussion on heat treatment legislation is provided by Komorowski (2006) and most recently by Hickey (2009).

8.4 Quality Assurance/Commercial Sterility: The Current Approach

8.4.1 Introduction

It is the aim of this section to provide a fuller understanding of the factors affecting the safety and quality of UHT treated products. In the past, quality control was based mainly upon end-product testing for detecting faulty products, or worse still from consumer complaints. Once a fault was reported, investigations were undertaken to find the cause of the fault and provide corrective action. However, if it was not found quickly, then a large amount of product was often wasted. However, this approach has moved to a philosophy which is based on a more thorough scientific understanding of the process. This leads to implementing procedures to control the process more effectively, thereby trying to prevent problems occurring in the first place. This approach has been very useful in improving both the safety and quality of heat-processed foods.

There are various publications which provide the guidelines and framework for an effective control mechanism. These include Good Manufacturing Practice (GMP) (IFST, 2013), HACCP Publications, Campden Guidelines (e.g., Gaze, 2009), UK Department of Health guidelines (Department of Health, 1994) and Food Standards Agency (2015). The general principles and practices for the safe processing of foods are covered in more detail by Shapton and Shapton (1991). These guidelines encourage each manufacturer to have, and to be able to demonstrate, an effective system of quality management, which is appropriate to its individual circumstances, and which implements the underlying principles of the guidelines. Attention must be paid to all products, but the utmost diligence is essential for specialised products such as infant food and pharmaceutical products. It is a strict adherence to these principles which will continue to ensure that thermal processing is one of the safest means of processing.

It is recognised that UHT processing is more complex than conventional in-container sterilisation (IFST, 2013). Therefore ensuring the safety of UHT-processed foods should be based upon preventing and reducing microbial spoilage by understanding and controlling the process, as discussed earlier. One way of achieving this is by using the principles of Hazard Analysis Critical Control Points (HACCP) (ICMSF, 1988; von

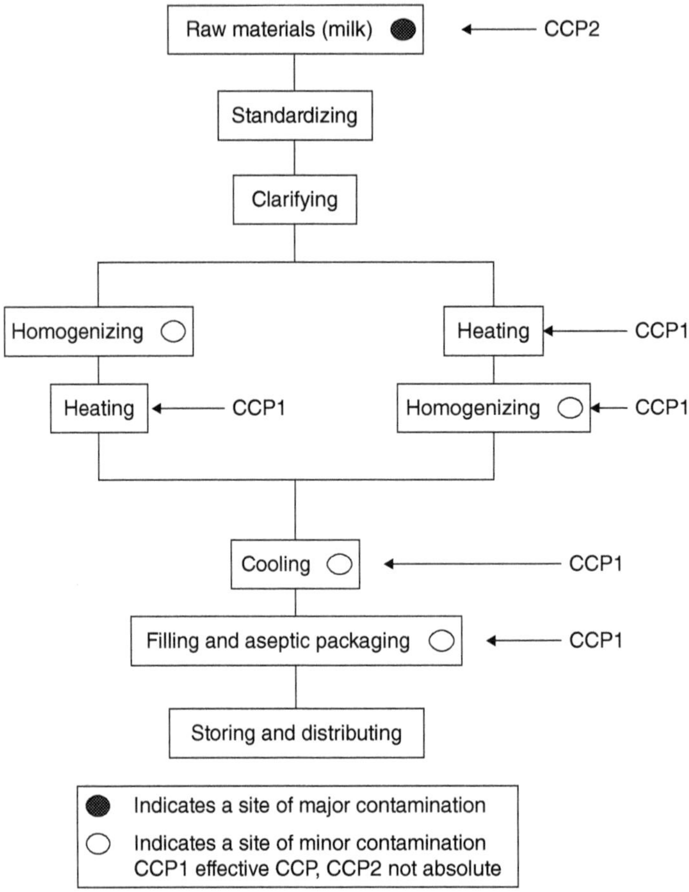

Figure 8.1 HACCP flow chart with Critical Control Points (CCPs) for a UHT process. (Source: Varnam & Sutherland, 2001. Reproduced with permission of John Wiley & Sons.)

Bockelmann & von Bockelmann, 1998; Jervis, 2002). This involves identification of the hazards involved, determination of the critical control points (CCPs), specifying the control criteria, monitoring the process and taking any necessary corrective action. A flow chart of a UHT process with the associated CCPs in shown in Figure 8.1. Critical control points are identified as raw milk quality, bulk storage, processing conditions and post-processing contamination; where the homogeniser is placed downstream, this becomes a major control point. It is important to be able to provide documented records of all the procedures involved.

An initial spoilage rate of 1 in 10,000 and ideally 1 in 100,000 should be the target. In order to verify that this is being achieved, very large numbers of samples need to be taken and examined when the process is being commissioned. The numbers of samples to be taken to achieve the target criterion of 1 in 1,000 with 95% confidence are given in Table 5.16. Accordingly, if 3,000 samples are tested, then the number of allowable unsterile packs is zero. However, the number of samples recommended by Tetra Pak is 7,720 from which a maximum of three unsterile packs is allowed for the sterility verification. The numbers of samples must be accumulated from the three different trials. It should be noted that if the target criterion is changed from 1 in 10,000 to 1 in 100,000, the number of samples to be tested increases 10-fold. Sampling plans are discussed further in Sections 8.4.3 and 8.4.4.

8.4.2 Commercially Sterile Products

For food sterilisation procedures, the aim is to produce a product which is deemed to be commercially sterile. According to first order reaction kinetics it is not possible to reduce the microbial (spore) population to zero and achieve **absolute sterility**. If sufficient samples are analysed, some surviving microorganisms may be found. However, providing these do not proliferate during storage and affect either the safety or the acceptability of the product, they are not considered to be a problem, although it is possible that a small percentage may cause spoilage. For this reason a distinction is made between absolute sterility and **commercial sterility**. Although the ideal situation is to achieve zero spoilage, most commercial producers of sterilised food products expect UHT products would be satisfied to achieve a spoilage rate of 1 in 10,000 units and preferably 1 in 100,000. In practice, 1 in 5000 may be acceptable, whereas rates lower than 1 in 1000 would be cause for some concern.

Thus achieving commercial sterility involves inactivating the most heat-resistant spoilage bacterial spores and then ensuring that the product does not become recontaminated. *G. stearothermophilus* produces one of the most heat-resistant bacterial spores that might be encountered. A process which would achieve about two decimal reductions of this spore would produce at least 24 decimal reductions of *Cl. botulinum*. Thus if commercial sterility is achieved, there should be no food safety risk arising from this organism. This is borne out by the rarity of botulism cases caused by commercially sterilised foods, of which many millions of units are processed each day. Although *G. stearothermophilus* is very heat resistant, it will not grow below about 40°C, so it is not considered to be of major concern in milk destined for temperate climates. It is of concern for milk that is to be transported through hot regions under non-refrigerated conditions. Some special concerns regarding milk destined for hot climates were first discussed in IDF bulletins (IDF, 1972, 1981).

It is a challenge to initially verify that a sterilisation process is performing as desired and under control. As indicated above, this involves analysing a large number of samples when first commissioning a process. However, even if a large number of samples are analysed, it does not guarantee complete safety (absolute sterility). For this reason, some of the problems related to sampling frequency with commercially sterile foods will be discussed in more detail. As for pasteurisation, the role of PPC on the keeping quality of UHT products cannot be overemphasised (see Section 4.4.3).

8.4.3 Sampling Theories and Probabilities

This section is an updated version of the discussion in Lewis and Heppell (2000). For any production run the ideal situation is to achieve zero defective items. Unfortunately, this cannot be achieved in practice and we have to resort to probability theory to gain a better understanding of why this is the case.

The probability of finding no defective items in a sample of n items, drawn at random from a batch whose fractional proportion of defectives is p [the proportion satisfactory is q, where $q = (1 - p)$], is given by the first term of the binomial expansion $(q + p)^n$, which is q^n.

As an example, if a batch contains ten defective items in a batch size of 10,000, then $p = 0.001$ and $q = 0.999$. If a sample of 1000 is analysed, then the probability of finding no defectives, that is, prob (0), is $0.999^{1000} = 0.368$. That is, there is a 36.8% chance of finding no defective samples. Therefore, the chance of finding 1 or more defectives is given by 1 - prob(0), which in this case would equal 1 - 0.368 = 0.632. Therefore there is a 63.2 % chance of finding one or more defective items. Thus the expression q^n can be used to evaluate the probability of finding zero defectives for combinations of different sample sizes and fraction defectives in the batch.

Some results are summarised in Table 8.4 for sample sizes from 10 to 10,000 and for percentage defectives in the batch ranging from a high level of 1 in 10, which would be considered to be a gross failure of sterilisation, up to an acceptable level for commercial sterility of 1 in 10,000. This corresponds to values of p from 0.1 to 0.0001. As an example, if 10,000 samples are analysed at this latter target level of spoilage, the chance of finding no defective items is 36.8%. Thus there will be a 63.2% chance of finding one or more defective items in 10,000 units of what is an acceptable batch.

The data in Table 8.4 highlight some of the inherent problems arising from this sampling theory. If only 10 samples are analysed, for a batch with a spoilage rate of 1 in 10, there would be a 36.8% chance of finding no defective items. In other words, a faulty batch may in fact be accepted for release. If the batch spoilage rate were 1 in 1000, there would be a 99% chance of finding no defective items and only a 1% chance of finding a defective item. *Therefore in circumstances where sample sizes are small, the main conclusion to be drawn if one or more defective items are found is that the spoilage rate is most probably high and that there is a major problem with the sterilisation process.* However there would be a very small probability that the batch is satisfactory. If the batch spoilage rate was 1 in 10 and 50 samples were inspected, there would be a ~99% chance of detecting it. However, for a batch spoilage rate of 1 in 1000, there would be only a ~5% chance of detecting it. Therefore, there is a small risk that a batch that is unacceptable would be accepted, that is, a 1% chance that a batch with 1 in 10 defectives, or a 95% chance that a batch with 1 in 1000 defectives would be accepted. This risk is discussed in more detail in the next section.

However, if 2000 samples are analysed and the process is well under control (i.e., a batch spoilage of 1 in 10,000), the chance of finding zero defectives is 81.9 %, or a 18.1% chance of finding one or more defectives, that is, the chance of an acceptable batch being rejected is high if the criterion for acceptance is set at zero defectives being found in 2000 samples. This gives rise to a second risk, which is that of an acceptable batch being rejected. This is perhaps a small price to pay compared to the costs involved in having to resort to a product recall.

Therefore, interpreting the results from sampling plans has to be done with care and nothing can ever be completely certain - or uncertain. Increasing the number of samples analysed decreases the level of uncertainty, but increases the time, effort, analysis costs and amount of wasted product.

8.4.4 Characteristic Curves

An alternative way of presenting the data in Table 8.4 is in the form of a series of characteristic curves (Figure 8.2). Here the probability of acceptance is plotted against the real defective fraction of a batch (% of defective units) for different sample sizes (n = 10

Table 8.4 The probability of finding no defective items for different sample sizes and spoilage rates.

			Probability of finding no defective items									
% defective	**p**	**q**	**Sample size (n)**	**10**	**50**	**100**	**200**	**500**	**1000**	**2000**	**5000**	**10000**
10	0.1	0.9		0.349	0.005	0.000	0.000	0.000	0.000	0.000	0.000	0.000
5	0.05	0.95		0.599	0.077	0.006	0.000	0.000	0.000	0.000	0.000	0.000
2	0.02	0.98		0.817	0.364	0.133	0.018	0.000	0.000	0.000	0.000	0.000
1	0.01	0.99		0.904	0.605	0.366	0.134	0.007	0.000	0.000	0.000	0.000
0.5	0.005	0.995		0.951	0.778	0.606	0.367	0.082	0.007	0.000	0.000	0.000
0.2	0.002	0.998		0.980	0.905	0.819	0.670	0.368	0.135	0.018	0.000	0.000
0.1	0.001	0.999		0.990	0.951	0.905	0.819	0.606	0.368	0.135	0.007	0.000
0.05	0.0005	0.9995		0.995	0.975	0.951	0.905	0.779	0.606	0.368	0.082	0.007
0.01	0.0001	0.9999		0.999	0.995	0.990	0.980	0.951	0.905	0.819	0.607	0.368
0.001	0.00001	0.99999		1.000	1.000	0.999	0.998	0.995	0.990	0.980	0.951	0.905

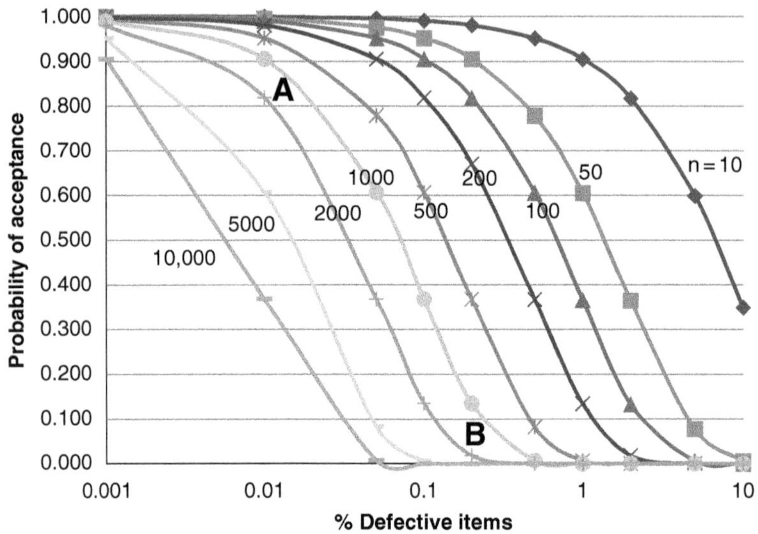

Figure 8.2 Characteristic curves of the probability of acceptance for different sample sizes and spoilage rates. Region A represents high probability of acceptance, region B represents low probability of acceptance (see Section 8.4.4 for further explanation).

to 10,000). It is useful to present it in this form as it highlights two interesting regions that should be briefly discussed.

Region A is that of a high probability of acceptance. This deals with the probabilty that a batch which is satisfactory may get rejected. This is termed the producer risk by Cerf (1989), as it is the chance that a perfectly acceptable batch will get rejected.

Region B is where a batch which is not satisfactory gets accepted. This is termed the consumer risk by Cerf (1989), that is, the risk that the consumer buys a faulty batch, although very unlikely that it will lead to food poisoning (see Section 8.6). However, it could be argued that this is also a potential risk to the producer, as it will be this batch that might be subjected to a product recall.

Thus region A deals with the situation where a satisfactory batch gets rejected. There will be a cost implication to the company as the batch is unsaleable. The upside is that the company has acted responsibly and not released the batch and the information is kept within the company. In contrast, a batch coming from region B is very likely to be subjected to a product recall, which will most probably have much costlier outcomes, both in terms of the product recall costs and the associated adverse media coverage.

The two terms producer risk and consumer risk can be confusing and should be avoided without a full explanation being offered. For example, as mentioned earlier, Cerf (1989) calls region A the producer risk and region B the consumer risk. In contrast, von Bocklemann and von Bockelmann (1998) describe a producer's risk as a substandard production being released (region B) and a buyer's risk (consumer risk) as claims being wrongly made when the product is good. Perhaps for these reasons, these terms are best avoided.

Sampling plans are normally discussed in terms of avoiding releasing a batch which is faulty (von Bocklemann & von Bocklemann, 1998). This is a challenge as large numbers of samples are required to ensure with a high probability that this is being achieved and thus this risk is reduced. It is also reduced by the process of sample pre-incubation (see

Section 8.7.2). It is consoling that the providers of aseptic processing systems have amounted a great deal of practical experience both in terms of devising sampling plans for commissioning new processes and for day-to-day production and should be consulted for further guidance. In this respect, cummulative data collected over a period of production can be very useful in terms of providing reassurance. The publications of Tetra Pak (undated2) and von Bockelmann and von Bockelmann (1998) are worth consulting.

8.4.5 Sampling for Process Verification

It has been shown that large numbers of samples need to be analysed in order to verify that a process is under control. This may be necessary when commissioning a new process or product. Another relevant question is what is the probable spoilage rate in the batch if a certain number of samples (n) are analysed and no defective samples are found? This can be determined at a 95% confidence limit ($c = 0.95$) by determining the fractional proportion satisfactory (q) which would result in a fractional probability of 0.05 (1 - c) (5% chance) of finding one or more defective items, using the following equation.

$$q = (1 - c)^{1/n}$$

q = fraction proportion satisfactory; c = confidence limit (expressed as a fraction), n = sample number.

If 500 samples are analysed and no defective items are found ($c = 0.95$), then $q = 0.9942$ and p, the fractional proportion defective = 1 - 0.9942 = 0.0058. Thus, the best estimate of the number of defectives in the population is 0.0058 or 5.8 in every 1000. *It is important to appreciate that this is nowhere near the 1 in 10,000 target, despite the large number of samples that were analysed.* Data in Table 8.5 have been derived using the equation above and show the most likely percentage level of defective samples in the batch for different sample sizes, when no defective items are found. The calculations are performed for three different confidence levels.

Table 8.5 Estimate for the percentage of defective items in the population for different samples sizes (when no defective items are found in the sample).

Sample size	Percentage defective in population (c = 0.9)	Percentage defective in population (c = 0.95)	Percentage defective in population (c = 0.99)
50	4.501	5.816	8.799
100	2.276	2.951	4.501
500	0.459	0.597	0.917
1000	0.230	0.299	0.459
5000	0.046	0.060	0.092
10000	0.023	0.030	0.046

c = level of confidence

Table 8.6 Estimate for the percentage maximum number of defective items in a population for different samples sizes, when different numbers of defective items are found in the sample (n= 0 to 2).

Sample size	n =	0	1	2
100		2.95	4.66	6.15
500		0.60	0.94	1.26
1000		0.30	0.47	0.63
2000		0.15	0.24	0.31

adapted from data in Cerf (1989)

In fact, the number of samples that need to be analysed, **without any defects being found**, to verify a target spoilage rate of 1 in 10,000 is just under 30,000, at the 95% confidence level. Note that the confidence limit is the expression of the likelihood that the conclusion drawn is correct. If confidence limits of 90% and 99% are selected, then these sample numbers are around 23,000 and 46,000 respectively. Thus more samples must be analysed to ensure a higher level of confidence in the result. Increasing the sample size would increase the reliability of the determination. However, it would also increase the workload, as well as the amount of product wasted.

It is also possible to determine the likely spoilage rate in the population if one or more defectives are found in the batch. Table 8.6 shows that the estimate for the number of defective items in a batch increases as the number of defective items in the sample increases. If three or more defective items are found in a sample size of 2000, the % defective will be 0.39%. Thus, it is highly likely that finding any more than two defective items in a sample will demonstrate that the batch is not commercially sterile.

Overall, the ideal situation is to aim to achieve zero spoilage. Although this is not realistic with food products, the philosophy of wishing to achieve zero defects (or the goal of production) can be adopted by seeking to identify and control all hazards that exist in the food and food production process (Harrigan & Park, 1991).

An ideal solution would be to analyse all samples by a non-destructive method. There has been a need for a long time for such an on-line, non-invasive sampling system, which would check every item and in principle be able to identify (and remove) every defective item. However, such a system would need to be able to detect defective packs due to post-processing contamination. This is a major challenge for testing products immediately after processing as only one bacterial cell per pack can lead to spoilage during storage. In terms of controlling the UHT process, the aspects discussed in Section 8.5 merit serious attention.

8.4.6 Sampling Plans for Refrigerated Products

For pasteurised and ESL products which are stored refrigerated, sampling plans are now commonplace for interpreting microbiological data. Harrigan and Park (1991) summarised what a correctly prepared microbial specification for any product would include:

1) A statement of the microorganisms or microbial toxins of concern.
2) A description of the sampling plan to be applied to obtain the samples to be examined.

3) The microbiological limits (number of microorganisms or concentration of toxin) appropriate to the food and its intended market.
4) The precise analytical method to be employed, including, for example, the methods of sub-sampling and preparation of dilutions, the medium to be employed, the source of the medium or constituents, the incubation temperature and the time.

Two-class and three-class attributes sampling plans are often used; their merits are discussed by Harrigan and Park (1991) and the IFST (1999). These plans are described in terms of four quantities, m, c, n and M, where 'm' is the microbiological limit (e.g., <1 per 25 g or 100 cfu/mL), 'n' is the number of samples from the lot that must be tested, 'c' is the number of samples that may exceed 'm' and 'M' is a level higher than m which must never be exceeded. For two-class plans the criterion is absolute, sample units either "pass" or "fail" and is described by three quantities only, m, c and n, while for three-class plans, the criterion includes the category of "marginally acceptable", 'M', and so is described by m, c, n, and M. In industry, two-class sampling plans are applied mainly when testing for specific pathogens using detection (presence/absence) tests, whereas three-class sampling plans are applied when using enumeration tests. One example of the use of such two-class and three-class sampling plans are in The UK Dairy Products Hygiene Regulations (SI, 1995) for pasteurised milk. To comply, products should meet the following microbiological standards. Plate count at 21 °C after incubation at 6 °C for 5 days: n = 5, c = 1, m = 5 x 10^4 cfu/g, M = 5 x 10^5 cfu/g. Thus, five samples should be analysed and all samples should give counts below 5 x 10^4 cfu/g. However, if one sample is higher than 5 x 10^4 cfu/g, but below 5 x 10^5 cfu/g, then the milk would also be compliant.

There are specific regulations for pathogens based on a two-class sampling plan: absence in 25 g, n = 5, c = 0. This criterion applies to *Listeria* spp in the current EU regulations (Hickey, 2009). A three-class class plan applies to Enterobacteriaceae: n = 5, c = 2, m = 1 cfu/mL, M = 5 cfu/mL.

It is also recommended that such microbiological data be used by individual producers to establish their own control limits and for following trends in production behaviour on a regular basis. Such monitoring should allow early identification or anticipation of problem areas, and allow faults to be investigated and rectified before they become too serious. *The role of post-processing contamination (PPC) and its effects on keeping quality for these products cannot be overemphasised.* Methods for measuring PPC have been reviewed by IDF (1993). Poor control of PPC is the main cause of poor keeping quality of pasteurised and ESL milk.

There has been some concern that the presence of bacterial clumps may compromise the effectiveness of heat treatments such as pasteurisation. However, Hasting *et al.* (2001) concluded that the thermal process delivered during milk pasteurisation is relatively unaffected by the diameter of any clumps of bacteria which are formed. Thus, in practical terms, clumping of cells of *Mycobacterium avium* subsp. *paratuberculosis*, previously suggested as a mechanism by which the organism survives pasteurisation, is unlikely to reduce the effectiveness of the normal pasteurisation process.

8.5 Important Quality Considerations for UHT Processing

This section provides an overview of the main areas that need to be understood and controlled.

8.5.1 Raw Material Quality

It is crucial that raw materials that are selected are fit for purpose. The focus here is on raw milk quality and also the other components that are added to milk and those used in the manufacture of non-dairy UHT products (see Section 9.17). The essential requirements are that the product can be processed at the high temperature and that the heat-treated product will not change substantially during storage.

Milk is the major raw material in many UHT products. There are many sources of data for composition of milk and milk products (McCance & Widdowson, 2015). However, these report average values and provide no information on the variability of raw milk due to species, breed, diet, climate and stage of lactation. Some recent data on variations over one year in bulk milk destined for processing are shown in Table 1.6 (Chen *et al.*, 2014) (see also Table 9.9).

Several aspects relating to raw material quality deserve attention, from an understanding of the physical properties described earlier, through to spore loadings, chemical composition and presence of heat-resistant enzymes. For some products, there are microbiological standards for raw materials; these vary between countries but those for milk intended for heat-treated drinks in the UK are as follows (SI, 1995): cow's milk <100,000 cfu/mL; goats, ewes or buffaloes milk <3,000,000 cfu/mL. An excellent guide to milk producers on "Hygiene on the farm" and on the microbial standards for raw milk for processing and raw milk for consumption was published by the Food Standards Agency (2016). According to the guide, raw milk for processing should contain <100,000 cfu/mL and have a somatic cell count of less than <400,000/mL. For other species the microbial count should be <1,500,000 cfu/mL. The standards for raw milk for grade "A" raw milk and milk products for pasteurisation, ultra-pasteurisation or aseptic processing in the US are <100,000 (single producer supply) and <300,000 cfu/mL (comingled milk) and a bulk somatic cell count of <750,000/mL (FDA 2009). The microbiological requirement for raw milk in China is currently 2,000,000 cfu/mL (Chinese Ministry of Health, 2010a). It should be noted that individual processing companies may set limits for bacterial and bulk somatic cell counts for farm supplies and pay either a premium or impose a penalty for low or high counts respectively. Target total bacterial and somatic cell counts for bulk farm milk supplies of 10,000 cfu/mL and <200,000/mL respectively are not uncommon. Higher standards apply to raw milk for direct consumption. For example, in the UK, it should have a microbial count of <20,000 cfu/mL and a coliform count of <100 cfu/mL (Food Standards Agency, 2016) while in Australia the requirements are for a standard plate count of ≤25,000 cfu/mL, coliform count of <10 cfu/mL, *E. coli* <3/mL, and *Salmonella* and *Campylobacter* to be absent in 25 mL (FSANZ, 2009).

In addition, raw materials for UHT-processed products should not contain high counts of heat-resistant spores and heat-resistant enzymes, as these could lead to increased spoilage and stability problems during storage. Spore counts are usually not specified in official standards but are included in the target specifications by some processors. In conditions where animals are not housed, spore counts in milk of <100 cfu/mL are reasonable but where animals are housed much higher levels can be expected. Burton (1988) pointed out that attention should be paid to how spore counts are determined. For example, spores are often determined by enumerating those microorganisms which survive 80 °C for 10 min, whereas other protocols use 100 °C for 10 min and even 100 °C for 30 min. These different conditions may give rise to a 100-fold difference in reported estimated spore population. For enumeration of spores in raw materials for UHT products, it is recommended that the more stringent conditions are used since non-sporeforming bacteria such as coryneforms can survive heating at 80 °C for 10 min. Dried products such as milk powder, other dairy powders, cocoa powder, rice and other functional powders need to be carefully monitored. Spices may contain high loads which may be effectively reduced by irradiation. Ingredients should have spore counts of ≤1000/g although for some milk powders a realistic limit may be 10,000/g. Thus, procedures are required for monitoring spore counts of all raw materials.

Ingredients with high activities of heat resistant enzymes should not be used in UHT applications. The level of heat-resistant enzymes in milk is never specified as there are no suitable test methods available for routine use (see Sections 11.2.14 and 11.2.19). Absence or a low level of these enzymes in milk is facilitated by ensuring that it has a low count of pyschrotrophic micro-organisms, which will ensure that bacterial protease and lipase activities are minimised (see Section 6.1.3.5). The total bacterial count of cold-stored milk is a reasonable guide to the probability of these enzymes being present, since psychrotrophic bacteria such as *Pseudomonas* species constitute the majority of the total bacteria present in such milk. Similarly, milk with a high plasmin activity should be avoided; the major strategies for achieving this are to avoid milk with high somatic cell count and milk from cows in late lactation. Further information on plasmin and measurement of its activity is given in Sections 5.2.1.3, 6.1.3.5 and 11.2.19.1. Several methods are available for measuring plasmin activity in raw milk but to our knowledge none is used routinely. Incubating raw milk at ~35 °C for 2 h and observing increases in NPN is one relatively simple method which could be used. Furthermore, development of a sensor for plasmin activity in raw milk may enable the activity to be routinely and conveniently measured.

A high titratable acidity in milk is indicative of microbial growth and may be a risk if used for ESL/UHT milk. Such milk may be susceptible to fouling and sedimentation, so measurement of pH, alcohol stability and titratable acidity is recommended to avoid these problems. Milk should have an alcohol stability greater than 74 % (Section 6.2.1.4) and a titratable acidity of less than 0.17% lactic acid (Burton, 1988). These may also be useful criteria for other milk-based products (Chapter 9).

Milk with a high fat acidity should also be avoided. This acidity is caused by free fatty acids produced by lipolysis catalysed by lipase. The free fatty acids cause rancid flavours in milk and milk products and also affect the functional properties of milk, particularly foaming. Lipolysis is raw milk is almost always caused by the natural milk lipase either on the farm or in the factory (Deeth & Fitz-Gerald, 2006).

For milk-based beverages, all the components in the formulation are important. These include the principal ingredients and their sources: sugar, starch and salts, and how they influence the final pH and ionic calcium of the formulation, particularly if it contains appreciable amounts of protein. Attention should be paid to the quality of any added water, particularly its mineral content and microbial status. Calcium and magnesium in water may reduce the heat stability of milk proteins and make them more susceptible to fouling. In contrast, using water which has been softened by ion exchange may increase the sodium content. In addition, the water condensed in the product during direct processing must also be of the highest quality; contamination from carryover in the steam can cause flavour and other problems in directly processed products.

For products requiring addition of powders, reproducibility in metering or weighing added ingredients is crucial. Routine measurement of freezing point depression can establish how consistently the formulation is being made up (see Section 11.2.8). It is also important to ensure that powdered materials are properly hydrated, dissolved or dispersed and that there are no clumps. Centrifugation or filtering may be useful to establish that this has been done properly (see Section 11.2.24). If clumps of powder are present in the formulation, it takes longer for heat to penetrate into these than into the surrounding liquid. The clumps may also offer some protection for spores located toward the centre of the clump. Poorly dissolved formulations may also block filters and small gaps between plates and homogeniser valves, and generally adversely interfere with the flow of the product through the process.

Care should be taken to avoid too much air being entrained into the product during mixing. This may cause foaming and decrease the density and heat conductivity of the fluid. Entrained air will come out of solution when the product is heated and form bubbles or air-locks in the heat exchanger; this effect will be reduced or eliminated when processing occurs under pressure. Excessive agitation during the mixing or blending period or holding the formulation at excessively high temperatures prior to thermal processing may cause some other unwanted reactions to take place. With raw milk, it may damage the milk fat globule membrane, promoting lipolysis and the development of a pronounced soapy flavour. Other undesirable reactions include proteolysis and microbial growth. If mixing takes place between 30 and 50 °C, which is commonplace, the amount of time the product spends at those temperatures is important. For example, for raw milk, this is the optimum temperature for reactions mediated by many of its enzymes to take place, and for microbial growth. These kinds of reaction can be avoided by mixing at temperatures above 50 °C, but other reactions such as precipitation of calcium phosphate and initiation of whey protein denaturation might come into play. Whatever the mixing temperature, an intermediate cooling stage following mixing may be required if the product is not to be processed within a short time of mixing. Even the order of mixing of ingredients can be important where milk is involved. With flavoured milk, excessively reducing the pH of even a small portion of the milk should be avoided as this can affect its heat stability. For example, blending in an acidic ingredient to milk is best done by adding it last to the bulk of the milk, thereby ensuring that the pH does not fall excessively. A poorer heat stability may result when the acidic ingredient is blended with a small fraction of the milk and this mixture is added to the bulk.

8.5.2 Processing Aspects

It has been previously discussed that requirements for product safety and food acceptability are often conflicting, as a certain amount of chemical change will occur during adequate sterilisation of the food. Therefore it is important to consider what is the meant by quality and what is the scope for improving or maximising the quality.

Achieving a safe product relies on eliminating pathogenic bacteria, and achieving a product which is commercially sterile by eliminating spoilage bacteria which could grow at the storage conditions of the product. A second important quality aspect is minimising chemical reactions and reducing nutrient loss. In this aspect, UHT processing offers some distinct advantages over in-container sterilisation for reasons discussed in Section 3.4.

To better understand a specific UHT process, the temperature – time profile for the product should be known. It is a worthwhile exercise for any UHT processor to calculate B* C* and F_0 values for their particular plant configuration, as they provide an indication of the severity of the process being used. If desired, the values can be compared with those of other UHT plants such as those reported by Tran *et al.* (2008). This can then be related to other product quality parameters. Similar calculations should be done for the same plant when it is operating below (and also above) its specified flow rate and for situations where the product temperature falls below the normal set-point. It is also wise to determine the temperature – time profile of any pilot plants being used for process and product development in order to simulate full-scale production as closely as possible (Tran *et al.,* 2008) (see also Section 5.2.1.7).

If the objective is to minimise chemical change but maintain the same bactericidal effect, this can be achieved by using high temperatures, e.g., >145 °C, for a short time rather than a lower temperature, say 137 °C, for a longer time. A means of achieving this is by using a direct process whereby heating and cooling between 70 and 150 °C are very fast, ≤0.5 s. The flash cooling in direct processes removes some volatile components and also some of the dissolved oxygen. There are claims that products produced by direct UHT cannot be distinguished from pasteurised products. In our opinion these are slightly exaggerated, although the cooked flavour intensity and level of whey protein denaturation in direct UHT milk are lower than those found for indirect UHT milk. Further information on direct and indirect UHT processes is given in Section 5.2.1. and Table 5.2.

A problem that arises when operating at higher temperature is the very short holding times required, and the control and validation of such short holding times. This is particularly important in jurisdictions such as USA where UHT plants are approved on the basis of the conditions in the holding tube only. In theory it should be possible to obtain products with very high B* and low C* values, at holding times of ≤1 s. While this is logistically difficult, some equipment manufacturers are now offering plants which have this capability. For indirect processes, the use of higher temperatures is limited by fouling and chemical changes, so it is important to ensure that the heat stability of the formulation is optimised. Some tests for assessing heat stability are discussed in Sections 6.2.1 and 11.2.25.

Thus holding time and temperature are the two most critical process parameters. Recording thermometers should be checked and calibrated regularly, and accurate flow control is crucial, as it is for pasteurisation.

The sterilisation and cleaning procedures are extremely important (see Section 5.6). The plant downstream of the holding tube, that is, all pipework and valves, aseptic tank and filling machine, should be sterilised at 130 °C for 30 min. Cleaning should be adequate (detergent concentrations and temperatures) to remove accumulated deposits and the extent of fouling should be monitored where possible. Steam barriers should be incorporated if some parts of the equipment are to be maintained sterile, whilst other parts are being cleaned.

8.5.3 Other Factors

It is important to record and report all important operational parameters. In the event of problems occurring, this will help to ensure that they can be properly investigated. Regular inspection and maintenance of equipment, particularly for eliminating leaks, is essential. All staff involved with the process should be educated about the principles underlying the process in order to understand the operation and be encouraged to be diligent and observant with respect to changes from the norm. This should be combined with monitoring plant and processing conditions. Ensuring that records are kept of critical processing conditions is crucial. Temperature measurement and recording is fundamental but pressure readings can also be good indicators. For example on steam lines and product lines they can give advanced warning of fouling (see Section 6.2.2). Product flow rates should also be monitored.

The layout of the processing units within the factory is important for ensuring that there is no cross-contamination between raw materials and finished products. This might seem unnecessary for heat-treated products as it is apparently only a remote possibility. However, it should not be forgotten that some of the most serious food poisoning outbreaks in heat-treated foods have occurred as a result of post-process contamination.

With experience, further hazards will become apparent and methods for controlling them introduced. The overall aim should be to minimise spoilage rates and maximise the quality of the product.

8.6 Some Practical Aspects

The ideal situation when manufacturing sterile products is where commercial sterility is being achieved and the process is under control. Unfortunately this is not always the case and spoilage of the product occurs. It has been shown to be possible to reduce the level of complaints about a product down to less than 1 in 100,000 and not all these are related to bacterial spoilage (Burton, 1988).

The other extreme is where gross spoilage occurs. This manifests itself as gas production, acid production or coagulation. In almost all such cases, bacterial growth is the cause. Ideally such batches would be identified before being released for sale and the source of the problem identified and corrected.

One challenge to the UHT milk producer is to work out how much product to inspect to identify microbiological issues. One suggestion is to look at this in two ways: firstly to ensure that any batch (or production run) of UHT milk is commercially sterile, and secondly to identify the cause of situations where commercial sterility has not been achieved. One comforting fact about UHT products is that they are very unlikely to contain pathogens, provided they have been properly processed and aseptically packaged. Thus many of the emerging methods for detecting pathogens will not be appropriate for these products.

However, if microbial activity is found, another challenge is to pinpoint a cut-off level, above which action is to be taken, that is, what level of microbial activity constitutes a failure for a product which is required to be commercially sterile. For example, UHT milk heat treatment regulations at one time stipulated that the product would be deemed to have failed if the count was greater than 100 cfu/mL after incubation for 15 days at 30 °C and this is still included in some advisory material. Although this may appear to be a generous standard for a product which is intended to be commercially sterile, a UHT product with this count, or even up to two orders of magnitude higher, is unlikely to be a health hazard. Tetra Pak (undated2) considers a product to be commercially sterile if a streak (from a 10 µL loop) contains less than 10 colonies, which equates to 1000 colonies/mL) and the product shows no sign of deviation and there is no deviation in pH. Milk will normally contain in excess of 10^6 cfu/mL before off-flavours become noticeable.

Where microbial activity is found, it is very important to ascertain the cause of that occurrence. For heat-treated products such failures fall into three categories: gas production observed by blown containers, and acid production without gas, the so-called flat-sour problem, and coagulation caused by proteolysis. Blown containers can be caused by anaerobic bacteria, primarily clostridia, but in the majority of cases they are associated with some form of post-processing contamination or packaging fault, such as a defective seal or pin hole. These are usually easy to identify and isolate due to bulging of the pack, and defective items can be removed before the product is dispatched.

In contrast, flat-sour spoilage shows no such visible signs and it is almost impossible to remove any spoiled items, without being able to see the product, which usually involves opening the container. Acid development may cause coagulation or gelation of milk-based products. Many of the heat resistant aerobic spore forming bacteria can cause flat-sour spoilage. *G. stearothermophilus* is in this category, but is unlikely to be a cause in temperate climates. Flat-sour problems have been experienced in UHT milk transported across the equator on container ships, when stored in hot warehouses and when stored under ambient conditions in tropical regions. Non-acid coagulation can occur in UHT milk contaminated with highly proteolytic bacteria such as some *Bacillus* species.

Problems from UHT plant failures are rare and easily monitored. More common are contamination of transfer lines or aseptic tanks and most frequent are failures arising from aseptic filling (Williams, 1996).

Usually, extensive microbial activity will noticeably change both the appearance and flavour of UHT milk. However, Burton (1988) reported that several organisms have been found which gave no change in appearance, smell or taste, even at counts of 10^8 cfu/mL. One UHT product encountered by the authors was a sour product with a pH of 5.4 which was curdled, but no viable bacteria could be found using plating methods. One explanation was contamination by an acid-producing bacterium which had been killed but left behind lactose-fermenting enzymes in the product; many degradative enzymes have been shown to be produced by *Bacillus* species isolated from flat-sour evaporated milk (Kalogridou-Vassiliadou, 1992). The product may have contained many dead bacteria but this was not investigated at the time, so the cause of this spoilage remains unknown. However, in these cases, the milk would have been rejected because it was sour.

About 20 years ago attention focused on an emerging highly heat-resistant mesophilic sporeformer (HHRS). This was isolated from UHT milk in various parts of Europe and caused them to fail the then current standard (>100 cfu/mL after incubation for 15 days at 30 °C) (Hammer *et al.*, 1996). It was characterised as Gram-positive to Gram-labile rods, with terminal endospores. It was observed to grow at temperatures between 20 and 52 °C, taking two days to reach a count of 10^5 cfu/mL in UHT milk at 30 °C. Dissolved oxygen was required for growth. However, microbial counts rarely increased much above this level and there were no noticeable changes in the organoleptic characteristics of the milk, except in some cases where a pink colouration was observed when milk was stored in plastic bottles with a low oxygen barrier; oxygen depletion may have limited

Table 8.7 Reported heat resistance data for of *Bacillus sporothermodurans*.

Institute	Result
Canning Research Institute, Campden (UK)	$D_{100} = 5.09$ min
Institute for Food Technology, Weihenstephan (Ger)	$D_{121} = 8.3$-34 s
Tetra Pak, Lund (S)	$F_0 > 68$ min (pilot plant, production unit) (for “satisfactory sterilising effect”)
Tetra Pak Research, Stuttgart (Ger)	$D_{98} = 60$ min; $D_{120} = 10$ min
Netherlands Institute for Dairy Research, Ede (NL)	$D_{126} = 1$ min; $D_{147} = 5$ s $D_{140} = 3.4$-7.9 s

[(from Hammer *et al.* (1996) with permission from International Dairy Federation, and Huemer *et al.* (1998)]

its further growth. This organism was classified as *Bacillus sporothermodurans* (Pettersson *et al.*, 1966). The heat resistance of its spores was investigated by several laboratories and wide variations were observed (Table 8.7). In fact some of the laboratories determined the spores to be very heat-resistant. One suggested reason for these differences between laboratories was the difference in heat resistance of spores in naturally contaminated milk and laboratory-cultured strains. From toxicity tests, there is no evidence to suggest this organism has any pathogenic or toxic properties. See Sections 4.2.2 and 4.4.3.1 for further information on *B. sporothermodurans.*

One debating point relates to what should be done with UHT milk that does not comply with microbiological criteria, but may still be both acceptable and safe for consumption. Generally such products are discarded. This is a source of frustration for the processor, but can lead to heightened vigilance in day-to-day practices. It further raises the question as to at what microbiological count should a UHT product be deemed to be spoiled. It is fortunate that milk has its own built-in warning system, in that UHT milk would generally, but not always, be deemed by the consumer to be to be unfit to drink well before it becomes unsafe.

For trouble shooting, it is important to be able to identify the source of the contaminating microorganisms. If large numbers of vegetative bacteria are found, it strongly suggests that the product has been subject to post-processing contamination. If heat-resistant spores are found, one possible explanation is that the product has been under-processed or that the raw materials contained unusually high counts of these spores. However, post-process contamination with spores, coming from biofilms for example (see Section 6.2.2.8), cannot be ruled out since PPC is the most common source of contamination in UHT products. Until on-line methods become available using non-destructive methods, which will allow for 100% testing, it will not be possible to achieve 100% reduction of micro-organisms, and a low level of microbial spoilage is inevitable.

A spoilage rate of less than 1 in 10,000 is the long-standing accepted criterion for canned food production and they have an excellent safety record. At this level of spoilage, if 1 million units were produced every day, 100 spoiled units could be released onto the market. This may be a difficult point to argue convincingly in a court of law, especially when involving infants or vulnerable adults, an articulate prosecuting advocate and a hostile press. However von Bocklemann and von Bocklemann (1998) point out that this acceptable quality limit (AQL) (i.e., spoilage rates of less than 1 in 10^4) should not be confused with "goal of production". Though a zero defect rate (in an unlimited volume of product) is not achievable, this nevertheless should be the goal of production. To our knowledge, there has not been one serious case of food poisoning resulting from consumption of UHT or sterilised milk, so one might argue that the quality assurance systems currently in place are working well. Even though spoiled products will occasionally be found from a reputable processor, it is highly unlikely that they would contain viable pathogens. However, the UHT producer must never become complacent and continue to be attentive to identifying and isolating any problematic batches. Some producers of UHT milk experience complaint rates of less than 1 in 100,000, which is a very creditable performance. Burton (1988) also reported that these very low spoilage levels can be obtained. Thus, over 20 years ago it had been concluded that problems from UHT plant failures were rare and easily monitored. More common then were contamination of transfer lines or aseptic tanks and failures arising from aseptic filling. It is our contention that further improvements have been made in reducing spoilage incidents in the sterile sections downstream of the heat exchanger.

8.7 Microbiological Examination of Heat-Treated Foods

8.7.1 Introduction

Microbiological standards or specifications for foods are commonplace for quality assurance purposes and in some cases are incorporated into food regulations (see Section 8.3). The requirements for microbiological testing of refrigerated (pasteurised and ESL) and shelf-stable (UHT and sterilised) products are quite different. For refrigerated products, the main aims are for PPC to have been minimised and the products to be free of pathogens. There is a wide range of conventional and rapid methods available for detecting pathogens; these are considered below (see Section 8.7.6).

One of the most serious problems for UHT products is caused by the presence of viable bacteria. Burton (1988) stated that it is important to establish the most reliable method of detecting spoilage in an individual pack, which is economic in terms of time and equipment. Most methods for detecting spoilage involve the destruction of the sealed pack and therefore the loss of saleable product, so some loss in this respect is inevitable. Another important concern is to be able to release product as soon as possible after production, without running the risk of it being recalled. Some of the rapid methods discussed below in Section 8.7.5 have helped to achieve these two goals and this is one of their advantages.

Microbiological sampling plays an important role in quality assurance. It allows easy detection of a gross failure of heat treatment (see Table 8.4). Thus two important decisions for the processor is their target level of spoilage (usually 1 in 10,000) and the confidence level (often 95%), as these will influence the number of samples to be taken.

A procedure used in France is described by von Bocklemann and von Bocklemann (1998). One per cent of the packages produced in a production run are sampled and incubated at 30 °C for 7 days. The remainder is stored and examined daily for blown packages. After incubation at 30 °C for 7 days, 25% of the incubated samples are examined for microbial activity. If no defects are found, the rest of the incubated sample (0.75% of the production run) is released together with the rest of the run. If defects are detected, the 75% remaining from the incubated packages are opened and checked for sterility. Again, if no further defects are found, the production run is released. Detection of additional unsterile units leads to further troubleshooting. von Bocklemann and von Bocklemann (1998) reported that in Germany, the sampling rate varied from 0.03% to 0.3% of the production, with only part of the sampled packages being analysed for defects.

A sampling procedure recommended by Tetra Pak for plant sterility verification is shown in Table 5.16 and described in Section 5.5.5.

von Bockelmann and von Bockelmann (1998) provide an example of testing 100 samples from every batch on a regular basis. Even if all the samples are sterile, at a probability of 90% the defect rate of this batch is less than 2.3%, so there is a 10% chance that it may be higher than this. This procedure will not be effective in preventing substandard product being released in the early stages. However, with time the circumstances will become better if one looks at sampling plans on a continuous basis. In the example above, 100 samples taken on consecutive days will allow failure rates of 1% to be detected between 1 and 3 days (average 1.6 days); failure rates of 0.1% to be detected between 1 and 23 days (average 10.5 days) and failure rates of 0.01% to be detected between 6 and 300 days (average 100 days). One conclusion is that it is more difficult to detect those samples which may lead to product recalls in the early stage of production, but with time this becomes easier. This is why a full-scale validation should be done on any new facility, to ensure that it is capable of producing target spoilage rates of 1 in 10,000.

Whatever level of sampling is used, detecting even one spoiled sample would indicate that a significant failure has most likely occurred at some point in the process. The more challenging scenario is to identify those situations where low levels of spoilage have occurred. This could arise because very low numbers of heat-resistant bacterial spores have survived the heat treatment or a low level of post-processing contamination has occurred, perhaps even for a very short period of the production run. In both of these situations, there will be a small number of viable bacteria amongst a sizeable population of dead bacteria. Sample pre-incubation is usually required in these circumstances.

8.7.2 Sample Pre-Incubation

Since the number of surviving organisms is likely to be very low immediately after production, some form of sample incubation is required to increase their numbers and make them easier to detect in any subsequent analyses. Burton (1988) reviewed some of the recommendations on sampling plans and the literature on determining optimum incubation conditions. He concluded that the optimum incubation conditions are 5 days at 30-35 °C. If time is at a premium, it could be reduced to three days at the cost of missing some potential spoilage. He suggested that in most cases there was little point in incubation at more than one temperature, as normally little additional information is obtained. However, von Bockelmann and von Bockelmann (1998) recommended incubation at ~30 °C for 30 days or at 30 °C for at least 7 days (Tetra Pak's recommendation). The latter accords with the findings of a detailed study by Langeveld and Bolle (1979) that the optimum conditions were 30 °C for at 7-9 days followed by incubation of the agar plates at 30 °C for 4 days. They found that pre-incubation for less than 5 days missed certain types of bacteria and that pre-incubation for longer than 9 days caused some microorganisms to die off.

The incubation times and temperatures may be specified in the recommended test methods and also in some legislation. Some common conditions are 30 °C for 14 days or 55 °C for 7 days. For most conditions under which UHT products are likely to be stored, the former incubation periods are more appropriate. The latter conditions would rarely be encountered 20 years ago, but now where milk is transported through or kept within hot climatic zones, they have become more pertinent. These high temperatures only allow the growth of thermophilic bacteria. Some browning will inevitably occur at these temperatures (see Section 7.2.5).

Some microbiological specifications for sterilised and UHT milk are that after 14 days in a sealed container at 30 °C, it should have a microbiological count at 30 °C of less than 100 cfu/mL. In addition, it should be organoleptically normal (i.e., no off-flavours), and not show any sign of deterioration. However, this would still mean that stock would not be released for at least two weeks based on results from incubation at 30 °C. This long incubation time is necessary for conventional microbiological methods since as von Bocklemann and von Bocklemann (1998) pointed out, the chances of finding any surviving bacteria increases as the incubation time increases. However, with the advent of more rapid microbial methods, shorter incubation periods can now be used. For example, Davenport (2009) found 24-48 h was suitable for testing dairy and non-dairy beverages with the 3M MLS ATP testing system (see Section 8.7.5).

8.7.3 Testing for Microbial Activity

The following sections discuss some methods available for assessing the microbiological status of heat-treated products. An overview of methods used in the dairy industry was published by Anand (2011). The analyses may be performed at the production site or by an off-site accredited provider of analytical services. In fact, there is a strong argument against conducting traditional plating methods (or any methods involving enrichment of microorganisms) at a heat treatment site, or if it is practised, taking precautions to avoid any cross contamination. Another important question is what kind of analyses should be performed, as microbiological methods have undergone some major changes over the past two decades.

8.7.4 Plate Counting and Microscopy

Traditional colony counting procedures are still widely used for heat-treated products. Total counts can be obtained by a number of procedures (Mostert & Jooste, 2002). These include conventional plate counts, surface counts, plate loop technique, roll tube method, spiral plate count, dry rehydratable film technique and a hydrophobic grid membrane technique. Such counting procedures usually need 2-3 days, incubation and therefore are suited to quality control and monitoring. Also, if there is considerable clumping of bacteria, a clump will grow as a single colony. This accounts for differences that may be found in results from plate counting (Deibel & Banwart, 1982) and the direct epifluorescent technique (DEFT) (McMath *et al.*, 1998).

Table 8.8 provides information for reference microbiological methods for milk and milk products (taken from Bintis *et al.*, 2008). Plating methods are used as the reference standard for comparisons where innovative methods are being developed.

One plating method which used to be stipulated in UHT regulations in the UK involves spreading a calibrated loopful of sample (typically 0.01 mL) onto agar and

Table 8.8 Reference microbiological methods for milk and milk products. (Source: Bintsis, 2008. Reproduced with permission of John Wiley & Sons.)

Microorganism	Method	Standard
Total plate count	Colony count technique at 30 °C	IDF 100B.1991
	Plate loop at 30 °C	IDF 131.2004, ISO 8553.2004
Coliforms	Most probable number (MPN) technique	IDF 73B.1998
E. coli	Part 1. MPN technique	IDF 170A.1999
	Part 2. MPN technique using 4-methylumbelliferyl β-D-glucuronide	
	Part 3. Colony count technique at 44 °C using membranes	
Coagulase-positive Stapylococci	MPN technique	IDF 60.2002, ISO 5944.2002
	Colony count technique	IDF 145A.1997
Salmonella spp.		IDF 93.2001, ISO 6785.2001
Yeasts and moulds	Colony count technique	IDF94.2004, ISO 6611.2004

(from Bintsis *et al.*, 2008)

performing counts after an appropriate incubation period (usually at 30 °C and 55 °C). The streaking method is the one favoured by Tetra Pak (undated2).

Alternatively, a pour plate method can be used. More information on these methods is provided by Mostert and Jooste (2002).

Another method is direct microscopical examination, using simple staining methods. However, for this method, the microbial counts need to be higher than 3 x10^5 cfu/mL and, to obtain accurate results, many fields have to be counted. Therefore, this is a labour-intensive method which would not be suitable for large numbers of UHT samples. However, it does provide an opportunity to examine samples if conventional microbiological laboratory facilities are not available. DEFT is a refinement of this, which involves filtering the sample and concentrating the microorganisms on the surface of a membrane, followed by the use of a fluorescent dye to stain and make the bacteria visible. It is capable of detecting lower levels than direct microscopy, with a range of 6 x 10^3 to 10^7 cfu/mL and it can distinguish between live and dead bacteria. Specific probes, such as fluorescent antibodies and oligonucleotides, have been applied to DEFT to allow counting of specific microbial populations. Digital image analysis can reduce the labour associated with manual counting of microbial cells. Therefore it would be most suitable for UHT milk which has been incubated, and it may also be useful for gauging the quality of raw products to be used for UHT processing (IDF, 1991). However, it is time consuming and milk needs to be treated to ensure that fat globules and somatic cells do not interfere with the procedure. A solution to these problems lies in the EZ-Fluo™ Rapid Detection System (Merck Millipore) discussed in Section 8.7.5.1.

8.7.5 Rapid Instrumental Methods for Total Bacteria

There are a number of rapid microbiological instrumental methods available which furnish results more quickly than manual procedures. Commercially available rapid testing methods and the principles on which they are based are listed in Table 8.9.

8.7.5.1 Based on DEFT Method

The *EZ-Fluo*™ method (Merck Millipore) is based on the DEFT method described above. After a pre-incubation period, milk is filtered through a membrane, a fluorescent staining reagent is applied, and viable and culturable microorganisms are stained. After a short incubation period, 30 min, the fluorescent colonies on the membrane are counted by the EZ-Fluo™ reader. The reader allows much smaller colonies to be counted than the naked eye can see.

8.7.5.2 Based on Impedance Measurement

Another approach for heat-treated products products is based on providing a suitable growth environment for any viable bacteria and then detecting physiological changes taking place as a result of growth. The *Bactometer* (BioMerieux, USA) detects bacterial contamination within a few hours, based on measuring the early stages of breakdown of nutrients by the bacteria through changes in the electrical impedance of a suitable

Table 8.9 Commercial rapid microbiological testing methods.

Method tradename	Supplier	Principle of method
Innovate® luminometer and RapiScreen reagent kits	Celsis Rapid Detection	ATP bioluminescence
Microbial Luminescence System (MLS) reagents and instruments	3M Food Safety	ATP bioluminescence
EPIC™ luminometer	Charm Sciences	ATP bioluminescence
PROMILITE-M1®	Promicol B.V.	ATP bioluminescence
EZ-Fluo™ Rapid Detection System	Merck Millipore	Based on DEFT method. Counts fluorescent bacteria on membrane filter used to filter milk
Soleris® Rapid Microbiological System	Neogen Corp	Colour change of indicators
GreenLight System	MOCON	Oxygen depletion
Bactoscan	Foss	Flow cytometry with fluorometric detection
Chemunex Bactiflow ALS® and D-Count®	bioMerieux	Flow cytometry with fluorometric detection
BD Accuri C6	BD Biosciences	Flow cytometry with fluorometric detection using 4 flourescent labels
Bactometer	bioMerieux	Impedance
RABIT	Don Whitely Scientific	Impedance and carbon dioxide sensing
BioLumix Rapid Microbiology	Neogen Corp	Carbon dioxide sensing

growth medium. In principle, a single viable bacterium will grow and any resulting metabolic reaction products will then be detected. A similar instrument is the *Rapid Automated Bacterial Impedance Technique (RABIT)* (Don Whitely Scientific). It is a rapid bacterial detection system which, in one mode, is based on impedance measurement of bacterial growth in a culture medium.

These systems are designed to allow tests to be carried out over a range of incubation temperatures to provide maximum flexibility for microbiological testing. They allow many samples to be measured simultaneously, making them ideal for UHT products, since large numbers of samples can be checked and results obtained within 24 h for initial counts of less than 100 cfu/mL. Drawbacks of these methods are that they are prone to contamination and may fail to detect some organisms, those with very slow growth rates and those that do not cause a change in impedence during the incubation. Despite the claimed advantages for UHT products, these instruments do not appear to be widely used, perhaps because of the emergence methods of other methods, particularly those based on ATP (see Section 8.7.5.6).

8.7.5.3 Based on Carbon Dioxide Detection

A second mode in which the *RABIT* (discussed in Section 8.7.5.2) operates is an indirect mode which monitors carbon dioxide produced by growing organisms. This mode is suitable for organisms that do not produce highly charged metabolites during growth. The CO_2 reacts with a sensor which changes colour to provide an indication of the bacterial growth. The recommended sample pre-incubation period is 30 h at 30 °C.

BioLumix (Neogen® Corporation) is another instrument based on the detection of CO_2 produced during metabolic activity. The BioLumix system is capable of quickly distinguishing 'clean' UHT samples from samples containing bacteria. In one study, no activity was detected in 60 commercial UHT samples but when the UHT samples were injected with a selection of live bacteria (~10 to 100 cells per carton) all were detected.

8.7.5.4 Based on Dissolved Oxygen Depletion

Microbial activity consumes, and hence reduces the level of, dissolved oxygen present in heat-treated milk (Section 7.1.2). Dissolved oxygen in milk can be measured with an oxygen electrode (Section 11.2.4). However, a more sensitive method for detecting the change in dissolved oxygen due to bacterial growth has been developed by Luxel and is known as the *GreenLight®* system. Samples are added directly to vials containing an oxygen sensor and respiration of any viable organisms present causes depletion of oxygen which is monitored by the instrument. It is particularly useful for pasteurised and ESL milk and is claimed to allow earlier detection of pyschrotrophic and thermoduric bacteria than other methods and to detect contaminants that might be missed by plating methods. However, it has not been advocated for use for detecting non-sterility in UHT products.

8.7.5.5 Based on Flow Cytometry (FCM)

Flow cytometry analyses the physical and chemical characteristics of cells or particles by streaming them past a laser. Information about the cell or particle may be deduced from the scattered laser light in the direction of the laser path (forward scatter) and at ninety degrees to the incident laser light (side scatter). It is especially useful for counting different types of cells within a mixed population and can count thousands of units per second.

FCM is particularly useful for counting cells and instruments based on this principle are now widely used in the dairy industry. FCM is extremely sensitive, avoids the need for culturing or enrichment procedures, and can be both qualitative and quantitative. Use of fluorescent stains or fluorogenic substrates in combination with FCM allows the detection and discrimination of viable culturable, viable non-culturable, and non-viable organisms. There is potential for detecting specific bacteria of interest against a background of other bacteria or non-bacterial particles by combining FCM and specific fluorescently labelled antibodies or oligonucleotide probes. FCM can distinguish between living and dead bacteria. As well as counting bacteria, it can also size them.

Gunasekera *et al.* (2000) developed a rapid method for detecting total bacteria in milk to demonstrate the potential of flow cytometry in this medium. They used proteases (proteinase K from Sigma-Aldrich or Savinase from Novo Nordisk) to clarify the milk to remove interference by milk proteins. There was a good correlation between the results of the FCM assay and conventional methods, plating and direct microscopic counting. The limit of detection was $\leq 10^4$ cfu/mL.

One instrument based on FCM is the *BactoScan* (Foss, 2013) which is widely used throughout the world for monitoring the microbial quality of raw milk. It counts all bacteria as single cells and not clusters. Instruments are now available which can analyse 200 samples per hour and results are available after a short time. Prior to measurement, all components in the milk, except the bacteria, are broken down during an incubation period. The bacteria are then stained with ethidium bromide and detected fluorometrically. Counts may be higher than those found by standard plate counting methods, especially for samples in which the bacteria form clusters or clumps which are counted as single cells in standard plate counting methods.

The BactoScan is widely used throughout the UK. In 2014, the average BactoScan count for UK raw milk was 2.6 x 10^4 cells/mL. A troubleshooting service is offered to farmers where high BactoScan counts are found in order to determine the source of the high count.

Other instruments based on FCM and fluorometric detection of living cells is the *Chemunex* range (bioMérieux) which has models with throughputs of 15 to 50 samples per hour. These instruments are now being used for sterility testing of UHT milk products and are claimed to be able to detect unsterility, even 1 bacterial cell/L, after a pre-incubation of 24 h. According to a testimonial on the bioMérieux website, one company (China Mengniu Dairy Company) reported that use of a Chemunex analyser enabled their UHT product release time to be reduced from 7-10 days to 3 days. The company claimed that the method's accuracy is greater than 99.5%; false negatives are at 0% and reproducibility is >99.5% against a traditional plate culture method.

A third FCM instrument is the *BD Accuri C6* (BD Biosciences) which is a dual-laser flow cytometer which allows simultaneous multiparametric analysis based on the use of four fluorescent labels. It has been used with some success to determine live-dead counts using thiazole orange (TO) and propidium iodide (PI). All cells take up the TO but only dead cells take up the PI, allowing their differentiation. This instrument is suitable for analysing bacteria in food and protocols are available for testing the sterility of UHT milk and detection of pathogens in raw milk. It is of interest that this FCM instrument is claimed to be 100-1000 times more sensitive than ATP methods in detecting the sterility of UHT milk. It requires a 24-h pre-incubation followed by a 1-hour analysis period.

8.7.5.6 Based on ATP of Viable Cells

Methods based on ATP detection have now come to prominence, especially for analysing UHT products for sterility. ATP bioluminescence methods are now widely used by UHT processors. They are rapid and sensitive and are available in easy-to-use kit formats. Bioluminescence is the emission of light from viable cells and is a widespread phenomena. The ATP-specific luciferein/luciferase reaction produces luminescence, with the amount of light released being proportional to amount of ATP. Since ATP degrades rapidly when a cell dies, this method provides a good approach for detecting low numbers of viable cells, which is the most important requirement for UHT products. Dormant spores do not contain detectable amounts of ATP (Pedraza-Reyes *et al.*, 2012) and sub-lethally stressed cells contain less than unstressed cells. The method gives an almost immediate response and the procedures and reagents used remove any interfering non-microbial ATP, particularly from somatic cells. The cost of these tests is now reasonable (less than 1£ per sample) and they have been adopted by many UHT processers.

The output of the ATP methods is a light reading (expressed as relative light units, rlu) and guidelines are provided by the suppliers about the cut-off value. Some users have reported occasional false positive results especially for products treated by direct heat treatment. The reason for this is not clear and plate counting methods were not able to detect any spoilage. False negative results are not an issue. A pre-incubation period is generally recommended.

ATP kit suppliers are *Celsis, 3M, Charm* and *Promicol.* All these kits contain reagents which remove non-microbial sources of ATP. ATP test kits are also useful for testing the microbiological status and cleanliness of food processing surfaces and for monitoring hygiene procedures in food factories. A wide variety of test kits are available for this and these have been recently compared for surfaces contaminated with orange juice (Anonymous, 2015).

8.7.5.7 Based on Colour Indicators

Other systems make use of colour indicators to give a visual indication of defective products. One such instrument is the automated *Soleris*™ system (Neogen® Corporation). It uses an LED and a photo detector to monitor colour changes caused by bacterial growth in a medium containing colour indicators. A protocol for sterility testing of UHT milk has been developed.

8.7.6 Analyses of Specific Bacteria

8.7.6.1 Molecular and Immunological Methods

These methods provide alternative approaches for identifying pathogenic and spoilage bacteria and dealing with stressed and damaged cells, and toxins produced by pathogens. Their applications in detecting foodborne pathogenic bacteria were recently reviewed by Law *et al.* (2015). Many of these methods are in use and available in kit form for detection of specific pathogens, so are most useful for monitoring pasteurised and ESL products and, in a UHT context, raw materials. Test-kits for detecting specific spoilage bacteria and heat-resistant spores are less readily available. No doubt considerable improvements have been made and as they become more widely used, they will also become cheaper.

8.7.6.2 Antibody-Based Methods

Antibody-based methods rely on specific binding of an antibody (animal-derived protein) to a target antigen, followed by detection of the antigen-antibody complex. The success of this approach depends upon the stable expression of target antigens in a microorganism, which is often influenced by temperature and the composition of the media. These assays have been developed for microbial detection using different labels to generate the signal. They are sensitive, rapid and simple; testing can be done directly from enrichment media. Immunochemical techniques applied to dairy products include the Enzyme-linked Immunosorbent Assay (ELISA), immunoprecipitation in gel, immunoblotting, immunosensors and antibody array. The best known of these is ELISA whose detection range is 10^4 to 10^7 cells/mL, so requires some form of enrichment. As mentioned, commercial kits are mainly for pathogens, rather than spoilage bacteria. High-throughput, automated ELISA systems such as VIDAS (bioMerieux), Assurance EIA (BioControl), GeneQuence® (Neogen) and Tecra™ (3M) are available for the detection of *Salmonella, E. coli,* O157:H7, *Listeria monocytogenes* and *Campylobacter* (Anand, 2011, Law *et al.,* 2015). These are listed in Table 8.10.

In addition to microorganisms, many other antigens can be detected by ELISA methods. These include individual milk proteins and peptides, toxins, hormones, antibiotics, indicators of heat treatment such as HMF, denatured α-lactalbumin and carboxymethyllysine, adulterants such as melamine and extraneous proteins (e.g., cow's milk proteins in goat's milk or soy milk), and bacterial enzymes such as protease and lipase. Therefore

Table 8.10 Commercially available immunological test kits for bacterial pathogens.

Method tradename	Supplier	Target bacterium or toxin	Principle of method
Assurance GDS	Biocontrol	*E. coli* 0157:H7 *Listeria* *Salmonella* *Campylobacter*	Enzyme-linked Immunosorbent Assay (ELISA)
TECRA	3M	*E. coli* 0157:H7, *Listeria* *Salmonella* *Staphylococcus aureus* *Pseudomonas*	ELISA
SELECTA	Bioline	*Salmonella*	ELISA
GeneQuence®	Neogen	*E. coli* 0157:H7	ELISA
VIDAS	bioMerieux	*Listeria* spp *L. monocytogenes* *Salmonella* Staphylococcal enterotoxins	Enzyme-linked Fluorescent Assay (ELFA)
Salmonella/Screen, Salmonella/Verify ListerScreen, Lister/Screen	Vicam	*Salmonella* *Listeria*	Immunomagnetic capture
LATEX TEST	Oxoid	*Salmonella* *E. coli* 0157:H7	Latex agglutination

Based on Bintsis *et al.* (2008) and http:/neogen.com/FoodSafety/pdf/QG_Catalog.pdf there are many applications of ELISA methods relevant to UHT milk products. An example of the application for detecting bacterial enzymes is the work of Clements *et al.* (1990) who demonstrated that an ELISA method could detect low levels of proteases produced by four strains of *Ps. fluorescens*. It was able to detect 0.24-7.8 ng protease per mL in UHT milk and could be used for quantifying spoilage proteases in dairy products. It should be noted however, that ELISA methods can determine the amount of enzyme protein present but cannot determine the enzyme activity as enzymes vary in their specific activity, that is, enzymic activity per weight of protein. For this reason they have limited practical value in detecting bacterial protease in UHT milk.

8.7.6.3 Nucleic Acid-Based Methods

Nucleic acid-based methods for microorganisms depend directly on their genotypic characteristics, which are far more stable than other methods which rely on phenotypic expression of the genotypic characteristics of the microorganism. They are based on hybridization of a characterised nucleic acid probe to a specific nucleic acid sequence in the test sample, followed by detection of the test hybrid. In this context, a "probe" is a nucleic acid sequence which is unique to the organism of interest and which is used to detect homologous DNA or RNA sequences in the target organism. RNA as a target sequence has an advantage of having far more copies per cell than DNA. Ribosomal RNA is a very important biomolecule that can be used to discriminate between different genera, species or even sub-species.

Several variations of DNA-based assays are used. These include Polymerase Chain Reaction (PCR) (several types), ribotyping, pulsed field electrophoresis and DNA microarray (Naum & Lampel, 2011). A PCR method which was recently developed and applied to *B. licheniformis* and *Geobacillus* spp is PCR – high-resolution melt analysis (HRMA) (Seale *et al.*, 2012, Dhakal *et al.*, 2013).

Although these tests may not be applicable for routine quality assurance procedures, they are valuable in situations where spoilage has occurred and the identity of the spoilage microorganism needs to be determined. The starting point may be a pure colony, an enrichment culture or, in some cases, the food itself. Bintsis *et al.* (2008) stated that genetic methods exist which can identify and characterise to the genus, species and sub-species level, by using a pure colony of the organism. A useful extension of this is proving whether two isolates of the same named bacterium are identical. A classic example of the use of these tests for diagnosis of UHT contamination issues is the case of a recombined UHT product contaminated with *Geobacillus* spores. The skim milk powder being used was implicated as the source of the spores but PCR-HRMA showed that the spores in the final UHT product were genetically distinct from all *Geobacillus* isolates from the powder (Dr M. Weeks, Pers Com).

One issue with molecular methods of detection is that they cannot distinguish the DNA of viable cells and from that of dead cells. This can lead to an overestimation of the level of contamination. This problem was overcome by Cattani *et al.* (2013) for assaying *B. sporothermodurans* in UHT milk. They blocked the availability of DNA from dead cells before using a PCR assay. They achieved this by using the dye, propidium monoazide which penetrates only membrane-damaged cells and intercalates the DNA.

There are several commercial nucleic acid-based kits available for use with dairy products. A selection of these together with their test principles are shown in Table 8.11.

Table 8.11 Commercially available nucleic acid-based test kits for bacterial pathogens.

Method tradename	Supplier	Target bacterium	Principle of method
Assurance GDS	Biocontrol	*E. coli* 0157:H7	DNA amplification
Probelia®	Biocontrol	*Clostridium botulinum,* *E. coli* 0157:H7 *Salmonella* *Listeria*	DNA amplification, PCR
BAX®	DuPont Qualicon	*L. monocytogenes* *E. coli*	PCR
GENE-TRACK®	Neogen	*Campylobacter* *L. monocytogenes* *Salmonella* *Yersinia enterocolitica*	DNA probe
AccuProbe	Holistic)formerly Gen-probe)	*L. monocytogenes* *Staphylococcus aureus* *Mycobacterium tuberculosis* *M. avium*	DNA probe
GeneQuence®	Neogen	*L. monocytogenes, Salmonella* *Listeria*	Nucleic acid hybridisation

Based on Bintsis *et al.* (2008) and manufacturers' websites

Table 8.12 Comparison of some available diagnostic methodologies for detecting *Salmonella.*

Point of comparison	GeneQuence®	PCR	ELFA	ELISA microwell
Enrichment time (h)	24	24-26	24 or 48	48-72
Pre-assay setup time (min)	10	45	15	15
Assay time (h)	1.8	3	1.25	1.5
Technology	Nucleic Acid hybridization	PCR	Fluorescent antibody	Enzyme-linked antibody
Throughput (per h)	79	31	24	62
Specificity	>99.7%	>98%	>98%	>96%
Sensitivity	>98.9%	>98%	>98%	>98%
Average total time to results (h)	27	28	26	50

Based on Pathogen Detection Solutions brochure (http://www.neogen.com/FoodSafety/pdf/GQ_Catalog.pdf)

Table 8.12 shows a comparison of two nucleic acid-based test kits and two immunological test kits available for *Salmonella.* This gives important information about the relative times required for particular tests as well as throughput rates, sensitivity and specificity. The high levels of both sensitivity and specificity are particularly significant. [Note that this table is based on one provided by Neogen, manufacturers of GeneQuence®; its inclusion here does not imply endorsement of this product].

8.7.7 Indirect Methods Based on the Metabolic Activity of Microorganisms

The microbial activity associated with viable bacteria leads to changes in some chemical and physical properties of the product. In many cases, microbial activity results in the production of acid and leads to a decrease in pH. Thus pH measurement is a simple means of monitoring microbial activity. For milk products a reduction in pH also leads to an increase in ionic calcium and a decrease in ethanol stability. It is also likely to increase freezing point depression, due to conversion of lactose to lactic acid. Such changes occur rapidly in raw milk, more slowly in pasteurised and ESL milk and not at all in sound UHT milk, and at variable rates in spoiled UHT milk. If pH is used for assessing sterilised products, reductions in pH resulting from microbial activity should not be confused with those resulting from storage at high temperature. pH reductions under these conditions are due to the Maillard reaction which results in production of formic and acetic acids.

The reduction in pH may lead to changes in the rheological characteristics of the product such as gelation or coagulation. Such changes are readily visible in transparent containers but not in opaque sealed containers. Measurement of such changes by using ultrasound has been investigated as a non-destructive test for product spoilage (see Section 8.8).

Dye reduction tests have in the past been commonly used to monitor raw milk quality and evaluate the keeping quality of pasteurised milk. Their main advantage is that they are simple to perform. The most commonly used dyes are resazurin and methylene blue. The dyes effectively measure redox potential. Davis (1955) considered that resazurin was a very useful test for detecting souring and reducing organisms in raw milk. The basis of the methods is that dyes are reduced at a rate which depends upon the extent of microbial activity (see Section 11.2.4).

Reinheimer and Demkow (1990) investigated a range of tests for detecting contaminated UHT samples and found resazurin to be the most sensitive. Other tests investigated were colony count, titratable acidity and pH modification, organoleptic assessment, stability towards ethanol (68, 80 and 88% v/v) and nitrate reduction. The resazurin test was recently compared with the *Soleris* method (see Section 8.7.5.7) and the standard method using a plating procedure, all with appropriate incubation periods, for assessing the sterility of UHT milk. Both the resazurin and Soleris methods were able to detect non-sterility but results were obtained after 2-3 days with the Soleris method but after 5 days with the resazurin method (Laboratoire-de-Microbiologie-d'Actalia-Cecalait, 2015). The limit of detection for the Soleris method was claimed to be ~1 cfu/mL.

Ahmad and Jindal (2006) reported an automated test for raw milk quality based on the methylene blue dye reduction test in which the change in dye colour was measured by a light sensing probe for about 40 min. The predictions of raw milk quality grades based on the light sensing probe output were in good agreement with those based on standard plate count. The authors suggested it could be used as an inexpensive and rapid alternative to methods such as *Bactoscan* (see Section 8.7.5.5) for assessing milk quality at collection centres in some countries.

8.8 Non-Invasive Methods

The use of non-invasive methods to assess UHT milk avoids sample waste and, in theory, enables 100% sampling. Two methods have been evaluated for detecting spoiled samples. The earliest method was the ElecTester (Elecster Oyj, Finland) which detects changes in the hydrodynamic behaviour of the product caused by changes in viscosity, coagulation and gelation. This is achieved by oscillating the package and measuring the 'damping' value (reduction in amplitude). It has been reported that The ElecTester MK V can test UHT brick-type cartons at speeds of up to 1,000 per hour. Cartons are not opened or damaged during testing; this enables tested samples to be sold as part of the normal batch. According to the manufacturer, ElecTester quality control systems are now used in more than 40 countries (Electster, undated).

The other approach to detect spoiled products is using ultrasound methods. Ahvenainen *et al.* (1989) found that ultrasound imaging was an effective non-invasive method for monitoring microbial growth. They suggested that bacterial loads in UHT processed foods, such as soft ice-cream and processed vanilla sauce, could be detected at levels of 10^5 cfu/g. It was shown to be effective for detecting growth of various bacteria including *B. cereus, St. aureus, Cl. perfringens* and *E. coli*; however, it is not suited to detecting low levels of spoilage. It was also shown to be effective in detecting gelation caused by proteolytic enzymes (Ahvenainen *et al.*, 1991). Wirtanen *et al.* (1992) examined the effect of frequency, probe area and other technical factors on the sensitivity of ultrasound imaging with respect to UHT milk and UHT soft ice cream base packed in Tetra Brik cartons. They found that the accuracy and rapidity of the method was affected by the frequency (3.75-7.5 MHz), dynamic range, echo enhancement and gamma-compensation. Gestrelius (1994) observed that a decrease in the streaming velocity occurred as the product was increasingly spoiled by *St. epidermidis* and *B. subtilis*, suggesting that ultrasound methods are capable of detecting contaminated packs.

Elvira *et al.* (2005) developed an eight-channel ultrasonic device which measured the amplitude and the delay of an ultrasonic pulse passing through packaged UHT-processed milk. Changes in these parameters produced by different microbes were detected even when other physicochemical parameters still remained within the sterility margins. Three different species (*B. cereus, Proteus vulgaris* and *B. pumilus*) were inoculated at different concentrations into UHT milk packs. Growth was detected between 7 and 48 h, depending on the number and type of bacteria inoculated. Elvira *et al.* (2014) used similar techniques to detect microbial contamination of fruit juice, inoculated with acidolactic bacteria, and yeasts and moulds, with a similar degree of success.

The drawbacks of these methods are that they do not detect low levels of spoilage and are usually only applicable after the sample has been incubated. Thus it becomes difficult to apply them in a QA role immediately after production. However, it might be worthwhile using them after the normal period of incubation at the factory, prior to release of the product. However, one has to conclude that although ultrasound shows promise for detecting spoilage in sealed containers, it has not been developed commercially for doing so in UHT products.

8.9 The Milk Microbiome

There are now well established procedures for studying the milk microbiome, which is the body of bacteria that inhabit a specific environment (in this case, milk) at a given time. Methods used are based mainly on molecular biology. It is possible to identify over 1000 different strains of bacteria in raw milk.

The strength of this approach is that in situations where the microbiology is complex, it is possible to obtain a snapshot of the diversity of the microbial flora. To date, considerable research has been carried out on human milk (e.g., Cabrera-Rubio *et al.*, 2012) and on some aspects of cow's milk such as the effect of mastitis (e.g., Bhatt *et al.*, 2012). We are unaware of studies on the microbiome of heat-treated milk. The viable microbial flora associated with pasteurised and ESL products under normal circumstances will be much simpler than that of raw milk and it may be possible to identify specific strains which are predominant at the end of shelf life. Thus, monitoring the changing microbial flora in heat-treated milk should lead to a better understanding of factors affecting shelf-life. The microbial flora associated with a spoiled UHT milk sample is generally even simpler and hence an examination of its microbiome should facilitate determination of the source of the spoilage bacteria. Major advances in this field which will be of benefit to the dairy industry are expected within the next ten years.

8.10 Use of Modelling Procedures

Reaction kinetics and computer modelling have been discussed in Section 6.3. In the context of quality assurance, Hotrum *et al.* (2010) discussed the use of deterministic and stochastic modelling procedures in thermal processing operations. Deterministic models produce a point estimate of the parameter of interest, for example, B^* or C^*, which is valid for that particular set of experimental conditions. Examples are given in Section 6.3. Deterministic models do not account for random variability that is inherent in food systems and thermal processing. An alternative approach is stochastic modelling. Here variability in different processing parameters (time and temperature), product properties (temperature, pH, water activity, salt content) and microbial properties (species, inactivation rate and dependency on product properties) can be taken into account. The final result is a probability function describing the likelihood and extent of an outcome of interest (Hotrum *et al.*, 2010).

Pujol *et al.* (2015) made some pertinent observations on UHT processing with respect to inactivation of *B. cereus*, *Cl. botulinum* and *G. stearothermophilus.* Their models predicted a sterility failure rate (SFR) of 3 bottles in a total of 10^{11} bottles for *Cl. botulinum*, 17,000 bottles for *B. cereus* and 380,000 bottles for *G. stearothermophilus.* For inactivation of G. *stearothermophilus*, the key processing step was the UHT-treatment. This was not surprising because of its known high heat resistance. For *B. cereus*, the operational key process step was the intermediate storage just after the UHT treatment. It was shown that *B. cereus* failure rate was mainly due to airborne recontamination (49%) and to the packaging (33%). Although most *B. cereus* strains do not resist UHT treatment, it was estimated that the sterility failure rate due to this microorganism was not negligible, at 5% of the total failure rate. This finding is supported by *B. cereus* being reported as a cause of UHT food product failures (European Commission, 2016). With the identification of the contamination pathways, microbiological criteria could be targeted to control or reduce the SFR. For instance, to control *B. cereus* by air re-contamination, a microbiological criterion associated with the air quality check of the aseptic tank room could be implemented. Interestingly, the storage of the sterilised packaging before filling did not have any impact on the SFR of the three bacteria. The authors considered that this probabilistic exposure assessment model offered the possibility to firstly implement process settings in a transparent and scientific manner and secondly to identify and quantify where sterility failures in the aseptic-UHT process are most likely to occur.

The future for this type of model could be as a generic tool for assessing the risks of sterility failure, and to assess the relevance and impact of different management options between aseptic-UHT process lines from various factory plants. However, the validity of any such models is wholly dependent on the accuracy of the data input into the models.

8.11 UHT Product Alerts and Recalls

There are mechanisms in place to alert consumers about any safety and quality problems related to foods. For example, the EU has a product alert system in place: https://webgate.ec.europa.eu/rasff-window/portal. Over the period August 2002 to September 2013 there were 22 notifications (not product recalls) related to UHT products and none was considered to be serious. Examples were inadequate thermal processing, coagulation, poor smell and taste. There were two *B. cereus* notifications, but with low counts, and several migrations of isopropyl thioxanthone from packaging around 2005/6. Overall, the fact that there were so few incidents and that they were considered to be a low safety risk pays testimony to the quality assurance systems that are in place. As discussed in Section 8.4, the main aim for the UHT processor is to be able to detect any faulty batches before they are released to the public.

UHT product recalls are not a common occurrence and when they occur information is disseminated either by the company concerned or by the regulatory authorities, so information is made available in the public domain. Some recalls are the result of consumer complaints and others result from in-house quality assurance procedures used by the manufacturer. The very low incidence of product recalls suggests that current quality assurance procedures are working well and overall spoilage rates are very low. If spoilage rates are in reality slightly higher than suggested by this evidence, the manufacturers must have in place effective procedures for detecting spoiled batches before they are released for sale. Where recalls due to bacterial contamination have been necessary, companies have used a precautionary approach and taken appropriate steps, issued advisory information and may have even identified the offending bacterium.

Some recent reported UHT product recall incidents were as follows. The first identified *B. subtilis* as the offending bacteria, with counts of up to 3000 cfu/g. This problem was identified as a result of complaints from customers. It is unlikely that the microbial counts that were found would not have been sufficient to cause any changes in sensory characteristics. One conclusion is that the company acted very responsibly, despite the high costs involved in a product recall. It is also interesting that *B. subtilis* is not usually considered to be a pathogenic bacteria, despite some indications of this in the media literature. The second case involved a UK supermarket recalling its own-label UHT soy drink because a manufacturing fault had caused curdling in some products. The third case was the recall of UHT skim milk, due spoilage by *B. circulans* in some packs resulting in the development of an objectionable flavour and odour. The fourth case was a UHT chocolate milk which was found to have a colony count exceeding the legal limit of the importing country, Hong Kong. Other recalls of UHT products were UHT milk contaminated with melamine (Hong Kong) and UHT milk to which sterile water had been added.

Other recently documented recalls involved a pre-germinated organic soybean drink with brown rice and an organic black bean drink in Hong Kong. These were contaminated with *B. cereus* to levels between 210,000 and 600,000 cfu/mL. It was reported that the guidelines in Hong Kong were that a food is potentially injurious to health or unfit for consumption if levels of *B.* cereus exceed 100,000 cfu/mL.

8.12 Time–Temperature Indicators

Time–temperature indicators ideally are small inexpensive devices that show a time–temperature dependent change which can be easily measured and which mimics the change in a target parameter in a food subjected to temperature change (http://timetempindicators.com/). The ultimate extension of these is an indicator incorporated into the packaging, either to indicate that the food has been adequately sterilised or has not been subject to high temperatures during ambient storage. The latter is probably of most use for UHT products, although the product will normally show noticeable browning once it is opened if it has been temperature-abused.

One application of time–temperature indicators is for storage of refrigerated products. A selected temperature-sensitive ink travels along a channel created in a paper laminate at a rate determined by the environmental temperature. Also simulated within the indicator ink are Arrenhius time–temperature growth profiles of some food poisoning bacteria, such as *Salmonella, Listeria, E. coli* 0157:H7, *Cl. perfringens* and *Campylobacter.* A white window mask is superimposed with selected action points for different foods contained in the multipoint cold chain. This allows accurate display of the potential for bacterial growth in the food product. Most bacteria demonstrate little or no growth from 2 to 8 °C. Temperature excesses are identified by indicator ink flowing into precisely positioned white windows. Depending upon the food and its preparation, these flow rates can be from minutes to days to months.

Timestrip® smart labels or time–temperature indicators help consumers recognise when food is at its nutritional best and is safe to eat and when it should be disposed of to avoid illness (http://timestrip.com/time-temperature-indicators/). Another use is to indicate temperature abuse during storage of UHT products, for example, when temperatures exceed 50 °C or fall below 0 °C. Note that freezing of UHT milk has been shown to have an adverse effect on UHT product quality (see Section 7.5.2).

8.13 Conclusions

UHT processes are now commonplace around the world and the factors affecting safety and quality are mostly well understood. It is strict adherence to understanding and controlling these factors which will continue to ensure the long-term success of this process. Questions which are often asked are what sampling rates and what pre-incubation periods should be used. There are no simple answers to these questions but the information in this chapter may alert the reader to some of the factors involved that might assist in making those decisions. There will always be a conflict between having products released as quickly as possible and further improving the AQL, for example, from 1 in 10^4 to 1 in 10^5, or better, which is a commendable goal.

Many UHT products now travel great distances before they are consumed. There may be statutory regulations for some products, which lay down conditions under which heat treatment should be performed and the quality standards of the final product. These vary around the world so it is important that UHT products are compliant with these regulations wherever they may be sold, and up to 12 months after they are processed.

References

Ahmad, I. & Jindal, V.K. (2006) An automatic procedure for rapid online estimation of raw milk quality *LWT - Food Science and Technology* **39**(4), 432–436.

Ahvenainen, R., Wirtanen, G. & Manninen, M. (1989) Ultrasound imaging - a non-destructive method for control of the microbiological quality of aseptically packed foodstuffs. *Lebensmittel Wissenschaft und Technologie* **22**(5), 273–278.

Ahvenainen, R., Wirtanen, G. & Mattila-Sandholm, T. (1991) Ultrasound imaging - a non-destructive method for monitoring changes caused by microbial enzymes in aseptically-packed milk and soft ice-cream base material. *Lebensmittel Wissenschaft und Technologie* **24**(5), 397–403.

Anand, S. (2011) Analytical methods: Microbiological. In: *Encyclopedia of Dairy Sciences.* 2nd edn. vol. 1, (eds J.W. Fuquay, P.F. Fox & P.L.H. McSweeney) pp. 215–220. Academic Press, San Diego.

Anonomous (2015) Assessing sanitation: a comparison performance testing of ATP devices. *New Food Magazine* **18**, 41–45.

ANZDAC (2007) *Guidelines for Food Safety. Validation and Verification of Heat Treatment Equipment and Processes* (http://www.agriculture.gov.au/SiteCollectionDocuments/aqis/exporting/dairy/publications/anzdac-validation-heat-treatment.pdf)

Bhatt, V.D., Ahir, V. B., Koringa, P., Jakhesara, S.J., Rank, D.N., Nauriyal, D.S., Kunjadia, A.P. & Joshi, C.G. (2012) Milk microbiome signatures of subclinical mastitis-affected cattle analysed by shotgun sequencing. *Journal of Applied Microbiology* **112**(4), 639–650.

Bintsis. T., Angelides A.S. & Psoni, L. (2008) Modern Laboratory Practices- Analysis of Dairy Products. In: *Advanced Dairy Science and Technology.* (eds T.J. Britz & R.K. Robinson), Blackwell Publishing Ltd, Oxford.

Burton, H. (1988) *Ultra High Temperature Processing of Milk and Milk Products.* Elsevier Applied Science, London.

Cabrera-Rubio, R., Carmen Collado, M., Laitinen, K., Salminen, S., Isolauri, E. & Mira, A. (2012) The human milk microbiome changes over lactation and is shaped by maternal weight and mode of delivery. *American Journal of Clinical Nutrition* **96**(3), 544–551.

Cattani, F., Ferreira, C. A. S. & Oliveira, S. D. (2013) The detection of viable vegetative cells of *Bacillus sporothermodurans* using propidium monoazide with semi-nested PCR. *Food Microbiology* **34**(1), 196–201.

CFIA (Canadian Food Inspection Agency) (2016) *Milk and Regulated Milk products.* AIED-DIEA-IE-2001-4-7 (http://www.inspection.gc.ca/animals/terrestrial-animals/imports/policies/animal-products-and-by-products/2001-4/eng/1321120021415/1321120121634).

Cerf, O. (1989) Statistical control of UHT milk. In: *Aseptic Packaging of Foods.* (ed. H. Reuter), Technomic, Lancaster, Pensylvania.

Chen, B., Lewis, M.J. & Grandison, A.S. (2014) Effect of seasonal variation on the composition and properties of raw milk destined for processing in the UK. *Food Chemistry* **158**, 216–223.

Chinese Ministry of Agriculture (2005) Identification of reconstituted milk in pasteurised and UHT milk. Agriculture Professional Standard of the Peoples' Republic of China. NY/T 939-2005.

Chinese Ministry of Health (2010a) National food safety standard. Raw milk. GB 25190-2010. The National Standard of the People's Republic of China. eagri.cz/public/web/file/294859/GB19301_2010_raw_milk.pdf.

Chinese Ministry of Health (2010b) National food safety standard. Sterilised milk. GB 19301-2010. The National Standard of the People's Republic of China. eagri.cz/public/web/file/294863/GB25190_2010_sterilized_milk.pdf.

Clements, R.S., Wyatt, D.M., Symons, M.H. & Ewings, K.N. (1990) Inhibition enzyme-linked immunosorbent assay for detection of *Pseudomonas fluorescens* proteases in ultra high-temperature-treated milk. *Applied and Environmental Microbiology* **56**(4), 1188–1190.

Codex (2000) *Heat Treatment Definitions.* Joint FAO/WHO Food Standards Programme, CX/MMP 00/15, Codex Committee on Milk and Milk Products.

Codex (2004) *Draft Code of Hygienic Practice for Milk and Milk Products at Step 8,* ALINORM 04/27/13, Appendix II.

Davenport, K. (2009) Rapid testing of non-dairy and mixed dairy beverages using the 3M™ Microbial Luminescence System (MLS, formerly Cogent) (http://multimedia.3m.com/mws/media/593593O/iafp09-poster-3m-mls-non-dairy-mixed-dairy-beverages.pdf).

Davis, J.G. (1955) *A Dictionary of Dairying.* 2nd edn. Leonard Hill, London.

Deeth, H.C. & Fitz-Gerald, C.H. (2006) Lipolytic enzymes and hydrolytic rancidity. Chapter 15 in *Advanced Dairy Chemistry, 2 Lipids.* (eds P.F. Fox and P. McSweeney) Springer, pp 481-556. New York.

Deibel, K.E. & Banwart, G.J. (1982) Comparison of the stomacher with other systems for breaking clumps and chains in the enumeration of bacteria. *Journal of Food Protection* **45**(10), 898–902.

Department of Health (UK) (1994) *Guidelines for the Safe Production of Heat Preserved Foods.* HMSO, London.

Dhakal, R., Chauhan, K., Seale, R.B., Deeth, H.C., Pillidge, C.J., Powell, I.B., Craven, H. & Turner, M.S (2013) Genotyping of dairy *Bacillus licheniformis* isolates by high resolution melt analysis of multiple variable number tandem repeat loci. *Food Microbiology* **34**, 344–351.

Elecster (Undated). Computerized ElecTester MK V Non-Destructive Quality Control Method for UHT Milk Products. (Brochure provided by Elecster February 2016).

Elvira, L., Duran, C.M., Urredja, J. & Montero de Espinosa, F.R. (2014) Detection of microbial contamination of fruit juices using non-evasive ultrasound. *Food Control* **40**, 145–150.

Elvira, L., Sampedro, L, Matesanz, J., Gomez-Ullate, Y., Resa, P., Iglesias, J.R., Echevarría, F.J. & Montero de Espinosa, F.R. (2005) Non-invasive and non-destructive ultrasonic technique for the detection of microbial contamination in packed UHT milk. *Food Research International* **38**, 631–638.

EU (1992) Directive 92/46, Laying down health rules related to the production of and placing on the market of raw milk, heat treated milk and milk-based products.

EU (1995) Directive No 95/2/EC Food additives other than colours and sweeteners. (http://ec.europa.eu/food/fs/sfp/addit_flavor/flav11_en.pdf).
EU (2005) EU regulations related to UHT heat treated milk (2005) Commission Regulation (EC) No 20732074. Annex VII, Amendments to Regulation (EC) No 853/2004. http://eur-lex.europa.eu/legal-content/EN/ALL/?uri=CELEX%3A32005R2074.
EU (2010) Commission Regulation (EC) No 605/2010.Laying down animal and public health and veterinary certification conditions for the introduction into the European Union of raw milk and dairy products intended for human consumption. L175/4. https://www.fsai.ie/uploadedfiles/legislation/food_legisation_links/milk_and_milk_products/reg605_2010.pdf.
European Commission (2016) Rapid Alert System for Food and Feed (RASFF) – RASFF Portal – online searchable database. (Available at https://webgate.ec.europa.eu/rasff-window/portal/(Date accessed: Jan 2016).
FAO/WHO (2004) *Code of Hygienic Practice for Milk and Milk Products* CAC/RCP 57-2004 (http://www.fao.org/fileadmin/user_upload/livestockgov/documents/CXP_057e.pdf).
Food Safety Info (undated) UHT Food and Beverage Microbiology http://www.rapidmicrobiology.com/test-method/microbial-detection-in-uht-food-and-beverage-products/(Accessed Jan 2016).
Food Standards Agency (2015) *Food Law Code of Practice*, FSA, London, UK.
Food Standards Agency (2016) *Milk. Hygiene on the Dairy. Farm. A Practical Guide for. Milk Producers to: The Food Safety and Hygiene (England) Regulations 2013.* (http://www.food.gov.uk/sites/default/files/multimedia/pdfs/milk-hygiene-guide-for-milk-producers.pdf).
Foss (2013) BactoScan™ FC+ The approved rapid method for determination of total bacteria in raw milk. (www.foss.dk).
FSANZ (Food Standards Agency Australia & New Zealand) (2009) *Microbiological Risk Assessment of Raw Cow Milk Risk Assessment Microbiology Section.*
Gaze, R. (2009) *HACCP: A Practical Guide.* 4th edn, Guideline 42 Campden and Chorleywood Food Research Association, Chipping Campden, UK.
Gestrelius, H., Mattila-Sandholm, T. & Ahvenainen, R. (1994) Methods for non-invasive sterility control in aseptically packaged foods. *Trends in Food Science & Technology* **5**(12) 379–383.
Griffiths, M.W. (2010) The microbiological safety of raw milk. In: *Improving the Safety and Quality of Milk,* Volume 1, *Milk Production and Processing. (*ed M.W. Griffiths), pp 27–63, Woodhead Publication, Cambridge.
Gunasekera, T.S., Attfield, P.V. & Veal, D.A. (2000) Flow cytometry method for rapid detection and enumeration of total bacteria in milk. *Applied Environonmental Microbiology* **66**(3), 1228–1232.
Hammer, P, Lembke, F., Suhren, G. & Heeschen, W. (1996) Characterisation of heat resistant mesophilic *Bacillus* species affecting the quality of UHT milk: In: *IDF Heat Treatment and Alternative Methods.* Document SI 9602, pp 9–16. International Dairy Federation, Brussels.
Harrigan, W.F. & Park, R.W.A. (1991) *Making Safe Food.* Academic Press, London.
Hasting, A.P.M., Blackburn, C.D.W. & Crowther, J.S. (2001) *Mycobacterium paratuberculosis* and the commercial pasteurization of milk. *Food and Bioproducts Processing* **79**, 83–88.

Hickey, M. (2009) Current legislation of market milks. In: *Milk Processing and Quality Management.* (ed A.Y. Tamime), pp101–138, Wiley Blackwell. Oxford.

Hotrum, N, Fox, M., van Lieverloo, H., Smit, E. & de Jong, P. (2010) Modelling heat processing of dairy products, In *Improving the Safety and Quality of Milk, Volume 1, Milk Production and Processing.* (ed. M.W. Griffiths), pp. 330–348, Woodhead Publishing Ltd, Cambridge.

Hubbell, Jr, B.E. & Collins, B.E. (1962) Ropiness in milk. Pyschrotrophic bacteria and California milk quality. *California Agriculture* **16**(10), 14–15.

Hubbert, W.T., Hagstad, H.V., Spangler, E., Hinton, M.H. & Hughes, K.L. (1996) *Food Safety and Quality Assurance.* 2nd edn, Iowa State University Press, Ames.

ICMSF (1988) *Micro-organisms in Foods 4, Application of the Hazard Analysis Critical Control Point (HACCP) System to Ensure Microbiological Safety.* Blackwell Scientific Publications, Oxford.

IDF (1972) *Monograph on UHT milk.* IDF Annual Bulletin, Part V. International Dairy Federation, Brussels.

IDF (1981) *New Monograph on UHT Milk.* IDF Doc. 133, International Dairy Federation, Brussels.

IDF (1991) *Methods for Assessing the Bacterial Quality of Raw Milk from the Farm,* IDF Bulletin 256. International Dairy Federation, Brussels.

IDF (1993) *Catalogue of Tests for the Detection of PPC of Milk.* Number 281. International Dairy Federation, Brussels.

IFST (1999) *Development and use of microbiological criteria for foods,* London, UK.

IFST (Institute of Food Science & Technology) (2013) *Food and Drink - Good Manufacturing Practice: A Guide to its Responsible Management,* 6th edn, IFST, London.

Jervis, D (2002) Application of process control. In: *Dairy Microbiology Handbook.* (ed. R.K. Robinson), pp. 593–654, Wiley Interscience, New York.

Kalogridou-Vassiliadou, D. (1992) Biochemical activities of *Bacillus* species isolated from flat sour evaporated milk. *Journal of Dairy Science* **75**(10), 2681–2686.

Kelly, A., Datta, N & Deeth, H.C. (2012) Thermal Processing of Dairy Products. In: *Thermal Food Processing: New Technologies and Quality Issues.* 2nd edn, (ed D-W. Sun) pp. 273–307, CRC Press, Taylor & Francis Group, Boca Raton, FL.

Komorowski, E.S. (2006) New dairy hygiene legislation. *International Journal of Dairy Technology,* **59**, 97–101.

Laboratoire de Microbiologie d'Actalia Cecalait (2015) Sterility testing of UHT milk. *New Food Magazine* **16**(6), 55–58.

Langeveld, L.P.M. & Bolle, A.C. (1979) Duration of the pre-incubation period in the sterility control of UHT-sterilised milk. *Netherlands Milk and Dairy Journal* **33**(4), 172–180.

Law, J.W.F., Ab Mutalib, N.S., Chan, K.G. & Lee, L.H. (2015) Rapid methods for the detection of foodborne bacterial pathogens: principles, applications, advantages and limitations. *Frontiers in Microbiology* **5**, 1–19.

Lewis, M.J. & Heppell, N. (2000) *Continuous Thermal Processing of Foods: Pasteurization and UHT Sterilization,* Aspen Publishers, Gaithersburg, MD.

MacMahon, S., Begley, T.H., Diachenko, G.W. & Stromgren, S.A. (2012) A liquid chromatography-tandem mass spectrometry method for the detection of economically motivated adulteration in protein-containing foods. *Journal of Chromatography A* **1220**, 101–107.

McCance, R.A. & Widdowson, E.M. (2015) *McCance and Widdowson's The Composition of Foods*, 7th edn, compiled by Institute of Food Research and Public Health England. Royal Society of Chemistry, Cambridge, UK.

McMath, S. M., Delanove, A., Holt, D. M., Chamberlain, A. H. L., & Lloyd, B. J. (1998) The optimization of an epifluorescence microscopy technique for the direct counting of clumped bacteria. *Journal of the Chartered Institution of Water and Environmental Management,* **12**(2), 113–116.

Mostert, J.F. and Jooste, P.J. (2002) Quality control in the dairy industry. In: *Dairy Microbiology Handbook.* 3rd edn, (ed. R.K. Robinson), InterScience.com.

NZMAF (New Zealand Ministry of Agriculture and Forestry) (2011) *Animal Products (Dairy Processing Specifications)* Notice 2011 (http://www.foodsafety.govt.nz/elibrary/industry/Animal_Products_Dairy-Requirements_Material-2011.pdf).

Naum, M. & Lampel, K.A. (2011) Analytical methods: DNA-based assays. In: *Encyclopedia of Dairy Sciences.* 2nd edn. vol. 1, (eds J.W. Fuquay, P.F. Fox & P.L.H. McSweeney) pp. 221–225. Academic Press, San Diego.

Pedraza-Reyes, M, Ramirez-Rameriz, N., Vidales-Rodriques, L.E. & Robleto, E.A. (2012) Mechanisms of bacterial spore survival. In: *Bacterial Spores: Current Research and Applications.* (ed. E. Abel-Santos). pp. 73–87. Horizon Scientific Press.

Pettersson, B., Lembke, F., Hammer, P., Stackebrandt, E. & Priest, F.G. (1996) *Bacillus sporothermodurans,* a new species producing highly heat-resistant endospores. *International Journal of Systematic Bacteriology* **46**(3), 759–764.

Pujol, L., Albert, L., Magras, C., Johnson, N.B. & Membré. J-M. (2015) Probabilistic exposure assessment model to estimate aseptic-UHT product failure rate. *International Journal of Food Microbiology* **192**, 124–141.

Reinheimer, J.A. & Demkow, M.R. (1990) Comparison of rapid tests for assessing UHT milk sterility. *Journal of Dairy Research* **57**, 239.

Rowe,.R. & Donaghy, J. (2011) Microbiological aspects of dairy ingredients, *In "Dairy Ingredients for Food Processing.* (eds. R.C. Chandan & A. Kilara), Wiley Blackwell, Iowa, USA.

Shapton, D.A. & Shapton, N.F. (1991) *Principles and Practices for the Safe Processing of Foods.* Butterworth Heinemann. Reprinted in 1998 by Woodhead Publishing Limited.

Seale, R.B., Dhakal, R., Chauhan, K., Craven, H., Deeth, H.C., Pillidge, C.J., Powell, I.B. & Turner, M.S. (2012) Genotyping of present-day and historical *Geobacillus* species isolates from milk powders by high-resolution melt analysis of multiple variable-number tandem-repeat loci. *Applied and Environmental Microbiology* **78**(19), 7090–7097.

SI (UK Statutory Instruments) (1995) *The Dairy Products (Hygiene) Regulations 1995*, No 1086, [legislation.gov.uk/uksi/1995/1086/contents/made].

Tetra Pak (undated) *The Role of Raw Milk Quality in UHT Production* available at: http://www.sdt.org/media/uploads/the_role_of_raw_milk_quality_in_uht_production_141215.pdf (4 May 2016).

Tetra Pak (undated2) *Procedures & Guidelines.Guideline for Microbiological Evaluation of Commercially Sterile Products.* Document number QAM – 588002-0107.

Tran, H., Datta, N., Lewis, M.J. & Deeth, H.C. (2008) Predictions of some product parameters based on the processing conditions of ultra-high-temperature milk plants. *International. Dairy Journal* **18**, 939–944.

von Bockelmann, B. & von Bockelmann, I.A. (1998) *Long-life Products: Heat-treated, Aseptically Packed: A Guide to Quality.* Tetra Pak Processing Systems AB, Lund, Sweden.

Wilbey, R.A. (1996) Estimating the degree of heat treatment given to milk. *Journal of Society of Dairy Technology* **49**(4), 109–112.

Williams T.W. (1996) UHT Course lecture notes, Leatherhead Food RA, UK.

Wirtanen, G., Ahvenainen, R. & Mattila-Sandholm, T. (1992) Non-destructive detection of spoilage of aseptically-packed milk products: effect of frequency and imaging parameters on the sensitivity of ultrasound imaging. *Lebensmittel Wissenschaft-und Technologie* **25**(2), 126–132.

9

Other Shelf-Stable Products

9.1 Introduction

In principle, UHT processing can be applied to any product that can be pumped through a heat exchanger and then be packaged aseptically. The most widely processed product is bovine milk, but there is now a wide variety of UHT milk-based products on the market. Tables 1.4 and 1.5 show a selection of the milk products that are available, or being developed. In addition, more non-dairy, low-acid UHT products are emerging, including a range of plant extracts and drinks such as tea and coffee. In addition to these, a large range of acidic beverages, which are aseptically packed and ambient-stable are available. This chapter discusses this range of products. One feature of the milk-based products is the important role of pH and ionic calcium in their heat stability.

9.2 Reconstituted and Recombined Milk

Where supplies of fresh milk are not available, milk can be reconstituted from powder. The issues for heated reconstituted milk products are similar to those posed by liquid milk products. Almost all milk powder is now produced by spray drying, which has superseded roller drying. The production of milk powder involves a number of processes, including centrifugal separation (for skim powder), forewarming, evaporation or concentration by membrane filtration, and drying. A good account of drying technology is provided by Kelly (2006) and the properties of milk powders have been reviewed in Tamime (2009).

One of the main characteristics for distinguishing between powders is the degree of heat treatment, the basis of the whey protein nitrogen index. Up to five different categories have been described for skim milk powder, but the most common are low-, medum- and high-heat. Low- or medium-heat powders are usually used for reconstituted milk destined for consumption as liquid milk. In all situations it is essential to ensure that a high-quality milk powder with no off flavours is used. In terms of minimizing problems during heat treatment, the powder should be well mixed and properly hydrated. There are a number of important properties of milk powders, such as wettability, ability to sink, dispersibility and solubility which influence its dispersion. Usually mixing is performed at 40 to 50 °C to achieve full rehydration of the powder. After mixing for about 15-20 min the milk is often left for another 20 min to expel occluded air, since

High Temperature Processing of Milk and Milk Products, First Edition. Hilton C. Deeth and Michael J. Lewis.

milk powder may contain up to 40% by volume of occluded air. Mixing can be done at 5 °C, but it takes longer than at 40 to 50 °C as the powder solubility/dispersibility is lower at this temperature. Effective air removal can be checked by measuring the density (see Section 11.2.3). Oxygen solubility is higher at low temperatures and the air is less easily removed. Air can also be removed by vacuum deaeration (see Section 5.4) which can be incorporated as part of the heat treatment, normally after regeneration and before the homogeniser and the sterilisation holding tube. Too much dissolved air may result in cavitation of the homogeniser and excessive oxidation of the product. Poorly dispersed powder can result in blockage of homogeniser valves and narrow gaps between plates of plate heat exchangers.

The quality of the water used for reconstitution is important. It should be good drinking quality, free of pathogenic microorganisms and an acceptably low hardness, less than 100 ppm of calcium carbonate (Tetra Pak, 2015). Addition of more minerals (from the added water) may jeopardise the salt balance, which may cause problems related to heat stability. Copper and iron contents are important as they promote oxidation and may have an adverse effect on flavour.

Kastanas (1996) investigated the fouling behaviour of skim milk powder, which was reconstituted with hard water and soft water and found that using soft water reduced its susceptibility to fouling. Faka (2008) evaluated the heat stability of skim milk powders subjected to in-container sterilisation and found the heat stability was not influenced by the water hardness, although some other properties were.

Heat stability is a very important property of any powder in formulations destined for sterilisation. It is influenced by variations in milk composition and the nature of protein and minerals, and interactions that take place during processing. Faka *et al.* (2009) evaluated methods to improve the heat stability of low-heat skim milk powder and found the ionic calcium level was a major factor; removal or reduction of ionic calcium improved the heat stability. Of special interest are stability issues occurring due to changes of season, feed and calving pattern (Chen *et al.*, 2017). The underlying causes of poor heat stability are not well understood.

In terms of heat stability, Newstead (1994) presented some useful guidelines to reduce fouling in recombined milk and reconstituted whole milk powder. Fouling was measured on a plate heat exchanger and the primary indicator was the rate of change of approach temperature (ΔT) with time. Experiments were performed for two hours and results were presented in terms of the run time possible. A wide range of fouling rates was obtained, corresponding to commercial production times of 20 h or more (a rise in ΔT of <0.2 °C/h) or less than 1 h (a rise in ΔT of >4 °C/h). Product pressure differentials were also measured but were seldom high enough to provide a useful differentiation between runs. The main results obtained are summarised below:

Process temperature had a considerable effect on fouling rate. Increasing it from 135 to 142 °C resulted in a 45% increase for each °C rise in temperature. A further increase from 142 to 145 °C resulted in an increase of 70% per °C change.

Pre-treatments: lower pre-treatment conditions for powders tended to favour lower fouling rates. However the heat treatment effects were reported to be complex and the "total" heat treatment, as indicated by the whey protein nitrogen index for example, was not a useful indicator.

Type of water: the minerals in water significantly influenced fouling, with harder waters producing more fouling.

Cold storage of the reconstituted milk for up to 48 h significantly reduced fouling rates.

Addition of polyphosphates such as SHMP reduced fouling significantly. The point of addition also had a significant effect, especially when hard water (300 ppm as calcium carbonate) was used for reconstitution. It was preferable to add it to the milk after reconstitution (or recombination).

Homogenisation during recombining significantly increased fouling. In this sense, recombined milks behave differently to fresh milks. It is probably best to avoid homogenisation prior to heat treatment.

Some other important properties of milk powders are bulk density, sensory characteristics, nutritional value and microbial population. The source and composition of the fat used in recombined or filled milks may also influence the sensory characteristics of the final product.

In countries where both fresh milk and reconstituted milks are available, there is an interest in being able to distinguish between them. It is relatively straight forward to distinguish between pasteurised fresh milk and pasteurised reconstituted milk because of the difference in chemical heat treatment indices (see Section 6.1.7). It is not straightforward to distinguish between UHT fresh and reconstituted milks as both types have increased values of heat indices such as whey protein denaturation, furosine, HMF and lactulose. However, UHT reconstituted milk has a higher ratio of furosine to lactulose values compared with UHT milk and this can be used for distinguishing between them (Corzo *et al.*, 1994; Montilla *et al.*, 1996). The relevant analytical procedures are discussed in Chapter 11.

9.3 Concentrated Milk Products

Concentrated milk products which can be produced by either in-container sterilisation or UHT processing are now available. The most popular are *evaporated* and *condensed milk* products. These can be made from fresh milk or by a recombination process using milk powder. Although evaporated milk is a traditional product in the UK, sales are low. Evaporated or condensed milk still remains popular in the Far East and also in parts of Africa, especially where fresh milk is not readily available. Traditionally, evaporated milk contained a minimum of 22% MSNF and 9% fat, giving 31% total solids and it was once common practice for milk to be standardized to this composition prior to evaporation. The English food regulation for condensed milk and dried milk (SI, 2015) states that condensed milk is partly dehydrated milk containing, by weight, not less than 7.5% fat, and not less than 25% total milk solids. There are other variants: condensed high-fat milk containing, by weight, not less than 15% fat, and not less than 26.5% total milk solids; condensed, partly skimmed milk, containing, by weight, not less than 1% and less than 7.5% fat, and not less than 20% total milk solids; and condensed skimmed milk partly dehydrated milk containing, by weight, not more than 1% fat, and not less than 20% total milk solids. These are similar to the Codex standard for evaporated milk (Codex, 2010).

Permitted additives can be present (2 g/kg singly or 3 g/kg combined, expressed as anhydrous materials). These are listed as firming agents (KCl, $CaCl_2$), stabilisers (citrates of sodium, potassium and calcium), acidity regulators including phosphates, thickeners and emulsifiers.

Evaporated milk is a term also used to describe concentrated milk. Hickey (2009) points out that this is a particularly English designation and refers to partly dehydrated milk containing, by weight, at least 9% fat and 31% total milk solids (SI, 2015). Some data for evaporated milk products in the UK are given in Table 9.1. It is interesting that the density and total solids contents are very close, whereas there is a considerable range in viscosity (measured by a capillary flow viscometer). It is also noteworthy that sediment is low in all products, suggesting that problems due to poor heat stability relate more to thickening and perhaps gelation of the product than to sediment formation.

Chen (2013) presented data for different forewarming conditions and stabiliser additions on the viscosity of evaporated milk, produced by in-container sterilisation at different times of the year. Samples with no forewarming had a viscosity in excess of 130 cSt, or formed a gel. Samples which were forewarmed but which had no added stabiliser were also excessively viscous but gelation was less common. Figure 9.1 shows a wide

Table 9.1 Some data for commercial evaporated milk products. (Source: Chen, 2013. Reproduced with permission.)

Source of samples	Viscosity (cSt)	Density (g/ml)	pH	Ca^{2+} (mM)	Total Solids %	Sediment %
Sainsbury	27.6	1.082	6.25	0.743	27.7	0.18
Tesco	36.1	1.083	6.26	0.76	27.9	0.06
Delicious Desserts	42.8	1.082	6.25	0.71	27.6	0.073
Milbona	45.5	1.081	6.33	0.72	27.7	0.05
ASDA	46.1	1.082	6.03	–	27.8	0.05
Morrison	30.0	1.081	6.28	0.71	27.7	0.08
Nestle	43.0	1.083	6.06	–	27.5	0.04

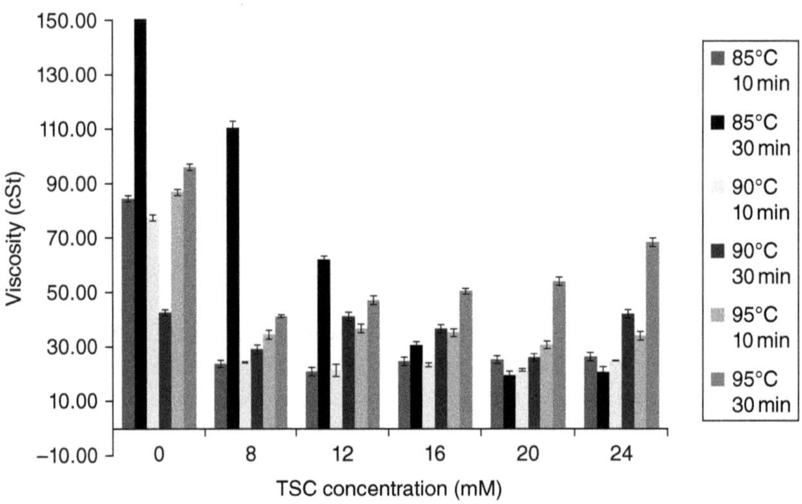

Figure 9.1 The influence of forewarming conditions and stabiliser concentration on the viscosity of evaporated milk for six batches of raw milk. (Source: Chen, 2013. Reproduced with permission.). (*See color plate section for the color representation of this figure.*)

range of viscosities found for six evaporated milk samples subjected to different levels of stabiliser addition (TSC at 8 to 24 mM) and different forewarming conditions (80 °C/10 min to 95 °C for 30 min).

In principle, the procedures for producing evaporated or condensed milk are straightforward. The aim is to produce a product which is pourable and which does not thicken or coagulate during the final sterilisation procedure. The target for viscosity is rarely recorded, but it should be in the range of 20 to 40 cP at 20 °C. Viscosity measurement is widely used for assessing heat stability. Chen (2013) recorded values for commercial products in the UK (Table 9.1).

Tarassuk and Tamsma (1956) proposed that the critical zone of viscosity that coincided with the first and partial appearance of gel formation was 90-100 cP (measured at 25 °C). The key steps in terms of controlling viscosity are forewarming prior to evaporation and ensuring the correct amount of stabiliser is added prior to sterilisation. It is then important to ensure that further thickening and gelation and crystal formation do not take place during storage. Although a small amount of sediment is usually formed, sediment formation is not reported to be a major issue (Table 9.1).

It was suggested that using high-temperature, short-time conditions results in good heat stability and that this could also be improved by further heat treatment of the concentrate prior to sterilisation. There is also a suggestion that applying correct forewarming conditions would eliminate the need for the addition of stabilisers. This is an interesting proposal that is worthy of further investigation. Even earlier, Benton and Aldberry (1926) measured heat stability of milk which had been concentrated to 18% solids-not-fat, by measuring the time required for the product to curdle at 120 °C. They reported considerable variations in heat stability of milk from individual cows (4 to 24 min). They also observed that addition of citrate or borate to improve heat stability was effective up to a point, beyond which the stability decreased. It was observed that adding these stabilisers increased alcohol stability (see Section 11.2.25.1). In the majority of the samples studied, the optimum heat stability lay between alcohol stabilities of 70 and 75%. They considered that the alcohol test provided a better means of avoiding over-stabilisation than pH adjustment, as there was no particular pH to which milk could be adjusted to provide uniform results. However, it was observed that samples which were naturally, or adjusted to be, above or below the pH range 6.55 to 6.65 prior to evaporation, were never heat stable.

Forewarming conditions initially were 80-95 °C for 10-30 min, although more recently temperatures over 100 °C for shorter time periods have been used. Augustin and Clark (1991) observed that 120 °C for 2 min (direct or indirect heating) resulted in lower viscosities than 85 °C for 30 min. Tarassuk and Tamsma (1956) reported that optimum conditions of pre-heating vary seasonally and with milk from different cows. All these factors contribute to the challenges faced by the processor to produce a consistent concentrated milk product.

There is relatively little increase in viscosity arising from the evaporation stage. Data in Figure 9.2 show how the viscosity changes as a result of pre-heating (forewarming), evaporation and sterilisation. Once the concentrate has been produced, it is homogenised prior to stabiliser addition and sterilisation. Cronshaw (1947) gives conditions of 2500 psi (~148 bar/14.8 MPa) at 140 °F (60 °C). The stabilisers most investigated are TSC and DSHP. Most of the early literature makes reference to TSC, but the stabiliser of choice in the UK is now DSHP. As discussed, the optimum amount to be added will be

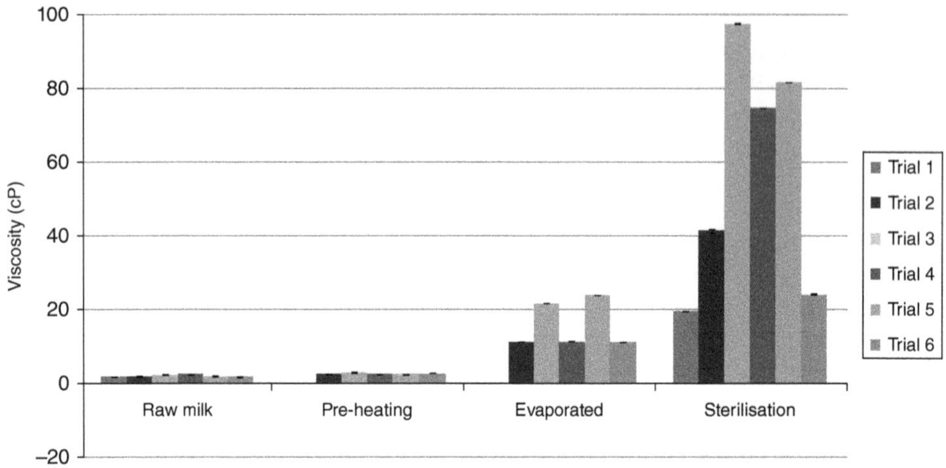

Figure 9.2 Effects of forewarming, evaporation and sterilisation on the viscosity of evaporated milk for six different milk samples. (Source: Chen, 2013. Reproduced with permission.). (*See color plate section for the color representation of this figure.*)

different for every batch as levels of phosphate and citrate naturally present in the milk are variable. Traditionally, the amount to be added has been determined experimentally, by sterilising a few cans with different amounts of salts under the conditions employed and determining which level does not cause coagulation (Cronshaw, 1947). This procedure is still widely practised today. Thus recommended levels of stabiliser addition are rarely reported. Cronshaw (1947) gives a range of 2 to 10 oz/1000 lb (0.013 to 0.063%) of TSC or DSHP. In our opinion, these values are on the low side.

Although the procedures seem relatively straightforward, ensuring a product of consistent viscosity is difficult because of the variability in heat stability of the starting milk. It is important to ensure that total solids content is controlled tightly as heat stability is influenced by even small changes in total solids. Viscosity data for concentrated milk of 26% to 34% TS during storage was provided by Tarassuk and Tamsma (1956). The time to reach an unacceptable viscosity increased as total solids was reduced. Marked retardation of gelation during storage was also achieved by a process that involved over-concentrating the milk (28 to 40% TS), pre-heating the concentrate and then diluting back to 26% TS. At higher concentrations, pre-heating conditions of less severity are required for optimum retardation of gelation during storage.

TSC is effective in preventing thickening and coagulation during sterilisation. Deysher and Webb (1952) reported that the crystals which formed during storage of evaporated milk were calcium citrate and that this could be retarded by addition of DSHP and by storage at low temperature. Thus, one drawback of TSC is that it appears to promote crystallisation of calcium citrate which adversely affects the texture of the product. This was found to be the case when citric acid was added to evaporated milk (Deysher & Webb, 1952).

Sterilisation results in a decrease in calcium ion activity and a decrease in pH. The decreases in calcium ion concentration caused by sterilisation were not influenced by the pre-heating treatment. Niewenhuijse *et al.* (1988) reported that calcium ion concentration appeared to be an essential factor in the heat stability of concentrated milk

(TS = 31.3%) at pH values below the HCT maximum. Generally, in the Northern Hemisphere, March milk and October milk are less stable than June and September milks.

Where fresh milk is not available, concentrated milk is made by reconstituting milk powder. High-heat powders are normally recommended for this purpose. Augustin and Clarke (1991) produced recombined milks from powders which had been subjected to different pre-heat treatments. They were sterilised at 120 °C for 13 min and heat stability was measured by viscosity. The ranges found for two experiments were 11.5 to 49 cP and 13 to >80 cP. Faka *et al.* (2009) found that low-heat skim milk powder reconstituted to 25% TS had poor heat stability. This could be improved by removal of some calcium by ion exchange or by addition of TSC prior to drying. Tsikritzi (2011) compared the heat stability of medium-heat SMP reconstituted to 20 and 25% TS. When sterilised at 115 °C for 15 min, the milk at 20 % TS was heat-stable, whereas that at 25% TS coagulated. Improved heat stability was best achieved by addition of DSHP (0.1% to 0.5%) or TSC (0.1%). Higher concentrations of TSC resulted in a higher viscosity but no coagulation. The disodium salt of EDTA and DHSP caused coagulation, whereas the tetrasodium salt of EDTA did not cause coagulation and viscosity decreased as its concentration increased from 0.1% to 0.4%. SHMP caused coagulation at about 6.6 mM (0.4%) and was the only salt to increase viscosity prior to sterilisation.

9.3.1 UHT Evaporated Milk

UHT evaporated milk is less common than its in-container sterilised equivalent. The following guidelines are available for its production (Alfa Laval, undated). Prior to evaporation, the milk is deaerated and then heated at 120 °C and held for 4 min for stabilisation (prewarming) and then flash cooled to 93 to 95 °C. If the product has to be stored for a long period before UHT treatment, it should be cooled to 5 °C. UHT processing involves heating to 135 to 140 °C for 1 min. Homogenisation should be downstream at 75 °C, in two-stages at a total of 190 bar (19 MPa). The amount of sediment can be determined by centrifugation (see Section 11.2.22). It is expected that it would normally be low, as shown for products in Table 9.1. The desired quality attributes of the end product are influenced by its intended end use. If it is to be diluted to normal concentration, it should have a low viscosity and low sediment and not be too cooked. For use as a coffee cream, a higher viscosity and more cooked flavour may be appropriate. It must also have good thermal stability when added to hot (acidic) coffee (Geyer & Kessler, 1989a,b).

Sediment formation and excessive product thickening, which may result in gelation, are the two most common problems encountered. As mentioned, sediment will be evident soon after heat treatment. It can be prevented to some extent by increasing homogenisation pressure but the most likely cause is poor heat stability. It is not a common occurrence.

Excessive thickening and gelation of UHT evaporated milk can be prevented by addition of stabiliser. It is claimed that SHMP and other sodium polyphosphates are more effective than sodium hydrogen phosphate, but it was not clear whether this was the dihydrogen or disodium salt (Alpha Laval, undated).

In 2015, 1.6 billion tonnes of evaporated milk were produced worldwide, as opposed to 216.8 billion litres of unconcentrated white milk (Table 1.6). Evaporated milk consumption is greatest in USA, Germany, Nigeria and South East Asia. About 30% is in cartons and the most popular size is in the range 250-360 mL. Another common form of packaging for UHT evaporated milk is the 10 L bag-in-box system, designed for the catering and food services industry. There is a trend towards aseptic processing as illustrated by the situation in Brazil, where 82% of evaporated milk was produced in cans in 2005, whereas in 2013, 72% was in cartons. Where evaporated milk is aseptically processed, it may be transported from where it is evaporated to the UHT plant or it may be processed on the same site.

Overall, there is relatively little published information on issues related to heat stability and sediment formation in UHT evaporated milk and whether it is more or less susceptible to age gelation than that produced in the can. Thus, there are no clear guidelines for selecting stabiliser type or levels for UHT evaporated milk.

Another concentrated milk product is *sweetened condensed milk*, which is partly dehydrated milk with added sucrose and containing, by weight, not less than 8% fat and not less than 28% total milk solids. The milk prior to evaporation is heat treated, typically at 82 °C for 10 min if a high-viscosity product is required or 116 °C for 30 s for a low-viscosity product. It is not given any additional heat treatment after the evaporation process, so hygienic handling is essential after evaporation. The high sucrose level confers high microbial stability on the product because of its low water activity. Data in Table 1.6 show that more sweetened condensed milk is produced compared to evaporated milk, although current literature and research activity does not reflect this. As for evaporated milk, there is a trend away from packaging this product in cans to aseptic packaging in cartons and pouches.

9.3.2 Concentration by Membrane Filtration

Milk can also be concentrated by membrane processes, although at the moment UHT treatment of such milk concentrates is not widely practiced. An overview of the membrane processes used for dairy products and other beverages was given by Tamime (2013) (see a summary in Table 10.2). A brief overview follows:

Reverse osmosis allows all components in milk to be concentrated. Thus, in this respect, it is similar to evaporation. It causes a reduction in pH, a slight increase in ionic calcium and a reduction in ethanol stability (Lin *et al.*, 2015).

Nanofiltration allows some reduction in minerals and may be useful for improving heat stability, especially as some calcium may be lost in the permeate.

Ultrafiltration provides a means of concentrating protein and fat in milk. The ionic environment is not substantially changed, neither is the alcohol stability nor the ionic calcium content of the concentrate. Total calcium in the retentate increases as that fraction which is associated with the casein micelle is concentrated. On-Nom (2012) performed ultrafiltration of milk up to 140 °C by positioning a UF module in the holding tube of a UHT plant. Some data on pH and ionic calcium in permeates at different temperatures is presented in Table 11.6.

UF is widely used to concentrate whey proteins for production of whey protein concentrates and isolates, but these do not withstand UHT processing temperatures without considerable pH adjustment. More recently there has been more interest in the properties of milk protein concentrates (MPCs) as these are now more widely available as functional proteins. Crowley *et al.* (2014a) have described the effects of adding calcium salts to MCPs (see Section 9.5.1).

Diafiltration reduces the concentration of soluble components in milk concentrates. For example, when water is added to UF milk retentate to return the protein content to its original level and ultrafiltration is repeated, there is a reduction in lactose, and, because it is not selective, a reduction in salts and soluble vitamins also occurs.

Protein standardization of milk is practiced in some countries using UF permeate as a diluent. This should be appropriately labelled and it would be prudent for the manufacturer to establish that the FPD has not changed substantially (see Section 11.2.8).

Microfiltration is used for removing bacteria and spores from milk (see Section 10.2). Microfiltered milk is normally sold as an extended shelf-life product (see Section 3.3.2). MF also offers the potential for fractionating whey proteins and caseins. It can also be used to concentrate micellar casein, by removing the whey proteins in the permeate, to produce milk protein concentrate.

Syrios *et al.* (2011) compared the heat stability of powders produced by RO, UF and NF. These were reconstituted to 25% TS and in-container sterilised. NF was shown to offer opportunities over RO for improving the heat stability of powders destined for reconstitution and sterilisation, possibly due to the loss of small amounts of calcium.

9.4 Lactose-Reduced Milk (LRM)

LRM or lactose-free milk is now widely available and consumed by people with lactose intolerance. The most common method for removing the lactose is to hydrolyse it with β-galactosidase. This converts lactose to glucose and galactose. Although it reduces the amount of lactose, there may be some residual lactose. The amount of residual lactose in the product determines whether it can be labelled as lactose-free.

Hydrolysis of lactose results in a number of changes to the milk. Firstly it makes the milk taste sweeter as the hydrolysis products are sweeter than lactose. Secondly, the glucose and galactose make the product more susceptible to browning, as they have increased reducing capacities compared to lactose. This is unlikely to be noticed immediately after UHT processing but it will become apparent if the product is stored for a period of time at higher temperatures (i.e., greater than ~30 °C). Some data are provided for LRM stored at 4 to 50 °C for four months in Table 7.3. Colour differences for LRM stored at 35 °C and 50 °C for 4 months were greater than those found for normal cow's milk and goat's milk (Table 7.4) (DIAL, 2014).

The hydrolysis procedure increases the osmotic pressure of the milk and hence decreases its freezing point. The freezing point of LRM is substantially lower than normal milk and is about -0.760 °C. As a result, FPD measurement is a convenient means of determining the extent of lactose hydrolysis and whether the conversion is complete. Data for one storage trial shows that UHT treatment of LRM hardly affected the FPD; it changed from 761 m°C to 756 m°C (see Section 7.5.1 and Table 7.7). In contrast to the effects on FPD, the hydrolysis process hardly affects the density of the milk or causes significant changes in pH. It does not alter the ionic balance and should not change the heat stability of the milk. Thus if the starting milk has good heat stability, then the resulting hydrolysed milk should also have good heat stability and be suitable for UHT processing. If it had poor heat stability prior to hydrolysis, then it would remain poor.

If the increased sweetness of LRM is thought to be problematic, some of the sugars can be removed by membrane filtration, for example, by diafiltration. One downside of this process which is not highlighted is that other low molecular weight compounds such as vitamins and soluble minerals are also reduced.

There are two options for manufacture of UHT lactose-reduced milk. The first is to hydrolyse the lactose prior to UHT processing and the second is to add a sterile enzyme solution to the milk after UHT processing. Both processes are practised. Downstream addition needs to be done aseptically and the enzyme remains active in the product. The enzyme preparation is usually sterilised by filtration through a MF membrane with a pore size of ~0.2 μm.

As well as some humans, many cats are also lactose intolerant. Thus specialty milk products for cats or kittens are usually based on lactose-reduced milk.

One might speculate how much residual lactose is considered to be permissible in these products (lactose-reduced or lactose-free). The answer relates to ensuring that it is below the level of detection, which varies according to the method used and can be as low as <5 ng/L for detection by GC-MS.

Enzymic hydrolysis of lactose can lead to the development of bitterness during storage of the lactose-reduced milk. The bitterness arises from proteolysis by contaminating proteases in the β-galactosidase preparation. A typical case was presented by Jansson *et al.* (2014).

9.5 Mineral-Fortified Milk

9.5.1 Calcium

Calcium is the main mineral to be fortified in milk. Cow's milk contains about 30 mM total calcium. Figure 11.4 shows the calcium content of different milk samples, as well as the phosphorus content. There is a wide variation in the calcium content in the milk of different species. The range is from about 7 mM in human milk to over 100 mM in milk from some bears. It is not often mentioned, but human milk contains only 25-30% of the calcium present in cow's milk. This reflects the much lower casein content in human milk.

The daily requirement for calcium is in the range 800-1500 mg/day. The maximum recommended daily allowance for any group is 1500 mg/day. Wark and Nowson (2003) reported that 60% of dietary calcium comes from dairy foods (another estimate is 74%) and that people who avoid dairy foods usually have an inadequate calcium intake, unless they have an adequate intake of other high-calcium foods. It is noteworthy that most non-dairy milks are naturally low in calcium and are therefore usually fortified with it (see Section 9.17). More detail on calcium fortification of milk is provided in a review by Deeth and Lewis (2014) and of soy milk by Pathomrungsiyounggul *et al.* (2013).

Although milk is a rich source of dietary calcium, there is interest in increasing its levels and hence calcium-fortified milks are available. A commonly used source of calcium for dietary supplementation is calcium carbonate. The main problem in using this is that it is not soluble. A list of calcium salts which are available for adding to milk is given in Table 9.2. Also listed in Table 9.2 are the calcium contents and number of molecules of water of crystallization for common forms of each salt which need to be taken into account in fortification. It is elemental calcium that matters, with calcium carbonate providing 40% whereas others provide less e.g. calcium citrate, 21%, calcium lactate, 13%,

Table 9.2 Properties of some available calcium salts and their effects when added to milk.

Calcium salt	E number	Waters of crystal-lisation[1]	Calcium (%)	Effect on Ca^{2+} when added to milk	Effect on pH when added to milk	Solubility (g/100 mL)
Soluble salts						
Ca chloride	E509	2	36	Increases	Decreases	75
Ca acetate	E263	2	40	Increases	Decreases	35
CaKcitrate (Gadocal-K)	E333 + E332		15	No change	No change	53
Ca lactate gluconate	E578 + E327		13	Increases	Decreases	40
Ca lactobionate	E399		5	Increases	Decreases	47.3[2]
$Na_2CaEDTA$	E385	2	10	Not known	Not known	
Insoluble salts						
Phosphate (tricalcium, $Ca_3(PO_4)_2$)	E341		39	No change	No change	0.002
Phosphate (dicalcium, $CaHPO_4$)	E341	2	23	No change	No change	0.004
Citrate	E333	4	21	No change	No change	0.095
Carbonate	E170		40	No change	No change	0.0007
Sparingly soluble salts						
Lactate	E327	5	13	Increases	Decreases	6.6
Gluconate	E578		9-13	Increases	Decreases	3.5
Hydroxide	E526		54	Decreases	Increases	0.17

[1] may vary; some salts are available in anhydrous as well as crystalline form
[2] solubility of anhydrous salt in 1 m NaCl (Vavrusova *et al.*, 2013)

and calcium lactate gluconate, 9-13%, and they may be more expensive. Calcium hydroxide (CHyd) is an interesting choice, as it is sparingly soluble and leads to an increase in pH and contains more than 54% calcium.

Table 9.2 categorises the salts of calcium into soluble, such as calcium chloride, sparingly soluble such as calcium gluconate and calcium lactate, and insoluble such as calcium carbonate, calcium phosphate and calcium citrate. Also included in Table 9.2 are the effects of additions of the different calcium salts to milk, some of which have been described by Omarukhe *et al.* (2010). Soluble calcium salts generally cause a decrease in pH and an increase in ionic calcium; CHyd is an exception as is calcium potasssium citrate (Gadocal-K®) (Ramasubramanian *et al.*, 2008). It only takes slightly over 2 mM addition of calcium chloride ($CaCl_2$) to make milk unstable to UHT treatment and it takes a slightly larger amount to make it unstable to in-container sterilisation. On-Nom *et al.* (2012) investigated the addition of calcium chloride on pH and ionic calcium, both before sterilisation and during the sterilisation procedure. The relationship between pH and ionic calcium, measured both at 20 °C and 120 °C is shown in Figure 11.8. The results are discussed in more detail in Chapter 6.

Some observations on the use of sparingly soluble salts, for example calcium gluconate and lactate are provided for milk by Omarukhe *et al.* (2010) and for soy milk by Pathomrungsiyounggul *et al.* (2013). These salts give rise to a decrease in both pH and an increase in ionic calcium, but to a lesser extent than calcium chloride. In contrast, insoluble salts have hardly any effect on pH and ionic calcium. One important aspect to appreciate is that although they are insoluble in water or milk, after ingestion they enter the stomach, where pH is much lower and their solubility will increase dramatically.

Crowley *et al.* (2014a) used a LUMisizer (see Section 11.2.25.3) to examine the effects of addition of different calcium salts to reconstituted skim milk powder (RSMP). Fortification of RSMP with soluble Ca salts caused significant ($P < 0.05$) increases in sedimentation rates, with CHyd-fortified samples having a significantly higher ($P < 0.05$) sedimentation rate than all other samples.

It was estimated that approximately 80% of Ca, added at levels between 4.5 and 13.5 mM as $CaCl_2$, became associated with casein micelles, regardless of addition level (Philippe *et al.*, 2003). It was suggested that increased CCP levels, combined with reduced casein micelle solvation in RSMP fortified with soluble Ca salts, created denser casein micelles which sedimented more rapidly. As CHyd resulted in the smallest increase in $[Ca^{2+}]$, a greater proportion of CCP may have formed on its addition, which, when combined with decreased micellar solvation and larger casein micelles, may have contributed to the high sedimentation rate.

9.5.2 Other Minerals

A significant proportion of the world's population has a diet deficient in minerals other than calcium. This is particularly applicable to developing countries but people in developed countries may also be deficient in some minerals; according to USDA (2009), 10%, 57% and 29% of Americans may have inadequate intake of iron, magnesium and zinc, respectively. Therefore, there is a need for fortification of foods with these minerals to alleviate these deficiencies. The long shelf-life and increasing worldwide consumption of UHT milk, together with its relatively low natural levels of these elements, makes it an ideal product for fortification. Abdulghani *et al.* (2014) produced UHT milk fortified with sulfates of iron, magnesium and zinc to give total concentrations (natural plus added) of approximately 25, 50, 75 and 100% of the recommended daily intake (RDI) per litre of the UHT milk. The highest concentrations (100% RDI) were 8, 320 and 16 mg/L for iron, magnesium and zinc, respectively. For the magnesium fortification, co-addition of trisodium citrate was necessary to prevent destabilisation of the protein. Zinc and iron preferentially partitioned into the casein micelles whereas most of the magnesium was found in the serum phase. The distribution of magnesium differs from that of calcium which preferentially partitions with the casein micelles. Iron fortification of milk is challenging as it a strong pro-oxidant. Abdulghani *et al.* (2014) found that UHT milk fortified with ferrous sulfate to 50% of RDI per litre developed a flavour defect after 30 days which correlated with a chemical measure of oxidation. Use of other iron supplements such as iron–milk protein complexes (Ellis *et al.*, 2012) or inclusion of ascorbic acid may alleviate this problem.

Milk may also be fortified with minerals such as selenium and iodine. This can be done by addition to the milk or through supplementation of the diet fed to cows (Harstad & Steinshamn, 2010). Cows in high-selenium locations are known to produce milk with much higher selenium contents (160-3000 μg/L) than cows in low-selenium locations (5-30 μg/L). With iodine, since ~25% of a cow's dietary iodine is excreted in the milk, the level in milk can be readily manipulated through the cow's diet.

9.6 Flavoured Milk

Many flavoured UHT milk products are now available. These may contain a number of ingredients, most of which must be declared on the package in most jurisdictions. A selection of ingredients used is given in Table 9.3. Hence, most formulated milk-based products are much more complex than white milk. Discerning consumers may show a preference for products with fewer additives, particularly avoiding those with E numbers or equivalent.

When developing a UHT formulation for a flavoured milk drink, it is crucial to ensure it has good heat stability and that it will remain stable during prolonged storage. It is worthwhile establishing how added ingredients change pH and ionic calcium as these are considered to be key determinants of heat stability. Tables 6.1 and 6.2 illustrate changes that take place in ionic calcium and pH when different components are added to milk. These additives may also affect the viscosity of the product and hence its mouthfeel and visual appeal.

Another important consideration is to ensure that added ingredients do not contribute to an increased microbial count, especially spore count (see Sections 4.4.3.2 and 4.4.3.3). This aspect was addressed well by Witthuhn *et al.* (2011) who provided

Table 9.3 Some ingredients used in flavoured milk drinks.

Ingredient class	Ingredient
Liquid base	Milk, water,
Sweetener	Sucrose, glucose syrup, lactose, fructose, cane molasses, artificial sweeteners, stevia
Fat/oil	Milk fat, palm oil, rapeseed (canola) oil
Protein powder	Skim milk powder, whey powder
Flavour	Cocoa powder (full fat and fat reduced), vanilla extract, caramel, malt extract, fruit purees
Mineral salt	Phosphates, sodium chloride, sodium bicarbonate, calcium salts, iron salts
Stabiliser	Carrageenan, xanthan, guar gum, alginate, pectin, modified starch, microcrystalline cellulose, carboxymethylcellulose
Emulsifier	Soy lecithin, mono- and diglycerides
Vitamin	Niacin, riboflavin, thiamine, folic acid, ascorbic acid
Colour	Carmine, tartazine, sunset yellow
Dietary fibre	Soluble wheat fibre, beta-glucan

information on the spore loadings that might be found in different foods and ingredients (Table 9.4) and an interesting chart showing spore inactivation by a heat treatment at 120 °C for 30 min in phosphate buffer (Figure 9.3).

Flavoured milk drinks fall into two broad categories: fruit flavoured and confectionary drinks such as chocolate, coffee and malt flavours. Many of these are sterilised by UHT treatment but some in-container sterilised varieties are available, some of which are heated in glass bottles and some in plastic bottles.

9.6.1 Fruit-Flavoured Milk

Fruit-flavoured milk drinks and shakes have been popular for a long time. However, not all fruit flavours are compatible with milk. Two of the most popular flavours are

Table 9.4 Examples of spore loads and types in ingredients commonly used in dairy processing. (Source: Witthuhn, 2011. Reproduced with permission of New Food.)

Ingredient	Spore count [cfu/g or cfu/mL]	Genera	Source
Milk powder	Up to 10^4	*Bacillus, Anoxybacillus, Geobacillus*	Rückert *et al.* (2004)
Cocoa powder	Up to 10^4	*Bacillus, Geobacillus*	Lima *et al.* (2011)
Herbs	Up to 10^5	Not determined	Witkowska *et al.* (2011)
Spices	Up to 10^7	*Bacillus, Geobacillus*	Oomes *et al.* (2007)
Raw milk	Up to 10^2	Different genera, including *Bacillus*	Coorevits *et al.* (2008)

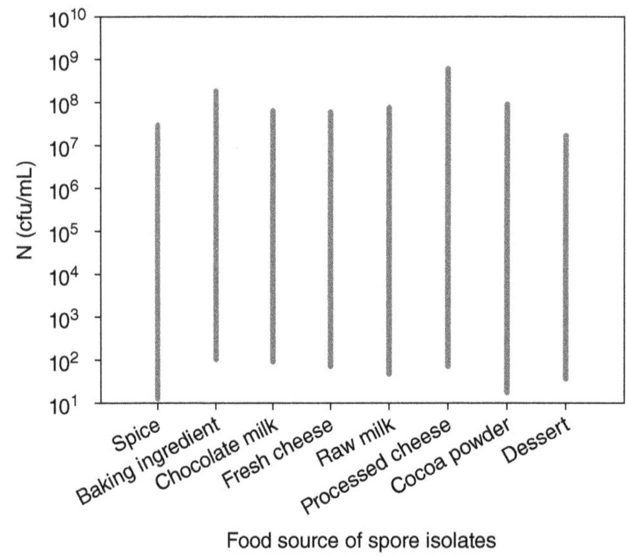

Figure 9.3 Spore inactivation in some foods and ingredients after heating at 120 °C for 30 min in phosphate buffer. The top end of the bar signifies the spore count before the heat treatment and the lower end the spore count after the heat treatment. (Source: Witthuhn, 2011. Reproduced with permission of New Food.)

strawberry and banana. A very important consideration is that the flavouring may be more acidic than milk, so care has to be taken not to reduce the pH of the formulation sufficiently to impair its heat stability. Again, pH and ionic calcium are key product parameters and acidity regulators may be required to control these. A cautionary note is to avoid lowering the pH of even a small fraction of the milk by too much during formulation. Even though pH may revert back to close to its original value, this action may lead to a product with a poorer stability (Ezeh & Lewis, 2011). Therefore, it is advisable to add any ingredients that lower pH towards the end of the mixing procedure to the bulk of the product and not to a portion of it.

Different approaches, such as use of real fruit, natural fruit flavours or synthetic fruit flavours are used to produce fruit-flavoured milk products, as can be observed by reference to product ingredient lists. However, the ingredients list does not provide a complete picture of the appearance of the product, how it pours, what it tastes like, and how stable it is during storage.

9.6.2 Chocolate and Other Confectionery Milk

Chocolate milk is one of the most popular flavoured milk drinks. A selection of those on the market is shown in Table 9.5. Included in the table are some products which have protein levels higher than that of milk, and breakfast milks which contain added fibre (see also Sections 9.7 and 9.8).

Chocolate milk has attracted considerable attention because of its reported benefits in post-exercise recovery and rehydration (Karp *et al.*, 2006; Mansfield, 2009). Conick (2016) stated that "consumers are now starting to realize the *refueling benefits* of chocolate milk". Ingredients added include cocoa powder, sugar and gums of which a wide range are used, as illustrated in Tables 9.3 and 9.5. Most chocolate milk drinks are made from liquid milk, but some are made from powders and some from milk fortified with powders. From Table 9.5, the protein content varies from 2.3 to 9.1% and the carbohydrate from 4.4 to 16.5%; in some cases low-calorie sweeteners are included to enable a reduced carbohydrate content. They can have a range of fat contents from 0.2 to 3.8%; reduced-fat levels of ≤2% are common. The amounts of fat and sugar are often targeted to specific markets, for example, children and health-conscious.

In some countries, chocolate milk has been introduced into school milk programs in order to reverse the decline in milk consumption amongst children. However, there is concern about the high sugar content of chocolate milk which as shown in Table 9.5 can be up to 16.5%. For this reason it has been withdrawn from some school lunch programs. Furthermore, there is also consumer concern about the use of high-fructose syrups for sweetening these milks. Hence it is considered important to reduce the sugar (sucrose and fructose) content of chocolate milk but maintain its sensory acceptability. Consequently, alternative natural sweetening agents such as stevia leaf and monk fruit have been investigated (Li *et al.*, 2015); as shown in Table 9.5, stevia is being included in some products. It has been found that stevia together with a reduced sugar level is preferable to use of stevia alone.

Another approach to reducing the sugar level in flavoured milk is to hydrolyse the lactose to glucose and galactose, both of which are sweeter than lactose (see Section 9.4). This has the added advantage of making the product suitable for lactose-intolerant consumers. With the increasing consumption of milk products in countries such as China,

Table 9.5 Composition of commercial flavoured milk drinks.

	Ingredient						
	Protein (%)	Fat (%)	Flavour	Stabiliser	Emulsifier	Carbohydrate (%)	Other ingredients
Commercial chocolate milk drinks (<5% protein)							
1	2.3	3.8	Chocolate (0.8% cocoa)	Carrageenan	Sorbitan monostearate	11.2	Malt
2	3.2	3.4	Chocolate (1.0% cocoa)			10	Colours (Brilliant Blue FCF, chocolate brown)
3	3.6	3.7	Chocolate	Carrageenan, euchuema seaweed			Colours (Brilliant Blue FCF, chocolate brown)
4	3.0	1.7		Carrageenan			Colours (Brilliant Blue FCF, chocolate brown) Vitamin D3
5	2.9	3.3	Chocolate	Carrageenan, guar gum	Mono- & diglycerides	11.5	
6	3.3	2.1	Chocolate (<2% alkalised cocoa)	Carrageenan, guar gum		12.1 (includes high-fructose corn syrup)	NaCl, $CaCO_3$, vitamins A & D
7	2.3	3.8	Chocolate	Carrageenan	Sorbitan monostearate	10.1	Malt, guarana extract
8	3.4	2.0	Chocolate (1.4% cocoa)	Carrageenan		12.7	Vitamin D
9	2.9	1.9	Drinking chocolate (3.4%)	Carrageenan, guar gum	Mono- & diglycerides	12.8 (includes glucose & molasses)	Malt
10	3.5	1.8	Cocoa & caramel	Alginate, carrageenan		14.4 (includes glucose)	
11	3.4	2.1	Organic cocoa	Carrageenan		11.4 (includes evaporated cane juice)	Vitamin A palmitate; Vitamin D3; fibre <0.4%
12	3.2	1.0		MCC, CMC, carrageenan		10.1	
13	3.4	2.2	Cocoa (1%)	CMC, carrageenan	Mono- & diglycerides	10.2 (includes stevia)	
14	3.4	1.9	Cocoa (1.5%)	Carrageenan, xanthan gum		11.1	NaCl (0.14%); fibre (0.4%)
15	3.4	2.0	Fat-reduced cocoa (1.5%)	Nothing else		10.0	Sterilised in the bottle

Breakfast milk							
a	3.5	0.5	Cocoa (1%)	MCC, carrageenan	Mono- & diglycerides	12	Corn fibre & beta-glucan (total fibre 2%) potassium phosphate, salt, WPC% (Ca, 106 mg/100 mL)
b	4.0	0.5	Cocoa (0.7%)	MCC, CMC, carrageenan		11.8	Malt extract, maltodextrin, oat fibre 1.25%, (Ca, 146 mg/100 mL)
c	5.9	1.5.	Cocoa (0.5%)	Carrageenan, CMC, MCC		11.3 including corn syrup	SMP, WPI, soy protein, maltodextrin, dietary fibre (1.5%), potassium citrate
High-protein (>5%) flavoured milk drinks							
A	5	1.5	Chocolate (0.5% cocoa)	Carrageenan	Soy lecithin	16.5 including glucose	Magnesium, calcium
B	5	0.5	Chocolate (0.7% cocoa)	CMC, MCC,	MG/DG	11.8 (including fibre)	Dietary fibre (1.25%) malt extract
C	5	0.5	Chocolate (0.7% cocoa)	Carrageenan, MCC	MG/DG	12.0 (including fibre)	Dietary fibre (2.0%), WPC, potassium phosphates
D	5.6	1.3	Cocoa (alkalined)	Carrageenan		Fructose, sucrose	WPC, NaCl, vitamin A palmitate, vitamin D3; fibre 0.3%
E	7.0	0.5	Coffee or vanilla	Caarageenan, guar gum, alginate	MG/DG	Fructose	Ca caseinate, WPC, SMP, Na citrate
F	7.3	1.8	Decaffeinated coffee	Carrageenan, xanthan gum, CMC, MCC	MG/DG	5.7 (including sucralose)	MPC, WPC, phosphates of Ca, K & Na, soluble fibre
G	8.0	0.3	Vanilla or strawberry	Carrageenan, guar gum	Mono- & diglycerides of fatty acids (MG/DG)	7.5 (including sucralose)	SMP, calcium caseinate, WPI, maltodextrin, strawberry colour
H	8	0.4	Chocolate (0.5% cocoa)	Carrageenan, guar gum	MG/DG Lecithin	7.6 (Includes fructose & sucralose)	Soy protein, WPI, WPC, SMP, calcium citrate
I	8	2.3		Carrageenan, guar gum			WPC, sodium caseinate, SMP
J	8	0.2	Vanilla & caramel			4.4	Na 60 mg/100 mL
K	9.1	0.2	Strawberry			5	

Figure 9.4 Some chocolate milk samples after centrifugation. (*See color plate section for the color representation of this figure.*)

this is an important consideration. According to DSM (2015), by including a lactose-hydrolysing enzyme such as Maxilact® in the production process, sugar content can be reduced by up to 20% without affecting the flavour. If lactose hydrolysis is combined with the use of a natural low-calorie sweetener such as stevia, the sugar content can be reduced by up to 50% and the product will have a better flavour profile than when these sweeteners are used alone.

Most contain added stabilisers such as carrageenan and those without stabilisers are susceptible to cocoa separation. This is mostly considered to be undesirable but some consumers now like to see sedimented cocoa, as it shows that it is present. Cocoa addition is usually about 1.0% but the range is quite wide (0.5 to 3.4% in a selection of commercial products, Table 9.5); the cocoa powder is sometimes fat-free and/or chemically treated. It is worthwhile for manufacturers to centrifuge chocolate milk products soon after UHT processing, to gauge their stability during short term storage and to predict any potential customer complaints. Some examples of centrifuged samples are shown in Figure 9.4. Ideally, the centrifuged samples should show a small cocoa sediment but no other phase separation (like samples 3 & 4 in Figure 9.4).

The selection of cocoa powder and its level of addition are factors to be considered. It is essential that the spore count of added cocoa powder is low (see Section 4.4.3.3). Difficult climatic conditions at the time of cocoa production and harvesting have in the past contributed toward high spore counts. Low-spore-count cocoa preparations designed for use in long-shelf-life chocolate milk drinks are available from some suppliers. Furthermore, cocoa suppliers can provide cocoa powder which has low alkalinity; high-alkalinity/pH powder has a tendency to cause protein agglomeration.

Chocolate milk drinks are seen as ideal products for fortification with minerals and vitamins, especially for children. Calcium and vitamin D are two of the most common additives. Adding value to milk through fortification may also be beneficial to the diary industry (Stones, 2015).

The FPD values of flavoured milks are higher than for white milk. Typical values fall within the range 800 to 1200 m°C. However, each individual product should show a constant FPD value from batch to batch if it is being produced consistently.

There are chilled and long-life variants of most brands. Some long-life products are produced by in-container sterilisation. One such drink is *Cocio,* which is item 15 in Table 9.5. This is unusual as it is sterilised in glass bottles, allowing the consumer to see the product; however, it does have a warning to store between 2 and 25 °C and to avoid direct sunlight. It has a shelf-life of 1 year. Another unusual feature is that it contains only one additive, a sweetener, so some cocoa separation occurs during storage. A low-calorie version is available, with reduced sugar and a sweetener, and a calorific value, of below 50 kcal/100 mL, which will appeal to calorie-conscious consumers.

Chocolate flavoured milk is more prone to fouling during UHT processing than is white (unflavoured) milk. Fouling is promoted by low or high levels of stabiliser gum (for example, for κ-carrageenan ~0.03% is optimal, lower and higher levels increase fouling) and high sucrose levels (> ~7%) (Prakash *et al.,* 2010).

There are many variants on chocolate milk, for example milk with added body (shakes), and combination flavours, with caramel and coffee being two of popular ones. They provide an attractive flavoured alternative to white milk. Some may also be fortified with minerals and vitamins and some are thickened with starch. Some are sold in transparent containers and some in opaque containers. Most of them come with an instruction to shake before use.

UHT chocolate sugar syrup concentrates are also available, for dilution with milk (about 4:1 to 5:1). They contain about 30% sugar, 1% protein and 0.5% fat and have a shelf life of 1 year. Additives include lactic acid, trisodium citrate and xanthan gum, and potassium sorbate as preservative. Syrups are also available in fruit flavours, such as strawberry and raspberry.

9.7 High-Protein Milk Drinks

Drinks are now available for sport and performance nutrition, weight management and managing a wide range of clinical conditions. UHT variants of these are available and represent a developing product category. Drinks for sport and performance nutrition are usually high in protein ("high protein" according to the US Food and Drug Administration is >4.2 % protein). Information obtained from the labels of some commercial high-protein milk products is shown in Table 9.5, products A-K.

High-protein dairy drinks often contain a balance of casein and whey protein as this is beneficial nutritionally. The caseins are slowly digested while the whey proteins are rapidly digested (Lacroix *et al.,* 2006; Luhovyy *et al.,* 2007). Ingestion of 'fast' whey protein induces intense, but transient, release of amino acids which stimulate anabolic muscle protein synthesis, and the slow digestion of casein may reduce muscle protein catabolism. A combination of casein and whey protein can benefit muscle building, maintenance and repair (Crittenden *et al.,* 2009). The caseins can be in the form of caseinate, micellar casein or milk protein concentrate (MPC). In the authors' experience however, care needs to be taken with the last two forms to ensure the level of ionic calcium is not high enough to cause rapid fouling. MPCs with high-protein contents have high levels of ionic calcium and are quite unstable to heat. For example, a 3.5%

solution of MPC80 had no stability (HCT = 0 s) at 140 °C and hence would be unsuitable for UHT processing (Crowley *et al.*, 2014b). The problem can be overcome by adding a chelating agent such as trisodium citrate.

Drinks containing high levels of whey protein are difficult to UHT process as they foul readily. In the authors' experience, this can be alleviated by the addition of sodium caseinate because of the chaperoning effect of the caseins on the whey proteins (O'Kennedy & Mounsey, 2006). The efficacy of this approach is illustrated by the following example in which a bench-top UHT plant was used to process mixtures of WPC (80% protein) and sodium caseinate. UHT processing trials were first attempted on 2% WPC alone; on each occasion, this caused severe fouling after short running times. Addition of 2% sodium caseinate to the 2% WPC enabled it to be successfully UHT processed. In fact, the following combinations were also successfully UHT processed: sodium caseinate + WPC at 4% + 4%, and 5% + 5%.

While whey proteins are unstable to heat at the normal pH of milk, they are stable at low pH (≤3). Hence it is possible to produce shelf-stable, high-protein acidic beverages based on WPI. These have become popular with athletes and health-conscious consumers. They have a shelf-life of up to 6 months at ambient temperature. When these formulations (containing ~7% WPI) are processed at pH ≤3, they remain clear. Processing at 120 °C for 20 s has been reported to be suitable for long term stability (Villumsen *et al.*, 2015a). In particular, these conditions prevent the formation of visible aggregates during storage which are largely caused by the presence of caseinomacropeptide (CMP) in WPI made from cheese whey (Villumsen *et al.*, 2015b). Processing at 120 °C for 20 s causes some hydrolysis and deamidation of the CMP (Le *et al.*, 2016). Another acidic whey protein drink, *Upbeat*, contains 8% protein and is based on microparticulated whey protein (see Section 9.14).

9.8 Breakfast Milk Products

The "breakfast milk" concept provides a convenient drink that has a similar nutrient content to a bowl of breakfast cereal and milk, in other words milk plus fibre in the range 2-3% (see Table 9.5, products a-c). These have been popular in Australia for over 10 years, supplied by Sanitarium, and are now more widespread, and in 2013 by Weetabix in the UK. The main component is usually milk and the difference is provided by the addition of some sugar, fibre and flavour. Fibres include inulin, oat fibre, corn fibre and β-glucan. Except for the added fibre, many of these products are similar to flavoured milk products, although variants, for example, with extra protein and addition of "super fruits" such as blueberries and blackberries are emerging. One observation is that many of these products do not taste like a bowl of cereal and milk.

Breakfast drinks with added cereals and fruit juice are available as short shelf-life products, for example, Nosh drink products are processed by high pressure processing (HPP) (see Section 10.3), but long-life variants using HPP are not yet available.

Alqahtani *et al.* (2014) investigated the consistency of beverages with a variety of fibres incorporated into UHT beverages. A variety of insoluble fibres were used including orange fibre, kibbled wheat fibre and oat fibre, whose particle size distribution ranged over an order of magnitude from 35 to 372 μm. The results showed that the acceptability of UHT beverages is significantly affected by the particle size distribution and level of fibre addition which varies from 0.5 to 5% in the formulations. The colour appearance of UHT beverages is also influenced by the incorporation of insoluble fibre where preparations with wheat fibre appear darker than their counterparts. Results indicate a positive relationship between fibre characteristics and beverage consistency, with increasing concentration or wider size distribution in particles yielding higher steady-shear viscosity. During storage for 12 weeks at 22 °C, the beverages exhibited increasing viscosity at a given concentration and type of fibre, whereas storage at 30 °C resulted in increasing flow, which is an interesting outcome for the commercialisation of the industrial formulations.

9.9 Starch-Based and Thickened Desserts

A wide variety of these products is now available. Waxy maize starches are widely used as the thickening agents. Some product development and pilot plant trials may be needed to select the best starches and their concentrations. Waxy maize starches are rich in amylopectin and several types are available. Again, stabilisers such as locust bean gum, carrageenan, agar, cellulose gum (CMC), cellulose gel (MCC) and xanthan gum need to be added. Sugars and sweetening agents are also used. In the case of milk-based drinks, non-dairy fat such as rapeseed (canola) oil or palm oil may be added. A raising agent such as sodium bicarbonate may be used. Some of these products, referred to as puddings in the Alfa Laval manual (Alfa Laval, undated), may even solidify in the pack. Instructions such as "shake well before use" and "keep chilled after opening" may be included on the package.

UHT custard is a starch-based milk dessert. One example contains thickener E442 (ammonium phosphatides), vegetable gum (carrageenan) and colours E102 (tartazine) and E110 (Sunset yellow). The declared composition of one commercial product is protein 3.0%, fat 3.3% and total carbohydrate 14.2%. Another declares 2.8% protein, 2.6% fat, and "glycaemic carbohydrate". ESL and in-container sterilised custards are also produced.

One manufacturing feature of these products is that they may be packed slightly hotter than for milk, at ~50 °C to avoid excessive increase in viscosity in the cooling section of the UHT plant, especially plate types. In fact, their production is better suited to indirect processing on tubular UHT plants. Normal UHT sterilisation conditions are recommended, for example, 140 °C for 4 s, combined with downstream homogenisation, either single- or double-stage. In some cases homogenisation may be dispensed with but in many cases it results in a smoother texture and glossy appearance.

Product thinning has been known to occur due to residual amylase activity (see Section 7.2.3). The source of the amylase is not always clear but xanthan gum is one ingredient which has been shown to contain amylase. Commercial amylase-free xanthan gums are available for use in starch-thickened products.

UHT savoury sauces, including gravies and spicy variants are also available. Another potential product is a UHT batter, which would be convenient for making Yorkshire puddings, pancakes and other food products. The high level of starch in this product makes its manufacture very challenging. To our knowledge this has not been achieved satisfactorily and requires further development. The product may be more suited as an ESL or short-shelf-life chilled product as these require less heat input during manufacture.

9.10 UHT Cream

Cream products have a much higher fat content and a higher total solids content than milk, but the concentrations of protein, lactose and minerals are slightly lower than in milk and decrease as the fat content increases. Accordingly, cream products do not pose any special problems related to heat stability, provided good quality milk is used for producing the cream. Information on cream production methods is provided in Rothwell (1989). Reviews on UHT cream have been published by Lewis (1989) and IDF (1996).

Typical fat contents for different creams are: coffee cream, 12%; single cream, 18% fat, whipping cream: 35% and double cream: 48% fat. However, the designation of creams and their fat contents vary from country to country. Usually no additives are used and if they are used, they must be declared according to local regulations. The main exception to this is for whipping cream, where additives such as sodium alginate, sodium carboxymethyl cellulose, carrageenan and gelatin are used to increase viscosity and retard fat separation. Also, emulsifiers such as mono- and diglycerides are added to stabilise the emulsion. Many ingredient suppliers market proprietory mixes of stabilisers and emulsifiers designed for UHT whipping cream.

A key process in the production of UHT cream is homogenisation which reduces fat separation during storage. However, incorrect conditions can cause excessive increases in product viscosity, or give rise to fat separation during storage, due to partial coalescence of the fat globules. Homogenisation is usually performed at relatively low pressures, 3.5-4.5 MPa at 65-78 °C (IDF, 1996). Downstream homogenisation is more effective than upstream homogenisation. Because cream has limited protein available for coverage of the smaller fat globules formed during homogenisation, milk proteins such as sodium caseinate or MPC at around 1% are sometimes added to improve stability. Addition of caseinate has also been shown to also reduce feathering when the cream is added to (acidic) coffee by increasing the buffering capacity of the cream.

Calcium is significant in UHT cream. Addition of chelating agents such as trisodium citrate (~0.15%) reduces ionic calcium and reduces feathering. It also increases the alcohol stability of the cream which should be ~80%. However, addition of calcium, as calcium lactate, increases the hardness of the cream whip.

The temperature of storage is an important factor in the stability of cream. While it is desirable for convenience to store the UHT cream at room temperature, storage under refrigeration is preferable for maintaining emulsion stability and is widely practiced. Frequent changes of temperature during storage reduces emulsion stability.

A most important feature is that UHT treatment should not impair the whipping performance of the cream. Unfortunately, some steps which can be taken to improve the stability of the cream during storage adversely affect the whipping characteristics. For example, homogenisation, reduction in calcium and addition of stabilisers which increase viscosity improve stability but reduce whipping capacity. Muir and Kjaerbye (1996) described a simple whipping test to evaluate the functionality of whipping and double cream. They also discuss how whipping time and overrun are affected by seasonal variations of milk.

Aerosol cream products are also popular and are produced by UHT treatment (Smiddy *et al.*, 2009). The usual propellant is nitrous oxide (E942). Other additives are sugar or glucose syrups, emulsifiers, for example, mono- and diglycerides of fatty acids or lactic acid esters of these mono- and diglycerides, and stabilisers such as carrageenan. There is little residual cooked flavour in these products, which is surprising as they have been UHT processed. One feature of aerosol creams is the high overrun of about 400 to 500%, as opposed to about 100% for normal whipped cream. The downside of this high overrun is poor longer term stability of the whipped cream. It collapses in a much shorter time than normal whipped cream and it not suitable as a filling for cakes and pastries. The high overrun may also contribute to their low cooked flavour intensity.

The conditions of separation affect the properties of the cream. While warm separation at 50-55 °C is faster (due to the higher viscosity) and more efficient than cold separation at ~12 °C, cold separation is preferred by some processors because of the lower risk of contamination of the final product with thermophilic organisms which can grow during long separation runs at 50-55 °C. Cold separation at 12 °C produces cream with higher viscosity (and longer whipping time) than does warm (50 °C) separation (Hillbrick *et al.*, 2000). IDF (1996) recommends separation at 55 °C. Importantly, separation and pumping of cream should not occur in the temperature range 15-40 °C in which the fat globules are most sensitive to damage.

One might expect some fat separation to take place during storage. Some does occur but not as much as one might expect, most probably because of the high viscosity of the product. The viscosity depends on numerous factors including fat content, stabilisers added, homogenisation conditions and cream separation conditions. Regretfully, there appears to be no information on optimum levels of viscosity in order to minimise fat separation during processing without unduly affecting whipping properties.

Creams containing vegetable oils in varying proportions are available, as alternatives to cream, although some of these also contain buttermilk and lactose, and thus would not be classified as non-dairy. In addition, a wide range of UHT coffee cream products are available, especially in catering portions of 10-12 mL. Arla have reinvented the tetradedron-shaped individual portions, at a more generous 20 mL size. A feature of a coffee cream is that they should show good stability in hot acidic conditions found for coffee. Such a test, sometimes referred to as a feathering test, is described in Section 11.2.25.2.

9.11 UHT Ice Cream Mix

A considerable amount of ice cream mix is produced by UHT processing. This is a convenient product which can be poured directly into an ice cream freezer. As such it is a popular choice for ice cream vendors, especially in holiday locations.

There are differences in what comprises ice cream in different countries, so national standards should always be consulted. For example, in several countries, ice cream must contain dairy fat; if it contains other oils/fats, it has a different designation such as ice confection. In contrast, in the UK and other countries ice cream can contain either vegetable fat and/or dairy fat. With the demise of the UK national standard for ice cream, an industry trade association (Ice Cream Alliance) introduced its own national quality standard to uphold quality and protect the public from 'artificial' ice cream. This standard states that ice cream should consist of a minimum of 5% fat and 2.5% milk protein. For it to be labelled dairy ice cream it should contain a minimum of 5% fat and contain no other fats than milk fat (Pearman, 2015).

Non-dairy ice cream, which is prevalent in the UK, contains coconut oil, palm oil or palm kernel oil, or combinations of these as the main source of fat. Other ingredients are milk solids-not-fat, which can be sourced from skim milk powder, liquid milk or cream, or from combinations of these. The other main ingredient is sugar or sugar derivatives. Ice cream mixes range from about 25% to >30% total solids. In terms of their gross composition, it is similar to evaporated milk, although most of the sugar is sucrose (or others) rather than lactose. Minor but very important components are stabilisers, flavours and colours.

The fat principally affects the texture, giving it the desired consistency and a smooth-bodied texture. Ice cream with a high fat content melts more slowly than ice cream with a lower fat content. The milk solids-non-fat (MSNF) also contributes to texture and flavour. Addition of these solids is limited by their depression of freezing point, making the ice cream melt at a lower temperature; it is also limited by the lactose content as too much lactose gives ice cream a sandy texture. The protein fraction has water-binding capability which must be also supplemented by addition of stabilisers. UHT treatment increases the water-binding capacity of the proteins, which allows a reduction in stabiliser addition when making a mix for UHT treatment (Alfa Laval, undated).

Stabilisers are added to bind the water phase in order to prevent the formation of large ice crystals when the product is subjected to changes in temperature. This is less of a problem for soft-scoop ice cream mixes. It also contributes to texture and melting characteristics. Commonly used stabilisers are sodium alginate, carragenan, CMC and gelatin. Emulsifiers are added to aid incorporation of air during freezing and improve the stability of the foam. They arrange themselves on the interface between fat, water and air. The traditional emulsifier is egg yolk, but now mono- and diglycerides, and sorbitan esters are used.

The quantities of fat and MSNF should be balanced for optimum consistency. Some suggestions for fat (%)/MSNF (%) combinations are: 10/11.5 to 12; 12/11 to 11.5; 14/10 to 10.5 and 16/9.5 to 10. If MSNF content is too low, it may lead to fat churning when the mix is whipped in the freezer. Sugar or sweeteners may be used. Sucrose is a cheap source of sugar but its addition may be limited because it sharply lowers the freezing point.

Making a good ice cream mix is not as simple as it seems and success will result from making sure that the formulation is properly balanced and that the water is correctly frozen. This balance is important to ensure a good consistency and texture throughout its storage.

A common defect of UHT ice cream mixes, which develops after a period of storage, is described as lumpy or granular, and is due to formation of a weak thixotropic gel. This breaks down once the pack is opened and the mixture is worked, which restores the original texture and allows it to be frozen without any problems (Alfa Laval, undated).

It is a legal requirement in many countries to heat-treat ice cream mix. Pasteurisation is widely used and typical HTST conditions are 79 °C for 15 s, which is slightly more severe than for liquid milk (see Table 2.5). UHT treatment of ice cream mix at one time involved conditions of 149 °C for 2 s. Even excluding heating and cooling periods, this corresponded to an F_0 value of 20.6, a B* value of 4.3 and a C* value of 0.2. If such a product was heated in a UHT indirect plant, arguably the product would be well over-processed, but not in a direct UHT plant. The conditions recommended by Alfa Laval (undated) are 140 °C for 4 s for indirect UHT processing. The approximate values of the parameters would be F_0, 17; B*, 4: and C* 1.5.

Despite their high total solids, ice cream mix formulations are available which are stable to UHT processing. Care should be taken to ensure that any added ingredients do not cause an excessive reduction in pH. The authors are not aware of any studies on the effects of pH and ionic calcium on heat stability of ice cream mix. Ice cream mix is well suited to processing by direct UHT processes, as this produces a mix with a very clean flavour. UHT indirect treatment will produce an ice cream with a slightly cooked flavour, which can easily be masked by addition of caramel and other such flavours, so there is no suggestion that UHT processing results in any unpleasant flavours.

Guidelines for providing the correct homogenisation conditions are provided by Alfa Laval (undated). The homogenisation pressure should range between 95 and 225 bar (9.5 to 22.5 MPa). Table 9.6 shows some data for single-stage homogenisation of ice cream mixes of different fat contents for fat derived from butterfat, cream and a vegetable source. It is apparent that homogenisation conditions need to be optimised for each individual product.

Some commercial products available in the UK are shown in Table 9.7. Most of these companies supply both dairy and non-dairy formulations.

Table 9.6 Recommended homogenisation pressures (bar) for UHT ice-cream mix. (Source: Reproduced with permission of Tetra Pak.)

Fat content (%)	Milk fat	Fat from cream	Vegetable fat
5	210	225	190
7.5	195	210	180
10	155	175	140
12.5	110	120	95

(from Alfa Laval, undated)

Table 9.7 Some commercial UHT ice-cream mixes available in UK and their composition.

	Fat (%)	Sugar (%)	Protein (%)	Stabiliser type	Colour	Total carbohydrate (%)
Kerry Maid	6.8	19.9	2.9	E412, E451, E407, E471		21.1
Jersey Luxury dairy	6.2	16.3	4.1	E401, E407, E410, E412, E466, E471	annatto	18.5
Spring Cool, Lakeland	6.2		4.0			18.8
Comelle	6.5*		3.9			18.8
	6.0**		4.3			19.3

* non-dairy;
** dairy

9.12 Infant Formulae

A small selection of infant formulae on the market is shown in Table 9.8. Companies producing these products usually supply a first infant formula and follow-on formulae. Infant formulae are produced in large quantities in dairy-intensive countries such as Ireland, New Zealand and Australia and currently command a high price of about 40 euros per kg. In recent years, China has become a major market for this product and with the relaxation of the one-child policy, sales of infant formulae in that country are expected to increase considerably.

Infant formulae are mostly provided in powdered form (Montagne *et al.*, 2009). During the production of these powders, heat is used in the evaporation and drying stages, but this is not sufficient to produce a sterile powder. One important criterion is that they do not contain anaerobic spore-forming bacteria, as one company found out to their misfortune when a *Clostridium* (not *C. botulinum*) was found in infant formula containing their whey powder (Lewis, 2013).

Infant formulae should mimic the composition of human milk which is very different to that of cow's milk. One important difference is the ratio of whey protein to casein. In cow's milk it is ~20:80 but in human milk it is 60:40. Therefore infant formulae contain increased amounts of whey protein. This may reduce the heat stability of the formulae. It should be noted that there is also no β-lactoglobulin in human milk so the protein in infant formulae containing unmodified cow's milk whey proteins does not match that of human milk. Also, the overall protein level is also much lower (~1.3%) in human milk than in cow's milk (~3.2%), and the lactose content is much higher (~7% compared with ~5%). The levels of oligosaccharides in human milk, ~0.8%, are higher than those in bovine milk, 30 to 60 mg/L. The regulations governing the essential nutrient compositions of infant formulae were reviewed by Montagne *et al.* (2009). A modification to the English infant and follow-on formula regulation in 2007 allowed for the first time the use of goat's milk protein in the manufacture of infant and follow-on formulae. Strict limits are placed on the levels of toxins; the permitted levels for aflatoxins in infant formulations have been set at <25 parts per trillion.

Recently UHT infant formulae have become available. Advantages of these ready-to-drink formulae are their freedom from vegetative pathogens, their ease of use, guaranteed chemical composition and absence of anaerobic spore-forming bacteria

Table 9.8 Some commercial sterilised and UHT-processed infant formulae.

SMA first infant milk (breast milk substitute) – glass bottles (sterilized) – 100 mL – 11 lines of ingredients: protein 1.3%, fat 3.6%, carbohydrate, 7.3%, calcium 42 mg
SMA follow-on milk – plastic bottle – UHT 250 mL – protein 1.5% (whey 0.9, casein 0.6), fat 3.6%, carbohydrate 7.2%, calcium 50 mg
SMA first infant milk – UHT 250 mL – protein 1.3%, fat 3.6%, carbohydrate 7.3%, calcium 42 mg
Aptamil first milk from birth, with GOS/FOS, ~100 mL – minimal ingredients on bottle (perhaps on sleeve)- contains milk, soya, fish protein
Aptamil toddler milk (Milupa), with GOS/FOS 200 mL – plastic bottle – 1.5 g protein (whey 1 g, casein 0.5 g), fat 2.8%, carbohydrate 8.5% calcium 84 mg, other – choline, taurine

such as clostridia (Lonnerdal & Hernell, 1998). In addition, UHT treatment may confer some nutritional advantages such as improving protein digestibility (Deeth & Lewis, 2016).

However, there are a number of challenges in producing UHT infant formulae. These lie not only in the formulation of the product, but also in being able to account for changes to nutrients during processing and, probably more importantly, during storage. Fortunately, UHT processing is a milder process than in-container sterilisation in producing a commercially sterile product. Montagne *et al.* (2009) reported that direct rather than indirect UHT methods were better suited to production of liquid infant formulae as they minimised fouling during processing, and browning and phase separation during storage. However, analysis of commercial infant formulae suggests the use of quite intense heat treatments in production of some of them.

Maillard browning reactions were investigated in a model infant formulation by Roux *et al.* (2009) during UHT processing. Early Maillard reaction products such as lactulosyl-lysine could be detected (as furosine) from the first seconds of the UHT treatment, suggesting that there was a significant contribution from the heating step. The Maillard reaction continues during storage. In a review of blocked lysine in milk and milk products, Mehta and Deeth (2016) reported furosine value ranges for liquid and powdered infant formulae of 730-1250 and 660-1890 mg/100 g protein, respectively; for comparison, the ranges for direct and indirect UHT milks were 35-170 and 54-300 mg/100 g protein respectively. Correspondingly, the percentage of blocked (that is nutritionally unavailable) lysine in infant formulae is much higher than in UHT milk. Martysiak-Zurowska and Stolyhwo (2007) suggested there should be a maximum allowable furosine level of 700 mg/100 g protein in infant and follow-up formulae.

UHT processing can also be used as a pre-treatment in the production of spray-dried formulations in order to minimise spore counts in the final product. Such a process involves pre-heating to 55 to 73 °C, followed by rapid heating to 133 to 148 °C using direct steam infusion or injection, and then flash cooling.

9.13 UF Permeate

In some situations, milk may be standardized with UF permeate, prior to heat treatment (Hardham, 1998). UF permeate contains all the low-molecular-weight components in milk and about one third of the calcium (~10 mM). However, the composition varies according to whether the starting material is milk/sweet whey or acid whey, and to the temperature (e.g., 10 or 50 °C) of the UF process. Permeates from acid whey and from low-temperature UF processes have higher levels of calcium and phosphorus. Rattray and Jelen (1996) found that addition of UF permeate from milk increased the heat stability of milk while the UF permeate from acid whey decreased heat stability. Hardham (1998) reported that UHT-processed milk with added permeate had a slightly longer storage time before gelation which was attributed to the decreased solids content; UF permeate from milk ultrafiltered at 50 °C extended the gelation time slightly more than 10 °C UF filtrate. The calcium content was the only discernible difference between the two permeates.

When UF permeate is heated alone, it becomes cloudy at a temperature slightly in excess of that of the UF process, due to calcium phosphate precipitation. Kastanas *et al.*

(1995) showed that on a tubular UHT heat exchanger, permeate caused rapid fouling. However, on a UHT plate pilot plant, no fall in the outlet temperature occurred over 20 min, although the pressure drop increased. This is consistent with the report of Lund and Blixby (1975) that pre-fouling of a heat exchanger with calcium phosphate did not affect its performance over a three-hour period, although the OHTC of the pre-fouled heat exchanger was half that of the clean-plate heat exchanger.

9.14 Whey Proteins

Cheese whey contains 20-25% of the protein present in milk. Whey proteins start to denature irreversibly and coagulate at temperatures in excess of ~70 °C. It is not possible to UHT process sweet whey or acid whey, as it results in almost complete blockage of the UHT plant in a very short time period. Barlow *et al.* (1984) also found that whey protein concentrates were unstable to UHT processing, with plant blockage and sedimentation being observed. Stability to UHT processing was improved by addition of 0.25% sodium dihydrogen phosphate, combined with pre-heat treatment at 85 °C for 5 min.

Two types of approach to preventing destabilisation of whey proteins during heating have been investigated: preventing their denaturation and/or aggregation, and rendering them stable to heat. These have been reviewed by Wijayanti *et al.* (2014). Approaches in the first category include use of caseins as chaperone proteins (see Section 9.7), addition of and conjugation with carbohydrates, chemical blocking the free –SH group of β-lactoglobulin, cross-linking with transglutaminase and chelation of minerals. Approaches in the second category include enzymic hydrolysis of the proteins, formation of soluble aggregates, ultrasonication, microencapsulation and microparticulation. Of these, microparticulation has been widely used and commercial microparticulated whey proteins products are available, for example, Simplesse® (CP Kelco) and LeanCreme™ (SPX).

The process of microparticulation involves controlled heating and shearing to promote controlled aggregation of the whey protein. Solutions of these aggregated whey proteins have been used as cream replacers and as the basis of high-protein beverages, as they have enhanced heat stability. One example of such a whey protein beverage is *Upbeat*, which contains 8% protein, although only acidic variants of these products have been so far produced commercially. Microparticulation can also be achieved by extrusion cooking and high-pressure homogenisation (Bansal & Bhandari, 2016).

9.15 Yogurt and Cheese

While many products are produced directly by UHT or in-container sterilization, others benefit, or have the potential to benefit, from such processes. Yogurt and cheese are in this category.

9.15.1 Yogurt

9.15.1.1 Yogurt Produced from UHT Milk

Manufacture of yogurt from UHT milk was reviewed by Krasaekoopt *et al.* (2005). In the conventional manufacture of yogurt, the yogurt milk mix is heated at various temperature–time combinations such as 85 °C for 30 min and 95 °C for 5-10 min in a batch operation. Although such as a treatment destroys pathogenic and most spoilage organisms, the major reason for its use is to increase the viscosity of the final yogurt product (Tamime & Deeth, 1980). The viscosity increase is due to denaturation of whey proteins and their attachment to the casein micelle. Since whey proteins are similarly denatured during UHT processing, continuous heating in a UHT plant has been investigated as an alternative to the batch process. Such a heating process, at a higher temperature for a shorter time than that used conventionally, if successful, would add considerable convenience, cost saving, and quality control to yogurt manufacture. Unfortunately attempts to produce yogurt from UHT-treated milk have not been very successful and hence it has not been adopted by industry. A major reason is that the viscosity of yogurt made from UHT-treated milk is lower than that of yogurt manufactured using the conventional method of heating (Labropoulos *et al.*, 1981a; Mottar *et al.*, 1989). This is illustrated in Figure 9.5 showing the development of viscosity of yogurt during manufacture from conventionally heated and UHT-heated milk (Parnell-Clunies *et al.*, 1988). The reduced viscosity in yogurt from UHT milk has led to the suggestion that UHT processing may be suitable for making drinking yogurt which has lower viscosity than normal yogurt (Labropoulos, 1980).

The effect of the type of heat treatment on the gel strength or firmness of yogurt is similar to that on the viscosity as yogurt made from UHT milk has a weaker gel than conventionally made yogurt (Labropoulos *et al.*, 1981a; Parnell-Clunies *et al.*, 1986a; Mottar *et al.*, 1989; Savello & Dargan, 1995). The firmness of yogurt is related to the cumulative effects of covalent and non-covalent interactions between proteins in the gel matrix (Kinsella, 1984). So the weak gel of yogurt made from UHT milk indicates it has less effective interactions between proteins than those in conventionally made yogurt. This is supported by microstructural studies which show the casein micelles in UHT-treated product exhibit more superficial filaments than the micelles from conventionally heated (90 °C for 10 min) milk with a similar degree of whey protein denaturation. It has been suggested that these filamentous appendages inhibit the fusion of the casein particles into a tight gel network and instead form a loose microstructure which results in low viscosity and gel strength. On the other hand, in yogurt made conventionally, the micelles fuse and form a dense network, which results in high gel strength and viscosity (Mottar *et al.*, 1989).

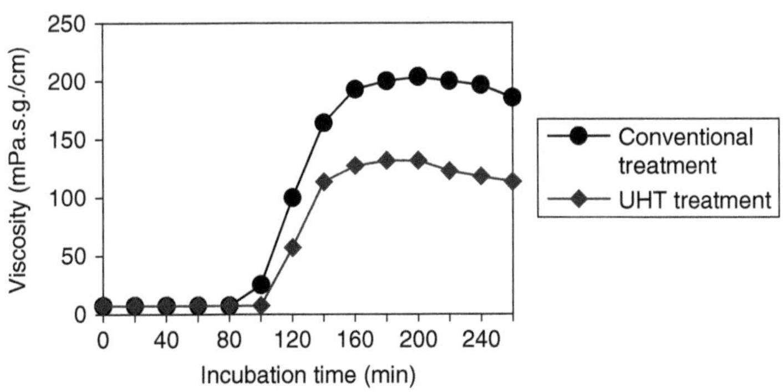

Figure 9.5 Viscosity change during fermentation of yogurt made with conventional heat treatment and UHT. (Source: Krasaekoopt *et al.*, 2003. Reproduced with permission of Dairy Industry Association of Australia.)

Curiously, yogurt made using UHT heat treatment displays less syneresis than conventional yogurt. (Schmidt *et al.*, 1985; Savello & Dargan, 1997). This has been attributed to the greater casein micelle coalescence which results in reduced hydration and increased water expulsion from the tight gel in conventional yogurt. This is consistent with the report by Parnell-Clunies *et al.* (1986b) that yogurt made from UHT-treated milk had higher water holding capacity and protein hydration indices than yogurt made by the conventional method.

A further difference between yogurt made with UHT heating and conventional heating is in the extent of nodulation or graininess. Stirred yogurt made from UHT-treated milk has been shown to have less nodulation than yogurt made from conventionally heated milk. This was observed when the levels of whey protein denaturation in the conventionally heated and UHT milk were similar, indicating that the extent of whey protein denaturation alone is not a major factor in the development of nodules (Weeragul, 2008).

There are conflicting reports on the effects of the heat treatment on the growth of yogurt cultures. Dougherty *et al.* (1980) and Smith *et al.* (1982) reported that the activity of yogurt cultures was significantly increased in UHT and De Brabandere and De Baerdemaeker (1999) showed that UHT treatment led to a shorter lag time before the pH decrease. In contrast, Baril and Cardwell (1984) and Prasad (1990) found no improvement in the performance of yogurt starters in UHT milk.

The firmness of yogurt gels is known to increase with the heat treatment normally applied during yogurt manufacture. This has been attributed to whey protein denaturation and the formation of whey protein–casein complexes through disulfide bonds with kappa-casein. Therefore it is not surprising that researchers have considered that the level of whey protein denaturation caused by UHT and conventional heat treatment may in some way explain the different properties of the yogurt gels produced. However, the reports have been conflicting. Labropoulos *et al.* (1981b) reported that almost complete denaturation of whey proteins occurs by heating milk at 85°C for 30 min but that denaturation never exceeded about 92% in UHT processes. In contrast, similar levels of whey protein denaturation in yogurt from UHT and conventionally heated milks were reported by Mottar *et al.* (1989) and Dargan and Savello (1990). It has been suggested that the lower firmness of yogurt from UHT milks is not related to degree of denaturation but to other differences in protein structure caused by the different heat processes (Labropoulos, 1980; Labropoulos *et al.*, 1981b). This is consistent with the reports that firmness, viscosity, water holding capacity and nodulation of yogurts did not follow the same pattern as the percentage whey protein denaturation (Labropoulos *et al.*, 1981b; Parnell-Clunies *et al.*, 1986b; Savello & Dargan, 1997).

The key to the different effects of conventional batch heating and UHT heating in yogurt manufacture may lie in the different interactions of the denatured whey proteins and casein during the different heating methods rather than in the different levels of whey protein denaturation. Denaturation of whey proteins by slow heating over a long time, as in conventional yogurt manufacture, leads to formation of active β-lactoglobulin monomers and small aggregated whey protein species while more rapid heating to the same denaturation level results in large whey protein aggregates. The active monomers and small aggregates are better able to penetrate the κ-casein hairy layer on the casein micelle and associate with the micelles than the larger whey protein aggregates. Thus, the amount of interaction between whey proteins and casein micelles is greater under slow heating conditions (Oldfield *et al.*, 1998a). The consequence of this appears to be the formation of filamentous appendages consisting of β-Lg attached to the casein micelles in yogurt from UHT treated milk. As a result, the tendency of casein micelles to aggregate is reduced (Dannenberg & Kessler, 1988a,b). This is consistent with the microstructure results of Mottar *et al.* (1989).

A further explanation may relate to the interaction of the different whey proteins with themselves and with casein micelles. During heating, β-lactoglobulin and α-lactalbumin associate with the casein micelle and the nature of this association depends on the severity of heating (Smits & Brouwershaven, 1980). During the least severe heating, β-Lg, but little α-La, attaches to the surface of the casein micelles while with more severe heating, more α-La becomes associated with the micelle surface, partly overlaying the attached β-Lg (Mottar *et al.*, 1989). The ratio of β-Lg to α-La attached to the micelle influences the hydrophilicity of the casein and consequently its physical properties such as water-holding capacity; the lower the ratio, the higher the surface hydrophilicity and the higher the water holding capacity (Mottar *et al.*, 1989).

9.15.1.2 Ambient Yogurt

According to Tetra Pak (2016), ambient, or shelf-stable, yogurt has become immensely popular in recent years. It is produced in a similar manner to conventional yogurt but with three major differences: it contains stabilizers; is heated after fermentation; and is packaged aseptically. The stabilisers are added because the post-manufacture heat treatment reduces the viscosity and causes whey to separate out from the yogurt. Ambient yogurt has a shelf-life of 4-12 months.

Two types of the product are recognized: low viscosity and high viscosity. The former is usually stabilised with starch, gelatin or pectin while the latter uses pectin with or without another stabiliser. The high-viscosity ambient yogurt is heated at 75 °C for 20 s while the low-viscosity yogurt is heated at a higher temperature, between 95 and 110 °C. The low-viscosity yogurt is usually homogenised at 150-200 bar (15-20 MPa) but the high-viscosity yogurt is not normally homogenised as this would reduce its viscosity.

In some jurisdictions, yogurt is defined as containing live yogurt bacteria. Hence the product described here would not be able to be labeled as yogurt.

9.15.2 Cheese made from UHT Milk

There are two major incentives to use UHT milk in cheesemaking. One is that it may increase the yield of cheese by incorporation of whey proteins in the curd which are lost during traditional cheesemaking and the other is destruction of sporeformers such as *Clostridium tyrobutyricum* which cause late fermentation and blowing in cheese (Schreiber, 2001). Unfortunately, high-heat treatment of milk impairs coagulation of milk which makes it unsuitable for cheese manufacture. For this reason, cheese is not generally made from high-heat-treated milk.

In attempts to make cheese from UHT milk, it was found that rennet clotting (RCT), curd formation times (CFT) and cheddaring times were increased significantly, with the effect increasing as the level of denatured whey proteins produced by the high-temperature treatment increased. Curd strength was negatively correlated with milk heating temperatures as was the total amount of whey collected from the curds. As a consequence of the latter, cheeses obtained from milk heated at 135-145 °C had significantly higher moisture contents. The cheese yields, unadjusted for moisture content, were significantly higher for cheeses made from milk heated at 135-145 °C than control cheese made from pasteurised milk; however, moisture-adjusted cheese yields were similar (Amenu & Deeth unpublished data).

In order to address the main problems encountered with the use of UHT milk for making Cheddar cheese, that is, long clotting times and soft curds with high moisture retention, some adjustments which have been reported to alleviate these problems were investigated. They included addition of calcium chloride, pH adjustment and pH cycling. These adjustments were partially successful. The coagulation, cooking and cheddaring times were reduced by pH adjustment to 6.2, addition of 0.02% calcium chloride and by a combination of pH cycling to pH 6.2 or 6.4 and addition of 0.02% calcium chloride. However, the resulting cheeses were dry and crumbly, probably due to excess acid produced during manufacture. Cheese yield appeared to be increased by a combination of pH cycling to pH 6.2 or 6.4 with addition of 0.02% calcium chloride (Amenu & Deeth, unpublished data).

Mechanistically, high-heat treatment is generally agreed to have a minor effect on the first stage of coagulation, hydrolysis of κ-casein by rennet, but it severely impedes the subsequent protein aggregation/coagulation step (Singh *et al.*, 1988). The effect on casein aggregation has been attributed to denaturation of whey proteins leading to formation of whey protein aggregates, complexes between β-lactoglobulin and κ-casein and whey-protein-coated casein micelles (Smits & van Brouwershaven, 1980; Oldfield *et al.*, 1998a,b; Vasbinder & de Kruif, 2003). In addition, heat-induced binding of whey proteins to the milk fat globule membrane (Corredig & Dalgleish, 1996) and changes in calcium phosphate equilibrium (van Hooydonk *et al.*, 1987; Singh *et al.*, 1988; Lucey & Fox, 1993) have been reported to adversely affect the processability of UHT milk for cheesemaking. Schreiber (2001) considered heat-induced changes in the calcium distribution between the micellar and serum phases to be the main factor impairing rennet coagulation. This conclusion was reached from experiments showing that heat treatments at temperatures up to 140 °C caused a decrease in the gel strength of casein dispersions even in the absence of whey protein.

Calcium has an important role in the rennet coagulation of milk and in the structure and buffering of cheese. Addition of ionic calcium reduces the rennet coagulation time of milk due to the neutralization of negatively charged residues on casein, which facilitates aggregation of renneted micelles and also increases gel firmness (Singh *et al.*, 1988). Cheese yield and rheological characteristics are strongly affected by the coagulation process which in turn is controlled by factors such as cheese milk processing temperature, pH and concentration of Ca^{2+}. These factors have been widely studied with pasteurised and high-heat-treated milks in which pH cycling or adding calcium chloride were used to enhance their processability (Singh *et al.*, 1988; Imafidon & Farkye, 1993); however, there is little information available related specifically to UHT milk.

9.16 Milk from Species other than Cows

Cow's milk accounts for 83% of world milk production, followed by buffalo's milk 13%, goat's milk 2%, sheep milk 1% and camel's milk 0.3%. A relatively small amount is produced from other species such as yaks, horses, reindeers and donkeys (FAO, 2016). The composition of the milk of these and other non-commercial milk-producing animals

Table 9.9 Composition (%) of milk from various milk-producing animals.

Species	Water	Fat	Protein	Lactose	Ash
Ass	89.8	1.2	1.7	6.9	0.4
Buffalo	82.5	7.7	4.3	4.7	0.8
Camel	85.6	4.9	3.7	5.1	0.7
Cow	86.9	4.1	3.6	4.7	0.7
Goat	87.0	3.5	3.1	4.6	0.8
Horse	89.0	1.6	2.7	6.1	0.5
Reindeer	70.3	15.5	10.3	2.5	1.4
Sheep	83.7	5.3	5.5	4.6	0.9
Yak	82.7	7.0	5.2	4.6	0.9
Zebra	86.5	4.8	3.0	5.3	0.7

(based on Park & Haenlein, 2006, Uniacke-Lowe *et al.*, 2010)

varies considerably as indicated for some species in Table 9.9 (Park & Haenlein, 2006, Uniacke-Lowe *et al.*, 2010). An indication of the variability can be seen in the ranges for components of milk from various animals:

- Fat: 1.2% (ass) to 41.5% (dolphin);
- Protein: 1.1% (human) to 13.6% (whale and sea lion);
- Lactose: 0.5% (polar bear) to 6.9% (mink);
- Minerals: 0.2% (chimpanzee) to 1.6% (moose);
- Total solids: 9.5% (kangaroo) to 67.7% (gray seal).

Information on the properties of milk from non-bovine species is contained in a Handbook of Milk of Non-bovine Mammals (Park & Haenlein, 2006).

Much less information is available on high-heated treated milk from these other species than there is on high-heat treated cow's milk. Discussed below are the milks of three animals, buffalo, goat and camel, for which some information is available, particularly on UHT-treated milk. While the milk from other species are UHT-treated, for example, sheep milk (http://www.beurrespa.es/en/oveja_en.html; Alvarez 2009) and donkey milk (Astley, 2015), insufficient information is currently available to warrant their inclusion here.

9.16.1 Buffalo's Milk

Worldwide buffalo's milk production in 2012 was 98 million tonnes (FAO, 2016). Buffalo's milk now accounts for about 65% of milk production in Pakistan, 55% in India, 20% in Egypt and 7% in China (IDF, 2015). Buffalo's milk has a higher level of total solids than cow's milk (~16.7% as opposed to 13.3%) and is about twice as rich in milk fat. Divalent cations are about 50% higher than in cow's milk and FPD is in the range -0.552 to -0.558 °C. Compared to cow's milk, there are more large fat globules and a lower urea content (17-22 mM as opposed to 37-40 mM).

UHT processing of buffalo's milk commenced in the 1980s (Pandya & Khan, 2006). UHT milk in India has increased from approximately 30 ML in 2000 to about 230 ML in 2014. This milk would have contained both buffalo's milk and cow's milk since buffalo's milk and cow's milk are not segregated in much of the milk processed in India. Kaushik *et al.* (2015) reported that all mixtures of cow's and buffalo's milk that they tested were stable in 75% ethanol, indicating their suitability for UHT processing. Buffalo's milk has a lower ethanol stability than cow's milk but is still stable to UHT treatment.

Sediment is not considered to be a major problem but oxidative and hydrolytic deterioration of fat are considered to be problems. Singh and Patil (1989a,b) investigated the storage stability of UHT buffalo's milk (6% fat) over 6 weeks at 30 °C. They found a substantial increase in acid degree value, a measure of lipolysis, from 3.75 to 5.87 mmol/100 g fat, an increase in TBA value, a measure of oxidation, from 0.047 to 0.074 µmol/L of malondialdehyde, and an increase in HMF, a measure of the Maillard reaction, from 8.89 to 22.38 µmol/L. The authors reported that UHT buffalo's milk deteriorated faster during storage than cow's milk. However, given the observed lipolysis during storage, it is apparent that heat-resistant bacterial lipases were present, implicating low raw milk quality. Hence the comparison between cow's and buffalo's milk in this case may be misleading.

Malmgren (2015) showed that gelation may be an issue in UHT-treated buffalo's milk, particularly if it is produced by direct heating. Indirect heating at 137 °C for 4 s was shown to be effective in preventing gelation during 6 months of storage while milk treated by direct heating at 142 °C for 6 or 12 s gelled in 5 months. From an analysis of the peptides in pH 4.6 and TCA filtrates (see method in Section 11.2.21.1), the cause of the gelation was determined to be plasmin. For direct heating, two strategies were devised to reduce the risk of gelation: increase the UHT holding time to 24 s or introduce a low-temperature inactivation treatment to inactivate the plasmin (see Section 6.1.3.5.1).

9.16.2 Goat's Milk

UHT goat's milk is available in many countries. It is whiter than cow's milk because of its lack of β-carotene, and may have a distinct goaty flavour. In terms of its major chemical components, goat's milk is similar to cow's milk (see Table 9.9), but there are some interesting differences. It has a lower α_{s1}-casein (about 5.6% compared to 38% in cow's milk) and usually more α_{s2}-casein. It is believed that the lower level of α_{s1}-casein may be responsible for the weak texture of goat's milk yogurt. In contrast, β-casein is higher: 54.8% as opposed to 36%. One consequence of this is that there is more soluble casein in goat's milk (about 25% at 5 °C and 10% at 25 °C, compared to 10% at 4 °C and 1% at 25 °C for cow's milk). This may be one of the factors contributing to the poorer heat stability of goat's milk toward UHT processing. Heat stability is lower than for cow's milk. The heat coagulation time is in the range 0.5 to 23.5 min at 140 °C and heat coagulation temperature (1 min) is in the range 118 to 140 °C (Malmgren, 2015).

The total calcium in goat's milk is similar to that in cow's milk. However, it has a higher content of ionic calcium and a much lower ethanol stability compared to cow's milk (Figure 9.6.), both of which are associated with low heat stability (see Section 6.2.1). This is considered to be due partly to a lower level of citrate than in cow's milk. A low micellar solvation may also contribute to poor heat stability.

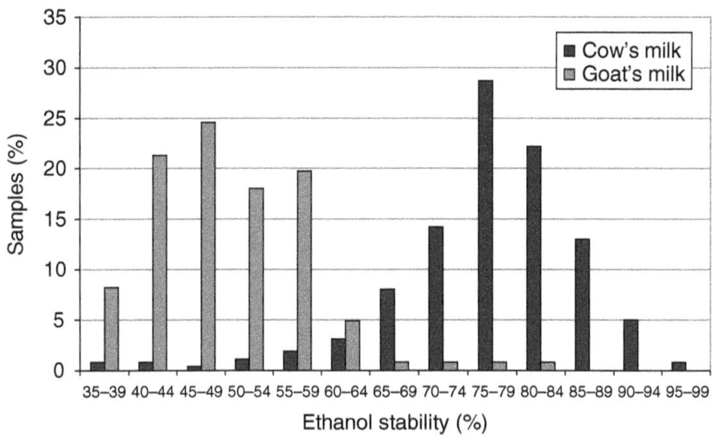

Figure 9.6 Variations in ionic calcium and ethanol stability in milk from individual cows and goats. (Source: Lin, 2002. Reproduced with permission.)

Since goat's milk has poor heat stability, it needs to be stabilised to UHT processing by addition of either TSC or disodium hydrogen phosphate (DSHP). Adding 0.1% DSHP would increase its pH from 6.65 to 6.99 and increase its alcohol stability from below 60% to between 70 and 75 %, the lower end of the desirable range for UHT processing.

Fouling of goat's milk during UHT processing has been shown to be reduced by addition of stabilisers or by removing some of the calcium by ion exchange (Prakash *et al.*, 2007). These processes reduce ionic calcium and increase pH slightly, which contribute to an increase in ethanol stability. Boumpa *et al.* (2008) found three added stabilisers (DSHP, TSC and SHMP) reduced ionic calcium in goat's milk. However at any specific reduced ionic calcium content, the amount of sediment following UHT processing was lowest for DSHP and highest for SHMP. It has also been recently observed that it is possible to add too much stabiliser and that attention should be paid to this (Chen *et al.*, 2012). Addition of stabiliser was also found to make the product much more susceptible to sediment formation during in-container sterilisation.

Our guideline for adding stabiliser for UHT processing is to add sufficient to increase the alcohol stability to about 75%. This procedure can be described as aiming to make goat's milk behave like cow's milk. It is almost the reverse of adding a small amount of calcium chloride to cow's milk to make it behave like goat's milk.

A batch of goat's milk was subjected to direct and indirect treatment in a UHT pilot plant trial after addition of various levels of TSC and DSHP. The results of the trial (Table 9.10) show clear differences between indirect and direct UHT samples for the same milk, as well as the beneficial effects of adding an appropriate amount of stabiliser on the level of sediment formed. Thus, it is possible to produce an acceptable UHT goat milk product, with appropriate additions of stabiliser.

It has not been reported whether UHT goat's milk is more susceptible to age-gelation than UHT cow's milk. One anecdote is that gelation was an issue 10-15 years ago, but this is not so much in evidence now as there is more awareness of milk quality issues and the need to avoid milk with high counts of pyschrotrophic microrganisms. Interestingly, inspection of some cartons of UHT goat's milk which were over 10 years old located at the University of Reading revealed no gelation but the product had a bitter flavour, indicating proteolysis.

Table 9.10 Some data from pilot plant trials on UHT goat's milk, comparing direct and indirect heating.

Stabilizer (%)	Ethanol stability (%)	pH	Ionic calcium (mM)	Sediment (% dwb)
Direct				
Control	60	6.54	1.59	2.31
0.15 TSC	70	6.68	0.97	0.58
0.20 TSC	70	6.69	0.82	0.56
0.25 TSC	70	6.73	0.73	0.56
0.15 DSHP	80	6.68	0.62	0.75
0.20 DSHP	80	6.77	0.46	1.46
0.25 DSHP	80	6.82	0.38	1.76
Indirect				
Control	60	6.53	1.62	4.44
0.15 TSC	80	6.67	0.98	0.60
0.20 TSC	80	6.71	0.82	0.55
0.25 TSC	80	6.74	0.70	0.55
0.15 DSHP	80	6.70	0.72	4.02
0.20 DSHP	80	6.76	0.59	5.44
0.25 DSHP	80	6.83	0.42	5.89

9.16.3 Camel's Milk

Camel's milk with an annual production of ~2.3 MT is a staple commodity for many Middle Eastern and North African countries. Its proximate composition is similar to that of cow's milk with fat of 3.2-5.2%, protein 2.7-4.5% (casein 1.9-2.3%, whey protein 0.7-1.0%), lactose 3.4-5.6% and ash 0.6-0.9%. However, it contains no β-lactoglobulin and has a lower casein level than cow's milk. Of the casein fractions, 65% is β-casein and only 4% is κ-casein (cf ~13% for cow's milk). The freezing point is -0.57 °C to -0.61 °C and it has a slightly better buffering capacity than cow's milk. It contains about 10 times more lactoferrin than cow's milk and has a higher content of iron. Its alcohol stability is reported to be below 60% (Malmgren, 2015), reminiscent of goat's milk (see Section 9.16.2).

The low alcohol stability suggests that it would be unstable to heat and form considerable sediment on UHT treatment, which has been found to be the case. Farah *et al.* (2004) evaluated direct heating (150 °C for 2 s) and indirect heating (138 °C for 4 s) of camel's milk. Excessive sediment was formed and the condition of the milk after 5 weeks storage at ambient temperature was not satisfactory. Malmgren (2015) showed that adding salts or increasing pH reduced sediment and suggested increasing the alcohol stability to about 80% to overcome the sediment problem. In contrast, we have found that sediment formation was not extensive when camel's milk (which had been stored frozen) was UHT processed. Perhaps the freezing process had improved its heat stability. However, the report of Zhao *et al.* (2010) of an alcohol stability of 75% for milk from 10 Alxa bactrian camels suggests that milk from this breed of camel at least would be stable to UHT.

Farah and Atkins (1992) found that the shape of the HCT–pH curve for camel's milk at low temperature was different from those at high temperatures. The milks heated at 130 and 120 °C were very unstable at all pH values and coagulated in 2-3 min. At 100 °C the HCT initially increased with pH, remained constant between pH 6.4 and 6.7 and then increased progressively with increasing pH. Kouniba *et al.* (2005) reported that the heat coagulation time (HCT) in the range 100-130 °C (<2 min) was too short for high-temperature processing and that heat preservation of camel's milk can be done only by pasteurisation. The exact cause of its poor stability at high temperatures has not been determined but the fact that reducing ionic calcium by addition of calcium chelating agents such as EDTA and citrate increases its heat stability (Emletan & Mohammad, 2003) suggests a high ionic calcium may be a cause. It has a lower levels of citrate compared to cow's milk but the level of ionic calcium in camel's milk does not appear to have been reported. The deficiency of κ-casein and β-lactoglobulin in camel's milk might be another cause of the low heat stability. Interestingly, the whey proteins of camel's milk have been shown to be considerably more heat stable than those of cow's milk (Farah, 1986). Clearly, further research is required on the heat stability of camel's milk.

Proteolytic activity was found in UHT camel's milk and was higher in the directly processed than indirectly processed milk (Malmgren, 2015). Proteolysis by both bacterial proteases and plasmin were observed in the pH 4.6 filtrates. In this regard, it appears that camel's milk behaves in a similar manner to cow's milk. Clearly, attention needs to be paid to raw milk quality (to eliminate proteolysis by bacterial proteases) for production of a product with a long shelf-life.

9.17 Non-Dairy Products

A large number of shelf-stable non-dairy products are now available. These are produced by both in-container sterilisation and UHT technologies. Table 9.11 gives a snapshot of some of these products that are available in UK and Australian supermarkets. Many have a best-before date of 9 months. Their composition varies widely as do the additives used in them. Emulsifiers and stabilisers are common ingredients and, more rarely, vitamins and minerals are added. Some are fortified with calcium up to the level of milk while others contain very little calcium. Typical pH and FPD values of the UK products are shown in Table 9.12. In general, the pH values are higher than those found for milk products which could make them more susceptible to browning during storage (see Section 7.2.5). The FPD values show considerable variation between the products which largely reflects the amount of sugar and mineral salts added.

A step in the preparation of some non-dairy milks is the use of enzymes, in much the same way as β-galactosidase is added to milk in the production of lactose-reduced milk (see Section 9.4). For example, papain is added in the preparation of peanut milk to break up the proteins (Rustom *et al.*, 1996b) and amylases (α and β) and glucoamylase are used in production of rice milk to convert starch into sugars (Mitchell, 1988; Ravagnani & Sambataro, 2004).

The term "milk" is commonly used as a product descriptor for many of these products, probably because they are made to resemble and, in many cases, substitute for milk. However, the proteins, fat and carbohydrates are different from those in milk, as are the sensory characteristics. Like milk, they are susceptible to spoilage by micro-organisms,

Table 9.11 The composition of some long-life non-dairy beverages in the UK and Australian markets (figures are g/100 mL unless otherwise specified).

Product	Protein	fat	Carbohydrate	Sugars	Energy kJ/100 mL	Stabilizers & emulsifiers	Fibre	Other additives & comments
UK								
Soy (unsweetened)	3.4	1.9	0.8	0.3	147	Gellan gum	0.6	Tricalcium phosphate, maltodextrin, vitamins B2, D, B12
Soy (sweetened)	3.3	1.9	2.4	2.4	180	Gellan gum,		Tricalcium phosphate Apple juice, salt, Vitamins B2, B12, D2, E
Oat	0.8	1.3		3.4	194	Carrageenan	0	Added iodine, vitamin D, vitamin B12
Coconut	0.2	1.9		2.0	118	Sucrose esters of fatty acids	1.0	Vitamin D, vitamin B12, Ca, 120 mg/100 mL
Rice 1	<0.1	0.8	11.0	3.8*	217	None	0.1	Iodine, vitamins D, B12
Rice 2	0.5	1.6		7.1	331			No calcium
Almond	0.5	1.2		3.3	120	Gellan gum, guar gum, sunflower lecithin	1.0	Ca, 120 mg/100 mL;
Hemp	0.5	2.5	3.4	1.6**	164	Sucrose ester, xanthan gum	<0.1%	Ca, 120 mg/100 mL; 50% omega-3 per glass
Australian								
Soy 1	3.0	1.8	3.0	2.0	169	Carrageenan, microcrystalline cellulose, carboxymethyl cellulose		Potassium phosphates, sodium citrate
Soy 2	3.2	3.5	5.1	2.0	273	Glycerol monostearate		Potassium citrate, sodium and potassium phosphates
Oat 1	2.1	1.8	7.7	3.7	237		0	Sunflower oil

Oat 2	1.0	1.0	11.8	1.5	256		0.9 (β-glucan, 0.6)	Ca, 120 mg/100 mL.
Coconut	0.2	3.4	0.7	0.6	142	Lecithin, gellan, guar, xanthan, carrageenan, starch	0.1	Coconut and sunflower oils; tricalcium phosphate, sodium bicarbonate, Ca, 75 mg/100 mL.
Rice 1	0.6	1.0	10.6	2.8	229		0	Sunflower oil, Ca, 120 mg/100 mL.
Rice 2	0.3	1.2	9.5	5.8	209		<1	Ca, 120 mg/100 mL.
Almond	0.7	1.8	0.3	0.1	84		0.3	Ca, 110 mg/100 mL.
Almond + cashew	0.4	1.2	0.6	0	63	Sunflower lecithin	<0.5	Ca, 80 mg/100 mL.
Almond + coconut	0.5	1.3	1.0	0.1	73	Tapioca starch	<0.2	Ca, 80 mg/100 mL.
Macadamia[1]	0.2	1.8	2.8	0.2	121	Locust bean gum, sodium alginate, guar gum		Monounsaturated fat 1.4 g/100 mL

[1] A native Australian nut; the oil has a high content (~70%) of monounsaturated fatty acids

* total carbohydrate = 11 g;

** total carbohydrate = 3.4 g

Table 9.12 Freezing point depression (FPD) and pH of some non-dairy milk products on the UK market.

Sample	pH	FPD
Soy	7.46	525
Coconut	8.02	335
Almond	7.82	258
Rice	7.69	516
Oat	7.80	450
Hemp	7.76	280
Semi-skim cow's milk	6.81	518

can undergo acid-curdling and are subject to proteolysis. These issues can be largely controlled by heating, either pasteurisation or sterilisation.

UHT processing of these products follows closely, and in most cases is based on, the experience gained in, UHT processing of milk. Interestingly, the issues faced during processing, such as fouling, and during storage, such as sedimentation, gelling and emulsion instability, are similar to those encountered with UHT cow's milk. Sedimentation is an issue with some of these products which Durand *et al.* (2003) concluded was related to particle size. They found, for example, that the particle sizes ($d_{3,2}$) (μm) of oat milk (1.55) and rice milk (2.52) were much higher than those of soy milk (0.42) and cow's milk (0.45). Furthermore, unstable milks (rice and oat) contained a relatively higher proportion of larger particles and clusters of particles. Colloid mill grinding, homogenisation, including high-pressure homogenisation (Valencia-Flores *et al.*, 2013; Bernat *et al.*, 2015), filtration or centrifugation, and addition of stabiliser gums are strategies used to minimise sedimentation problems.

Some of the more common non-dairy products are discussed below; a full discussion of all such products is beyond the scope of this book, as is a discussion of any nutritional or therapeutic claims associated with them. In contrast to the vast amount of information on shelf-stable milk products, the publicly available information on the non-dairy products is somewhat limited.

9.17.1 Soy Milk

Soy milk, which is a traditional drink in the Far East, has increased in popularity in other parts of the world in the last 20 years. UHT processing with aseptic packaging has contributed to its increasing popularity. Both unflavoured and flavoured varieties are available, with chocolate being a popular choice. As for cow's milk, soy milk is UHT processed at between 135 and 150 °C for few seconds, using indirect or direct processing methods.

Soy milk is an aqueous extract of ground whole soy beans or soy flour. The beans are soaked, ground, blanched and the soluble components extracted. The four main extraction techniques are the traditional method, the hot water grind, the Illinois process and the rapid rehydration hydrothermal cooking (RHHC) process (Figure 9.7). These have been compared by Kwok and Niranjan (1995). Soy milk can be made from whole beans

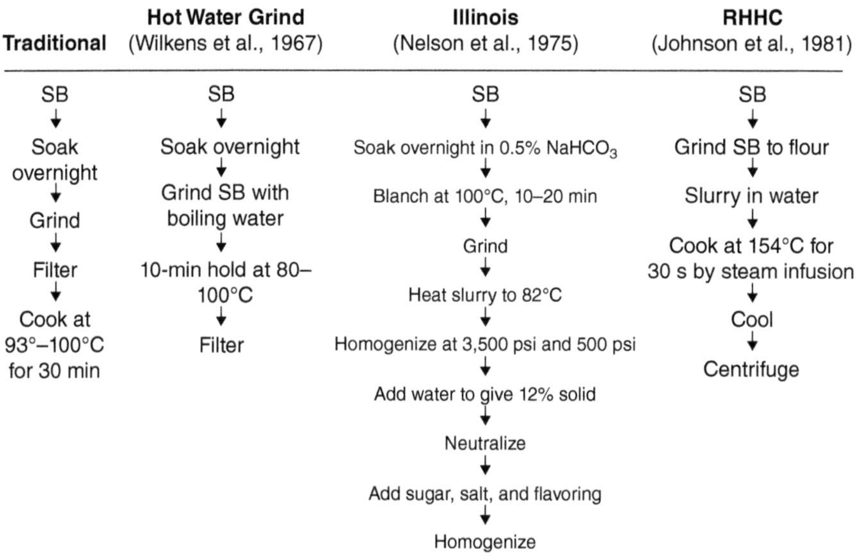

Note: SB = soy beans; RHHC = rapid hydration hydrothermal cooking.

Figure 9.7 Soy milk extraction processes. (Source: Kwok *et al.*, 1995. Reproduced with permission of John Wiley & Sons.)

Table 9.13 Soy milk composition compared to cow's milk composition (from McCance *et al.*, 1991).

Constituent	Soy milk	Cow's milk (whole)
Protein (g/100 mL)	2.9	3.2
Fat (g/100 mL)	1.9	3.9
Carbohydrate (g/100 mL)	0.8	4.8
Calcium (mg/100 mL)	13	115
Magnesium (mg/100 mL)	15	11
Sodium (mg/100 mL)	32	55
Iron (mg/100 mL)	0.40	0.05
Vitamin C (mg/100 mL)	0	1

or from soy flour, concentrate or isolate. Sugar may be added. Similar options exist for the other plant-based products, provided suitable flours, concentrates and isolates are available. Thus these "milks" vary much more widely in composition than bovine and non-bovine milk products. Soy milk is neutral in pH and a typical analysis is shown in Table 9.13 together with the composition of cow's milk for comparison. Soy products can be formulated in a variety of ways. This is best illustrated by the wide range of compositions (see Table 9.11).

Heating is involved at several stages in the production of soy milk, including pre-treatment of the beans, extraction to produce the soy milk, followed by either pasteurisation or sterilisation to increase its shelf-life. The final quality depends on the sum of the effects of all these heat treatments. For UHT products, it is important to inactivate the heat-resistant spores which may be present, including the mesophilic spore-formers (*Cl. botulinum* and *Cl. sporogenes*) and at least some of the thermophilic sporeformers such as *Geobacillus stearothermophilis.* This last organism has been isolated from soy milk and its D-value determined as 2.76 min (121 °C) and 1.39 min (125 °C), with a z-value of 8.36 °C (Chung *et al.*, 1988).

One problem in the early days of soy milk production was a "beany flavour". This arises from the action of lipoxgenase on linoleic and linolenic acids, resulting in the formation of aldehydes, ketones and alcohols. This is most likely to occur at the grinding and extraction stage and can be controlled by appropriate heat treatment at that stage. Lipoxygenase inactivation has been reviewed by Kwok and Niranjan (1995). Another issue is inactivation of trypsin inhibitors which are growth inhibitors present in raw soy beans. The aim is to reduce the original activity by at least 90%, as it should not then interfere with the biological value of the protein. The UHT heating conditions to achieve this are more severe than those normally used for UHT cow's milk and may result in adverse heat-induced changes. Kwok *et al.* (1993) heated soy milk at high temperatures at three different pH values, namely 2, 6.5 and 7.5. At 95, 121 and 132 °C, trypsin inhibitor activity was considerably reduced as the pH increased. However, at 143 °C and 154 °C, the effect of pH on thermal inactivation was less pronounced. Johnson *et al.* (1980a,b) reported that steam infusion heating at 154 °C for 40 s at pH 6.7 resulted in 92.4% reduction in trypsin inhibitor activity.

The kinetic parameters for inactivation of trypsin inhibitors have been reviewed by Kwok and Niranjan (1995). There are at least two fractions, one which is heat-labile and the other heat-resistant. Table 9.14 summarises the data on heat inactivation of trypsin inhibitors in soy milk. A contour plot, showing the effects of time and temperature on trypsin inhibitor inactivation is shown in Figure 9.8.

Table 9.14 Heat inactivation data for soy milk trypsin inhibitors. (Source: Kwok *et al.*, 1995. Reproduced with permission of John Wiley & Sons.)

pH	Temperature Range (°C)	E_a (kJ mol^{-1})	D-value (s)	Reference
6.8	93–121	77.4	4,200 (83°C)	Hackler et al. (1965)
6.7	99–154	48.4	3,600 (99°C)	Johnson et al. (1980)
			280 (121°C)	
			165 (132°C)	
			100 (143°C)	
			40 (154°C)	
6.5	93–154	60.7	3,600 (93°C)	Kwok et al. (1993)
			360 (121°C)	
			150 (132°C)	
			56 (143°C)	
			23 (154°C)	

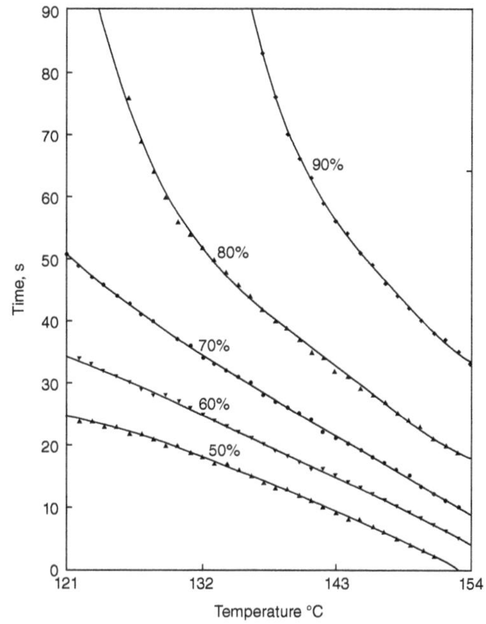

Figure 9.8 Contour changes for soy milk products for inactivation of trypsin inhibitor. (Source: Kwok, 1997. Reproduced with permission.)

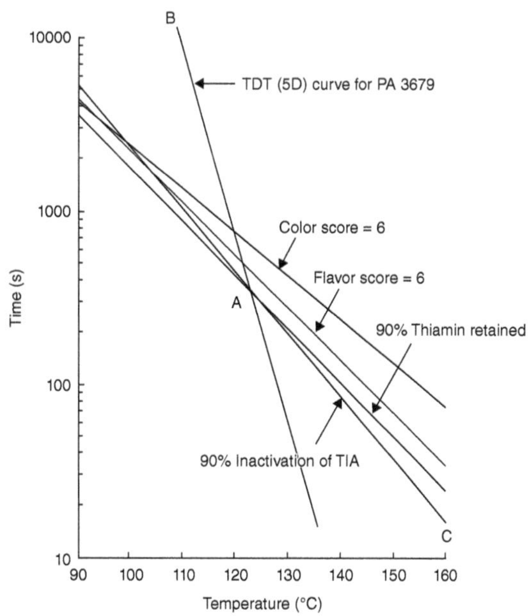

Figure 9.9 Soy milk quality changes with temperature and time of heating. (Source: Kwok, 1997. Reproduced with permission.)

Kwok (1997) presented data for quality changes in heat-treated soy milk (Figure 9.9), which showed that a wide range of conditions can be used to ensure good colour and flavour, over 90% retention of thiamine and 90% inactivation of trypsin inhibitor.

The pH of soy milk can vary from about 6.7 to over 7.5 and tends to increase after UHT treatment (by 0.4 to 0.5). This is in marked contrast to the effect on white milk and flavoured milk. It is not clear what reactions are responsible for this increase in pH.

Most natural soy milk products contain much less calcium than cow's milk and thus many are fortified, using a variety of calcium salts. One problem that this creates is that addition of soluble calcium salts lower pH and increase ionic calcium and both of these changes reduce the heat stability of the product. In fact, soy bean curd (tofu) is manufactured by using calcium chloride and/or magnesium chloride with heat to coagulate the soy protein. Calcium fortification of soy milk and its effects on heat stability and other physical properties have been reported by Pathomrungsiyounggul *et al.* (2007, 2009, 2010) and reviewed by Pathomrungsiyounggul *et al.* (2013).

One interesting property which shows wide variation is freezing point depression. Abdul Rahman (2015) reported that the freezing point depression of a selection of UHT treated soy milk products ranged from <100 m °C to >500 m °C. This is a reflection of the manufacturing process and also whether sugar has also been added. Furthermore, commercial soy milk products in the UK have been shown to have a wide range of pH values, FPD values and sediment levels. Tables 9.15 and 9.16 show some values for products on the UK market (Abdul Rahman, 2015). FPD is useful because it helps to identify whether the product is made from soy flour, isolate or concentrate, especially if it is not sweetened. Other uses for FPD measurement are discussed in Section 11.2.8. Table 9.16 shows the results obtained from dialysates of these soy milk samples, taken at 115 °C for 15 min. As for cow's milk, the pH and ionic calcium of these dialysates are lower than those found in their corresponding milk samples at 20 °C, indicating that both pH and ionic calcium decrease as temperature increases. FPD is lower in dialysates than in milk, suggesting that some of the soluble components may interact with the protein at high temperature. These results should also be compared with those of Pathomrungsiyounggul *et al.* (2012) on soy milk fortified with different calcium salts.

9.17.2 Peanut Milk

Peanut milk and peanut milk products have been known for a long time as nutritious foods. They have been principally produced and consumed in countries where mammalian milk has not been available or has been too expensive, especially for low-income people. However, peanut milk has become more common along with the other plant-based "milks". In India and some developing countries, a combination product made from buffalo's milk and peanut milk called "Miltone" has been developed (Diarra *et al.*, 2005).

Table 9.15 Range of values of some properties for soy milk samples available in UK supermarkets (Abdul Rahman, 2015. Reproduced with permission.)

Property	Pasteurised (n = 4)	UHT treated (n = 6)
pH	7.04–7.75	7.05–7.92
Total calcium (mg/100 mL)	102–109	38–115
Ionic calcium (mM)	0.01–0.35	0.04–0.21
Total solids (g/100 mL)	5.96–8.6	4.3–9.1
Sediment* (g/100 mL)	0.54–0.96	0.2–0.89
Freezing point depression (m °C)	142–470	99–539

* dry weight basis; measured after centrifugation of the sample

Table 9.16 Some commercial soy milk properties which were measured at 20 °C and by dialysis at 115 °C for 30 min. (Source: Abdul Rahman, 2015. Reproduced with permission.)

		pH		Ca^{2+} (mM)		Freezing point depression (m°C)	
Sample		**NT***	**Dialysate****	**NT**	**Dialysate**	**NT**	**Dialysate**
Pasteurised	A	7.62	6.83	0.01	0.01	246	204
	B	7.57	7.11	0.02	0.01	470	378
	C	7.04	6.34	0.07	0.07	293	228
	D	7.75	7.17	0.35	0.04	142	121
UHT	E	7.06	6.37	0.06	0.03	99	69
	F	7.72	7.18	0.04	0.04	145	103
	G	7.85	7.33	0.04	0.03	260	200
	H	7.05	6.62	0.20	0.06	539	440
	I	7.92	6.82	0.21	0.11	355	289
	J	7.15	6.90	0.15	0.07	207	172

* NT = Non-treated, measured at 20 °C.
** Dialysate obtained from soy milk heated at 115 °C for 30 min. Results are means of three replicates.

Peanut milk is produced in several ways but a general process involves the following steps: peanut kernels (deskinned, full-fat or defatted, raw or roasted) are hydrated in 0.5-1.0% sodium bicarbonate for up to 18 h, drained, washed and ground with water (5-10 parts), filtered, and the filtrate homogenised and heated. The heat treatment can be at sterilisation conditions of 121 °C for 10-15 min or sub-sterilisation at 72 to 111 °C. The stability of the product can be improved by adjusting the pH from the normal value of 6.0-6.2 to up to 6.8, which increases the solubility of the proteins, and adding emulsifying agents, for example, Tween 80 and mono- and diglycerides, and stabilisers such as alginates, gelatin or vegetable gums. Sugar (3-6%) may be added to sweeten the product (Bucker *et al.*, 1979; Hinds *et al.*, 1997; Diarra *et al.*, 2005). The composition of an unsweetened peanut milk prepared using a peanut:water ratio of 1:9 and non-defatted peanuts, was as follows: total solids, 9.4%; fat, 4.4%; protein, 2.8%; and carbohydrate, 2% (Bucker *et al.*, 1979).

Hinds *et al.* (1997) compared heat treatments of 72-82 °C for 2 min with 111 °C for 8 min and found the higher temperature treatment increased the viscosity to 17.4 cP compared with 2.1-8.4 cP for the lower-temperature-treated samples. They also found the higher-temperature treatment to impart a chalky mouthfeel and slightly bitter and beany aftertastes. On this basis, the authors concluded that the higher heat treatment would be unsuitable for producing a beverage with acceptable sensory attributes.

Rustom *et al.* (1995, 1996a, 1996b) subjected chocolate- and strawberry-flavoured peanut beverages to UHT processes of 137 °C for 4 and 20 s. The composition of the products was: total solids, 18.0%; fat, 5.2%; protein, 3.0%; and carbohydrate 9.6%. In contrast to the above findings of Hinds *et al.* (1997), these UHT-treated products were not bitter and the products subjected to the most intense heat treatment (137 °C for

20 s) had the lower viscosity (7.7 compared with 12.5 cP). The sample treated for the longer holding time was slightly darker and had slightly more sediment than the one treated for the shorter time. pH decreased and sediment increased in all samples during storage. A decrease in overall acceptability of the beverages was highly correlated with the decrease in pH, increase in sedimentation and reduced emulsion stability. The viscosity of the strawberry sample remained constant during storage whereas the viscosity of the chocolate samples remained constant for 19 weeks and then increased and the samples gelled. The time of gelation was not affected by either the UHT treatment time or the temperature of storage (4, 20 or 35 °C).

9.17.3 Coconut Milk

Several different sterilised liquid coconut products are now available. These include coconut cream, coconut milk, coconut drink and coconut water. The first two are mostly canned and used in cooking while the last two are UHT processed and sold in aseptic packages for direct consumption as drinks (IDF, 1981).

Coconut cream is produced by pressing the flesh of the coconut and filtering to remove most of the fibre. This product typically contains around 30% to 40% fat (Seow & Gwee, 1997; Saikhwan *et al.*, 2015). *Coconut milk* is made by diluting the coconut cream with water with the final product composition depending largely on the dilution factor. Seow and Gwee (1997) reported the composition of samples of coconut milk produced in four countries (Malaysia, Singapore, Thailand and Western Samoa) to be: total solids, 15.6 to 24%; fat, 11 to 18.4 %; protein, 0.3 to 0.9%; and carbohydrate, 3.5 to 8.1%. The pH value (measured on two only) was 6.2 and 6.3; Mepba (2002) reported pH values of 5.4-6.2 for coconut milks. *Coconut drinks* have lower solids contents than coconut milk with a typical composition being: total solids, ≤5%; fat, ≤3.5%; protein, ≤0.3%; and carbohydrate, ≤2%. *Coconut water* is produced from the liquid inside the coconut. It is very low in fat (~0%) and protein (≤0.1%), has ~5% carbohydrate and is relative high in some minerals, particularly potassium (up to 200 mg/100 mL).

Problems arising from sterilisation relate to instability during heating and cooling; instability can manifest as curdling or separation which can be mitigated by selection of appropriate emulsifiers and stabilisers, by (two-stage) homogenisation and pre-heating. For in-container sterilised products, use of agitating retorts reduces stability problems. During UHT processing of these products, heat exchanger fouling is an issue (Seow & Gwee, 1997; Saikhwan *et al.*, 2015). This is attributable to proteins which are denatured in a similar manner to whey proteins in cow's milk. Pre-heating at 90-95 °C is recommended to alleviate this problem. Coconut flesh contains lipases which hydrolyse the lipid component and produce unpleasant odours due to the release of short-chain fatty acids (C4 to C10, butyric to capric) and soapy tastes due to release of longer-chain fatty acids (C12 and C14, lauric and myristic). A pre-heat treatment of 80 °C for 10 min inactivates the lipase. Oxidative deterioration can also occur through the action of the enzymes lipoxidase and polyphenoloxidase; however, sterilisation heating conditions inactivate these enzymes and protect the product from oxidation off-odours and off-flavours (Seow & Gwee, 1997). Sucupira *et al.* (2015) reported that polyphenoloxidase in coconut water was inactivated by UHT processing.

For long-shelf-life coconut cream, milk and drink products, it is desirable for the emulsion to remain stable during storage and to not separate into layers. For this reason, emulsifiers, such as soy lecithin and mono- and diglycerides, and stabilisers, which increase viscosity and reduce the rate of phase separation, are often, but not always, added. A large range of stabilisers are used including carrageenan, alginate, CMC, guar gum, gellan gum, xanthan gum, locust bean gum and modified starch.

9.17.4 Almond Milk

The production of almond milk follows a similar pattern to that of peanut milk. Almond powder in water (4-8%) is mixed at 60-90 °C and ground in a colloid mill. The resulting suspension is filtered (100 µm steel sieve) or centrifuged to remove large particles. An emulsifier such as lecithin and a stabilising hydrocolloid may be added to the liquid phase. The emulsion can then be UHT processed (e.g. at 142 °C for 6 s) with downstream homogenisation during the cooling process (Berger *et al.*, 1997; Valencia-Flores *et al.*, 2013). Valencia-Flores *et al.* (2013) also used ultra-high-pressure homogenisation (300 MPa at 65 or 75 °C) which produced a sterile product (the temperatures reached were 127.7 ± 9.7 and 129.3 ± 12.6 °C, respectively, for ~0.7 s) with a very stable emulsion. The composition of the almond milk with lecithin (0.03%) added was: total solids, 3.4%: fat, 2.0%; protein, 1.2%; and carbohydrate, 0.12%.

9.18 Other Non-Dairy Beverages

9.18.1 Tea and Coffee

Ready-to-drink beverages such as tea and coffee are UHT processed. The pH of black tea is 6.0 to 6.6 but it can be as low as 4.9. The pH of green tea is around 6.7. Coffee is in the range of 5.2 to 6.9 and the feathering test is useful for cream which may be added to coffee (see Section 11.2.25.2). The influence of a UHT treatment on the chemical composition and sensory properties of Arabica coffee brews has been studied by Sopelana *et al.* (2013). Treatment at 120 °C for 2 s produced a microbiologically safe coffee brew with good sensory characteristics. The behavior of the UHT-treated coffee compared to non-UHT-treated coffee was followed throughout 120 days of storage at 4 °C. The UHT treatment retained the typical acidity of the brews longer, delaying and reducing the pH decrease and the development of sourness, which is one of the main causes of the rejection of stored coffee brews. The UHT treatment hardly affects the concentrations of caffeine and trigonelline, and of some phenolic compounds such as 5-caffeoylquinic (5-CQA), caffeic and ferulic acids. Sixteen key odorants and staling volatiles were analyzed by headspace GC–MS and smaller changes were observed in the UHT-treated coffee brew throughout storage. Therefore UHT treatment is a viable process for extending the shelf-life (up to 60 days) of coffee brews. It is also used for green tea, black tea, green tea/soy milk combinations, jasmin, iced lemon and chrysanthemum tea.

Another UHT beverage which is produced commercially is tea with milk. After water, tea is the most widely drunk beverage in the world. It has reputed beneficial health effects based largely on its tannins or polyphenols, namely epigallocatechin gallate (EGCG), epigallocatechin (EGC), epicatechin gallate (ECG), epicatechin (EC), gallocatechin (GC) and catechin. Black tea is drunk either without or with milk. Components of tea have some interesting effects in milk. Extracts of black tea have been shown to increase the heat stability of milk (including concentrated milk) through the catechins binding strongly to the casein micelle and preventing aggregation, and by chelating ionic calcium. They also increase rennet coagulation time and improve alcohol stability (O'Connell & Fox, 2001; Shukla *et al.*, 2009). Tea polyphenols have also been shown to prevent Maillard browning in UHT milk during storage (Schamberger & Labuza, 2007).

9.18.2 Fruit Juices, Purees and Drinks

There has been a large increase in recent years in fruit-juice consumption worldwide. In the UK, some juice, such as apple and blackcurrant, is produced locally but from fruit of Mediterranean and tropical origin, it is usually imported as a frozen concentrate. It is then thawed, diluted back to near its original solids content, pasteurised and packaged. The consumer pays a premium for juice which is designated "not from concentrate".

The key stages in fruit juice technology have been described by Veal (1987) and for soft drink technology by Houghton (1987). In general they involve cleaning and preparation of the raw materials, and juice extraction, usually by pressing, although other extraction processes have been investigated.

Almost all fruit fall into the acidic product category (pH <4.5), with many having a pH below 4.0. Some pH values are given in Table 9.17. Sound fruit quality provides the starting point for the production of high-quality juices, purees, nectars and fruit-based drinks. Most of these are heat-treated, the severity depending on their pH, initial microbial count and whether the juice is fresh or reconstituted.

Continuous heat treatment may be used at several stages in the production process: a) immediately after juice or puree is produced, for example, for apples; flash heating at 82.2 to 85 °C is used to help to coagulate those components which make juice difficult to filter with final pasteurisation at 85 to 90 °C; b) as a pre-treatment to hot-filling, or after cooling for the production of frozen concentrates, or a pre-treatment for evaporation or drying. In addition to heat treatment, preservatives such as benzoate and sorbate may be added to prevent spoilage by yeasts and moulds (Spittstoesser & Churey, 1989).

Table 9.17 pH values of various fruit juices.

Fruit juice	pH
Orange	3.30 to 4.19
Lemon	2.75
Grapefruit	3.65
Lime	2.88
Apple	3.35 to 4.00
Pineapple juice	3.30 to 3.60
Prune juice	3.95 to 3.97
Grape juice	2.92 to 3.53
Cranberry	2.3 to 2.5
Tomatoes (Mexican)	3.96 to 4.75

compiled from various sources

The main reasons for heat treatment are to inactivate the spoilage micro-organisms including yeasts, moulds, and lactic acid and acetic acid bacteria, as well as some of the enzymes that may pose a problem during processing and storage. The main enzymes are those responsible for browning and cloud stability. Salunkhe and Kadam (1995) reviewed the production of juices from a wide variety of fruits. In general, acidic juices are heated within the range 85 °C for 15 s to 95 °C for 2 s. Aseptic packaging is required if the product is to have a shelf life of six months or longer. In contrast to milk products, there are no heat treatment regulations for these to comply with and the conditions used by manufacturers are not usually disclosed.

The following decimal reduction times (D-values) at 75 °C have been reported for some organisms relevant to juices: 0.004 s for *Saccharomyces cerevisiae,* 0.053 s for *Lactobacillus fermentum* and 0.02 s for *Aspergillus niger* spores (Hasselbeck *et al.*, 1992). It can be clearly seen that these organisms would be inactivated by normal juice pasteurisation conditions. Acetic acid bacteria grow well in single-strength juice; temperatures of about 74 °C destroy these major spoilage flora. Although pathogenic bacteria will not grow at the low pH values of fruit juice, they may be present in the raw fruit, so heat treatment will inactivate them. The conditions required for ensuring cloud stability are more severe; it is necessary to use temperatures in the region 85 to 99 °C. The following combinations have been shown to be successful in controlling cloud: 99 °C for 1.75 s; 91 °C for 12.75 s; and 85 to 88 °C for 42.6 s (Nelson & Tressler, 1980). Browning may be a problem for some juices. Most of the browning which takes place during juice production is enzymatic whereas most which takes place after heat treatment and throughout storage is non-enzymatic (see Section 7.2.5). Enzymatic browning can be minimised by controlling heat treatment conditions, by ascorbic acid addition or by use of sulphur dioxide.

Pectin, which is naturally present, stabilises the colloidal material responsible for the cloudy appearance of many juices. If a cloudy juice is required, it may be necessary to inactivate pectic enzymes or add some additional pectin. For the production of clear juices, pectic enzymes are added to the juice and later the resulting clear juice is filtered from the deposited sediment. If juice is pasteurised at only 65 °C, pectinesterases survive and cause the juice to separate into a clear liquid and a sediment. Therefore higher temperatures, up to 90 °C, are required to confer additional stability. Apple juice has been reported to be more affected by higher processing temperatures than orange juice (Hasslebeck *et al.*, 1992).

Production of citrus juice involves extraction and screening. The juice may also be deoiled using a vacuum evaporator which also deaerates the product. For orange juice, deaeration improves flavour stability and nutrient retention and also improves heat transfer in the heat exchanger and filling operations. A certain amount of oil is essential but some needs to be removed, as excessive amounts lead to objectionable flavours. The heating conditions required to prevent sediment formation, by destabilising pectic enzymes, or loss of cloud are 86 - 99 °C for 1 to 40 s. For lemons and limes, cloud stabilisation is achieved by inactivation of pectinesterase activity between 69 and 74 °C; a higher temperature of 77 °C for 30 s followed by immediate cooling by flashing in a vacuum chamber provides an adequate margin of safety.

Some tropical fruit are sold as either juices or purees and may be heated in either tubular or scraped surface heat exchangers (Jagtiani *et al.,* 2012). With mango, heat resistant enzymes are a problem. For puree which is to be frozen, a heat treatment of 90.5 - 93.3 °C for 2 min followed by cooling to 29.4-37.8 °C is recommended. Kalra *et al.*

(1995) reported that mango pulp showed no sensory or microbial spoilage when stored at 18-20 °C for 4 months. This was achieved by pasteurising for 1 min at 88-92 °C with a scraped-surface heat exchanger, aseptically filling into sterilised pouches at 88-92 °C, evacuating, sealing and cooling. Argaiz and Lopez-Malo (1996) showed that pectinesterase was more temperature-resistant in mango and papaya nectars than in purees and that this enzyme required more intensive heat than the destruction of deteriorative microorganisms. Deaeration of mango juice was found to reduce dissolved oxygen from 6.5 to 0.8 ppm and minimise vitamin C loss during processing (Shyu *et al.*, 1996).

Guava puree is processed at 90.5 °C for 1 min, followed by hot filling, using a scraped-surface heat exchanger or at 93 °C for 38 s, followed by aseptic packaging. For papaya puree, gelling is prevented by inactivation of pectinesterase enzyme at 96 °C for 2 min, followed by cooling to below 29 °C. An aseptic puree is described, which is acidified to pH 3.5 by addition of citric acid, followed by heating at 93 °C for 60 s using a scraped surface heat exchanger. Note that its normal pH is 5.1 to 5.3.

Passion fruit has an extremely sensitive flavour, some of which is lost during pasteurisation. It also contains considerable starch, which produces a gelatinous deposit which is prone to foul heat exchangers. Because of these problems, the usual method or preservation is freezing.

Tomatoes are the most abundantly grown fruit in the world and have a moisture content of 94 to 95%. They are processed into a wide range of products, including juices (up to 12% Brix), purees with 12-22% Brix and pastes with greater than 21% but usually not exceeding 31% Brix. The pH of tomatoes is usually 4.3 to 4.5 and in general they are treated as an acidic product. Heat treatment is an integral part of these processes, not only in terms of controlling microbial activity but also in terms of its effects on the various pectin enzymes which control the viscosity of the product. The desirable characteristics of tomatoes for processing are full body, tough to eat (to avoid damage during transportation), high soluble solids content and good colour; flavour is also important for juices, but less so for sauces.

The tomatoes are cleaned and crudely chopped and then subjected to either a hot-break or a cold-break heating process. The hot break involves heating at 95 °C for 5 to 8 min to deactivate the pectic enzymes; this results in a product with better viscosity, but flavour loss is higher than for cold-break heating. The cold-break process involves heating at 65 °C for 10 min and results in a fresh, flavoursome (tomatory) product. The material is then screened while hot to remove skins and seeds, typical screens being 1.5 to 2.5 mm. Hot-break juice would be at 85 to 100 °C and have a pH of about 4.2 and 4-4.5 Brix. At this stage the titratable acidity and serum viscosity may be measured and citric acid and salt may be added.

Much juice is also evaporated. Tomato evaporation may operate almost continuously throughout the growing season. Following evaporation, the concentrate is subject to a final "sterilisation" process of 95 °C for 2-3 min, before being packaged. Direct steam injection processes or tubular heat exchangers are suitable. Bulk aseptic packaging systems using 5 L to 1 metric tonne containers and larger are becoming more widespread (Nelson, 2014).

Some tomato concentrate may be used for the production of tomato-based sauces; these also require a final heat treatment before being packaged. In addition, chopped tomatoes in tomato juice are now available as a retail product in an aseptic carton, with a shelf-life of one year; this product contains 4.2% sugar, 1% protein and 0.5% fibre.

Adams *et al.* (1955) found that addition of pectic enzymes to tomato juice completely inhibited fouling, suggesting that pectin contributes to fouling as well as protein. Tomato juices which had not been heated during the break process and so contained natural pectic enzymes, when concentrated produced a reduced amount of fouling. However, this method of controlling fouling produces a product with a greatly decreased viscosity. Morgan and Carlson (1960) found that temperature was the most important factor which affected fouling of tomato products.

A wide variety of commercial vegetable drinks are available, with some as a UHT product. Popular choices are carrot and beetroot, although both of these are low-acid products. One option is to blend these with tomato juice, thereby keeping the pH low and processing them as acid-products. Long-life products of this type are available with up to 90% tomato juice.

References

Abdulghani, A.H., Ali, M.Y., Prakash, S. & Deeth, H.C. (2015) Sensory evaluation and storage stability of UHT milk fortified with iron, magnesium and zinc. *Dairy Science and Technology* **95**(1), 33–46.

Abdul Rahman, R.H. (2015) *Partitioning and Physicochemical Properties of Calcium-fortified Soymilk at High Temperature.* PhD thesis, University of Reading, UK.

Adams, H.W., Nelson, A.I. & Legault, R.R. (1955) Film deposition on heat exchanger coils. *Food Technology* **9**(7), 354–357.

Alfa Laval (undated) *UHT Treatment - Processing Guidelines.*

Alqahtani, N.K., Ashton, J., Katopo, L, Haque, E, Jones, O.A.H. & Kaspasis, S. (2014) Consistency of UHT beverages enriched with insoluble fibre during storage. *Bioactive Carbohydrate and Dietary Fibre* **4**, 84–92.

Alvarez, J.B. (2009) Method for UHT treatment of milk with a high protein content. Patent Application EP 2092832 A1.

Argaiz, A. & Lopez-Malo, A. (1996) Kinetics of first change on flavour, cooked flavour development and pectinesterase inactivation on mango and papaya nectars and purees. *Fruit Processing* **6**(4) 145, 148–150.

Astley, M (2015) Eurolactis launches world's first carton-packed donkey milk. http://www.dairyreporter.com/Manufacturers/Eurolactis-launches-world-s-first-carton-packed-donkey-milk.

Augustin, M.-A. & Clark, P.T. (1991) Heat stability of recombined concentrated milk: changes in calcium activity and pH on sterilization. *Journal of Dairy Research* **58**, 67–74.

Bansal, N. & Bhandari, B. (2016) Functional milk proteins: production and utilisation – Whey-based ingredients. In: *Advanced Dairy Chemistry.* Volume 2. *Proteins.* (eds P.L.H. McSweeney & J.A. O'Mahony), pp 67–98, Springer, New York.

Baril, T.A. & J.T. Cardwell (1984) The effect of UHT milk on viable cell count and organoleptic qualities of unflavoured yogurt. *Journal of Dairy Science* **67**(Suppl. 1), 227–228.

Barlow, I.E., Hardham, J.F & Zadow, J.G. (1984) Stability of reconstituted whey protein concentrates to UHT processing. *Journal of Food Science* **49**, 32-33, 39.

Benton, A.G. & Aldbery, H.G. (1926) Studies on the stability of evaporated milk during sterilization with reference to hydrogen ion concentration, alcohol test and the addition of specific buffers. *Journal of Biological Chemistry* **68**, 251–263.

Berger. J, Bravay, G & Berger, M. (1997) Method for preparing almond milk preparation and products obtained. US patent 5656321 A.

Bernat, N., Chafer, M., Chiralt, A. & Gonzalez-Martinez, C. (2015) Development of a non-dairy probiotic fermented product based on almond milk and inulin. *Food Science and Technology International* **21**(6), 440–453.

Boumpa, T., Tsioulpas, A., Grandison, A. & Lewis, M. (2008) Effects of phosphate and citrates on sediment formation in UHT goat's milk. *Journal Dairy Research* **75**, 160–166.

Bucker, E.R., Mitchell, J.H. & Johnson, M.G. (1979) Lactic fermentation of peanut milk. *Journal of Food Science* **44**(5), 1534–1538.

Chen B.Y. (2013) *Best Use for Milk and Some Associated Issues*. PhD thesis, University of Reading, UK.

Chen, B.Y., Grandison, A.S & Lewis, M.J. (2012) Comparison of heat stability of goat's milk subjected to UHT and in-container sterilisation. *Journal of Dairy Science* **95**, 1057–1063.

Chen, B.Y., Grandison, A.S. & Lewis, M.J. (2017) Best use for milk- A review I Effect of breed variations on the physico-chemical properties of bovine milk. *International Journal of Dairy Technology.* **70**, 3–15.

Chung, J.B., Lee, K.H., John, H.S. & Kim, S.M. (1988) Studies on thermophilic flat-sour bacteria in soymilk: isolation, identification and determination of, heat resistance. *Korean Journal of Food Science and Technology* **20** (2), 218–224.

CODEX (2010) *CODEX Standard for Evaporated Milk* 281–1971, formerly A-3-1971, Revision 1999. Amendment 2010.

Conick, H. (2016) Dean Foods releases new calcium-enriched chocolate milk. *Dairy Reporter* 19 January2016. http://www.dairyreporter.com/Markets/How-to-make-more-from-milk-fortify-it.

Coorevits, A., De Jonghe, V., Vandroemme, J., Reekmans, R., Heyrman, J., Messens, W., De Vos, P. & Heyndrickx, M. (2008) Comparative analysis of the diversity of aerobic spore-forming bacteria in raw milk from organic and conventional dairy farms. *Systematic and Applied Microbiology* **31**(2), 126–140.

Corredig, M. & Dalgleish D.G. (1996) Effect of different heat treatments on the strong binding interactions between whey proteins and milk fat globules in whole milk. *Journal of Dairy Research* **63**, 441–449.

Corzo, N., Delgado, T., Troyano, E. & Olano, A. (1994) Ratio of lactulose to furosine as indicator of quality of commercial milks. *Journal of Food Protection* **57**(8), 737–739.

Crittenden., R., Buckley., J., Cameron-Smith., D., Brown., A., Thomas., K., Davey., S. & Hobman., P. (2009) Functional dairy protein supplements for elite athletes. *Australian Journal of Dairy Technology* **64**, 133–137.

Cronshaw, H.B. (1947) *Dairy Information*. Dairy Industries Ltd, London.

Crowley, S.V., Kelly, A.L. and O'Mahony, J.A. (2014a) Fortification of reconstituted skim milk powder with different calcium salts: Impact of physicochemical changes on stability to processing. *International Journal of Dairy Technology* **67**, 474–482.

Crowley, S.V., Megemont, M., Gazi, I., Kelly, A.L., Huppertz, T. & O'Mahony, J. A. (2014b) Heat stability of reconstituted milk protein concentrate powders. *International Dairy Journal* **37** 104–110.

Dannenberg, F. & Kessler, H.G. (1988a) Effect of denaturation of β-lactoglobulin on texture properties of set-style nonfat yoghurt. 1. Syneresis. *Milchwissenschaft* **43**, 632–635.

Dannenberg, F. & Kessler, H.G. (1988b) Effect of denaturation of β-lactoglobulin on texture properties of set-style nonfat yoghurt. 2. Firmness and flow properties. *Milchwissenschaft* **43**, 700–704.
Dargan, R.A. & Savello, P.A. (1990) Yogurt with improved physical properties from ultrafiltered and ultra-high temperature treated skim milk. *Journal of Dairy Science* **73**(Suppl. 1), 94.
De Brabandere, A.G. & De Baerdemaeker, J.G. (1999) Effects of process conditions on the pH development during yoghurt fermentation. *Journal of Food Engineering* **41**, 221–227.
Deeth, H.C. & Lewis, M.J. (2014) Practical consequences of calcium addition to and removal from milk and milk products. *International Journal of Dairy Technology* **68**, 1–10.
Deeth, H.C. & Lewis, M.J. (2016) Protein stability in sterilised milk and milk products. In: *Advanced Dairy Chemistry.* Volume 2. *Proteins.* (eds P.L.H. McSweeney & J.A. O'Mahony) pp 247–286, Springer, New York.
Deysher, E.F. & Webb, B.H. (1952) Factors that affect the formation of a crystalline deposit in evaporated milk. *Journal of Dairy Science* **35**, 106–115.
DIAL (2014) UHT and thermal processing – Results of trial performed in association with a Dairy Innovation Australia Ltd (DIAL) workshop. May 21-22, 2014, Werribee, Australia.
Diarra, K., Nong, Z.G. & Jie, C. (2005) Peanut milk and peanut milk based products production: A review. *Critical Reviews in Food Science and Nutrition* **45**(5), 405–423.
Dougherty, S.L., Feldstein, F.J., Ousley, R.S. & Westhoff, D.C. (1980) The behaviour of lactic starter cultures in ultra-high temperature processed milk. *Journal of Dairy Science* **62**(Suppl. 1), 67.
DSM (2015) Understanding consumer preference on sugar reduction: flavoured milk. [http://www.dairyreporter.com/Product-innovations/DSM-REPORT-POTENTIAL-FOR-FLAVOURED-MILK/(result)/go].
Durand, A., Franks, G.V. & Hosken, R.W. (2003) Particle sizes and stability of UHT bovine, cereal and grain milks. *Food Hydrocolloids* **17**(5), 671–678.
Ellis, A., Mittal, V. & Sugiarto, M. (2012) Innovation in iron fortification. Is the future in iron-binding milk proteins. In: Innovation in Healthy and Functional Foods. (eds D. Ghosh, S. Das, D. & Bagshi, R.B. Smarta) pp. 249–268, CRC Press, Boca Raton.
Emletan, A.M. & Mohammad, K.S. (2003) Heat stability of camel milk. *Egyptian Journal of Dairy Science* **31**(1), 81–86.
Ezeh, V.N. & Lewis, M.J. (2011) Milk reversibility following reduction and restoration of pH. *International Journal of Dairy Technology* **64**, 179–187.
Faka, M. (2008) *Heat Stability of Skim Milk Powder.* PhD thesis, University of Reading, UK.
Faka, M., Lewis, M.J., Grandison, A.S. & Deeth, H.C. (2009) The effect of free Ca^{2+} on the heat stability of low-heat skim milk powder. *International Dairy Journal* **19**, 386–392.
Farah, Z. (1986) Effect of heat-treatment on whey proteins of camel milk. *Milchwissenschaft* **41**(12), 763–765.
Farah, Z. & Atkins, D (1992) Heat coagulation of camel milk. *Journal of Dairy Research* **59**, 229–231.
Farah, Z., Eberhard, P., Meyer, J., Rehberger, B., Thomet, A. & Gallmann, P.U. (2004) UHT processing of camel milk. UBISI symposium (IDF) Cape Town 1.-5. March 2004 Poster (http://www.db-alp.admin.ch/de/publikationen/docs/pub_FarahZ_2004_15718.pdf).
FAO (2016) Dairy production and products: Dairy animals [http://www.fao.org/agriculture/dairy-gateway/milk-production/dairy-animals/en/#.VpIeLvl96M8].

Geyer, S. & Kessler H.G. (1989a) Influence of individual milk constituents on coffee cream feathering in hot coffee. *Milchwissenschaft* **44**(5), 284–288.

Geyer, S. & Kessler H.G. (1989b) Effect of manufacturing methods on the stability to feathering of homogenized UHT coffee cream. *Milchwissenschaft* **44**(7), 423–427.

Hackler, L., Van Buren, J.P., Steinkraus, K.H., El Rawi, I. & Hand, D.B. (1965) Effect of heat treatment on nutritive value of soymilk protein fed to weanling rats. *Journal of Food Science* **30**(4), 723–728.

Hardham, J.F. (1998) Effect of protein standardisation of milk by addition of UF milk permeate on the composition and storage stability of UHT processed milk. *Australian Journal of Dairy Technology* **53**(1), 22–27.

Harstad, O.M. & Steinshamn, H. (2010) Cows' diet and milk composition. In: *Improving the Safety and Quality of Milk,* Volume 1*: Milk Production and Processing.* (ed. M.W. Griffiths), pp. 223–245, Woodhead Publishing Ltd, Cambridge.

Hasselback, U., Ruholl, T & Knorr, D. (1992) Fruit juice pasteurisation under reduced thermal load. *Fluessiges-Obst* **59**(10), 592–593.

Hillbrick, G., Shen, Z., Aisworth, S., Poivesan, D. & Augustin, M.A. (2000) Influence of cream separation conditions on the physical properties of creams. *Australian Journal of Dairy Technology* **55**, 84.

Hickey, M. (2009) Current legislation on concentrated and dried milk products. In: *Dairy Powders and Concentrated Products.* (ed A.Y. Tamime), Wiley-Blackwell, Chichester, UK.

Hinds. M.J., Beuchat, L.R. & Chinnan, M.S. (1997) Properties of a thermal processed beverage prepared from roasted partially defatted peanuts, *International Journal of Food Science and Technology* **32**, 203–211.

Houghton, H.W. (1987) How soft drinks have benefited from new technology. *Food Technology International Europe* 139–141.

IDF (1981) *The Composition of Ewe's and Goat's Milk.* IDF Bulletin 130, International Dairy Federation, Brussels.

IDF (1996) *UHT Cream.* IDF Bulletin 315, pp. 3-34. International Dairy Federation, Brussels.

IDF (2015) https://dairyindustrynewsletter.co.uk/wp-content/uploads/World-milk-production-IDF-2015.pdf.

Imafidon, G.I. & Farkye, N.Y. (1993) Rennet coagulability of high-heat treated milk influenced by time of pH adjustment. *Journal of Food Science* **58**, 1300–1302.

Jagtiani, J., Chan, H. & Sakai, W. (2012) *Tropical Fruit Processing.* Academic Press, San Diego.

Jansson, T., Jensen, H.B., Sundekilde, U.K., Clausen, M. R., Eggers, N., Larsen, L.B., Ray, C., Andersen, H.J. & Bertram, H.C. (2014) Chemical and proteolysis-derived changes during long-term storage of lactose-hydrolyzed ultrahigh-temperature (UHT) milk. *Journal of Agricultural and Food Chemistry* **62**(46), 11270–11278.

Johnson, L.A., Hoover, W.J., Deyoe, C.W. & Erickson, L.E. (1980) Modeling the kinetics of heat inactivation of trypsin inhibitors during steam-infusion cooking of soymilk. *Transactions of ASAE* **3**, 1326–1329.

Johnson, L.A., Deyoe, C.W., Hoover, W.J. Schwenke, L.R. (1980) Inactivation of trypsin inhibitors in aqueous soybean extracts by direct steam infusion. *Cereal Chemistry* **57**(6), 376–379.

Kalra, S.K.,Tandon, D.K. & Singh, B.P. (1995) Mango. In: *Handbook of Fruit Science and Technology,* Marcel Dekker, New York.

Karp, J.R., Johnston, J.D., Tecklenburg, S., Mickleborough, T.D., Fly, A.D. & Stager, J.M. (2006) Chocolate milk as a post-exercise recovery aid. *International Journal of Sport Nutrition and Exercise Metabolism* **16**(1), 78–91.

Kastanas, P. (1996) *Studies on the Fouling of Heat Exchangers during the Sterilisation of Cows' and Goats' Milks*. PhD thesis, the University of Reading, UK.

Kastanas, P., Lewis, M.J. & Grandison, A.S. (1995) Design and development of a miniature UHT plant for fouling studies and its evaluation using milk adjusted to different pH values, goats' milk and milk permeate. *Transactions Institute of Chemical Engineers* **73**, Part C, 83–92.

Kaushik, R., Sachdeva, B., Arora, S. & Gupta, C. (2015) Effect of fat content on sensory and physico-chemical properties of laboratory-pasteurised calcium- and vitamin D-fortified mixture of cow and buffalo milk. *International Journal of Dairy Technology* **68**, 135–143.

Kelly, P.M. (2006) Innovation in milk powder technology. *International Journal of Dairy Technology* **59**, 70–75.

Kinsella, H. (1984) Milk proteins, physio-chemical and functional properties. *CRC Critical Reviews in Food Science and Nutrition* **21**(3), 197–263.

Kouniba, A., Berrada, M., Zahar, M. & Bengoumi, M. (2005) Composition and heat stability of Moroccan camel milk. *Journal of Camel Practice and Research* **12**(2), 105–110.

Krasaekoopt, W., Bhandari, B. & Deeth, H.C. (2003) Yogurt from UHT milk: A review *Australian Journal of Dairy Technology* **58**, 26–29.

Kwok K.C. & Niranjan, K. (1995) Review: Effect of thermal processing on soymilk. *International Journal of Food Science and Technology* **30**, 263–295.

Kwok, K.C., Qin W H & Tsang J C (1993) Heat inactivation of trypsin inhibitors in soymilk at ultra high temperatures. *Journal Food Science* **58**(4), 859–862.

Kwok, K.C. (1997) *Reaction Kinetics of Heat-induced Changes in Soymilk*, PhD thesis, The University of Reading, UK.

Labropoulos, A.E. (1980) Process variables in a UHT system and their effects on psychophysics and rheology of yoghurt. *Dissertation Abstracts International* **41**(3), 887B–888B.

Labropoulos, A.E., Lopez, A. & Palmer, J.K. (1981a) Apparent viscosity of milk and cultured yoghurt thermally treated by UHT and vat systems. *Journal of Food Protection* **44**, 847–846.

Labropoulos, A.E., Palmer, J.K. & Lopez, A. (1981b) Whey protein denaturation of UHT processed milk and its effect on rheology of yoghurt. *Journal of Texture Studies* **12**, 365–374.

Lacroix, M., Bos, C., Leonil, J., Airinei, G., Luengo, C., Dare, S., Benamouzig, R., Fouillet, H., Fauquant, J., Tome, D. & Gaudichon, C. (2006) Compared with casein or total milk protein, digestion of soluble proteins is too rapid to sustain the anabolic postprandial amino acid requirement. *American Journal of Clinical Nutrition* **84**, 1070–1079.

Le, T.T., Nielsen, S.D., Villumsen, N.S., Kristiansen, G.H., Nielsen, L.R., Nielsen, S.B., Hammershøj, M. & Larsen, L.B. (2016) Using proteomics to characterise storage-induced aggregates in acidic whey protein isolate drinks. *International Dairy Journal* **60**, 39–46.

Lewis, M. (2013) The mythical *Clostridium botulinum*. *Food Management Today* Nov/Dec, 12–17.

Lewis, M.J. (1989) UHT cream processing. In: *Cream Processing Manual.* (ed J. Rothwell) Society of Dairy Technology, Huntingdon, England.

Li, X.E., Lopetcharat, K. & Drake, M.A. (2015) Parents' and children's acceptance of skim chocolate milks sweetened by monk fruit and stevia leaf extracts. *Journal of Food Science* **80**(5), S1083–S1092.

Lima, L.J.R., Kamphuis, H.J., Nout, M.J.R. & Zwietering, M.H. (2011) Microbiota of cocoa powder with particular reference to aerobic thermoresistant spore-formers. *Food Microbiology* **28**(3), 573–582.

Lin, M.-J. (2002) *Measurement of Ionic Calcium and its Role in Milk Stability.* PhD thesis, University of Reading, UK.

Lin,M.-J., Grandison, A.S. & Lewis, M.J. (2015) Partitioning of calcium and magnesium (total divalent cations) during membrane filtration of milk. *Journal of Food Engineering* **149**, 153–158.

Lonnerdal, B. & Hernell, O. (1998) Effects of feeding ultrahigh-temperature (UHT)-treated infant formula with different protein concentrations or powdered formula, as compared with breast-feeding, on plasma amino acids, hematology, and trace element status. *American. Journal. Clinical. Nutrition.* **68**, 350–356.

Lucey, J.A. & Fox, P.F. (1993) Importance of calcium and phosphate in cheese manufacture - a review. *Journal of Dairy Science* **76**, 1714–1724.

Luhovyy, B. L., Akhavan, T. & Anderson, H. (2007) Whey proteins in the regulation of food intake and satiety. *Journal of the American College of Nutrition* **26**, 704S–712S.

Lund, D.B. & Blixby, D. (1974) Fouling of heat exchangers by biological fluids. *Proceedings of the International Congress of Food Science and Technology* Vol. **IV**, 434–442.

Malmgren, B. (2015) Challenges in processing of goat, buffalo and camel milk for longer shelf life. Presentation to 7th International Symposium on Sheep, Goat and other Non-cow Milk. Limassol, Cyprus, March, 2015.

Mansfield, E. (2009) Chocolate milk: an ideal choice for rehydration and recovery post exercise *Canadian Journal of Dietetic Practice and Research* **70**(3), 109–109.

Martysiak-Zurowska D. & Stolyhwo A. (2007) Content of furosine in infant formulae and follow-on formulae. *Polish Journal of Food & Nutrition Sciences* **57**(2), 185–190.

McCance, R.A., Widdowson, E.M. & Holland, B (1991) McCance and Widdowson's, The Composition of Foods (5th Edition), 1991, Royal Society of Chemistry, MAFF, Cambridge, UK.

Mehta B.M. & Deeth H.C. (2016) Blocked lysine in dairy products – formation, occurrence, analysis and nutritional implications. *Comprehensive Reviews in Food Science and Food Safety* **15**, 206–218.

Mepba, H.D. (2002) Effects of processing methods and storage on the qualities of coconut milk. *Discovery and Innovation* **14**(3-4), 179–185.

Mitchell, C.R. (1988) Nutritional rice milk production. United States Patent 4744992.

Montagne, D-H., Van Dael, P. Skanderby, M. & Hugelshofer, W (2009) Infant formulae - powders and liquids. In: *Dairy Powders and Concentrated Products,* (ed A.Y. Tamime), pp. 294 -331. Wiley-Blackwell, Chichester, UK.

Mottar, J., Bassier, A., Joniau. & Baert, J. (1989) Effect of heat-induced association of whey proteins and casein micelles on yoghurt texture. *Journal of Dairy Science* **72**, 2247–2256.

Montilla, A., Calvo, M. M., Santa Maria, G., Corzo, N. & Olano, A. (1996) Correlation between lactulose and furosine in UHT-heated milk. *Journal of Food Protection* **59**(10), 1061–1064.

Morgan, A.I. and Carlson, R.A. (1960) Fouling inside heat exchanger tubes. *Food Technology*, **14**, 594–596.

Muir, D.D. & Kjaerbye, H. (1996) Quality aspect of UHT cream. In: *UHT Cream*, IDF Bulletin, 315, International Dairy Federation, Brussels.

Nelson, P. E. (2014) From tomato king to world food prize laureate. *Annual Review of Food Science and Technology* **5**, 1–20.

Nelson, P.E. & Tressler, D.K. (1980) *Fruit and Vegetable Juice Processing Technology*, AVI Publishers, Westport, Connecticut.

Newstead, D. (1994) Observations on UHT plant fouling by recombined milk, AP270-1, March, 1–10

Nieuwenhuijse, J.A., Timmermans, W. & Walstra, P. (1988) Calcium and phosphate partitions during the manufacture of sterilised milk and their relations to the heat stability. *Netherlands Milk and Dairy Journal* **42**, 387–421.

O'Kennedy, B.T. & Mounsey, J.S. (2006) Control of heat-induced aggregation of whey proteins using casein. *Journal of Agricultural & Food Chemistry* **54**(15), 5637–5642.

O'Connell, J.E. & Fox, P.F. (2001) Significance and applications of phenolic compounds in the production and quality of milk and dairy products: a review. *International Dairy Journal* **11**, 103–120.

Oldfield, D.J., Singh, H. & Taylor, M.W. (1998a) Association of β-lactoglobulin and α-lactalbumin with the casein micelles in skim milk heated in an ultra-high temperature plant. *International Dairy Journal* **8**, 765–770.

Oldfield, D.J., Singh, H., Taylor, M.W. & Pearce, K.N. (1998b) Kinetics of denaturation and aggregation of whey proteins in skim milk heated in an ultra-high temperature (UHT) pilot plant. *International Dairy Journal* **8**, 311–318.

Omoarukhe, E.D., On-Nom, N., Grandison, A.S. & Lewis. M.J. (2010) Effects of different calcium salts on properties of milk related to heat stability. *International Journal of Dairy Technology* **63**, 504–511.

On-Nom, N. (2012) *Partitioning of Milk at High Temperature*, PhD thesis, University of Reading, UK.

On-Nom, N., Grandison, A.S. & Lewis, M.J. (2012) Heat stability of milk supplemented with calcium chloride. *Journal of Dairy Science* **95**, 1623–1631.

Oomes, S.J.C.M., van Zuijlen, A.C.M., Hehenkamp, J.O., Witsenboer, H., van der Vossen, J.M.B.M. & Brul, S. (2007) The characterisation of *Bacillus* spores occurring in the manufacturing of (low acid) canned products. *International Journal of Food Microbiology* **120**(1-2), 85–94.

Pandya, A. J. & Khan, M. M. H. (2006) Buffalo milk utlilization for dairy products. In: *Handbook of Milk of Non-Bovine Mammals*. (eds Y.W. Park and & G.F.W. Haenlein), pp. 215–256, Blackwell Publishing, Iowa, USA.

Park, Y.W. & Haenlein G F W (eds) (2006) *Handbook of Milk of Non-Bovine Mammals*. Blackwell Publishing, Iowa, USA.

Parnell-Clunies, E., Kakuda, Y., deMan, J.M., & Cazzola, F. (1988) Gelation profiles of yoghurt as affected by heat treatment of milk. *Journal of Dairy Science* **71**, 582–588.

Parnell-Clunies, E.M., Kakuda, Y. & Deman, J.M. (1986a) Influence of heat treatment of milk on the flow properties of yoghurt. *Journal of Food Science* **51**, 1459–1462.

Parnell-Clunies, E.M., Kakuda, Y., Mullen, K., Arnott, D.R. & deMan, J.M. (1986b) Physical properties of yoghurt, A comparison of vat versus continuous heating systems of milk. *Journal of Dairy Science* **69**, 2593–2603.

Pathomrungsiyounggul, P., Grandison, A.S. & Lewis, M.J. (2007) Effects of calcium chloride and sodium hexametaphosphate on certain chemical and physical properties of soymilk. *Journal of Food Science* **72**, E428–E434.

Pathomrungsiyounggul, P., Grandison, A.S. & Lewis, M.J. (2010) Effect of calcium carbonate, calcium citrate, tricalcium phosphate, calcium gluconate and calcium lactate on certain properties of soymilk. *International Journal of Food Science and Technology* **45**(11), 2234–2240.

Pathomrungsiyounggul, P., Grandison, A.S. & Lewis, M.J. (2012) Feasibility of using dialysis for determining calcium ion concentration and pH in calcium-fortified soymilk at high temperature. *Journal of Food Science* **77**, E10–E16.

Pathomrungsiyounggul, P., Lewis, M.J. & Grandison, A.S. (2009) Effects of calcium chelating agents and pasteurisation on certain properties of calcium-fortified soy milk. *Food Chemistry* **118**, 808–814.

Pathomrungsiyounggul, P., Lewis, M.J. & Grandison, A.S. (2013) Calcium-fortified soymilk. In: *Handbook of Food fortification and Health; from Concepts to Public Health Applications* Volume 1 *Nutrition and Health*, (eds V.R Preedy, R. Srirajaskanthan & V.B. Patel), Springer Science and Business Media, New York.

Pearman, P. (2014) The ice cream compositional requirements laid down by the Ice Cream Alliance. http://www.ice-cream.org/sites/default/files/uploads/news/code_of_practice.pdf (accessed Jan 2016).

Philippe, M., Gaucheron F, Le Graet Y, Michel F & Garem A (2003) Physicochemical characterization of calcium-supplemented skim milk. *Lait* **83**, 45–59.

Prakash, S., Datta, N., Lewis M. J. & Deeth H.C. (2007) Reducing fouling during UHT treatment of goat's milk *Milchwissenschaft* **62**, 16–19.

Prakash, S, Huppertz, T., Kravchuk, O. & Deeth, H.C. (2010) Ultra-high temperature processing of chocolate-flavoured milk. *Journal of Food Engneering* **96**(2), 174–184.

Prasad, D.N. (1990) Performance of yoghurt starters in UHT milk. *23th International Dairy Congress - Brief Communications* **532**.

Ramasubramanian, L., Restuccia, C. & Deeth H. C. (2008) Effect of calcium on the physical properties of stirred probiotic yogurt. *Journal of Dairy Science* **91**, 4164–4175.

Rattray, W. & Jelen, P. (1996) Thermal stability of skim milk with protein content standardisation by the addition of ultrafiltration permeates. *International Dairy Journal* **6**, 157–170.

Ravagnani, V. & Sambataro, D. (2004) Method for preparing a rice milk. Patent application EP1400177 (http://www.google.com/patents/EP1400177A1?cl=enA1).

Rothwell, J. (1989) *Cream Processing Manual.* Society of Dairy Technology, Huntingdon, England.

Roux, S., Courel, M., Ait-Ameur, L. Birlouez-Aragon, I. & Pain, J-P. (2009) Kinetics of Maillard reactions in model infant formula during UHT treatment using a static batch ohmic heater. *Dairy Science and Technology* **89**, 349–362.

Rückert, A., Ronimus, R.S. & Morgan, H.W. (2004) A RAPD-based survey of thermophilic bacilli in milk powders from different countries. *International Journal of Food Microbiology* **96**(3), 263–272.

Rustom, I.Y. S., Lopez-Leiva, M.H. & Nair, B.M. (1996) Nutritional, sensory and physicochemical properties of peanut beverage sterilised under two different UHT conditions. *Food Chemistry* **56** (1), 45–53.

Rustom, I.Y.S., Lopezleiva, M.H. & Nair, B.M. (1995) UHT-sterilized peanut beverages - changes in physicochemical properties during storage. *Journal of Food Science* **60**(2), 378–383.

Rustom, I.Y.S., Lopezleiva, M.H. & Nair, B.M. (1996b) UHT-Sterilized peanut beverages: Kinetics of physicochemical changes during storage and shelf-life prediction modeling. *Journal of Food Science* **61**(1), 198–203, 208.

Saikhwan, P., Thongchan, S., Jumwan, N., Thungsiabyuan, P., Sakdanuphap, J., Boonsom, S., Kraitong, P. & Danwanichakul, P. (2015) Cleaning studies of coconut milk foulants formed during heat treatment process. *Food and Bioproducts Processing* **93**, 166–175.

Salunkhe, D.K. and Kadam, S.S. (eds) (1995) *Handbook of Fruit Science and Technology*, Marcel Dekker, New York.

Savello, P.A. & Dargan, R.A. (1995) Improved yogurt physical properties using ultrafiltration and very-high temperature heating. *Milchwissenschaft* **50**, 86–90.

Savello, P.A. & Dargan, R.A. (1997) Reduced yogurt syneresis using ultrafiltration and very-high temperature heating. *Milchwissenschaft* **52**, 573–577.

Schamberger, G.P. & Labuza, T.P. (2007) Effect of flavanoids on Maillard browning in UHT milk. *LWT-Food Science and Technology* **40**, 1410–1417.

Schmidt, R.H., Vargas, M.M., Smith, K.L. & Jezeski, J.J. (1985) The effect of ultra-high temperature milk processing on yoghurt texture. *Journal of Food Processing and Preservation* **9**, 235–240.

Schreiber, R. (2001) Heat-induced modifications in casein dispersions affecting their rennetability. *International Dairy Journal* **11**, 553–558.

Seow, C.C. & Gwee, C.N. (1997) Coconut milk: chemistry and technology. *International Journal of Food Science and Technology* **32**, 189–201.

Shukla, A., Narayanan, T. & Zanchi, D. (2009) Structure of casein micelles and their complexation with tannins. *Soft Matter* **5**, 2884–2888.

Shyu, Y.T., Hsu, W.Y. & Hwang, H.S. (1996) Study on the improvements of mango juice processing. *Food Science Taiwan* **23**(4), 469–482.

Singh, H., Shalabi, S. I., Fox, P. F., Flynn, A. & Barry, A. (1988) Rennet coagulation of heated milk - influence of pH adjustment before or after heating. *Journal of Dairy Research* **55**, 205–215.

Singh, R. R. B. & Patil, G. R. (1989) Storage stability of UHT buffalo milk. I. Change in physico-chemical properties. *Indian Journal of Dairy Science* **42**(2), 150–154.

Singh, R. R. B. & Patil, G. R. (1989) Storage stability of UHT buffalo milk. II. Sensory changes and their relationship with physico-chemical properties. *Indian Journal of Dairy Science* **42**(2), 384–387.

Smiddy, M.A., Kelly, A.L. & Huppertz, T. (2009) Cream and cream related products. In: *Dairy Fats and Related products*. (ed. A.Y. Tamime), pp. 61–85. Wiley-Blackwell, Chichester, UK.

Smith, K.L., Schmidt, R.H. & Adams, J.P. (1982) Growth of yoghurt cultures in UHT treated milk. *Journal of Dairy Science* **65**(Suppl. 1), 221.

Smits, P. & van Brouwershaven J.H. (1980) Heat-induced association of β-lactoglobulin and casein micelles. *Journal of Dairy Research* **47**, 313–325.

Sopelana, P., Pérez-Martínez, M., López-Galilea, I., Paz de Peña, M. & Cid, C. (2013) Effect of ultra high temperature (UHT) treatment on coffee brew stability. *Food Research International* **50**, 682–690.

Splittstoesser, D.F. & Churey, J.J. (1989) Effect of low concentrations of sorbic acid on the heat resistance and viable recovery of *Neosartorya fischeri* ascospores, *Journal of Food Protection* **52**(11), 821–822.

SI (Statutory Instruments). (2015) *The Condensed Milk and Dried Milk (England) Regulations 2015.* UK Food Regulation 675.

Stones, M. (2015) How food industry can boost milk's value: fortify it. *Dairy Reporter* **17** June 2015.

Sucupira, N.R., Wurlitzer, N.J., Dionisio, A.P., de Abreu, F.A.P. & de Sousa, P.H.M. (2015) Effect of sulphite and UHT sterilisation parameters in the coconut water quality. In: *Frutic Italy 2015: 9th Nut and Vegetable Production Engineering Symposium.* (eds R. Guidetti, L. Bodria, & S. Best). Vol. **44**, pp. 79–84.

Syrios, A., Faka, M., Grandison, A.S. & Lewis, M. J. (2011) Comparison of reverse osmosis, nanofiltration and ultrafiltration as concentration processes prior to spray drying of skim milk. *International Journal of Dairy Technology*, **64**, 467–472.

Tamime, A.Y. & Deeth, H.C. (1980) Yogurt, technology and biochemistry. *Journal of Food Protection* **43**, 939–977.

Tamime, A.Y. (ed) (2009) *Dairy Powders and Concentrated Products.* Wiley-Blackwell, Chichester, UK.

Tamime, A.Y. (ed) (2013) *Membrane Processing: Dairy and Beverage Applications.* Wiley-Blackwell, Chichester, UK.

Tarassuk, N.P. & Tamsma, A.F. (1956) Control of gelation in evaporated milk *Journal of Agricultural & Food Chemistry* **4**, 1033–1035.

Tetra Pak (2015) *Dairy Processing Handbook.* Tetra Pak Processing Systems, Lund, Sweden.

Tetra Pak (2016) *Understanding Ambient Yogurt - Challenges and Opportunities* Tetra Pak Processing Systems, Lund, Sweden.

Tsikritzi, R. (2011) *Mineral Partitioning of Milk under various Conditions in relation to Heat Stability.* PhD thesis, University of Reading, UK.

Uniacke-Lowe, T., Huppertz, T. & Fox, P.F. (2010) Equine milk proteins: Chemistry, structure and nutritional significance. *International Dairy Journal* **20**, 609–620.

USDA (United States Department of Agriculture, Agricultural Research Service) (2009) Nutrient intakes. Percent of population 2 years old and over with adequate intakes based on average requirement (http://www.ars.usda.gov/Services/docs.htm?docid=15672, accessed 23 February 2014).

Valencia-Flores, D.C., Hernandez-Herrero, M., Guamis, B. & Ferragut, V. (2013) Comparing the effects of ultra-high-pressure homogenisation and conventional thermal treatments on the microbiological, physical, and chemical quality of almond beverages. *Journal of Food Science* **78**(2), E199–E205.

Van Hooydonk, A.C.M., De Koster, P.G. & Boerrioter, I.J. (1987) The renneting properties of heated milk. *Netherlands Milk and Dairy Journal* **41**, 3–18.

Vasbinder, A.J. & De Kruif, C.G. (2003) Casein-whey protein interactions in heated milk: the influence of pH. *International Dairy Journal* **13**, 669–677.

Vavrusova, M., Munk, M.B. & Skibsted, L.H. (2013) Aqueous solubility of calcium L-lactate, calcium D-gluconate, and calcium D-lactobionate: importance of complex formation for solubility increase by hydroxycarboxylate mixtures. *Journal of Agricultural & Food Chemistry* **61**(34), 8207–8214.

Veal, K. (1987) Key stages in fruit juice technology. *Food Technology International Europe* 136–138.

Villumsen, N.S., Hammershøj, M., Nielsen, L.R., Poulsen, K.R., Sørensen, J. & Larsen, L.B. (2015) Control of heat treatment and storage temperature prevents the formation of visible aggregates in acidic whey dispersions over a 6-month storage period. *LWT - Food Science and Technology* **64**, 164–170.

Villumsen, N.S., Jensen, H.B., Le, T.T.T., Møller, H.S., Nordvang, R.T., Nielsen, L.R., Nielsen, S.B., Sørensen, J., Hammershøj, M. & Larsen, L.B. (2015) Self-assembly of caseinomacropeptide as a potential key mechanism in the formation of visible storage induced aggregates in acidic whey protein isolate dispersions. *International Dairy Journal* **49**, 8–15.

Wark, J.D. & Nowson, C. (2003) Calcium supplementation: the bare bones. *Australian Prescriber* **26**,126–127.

Weeragul. K (2008) *The Effect of Manufacturing Factors on Stirred Yogurt Properties*. PhD thesis, University of Queensland, Brisbane, Australia.

Wijayanti, H.B., Bansal. N. & Deeth, H.C. (2014) Stability of whey proteins during thermal processing: a review. *Comprehensive Reviews in Food Science and Food Safety* **13**, 1235–1251.

Witkowska, A.M., Hickey, D.K., Alonso-Gomez, M. & Wilkinson, M.G.(2011) The microbiological quality of commercial herb and spice preparations used in the formulation of a chicken supreme ready meal and microbial survival following a simulated industrial heating process. *Food Control* **22**(3-4), 616–625.

Witthuhn, M., Hinrichs, J. & Atamer, Z. (2011) Emerging challenges and trends in UHT processing of milk and other dairy products. *New Food* **14**, 30–34.

Zhao, D.B., Bai, Y.H, Mutu, J. & Zhang, H.P. (2010) Ethanol stability of Alxa bactrian camel milk. *Journal of Camel Practice and Research* **17**(2), 189–193.

10

Non-Thermal Technologies

10.1 Introduction

There have been major developments in non-thermal technologies over the last few decades as replacements for, or adjuncts to, thermal technologies. These have been largely driven by consumer demand for minimally processed food products with the flavour and nutritive properties of fresh foods. Other drivers have been the need to process heat-sensitive food products and ingredients, the ability of the new technologies to give rise to new product concepts, and the availability of equipment for use with the new technologies, often developed and used originally for non-food applications. In relation to their use with milk and dairy products, some non-thermal technologies are well established commercially, for example, microfiltration, some are now commonly used commercially in other food industries but very little in the dairy industry, for example, high-pressure processing, some have been developed in the laboratory and have found limited applications in other food industries but none in dairy, for example, pulsed electric field technology, whereas others have shown great potential in the laboratory and limited small-scale use in some industries but none in the dairy industry, for example, ultrasonication and high-pressure homogenisation.

In this chapter, non-thermal technologies which have potential for producing extended shelf life (ESL) or shelf-stable (commercially sterile) milk and dairy products, alone or with some thermal processing, are discussed. Table 10.1 summarises the advantages, limitations and commercialization status of these technologies. General dairy applications of three of these technologies, high pressure processing, pulsed electric field technology and high pressure homogenisation, have been reviewed by Deeth *et al.* (2013). Several other non-thermal technologies exist including ultrasonication, cold plasma and high-intensity light (Deeth & Datta, 2011) which may in the future find application in extending the shelf life of milk but are not considered to be viable for this purpose in the forseeable future.

10.2 Microfiltration

Microfiltration (MF) is a mature technology in the dairy industry, having been used for many years for a range of applications including pre-treatment of milk for cheesemaking to remove *Clostridium tyrobutryicum* and prevent gas formation during cheese

High Temperature Processing of Milk and Milk Products, First Edition. Hilton C. Deeth and Michael J. Lewis.

Table 10.1 Summary of the potential of various non-thermal processes for production of ESL or shelf-stable milk.

Non-thermal technology	Advantages	Limitations	Commercialisation status
Microfiltration	Continuous process. Removes bacteria, both vegetative cells and spores, and somatic cells at mild temperature (30-50 °C). Causes little noticeable change to colour, flavour and nutritive value	Unable to process fat-containing milk. Cream portion has to be high-heat treated. Maximum reduction of bacteria insufficient for producing shelf-stable products. Final product has to be pasteurised (in most countries).	Well established in the dairy industry for several applications, including production of ESL milk
High pressure	Inactivates a large range of bacteria and enzymes. Causes little noticeable change to colour, flavour and nutritive value. Suitable for heat-sensitive products/ingredients.	A non-continuous process. Maximum capacity of processing vessels makes it unsuitable for large-volume products. Destroys bacterial spores only under severe conditions – very high pressure with heat. Does not destroy some baratolerant bacteria.	Well established in other food industries such as meat, horticultural products, seafood and prepared meals but not in the dairy industry. Patents exist for several dairy applications
Pulsed electric field technology	Continuous process. Inactivates a large range of vegetative bacteria and some enzymes. Can be used for producing ESL milk with some thermal treatment.	Does not destroy bacterial spores. Limited capacity of commercial equipment.	Has been used in other food industries with mixed success. Not commercialised in the dairy industry
High-pressure homogenisation	Continuous process. Homogenises the product. Heat generated by homogenisation can be used for inactivation of bacteria and enzymes.	Bacterial and enzyme inactivation is largely by heat. Small capacity of available equipment.	Used in the pharmaceutical industry. Not used in the dairy industry
Bactofugation	Continuous process. Reduces spore levels by up to 2 logs and total bacteria by around 1 log.	Insufficient removal of bacteria for "pasteurisation' or sterilisation when used alone	Used extensively for reducing clostridia in cheese milk to prevent late blowing defect. Also used to reduce spore levels in powders, infant formulae, and ESL and UHT milk.

UV irradiation	Continuous process, capable of killing most bacteria, including spores. Energy efficient.	Effectiveness is greatly reduced by opacity/turbidity. Causes production of unclean flavours due to light-induced oxidation.	Has been approved in EU for extending the shelf-life of pasteurised milk
Gamma irradiation	At low doses (≤ ~1 kGy) extends the shelf-life of pasteurised milk with little off flavour production	Higher doses produce unacceptable off-flavours. Spores are only inactivated at high doses (≥10 kGy). Negative consumer attitude	Used for several foods in >60 countries but no current commercial dairy applications
Carbon dioxide	Can be continuously added to milk by in-line injection at pressures ≤1 bar (0.1 MPa). Bacterial growth in inhibited and the shelf-life of raw and pasteurised milk is extended by several days	Does not kill bacteria. At concentrations above ~12 mM, carbon dioxide imparts a taste to milk. Removal is usually necessary	Most beneficial for inhibiting growth of psychrotrophic bacteria in milk to prevent production of proteases and lipase which can cause defects in UHT milk
High pressure carbon dioxide	Inactivates both vegetative bacteria and spores	Causes destabilization of milk proteins. Equipment limitations for high-volume products.	No current commercial dairy applications

maturation, and production of milk protein concentrate and ESL milk. In the context of this book, its major application of interest is in producing ESL milk (see Section 3.3.2). MF has not been used commercially for producing "commercially sterile" milk although Tetra Pak have patented a system capable of producing such a product. This system involves the use of membranes with a pore size of 0.5 μm combined with a heat treatment of 100 °C for 2 s. It is claimed to produce a milk which is stable at room temperature for 6-9 months.

Use of MF in conjunction with UHT has been proposed for improving the keeping quality of UHT milk. Kong *et al.* (2011) used a 1.4 μm ceramic membrane to microfilter high-somatic-cell-count (SCC) and low-SCC milk prior to UHT processing and found that the resulting MF + UHT milks had almost no residual plasmin while the milks without MF retained 3-4% of the original plasmin. The MF + UHT milks showed less change in non-casein nitrogen and pH during storage than the control UHT milks. The effect of MF was greater for the high-SCC milk than for the low-SCC milk and this led the authors to suggest that the use of MF could be particularly beneficial when producing UHT milk from high-SCC milk.

The technology used in the dairy industry is a continuous, cross-flow or tangential-flow process in which the liquid feed flows under pressure parallel to a membrane to produce a retentate and a permeate. The membranes used have pore sizes of 0.1 to 10 μm and are designed to retain particulate matter and allow all soluble material to permeate through the membrane. When used with milk, MF membranes retain most bacteria, somatic cells and milk fat globules but allow proteins, including most of the colloidal casein micelles, and other smaller molecules to pass through. Tighter membranes, that is, those with smaller pore sizes, are used in other membrane technologies to separate soluble components according to their molecular size. Thus ultrafiltration retains large molecules such as proteins, nanofiltration retains most soluble material except some mineral salts and reverse osmosis retains virtually everything allowing only water to pass through the membrane (Table 10.2).

Table 10.2 Characteristics of membrane technologies used in the dairy industry.

Process	Pore size (μm)	Milk components retained by membrane	Milk components that permeate the membrane
Microfiltration	0.1-10	Bacteria including spores, somatic cells, fat globules, large casein micelles	Soluble compounds and most casein micelles, water
Ultrafiltration	0.001- 0.1	Particulate matter as above plus micellar and non-micellar proteins, large peptides	Lactose, mineral salts, small organic molecules including vitamins, small peptides and amino acids, water
Nanofiltration	0.0001-0.001	Particulate matter and large molecules as above plus lactose, most mineral salts, small organic molecules including vitamins, small peptides and amino acids	Some monovalent ions, water
Reverse osmosis	<0.0001	All particulate matter and solutes	Water

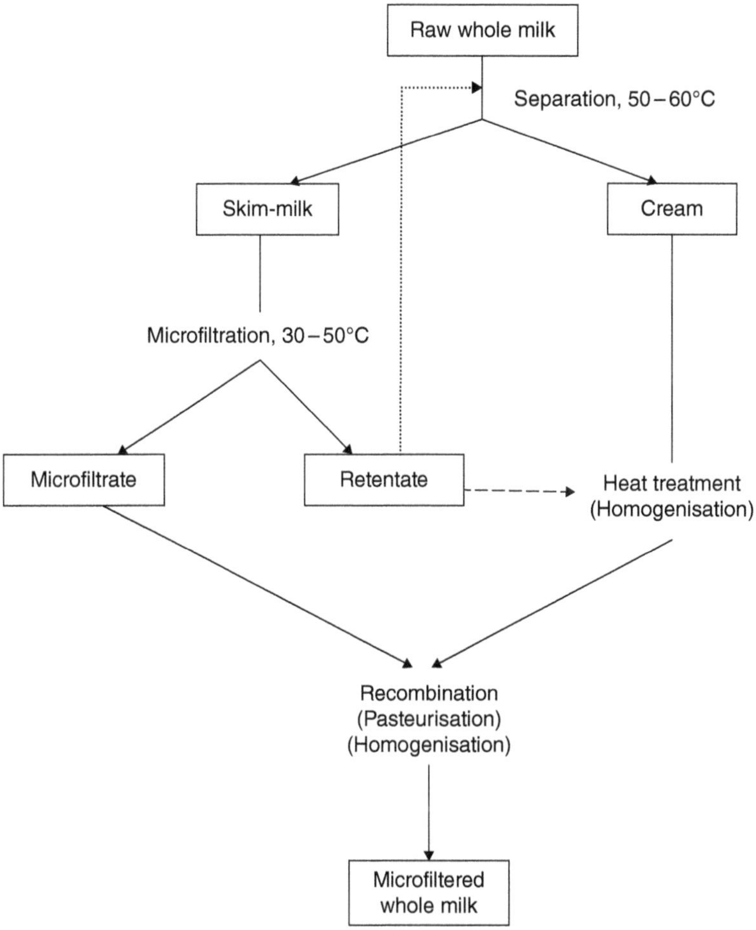

Figure 10.1 Process for producing ESL milk by microfiltration.

Most modern MF membranes are made of ceramic material and are relatively robust to mechanical stresses such as those encountered in recirculation of product and cleaning operations, such as back-flushing, and to chemicals used in sanitising and cleaning the membranes; they have a pH tolerance from 0.5 to 13.5. The membranes are 3-5 μm thick and are usually backed by a porous support which gives them some physical rigidity.

MF using ceramic membranes has been used commercially for producing ESL milk since the 1990s using the Bactocatch® process. An outline of such a process is shown in Figure 10.1. Because MF membranes retain milk fat globules, the cream has to be removed from milk before the skim milk can be microfiltered. The cream, with or without the MF retentate, is heated at a high temperature for a short time (120-130 °C for 2-4 s) to destroy bacterial contaminants and then recombined with the microfiltered skim milk. In most cases the recombined milk is HTST pasteurised to inactivate the native milk lipase and to destroy any pathogens that may have permeated the membrane. Pasteurisation is usually performed to meet regulatory requirements. In jurisdictions where consumption of unpasteurised milk is permitted, the pasteurisation step may be omitted. The shelf-life of commercial pasteurised ultrafiltered ESL milk is 30-45 days.

In a variation of the above process, the retentate, ~5% of the feed volume, which contains most of the bacteria and somatic cells as well as large casein micelles, is continuously recycled back into the raw whole milk entering the separator. A large proportion of the accumulated bacteria and somatic cells in the MF retentate are removed in the separator sludge and not added back to the milk. A further modification made to the Bactocatch® process in Germany is to not include the retentate, which contains the microfiltered bacteria and somatic cells, in the final product (Hoffmann *et al.*, 2006).

In a trial using a system which did not include the retentate in the final (pasteurised) ESL product, Hoffmann *et al.* (2006) examined the effect of MF combined with a high-heat treatment (125 °C for 4.5 s) of the cream on some milk components. They found the final pasteurised ESL milk contained 0.2-0.3 g/kg less protein than the raw milk, 20% less acid-soluble (undenatured) β-lactoglobulin than a control HTST-pasteurised milk, 21% less immunoglobulin than a control HTST-pasteurised milk (40% loss vs 24% loss), 18 mg/kg of lactulose (detection limit was 10-30 mg/kg) and 12 mg furosine/100 g protein (compared with 5-6 mg/100 g protein in raw milk). It is assumed that the lower protein content in the ESL milk is due to removal of very large casein micelles in the retentate.

One measure of the extent of heat treatment a milk receives is whether it is lactoperoxidase positive. HTST pasteurised milk is lactoperoxidase-positive but ESL milk produced by a thermal process (123-145 °C for <1-5 s) is not. ESL skim milk produced by MF and pasteurisation is lactoperoxidase-positive but when microfiltered skim milk is combined with heated cream to produce a whole milk or fat-reduced milk, the amount of heat treatment given to the cream will affect its lactoperoxidase status. In the work by Hoffmann *et al.* (2006) described above, the concentration of acid-soluble β-lactoglobulin in the ESL milk was 11.1% of the total protein which, according to the authors exceeded lower the limit of 10% for a lactoperoxidase-positive milk. However, the furosine level of 12 mg/100 g protein was higher than the limit of 8.5 mg/100 g protein set for lactoperoxidase-positive milk. The authors considered the heat load during MF at 48-50 °C may have contributed to the furosine content. It is noted, however, that Hoffmann *et al.* (2006) diluted their 31.5% fat cream to 15% fat with UF permeate before heating it at 120-130 °C for 2-4 s. Hence the permeate added also received the high-heat treatment which it would not have received if the cream had been heat treated without dilution. This suggests that lactoperoxidase-positive ESL milk is able to be produced by MF.

MF membranes with a pore size of 1.4 μm used for production of ESL milk cause a 3-4 log reduction in bacteria and virtually complete removal of somatic cells. Membranes with smaller pores of 0.5-0.8 μm result in a greater reduction of bacterial count of 5-6 logs. The Tetra Therm ESL™ process using 1.4 μm membranes is claimed to give a 3-5 log reduction in milk with an initial count of 30,000 cfu/mL (Kelly, 2011). Elwell and Barbano (2006) reported an average reduction in bacterial count of skim milk after MF of 3.8 log using a 1.4 μm ceramic membrane at 50 °C. The average initial bacterial count was 2,500 cfu/mL. When the MF process was followed by a pasteurisation step at 72 °C for 15 s, a 5.6 log reduction was observed. The shelf-life of the pasteurised MF milk, taken as the time at which the count reached >20,000 cfu/mL, was 16 days when stored at 6.1 °C and 66 days when stored at 0.1 °C.

The major reasons for the longer shelf-life of MF-generated ESL milk compared with HTST pasteurised milk (shelf-life of 10-16 days) is the almost complete absence of thermoduric bacteria, both non-spore-forming and spores. However, unless ESL milk is packaged aseptically, post-processing contamination remains the major cause of bacterial growth during storage (see Sections 3.3 & 4.4.2).

The major differences between ESL milk produced by MF and by thermal processing, as discussed in Sections 3.3.1 and 3.3.2, are the reduced heat-induced flavour changes and the removal of a high proportion of all spores, including those of psychrotrophic spore-formers which can limit the shelf-life of ESL milk produced by heat treatment.

An advance which improved the efficiency of microfiltration for producing ESL milk was the development of a Uniform Trans-membrane Pressure (UTP) system. This overcomes the problem with conventional microfiltration of a drop in hydrodynamic pressure between the inlet and outlet caused by recirculating flow. It results in a decrease in the trans-membrane pressure which enhances formation of a fouling layer which reduces permeate flux, separation efficiency and run time. The UTP system creates a pressure gradient on the permeate side to match that on the retentate side so that a uniform trans-membrane pressure exists along the length of the flow channel (Maubois, 2002; Kelly, 2011).

10.3 High-Pressure Processing

High-pressure processing involves the application of pressures of 300 to 1000 MPa for several minutes. Treatments of ~600 MPa for 10-30 min are capable of inactivating most vegetative cells with little loss of nutrients or change in flavour or colour. The pressure is usually applied in a batch system in volumes up to ~400 L. This maximum batch volume currently limits the application of this technology to low-volume, high-value products and makes it impractical and uneconomic for use with large-volume products such as ESL or long-shelf-life milk.

The application of high pressure to milk is not new, having been first reported by Hite (1899, 1914) who achieved a 5-6 log reduction in the number of bacteria in milk when it was treated at 680 MPa for 10 min at room temperature. Treatment at 680 MPa for 7 days achieved complete sterilisation although this result does not appear to have been confirmed by other researchers. High-pressure processing of milk and dairy products has been reviewed by Trujillo *et al.* (1997, 2002), Datta and Deeth (1999), Messens *et al.* (2003), Huppertz *et al.* (2006), López-Fandiño (2006) Considine *et al.* (2007), Deeth *et al.* (2013) and Datta and Tomasula (2015).

10.3.1 Effect on Bacteria and Potential for Producing ESL and Shelf-Stable Milk

The effect of high pressure on spores is of most relevance in this book. Several researchers have reported the inability of high pressure to inactivate bacterial spores when applied at or below room temperature. For example, Nakayama *et al.* (1996) could not inactivate spores of *Bacillus stearothermophilus* (now *Geobacillus stearothermophilus*), *B. subtilis and B. licheniformis* by treatment at 981 MPa for 40 min or 588 MPa for 120 min at 5 and 10 °C. Even at pressures up to 1033 MPa, Timson and Short (1965) were unable to inactivate bacterial spores.

However, treatment at high pressure and elevated temperature is capable of inactivating spores. Reddy *et al.* (1999) and Hayakawa *et al.* (1994b) found little inactivation of spores by pressures of ≥800 MPa at 20-35 °C but obtained log reductions of ~4.5 at 55 °C and 70 °C. The sporicidal effect increased with temperature, pressure and treatment time. Fornari *et al.* (1995) reported complete inactivation of *G. stearothermophilus* and *B. licheniformis* by treatment for 5 min at 700 MPa at 70 °C and 800 MPa at 70 °C, respectively. Such treatments are now commonly known as Pressure-Assisted Thermal Processing (PATP) or Pressure-Assisted Thermal Sterilisation (PATS).

PATS uses pressures of ≥600 MPa and initial temperatures of 60 to 90 °C. Due to the adiabatic heat of compression (~5 °C/100 MPa) which occurs during pressurisation, the temperatures rises to ~90 to 130 °C (Barbosa-Cánovas & Juliano, 2008) which puts PATS into the temperature range for thermally processed ESL milk. When the pressure is released, the temperature immediately reduces to its initial temperature, thereby minimizing heat-induced changes to the product (Fryer & Versteeg, 2008).

Tovar-Hernandez *et al.* (2005) reported that treatment of milk at 586 MPa at 55 °C for 3 or 5 min was equivalent to a thermal ESL treatment and the milk had a shelf-life of >45 days under refrigeration. Production of shelf-stable milk equivalent to UHT milk but of higher quality has been reported to be achievable with PATS (Ramirez *et al.*, 2009; Lamela & Torres, 2008).

Several variations of the normal pressure treatment have been developed which have been claimed to increase its bactericidal effectiveness. These include *pulse pressurisation* (Meyer, 2000), *rapid decompression* (Hayakawa *et al.*, 1998), and *oscillatory pressurisation* (Hayakawa *et al.* (1994a). For example, pulse pressurisation is a two-pulse treatment at 690 MPa at 90 °C for 1 min with a holding time at atmospheric pressure of 1 min between cycles.

Using this procedure, Meyer *et al.* (2000) investigated the sporicidal effects of pressures from 483 to 1931 MPa and (initial) temperatures of 50 to 100 °C and developed a pressure–temperature matrix chart which showed certain combinations such as 483 MPa at 100 °C, 621 MPa at 90 °C, 621 MPa at 80 °C, 900 MPa at 70 °C and 1724 MPa at 60 °C which were capable of causing sterilisation. The corresponding final treatment temperatures, which reflect initial temperatures plus adiabatic heating during pressurisation, were 105 to 127 °C. Care should be exercised in the acceptance of such data in relation to sterilisation as the authors did not test the effectiveness of these conditions against *Cl. botulinum.*

The effect of PATS on spores is variable. For example, Scurrah *et al.* (2006) found inactivation levels of <0 to 6 logs of 40 isolates of eight *Bacillus* species treated at 600 MPa at 72 to 95 °C for 1 min; one isolate was actually activated. Higher temperatures caused greater inactivation with many isolates achieving high log reductions, for example, 5-6 log reductions of *B. licheniformis* and *B. cereus* spores. However, such log reductions are less than that required for producing shelf-stable milk.

10.3.2 Effect on Milk Components

High pressure has different effects on casein micelles according to the pressure applied. There is little effect at pressures <200 MPa, an increase in size after treatment at ~250 MPa due to aggregation, and a marked reduction at pressures of ≥400 MPa such that skim milk loses most of its opacity to appear translucent and almost clear (Gaucheron *et al.*, 1997). The milk will remain in this state for several days under refrigeration (Considine *et al.*, 2007) but regains most of its opacity on heating. The reduction in casein micelle size is due to solubilisation of colloidal calcium phosphate and rupture of hydrophobic bonds between caseins in the micelle (Huppertz *et al.*, 2006). The caseins are solubilised in the order: β-casein > κ-casein > α_{s1}-casein > α_{s2}-casein (López-Fandiño *et al.*, 1998). The order is related to both their content of serine phosphate, which binds to calcium, and their hydrophobicity.

At ~150 MPa, β-lactoglobulin commences to denature and the extent of denaturation increases with increasing pressure and time. Aggregation of whey proteins continues during the holding phase in HPP. β-Lactoglobulin is almost completely denatured by treatment at 750 MPa at 30 °C for 30 min, or 450 MPa at 60 °C for 15 min (Rademacher *et al.*, 1997). α-Lactalbumin (α-La) and bovine serum albumin (BSA) are more resistant to pressure than β-lactoglobulin. The pressure sensitivity of the whey proteins is: lactoferrin > β-Lg > immunoglobulin > BSA > α-La (Patel *et al.*, 2006). The sensitivity of β-lactoglobulin to pressure is evident in the increase in viscosity and turbidity of whey protein solutions when subjected to pressure. These changes are largely due to polymerisation and insolubilisation of β-lactoglobulin, the major whey protein.

Gel formation occurs when whey protein isolate (WPI) at ≥10% or WPC (~75% protein) at >18%) is subjected to pressures of ≥400 MPa. At lower concentrations, 6% WPI and >12% WPC, viscosity increased but no gel formed (Hinrichs & Kessler, 1997; Kanno *et al.*, 1998).

The enzymes in milk are less affected by high-pressure treatment than by heat treatment. *Alkaline phosphatase* (ALP) is quite resistant to high pressure inactivation and hence is not a suitable indicator of the effectiveness of (pressure-induced) pasteurisation of milk, as it is for thermal pasteurisation. Treatment at 800 MPa for 8 min is required for its complete inactivation (Rademacher *et al.*, 1998). *Plasmin* is reduced by only 30% and 75% at 400 MPa and 600 MPa respectively for 30 min at 20 °C (Huppertz *et al.*, 2004c); higher inactivation levels are achieved at higher temperatures, for example, 86.5% at 400 MPa for 15 min at 60 °C (Garćia-Risco *et al.*, 1998). *Lactoperoxidase* (LPO) is very stable to pressure with little or no inactivation by treatments up to 700 MPa at 20 to 65 °C (Ludikhuyze *et al.*, 2001). *Milk lipoprotein lipase* activity is reported to be not decreased but actually increased by pressure pulse treatment (with zero holding time) at 350 and 400 MPa (Pandey & Ramaswamy, 2004). Milk *lysozyme* is stable to high pressure (Trujillo *et al.*, 1997; Viazis *et al.*, 2007).

The major effect on minerals of high pressure treatment of milk is the increase in the levels of soluble calcium and phosphate of milk due to solubilisation of micellar calcium phosphate. Kiełczewska *et al.* (2009) reported that the percentage increases at 350 MPa were 42 and 63% respectively. Despite the increase in soluble calcium, we have found that ionic calcium does not change.

10.4 Pulsed Electric Field (PEF) Technology

Pulsed electric field (PEF) technology is a continuous process which uses very short pulses (microseconds) at very high electric field strengths. It is generally performed at room temperature but can be performed at temperatures up to ~60 °C. Several factors affect the outcome of PEF treatments which make it difficult to make blanket statements about its effects on milk and to compare the results from different studies. These factors include the electric field strength, the pulse characteristics (width, frequency, total number, shape of the pulse wave), product flow rate and flow conditions (laminar or turbulent), and product characteristics (electrical conductivity, temperature, viscosity, pH and composition). The total specific energy input (in kJ/kg or kJ/L), electric field strength, and treatment times and temperatures are essential parameters which should be reported in order to be able to compare treatments from different laboratories (Alvarez *et al.*, 2003) but this has been seldom done in the past. Electrical conductivity is important as PEF is more effective in liquids with low conductivity (Jayaram *et al.*, 1993) such as milk and fruit juices which have conductivities in the range 4-6 mS/cm.

Some authors have reported that PEF treatment is a low-energy process compared with thermal treatments (Ho *et al.*, 1995: Qin *et al.*, 1995). However, this varies according to the use of PEF. Some applications, such as permeabilisation of plant cells, require very little energy, ~1.5 kJ/kg, but other applications such as microbial inactivation for pasteurisation purposes require quite high energy levels of 40-1000 kJ/kg (Toepfl *et al.*, 2006). This can be reduced considerably if the PEF treatment is performed at temperatures higher than ambient, such as 50 °C (Sepulveda *et al.*, 2009) and the process is optimised. Under these conditions, the energy input can be reduced to close to that of thermal pasteurisation, ~20 kJ/kg.

Several reviews on PEF treatment of liquid foods have appeared in the last 20 years. Some devoted specifically to milk include Bendicho *et al.* (2002), Alvarez *et al.* (2003), Sampedro *et al.* (2005), Deeth *et al.* (2007, 2013), Deeth and Datta (2011) and Shamsi (2010).

PEF treatments are generally carried out at an electric field strength of 10-50 kV/cm in short pulses of 1-5 μs duration at frequencies of 200-400 Hz for a total time of <1 s (Wan *et al.*, 2009). The field strength is calculated from the voltage produced by the power supply and the distance between the electrodes in the treatment chamber. A voltage of 20 kV across a distance of 0.5 cm yields a field strength of 40 kV/cm. The temperature increases during PEF treatment; for example, milk processed at a field strength of 28 kV/cm and a specific energy input of 111.6 kJ/L reached 15, 40, 50 or 55 °C from inlet temperatures of 10.5, 30.5, 40.5 and 45.5 °C (Craven *et al.*, 2008).

10.4.1 Effect on Bacteria and Potential for Producing ESL and Shelf-Stable Milk

PEF technology can inactivate vegetative cells of most pathogenic and spoilage microorganisms by up to 5-6 logs after short treatment times under appropriate conditions (Sampedro *et al.*, 2005). Thus the treatment conditions can be designed to produce bactericidal effects in milk equivalent to extended shelf-life processing and, possibly, UHT sterilisation. Treatments at 65 °C (Sepulveda *et al.*, 2009) or PEF combined with a separate HTST pasteurisation step (Sepulveda *et al.*, 2005) produced extended-shelf-life milk while PEF treatment followed by heat treatment at 105-112 °C yielded a shelf-stable product (Evrendilek *et al.*, 2001).

The temperature at which PEF treatments are performed has a major effect on the bactericidal effect of the treatment and care must be taken to separate the two effects. For example, PEF treatment (electric field strength, 31 kV/cm and specific energy input, 139.4 kJ/L) of sterile milk inoculated with *Pseudomonas* was much more bactericidally effective at 50 and 55 °C than at 15 and 40 °C with the highest inactivation of >5 logs being achieved at 55 °C. A control thermal treatment at 55 °C caused only a 0.2-0.3 log reduction of *Pseudomonas* counts (Craven *et al.*, 2008). In contrast, Shamsi *et al.* (2008)

using similar PEF conditions to those of Craven *et al.* (2008) but conducting the treatments at a final temperature of 60 °C achieved a reduction in *Pseudomonas* of 5.9 logs but found that the thermal effect of the treatment accounted for a 2.4 log reduction in *Pseudomonas.* This suggests that, for this organism, thermal effects become significant between 55 and 60 °C.

Sepulveda *et al.* (2009) used PEF (35 kV/cm, total time 11.5 μs) at a treatment temperature of 65 °C with a residence time of 10 seconds and achieved a shelf life extension of milk of at least 24 days. They concluded that, with use of a thermal regeneration system, the energy efficiency of such a PEF treatment process would be highly competitive with thermal pasteurisation, using the same PEF treatment conditions.

Sepulveda *et al.* (2005) used two processes for producing ESL milk using PEF. These involved applying a PEF treatment (35 kV/cm, total treatment time 11.5 μs at 65 °C) directly after HTST pasteurisation or 8 days after pasteurisation. They resulted in shelf-life extensions of 60 and 78 days respectively. The advantage gained by performing the PEF treatment 8 days after pasteurisation may be due to inactivation of vegetative cells of spore-forming bacteria which had begun to grow during storage. These shelf-lives are similar to those achieved by thermal treatments of 120-130 °C for 1-4 seconds (see Section 3.3.1). Evrendilek *et al.* (2001) used a PEF treatment (35 kV/cm, total treatment time μs at 11.9 to 24 °C) followed by heat treatment at 112 °C for 31.5 s on chocolate milk and obtained a shelf-life of 119 days when stored at 4, 22 or 37 °C.

Another hurdle approach to enhancing the bactericidal effectiveness of PEF is to combine it with antibacterial agents (Ross *et al.*, 2003). Both nisin and lysozyme exert a synergistic bactericidal effect with PEF achieving up to 7 log reductions in milk (Smith *et al.*, 2002; Sobrino-Lopez & Martin-Belloso, 2008). Since PEF has a destructive effect on nisin, it must be added after the PEF treatment (Terebiznik *et al.*, 2000).

10.4.2 Effect on Milk Components

PEF has little effect on the fat or proteins in milk. The size of casein micelles may be decreased slightly (Floury *et al.*, 2006) but some authors have reported no decrease (Shamsi, 2010). No change in β-lactoglobulin, α-lactalbumin, lactoperoxidase, IgG and lactoferrin after PEF treatments of skim milk or whey has been reported by some authors (de Luis *et al.*, 2009; Barsotti *et al.*, 2002; Li *et al.*, 2003, 2005), whereas Odriozola-Serrano *et al.* (2006) reported denaturation levels of serum albumin, β-lactoglobulin and α-lactalbumin of 24.5%, 20.1% and 40% respectively. Floury *et al.* (2006) observed decreased viscosity and enhanced coagulation properties of milk, suggestive of an effect on the proteins. PEF treatment of whey protein isolate had no effect on protein aggregation, surface hydrophobicity, sulfhydryl groups or thermal stability but increased gelation times and decreased the strength of heat-induced gels made from the treated whey protein isolates (Sui *et al.*, 2010). Overall, although there are some apparently conflicting reports, which may be due to different analytical methodologies and PEF conditions used, PEF appears to have a relatively small effect on milk proteins compared with other technologies such as high-temperature and high-pressure processing.

Inactivation of *alkaline phosphatase* is not a suitable indicator of "pasteurisation" by PEF as the highest reported percentage inactivation, without thermal input, appears to be 42% (Shamsi *et al.*, 2008). Higher reported inactivation rates have been due to PEF plus some heat (van Loey *et al.*, 2001). Significantly, Shamsi *et al.* (2008) reported that the percentage inactivation of alkaline phosphatase under PEF conditions which caused a 5.9-log reduction (equivalent to or greater than the effect of thermal pasteurisation) of *Pseudomonas* was 67% while the percentage for HTST pasteurisation in the same study was 98%. Under similar PEF conditions, milk *plasmin* was inactivated by 42% (Shamsi, 2010). However, Vega-Mercado *et al.* (1995) reported a 90% inactivation of plasmin with more severe PEF treatment but at a lower temperature. While a direct comparison is not possible, it is apparent that plasmin is susceptible to PEF treatment. This is significant because plasmin is quite heat resistant and can cause quality defects in stored UHT milk (see Sections 7.1.3.4 and 7.1.4). Under some conditions, bacterial proteases are inactivated. *B. subtilis* protease was inactivated by >80% (Bendicho *et al.*, 2003) and *Ps. fluorescens* protease in tryptic soy broth was reduced by 80% (Vega-Mercado *et al.*, 2001). However, no inactivation of *Ps. fluorescens* protease occurred when it was treated with PEF in a casein-Tris buffer, indicating the importance of the treatment medium in PEF treatments. *Lactoperoxidase* is not affected by PEF (de Luis *et al.*, 2009; van Loey *et al.*, 2002; Grahl & Markl, 1996) and milk *lipoprotein lipase* and *xanthine oxidase* are inactivated to a small extent by PEF. Shamsi (2010) reported inactivation rates of 33 and 23%, respectively, under conditions which did not include a thermal component. As the temperature of PEF treatment is increased, higher levels of inactivation occur but the contribution of thermal inactivation became increasingly significant.

10.5 High-Pressure Homogenisation

The dairy industry is very familiar with homogenisation at pressures up to about 40 MPa. At these pressures, homogenisers break up milk fat globules with an average diameter of ~3 μm into much smaller globules with an average diameter of ≤1 μm and coat the small globules with milk proteins such that their density is increased and they remain dispersed in the milk rather than rise to the surface. Little other change occurs to the milk's chemical constituents or bacteria (further details on homogenisers are given in Section 5.3). However, homogenisation at much higher pressures of 100-400 MPa causes considerable change to both the chemical components and bacteria. This type of homogenisation is known as high pressure homogenisation although some authors use the term "Ultra High Pressure Homogenisation", especially for homogenisation at ≥200 MPa. In this chapter, the term High Pressure Homogenisation or HPH is used throughout. HPH should not be confused with high pressure processing discussed in Section 10.3 which is a completely different technology. The pressures referred to in HPH are reached and experienced over an extremely short time period, the time it takes to pass through the homogeniser valve, while in high pressure processing, the pressure is held for 1 to ~30 min, and high pressure homogenisers operate at pressures ≤400 MPa while high pressure processing operates at up to ~800 MPa. Another major difference is that HPH is a continuous process while high pressure processing is mostly a batch operation. Sometimes HPH is referred to as Dynamic High Pressure Processing to distinguish it from high pressure processing which is a "static" technology. The two technologies are compared in Table 10.3.

HPH is believed to have considerable potential for processing of milk and dairy products and ingredients (Leadley, 2003; Hayes *et al.*, 2005). It is currently used in the chemical, pharmaceutical and biochemical industries (Floury *et al.*, 2000) but not in the

Table 10.3 Comparison of high pressure homogenisation and high pressure processing.

Parameter	High pressure homogenisation (HPH)	High pressure processing (HPP)
Mode of operation	Continuous	Batch and semi-continuous
Operating pressures	~100-350 MPa	100-1000 MPa
Residence time under pressure	Very short, ~10^{-4} second	1-30 min
Capacity	Up to 8,000 litres per hour at 100 MPa and 5,000 litres per hour at 150 MPa; smaller capacity at higher pressures	Pressure vessels up to 420 litres, 5-10 batches per hour
Temperature increase during processing	17-24 °C per 100 MPa	~4 °C per 100 MPa, same decrease on pressure release
Energy recovery	Possible if heat regeneration system used	Not possible in single batch mode of operation, some possible when multiple vessels are used in a semi-continuous mode
Filling and packaging	Product packaged after treatment; aseptically if necessary	Products in flexible packaging can be treated directly
Post-processing contamination	Possible if aseptic packaging not used.	Not possible when processing occurs in closed containers
Microorganism inactivation	Minimal inactivation at low temperature. Susceptibility order: Gram-negative vegetative > Gram-positive vegetative > spores Substantial inactivation at higher temperatures, mostly due to thermal effect.	Substantial inactivation of vegetative cells at 500-600 MPa. Spore inactivation by high pressure at high temperature.
Sub-lethal injury to microorganisms	No sub-lethal injury	Partial inactivation or sub-lethal injury possible
Physical forces	Shear, turbulence and cavitation	Compression and decompression
Mechanism of microorganism inactivation	Rupture of the cell membrane and release of cytoplasmic material	Protein denaturation and membrane damage
Emulsification effect	Yes	No
Effect on casein micelle	Size decrease up to 200 MPa, size increase at >250 MPa	At ~400 MPa micelle disintegrates.
Effect on whey proteins	Minimal effect at low temperature. Denaturation at higher temperatures mostly due to thermal effect.	Causes denaturation
Appearance of treated milk	Little effect on appearance	Skim milk becomes translucent due to casein micelle dissociation
Potential for producing ESL or UHT milk	Sterility or near sterility may be possible at high pressure (~300 MPa) with an initial temperature of ~75-80 °C	Sterility can be achieved with PATS but milk becomes translucent

Based on Deeth *et al.* (2013)

dairy industry. This may be due to the small capacity of the available equipment making it suitable only for small-volume products. Aspects of the technology have been reviewed by Paquin (1999), Hayes *et al.* (2005), Diels & Michiels (2006) and Deeth *et al.* (2013).

Two different types of high pressure homogeniser are available, valve-type homogenisers, which are similar to conventional homogenisers but operate at much higher pressures, and microfluidisers which operate on a completely different principle. High pressure valve homogenisers operate up to 400 MPa with most in the range 100-300 MPa. Like conventional homogenisers, they usually have two valves in series, a primary and a secondary, with the secondary operating at ~10% of the pressure in the primary valve. The pressures are maintained for a very short time (approx 10^{-4} s); the generated cavitation, shear and turbulence are responsible for the physical effects observed.

In a microfluidiser, the liquid feed stream is split into two streams which are collided head-on with each other at high velocity (up to 50 m/s) in an interaction chamber. The pressure achieved in microfluidisers is 100 to 200 MPa, although some operate at pressures up to 500 MPa (Paquin, 1999). However, because the principles of the valve-type high pressure homogeniser and the micofluidiser are very different, caution needs to be exercised in making comparisons between them based on the nominal pressures. Cavitation and the high force of impact of the two feed streams in the interaction chamber are responsible for the effects exerted on the treated liquid by microfluidisation (Paquin, 1999; Hardham *et al.*, 2000).

High-pressure homogenisation by both types of homogeniser causes an almost instantaneous temperature rise in the treated product of ~17 to 23 °C per 100 MPa (Bouaouina *et al.*, 2006; Roach & Harte, 2008) which is exploited in some applications. The temperature rise depends on the composition of the product with the rise increasing with fat content (Hayes & Kelly, 2003a).

10.5.1 Effect on Microorganisms and Potential for Producing ESL and Shelf-Stable Milk

High-pressure homogenisers cause destruction to some extent of a wide range of organisms (Diels *et al.*, 2006). The effect is due to both the physical forces and the heat generated in the homogeniser. Greater bacterial destruction occurs at higher homogenisation pressures and higher temperatures. In microfluidisation, multiple passes of the product through the homogeniser increases the extent of inactivation (Picart *et al.*, 2006). The composition of the product can also affect inactivation which was reported to be greater for *E. coli* in whole milk than in skim milk (Diels *et al.*, 2006; Brinez *et al.*, 2006). Given the report of Hayes and Kelly (2003a) that the rise in temperature during high pressure homogenisation increases with product fat content, this may explain the greater bactericidal effect in whole milk than in skim milk.

HPH treatment using a valve-type homogeniser of milk at 300 MPa with an initial temperature of 30-40 °C has been found to reduce the bacterial count by ~3.5 logs (Perada *et al.*, 2007). The treated milk had a microbial shelf-life of 14-18 days at 4 °C, approximately equivalent to thermally pasteurised milk. Given the temperature rise of ~20 °C/100 MPa, this treatment would have resulted in a temperature rise of ~60 °C, giving temperatures at the homogenisation valve of 90-100 °C for a very short time.

Because of the instantaneous temperature rise, it has been suggested that HPH could be used to produce extended shelf-life milk (Pereda *et al.,* 2007) or shelf-stable (Asano *et al.*, 2000) products. These would have superior flavour profiles compared with those produced by thermal processes because of the very short holding time at the highest temperatures reached. In terms of the indices discussed in Section 3.2, ESL and shelf-stable milk produced by HPH would have lower C* values for the same B* value than corresponding thermally produced products.

The potential use of HPH for producing shelf-stable products is well illustrated in the work of Valencia-Flores *et al.* (2013), Poliseli-Scopel *et al.* (2014) and Amador-Espejo *et al* (2014).Valencia-Flores *et al.* (2013) treated an almond beverage at 300 MPa with a starting temperature of 75 °C which resulted in a final temperature of 129.3 ± 12.6 °C acting for <0.7 s. The final product showed high physical stability and no bacterial growth during storage at 30 °C for 20 days. Poliseli-Scopel *et al.* (2014) produced a shelf-stable soymilk by high pressure homogenisation at 300 MPa with a starting temperature of 80 °C, a temperature at the homogenisation valve of 144 °C with a holding time estimated to be 0.8 s. The aseptically packaged product had high colloidal stability and acceptable sensory properties, and did not show microbial growth during storage at room temperature for 6 months. Amador-Espejo *et al.* (2014) reported similar results for whole cow's milk. High-pressure homogenisation at 300 MPa and an initial temperature of 75 or 85 °C produced a sterile milk. The temperatures reached were 133 and 139 °C, respectively, and the residence time was estimated to be 0.5 s. The HPH- and UHT-treated milks were not significantly different in ethanol stability, colour, particle size, buffering capacity and plasmin proteolysis (determined by the increase in pH 4.6-soluble nitrogen during incubation for 7 days at 37 °C). However, they differed in viscosity with the HPH samples (2.1 and 2.39 mPa.s) being more viscous than the UHT milk (1.42 mPa.s). The sensory properties were similar except that the HPH samples exhibited less cooked flavour and slightly less salty flavour compared with the UHT samples. The authors have subsequently shown that the concentrations of ketones and dimethyl sulfide were lower in the HPH- than in the UHT-treated samples, which may account for the reduced cooked flavour. However, the concentration of aldehydes was higher in the HPH milk than in the UHT milk, suggesting more oxidation in the HPH milk (Amador-Espejo *et al.,* 2016).

The susceptibility of the different bacterial groups to HPH treatment is of the order Gram-negative vegetative >Gram-positive vegetative >spores. This is most significant at low temperatures where the effect of the homogenisation forces is more important than the heat effects. For example, HPH with a valve-type homogeniser of inoculated milk at 290 MPa and an initial temperature of 24 °C, resulted in log reductions of *Ps. fluorescens, Micrococcus luteus* and *Listeria innocua* of 4.0, 2.6 and 1.8 respectively (Picart *et al.*, 2006). HPH at low temperature has little effect on spores with ≤1 log reductions being reported (Feijoo *et al.*, 1997; Pereda *et al.*, 2007). However, as shown above by the work of Poliseli-Scopel *et al.* (2014), when the initial temperature is raised so that the final temperature is sufficient to inactivate spores, HPH can be very effective.

An important bacteriological aspect of HPH is that it does not cause sub-lethal injury to bacteria as does high pressure processing (Wuytack *et al.*, 2002; Brinez *et al.*, 2006; Jose-Brinez *et al.*, 2007). As sub-lethally injured bacterial can revive, grow and cause spoilage during storage of the product, this is a major advantage of HPH over high-pressure processing.

10.5.2 Effect on Milk Components

HPH with a valve-type homogeniser reduced the average size of milk fat globules: from ~3.2 μm to 0.47 μm ($d_{4,3}$) at 200 MPa (Hayes & Kelly, 2003a) and to 0.12-0.16 ($d_{3,2}$) at 100-200 MPa (Serra *et al.*, 2007). However, at higher pressures, fat globule aggregates form and the homogenisation efficiency is reduced. Serra *et al.* (2007) found that HPH at 230 to 330 MPa, resulted in polydisperse size distribution with some large particles present which were believed to be fat clusters (Thiebaud *et al.*, 2003).

Using a microfluidiser, Paquin (1999) found that the fat globules produced were finer and had a narrower size distribution than the globules produced with conventional homogenisation. This was supported by the work of Hardham *et al.* (2000) in which thermally processed UHT milk homogenised by microfluidisation was more stable during storage than UHT milk which had been conventionally homogenised. However, neither Paquin (1999) nor Hardham *et al.* (2000) made comparisons with milk homogenised by HPH using a valve-type homogeniser.

When using HPH on raw milk, care must be taken to ensure the temperature increases to ~80 °C to ensure the native milk lipase is inactivated. Otherwise lipolysis occurs and the milk will become rancid (Deeth, 2005; Lanciotti *et al.*, 2006; Serra *et al.*, 2008). HPH of raw milk using combinations of pressure and inlet temperature of 200 MPa/40 °C and 300 MPa/30 °C (outlet temperatures: 79.6 and 95.3 °C respectively) did not result in lipolysis, whereas HPH at 200 MPa/30 °C (outlet temperature 73.6 °C with a residence time of <0.7 s), resulted in considerable lipolysis (Serra *et al.*, 2008).

Both casein micelles and whey protein particles are altered by HPH. The size of casein micelles in milk is decreased by HPH at 200 MPa (Hayes & Kelly, 2003a) but increases again at higher pressures of 350 MPa (Roach & Harte, 2008). In a microfiltered casein micelle suspension, Roach and Harte (2008) found that HPH up to 250 MPa caused no change in the casein micelle size whereas it increased at 300 and 350 MPa reaching 363 nm. Using microfluidisation at 41-186 MPa on reconstituted skim milk powder, Sandra and Dalgleish (2005) observed a decrease in casein micelle size and an increase in non-sedimentable caseins with increasing pressure and number of passes.

HPH at temperatures lower than that which would cause thermal denaturation of whey proteins have been reported to cause no denaturation (Bouaouina *et al.*, 2006; Dissanayake & Vasilejvic, 2009) or some denaturation (Datta *et al.*, 2005; Hayes *et al.*, 2005). These results suggest that the homogenisation *per se* may have little effect. However, HPH has significant effects on the functional properties of whey proteins when a whey protein isolate (WPI) solution is homogenised. It markedly reduces the particle size as measured by a Malvern Mastersizer; at >200 MPa, HPH with a valve-type homogeniser reduced the size so that 90% were <1 μm compared with 9% of <1 μm before homogenisation (Bouaouina *et al.*, 2006). Microfluidisation at 140 MPa with 5 passes caused a similar reduction (Dissanayake & Vasilejvic, 2009). The microfluidisation treatment also caused an increase in heat stability, solubility and foaming capacity (Dissanayake & Vasilejvic, 2009). The improvement in foaming was also observed by Bouaouina *et al.* (2006) using a valve-type high-pressure homogeniser. Microfluidisation at 150 MPa substantially increased the solubility of whey protein which had been heated at 90 °C for 10 min by reducing the size of the protein aggregates (Iordache & Jelen, 2003). This has practical implications for improving the stability of products containing heat-denatured whey proteins.

HPH has little effect on the activity of *alkaline phosphatase* if the treatment is carried out such that the heat generated does not affect the enzyme (Picart *et al.*, 2006; Hayes & Kelly, 2003b). Under conditions where the heat generated can affect the enzyme, the inactivation by HPH is almost entirely due to thermal inactivation (Datta *et al.*, 2005). *Plasmin* has been reported to show no loss of activity by HPH (Iucci *et al.*, 2008). During HPH, plasmin associates with fat globules which are removed during analyses; Iucci *et al.* (2008) suggested this was why previous reports showed considerable inactivation of plasmin by HPH. Datta *et al.* (2005) reported that native milk *lipoprotein lipase* was activated by HPH when the outlet temperature was 58 °C but when HPH was carried out at 200 MPa with an outlet temperature of >71 °C, the lipase was completely inactivated. HPH decreases the activity of *lactoperoxidase* by up to 20% when the outlet temperature is lower than that which would cause thermal inactivation, ~70 °C. When the outlet temperature was 80 °C, lactoperoxidase was completely inactivated in both HPH-treated and thermal control milk samples (Datta *et al.*, 2005).

10.6 Bactofugation

Bactofugation is a process for removal of bacteria (and somatic cells) by centrifugation. It is commonly referred to by this name because the commercial equipment manufactured by Tetra Pak is marketed under the tradename of *Bactofuge*™. Bactofuges operate at a centrifugal force of about 9,000 g and the ideal operational temperature for milk is 55-60 °C.

Separation of the bacteria is based on specific gravity (SG). Bacterial spores have SGs of 1.30-1.32 g/mL and vegetative bacterial cells have SGs of 1.07-1.12 g/mL. Milk has an SG of 1.028-1.038 g/mL so it is difficult to remove vegetative bacterial cells from milk. Bactofugation reduces the total bacterial count of milk by ≤1 log in a single pass but a greater reduction can be achieved if the milk is passed through a second bactofuge.

A single pass removes around 98% of anaerobic spores and for this reason bactofugation is used extensively in some countries for cheese milk to reduce *Cl. tyrobutyricum and Cl. butyricum* which cause the late blowing defect in cheese (Stack & Sillen, 1998). It also removes around 90-95% of aerobic spores and can be used to pre-treat milk for manufacture of low-spore-count powder and UHT milk and to remove *B. cereus* (and reduce the level of the non-spore-former pathogen *Cronobacter sakazakii*) in preparation of infant formulae. Used with HTST pasteurisation, it can also extend the shelf-life of pasteurised milk by up to 5 days, depending on the quality of the raw milk. Recently, it has been reported to be used in conjunction with HTST pasteurisation and aseptic packaging for production of ESL milk (Mugadza & Buys, 2015).

Bactofugaton can be carried out on whole milk or skim milk after separation. Cream removal reduces the viscosity and increases the rate of removal of bacteria. However, some 30% of spores move with the cream phase and hence treatment of the whole milk is preferable for removing the bulk of these organisms.

During bactofugation, the centrifugate which contains most of the somatic cells and bacteria, is continuously discharged. This centrifugate, which represents about 0.2-0.3% of the milk volume, can be discarded, heat-treated to destroy the bacteria, including spores, and returned to the bactofuged milk, or continuously returned to the feed to be recirculated. Discarding the centrifugate is obviously the least economic and hence the last two options are preferred. However, under legislation in some countries, heat-treatment and return of the centrifugate is not permissible. The last option results in very little loss but does reduce the separation efficiency; Stack and Sillen (1998) reported a reduction in removal of anaerobic spores from 98 to 96% and from 96 to 90% for aerobic spores.

Another factor which affects the efficiency of the bactofuge is the capacity of the unit. High-capacity units have a shorter residence time which reduces the spore removal rate compared with lower-capacity units. Stack and Sillen (1998) cited a reduction in percentage spore removal with increasing capacity from 20,000 L/h to 40,000 L/h from 98-99 to 88-90% for anaerobic spores and from 96-97 to 80-82% for aerobic spores. The percentage removal by the 40,000 L/h unit was increased to 98-99% for anaerobic spores and 96-97% for aerobic spores when a second 40,000 L/h unit was placed in series with the first.

10.7 UV Irradiation

Irradiation of foods with UVC light (200-280 nm wavelength) is effective in killing most microorganisms. It commonly uses a mercury vapor lamp which transmits 85% of its energy at 254 nm. It has been used for many years for decontaminating air, surfaces of packaging materials and equipment, and water, and has been approved by the US FDA for commercial pasteurisation, without heat, of fruit juices (FDA, 2015), and by the European Food Safety Association for pasteurised milk (EFSA, 2016). UV treatment of foods has been reviewed by Bintsis *et al.* (2000), Guerrero-Beltrán and Barbosa-Cánovas (2004) and Koutchma *et al.* (2009).

UV light is most effective for clear liquids; its effectiveness is reduced by opacity and turbidity. Hence it is most suitable for water, fruit juices with a low level of insoluble solids, and clarified whey. Milk presents challenges because of its opacity. UV treatment equipment typically involves a UV-penetrable tube through which the liquid product is pumped. Ideally the flow in the tube is turbulent. In fact, the US FDA regulations (FDA 2015) specify that the minimum Reynolds number for fruit juice treatment should be 2,200. Turbulent flow continuously renews the surface and ensures all parts of the liquid come into contact with the UV light. This is essential for an opaque product such as milk. This has been achieved with the use of static mixers (Altic *et al.*, 2007) or a complex swirling flow through a corrugated spiral tube as in the commercial SurePure Turbulator™ system (Cilliers *et al.*, 2014). An alternative arrangement to turbulent flow is laminar flow in a very thin film.

Research on UV radiation of milk has been conducted over many decades. Burton, better known for his classic book on UHT processing (Burton, 1988), reviewed the early literature on this topic (Burton, 1951). He concluded that UV radiation could cause a 3-log reduction of total bacteria and the keeping quality of the UV-treated milk was poorer than that of thermally pasteurised milk. He also discussed the fact that UV irradiation increased the level of vitamin D in milk; this had been used commercially in Germany and North America in the 1020s, 1930s and 1940s. Increases in the level from ≤1 μg/L to as high as 31 μg/L were reported. More recent findings on UV treatment of milk have been reviewed by Reinemann *et al.* (2006) and Cilliers *et al.* (2014). Reinemann *et al.* (2006) showed that UV irradiation of 1.5 kJ/L caused a 3-log reduction in aerobic bacteria, yeasts and moulds, coliforms (including *E. coli*) and psychrotrophs in milk.

The major drawback of UV treatment of milk is the production of light-catalysed unpleasant flavours commonly described as "unclean". These flavours arise after doses of UV which are insufficient to destroy bacteria to the extent required for pasteurisation (that is, a 5-log reduction). Reinemann *et al.* (2006) observed cooked, barny, rancid and unclean off-flavours in milk treated with UV at 1.5 kJ/L which achieved a 3-log reduction in standard plate counts. They concluded that the treatment limit to retain sensory quality was 1 kJ/L. Cilliers *et al.* (2014) reported that milk treated with UV at a dosage of 1.045 kJ/L had more cardboard and tallowy off-flavours than a thermally pasteurised milk control. Similarly, Rossitto *et al.* (2012) noted sensory defects in milk treated with UV at 0.88 and 1.76 kJ/L, with the severity of the defects increasing as the UV dosage increased. Interestingly, Cilliers *et al.* (2014) found that the milk which had been thermally pasteurised after the UV treatment had less intense cardboard and tallowy off-flavours than a UV-treated raw milk; this led the authors to suggest that the off-flavours could be minimised by optimising the heat treatment used in combination with the UV treatment.

Cappozzo *et al.* (2015) treated raw and pasteurised milk at 1.045 and 2.090 kJ/L and found that the treatment caused no significant changes in proximate composition, fatty acid profile, lipid oxidation with respect to volatile flavour compounds produced, or protein profile. Similarly EFSA (2016) concluded that pasteurised milk UV treated at 1.045 kJ/L had a comparable nutrient composition to pasteurised milk. The EFSA Panel also concluded that UV treatment of milk does not cause significantly more lipid oxidation than thermal pasteurisation. However, neither Cappozzo *et al.* (2015) nor the EFSA panel addressed the sensory characteristics of the UV-irradiated milk; EFSA (2016) specifically stated that they were only concerned with the safety of the UV-treated pasteurised milk. One significant point where these two recent publications differ is with respect to vitamin D formation. EFSA (2016) concluded that vitamin D3 was increased significantly by the UV treatment, whereas Cappozzo *et al.* (2015) reported a 56% loss of vitamin D. The EFSA panel considered the Cappozzo *et al.* (2015) paper and suggested the loss of vitamin D may be due to the different UV treatment methodology (extensive recirculation) and that "no conclusions can be drawn from this publication".

Cilliers *et al.* (2014) carried out an extensive study of the microbiological, chemical and sensory aspects of milk irradiated in a SurePure Turbulator™, without recirculation, at a dosage of 1.045 kJ/L. They found that this UV treatment had a similar bactericidal efficacy to thermal pasteurisation. It had no significant effect on 7-ketocholesterol, riboflavin, vitamin B12, free fatty acids, the protein oxidation products methional, methionine, methionine sulfoxide and dimethyl disulfide, lipid oxidation (TBARS, TBA-reactive substances) or proteolysis (α-amino groups) levels and it had no effect on the activities of the enzymes plasmin, alkaline phosphatase and lactoperoxidase. However, it significantly reduced cholesterol levels and increased methionine sulfone (a protein oxidation product) content. Unfortunately, these authors did not report the effect of UV treatment on vitamin D levels.

Another potential application of UV irradiation in the dairy industry is treatment of cheese whey. This is very significant as whey often has to be stored for some time before being processed into WPC or WPI and cannot be thermally pasteurised. In its native state, cheese whey is turbid as it contains a high level of suspended solids. Hence it has a very poor transmittance to UV (Mahmoud & Ghaly, 2004). However, UV treatment of clarified whey can result in a high level of destruction of bacteria. Some denaturation of the whey proteins occurs but much less that by heat treatment to produce the same bactericidal effect (Kristo *et al.*, 2012). Cheese brine is another liquid where UV irradiation has been effective in markedly reducing the microbial load.

UV treatment using continuous turbulent flow technology flow has been shown to have potential for extending the shelf-life of pasteurised milk (Cappozzo *et al.*, 2015; EFSA, 2016). In fact, EFSA (2016) has approved the safety of the process for the general population in the EU, except for infants under 12 months of age.

10.8 Gamma Irradiation

Gamma rays, X-rays and high-energy electrons are forms of ionizing radiation. These have been researched for their bactericidal effects in food, with gamma irradiation attracting the most attention.

Irradiation with gamma-rays produced by a radionucleide such as cobalt-60, has been known to destroy microorganisms in food since the nineteenth century discovery of radioactivity by Becquerel. It is now approved for use in over 50 countries and used for treatment of a range of foods including spices, fruits and vegetables, seafood and meats (Lacroix, 2005; Barbosa-Cánovas & Bermúdez-Aguirre, 2010). The doses generally used for shelf-life extension and elimination of pathogens are 1-10 kGy and for sterilisation, 10-50 kGy. Gram-negative and Gram-positive vegetative bacteria have similar susceptibilities, being destroyed in the 1-10 kGy dose range, but spores are much more resistant.

Irradiation is not used commercially on milk although a considerable amount of research has been carried out on it. A major reason for this is the production of off-flavours from relatively low doses. Irradiation by either gamma rays or high-energy electrons produces similar off-flavours. Other adverse effects include cross-linking of proteins and reduction in vitamins and carotene (Barbosa-Cánovas & Bermúdez-Aguirre, 2010).

Sadoun *et al.* (1991) reported that the irradiation of milk at 4 °C could not be used at doses >0.5 kGy because of off-flavour production. Such a dose could only achieve a reduction in mesophiles of ~2 logs. Interestingly, they found that irradiation of pasteurised milk at room temperature with a dose of only 0.25 kGy resulted in a considerable extension of shelf-life without a change in organoleptic properties. The bacterial count after 18 days was $<10^4$/mL while the count in a control milk after 10 days was $>10^7$/mL. The degradation in vitamins B1, B2 and A was 10-20%. This suggests that irradiation could be used for producing ESL milk.

The conclusions of Sadoun *et al.* (1991) differ from those of de Oliveira Silva *et al.* (2015) who irradiated raw whole milk at 1, 2 and 3 kGy. Irradiation at all three doses eliminated psychrotrophic bacteria and *E. coli*. Mesophilic bacteria were reduced, with the reduction being dependent on the dose. During storage at 4 °C for 60 days, the total bacterial counts (log cfu/mL) remained constant at around 2.5, 1.5 and 1.2 for the 1, 2 and 3 kGy doses respectively. Titratable acidity also remained constant for 60 days. The irradiated samples could be distinguished from the non-irradiated samples by a consumer panel but the irradiated samples attracted both positive and negative comments. The results suggest that irradiation at 1 or even 2 kGy causes little adverse change in sensory properties and could be used to produce milk with an extended refrigerated shelf-life of ≥60 days.

Osaili *et al.* (2008) showed that *Enterobacter sakazaki* (now known as *C. sakazaki*, see Section 4.4.1), a pathogen which has been associated with powdered infant milk formulae, can be destroyed by irradiation. They recommended irradiating infant formulae with a dose of 5.13 kGy, a 3-log reduction of the most resistant strain investigated. Lee *et al.* (2007) found that irradiation at 5.0 kGy eliminated *C. sakazakii* inoculated at 8.0 to 9.0 log cfu/g of powdered infant formula. Unfortunately, neither of these papers reported organoleptic assessment of the irradiated powders.

Irradiation of foods is still an emotive issue in many countries. Although the safety of irradiated foods has been demonstrated (Crawford & Ruff, 1996), many consumers are still concerned about the possible presence of harmful radiolytic products in irradiated foods.

10.9 Carbon Dioxide

Carbon dioxide (CO_2) at ≤1 atmosphere (1 bar, 0.1 MPa) is an effective antimicrobial agent and can be used to inhibit the growth of bacteria, particularly Gram-negative bacteria, in raw or pasteurised milk (Hotchkiss *et al.*, 2006). This is particularly relevant to the quality and shelf-life of UHT milk as a reduction in growth of psychrotrophic bacteria, such as *Pseudomonas* species, in raw milk reduces the risk of residual heat-resistant proteases and lipases being present in the final product and causing bitterness, gelation and rancidity (see Section 6.1.3.5). Reviews on the use of CO_2 have been published by Hotchkiss *et al.* (2006), Singh *et al.* (2011) and Lee (2014).

CO_2 is a colourless, odourless gas which is naturally present in milk, albeit at low concentrations (<5% by volume) and is generally regarded as safe (GRAS). When added to milk and dairy products, it has little or no effect on taste, appearance or aroma if present at concentrations less than the threshold of 11.9 mM (Hotchkiss *et al.*, 1999). It is usually added to cold milk at atmospheric pressure where it reaches a saturation concentration of ~30 mM. This reduces the pH of milk from ~6.7 to 6.0-6.2 due to the formation of carbonic acid. This change is reversible through removal of the CO_2. Higher concentrations in milk can be achieved if the CO_2 is added under pressure but this can cause instability of the protein (King & Mabbit, 1982; Tomasula *et al.*, 1995).

The antimicrobial effect of CO_2 treatment increases with CO_2 concentration and treatment temperature but is also dependent on the microbial species in the product (Martin *et al.*, 2003). Gram-negatives such as *Pseudomonas* are more susceptible than Gram-positives such as *Lactobacillus* (Martin *et al.*, 2003). Spores are the least susceptible. The effect is mostly maifested in an extension of the lag phase of growth (Martin *et al.*, 2003; Dechemi *et al.*, 2005; Liang *et al.*, 2007).

At elevated temperatures, carbon dioxide has the additional bactericidal effect of reducing the heat resistance of the organisms. Loss and Hotchkiss (2002) reported significant reductions in the D-values of *Ps. fluorescens* and spores of *B. cereus* when heated in milk containing 33 mM CO_2: for *P. fluorescens* the D-value at 50 °C decreased from 14.4 to 7.2 min. while for *B. cereus* spores the D-value at 89 °C decreased from 5.56 to 5.29 min. This effect may be particularly significant for ESL and UHT processing as lower temperatures may be able to be used to cause the same bactericidal effect as when CO_2 is not present. In contrast, Guirguis *et al.* (1984) found that heating reconstituted skim milk, saturated with CO_2 and to which spores of *B. cereus, B. firmus, B. licheniformis,*

B. subtilis and *B. circulans* had been added, at 115 °C for 1 s caused activation of the spores. Furthermore, the spores had enhanced heat stability in the presence of the CO_2. These results suggest that CO_2 may have an adverse effect on the keeping quality of heat-treated milk and milk products.

Degassing or removal of CO_2 before heat treatment is recommended as it can enhance fouling of heat exchangers (Calvo & Derafael, 1995) and affect the taste of milk; it produces a tactile sensation if present above the flavour threshold of 11.9 mM (Hotchkiss *et al.* 1999). CO_2 can be removed from milk by vacuum degassing systems (Ruas Madiedo *et al.*, 1996; Hotchkiss *et al.*, 2006).

While the decrease in pH has an inhibitory effect on the growth of bacteria, it has been conclusively shown that this is not the major mechanism of the CO_2 action. The dissolved carbon dioxide has been shown to be more significant than the carbonic acid with which it is in equilibrium in solution. According to Loss and Hotchkiss (2002), there are four main explanations for the antimicrobial properties of CO_2:

- displacement of oxygen
- lowering of pH
- disruption of cell membrane
- ability to disturb the enzymes in the cytoplasm

While displacement of oxygen and lowering the pH may contribute to the antibacterial effect of CO_2, the effects on the cell metabolism through, inter alia, disruption of cell membranes and disturbing cytoplasmic enzymes appear to be the main factors involved. CO_2 is absorbed much faster than oxygen into cells. It is non-polar and is readily taken up by the lipid bilayer of the cell membranes. It has been observed to stress cells through the resulting effects on the membrane properties, including decreased ionic permeability (Sears & Eisenberg, 1961). Once the CO_2 has entered the cell it reduces the intracellular pH and interferes with certain cytoplasmic enzymes. For example, King and Nagel (1975) reported that it affected isocitrate dehydrogenase, malate dehydrogenase and decarboxylating enzymes in *Pseudomonas*.

Dechemi *et al.* (2005) investigated the effect of mixtures of CO_2 and N_2 on raw milk. The treated milks were stored at 7 °C for 10 days. More inhibition of psychrotrophic bacteria occurred when a mixture of 50% CO_2 and 50% N_2 was used than when 100% CO_2 was applied. No protease or lipase activities were detectable with pure CO_2 or a 50% CO_2 and 50% N_2 mixture after 8 days at 7 °C whereas both enzymes were detected in the untreated control after 4 days. The effect on protease and lipase production is consistent with the report by Ma *et al.* (2003) that high-quality raw milk treated with 1500 mg/L (34 mM) CO_2 could be stored at 4 °C for 14 d with minimal proteolysis and lipolysis.

Typical reported shelf-life extensions for raw milk after addition of CO_2 are: 4 days at 4 °C (Rajagopal *et al.*, 2005), 3.5 days at 7 °C (Mabbit, 1982) and 4 days at 2-10 °C (Hotchkiss *et al.*, 2006). Hence a CO_2 treatment can be viewed as an alternative to thermisation (see Section 2.2). Some authors have concluded that greater benefit is achieved by the addition of CO_2 to a high-quality raw milk as compared to poor quality milk (e.g., Mabbit, 1982).

Fewer studies have been carried out on addition of CO_2 to pasteurised milk because of the limitation of food regulations and the risk of causing an adverse sensory effect in the milk. However, from studies that have been done, beneficial microbiological and related shelf-life effects have been observed. For example, Duthie *et al.* (1985) found that whole pasteurised milk with CO_2 added at 1.81 to 3.18 mM and stored at 6 °C showed reduced psychrotrophic bacterial counts and superior keeping quality to control milk without added CO_2. After 14 days, the treated milk scored significantly higher than the untreated milk. Similarly, Hotchkiss *et al.* (1999) found that CO_2 inhibited the growth of psychrotrophic bacteria in pasteurised milk and caused a moderate extension of shelf-life.

Vianna *et al.* (2012) demonstrated the beneficial effect of CO_2 addition to raw milk on the quality of UHT milk. The UHT milk was produced from raw milk treated with CO_2 (to pH 6.2) and an unheated control after storage of the raw milks for 6 days at 4 °C. Less proteolysis and lipolysis occurred in the UHT from the untreated milk than in the treated milk during 120 days storage.

10.9.1 High Pressure Carbon Dioxide

Section 10.9 discussed the use of CO_2 at or below atmospheric pressure. At these pressures, bacterial growth is inhibited but the bacterial cells are not killed. There is now considerable interest in CO_2 application at elevated pressure [high-pressure carbon dioxide (HPCD)] because at these pressures microbial cells can be inactivated. HPCD is a collective term which includes liquid (subcritical) CO_2, supercritical CO_2 and dense-phase CO_2 (see review by Garcia-Gonzalez *et al.*, 2007). The pressures that have been used in HPCD treatment of milk range from 2 to 30 MPa at temperatures from 20 to 50 °C and for up to 300 min duration (Erkmen, 1997, 2001; Calvo & Balcones, 2001; Werner & Hotchkiss, 2006; Liao *et al.*, 2014). A good example of the effect of HPCD treatment is the report by Liao *et al.* (2014). They studied the effect of HPCD on the natural flora of raw milk and found a maximum reduction in total aerobic bacteria of ~5 log by a treatment at 25 MPa/50 °C for 70 min. The reduction increased with the treatment pressure, temperature and duration. Milk which had been treated showed no significant change over 15 days at 4 °C which suggests HPCD has the potential for extending the shelf-life of milk or milk products. Furukawa *et al.* (2009) reported that HPCD at 10 MPa/35 °C for 1 min caused a reduction of ~6 log of strains of the pathogens *Ps. aeruginosa, Aeromonas hydrophila, Salmonella enteritidis, S.* Typhimurium, *Yersinia enterocolitica, E. coli* O157 (two strains), *St. aureus* and *L. monocytogenes*.

Bacterial spores can also be inactivated by HPCD. Watanabe *et al.* (2003) studied the effect of HPCD treatment on spores of *B. coagulans, B. subtilis, B. cereus, B. licheniformis,* and *G. stearothermophilus*. While *G. stearothermophilus* is known to be the most heat-resistant of these organisms, it was not the most resistant to HPCD. Treatment of *G. stearothermophilus* spores at 30 MPa/95 °C for 120 min caused a 5-log reduction; in contrast, neither a heat treatment of 95 °C for 120 min nor a high-pressure treatment (without CO_2) of 30 MPa/95 °C for 120 min had a significant effect, indicating the substantial additional sporicidal effect of the CO_2. Ballestra and Cuq (1998) concluded that the inactivation of spores by HPCD treatment was highest when performed at a temperature above a certain threshold for each organism; this was 80 °C for *B. subtilis*. They maintained that this threshold temperature had to be sub-lethal or "low lethal" for maximal effect. The additional effect of the CO_2 over pure pressure treatment was also evident in a study on the germination of bacterial spores (Furukawa *et al.*, 2004).

Approximately 40% of *B. coagulans* and 70% of *B. licheniformis* spores were germinated after HPCD treatment at 6.5 MPa/35 °C for 120 min; in contrast, high-pressure treatment at 6.5 MPa/35 °C for 120 min resulted in no germination. Another spore-related effect of HPCD is the reduction in the heat tolerance of spores after treatment (Watanabe *et al.*, 2003). Spores of *B. coagulans* and *B. licheniformis* were inactivated completely in 30 min at 90 °C after HPCT at 6.5 MPa/35 °C.

HPCD causes coagulation of casein. Tomasula *et al.* (1995, 1997) developed a process for production of casein using HPCD. They found maximal coagulation at 2.7 to 5.52 MPa/38-49 °C for 5 min although the residence time was not significant. This is consistent with the report by Calvo and Balcones (2001) that HPCD treatment at 4 MPa/40 °C for 3 min caused coagulation of 85% of the casein in milk. It could be reasonably presumed that the coagulation is caused by a reduction in the pH of milk; however, Tomasula *et al.* (1995) found the pH of their whey to be 6.0 and Erkmen (2001) reported the pH of skim milk after HPCD treatment at 10 MPa/30 °C to be 5.84-5.96 (average 5.92). These pH values are considerably higher than the isoelectric point of ~4.6 at which casein is usually precipitated. Therefore, although the microcidal, especially sporicidal, effects of HPCD are interesting, the application of HPCD in the dairy industry may be limited by its destabilising effect on the milk proteins.

References

Altic, L.C., Rowe, M.T. & Grant, I.R. (2007) UV light inactivation of *Mycobacterium avium* subsp. *paratuberculosis* in milk as assessed by FASTPlaqueTB phage assay and culture. *Applied and Environmental Microbiology* **73**(11), 3728–3733.

Alvarez, V.B. & Ji, T.Y. (2003) Emerging processing and preservation technologies for milk and dairy products. In: *Food Science and Food Biotechnology*, pp. 313–327. CRC Press, Boca Raton.

Amador-Espejo, G.G., Gallardo-Chacón, J.J., Juan, B., & Trujillo, A.J. (2016) Effect of ultra-high pressure homogenization at moderate inlet temperatures on the volatile profile of milk. *Dairy Science and Technology* (in review March 2016).

Amador-Espejo, G.G., Suarez-Berencia, A., Juan, B., Barcenas, M.E. & Trujillo, A.J. (2014) Effect of moderate inlet temperatures in ultra-high-pressure homogenization treatments on physicochemical and sensory characteristics of milk. *Journal of Dairy Science* **97**, 659–671.

Asano, Y., Ihara, K., Mori, T., Mizota, T., Iwatsuki, K. & Sotoyama, K. (2000) Flash sterilization system using a high - pressure homogenizer. *Journal of the Japanese Society for Food Science and Technology (Nippon Shokuhin Kagaku Kogaku Kaishi)* **47**, 130–135.

Ballestra, P. & Cuq, J. L. (1998) Influence of pressurized carbon dioxide on the thermal inactivation of bacterial and fungal spores. *Food Science and Technology-Lebensmittel-Wissenschaft & Technologie* **31**(1), 84–88.

Barbosa-Cáaovas, G. & Bermúdez-Aguirre, D. (2010) Other novel milk preservation technologies: ultrasound, irradiation, microwave, radio frequency, ohmic heating, ultraviolet light and bacteriocins. In: *Improving the Safety and Quality of Milk*, Volume 1, *Milk Production and Processing*. (ed M.W. Griffiths), pp 420–450, Woodhead Publication, Cambridge.

Barbosa-Cánovas, G.V. & Juliano, P. (2008) Food sterilization by combining high pressure and thermal energy. In *Food Engineering Integrated Approaches.* (eds G. F. Gutierrez-Lopez, G.V. Barbosa-Cánovas, J.Welti-Chanes, & E. Parada-Arias). pp. 9–46, Springer, New York.

Barsotti, L., Dumay, E., Mu, T.H., Diaz, M.D.F. & Cheftel, J.C. (2002) Effects of high voltage electric pulses on protein-based food constituents and structures. *Trends in Food Science and Technology* **12**, 136–144.

Bendicho, S., Barbosa-Cánovas, G. V. & Martin, O. (2002) Milk processing by high intensity pulsed electric fields. *Trends in Food Science and Technology* **13**, 195–204.

Bendicho, S., Barbosa-Cánovas, G. V. & Martin, O. (2003) Reduction of protease activity in milk by continuous flow high-intensity pulsed electric field treatments. *Journal of Dairy Science* **86**, 697–703.

Bintsis T., Litopoulou-Tzanetaki, E. & Robinson R.K. (2000) Existing and potential applications of ultraviolet light in the food industry – a critical review. *Journal of the Science of Food and Agriculture* **80**, 637–645.

Bouaouina, H., Desrumaux, A., Loisel, C. & Legrand, J. (2006) Functional properties of whey proteins as affected by dynamic high-pressure treatment. *International Dairy Journal* **16**, 275–284.

Brinez, W.J., Roig-Sagues, A.X., Hernandez-Herrero, M.M. & Guamis-Lopez, B. (2006) Inactivation of two strains of *Escherichia coli* inoculated into whole and skim milk by ultrahigh-pressure homogenisation. *Lait* **86**, 241–249.

Burton, H. (1951) Ultra-violet irradiation of milk. *Dairy Science Abstracts* **13**, 229–244.

Burton, H. (1988) *Ultra-high Temperature Processing of Milk and Milk Products.* Elsevier Applied Science, London.

Calvo M.M. & Derafael, D. (1995) Deposition formation in a heat exchanger during pasteurisation of CO_2 acidified milk. *Journal of Dairy Research* **62**, 641–644.

Calvo, M.M. & Balcones, E. (2001) Inactivation of microorganisms and changes of proteins during treatment of milk with subcritical carbon dioxide. *Milchwissenschaft* **56**(7), 366–369.

Cappozzo, J.C., Koutchma, T. & Barnes G. (2015) Chemical characterization of milk after treatment with thermal (HTST and UHT) and non-thermal (turbulent flow ultraviolet) processing technologies. *Journal of Dairy Science* **98**, 5068–5079.

Cilliers, F.P., Gouws, P.A., Koutchma, T., Engelbrecht, Y., Adriaanse, C. & Swart, P. (2014) A microbiological, biochemical and sensory characterisation of bovine milk treated by heat and ultraviolet (UV) light for manufacturing Cheddar cheese. *Innovative Food Science & Emerging Technologies* **23**, 94–106.

Considine, T., Patel, H.A., Anema, S.G., Singh, H. & Creamer, L.K. (2007) Interactions of milk proteins during heat and high hydrostatic pressure treatments - A Review. *Innovative Food Science* and *Emerging Technologies* **8**, 1–23.

Craven, H.M., Swiergon, P., Ng, S., Midgely, J., Versteeg, C., Coventry, M. J. & Wan, J. (2008) Evaluation of pulsed electric field and minimal heat treatments for inactivation of pseudomonads and enhancement of milk shelf-life. *Innovative Food Science and Emerging Technologies* **9**, 211–216.

Crawford, L.M. & Ruff, E.H. (1996) A review of the safety of cold pasteurization through irradiation. *Food Control* **7**(2), 87–97.

Datta, N. & Deeth, H.C. (1999) High pressure processing of milk and milk products. *Australian Journal of Dairy Technology* **54**, 41–48.

Datta, N., Hayes, M.G., Deeth, H.C. & Kelly, A.L. (2005) Significance of frictional heating for effects of high pressure homogenisation on milk. *Journal of Dairy Research* **72**, 393–399.

Datta, N. & Tomasula, P.M. (2015) *Emerging Dairy Processing Technologies: Opportunities for the Dairy Industry*. John Wiley and Sons, New York.

de Luis, R., Arias, O., Puertolas, E., Benede, S., Sanchez, L., Calvo, M. & Perez, M. D. (2009) Effect of high-intensity pulse electric fields on denaturation of bovine whey proteins. *Milchwissenschaft* **64**, 422–426.

de Oliveira Silva, A.C., De Oliveira, L.A.T., De Jesus, E.F.O., Cortez, M.A.S., Alves, C. C. C., Monteiro, M. L. G., & Conte, C. A. (2015) Effect of gamma irradiation on the bacteriological and sensory analysis of raw whole milk under refrigeration. *Journal of Food Processing and Preservation* **39**(6), 2404–2411.

Dechemi, S., Benjelloun, H. & Lebeault, J.M. (2005) Effect of modified atmospheres on the growth and extracellular enzyme activities of psychrotrophs in raw milk. *Engineering in Life Sciences* **5**, 350–356.

Deeth, H.C. (2005) Lipoprotein lipase and lipolysis in milk. *International Dairy Journal* **16**, 555–562.

Deeth, H.C. and Datta, N. (2011) Heat treatment of milk | Non-thermal technologies: Introduction. In: *Encyclopedia of Dairy Sciences* (eds H. Roginski, J.W. Fuquay & P.F. Fox), 2nd edn, pp. 725–731. Academic Press, London.

Deeth, H.C. and Datta, N. (2011) Heat treatment of milk | Non-thermal technologies: Pulsed electric field technology and ultrasonication. In: *Encyclopedia of Dairy Sciences* (eds H. Roginski, J.W. Fuquay & P.F. Fox), 2nd edn, pp. 738–743. Academic Press, London.

Deeth, H.C., Datta, N & Versteeg, C. (2013) Non thermal technologies In *Advances in Dairy Ingredients* (eds G. Smithers & M.A. Augustin). pp 161–216. Wiley-Blackwell Publishing, Ames, Iowa.

Deeth, H.C., Datta, N. Ross, A.I.V. & Dam, X.T (2007) Pulsed electric field technology: Effect on milk and fruit juices. In: *Advances in Thermal and Non-thermal Food Preservation*.(eds: G. Tewari & V.K. Juneja), pp. 241–267. Blackwell Publishing, Ames.

Diels, A.M. J. & Michiels, C.W. (2006) High-pressure homogenization as a non-thermal technique for the inactivation of microorganisms. *Critical Reviews in Microbiology* **32**, 201–216.

Dissanayake, M. & Vasilejvic, T. (2009) Functional properties of whey proteins affected by heat treatment and hydrodynamic high-pressure shearing. *Journal of Dairy Science* **92**, 1387–1397.

Duthie, C., Shipe, W. & Hotchkiss, J. (1985) Effect of low level carbonation on the keeping quality of processed milk. *Journal of Dairy Science* **68**, 69.

EFSA (2016) Safety of UV-treated milk as a novel food pursuant to Regulation (EC) No 258/97. Report of Panel on Dietetic Products, Nutrition and Allergies. *EFSA Journal* **14**(1), 1–14.

Elwell, M.W & Barbano, D.M. (2006) Use of microfiltration to improve fluid milk quality. *Journal of Dairy Science* **89**(E. Suppl.), E10–E30.

Erkmen, O. (1997) Antimicrobial effect of pressurized carbon dioxide on *Staphylococcus aureus* in broth and milk. *LWT* **30**, 826–829.

Erkmen, O. (2001) Effects of high-pressure carbon dioxide on *Escherichia coli* in nutrient broth and milk. *International Journal of Food Microbiology* **65**(1-2), 131–135.

Evrendilek, G.A., Dantzer, W.R., Streaker, C.B., Ratanatriwong, P. & Zhang, Q.H. (2001) Shelf-life evaluations of liquid foods treated by pilot plant pulsed electric field system. *Journal of Food Processing and Preservation* **25**, 283–297.

FDA (2015) Ultraviolet radiation for the processing and treatment of food. In: *Irradiation in the Production, Processing, and Handling of Food.* Code of Federal Regulations Title 21, section 179.39. Office of the Federal Register. Government Printing Office Washington, DC, USA.

Feijoo, S.C., Hayes, W.W., Watson, C.E. & Martin, J.H. (1997) Effects of Microfluidizer technology on *Bacillus licheniformis* spores in ice cream mix. *Journal of Dairy Science* **80**, 2184–2187.

Floury, J., Grosset, N., Leconte, N., Pasco, M., Madec, M. N. & Jeantet, R. (2006) Continuous raw skim milk processing by pulsed electric field at non-lethal temperature: effect on microbial inactivation and functional properties. *Lait* **86**, 43–57.

Fornari, C., Maggi, A., Gola, S., Cassara, A. & Manachini, P.L. (1995) Inactivation of *Bacillus* endospores by high-pressure treatment. *Industria Conserve* **70**(3), 259–265.

Fryer, P.J. & Versteeg, C. (2008) Processing technology innovation in the food industry. *Innovation: Management, Policy and Practice* **10**, 74–90.

Furukawa, S., Watanabe, T., Koyama, T., Hirata, J., Narisawa, N., Ogihara, H. & Yamasaki, M. (2009) Inactivation of food poisoning bacteria and *Geobacillus stearothermophilus* spores by high pressure carbon dioxide treatment. *Food Control* **20**(1), 53–58.

Furukawa, S., Watanabe, T., Tai, T., Hirata, J., Narisawa, N., Kawarai, T., Ogihara, H. & Yamasaki, M. (2004) Effect of high pressure gaseous carbon dioxide on the germination of bacterial spores. *International Journal of Food Microbiology* **91**(2), 209–213.

Garcia-Gonzalez, L., Geeraerd, A. H., Spilimbergo, S., Elst, K., Van Ginneken, L., Debevere, J., Van Impe, J. F. & Devlieghere, F. (2007) High pressure carbon dioxide inactivation of microorganisms in foods: The past, the present and the future. *International Journal of Food Microbiology* **117**(1), 1–28.

García-Risco, M.R., Cortés, E., Carrascosa, A.V. & López-Fandiño, R. (1998) Microbiological and chemical changes in high -pressure-treated milk during refrigerated storage. *Journal of Food Protection* **61**,735–737.

Gaucheron, F., Famelart, M.H., Mariette, F., Raulot, K., Michel, F. & Le Graet, Y. (1997) Combined effects of temperature and high-pressure treatments on the physicochemical characteristics of skim milk. *Food Chemistry* **59**, 439–447.

Grahl, T. & Markl, H. (1996) Killing of microorganisms by pulsed electric fields. *Applied Microbiology and Biotechnology* **45**, 148–157.

Guerrero-Beltrán J.A. & Barbosa-Cánovas, G.V. (2004) Advantages and limitations on processing foods by UV light., *Food Science & Technology International* **10**, 137–147.

Guirguis, A.H., Griffiths, M.W. & Muir, D.D. (1984) Spore-forming bacteria in milk. II. Effect of carbon dioxide addition on heat activation of spores of *Bacillus* species. *Milchwissenschaft* **39**, 144–146.

Hardham, J.F., Imison, B.W. & Frenc, H.M. (2000) Effect of homogenisation and microfluidisation on the extent of fat separation during storage of UHT milk. *Australian Journal of Dairy Technology* **55**, 16–21.

Hayakawa, I., Furukawa, S., Midzunaga, A., Horiuchi, H., Nakashima, T., Fujio, Y., Yano, Y., Ishikura, T. & Sasaki, K. (1998) Mechanism of inactivation of heat-tolerant spores of *Bacillus stearothermophilus* IFO 12550 by rapid decompression *Journal of Food Science* **63**, 371–374.

Hayakawa I., Kanno, T., Yoshiyama, K. & Fujio, Y. (1994b) Oscillatory compared with continuous high pressure sterilization on *Bacillus stearothermophilus* spores. *Journal of Food Science* **59**, 164–167.

Hayakawa, I., Kanno, T., Tomita, M. & Fujio, Y. (1994a) Application of high pressure for spore inactivation and protein denaturation. *Journal of Food Science* **59**, 159–163.

Hayes, M.G. & Kelly, A.L. (2003a) High pressure homogenisation of raw whole bovine milk (a) effects on fat globule size and other properties. *Journal of Dairy Research* **70**, 297–305.

Hayes, M.G. & Kelly, A.L. (2003b) High pressure homogenisation of milk (b) effects on indigenous enzymatic activity. *Journal of Dairy Research* **70**, 307–313.

Hayes, M.G., Fox, P.F. & Kelly, A.L. (2005) Potential applications of high pressure homogenisation in processing of liquid milk. *Journal of Dairy Research* **72**, 25–33.

Hinrichs J. & Kessler H.G. (1997) Kinetics of pressure-induced denaturation of whey proteins at different temperatures and functional properties. In *High Pressure Research in the Biosciences and Biotechnology*, (ed K. Heremans). pp 407–410, Leuven University Press, Leuven, Belgium.

Hite, B.H. (1899) The effect of pressure in the preservation of milk. *Bulletin of the West Virginia Agricultural Experimental Station* **58**, 15–35.

Hite, B.H., Giddings, N.J. & Weakley, C.E. (1914) The effect of pressure on certain microorganisms encountered in the preservation of fruits and vegetables. *Bulletin of the West Virginia Agricultural Experimental Station* **146**, 1–67.

Ho S.Y., Mittal G.S., Cross J.D. & Griffiths, M.W. (1995) Inactivation of *Pseudomonas fluorescens* by high volatage electric fields. *Journal of Food Science* **60**, 1337–1343.

Hoffmann, W., Kiesner, C., Clawin-Radecker, I., Martin, D., Einhoff, K., Lorenzen, P. C., Meisel, H., Hammer, P., Suhren, G. & Teufel, P. (2006) Processing of extended shelf life milk using microfiltration. *International Journal of Dairy Technology* **59**, 229- 235.

Hotchkiss, J.H., Chen, J.H. & Lawless, H.T. (1999) Combined effects of addition of carbon dioxide and barrier films on microbial and sensory changes in pasteurised milk. *Journal of Dairy Science* **82**, 690–695.

Hotchkiss, J.H., Werner, B.G. & Lee, E.Y.C. (2006) Addition of carbon dioxide to dairy products to improve quality: a comprehensive review. *Comprehensive Reviews in Food Science and Food Safety* **5**, 158–168.

Huppertz, T., Fox, P.F. & Kelly, A.L. (2004c) Plasmin activity and proteolysis in high pressure-treated bovine milk. *Le Lait* **84**, 297–304.

Huppertz, T., Smiddy, M.A., Upadhyay, V.K. & Kelly, A.L. (2006) High-pressure-induced changes in bovine milk: a review. *International Journal of Dairy Technology* **59**, 58–66.

Iordache, M. & Jelen, P. (2003) High pressure microfluidization treatment of heat denatured whey proteins for improved functionality. *Innovative Food Science and Emerging Technologies* **4**, 367–376.

Iucci, L., Lanciotti, R., Kelly A. & Huppertz, T (2008) Plasmin activity in high-pressure-homogenised bovine milk. *Milchwissenschaft* **63**, 68–70.

Jayaram, S., Castle, G.S.P., and Margaritis, A. (1993) The effects of high field DC pulse and liquid medium conductivity on survivability of *Lactobacillus brevis*. *Applied Microbiology and Biotechnology* **40**, 117–122.

Jose-Brinez, W., Roig-Sagues, A.X., Hernandez-Herrero, M.M. & Guamis-Lopez, B. (2007) Inactivation of *Staphylococcus* spp. strains in whole milk and orange juice using ultra high pressure homogenisation at inlet temperatures of 6 and 20 °C. *Food Control* **18**, 1282–1288.

Kanno, C., Mu, T.-H., Hagiwara, T., Ametani, M. & Azuma, N. (1998) Gel formation from industrial milk whey proteins under hydrostatic pressure: Effects of hydrostatic pressure and protein concentration. *Journal of Agriculture and Food Chemistry* **46**, 417–424.

Kelly, P.M (2011) Membrane-based fractionation. *Encyclopedia of Dairy Sciences*. (eds J.W. Fuquay, P.F. Fox & P.L.H. McSweeney) 2nd edn, Volume 3, pp. 864-872. Elsevier Academic Press, Amsterdam.

Kiełczewska, K., Kruk, A., Czerniewicz, M. & Kopeć, M. (2009) Effect of pressure on constituents of the colloidal phase of milk. *Milchwissenschaft* **64**, 358–360.

King, A.D. & Nagel, C.W. (1975) Influence of carbon dioxide upon metabolism of *Pseudomonas aeruginosa*. *Journal of Food Science* **40**, 362–366.

King, J.S. & Mabbitt, L.A. (1982) Preservation of raw milk by the addition of carbon dioxide. *Journal of Dairy Research* **49**, 439–447.

Kong, F., Liu, L., Zhang, S. & LuJiaPing (2012) Microfiltration of ceramic membrane for improving UHT milk quality. *Transactions of the Chinese Society of Agricultural Engineering* **28**(17), 268–274.

Koutchma, T. (2009) Advances in ultraviolet light technology for non-thermal processing of liquid foods. *Food and Bioprocess Technology* **2**, 138–155.

Kristo, E., Hazizaj, A. & Corredig, M. (2012) Structural changes imposed on whey proteins by UV irradiation in a continuous UV light reactor. *Journal of Agricultural and Food Chemistry* **60**(24), 6204–6209.

Lacroix, M. (2005) Irradiation of foods. In: *Emerging Technologies in Food Processing*. (ed D.-W. Su), pp. 353–386. Elsevier Academic Press, San Diego.

Lamela, C. P. & Torres, J. A. (2008) Pressure processing of foods: an alternative for high flavour quality foods retaining health enhancing factors - Part 1. *AgroFood Industry hi-tech* **19**, 60–62.

Lanciotti, R., Vannini, L., Patrignani, F., Lucci, L., Vallicelli, M., Ndagijimana, M. & Guerzoni, M. E. (2006) Effect of high pressure homogenisation of milk on cheese yield and microbiology, lipolysis and proteolysis during ripening of Caciotta cheese. *Journal of Dairy Research* **73**, 216–226.

Leadley, C.E. & Williams, A. (2006) Pulsed electric field processing, power ultrasound and other emerging technologies in *Food Processing Handbook*, (ed J.G. Brennan), Wiley-VCH, Weinheim.

Lee, E.Y.C. (2014) *The Effect of Carbon Dioxide on Physico-chemical and Microbiological Properties of Skim Milk and Protein Concentrates*. PhD thesis, University of Queensland, Brisbane, Australia.

Lee, J.W., Oh, S.H., Byun, E.B., Kim, J.H., Woon, J.H. & Byun, M.W. (2007) Inactivation of *Enterobacter sakazakii* of dehydrated infant formula by gamma-irradiation. *Radiation Physics and Chemistry* **76**(11-12), 1858–1861.

Li, S.Q., Bomser, J.A. & Zhang, Q.H. (2005) Effects of pulsed electric fields and heat treatment on stability and secondary structure of bovine immunoglobulin G. *Journal of Agricultural and Food Chemistry* **53**, 663–670.

Li, S.Q., Zhang, Q.H., Lee, Y.Z. & Pham, T.V. (2003) Effects of pulsed electric fields and thermal processing on the stability of bovine immunoglobulin G (IgG) in enriched soymilk. *Journal of Food Science* **68**, 1201–1207.

Liang, Y., Li, X. & Bai, S. (2007) Research on CO_2 extending the shelf-life of raw milk and dairy products. *China Dairy Industry* **35**, 53–56.

Liao, H.M., Zhong, K., Liao, X.J. & Hu, X.S. (2014) Inactivation of microorganisms naturally present in raw bovine milk by high-pressure carbon dioxide. *International Journal of Food Science and Technology* **49**(3), 696–702.

López-Fandiño, R. (2006) High pressure-induced changes in milk proteins and possible applications in dairy technology. *International Dairy Journal* **16**, 1119–1131.

López-Fandiño, R., De La Fuente, M.A., Ramos, M. & Olano, A. (1998) Distribution of minerals and proteins between soluble and colloidal phases of pressurized milks from different species. *Journal of Dairy Research* **65**, 69–78.

Loss, C.R. & Hotchkiss, J.H. (2002) Effect of dissolved carbon dioxide on thermal inactivation of microorganisms in milk. *Journal of Food Protection* **65**, 1924–1929.

Ludikhuyze, L., Claeys, W. & Hendrickx, M. (2001) Effect of temperature and/or pressure on lactoperoxidase activity in bovine milk and acid whey. *Journal of Dairy Research* **68**, 625–637.

Ma, Y., Barbano, D.M. & Santos, M. (2003) Effect of CO_2 addition to raw milk on proteolysis and lipolysis at 4 °C. *Journal of Dairy Science* **86**, 1616–1631.

Mabbit, L.A. (1982) Preservation of refrigerated milk. *Kieler Milchwirtschaftliche Forschungsberichte* **34**, 28–31.

Mahmoud, N.S. & Ghaly, A.E. (2004) On-line sterilization of cheese whey using ultraviolet radiation. *Biotechnology Progress* **20**(2), 550–560.

Martin, J.D., Werner, B.G. & Hotchkiss, J.H. (2003) Effects of carbon dioxide on bacterial growth parameters in milk as measured by conductivity. *Journal of Dairy Science* **86**, 1932–1940.

Maubois, J.-L. (2002) Membrane microfiltration: a tool for a new approach in dairy technology. *Australian Journal of Dairy Technology* **57**, 92–96.

Messens, W., Van Camp, J. & Dewettinck, K. (2003) High pressure processing to improve dairy quality. In *Dairy Processing: improving quality*, (ed G. Smit). pp. 310–332. CRC Press, New York.

Meyer, R.S. (2000) Ultra high pressure, high temperature food preservation process. US patent 6,017,572.

Meyer, R.S., Cooper, K.L., Knorr, D. & Lelieveld, H.L.M. (2000) High-pressure sterilization of foods. *Food Technology* **54**(11), 67–68, 70, 72.

Mugadza, D. & Buys, E. (2015) Characterization of *B. cereus* in extended shelf-life (ESL) milk. *International Association for Food Protection Conference*. pp. 2–138. Portland, Oregon.

Nakayama, A, Yano, Y, Kobayashi, S, Ishikawa, M & Sakai, K. (1996) Comparison of pressure resistances of spores of six *Bacillus* strains with their heat resistances. *Applied and Environmental Microbiology* **62**, 3897–3900.

Odriozola-Serrano, I., Bendicho-Porta, S. & Martin-Belloso, O. (2006) Comparative study on shelf life of whole milk processed by high-intensity pulsed electric field or heat treatment. *Journal of Dairy Science* **89**, 905–911.

Osaili, T.M., Shaker, R.R., Abu Al-Hasan, A.S., Ayyash, M.M. & Martin, E.M. (2007) Inactivation of *Enterobacter sakazakii* in infant milk formula by gamma irradiation: Determination of D_{10}-value. *Journal of Food Science* **72**(3), M85–M88.

Pandey, P.K. & Ramaswamy, H.S. (2004) Effect of high-pressure treatment of milk on lipase and glutamyl transferase activity. *Journal of Food Biochemistry* **28**, 449–462.

Paquin, P. (1999) Technological properties of high pressure homogenizers: the effect of fat globules, milk proteins and polysaccharides. *International Dairy Journal* **9**, 329–335.

Patel, H.A., Carroll, T. & Kelly, A.L. (2008) Nonthermal preservation technologies for dairy applications. In: *Dairy processing and Quality Assurance*, (eds R.C. Chandan, A. Kilara & N.P. Shah). pp. 465–482, Wiley-Blackwell, New Jersey.

Pereda, J., Ferragut, V., Quevedo, J.M., Guamis, B. & Trujillo, A.J. (2007) Effects of ultra-high pressure homogenization on microbial and physicochemical shelf life of milk. *Journal of Dairy Science* **90**, 1081–1093.

Picart, L., Thiebaud, M., Rene, M., Guiraud, J.P., Cheftel, J.C. & Dumay, E. (2006) Effects of high pressure homogenisation of raw bovine milk on alkaline phosphatase and microbial inactivation. A comparison with continuous short-time thermal treatments. *Journal of Dairy Research* **73**, 454–463.

Poliseli-Scopel, F. H., Hernandez-Herrero, M., Guamis, B. & Ferragut, V. (2014) Sterilization and aseptic packaging of soymilk treated by ultra high pressure homogenization. *Innovative Food Science & Emerging Technologies* **22**, 81–88.

Qin, B.-L., Chang F.-J., Barbosa-Cánovas, G.V. & Swanson, B.G. (1995) Nonthermal inactivation of *Saccharomyces cerevisiae* in apple juice using pulsed electric fields. *Lebensmittel Wissenschaft und Technologie* **28**, 564–568.

Rademacher, B., Hinrichs J. & Kessler, H.G. (1997) New dairy products at 1000 MPa. *DMZ-Lebensmittelindustrie-und-Milchwirtschaft* **117**, 1035–1039.

Rajagopal, M., Werner, B.G. & Hotchkiss, J.H. (2005) Low pressure CO_2 storage of raw milk: microbiological effects. *Journal of Dairy Science* **88**, 3130–3138.

Ramirez, R., Saraiva, J., Pèrez, C., Lamela, J. & Torres, A. (2009) Reaction kinetics analysis of chemical changes in pressure-assisted thermal processing *Food Engineering Review* **1**, 16–30.

Reddy, N.R., Solomon, H.M., Fingerhut, G.A., Rhodehamel, E.J., Balasubramaniam, V.M. & Palaniappan, S. (1999) Inactivation of *Clostridium botulinum* type E spores by high pressure processing. *Journal of Food Safety* **19**, 277–288.

Reinemann, D.J., Gouws, P., Cilliers, T., Houck, K. & Bishop, J. R. (2006) New methods for UV treatment of milk for improved food safety and product quality. *ASABE (*American Society of Agricultural and Biological Engineers) *Paper Number: 066088* presented at 2006 ASABE Annual International Meeting, Portland, Oregon, USA.

Roach, A. & Harte, F. (2008) Disruption and sedimentation of casein micelles and casein micelle isolates under high-pressure homogenization. *Innovative Food Science and Emerging Technologies* **9**, 1–8.

Ross, A.I.V., Griffiths, M.W., Mittal, G.S. &. Deeth H.C. (2003) Combining non-thermal technologies to control foodborne microorganisms. *International Journal of Food Microbiology* **89**,125–138.

Rossitto, P. V., Cullor, J. S., Crook, J., Parko, J., Sechi, P. & Cenci-Goga, B. T. (2012) Effects of UV irradiation in a continuous turbulent flow UV reactor on microbiological and sensory characteristics of cow's milk. *Journal of Food Protection* **75**(12), 2197–2207.

Ruas Madiedo, P., Bada Gancedo, J.C., Fernandez Garcia, E., Gonzalez de Llano, D. & Reyes Gavilan, C.G. (1996) Preservation of the microbiological and biochemical quality of raw milk by carbon dioxide addition: a pilot-scale study. *Journal of Food Protection* **59**, 502–508.

Sadoun, D., Couvercelle, C., Strasser, A., Egler, A. & Hasselmann, C. (1991) Low-dose irradiation of liquid milk. *Milchwissenschaft* **46**(5), 295–299.

Sampedro, F., Rodrigo, M., Martinez, A., Rodrigo, D. & Barbosa-Cánovas, G.V. (2005) Quality and safety aspects of PEF application in milk and milk products. *Critical Reviews in Food Science and Nutrition* **45**, 25–47.

Sandra, S. & Dalgleish, D.G. (2005) Effects of ultra-high-pressure homogenization and heating on structural properties of casein micelles in reconstituted skim milk powder. *International Dairy Journal* **15**, 1095–1104.

Scurrah, K.J., Robertson, R.E., Craven, H.M., Pearce, L.E. & Szabo, E.A. (2006) Inactivation of *Bacillus* spores in reconstituted skim milk by combined high pressure and heat treatment. *Journal of Applied Microbiology* **101**, 172–180.

Sears, D. & Eisenberg, R. (1961) A model representing a physiological role of CO_2 at the cell membrane. *Journal of General Physiology* **44**, 869–887.

Sepulveda, D.R., Gongora-Nieto, M.M., Guerrero, J.A. & Barbosa-Cánovas, G.V. (2009) Shelf life of whole milk processed by pulsed electric fields in combination with PEF-generated heat. *LWT-Food Science and Technology* **42**, 735–739.

Sepulveda, D.R., Gongora-Nieto, M.M., Guerrero, J.A. & Barbosa-Cánovas, G.V. (2005) Production of extended-shelf life milk by processing pasteurized milk with pulsed electric fields. *Journal of Food Engineering* **67**, 81–86.

Serra, M., Trujillo, A.J., Pereda, J., Guamis, B. & Ferragut, V. (2008) Quantification of lipolysis and lipid oxidation during cold storage of yogurts produced from milk treated by ultra-high pressure homogenization. *Journal of Food Engineering* **89**, 99–104.

Serra, M., Trujillo, A.J., Quevedo, J.M., Guamis, B. & Ferragut, V. (2007) Acid coagulation properties and suitability for yogurt production of cows' milk treated by high-pressure homogenisation. *International Dairy Journal* **17**, 782–790.

Shamsi K. (2010) *Effects of Pulsed Electric Field on Milk Properties*. VDM Publishing House Ltd, Beau-Bassin, Mauritius.

Shamsi, K, Versteeg, C., Sherkat, F. & Wan, J. (2008) Alkaline phosphatase and microbial inactivation by pulsed electric field in bovine milk. *Innovative Food Science and Emerging Technologies* **9**, 217–223.

Singh, P., Wani, A.B., Karim, A.A. & Langowski, H.-C. (2011) The use of carbon dioxide in the processing and packaging of milk and dairy products: a review. *International Journal of Dairy Technology* **64**, 1–17.

Smith, K., Mittal, G.S. & Griffiths, M.W. (2002) Pasteurization of milk using pulsed electric field and antimicrobials. *Journal of Food Science* **67**, 2304–2308.

Sobrino-Lopez, A. & Martin-Belloso, O. (2008) Use of nisin and other bacteriocins for preservation of dairy products. *International Dairy Journal* **18**, 329–343.

Stack, A. & Sillen, G. (1998) Bactofugation of liquid milks. *Nutrition and Food Science* **98**, 280–282.

Sui, Q, Roginski, H., Williams, R.P.W., Versteeg, C. & Wan, J. (2010) Effect of pulsed electric field and thermal treatment on the physicochemical and functional properties of whey protein isolate. *International Dairy Journal* **21**, 206–213.

Terebiznik, M.R., Jagus, R.J., Cerrutti,P., de Huergo, M.S. & Pilosof, A.M.R. (2000) Combined effect of nisin and pulsed electric fields on the inactivation of *Escherichia coli*. *Journal of Food Protection* **63**, 741–746.

Thiebaud, M., Dumay, E., Picart, L., Guiraud, J. P. & Cheftel, J.C. (2003) High-pressure homogenisation of raw bovine milk. Effects on fat globule size distribution and microbial inactivation. *International Dairy Journal* **13**, 427–439.

Timson, W.J. & Short, A.J. (1965) Resistance of microorganisms to hydrostatic pressures. *Biotechnology and Bioengineering* 7, 139–159.

Toepfl, S., Mathys A., Heinz V. & Knorr D. (2006) Review: Potential of high hydrostatic pressure and pulsed electric fields for energy efficient and environmentally friendly food processing. *Food Reviews International* **22**, 405–423.

Tomasula, P. M., Craig, J. C. & Boswell, R. T. (1997) A continuous process for casein production using high-pressure carbon dioxide. *Journal of Food Engineering* **33**(3-4), 405–419.

Tomasula, P.M., Craig, J.C., Boswell, R.T., Cook, R.D., Kurantz, M.J. & Maxwell, M. (1995) Preparation of casein using carbon-dioxide. *Journal of Dairy Science* **78**, 506–514.

Tovar-Hernandez G., Peña, H. R., Velazquez, G., Ramirez, J.A. & Torres, J.A. (2005) Effect of combined thermal and high pressure processing on the microbial stability of milk during refrigerated storage. *IFT Annual Meeting New Orleans, Los Angeles,* Abstract 54F-12.

Trujillo, A.J., Capellas, M., Saldo, J., Gervilla, R. & Guamis, B. (2002) Applications of high-hydrostatic pressure on milk and dairy products: a review. *Innovative Food Science and Emerging Technologies* **3**, 295–307.

Trujillo, A.J., Ferragut, V., Gervilla, R., Capellas, M. & Guamis, B. (1997) High pressure effects on milk and milk products. *Recent Research Developments in Agricultural and Food Chemistry* **1**, 137–159.

van Loey, A., Verachtert, B. & Hendrikx, M. (2001) Effects of high electric fields pulses on enzymes. *Trends in Food Science and Technology* **12**, 94–102.

Valencia-Flores, D.C., Hernandez-Herrero, M., Guamis, B. & Ferragut, V. (2013) Comparing the effects of ultra-high-pressure homogenization and conventional thermal treatments on the microbiological, physical, and chemical quality of almond beverages. *Journal of Food Science* **78**(2), E199–E205.

Vega-Mercado, H., Powers, J. R., Barbosa-Cánovas, G.V. & Swanson, B.G. (1995) Plasmin inactivation with pulsed electric fields. *Journal of Food Science* **60**, 1143–1146.

Vega-Mercado, H., Powers, J., Barbosa-Cánovas, G.V., Luedecke, L. & Swanson, B.G. 2001) *Pulsed Electric Fields in Food Processing. Fundamental Aspects and Applications,* (eds G.V. Barbosa-Cánovas & Q.H. Zhang). pp. 105–120. Technomic Publishing Company Inc., Lancaster.

Vianna, P.C.B., Walter, E.H.M., Dias, M.E.F., Faria, J.A.F., Netto, F.M. & Gigante, M.L. (2012) Effect of addition of CO_2 to raw milk on quality of UHT-treated milk. *Journal of Dairy Science* **95**, 4256–4262.

Viazis, S., Farkas, B. E. & Allen, J. C. (2007) Effects of high-pressure processing on immunoglobulin A and lysozyme activity in human milk. *Journal of Human Lactation* **23**, 253–261.

Wan, J., Coventry, M. J., Swiergon, P., Sanguansri, P. & Versteeg, C. (2009) Advances in innovative processing technologies for microbial inactivation and enhancement of food safety – pulsed electric field and low-temperature plasma. *Trends in Food Science and Technology* **20**, 414–424.

Watanabe, T., Furukawa, S., Tai, T., Hirata, J., Narisawa, N., Ogihara, H. & Yamasaki, M. (2003) High pressure carbon dioxide decreases the heat tolerance of the bacterial spores. *Food Science and Technology Research* **9**(4), 342–344.

Werner, B.G. & Hotchkiss, J.H. (2006) Continuous flow nonthermal CO_2 processing: the lethal effects of subcritical and supercritical CO_2 on total microbial populations and bacterial spores in raw milk. *Journal of Dairy Science* **89**, 872–881.

Wuytack, E.Y., Diels, A.M.J. & Michiels, C.W. (2002) Bacterial inactivation by high-pressure homogenisation and high hydrostatic pressure. *International Journal of Food Microbiology* 77, 205–212.

11

Analytical Methods

11.1 Introduction

The analysis of raw materials and final products is crucial for understanding and thereby controlling the factors affecting the safety and quality of heat-treated dairy products. Having access to accurate analytical data is invaluable for ensuring that quality criteria are being met, and standards maintained. The information allows early detection of deviations from normal practice and allows problems to be detected at an early stage, thereby providing an opportunity to solve them before they become too serious. Overall, this will reduce the need for trouble-shooting and reduce the number of consumer complaints. The best control is achieved where problems are anticipated in advance and thereby prevented from occurring.

In a factory situation, the analytical methods used should be as quick and simple as possible. It may also help if they are fully automated, but this comes at a cost. However, where problems persist, one may need to resort to more sophisticated analyses.

This chapter discusses analytical procedures which are available for quality assurance and furthering knowledge of heat-treated products with particular focus on UHT products. It is included because textbooks on methods for examination of dairy products include little information relevant to UHT products (Sundekilde, 2012).

The methods given are not detailed recipes. For this detail the reader is referred to the numerous references provided.

The methods are arranged in alphabetical order in two broad categories: commonly used methods and advanced analytical techniques. In general, methods in the first category can be carried out in industrial laboratories while those in the second category would be available only in specialist laboratories. The methods in this chapter are mostly for chemical and physical analyses; microbiological methods are discussed in Section 8.7.

11.2 Commonly Used Analytical Methods

11.2.1 Amylase

Amylases can cause thinning of starch-based products such as dairy desserts and custard due to degradation of the starch (see Section 7.2.3). They have traditionally been determined by: 1. Reduction in viscosity of a starch paste; 2. Decrease in the

High Temperature Processing of Milk and Milk Products, First Edition. Hilton C. Deeth and Michael J. Lewis.

iodine-staining properties of a starch substrate (Robyt & Ackerman, 1968); and 3. Increase in the reducing power of a starch substrate (Bernfeld, 1955). However, there are several methods based on oligosaccharides as substrates which are purported to be more sensitive. A method based on maltohexaose and glucose determination has been used for bacterial amylase (Saraiva *et al.*, 1996) and reported to be convenient and quantitative, and correlate well with the iodine method of Somogyi (1952) (Ohnishi *et al.*, 1990).

A colourimetric assay method based on the substrate *p*-nitrophenylmaltoheptaoside in the presence of amyloglucosidase and alpha-glucosidase is reputed to be "simple, reliable, accurate and absolutely specific for α-amylase" (Sheehan & McCleary, 1988). It is available as a commercial kit. In a similar enzyme-linked assay, Teshima *et al.* (1985) employed the substrate β-2-chloro-4-nitrophenylmaltopentaoside. The *p*-nitrophenylmaltoheptaoside method has been used successfully for measuring α-amylase in enzyme-based detergent residues in dairy products (Tran *et al.*, 2003).

11.2.2 Browning

Non-enzymatic browning due to the Maillard reaction occurs in high-temperature-processed milk both during processing and during storage. The browning is visible by eye when it becomes quite intense (see Section 7.2.5) but can be detected much earlier with appropriate methodology (Omoarukhe *et al.*, 2010). Two major types of analysis of browning are performed: instrumental analysis using commercial colourmeters such as the Minolta Chromameter, which has become the most common method, and colourimetric methods in which the colour is measured spectroscopically.

11.2.2.1 Colour Meter Analysis

The milk or milk product is placed in a container such as a Petri dish and illuminated according to the manufacturer's instructions. The CIELAB values [L* (degree of lightness, 100 is perfect white and 0 is black), a* (red-green hues) and b* (yellow-blue hues)] are obtained using a colour meter such as a Minolta Chromameter. [An alternative colour measurement system is the Hunter L, a, b system. The difference between the two systems is explained in Hunterlab (2012). To a large extent, the CIELAB system has now replaced the earlier Hunter system.] Several (≥5) readings are recorded for each sample and the mean value calculated (Al-Saadi & Deeth, 2008). The L* and b* values are the most appropriate for detecting browning in milk. However, to measure the total colour difference between samples or between samples and a colour reference such as a white tile, the parameter delta E (ΔE, ΔE* or ΔE^*_{ab}) can be calculated by the formula in Equation 11.1:

$$\Delta E = [(L^*_2 - L^*_1)^2 + (a^*_2 - a^*_1)^2 + (b^*_2 - b^*_1)^2]^{0.5} \tag{11.1}$$

where L^*_2, a^*_2 and b^*_2 are values for one sample and L^*_1, a^*_1 and b^*_1 are values for a second sample or reference. According to the 1989 Handbook of Colour Science published by the Japanese Academy of Colour Science (cited in Yousif *et al.*, 2003), ΔE values between 0 and 0.5 are impossible to detect by eye, between 0.5 and 1.5 are difficult to detect by eye, between 1.5 and 3.0 are detectable by trained people, and between 3 and 6 are detectable by most people. A value of 2.3 has been reported by Sharma (2003) as the

JND (Just Noticeable Difference) while Pagliarini *et al.* (1990) showed that a minimum ΔE of 3.8 should be attained before there is a visual perception of milk browning (Nollet & Toldra, 2010).

11.2.2.2 Colourimetric Analysis

The brown colour which develops in milk is associated with both the casein and the serum phase. To measure the colour of free coloured compounds in the serum, the sample is deproteinised with 12% TCA, centrifuged, diluted and the colour read at 420 nm and 550 nm. An alternative method of producing the coloured extract is by dialysis (see Section 11.2.24.1). A browning index can be obtained from the difference in optical density readings at 420 and 550 nm. In order to measure the total brown colour, that is, of both the free and protein-bound colour compounds, the sample can be first enzymatically proteolysed with a protease such as pronase followed by centrifugation to obtain a clear solution. The colour in the solution is then read at 420 and 550 nm as above (Morales & van Boekel, 1998). As an alternative to the above enzymatic digestion of the protein to obtain a clear solution for colourimetric measurement, the sample can be clarified with a clarifying reagent (Morales *et al.*, 1997). These reagents are available commercially but can be readily prepared in the laboratory from published formulae (Humbert *et al.*, 2006). One such clarifying agent can be prepared with the following components: part a: 10 g SDS in 1 L of 0.1 M NaOH; part b: *n*-butanone + Triton X-100 (1:1); parts a and b are mixed in the ratio of 1:3 (Mottar & Moermans, 1988).

11.2.3 Density/Specific Gravity

Density or specific gravity can be quickly and cheaply measured by a hydrometer and more accurately by means of a specific gravity (SG) bottle. There are also more sophisticated instruments and also on-line instruments with claimed accuracies of +/- 0.1 kg m^{-3} to measure SG accurately to 4 or 5 decimal places. In a similar fashion to measurement of freezing point depression (see Section 11.2.8), it can be used to determine whether the correct amount of water has been removed in a direct UHT process, or for checking the consistency of flavoured milk and other formulations.

11.2.4 Dissolved Oxygen

The level of dissolved oxygen has a major effect on the storage stability of UHT mlk (see Section 7.1.2). The most common method of measuring dissolved oxygen is to use a dissolved oxygen meter fitted with a suitable oxygen electrode probe. The oxygen electrode is calibrated in air so that 100% dissolved oxygen corresponds to the concentration of oxygen in air (21%). A zero calibration using a 20 g/L sodium sulfite solution is conducted for each set of measurements (Perkins *et al.*, 2005b).

For some products it is useful to measure dissolved oxygen or redox potential. Dissolved oxygen disappears very quickly when there is excessive microbial activity and the redox potential also falls. Dissolved oxygen is the major contributor to the redox potential of milk; lesser contributors are ascorbate, riboflavin and the thiol – disulfide system. The redox status of milk can be estimated by dye-reduction tests, such as the resazurin test for raw milk quality and the methylene blue test for the keeping quality of pasteurised milk or cream, although the interpretation of the results is subject to debate (Burton, 1988). The principle is that enzyme activity in milk arising from bacterial growth can transfer hydrogen atoms from a substrate to the dye (biological acceptor) when added to the milk. The period of time required to decolourise the dye is an index of the bacteriological load of the milk. According to Bintsis *et al.* (2008) the incubation period for bulk milk with a plate count of approximately 10^5 cfu/mL was 6 h at 37 °C for methylene blue and 3 h for resazurin. These tests were once widely used but have lost popularity due to their poor correlation with other bacterial analyses, particularly for milk with a high psychrotrophic bacterial count. Hence they may have limited use on milk destined for UHT manufacture. However, Tetra Pak (undated) has included a simplified method for the methylene blue test in their booklet *The role of raw milk quality in UHT production.* Further information on dye reduction tests is given in Section 8.7.7.

11.2.5 Fat Separation and Fat Particle Size

Fat separation is a common occurrence in stored UHT milk (see Section 7.2.4). Although the milk is homogenised, the resulting emulsion does not always remain completely stable for the duration of storage. Because UHT milk is expected to remain stable at room temperature for six months or longer, even a very slow rate of rise of the fat over that time can result in a significant fat layer. The rate of rise is proportional to the size of the fat particles and hence measurement of the fat globule size, or more importantly, the fat globule size distribution, is often required.

11.2.5.1 Fat Separation

The simplest method of measuring the extent of fat separation is to measure the thickness of the fat layer of a sample left undisturbed for some time, for example, 48 h (Hillbrick *et al.,* 1999). This is suitable for clear containers but not for opaque containers. Alternatively, the milk can be transferred into a clear container for assessment. A second method is to measure the fat percentage in the milk at different levels. One variation is to leave the milk undisturbed for at least 48 h and carefully remove the top 10%. The fat contents of this portion and of the bottom 90% are measured. The milk is satisfactorily homogenised if 0.9 times the fat content of the top portion is less that the fat content of the bottom portion (Byland, 2003). The homogenisation efficiency or creaming index can be expressed as the difference in fat percentage of the two portions expressed as a percentage of the fat percentage of the top portion (Standards Association of Australia, 1980; Harding, 2013). For example, a milk with a fat content of 3.5% in the top layer (10%) and a 2.9% in the bottom layer (90%), the homogenisation creaming index would be 17%. The aim is to achieve an index of <10; a value of >30 indicates a poor emulsion. A third method is similar to the second but involves a mild centrifugation step (1000 g for 30 min). A 25 mL sample of the milk is centrifuged at 40 °C and the fat content of the bottom 20 mL is divided by the fat content of the whole sample and multiplied by 100. The result is termed the NIZO value which should be in the range 50-80% (Byland, 2003).

11.2.5.2 Fat Particle Size

This can be determined in several ways including light microscopy, spectroturbidimetry, electrical impedance and laser diffraction. Light microscopy is the simplest method but is a tedious manual operation in which the fat particles of different sizes are counted in a large number (50-100) of fields of view under a 1000 x magnification microscope.

The fat particles can be stained with a dye such as Sudan Black B to aid counting. A large number of fields is required because fat separation is largely due to large fat particles of which there may only be a small number unevenly dispersed in the different fields of view. Specifically, measuring particles between 2 and 5 μm and greater than 5 μm diameter is recommended. The determination of the size of the particles is facilitated by the use of an eyepiece containing a calibrated scale (Hillbrick *et al.*, 1998). Common rules of thumb in the dairy industry are that 90% of the fat particles should be ≤1 μm and that <5% should be greater than 2 μm and <2% should be greater than 5 μm.

Spectroturbidimetry involves measuring absorbance at 900 nm and reading the average fat particle size from a calibration chart relating the "absorbance index" with the fat content of the milk. Such charts are available from suppliers of commercial spectroturbidimetry equipment such as the Emulsion Quality Analyser (EQA™) (SPX, 2008).

Electrical impedance as employed by commercial Coulter counters is used to detect particles of different size as they pass through narrow-aperture tubes. Milk is normally diluted in a sodium chloride solution and a large number (>1 million) of particles are counted and sized. Hillbrick *et al.* (1998) used this method to determine the volume mean diameter, the volume mode diameter and the percentage of particles within certain size ranges (1.5-2 μm, 2-5 μm and 5-10 μm) in UHT milk homogenised at different pressures.

Particle size distribution is most commonly determined using laser diffraction instruments such as the Malvern Mastersizer™. "Laser diffraction measures particle size distributions by measuring the angular variation in intensity of light scattered as a laser beam passes through a dispersed particulate sample. Large particles scatter light at small angles relative to the laser beam and small particles scatter light at large angles. The angular scattering intensity data is then analysed to calculate the size of the particles responsible for creating the scattering pattern, using the Mie theory of light scattering. The particle size is reported as a volume equivalent sphere diameter" (Malvern Instruments Limited, 2012). The different size distributions of an homogenised milk sample and one that has not been homogenised are shown in Figures 11.1a and 11.1b.

The results from laser diffraction measurements can be expressed in several ways. The overall size of the particles can be expressed as the mean ('average' size of all the particles), median (size of which 50% of the particles are below/above) or mode (particles size with the highest frequency). The mean is most commonly used for fat particle measurement although mode has been preferred by some authors (e.g., Hardham *et al.*, 2000). Furthermore, there are several ways of expressing the mean size of the particle but the ones most commonly used for fat globules are the volume mean diameter, D[4,3], and the surface area mean diameter, D[3,2], with the former being preferred by most workers. Another means of characterising the size of fat particles in a sample is by percentiles which indicate the maximum particle size for a given percentage volume of the sample particles. Hence, D50 [(sometimes given as d (0.5)] designates the maximum particle diameter below which 50% of the volume of the particles exists (this is also the median particle size by volume). D90 {d (0.9)}, D50 {d (0.5)} and D10 {d (0.1)} are the most commonly used percentiles.

Hooi *et al.* (2004) recommended that the d(0.9) value for homogenised milk be <1.7 μm, whereas for raw milk it is 5 to 6 μm. They claimed that a new homogeniser should give a value of <1.3 μm. A d(0.9) value of 1.3 μm indicates that 90 % of the fat is in globules below that size, or 10% of the fat globules (by volume) are above that diameter. It is the fat

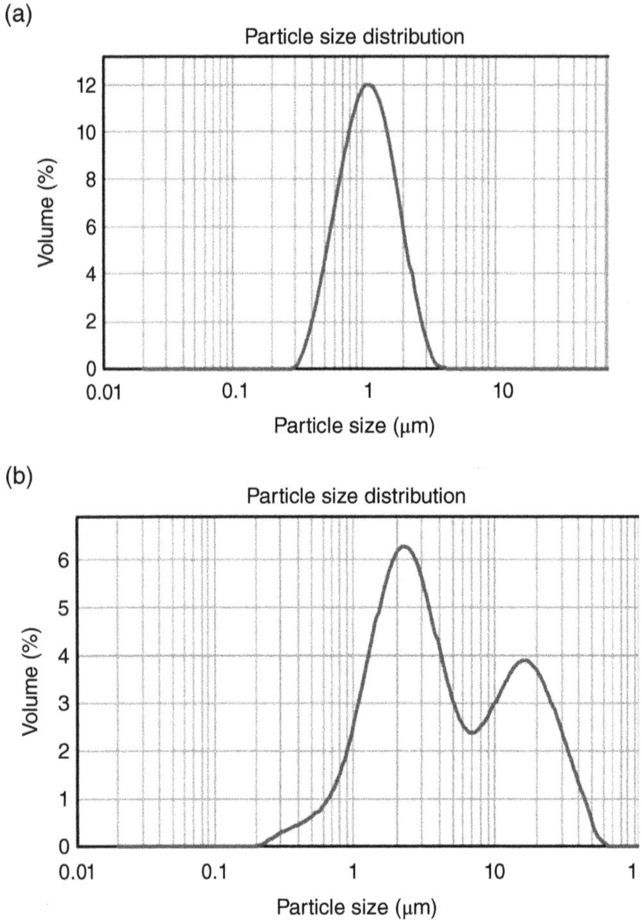

Figure 11.1 Fat particle size distribution in a. well-homogenised UHT milk and b. poorly homogenised UHT milk.

particles in this latter fraction that are most likely to separate (see Section 7.2.4). There are no precise guidelines for the fat globule size distribution in UHT milk products.

Hillbrick *et al.* (1999) compared the fat globule size measurements from electrical impedance (Coulter Counter) and laser light scattering (Masterisizer) of homogenised UHT milks processed by the three different UHT treatments (indirect heating with upstream homogenisation, indirect heating with downstream homogenisation and direct heating with downstream homogenisation). They found the fat particle sizes determined by the former to be much larger than those by the latter. At a homogenisation pressure of 17.3 MPa, the mean diameter by electrical impedance and the volume mean diameter, D(4,3), by laser light scattering of milk from the three treatments were 1.37, 1.35 and 1.46 μm, and 0.50, 0.50 and 0.51 μm, respectively. The d(0.9) values for these three milks determined by laser light scattering were 0.80, 0.80 and 0.83 μm, respectively. The authors concluded that laser light scattering underestimates the large fat globules, whereas electrical impedance analysis accurately determines the large fat globules in milk.

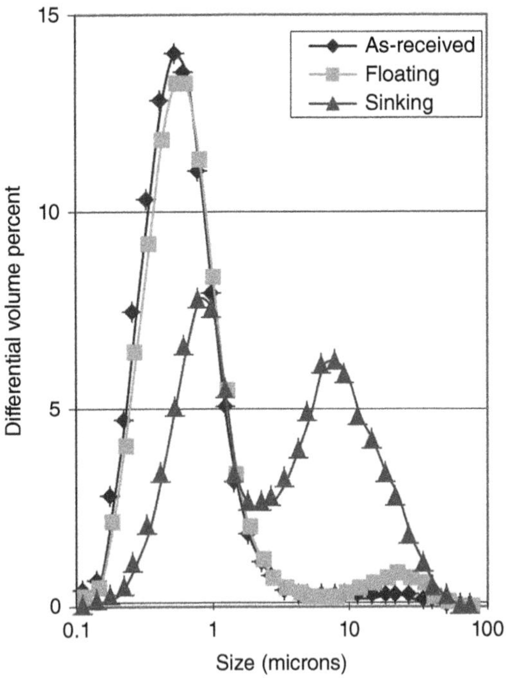

Figure 11.2 Particle size distribution in commercial UHT milk. (Source: Durand *et al.*, 2003. Reproduced with permission of Elsevier.)

The application of particle size distribution to commercial UHT milk is exemplified in the work of Durand *et al.* (2003). They examined bovine milk as well as soy, oat and rice milk. They analysed the entire milk samples as they were received and also the fractions which floated and sedimented after a low-speed (43 g) centrifugation for 30 min. Results for bovine milk are shown in Figure 11.2. The milk as received and the floating fraction exhibited single peaks at D[3,2] (D50) of 0.45 (0.43) and 0.5 (0.60) μm, respectively, while the sediment showed a bimodal distribution with overall D[3,2] (D50) of 1.21 (2.63) μm. The sediment would be expected to contain mostly protein aggregates with some fat.

11.2.6 Flavour Volatiles

The flavour of high-temperature-heated milk is a continuing issue for the dairy industry. UHT milk was introduced largely because the flavour of in-container sterilised milk was considered undesirable. Although the flavour of UHT milk has improved considerably since it was first introduced, many consumers still find it undesirable. It is therefore not surprising that the flavour of UHT milk has been researched extensively. There are five major chemical classes which contribute to this flavour, each of which requires its own chemical analytical approach: volatile sulfur compounds, stale/oxidised flavour compounds, heat-induced compounds, peptides resulting from proteolysis and free fatty acids resulting from lipolysis (see Sections 6.1.6 and 7.1.3). Methods for peptides and free fatty acids are covered in Sections 11.2.21 and 11.2.15.

The methods used for analysing the volatile sulfur compounds, stale/oxidised flavour compounds and Maillard reaction products generally include an extraction step followed by gas chromatography. Distillation has been used to extract the volatiles from milk but this is not suitable for all compounds because the heating generates increased quantities of some compounds, degrades others and forms artefacts. This has been overcome by the use of headspace methods, both dynamic and static. A static method which has gained acceptance is solid phase microextraction (SPME) which is considered better than other headspace techniques due to its sensitivity, selectivity and simplicity (Vazquez-Landaverde *et al.*, 2005). The sensitivity of the dynamic purge and trap technique is in parts per billion while for SPME it is in parts per trillion.

SPME is based on the absorption of the target volatiles on a small (1 or 2 cm long) fused silica fibre coated with a thin film of a polymer such as polydimethylsiloxane (PDMS) which is placed in the headspace of the product. The fibre is left for a certain time at a particular temperature to enable the volatiles to absorb into the polymer film. The fibre with the absorbed volatiles is then placed into the injector of the GC and the volatiles are released from the fibre and swept by an inert carrier gas, such as helium, onto the GC column where separation of compounds occurs. For each type of analysis, the nature of the fibre, and the temperature and time of extraction need to be optimised.

Volatile sulfur compounds in milk are difficult to analyse because of their volatility, susceptibility to oxidation and their presence in low concentrations. The methods used have been reviewed by Al-Attabi *et al.* (2009). SPME has become the extraction method of choice and has been used successfully in conjunction with sensitive GC detectors such as a pulsed flame photometric detector (PFPD). A typical extraction involves 5 mL milk and a carboxen/polydimethylsiloxane (CAR/PDMS) SPME fibre with an 85 μm thick film at 30 °C for 15 min (Vazquez-Landaverde *et al.*, 2006; Al-Attabi *et al.*, 2014). A typical GC method uses a capillary column with a 4 μm thick dimethylpolysiloxane film with pulsed flame photometric detection. Quantification is performed by reference to calibration curves for standards where possible. However, constructing calibration curves for H_2S and MeSH is difficult due to their high volatility and sensitivity to oxidation. For these compounds, the calibration curve for dimethyl sulfide can be used because the sulfur-selective PFPD has an equimolar response, meaning that compounds with the same number of sulfur atoms per molecule produce signals of equal size.

Stale flavours in UHT milk result from oxidation of lipids and tend to occur after a period of storage. They principally consist of aliphatic aldehydes and methyl ketones. They are conveniently analysed using SPME headspace extraction followed by gas chromatography with flame photometric detection. An optimised SPME method for extracting stale flavour volatiles from UHT milk was developed by Perkins *et al.* (2005a). It used a 2 cm SPME fibre with a 50/30 μm DVB/Carboxen/PDMS coating and extraction conditions of 15 min at 40 °C followed by chromatography on a fused silica capillary column coated with a 0.25 μm thick phenylpolysilaphenylene-siloxane film and FID detection. Calibration curves were constructed using pure samples of C_2 to C_{13} methyl ketones and C_4 to C_{10} aldehydes.

Similar conditions to those used for stale flavours can be used to analyse heat-induced flavours. For example, Scanlan *et al.* (1968) used a Carbowax™ 20 M polyethyleneglycol capillary column to analyse a wide range of volatiles, including aldehydes and ketones in UHT milk that were not present in raw milk. One compound they identified was diacetyl which was present in concentrations considerably greater than the threshold level.

11.2.7 Fouling of Heat Exchangers

Fouling of heat exchangers is the main influence on UHT run times (see Section 6.2.2). The changes that occur due to fouling include an increase in pressure and a reduced flow rate due to narrowing of the flow path of the product, a decrease in the temperature of the product at constant heat input or an increase in the temperature differential between the heating medium and product if the temperature of the product is kept constant due to the decrease in thermal conductivity of the heat exchanger caused by the deposit build-up. Each of these changes can be monitored to determine the extent of fouling and when it is necessary to stop the plant and perform either an intermediate or complete clean. An overview of the different reported methods of fouling detection was provided by Prakash *et al.* (2005).

Monitoring pressure as a fouling indicator is important commercially as it can provide a timely warning of the development of excessive pressures that can cause damage to heating equipment. It is not unlike other control systems used in dairy processing and hence is a convenient means of continuous real-time monitoring of fouling. A possible limitation is that it is a relatively insensitive measure of fouling as it is only changes when a critical thickness of deposit forms. It is more suitable to monitor fouling in plate heat exchangers than in tubular heat exchangers due to the narrowness of the flow channels in plate heat exchangers compared with the diameter of tubes in tubular systems.

Monitoring temperature of the heating medium in commercial plants where product temperature is kept constant to safeguard product safety and quality is an efficient method of continuous monitoring of fouling as in most cases it would not require additional instrumentation. Like pressure, the change in temperature may not be evident until a critical thickness of deposit develops. This means that the initial stages of fouling may not be detected; however, in commercial practice this is of little consequence. In UHT pilot plants which do not contain the facility to maintain constant product temperature, the decrease in temperature to less than a predetermined point, e.g., 120 °C (Prakash *et al.*, 2015), can be used to mark the end of the run. Of course such a criterion in commercial practice would be unacceptable.

The overall heat transfer coefficient (U) is a better parameter for monitoring deposit formation than temperature or pressure as it is takes into account changes in temperature and flow rate during build-up of the fouling deposit. U is calculated for milk by Equation 11.2 (Kastanas, 1996; Prakash *et al.*, 2007):

$$U = \frac{GC_p\Delta\theta}{A\Delta T_{lm}} \tag{11.2}$$

where G = mass flow rate of the milk in kg/s; Cp = specific heat of milk in J/(kg °C); $\Delta\theta$ = temperature difference between the inlet and outlet of the UHT section, in °C; A = surface area of heat exchanger in m^2; ΔT_{lm} = logarithm mean temperature difference (LMTD) in °C as calculated according to Equation 11.3:

$$\Delta T_{lm} = \frac{(T_0 - T_{m0}) - (T_0 - T_{mi})}{\ln[(T_0 - T_{m0}) - (T_0 - T_{mi})]} \tag{11.3}$$

where T_o is the temperature of the heating medium in the high-temperature section in °C; T_{mo} and T_{mi} are the temperatures of milk at the inlet and outlet of the high-temperature section respectively (Kastanas, 1996). OHTC has been used successfully in research on fouling using laboratory-scale UHT plants instrumented to obtain the required data and compute the OHTC in real time (Kastanas *et al.*, 1995; Prakash *et al.*, 2007).

11.2.8 Freezing Point Depression (FPD)

Freezing point depression measurement is a rapid method which is most commonly used for the detection of added water in raw milk. It can be measured very accurately by a cryoscope (to 1 m°C), by means of a very accurate thermistor. The manufacturers also supply accurate standards for calibrating the instrument. Results can be expressed as freezing point, for example, −0.520 °C or freezing point depression (FPD), for example, 520 m°C. More background information is provided by Harding (1995).

Of all the properties of milk, freezing point is the one that shows the least variation; milk should have an FPD above 515 m°C (0.530 H) for it to be deemed free of added water (Burgess, 2010). The freezing point corresponds to a solution of about 0.85% (w/w) of NaCl. The contributors to freezing point depression are lactose (54%), Na, K and Cl (31%) and other low molecular weight compounds (about 15%). In the early days of measuring freezing point, units of degree Hortvet (H) were used. These are still encountered in the literature, so conversion to degree Celsius are given in Table 11.1.

From our experience with pilot plant trials, FPD is not affected by indirect UHT processing, unless water somehow enters the milk. FPD measurement is very useful for direct UHT processes, where steam condenses and dilutes the product, for ensuring that the correct amount of water is removed in the flash cooling process. In all UHT processes, FPD measurement may help to establish whether plates or valves are leaking or whether extraneous water is added due to short changeover periods.

Some other applications of a cryoscope which have been investigated by the authors which UHT practitioners may wish to consider are as follows:

- determining the extent of the reaction during lactose hydrolysis, where FPD increases from 520 to approximately 760 m°C (see Section 9.4).
- checking the consistency of formulated milk drinks. Some examples are given in Chapter 9.
- measuring osmolarity for infant formulations and ensuring that it is similar to that of human milk
- measuring the extent of some chemical changes during storage of UHT milk
- indicating the possible addition of UF permeate to milk
- checking the purity of laboratory distilled or deionized water.

Table 11.1 Conversion from Hortvet (°H) to Celsius (°C).

°H	°C
−0.543	−0.528
−0.535	−0.515
−0.530	−0.510
−0.529	−0.509
−0.525	−0.505

11.2.9 Furosine

Furosine is a measure of early-stage Maillard reaction products and serves as a chemical index of the severity of heat treatment of milk and other milk products (see Section 6.1.7). It is also used to measure the extent of Maillard reactions during storage of long-shelf-life products such as UHT and in-container sterilised milks, and milk powders. Furosine does not occur in milk but is produced during the acid hydrolysis of the Maillard reaction products, lactulosyl-lysine, fructosyl-lysine and tagatosyl-lysine, produced by reaction of lysine with lactose, glucose, and galactose respectively (Finot & Mauron, 1972; Erbersdobler & Hupe, 1991). In the furosine analysis, the product is digested with concentrated acid, for example, 6-8 N HCl at ~110 °C for ~23 h (Elliott *et al.*, 2003).

Ion-pair, reversed-phase high-performance liquid chromatography (HPLC) using sodium-heptanosulfonate with UV detection at 280 nm is the most commonly used technique for determining furosine (Resmini *et al.*, 1990; Delgado *et al.*, 1992). Even low levels of furosine can be analysed by this method (Resmini *et al.* 1990). A calibration curve is prepared with a furosine dihydrochloride standard.

Furosine is expressed in units of mg/100 g protein and hence it is necessary to determine the protein content of the product using methods such as the Kjeldahl, Dumas, Lowry/Folin-Ciocalteu or bicinchoninic acid methods (see Section 11.2.20).

11.2.10 Hydrogen Peroxide

Hydrogen peroxide is the main sterilising agent used in aseptic packaging systems. After it has been applied, it is usually removed by hot air. Residual amounts, less than 0.5 ppm (0.5 mg/L), may remain, but this is not thought to be an issue in UHT products. The Food and Drug Administration (FDA) limits H_2O_2 residues to 0.5 mg/L in the finished food package (CFR, 2003). Hydrogen peroxide has been detected in UHT milk samples and believed to be due to the illegal practice of adding it as a preservative to raw milk (Souza *et al.*, 2011). Martin *et al.* (2014) commented that despite the recommendation that it should not be directly added to raw milk as a preservative, it is still used in some parts of the world.

Hydrogen peroxide, even at the low level of 10 mg/L is effective in reducing the levels of volatile sulfur compounds in UHT milk which are responsible for the sulfurous flavour in the milk. At this level of addition, either before or after UHT processing, the hydrogen peroxide could not be detected after one day's storage at 22 °C; at 50 mg/L addition, it could not detected after three days' storage (Al-Attabi, 2009).

A standard procedure for measuring hydrogen peroxide uses the reagent 1,4-phenylenediamine which is oxidised by any residual hydrogen peroxide to a blue colour. Detection strips are available which measure down to 0.5 ppm; these are very convenient to use and effective for semi-quantitative milk analysis (Al-Attabi, 2009; Martin *et al.* 2014). Marks *et al.* (2001) used such strips to determine whether lactoperoxidase was still active in pasteurised milk as it requires the presence of hydrogen peroxide. A batch injection analysis with amperometric detection has been described by Silva *et al.* (2012) for analysing hydrogen peroxide in milk. It is claimed to be a rapid, precise and selective method with a limit of detection of 0.38 ppm (10 μmol/L).

Hydrogen peroxide, together with thiocyanate (Marks *et al.*, 2001) or a halide (Fweja *et al.*, 2007), is also an active component of the natural antimicrobial lactoperoxidase (LP) system in raw milk (see Section 4.2). It can be generated in situ by addition of sodium percarbonate. However, the LPS cannot be stimulated in UHT milk because LP is inactivated by the heat treatment.

Hydrogen peroxide is generally recognised as safe and its presence in low concentrations is permitted in some foods in some countries. Its use as a processing aid is often permitted provided it is destroyed in the final product by the addition of catalase which converts it to water and oxygen. Hydrogen peroxide is an oxidant and oxidises some components of food such as folic acid and some amino acids. Thus treatment of milk with hydrogen peroxide could reduce its nutritive value.

11.2.11 Hydroxymethyl Furfural (HMF)

5-Hydroxymethylfurfural (HMF) is used as a chemical indicator of the intensity of heat treatment food products such as UHT milk receive (see Sections 6.1.7 and 7.1.7). Accordingly, it has been used to distinguish between milk samples subjected to different heat treatments (Fink & Kessler, 1986). It has also been proposed to be used in setting the maximum conditions for UHT processing, namely development of 10 μmol/L of HMF (Burton, 1988).

As indicated in Section 6.1.4.1, both total and free HMF are measured. For measurement of total HMF, a preliminary digestion with oxalic acid at ~100 °C is required to convert the Amadori products into HMF. For determination of free HMF, this step is omitted. HMF is usually analysed by reversed-phase HPLC with an acetic acid-acetate buffer solution at pH 3.6, and detected at 280 nm (Morales *et al.*, 2000).

HMF can also be detected colourimetrically after reaction with thiobarbituric acid (Keeney & Bassette, 1959). Park and Hong (1989) demonstrated the usefulness of this method on a range of domestic milk samples, reporting HMF values of 1.45 μmol/L for pasteurised milk, 2.54-3.54 μmol/L for directly sterilised UHT milks and 14.92 μmol/L for indirectly sterilised UHT milk. These values are similar to those reported by Elliott *et al.* (2003) obtained by reversed-phase HPLC which has now largely replaced the colourimetric method.

A consideration when measuring HMF is that even raw milk and pasteurised milks contain some HMF (Fink & Kessler, 1986) and hence this needs to be taken into account when considering the HMF resulting from the heat treatment or storage of the product.

11.2.12 Lactulose

Lactulose is not present in raw milk and is formed during heating (see Section 6.1.4.2). It is widely accepted as a chemical index of the severity of heating (see Section 6.1.7). It has also been proposed by the International Dairy Federation (IDF) and by the European Commission (EC) as an analytical index to distinguish UHT milk from sterilised milk. The IDF proposed that the lactulose content should be between 100-600 mg/L and 850-2000 mg/L in UHT and in container sterilised milk respectively.

Lactulose can be determined by the IDF-validated HPLC method (IDF, 1998) using a cation exchange column, operating at 75 °C, with refractive index detection. It can also be determined by gas chromatography of the silyl ethers on packed or capillary columns (Martinez-Castro *et al.*, 1987).

An alternative commercially available method is the enzymatic-spectrophotometric method in which the lactulose is converted to fructose and galactose and the lactose to glucose and galactose by β-D-galactosidase (Geier & Klostermeyer, 1980; Feier & Goetsch, 1990). Glucose is removed by glucose oxidase and the fructose is determined spectrophotometrically after a series of enzymatic reactions. This method has a detection limit of approximately 10 mg/L (De Block *et al.*, 1996). It has been adopted as a standard method by the IDF (IDF, 2004), ISO (ISO, 2004) and several countries.

11.2.13 Lysinoalanine (LAL)

Lysinoalanine (LAL) is an isodipeptide formed during protein cross-linking through the reaction of dehydroalanine with lysine residues (see Sections 6.1.3.3 and 7.1.5). It is caused by intense heating, especially at alkaline pH, and during storage of long-shelf-life products such as UHT milk (Fritsch *et al.*, 1983; Al-Saadi *et al.*, 2013). It can be analysed by ion-exchange chromatography (Friedman *et al.*, 1984) and GC/MS after derivatisation of both amino and carboxy groups (Hasegawa *et al.*, 1987). However, amino acid analysis and reversed-phase HPLC have become the methods of choice. The initial step in all methods is acid hydrolysis (6 M HCl at 110 °C for 23 h) of the protein to release amino acids and isodipetides including LAL, lanthionine and histidinoalanine. After neutralisation of the hydrolysate, Fritsch *et al.* (1983) determined LAL on an amino acid analyser with fluorometric detection. Pelligrino *et al.* (1996) derivatised the LAL in the hydrolysate with 9-fluorenylmethylchloroformate (FMOC), and, after a solid-phase extraction clean-up, determined the derivatised LAL by reversed-phase HPLC with fluorescence detection. Al-Saadi and Deeth (2008) reported a simple procedure based on the method of Faist *et al.* (2000) in which the LAL in the hydrolysate was derivatised with dansyl chloride and the dansyl derivative was analysed by reversed-phase HPLC with fluorescence detection.

11.2.14 Lipase

Because of the heat stability of bacterial lipases, some residual lipase activity can remain after high-temperature treatments such as UHT processing (see Section 6.1.3.5.2). This activity can cause lipolysis of the milk fat to free fatty acids and result in development of a rancid flavour in the milk or milk product (see Section 7.1.3.5). The level of activity which can cause the development of rancidity is very low, given the potentially long period of storage at room temperature. Consequently, sensitive lipase assay methods have been sought by the dairy industry to detect these low lipase activities before they cause off-flavours in the product.

Many different assays have been used for lipase (Stead, 1989; Deeth & Touch, 2000). In their review, Deeth and Touch (2000) listed 37 methods collated into eight different categories. Some assays use triacylglycerol substrates such as tributyrin, triolein and milk fat, some use colourogenic substrates such as *p*-nitrophenyl and β-naphthyl esters while others use fluorogenic substrates such as 4-methylumbelliferyl esters. While the colourigenic and fluorogenic methods are quite sensitive, the activity of the enzymes towards them does not necessarily match their activity against the true substrate in the product, that is, milk fat. Furthermore, the lipases tend to be more active against the short-chain esters, including short-chain triaglycerides such as tributyrin but this activity can be misleading as these substrates are also hydrolyed by esterases which do not have the ability to hydrolyse long-chain esters which are the true substrates of lipases. A further distinction between esterases and lipases is that esterases act on soluble substrates while lipases act at the interface between an insoluble substrate, normally an oil, and water (Fojan *et al.*, 2000). One long-chain fluorogenic substrate which has been shown to give results for lipases from psychrotrophic bacteria which correlate well with those using triacylglycerol substrates, such as milk fat, is 4-methylumbelliferyl oleate (Fitz-Gerald & Deeth, 1983; Stead, 1984a). It has been shown to be suitable for milk powders as well as fluid milk (Stead, 1984b,c).

Long-chain triacylglycerol substrates such as triolein are usually used in the form of emulsions in lipase assays. Gum acacia is commonly used as an emulsifier but several other materials such as xanthan gum can be used. A substrate which is very suitable for assaying lipase in milk is (homogenised) cream (Fitz-Gerald & Deeth, 1983). UHT cream, which is usually homogenised, is recommended because of its convenience, low-cost, emulsion stability, small size of the emulsion fat droplets, bacteriological sterility, and ideal chemical nature of the substrate for lipases in milk. In a typical assay, the lipase source, substrate emulsion and a suitable buffer such as Tris at pH of ~7.5 are mixed and incubated at 37-40 °C for a set time. If the lipase activity is quite high, an incubation time of 24 h may be sufficient but for very low levels of lipase activity, an extended incubation period of days can be used; in this case, the antibacterial agent, sodium azide (at 0.05%) should be added to prevent bacterial growth. After incubation the released free fatty acids are extracted with an organic solvent and quantified by one of the methods outlined in Section 11.2.15.

Button (2007) investigated the suitability of three lipase assay methods for quantifying very low levels of a *Pseudomomas* lipase in UHT milk. The methods were based on three substrates: triolein emulsified in gum acacia, 4 methylumbelliferyl oleate and β-naphthyl caprylate. A cell-free bacterial culture supernatant containing lipase was added to UHT milk aseptically at the addition rates of 0.00025%, 0.0005%, 0.001% and 0.002% and the milks were stored for up to 6 months. Aliquots of the milks were taken and assayed using the three substrates after incubation periods of up to 7 days. Using triolein, there were significant increases in FFA concentration after incubation for 4 h for the milk samples with 0.002% added crude lipase, after 24 h for the samples with 0.001% added lipase and after 7 days for the samples with 0.0005% added lipase. Even after incubation for 14 days there was no significant increase in the FFA concentration for the milk sample with 0.00025% added lipase. It was not possible to detect the low lipase activities with either 4 methylumbelliferyl oleate or β-naphthyl caprylate during long incubation times as the substrates were too unstable and spontaneously hydrolysed even in controls containing no added lipase.

The UHT milks containing the lowest added lipase concentration (0.00025%) did not show a significant increase in FFA during storage for 6 months while the milk containing the second lowest lipase concentration (0.0005%) showed significant increase after 28 days' storage and reached 1.9 mmol/L after 168 days, a level which would render the milk unacceptable to most people. Overall, the lipase assay results using triolein showed a high correlation with the FFA levels in the stored UHT milks. These results show that low levels of lipase which could cause rancid off-flavours in UHT milk during storage could be detected by a triolein assay with an incubation period of up to 7 days.

11.2.15 Lipolysis (Free Fatty Acids)

Several methods are used for determining free fatty acids in milk as a measure of the extent of lipolysis by lipases. These were summarised by Deeth and Fitz-Gerald (2006). They can be divided into two broad categories: those that estimate the total free fatty acid content, which are the most commonly used, and those that measure the concentrations of individual fatty acids chromatographically using either GC or HPLC. The latter can also be used to estimate the total FFA content by summation of the concentrations of the individual acids.

Most FFA methods include an initial solvent extraction step using solvent systems such as isopropanol-hexane-H_2SO_4 or ether-hexane-HCL to separate all the lipid material, including the free fatty acids, from the protein, carbohydrate and mineral components of the products. This is followed by a quantification step such as titration with methanolic alkali (Deeth *et al.*, 1975; Mouillet *et al.*, 1981). While small numbers of samples can be handled manually, autoanalysers are used for large numbers (Cartier *et al.* 1984). Alternative quantification methods include colourimetric assays of copper soaps or rhodamine dye adducts of the free fatty acids.

For the GC analyses, the total lipid extract is treated by solid phase adsorption using, for example, aminopropyl resin, ion exchange resins, neutralised silicic acid or deactivated alumina, to separate the free fatty acids from the other lipid material. GC can then be used to separate the free fatty acids (Deeth *et al.*, 1983) or their methyl esters (Spangelo *et al.*, 1986) using flame ionisation detection. Separation on a modified FFAP capillary column with mass spectrometric detection (de Jong & Badings, 1990) provides excellent chromatograms (see Figure 11.3). Alternatively, the free fatty acids can be derivatised to produce UV-absorbing esters such as *p*-bromophenacyl esters before HPLC separation (Reed *et al.*, 1983). HPLC can be used on the total lipid extract using UV detection or, better, Evaporative Light Scattering Detection (ELSD) (Mengesha & Bummer, 2010).

SPME can also be used to extract free fatty acids with 4 to 12 carbon atoms (González-Córdova & Vallejo-Cordoba, 2001). This method is difficult to quantify as SPME extraction depends on volatility and some fatty acids are not very volatile; however, it may be useful in some situations to monitor any increase in free fatty acids during storage (Perkins *et al.*, 2005b).

11.2.16 Lysine – Blocked and Reactive

11.2.16.1 Blocked Lysine

Lysine in milk and milk products becomes "blocked" when it reacts with reducing sugars, principally lactose but also glucose and galactose in lactose-hydrolysed products, in the first stage of the Maillard reaction (see Section 6.1.4.1). After lysine is glycated to lactosyl-lysine it rearranges in an Amadori reaction to produce the first stable Maillard reaction product, lactulosyl-lysine. In this form, lysine is rendered unavailable for metabolic functions and also the parent protein becomes less digestible by proteolytic enzymes. Lysine blockage is particularly significant for infant formulae because of the greater requirement of infants for lysine compared with adults, and because the composition (e.g., high lactose and whey protein contents), processing, which includes quite intense heat treatments including UHT, and storage all promote the Maillard reaction (Mehta & Deeth, 2016).

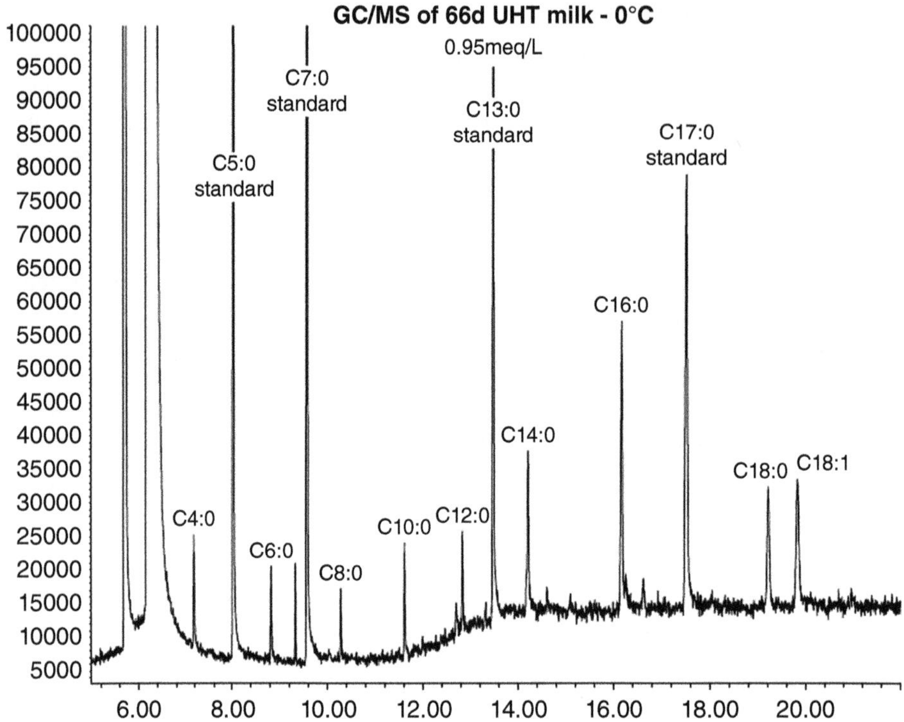

Figure 11.3 GC-MS chromatogram of free fatty acids extracted from UHT milk with a total free fatty acid concentration of 0.95 mmoles/L. (GC conditions: HP-FFAP nitroterephthalic acid modified polyethylene glycol capillary column; carrier gas, helium; split ratio, 20:1; injection temperature, 250 °C; oven temperature,. initial 65 °C increased at 15 °C/min to 240 °C and held for 13 min; detection, mass selective). (Source: Button, 2007. Reproduced with permission.)

Several methods are used to assess the extent of lysine blockage and hence the reduction in biological value. One of the most common methods involves the analysis of furosine (as discussed in Section 11.2.9) which is not present in milk but is formed by acid hydrolysis of lactulosyl-lysine. The acid hydrolysis yields approximately 32% furosine and 40% is converted back to lysine. Therefore, if lysine is independently analysed by, for example, HPLC or amino acid analysis, the percentage blocked lysine can be calculated using Equation 11.4 (Bujard & Finot, 1978; Erbersdobler & Somoza, 2007):

$$\text{Blocked lysine}(\%) = \frac{3.1\ \text{furosine}}{\text{Total lysine} + 1.87\ \text{furosine}} \times 100 \qquad (11.4)$$

Mehta and Deeth (2016) suggested an alternative approach to determining blocked lysine from furosine. Since the theoretical amount of lysine in the milk proteins is known from their amino acid sequences, the percentage blocked lysine can be estimated by Equation 11.5:

$$\text{Blocked lysine}(\%) = \frac{3.1\ \text{furosine}}{\text{Theoretical lysine in the milk proteins}} \times 100 \qquad (11.5)$$

Both furosine and theoretical lysine are expressed as mg/100 g protein. Since products such as infant formulae contain different casein:whey protein ratios, and casein and whey proteins contain different amounts of lysine, the theoretical lysine in the formula has to be calculated for each product. As an example, two products with a furosine value of 100 mg/100 g protein, and casein to whey protein ratios of 4:1 and 4:6 would have around 3.4% and 2.9% blocked lysine respectively (see also Section 6.1.4.1).

11.2.16.2 Chemically Reactive or Available Lysine

The alternative approach to assessing the effect of processing and storage on lysine is to measure the lysine which is not "blocked". If this is done chemically, the result is "chemically reactive lysine" which is an estimate of the amount of biologically available lysine. Measuring chemically reactive lysine is much easier and more convenient than measuring the true biologically available lysine using animals such as rats. Methods, available for determining the amount of chemically reactive and biologically available lysine have been reviewed by Hurrell and Carpenter (1981) and Rutherfurd and Moughan (2007). An overview of the chemical methods is given here. These methods include the 1-fluoro-2,4-dinitrobenzene (FDNB), trinitrobenzenesulfonic acid (TNBS), o-phthalaldehyde (OPA), guanidination, sodium borohydride, and dye binding methods.

FDNB and TNBS react with the free ε-amino group of lysine and, after acid hydrolysis, the yellow products dinitrophenyl (DNP)–lysine and trinitophenyl (TNP)–lysine, respectively, are formed. They can be measured colourimetrically or by HPLC (Carpenter, 1960; Kakade & Liener, 1969; Erbersdobler & Anderson, 1983).

OPA reacts with the ε- and α-NH_2 groups of proteins and forms a product which can be measured fluorimetrically almost immediately without any heating or hydrolysis (Goodno *et al.*, 1981). It is claimed to be rapid, simple and sensitive and suitable for routine analyses of reactive lysine. A correction can be made for the small contribution from the terminal α-NH_2 groups.

In the guanidination reaction, O-methylisourea reacts with the ε-amino group of lysine to produce the acid-stable homoarginine which can be determined chromatographically (Nair *et al.*, 1978).

The sodium borohydride method is based on a different principle from the above. Sodium borohydride reduces lactulosyl-lysine to compounds that do not include lysine so that a subsequent acid hydrolysis and lysine analysis gives a measure of only reactive lysine (Hurrell & Carpenter, 1974; Couch & Thomas, 1976).

The dye binding method uses a negatively charged dye which associates with the positively charged basic amino group of lysine. The reacted proteins precipitate and the amount of bound dye, which relates to the reactive lysine, can be estimated by determining the concentration of the remaining dye in solution. A correction can be made for the amount of dye which reacts with histidine and arginine (Hurrell *et al.*, 1979; Aalaei *et al.*, 2016).

11.2.17 Minerals and Salts

The minerals and salts are important components of the milk system and make a significant contribution to its heat stability and its nutritional value (see Section 6.2.1). The main minerals in milk are sodium, potassium, calcium and magnesium, and important salts are chloride, phosphates and citrate. From a nutritional standpoint, probably

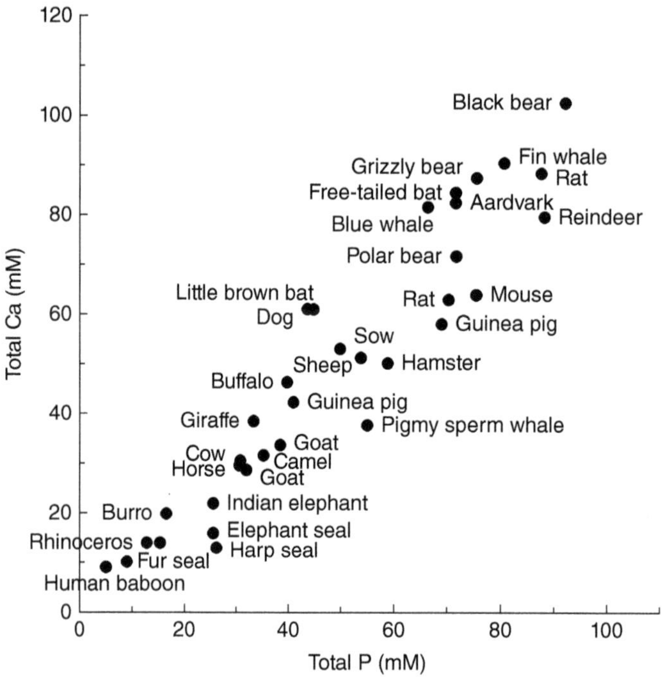

Figure 11.4 Total calcium and total phosphorus contents of the milk of various species (from Holt, 2011).

the most important mineral in milk is calcium. Calcium levels vary considerably in milk from different species: 7-8 mM in human milk, about 30 mM in cow's and goat's milk, 70 to 90 mM in mouse and rat milk and over 100 mM in milk from the black bear. Some data are shown in Figure 11.4.

Also of importance is magnesium which is present at about 15 to 20% of the amount of calcium in cow's milk. Both calcium and magnesium are distributed between the colloidal and soluble phases and at the normal pH of milk about half the magnesium and about two thirds of the calcium is associated with the casein micelle. As the pH of milk is reduced, more of both these minerals migrate to the soluble phase. However, as milk temperature increases, calcium (and probably magnesium) move from the soluble to the colloidal phase, but this is largely reversible on cooling. Migration in UHT milk continues slowly during storage (Geerts *et al.*, 1983). Sodium and potassium are almost completely ionized, whereas only small proportions of the calcium and magnesium are ionized at the normal pH of milk. For comparison, the concentrations ranges for K^+ and Na^+ in milk are 31-43 mM and 17-28 mM, respectively, while Ca^{2+} is about 2 mM and Mg^{2+} is about 0.8 mM (Walstra & Jenness, 1984). It is important not to change the K^+:Na^+ ratio substantially, as this may influence the flavour of milk (Ranjith, 1995).

Minerals are hardly affected by UHT treatment, although their distribution between the soluble and micellar phase might change, but only slightly. However, during the UHT process itself, when milk is at its high temperature, substantial movement of Ca, Mg and P calcium occurs from the soluble to the colloidal phase, but this is mostly reversible on cooling. This is reflected in the fact that there is little change in either pH and freezing point depression after UHT processing, despite the fact that the pH is in the region of 5.6 at 140 °C. Dialysis and ultrafiltration are methods for examining mineral partitioning in milk at high temperature (see Section 11.2.24.1).

Mineral analysis is not a routine quality assurance procedure for UHT milk. The equipment required for detailed mineral analysis is expensive and would not be found at most UHT production sites. However, data for minerals will be required for UHT products where specific claims are made about mineral contents, or where fortification with minerals such as calcium or iron is practised; some methods which could be adapted for a factory quality control laboratory are available for this purpose.

Minerals can be measured by a variety of methods. The total mineral content is determined by ashing at around 525 °C and is usually between 0.7 and 0.8%.

White and Davies (1958) undertook some detailed work on the partitioning of phosphorous, calcium and magnesium in milk. Calcium was measured by precipitating it as oxalate and titrating the oxalate with potassium permanganate solution. Chemical methods involving titration are available for some minerals. EDTA is a complexing agent which has been used to determine hardness in water and the method has been adapted for determining calcium and magnesium (total divalent cations) in milk and milk products and a number of different indicators have been evaluated. Davies and White (1962) investigated EDTA procedures for measuring calcium and magnesium in milk and milk diffusate, using Eriochrome Black T and calmagite. They concluded that EDTA methods can be speedy and convenient, but practice in recognising the end-points is necessary. Lin (2002) successfully applied the EDTA method using calmagite as the indicator and this gave an end point that was easy to detect by eye. With practice, it can give reproducible results. It provides a quick method in circumstances where products are fortified with calcium and magnesium for determining how much has been added. The same applies where small amounts are removed. For example, it was useful for rapid determination of calcium removal for improving the stability of goat's milk for UHT processing (Prakash *et al.*, 2007). Auto-titration systems are also available and remove the problem of visibly detecting the end-point. Note that the EDTA method, as used by Lin *et al.* (2015), measures total calcium and magnesium (total divalent cations). EGTA (ethylene glycol tetraacetic acid) can be used to measure calcium only as it has a very low affinity for magnesium.

Phosphorus can be determined by a colourimetric method, the reaction forming phosphomolybdate, which is blue (British Standard, 1992). This method has been widely used. Phosphorus can also be determined by Inductively Coupled Plasma (ICP) techniques and by ion chromatography (Gaucher *et al.*, 2008a).

Citrate has been measured traditionally by a colourimetric method. One widely cited method is the modified method of White and Davies (1963), using pyridine-acetic acid reagents. It can also be measured by HPLC (Garnsworthy *et al.*, 2006). It is also one of the numerous chemical species that can be detected by NMR, but isolating the citrate signal from all the others is not straightforward.

Phosphorus in milk is distributed between the micellar and soluble phases and this distribution is strongly influenced by pH (Walstra & Jenness, 1984). A lesser amount of citrate is associated with the micellar phase. Techniques are available for partitioning minerals, e.g. dialysis and UF (see Section 11.2.24.1) and ultracentrifugation (see Section 11.2.24.2).

Atomic absorption spectrophotometry (AAS) has been a standard method for minerals for a long time. The sample must be ashed and digested before measurement. Lamps with specific wavelengths can be selected for measuring a range of cationic minerals. More accurate procedures are ICP – mass spectrometry (ICP-MS) and ICP – atomic emission spectroscopy (ICP-AES) [sometimes called ICP – optical emission spectroscopy (ICP-OES)] which have become very popular since their introduction in the 1980s. They are more sensitive than AAS and have detection limits of 1-10 parts per billion (ICP-AES) to 1-10 parts per trillion (ICP-MS). They can be used to detect metals and some non-metallic elements (Tyler, undated).

Another technique for measuring specific minerals is ion chromatograpy (Gaucher *et al.*, 2008a). Columns are available for measuring either cations or anions in milk. It is useful as no special preparation is required for milk and whey, apart from selecting an appropriate level of sample dilution. Although the technique is called ion chromatography, it in fact measures total, ionic plus non-ionic, minerals such as calcium, magnesium, potassium and sodium in milk and dairy products.

In nutritional terms, the mineral content is important, with different products being enriched with specific minerals. Calcium supplementation of milk is widely practised (Deeth & Lewis, 2015) but there is also interest in magnesium, iron and zinc supplementation because of the prevalence of deficiencies of these minerals in diets of many people (Abdulghani *et al.*, 2015). Magnesium is also of interest technologically as it is a divalent cation which tends to destabilise heat-treated milk, in a similar fashion to calcium.

Elements such as iron, zinc, cadmium, copper, selenium and iodine are found in trace amounts and require sophisticated methods for their analysis. Overall, they play only a minor role in the quality of most UHT products.

11.2.17.1 Ionic Calcium

Ionic calcium is now considered to be an important determinant of heat stability and thus is discussed here in more detail. It is a challenge to measure Ca^{2+} in milk because almost anything that is done to milk is likely to change it. The equilibrium between insoluble calcium phosphate and Ca^{2+}can be represented in simplistic terms by Equation 11.6:

$$3\,Ca^{2+} + 2HPO_4{}^{2-} \Leftrightarrow Ca_3PO_4\downarrow + 2H^+ \qquad (11.6)$$

Raw milk is usually refrigerated immediately after milking, which increases its Ca^{2+} concentration. During prolonged cold storage, its pH usually decreases, due mainly to microbial activity, and this increases Ca^{2+} (Geerts *et al.*, 1983; Tsioulpas *et al.*, 2007). Methods for measuring Ca^{2+} in milk fall into three main categories: ion exchange resins; titration using murexide, which is a complexing agent, and ion selective electrodes (ISEs). These have been reviewed by Lewis (2010). The focus here is on *ion selective electrodes*. Some studies using ion selective electrodes are shown in Table 11.2.

All ion selective electrodes measure the thermodynamic function activity. According to the Nernst equation, there is a linear relationship between the electrical output (mV) from the electrode and the log of the activity. One important issue relates to how results are reported (i.e., as activities or concentrations) (see Comments column in Table 11.2). More detail of the relationship between activity and concentration is given by

Table 11.2 Use of calcium selective ion electrodes. (Source: Lewis, 2010. Reproduced with permission of John Wiley & Sons.)

Electrode type	Product	(Ca^{2+}) determined	Calibration procedure	Comments	Reference
Radiometer F3223 electrode	Milk and UF permeate from milk	Compared results with murexide and ion exchange methods	0–5 mM Ca standards	Also predicted Ca^{2+} in milk diffusate (2 mM)	Holt *et al.* (1981)
Orion	Raw, pasteurised and sterilised milk	Raw (2.24 mM0; past 1.94 mM, sterilised 2.24 mM	0–3 mM calcium salts, with added Na, K and Mg	Compared results with ion exchange method	Muldoon and Liska (1969)
Orion	Whole milk Skim milk (key paper)	Variations in bulked milk, Changes following heating and recovery	0.1, 1.0 and 10 mM: ionic strength = 0.08	Reported activities (activity coefficient = 0.4	Geerts *et al.* (1983)
Radiometer	Recombined concentrated milk	Recombined milks, subjected to different preheat treatments	As Geerts, but with ionic strength = 0.2 M	Expressed results as activities activity coefficient = 0.3	Augustin and Clarke (1991)
Calcium electrode, Ca 500 AT	Raw milk	Looked at changes during ageing	Not mentioned	Reported a slight increase in pH and decrease in Ca^{2+}	Lieske(1998)
SIE – Orion, 93–20 electrode	Skim milk Past skim milk	2.80 mM 2.50 mM	$CaCl_2$,10, 100 and 1000 ppm	Not clear; added ionic strength adjuster to sample	May and Smith (1998)
SIE Radiometer	Milk milk + SHMP	2.05 mM 1.05 mM		Looked at changes during storage	Mittal *et al.* (1990)
Radiometer 813D-12	Compared cow, goat and sheep milk	Obtained different results for different procedures	0.1–100 mM Ca: 0.1 M KCl	Considered most values are grossly underestimated	Silanikove *et al.* (2003)
Not mentioned	Milk and whey	Changes during heating storage and dilution	0.01–10 mM Ca: 70 mM KCl	Coefficient of variation = 13%: Looked at effects of dilution	Demott (1968)

(*Continued*)

Table 11.2 (Continued)

Electrode type	Product	(Ca^{2+}) determined	Calibration procedure	Comments	Reference
Pheonix electrode	89 bulk milk samples	1.56–2.59 mM (95% of samples)	Same as Geerts	Little experimental detail	Chavez *et al.* (2004)
Radiometer	Phosphate, citrate, EDTA addition	Combinations of P, citrate and EDTA and CaCl addition	$CaCl_2$ and 0.08 M KCl	Expressed results as activity; activity coefficient = 0.4	Udabage *et al.* (2000)
Orion	UHT calcium fortified milks	1.10–2.28 mM	Same as Geerts *et al.* (1983)	Conc. not mentioned	De la Fuente *et al.* (2004)
Radiometer	Calcium fortified milk samples		As for Augustin and Clarke (1991) (Geerts *et al.* 1983)	Report activities, but skim milk is cited as 2 mM	Williams *et al.* (2005)
Radiometer F2110	Measured fouling behaviour	Added and reduced Ca^{2+} (0.75–1.24 mM)	Not mentioned	Reported activities	Jeumink and DeKruif (1995)
Orion 93–20 electrode	Looked at partitioning Long paper, worthy of further investigation	Concentrated milks: Ca^{2+} decreased on forewarming but hardly changed on concentration	Same as Geerts *et al.* (1983)	Both activities and concentrations: activity coefficient = 0.4 for milk and 0.3 for concentrates	Nieuwenhuijse *et al.* (1988)
Orion	Changes during buffalo milk lactation	Overall, values are higher than most other studies. Also higher in buffalo milk compared with cows milk	10-fold changes in $CaCl_2$, adjusted with 4 M KCl	Results are also expressed in terms of μg/mg protein	Sodhi *et al.* (1996)
Ciba Coming*	Lactation studies	Wide variations	1.0–5.0 mM, ionic strength 81 mM		Lin (2002)
	Compared three electrodes		As above		Lin *et al.* (2006)
Ciba Coming*	253 milk samples from 36 different cows	Range was 1.05–5.29 mM (average = 1.88 mM)	1.0–5.0 mM – used five standards	Looked at effects of Ca^{2+} on ethanol stability and RCT.	Tsioulpas *et al.* (2007a,b)

Ciba Coming*	Colostrums and early lactation milks		1.0–5.0 mM – used five standards		Tsioulpas *et al.* (2008)
Ciba Coming*	Effects of stabiliser addition to goats' milk		1.0–5.0 mM – used five standards	Measured heat stability during UHT treatment by sediment	Boumpa *et al.* (2008)
Ciba Coming*	Changes during the Spring flush	Bulk milk from five farms over Spring flush period	1.0–5.0 mM – used five standards	Only small changes over the Spring flush period	Grimley *et al.* (2009)
Ciba Coming*	Measuring Ca^{2+} at high temperature	Ca^{2+} decreased as temp. increased	1.0–5.0 mM – used five standards	Compared dialysis and UF	On-Nom *et al.* (2010)
ISFET sensor	$CaCl_2$ solution with and without Mg and milk	Also measured milk, 1.33 mM	$CaCl_2$, with different electrolytes	Evaluated ion selective field effect transistor (ISFET) sensor	Bratov *et al.* (2000)

* Ciba Corning now trade under the name Siemens Healthcare Diagnostics Ltd, Frimley, Surrey, UK

Nieuwenhuijse and van Boekel (2003). For dilute solutions, concentrations and activities are the same, but in concentrated solutions they diverge. The relationship between concentration and activity is given by Equation 11.7:

$$\text{activity} = \text{activity coefficient x concentration} \tag{11.7}$$

For calcium (and other divalent ions) in milk the activity coefficient is usually taken as 0.4. In concentrated milk, it is 0.3 (Augustin & Clarke, 1981). Note that calibration procedures use standards of known concentration, not activity. Where activities are used, the activity coefficients discussed above are usually used, although concentration is easier to understand than activity. At the Ca^{2+} concentrations found in milk (~2 mM), milk is regarded as a dilute solution.

It is important to ensure that the calibration fluids resemble milk as closely as possible in terms of ionic strength, which is considered to be 81 mM for normal milk, but which increases in concentrated milk to about 200 mM (Augustin & Clarke, 1981). The pH of the standards should also match that of the product being tested; for normal milk this is about 6.7. Many studies adjust the ionic strength of standards, but few consider their pH.

The most important operating criterion is that the electrode should give a stable electrical output (mV). This should also comply with the Nernst equation, that is, for divalent cations, increasing the concentration ten-fold increases the mV response by about 29 mV (Geerts, 1983). For narrower ranges of concentration, doubling the concentration increases its response by 8.7 mV. If this is not found to be the case and the electrode does not give a stable reading within a reasonable period (no longer than 5 min), then the electrode needs to be cleaned, conditioned, topped up with electrode fluid, or replaced. If these procedures are adhered to, it is possible to obtain reproducible results with such electrodes. Some recent experimental work is discussed in Section 6.1.1.

11.2.18 pH and Titratable Acidity

11.2.18.1 pH

Milk is a complex fluid and its pH is influenced by a host of factors. The pH of raw milk should always be monitored, although the pH alone does not provide adequate information for assessing milk quality. The most convenient method is to use a pH probe, which is one example of a selective ion electrode. The electrode should always be calibrated prior to use using appropriate pH standards (4 and 7 are the norms for milk products, but 10 is also available). The status of the pH probe can be established by ensuring that a pH change of 1 unit gives rise to a 58-59 mV change in the millivolt response (sometimes called the slope). Note that this range is slightly temperature-dependent. If not within or close to this range, then the electrode is in need of cleaning or replacing.

As discussed in Section 6.1.1, it is difficult to predict the heat stability of a sample of milk solely from its pH value. However, for a specific sample or batch of milk (whatever its volume), any event which lowers its pH is likely to have a detrimental effect on heat stability, especially toward UHT processing (see Section 6.2.1). This could be due to microbial activity, addition of soluble calcium salts, SDHP, or any other acidic ingredient. Conversely, any substance that increases the pH, such as TSC and DSHP, will most likely initially lead to an improvement in heat stability. However, if it is increased too much, its heat stability may deteriorate (moreso to in-container sterilisation than to UHT treatment). Also, any changes in pH are accompanied by a change in ionic calcium (see Section 6.1.1).

Milk pH is temperature-dependent and it is important to appreciate this; pH decreases as temperature increases. It is possible to obtain pH probes for measuring fluids at high temperature. An alternative procedure is to use dialysis or ultrafiltration at high temperatures and measure the pH of the dialysate or permeate. These methods have been described for cow's milk by On-nom *et al.* (2010) and On-nom (2012) and for soymilk by Pathomrungsiyounggul *et al.* (2012). They are discussed further in Section 11.2.24.1. Using these procedures, it has been established that the pH of milk may fall to about 5.9 when it is heated to 120 °C and as low as 5.6 when it is heated to 140 °C. However, on cooling, the pH reverts to close to its original value following UHT processing. These changes have been compared for UHT and in-container processes by Chen *et al.* (2015). Much longer holding times at 140 °C, or extended heating and cooling periods may result in small changes in pH, compared to those resulting from in-container sterilisation.

11.2.18.2 Titratable Acidity (TA)

Titratable acidity (TA) is often used as a quality indicator of milk products, but the name is confusing as it is not measuring acidity, but rather buffering capacity, as discussed later.

TA is determined by titrating 10 mL milk (sample) with 0.11 M (0.11 N) caustic soda solution, using phenolphthalein as indicator. This, according to many texts, allows lactic acid (%) in the sample to be determined by dividing the titre value (mL) by 10. While TA is commonly expressed as % lactic acid (or % LA), it is also expressed in other ways as follows: Soxhlet Henkel degrees (°SH) which is % LA x 400/9; Therner degrees (°Th) which is % LA x 1000/9; and Dornic degrees (°D) which is 100 x % LA.

It might appear that TA is measuring the amount of lactic acid in the milk, but this is not the case. What is being measured in reality is the buffering capacity of the milk, i.e. the amount of alkali required to increase the pH of milk from its starting value to about 8.4, which is the pH at which the indicator colour changes from colourless to pink. Cronshaw (1947) makes some pertinent observations about titratable acidity, which are worth emphasising. It measures the total amount of acid, both ionized and unionized. He suggested the better and more explicit name of total absorbing capacity. He also stated that fresh normally expressed milk (normal milk) contains no lactic acid, although it is tacitly assumed that fresh milk has an acidity of approximately 0.14% LA. A similar assertion that fresh milk contains practically no lactic acid was made more recently by Skanderby *et al.* (2009). They gave the contributions to a TA value of 0.14% LA as follows: CO_2, 0.01; citrate, 0.01; casein, 0.07; whey P, 0.01 and phosphates, 0.3. Udder infections can cause TA to be lower and pH to be higher.

The initial TA of milk samples varies within wide limits, with ranges reported to be 0.12 to 0.20% (Cronshaw, 1947) and 0.13 to 0.20% (Bhandari & Singh, 2011). Values for herd milks vary less, but occasionally mixed milk has a TA up to 0.23%.

Some significant values for TA values (% LA) in relation to milk processing are:

- 0.12 to 0.15%- normal milk
- 0.18 -0.19% an arbitrary maximum between satisfactory and unsatisfactory milk
- 0.22% practical limit for milk pasteurisation
- 0.25% milk clots on boiling
- 0.3 to 0.4%: it is possible to detect a sour taste

- 0.32% milk begins to clot at pasteurisation temp
- 0.50% milk clots at room temperature

There is no information about limits for raw milk to be suitable for UHT processing. Decisions should not be made on the microbial quality of raw milk based on TA results alone. Although this is appreciated in many quarters, it still happens. In fact a reported definition of TA as the amount of titrant needed to react stoichiometrically with lactic acid in milk (Hooi *et al.*, 2004) is not helpful in this respect.

China has specifications for minimum and maximum levels for TA of imported UHT milk (Chinese Ministry of Health, 2010). The TA should range between 12 and 18°Th (0.108-0.162% LA). The lower limit may have been introduced to prevent adulteration of milk with alkali, which would decrease the TA value (but would also change the Na or K concentrations and most probably FPD). This is important for companies exporting milk to China as milk having TA values below the minimum can be rejected.

11.2.19 Protease

11.2.19.1 Plasmin

Plasmin is the major indigenous milk protease and is associated with the casein micelles in milk. As discussed in Sections 6.1.3.5.1, 7.1.3.4 and 7.1.4, plasmin has a high heat stability and withstands some UHT treatments and causes flavour (bitterness) and stability (gelation) defects in UHT milk. Plasmin is secreted with its inactive precursor plasminogen, which is present at 2 to 30 times the concentration of plasmin (Ismail & Nielsen, 2010). Plasminogen activators present in milk convert plasminogen into plasmin by hydrolysis of the Arg-Ile bond of plasminogen. This conversion occurs during storage of UHT milk and occurs more readily in high-somatic-cell-count milk (Kelly & Foley, 1997). Because of this conversion, it is not unusual to measure both plasmin and plasminogen. This is achieved by measuring the total plasmin after converting the plasminogen to plasmin with the enzyme urokinase (Korycka-Dahl *et al.*, 1983).

Plasmin hydrolyses proteins at the carboxyl side of Lys-X and Arg-X bonds. It has optimum activity at pH 7.5 and 37 °C. It degrades β-, α_{s1}- and α_{s2}-caseins to γ-caseins, proteose-peptones and λ-casein, respectively. It hydrolyses κ-casein very slowly and does not hydrolyse whey proteins. Plasmin activity in milk can be determined by measuring the concentration of the fluorescent product AMC (7-amido-4-methyl coumarin) released from the non-fluorescent coumarin peptide N-succinyl-L-alanyl-L-phenylalanyl-L-lysyl-7-amido-4-methyl coumarin (Richardson & Pearce, 1981; Saint-Denis *et al.*, 2001) or the coloured product 4-nitroaniline released from the chromogenic substrate H-D-valyl-L-leucyl-L-lysyl-4-nitroanilide (S-2251) (Korycka-Dahl *et al.*, 1983; Manji *et al.*, 1986).

Plasmin activity in UHT milk can also be detected by the rate of breakdown of β-casein into γ-casein. Newstead *et al.* (2006) achieved this by using densitometric quantification of the β-casein band on SDS-PAGE.

11.2.19.2 Bacterial Proteases

Several protease assay methods have been used for measuring bacterial protease activity in milk and milk products. Substrates used include casein, chromogenic substrates such as hide powder azure (HPA) (Cliffe & Law, 1982), azocoll (Chavira *et al.*, 1984) and azocasein (Christen & Marshall, 1984; Schokker & van Boekel, 1997), and fluorogenic substrates such as fluorescein isothiocyanate-labelled casein (FITC-casein) (Twining, 1984; Christen & Senica, 1987) and 4-methylumbelliferyl casein (Khalfan *et al.*, 1983). When casein is used as a substrate, the released peptides are quantified by one of the methods discussed in Section 11.2.21, for example, fluorescamine (Button *et al.*, 2011). Immunological methods using ELISA have also been developed (Stepaniak *et al.*, 1987); however, these methods measure the protein not its enzyme activity and hence have limited application in products like UHT milk.

Rollema *et al.* (1989) compared methods based on fluorescamine, trinitrobenzene sulfonic acid (TNBS), HPA, azocoll and ELISA for detection of bacterial proteinase in UHT milk. Assays based on fluorescamine, TNBS and azocoll showed similar sensitivities and were more sensitive than those based on HPA. The ELISA-based assay was very sensitive and rapid but the authors concluded it was too specific. The authors did not evaluate their assays on very low protease activity levels which are responsible for proteolysis and its associated problems of age gelation and bitterness in high-temperature heated products such as UHT milk. Therefore in order to be able to quantify the activity so that shelf-life can be predicted, very sensitive assays are required. Button *et al.* (2011) explored methods of detecting very low activities and found that the FITC-casein method was more sensitive and correlated better with proteolysis in UHT milk during normal storage than azocasein and casein–fluorescamine methods. The FITC-casein was very sensitive when used in accordance with the method of Christen and Senica (1987) modified to allow incubation times of up to a few days.

11.2.20 Protein

Determination of the protein content of milk is an every-day occurrence in the dairy industry. In most countries protein content is a major element of payment systems and is therefore routinely measured on raw milk by automatic machines. Such a machine is the Foss MilkoScan™FT1 based on Fourier Transform Infrared (FTIR) technology. It analyses total protein and casein as well as fat, lactose, total solids, solids-not-fat, freezing point depression, total acidity, density, free fatty acids, citric acid and urea (Foss, 2014).

A large number of protein analysis methods have been reported and used. Many of these are covered in the review by Tremblay *et al.* (2003). A brief discussion only of some of these is given here. Traditionally, protein has been determined from a Kjeldahl nitrogen analysis. While it has been superseded by instrumental methods for routine analyses, it remains the standard reference method (e.g. IDF, 1993). The factor used for conversion of nitrogen to protein is 6.38 for milk proteins. A consideration when using nitrogen methods to determine protein concentration is that the total nitrogen includes some non-protein nitrogen (NPN) which is due to nitrogenous compounds of which urea constitutes about 50%. The actual content of these NPN compounds contributes about 0.19% to the crude protein figure. Hence true protein is approximately equal to total protein percentage determined by nitrogen analysis multiplied by 6.38 minus 0.19.

In addition to total protein, casein and whey (or serum) proteins can also be determined. This is achieved by analysing the total protein, then precipitating the casein at pH 4.6 and measuring the remaining soluble non-casein protein which is largely the whey proteins. Subtracting the whey protein percentage from the total protein percentage gives the casein percentage. The non-protein nitrogen content can be measured by precipitating all but the NPN compounds in 5% phosphotungstic acid and determining the nitrogen in the supernatant (Jarrett *et al.*, 1991).

Several other methods are used for protein analysis in the laboratory. A widely used method is the Lowry method based on the colour produced with the Folin-Ciocalteu phenol reagent (Lowry *et al.*, 1951). This reagent and protocols for its use in protein determination are readily available from chemical companies such as Sigma-Aldrich. The colour formed is measured at either 550 or 750 nm. A variation of the Lowry method uses bicinchoninic acid (BCA) instead of the Folin-Ciocalteu phenol reagent (Stoscheck, 1990). The colour generated is measured at 562 nm. A BCA kit is available from the Pierce Chemical Company.

Another method commonly used is the Bradford dye binding method based on the interaction of proteins with Coomassie Brilliant Blue G-250 in acidic solution (Bradford, 1976). The colour produced is read at 595 nm. The sensitivity is similar to that of the Lowry method, being able to detect 5-100 μg of protein. As for the Lowry method, the reagents and protocol for the method are available from chemical companies such as the Bio-Rad Corporation.

Individual proteins can also be separated by electrophoresis and chromatography. 1D PAGE analysis has been the traditional method for separating proteins in milk. PAGE gels can be run in three different forms: native, SDS and SDS reducing. Native gels separate the proteins as they occur in milk, SDS gels incorporate sodium dodecylsulfate (SDS) which breaks down hydrophobic bonds between and within proteins and the SDS reducing form incorporates SDS as well as a reducing agent such as methyl mercaptan or dithiothreitol to reduce disulfide bonds between proteins. The SDS reducing form of PAGE is the most common form used. As well as PAGE, starch gel electrophoresis (SGE) is also used (e.g., Snoeren & Both, 1981; Kalogridou-Vassiliadou *et al.*, 1982).

2D PAGE, in which the proteins are separated in the first dimension according to charge by isoelectric focussing and then separated on a second dimension on the basis of size, has been increasingly used, especially in conjunction with mass spectrometry in proteomic analyses (Holland *et al.*, 2011) (see Section 11.3.3). In this case proteins (and peptides with molecular weights greater than ~10 Da) are shown as spots on large gels. The 2D PAGE analysis produces a large number of spots as proteins with different charge or molecular weight are separated (see Figure 7.3). It is a very powerful technique as a vast amount of information can be obtained from a gel.

Reversed-phase (RP) HPLC is being used increasingly for milk protein analyses. Whereas 1D SDS PAGE and SGE tend to show one band for each of the major proteins, HPLC separates some of the protein variants caused by genetic polymorphism and post-translational modifications of phosphorylation and glycosylation. The multiple forms for the various gene products can be seen in Figure 11.5.

11.2.21 Proteolysis (Peptides)

Because of the importance of proteolysis in development of bitter flavours and age gelation during storage of UHT milk (see Sections 7.1.3.4 and 7.1.4), many methods of measuring the proteolysis products, peptides, have been proposed and used (see, for example, AlKanal, 2000). Since all milk contains some endogenous peptides, these must be taken into account when determining increases during storage. For example, McKellar (1981) reported background TCA-soluble free amino groups to be ~0.82

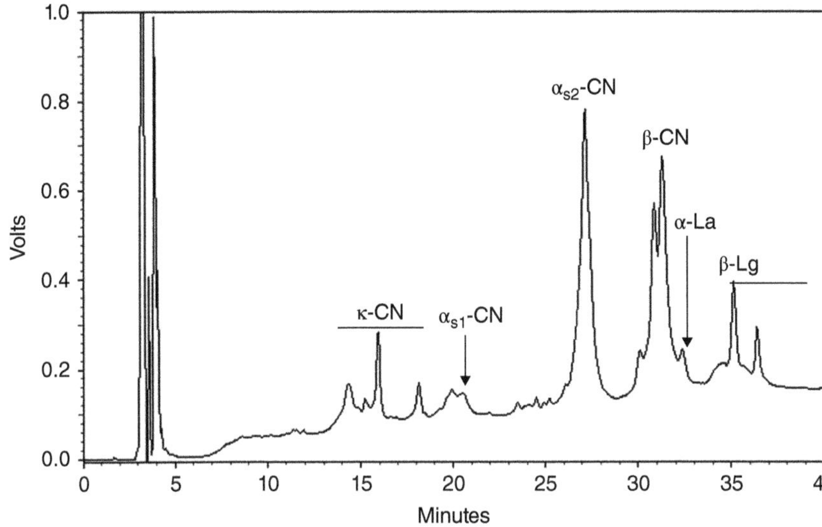

Figure 11.5 Reversed-phase HPLC of milk proteins in UHT milk soon after manufacture (HPLC conditions: 5 μ 250 mm × 4.6 mm C18 column run at 50 °C; binary solvent gradient system: solvent A 0.1 % TFA in water, solvent B was 0.1 % TFA in acetonitrile; gradient was: 0-2 min, 27% B; 2-37 min, 32% B; 37-40 min, 48.4% B; >40 min, 50% B. (Source: Al Saadi & Deeth, 2008. Reproduced with permission of Dairy Industry Association of Australia.)

μmol/mL. A further consideration is that the peptides can be formed from either plasmin or bacterial enzymes and these have different properties. Plasmin produces mostly large hydrophobic peptides from β-casein while most bacterial enzymes produce smaller, more hydrophilic peptides, principally from κ-casein. Most of the ones derived from plasmin are soluble at pH 4.6 but insoluble in TCA (4-12%) while most of the ones derived from bacterial proteases are soluble at pH 4.6 and in 4-12% TCA. The different properties of these peptides can be used to distinguish between them and hence provide information on the cause of the proteolysis (Datta & Deeth, 2003). It should be noted, however, that the bacterial proteases vary considerably in their specificity towards the different caseins and hence care needs to be exercised in interpreting the analytical results. For example, a competitive ELISA method was developed for detecting proteolysis in UHT milk by measuring the level of caseinomacropeptide (CMP), a peptide released by proteolysis of κ-casein (Picard *et al.*, 1996). Such an assay would be effective for bacterial proteases that are specific for κ-casein but would not be effective for assaying peptides released from other caseins by some bacterial proteases.

The peptide methods can be divided into three major categories: 1. measures of nitrogen in the non-casein nitrogen (NCN) and non-protein nitrogen (NPN) fractions of milk; 2. measures of primary amine groups; and 3. chromatographic and electrophoretic analyses of individual peptides. With the possible exception of the electrophoretic analysis, the methods include a preliminary fractionation step to produce a delipidiated, deproteinised extract of the product. These extracts are commonly pH 4.6 or TCA filtrates.

Category 1 methods include analysis of nitrogen by Kjeldahl and high-temperature combustion methods, for example, Dumas, for which standard methods are available

(Wiles *et al.*, 1998; Lynch & Barbano, 1999; Tremblay *et al.*, 2003); category 2 methods are based on the reaction with primary amino groups with reagents including fluorescamine (Chism *et al.*, 1979; Beeby, 1980; Kocak & Zadow, 1985), trinitrobenzene sulfonic acid (TNBS) (Spadaro *et al.*, 1979), Hull tyrosine (Hull, 1947) and o-phthalaldehyde (OPA) (Church *et al.*, 1983). Category 3 methods include reversed-phase HPLC and 1D- and 2-D PAGE. Proteomic techniques which combine either HPLC or 2D-electrophoresis with mass spectrometry are used for more detailed peptide analyses (e.g., Gaucher *et al.*, 2008b).

Category 1 and 2 methods are usually used for measuring the total amount of peptides in a sample while the category 3 methods are more suited to determining the different peptides formed. However, HPLC methods can be used to quantify the total peptides by integrating the peaks and using a suitable internal standard such as a dipeptide like Tyr-Leu or Leu-Gly (Le *et al.*, 2006). Similarly, 1D-PAGE bands can be quantified by densitometry although the sensitivity and accuracy are less than for HPLC methods.

11.2.21.1 Distinguishing Peptides Produced by Plasmin and Bacterial Proteases by Analysis of Primary Amine Groups

On hydrolysis, proteins are converted to peptides which have more primary amino groups than the parent proteins. Therefore a measure of the primary amine groups gives a measure of the extent of proteolysis. Fluorescamine, TNBS and OPA react with the primary amino groups to give fluorescent or coloured products. The fluorescence or absorbance is a measure of the level of peptides and amino acids present in the milk. By measuring the peptides in both the pH 4.6 and 4% TCA extracts, it is possible to determine the type of protease causing the proteolysis (Datta & Deeth, 2003). The results can be interpreted according to Table 11.3.

11.2.21.2 HPLC Analysis

RP-HPLC is a very sensitive method for detecting proteolysis in milk. This was shown by Haryani *et al.* (2003) who were able to detect low levels of proteolysis in cold-stored milk. It is also very useful for distinguishing between peptides produced by plasmin and bacterial proteases.

Table 11.3 Interpretation of the results from the primary amino group analysis on TCA- and pH 4.6 filtrates from UHT milk (based on use of fluorescamine by Datta *et al.*, 2003).

Peptide content		
TCA filtrate	**pH 4.6 filtrate**	**Interpretation**
Very low	Low	Good quality milk
Very low	Significant	Plasmin action only
Significant	Significant	Bacterial protease action only **or** both bacterial protease and plasmin action

Table 11.4 Interpretation of the results of peptide analysis on TCA- and pH 4.6 filtrates from UHT milk (based on the RP-HPLC method of Datta *et al.*, 2003).

Peptide peaks by HPLC		
TCA filtrate	**pH 4.6 filtrate**	**Interpretation**
Virtually no peaks after the initial solvent peak	Some sharp peaks after 20 min[1]	Good quality milk
Some small peaks after the initial solvent peak, especially with extensive proteolysis	Significant peaks after 20 min; peaks may be broad with extensive proteolysis	Plasmin action only
Significant peaks before 20 min but few if any after 20 min	Significant peaks before 20 min but few if any after 20 min	Bacterial protease action only
Significant peaks before 20 min but few, if any, after 20 min	Significant peaks both before and after 20 min	Both bacterial protease and plasmin action

[1] This time is based on the method used by Datta and Deeth (2003) and will vary according to the HPLC conditions used. The conditions used by Datta and Deeth (2003) were as follows: Column: 150 x 4.5 mm, 5 μm, C18 RP; Temperature: 40 °C; Solvent: Binary gradient system. Solvent A, 0.1% (v/v) trifluoroacetic acid (TFA) in water, Solvent B, 0.1% (v/v) TFA in acetonitrile. Gradient: Proportion of solvent B increased from 20 to 35% over 20 min, and after 5 min increased to 65% over 20 min and then increased to 100% over 5 min.

The peptides released by bacterial proteases (produced by psychrotrophic bacterial contaminants of milk) are relatively small and hydrophilic while those released by the native milk protease, plasmin, are relatively large and hydrophobic. On reversed-phase HPLC, the former elute before the latter. As discussed above, the peptides formed from plasmin hydrolysis remain soluble at pH 4.6 but precipitate in 4% TCA, while the latter are soluble at pH 4.6 and in 4% TCA. Therefore by measuring the peptides in both the pH 4.6 and TCA extracts, it is possible to analyse all the peptides and determine the type of protease(s) causing the proteolysis. This analysis complements the chemical analyses above (Datta & Deeth, 2003; Le *et al.* 2006). An interpretation of the reversed-phase HPLC results is given in Table 11.4.

11.2.21.3 Polyacrylamide Gel Electrophoresis (PAGE) Analysis

PAGE is widely used to separate milk proteins and peptides derived from them. In peptide analysis, PAGE can be used qualitatively for identifying the presence of a peptide or quantitatively with the use of a densitometer. The method is based on that of Laemmli (1970). Gels are commonly run with a protein-denaturant sodium dodecyl sulfate (SDS) or urea to break non-covalent bonding within and between proteins (Bansal *et al.*, 2010). They can also be run under non-reducing or reducing conditions. In the latter, disulfide bonds are cleaved to sulfhydryl groups by heating the sample with mercaptoethanol or dithiothreitol before being loaded onto the gel. After the gel is run, the protein bands are visualised by staining the gel in a solution of a protein-binding dye such Coomassie Blue and then destaining the background gel with dilute acetic acid.

For proteolysis of milk proteins the major bands of interest are those of para-κ-casein, which is a fair indication of bacterial proteolysis, and γ-caseins,

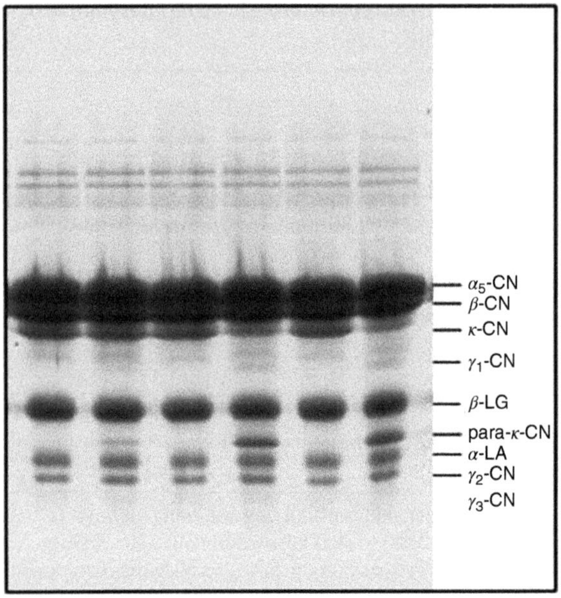

Figure 11.6 Reducing SDS-PAGE gel of protein degradation of UHT milk samples before and during storage at 25 °C. a) Lane 1: Day 0 - control, Lane 2: Day 0 - treatment, Lane 3: Day 3 - control; Lane 4: Day 3 - treatment, Lane 5: Day 6 - control, Lane 6: Day 6 - treatment. This chromatogram clearly shows the increase in the para-κ-casein band and decrease in the κ-casein band in lanes 1 to 2 to 4 to 6, indicating proteolysis by bacterial protease. (Source: Button, 2007. Reproduced with permission.)

proteose-peptones and λ-casein which are a good indication of plasmin-mediated proteolysis of β-, α_{s1}- and α_{s2}-caseins (these are illustrated in Figure 11.6). Concomitantly, one or more of the casein bands may show reduced intensity compared with a non-proteolysed sample.

11.2.22 Sediment

Sediment in heated milks is normally determined by centrifugation (see Section 11.2.24.2). The best centrifugation conditions should be established for each product. As a starting position, centrifugation at approximately 3000 g for 30 min with a sample size of 45 to 50 mL has been used successfully (Boumpa *et al.*, 2008; Chen *et al.*, 2012; Chen, 2013). This is equivalent to about 62.5 days storage at normal gravity (free standing). The centrifuge tubes should be transparent so that any separation can easily be observed. Note that the sample temperature will increase slightly as a result of centrifugation. The volume of wet sediment can be measured but for more quantitative studies, the amount of sediment should be measured on a dry weight basis. Some centrifuged products are shown in Figure 9.4.

Higher centrifugal forces may be required for evaluating the stability of more viscous products (see Stokes Law, equation 5.2). For these products, the higher viscosity hinders separation and potentially improves their stability. Note also that products stored at higher temperatures will have a lower viscosity and any fat creaming or sediment formation will take place more quickly.

11.2.23 Sensory Characteristics

The sensory characteristics of the final heat-treated milk or milk product are extremely important and are affected by changes which take during processing (Chapter 6) and especially throughout storage (Chapter 7). The major sensory characteristics relevant to UHT milk products are:

- the presence or absence of visual defects, for example, visible separation (creaming sedimentation), coagulation or gelation, change in colour,
- the aroma and taste which together constitute the flavour, for example, cooked, bitter, rancid, stale
- the texture and mouthfeel; defects may be described as watery, thin, coagulated, sandy, gritty, separated.

Sensory evaluation techniques may employ one or more of the following three methodologies:

- discrimination testing for determining whether there is a perceived difference between two products;
- acceptance or affective testing to determine the degree of acceptance or preference of a product. Often a standard product is used as a control or for benchmarking. Hedonic scales may be set up, typically with five, seven or nine points;
- descriptive analysis, which is a much more detailed approach, giving a total sensory description of a product, and which usually involves training a panel. Prasad *et al.* (1989) described the development of a vocabulary for heat-treated milks. This involved the use of 37 descriptive terms, covering its appearance, odour, flavour, mouthfeel and aftertaste. A list of terms used is given in Table 11.5, together with their descriptions. Clare *et al.* (2005) used descriptive analysis in their sensory analysis of UHT milks, produced by microwave and indirect heat exchanger methods, during storage. The descriptor terms they used were: cooked/caramelized, sweet aromatic/cake mix, fatty/stale, sweet, bitter, astringent, colour intensity and chocolate flavour (for chocolate milk only).

It is beyond the scope of this book to fully describe these methodologies. The reader is advised to consult the many excellent published texts and papers on sensory evaluation (e.g., Lawless & Heymann, 2010). Some of these have also related sensory characteristics with volatile compounds identified by chemical analysis (Jensen *et al.*, 2015).

In most countries, it is standard practice for samples of heat-treated products from every batch to be tasted by as many people as possible who work at the factory. These people soon become experts in the specific products and, in principle, should be able to detect any sensory defects and prevent sub-standard product being released for sale. However, this is not straightforward with some UHT products, especially milk, as the products have a high cooked flavour intensity immediately after production, which is not generally liked. If this approach is to be adopted, products should be tasted about one week after production or just prior to release from the factory. Retention samples should be tasted periodically throughout storage.

Table 11.5 Descriptive terms used for UHT milk. (Source: Prasad, 1989. Reproduced with permission.)

Term	Definition
Visual characteristics (by eyes)	
(1) Degree of creamy/yellowness (none intense)	A continuum exemplified by the lack of creamy/yellow color at one end to butter-like yellow at the other
(2) Degree of brownness (none intense)	A continuum exemplified by canned or in-bottle sterilized milks as the maximum
(3) Degree of grey/greenness (none intense)	A continuum exemplified by milk permeate or whey (maximum) and full cream milk (minimum)
(4) Transmittance (translucent opaque)	Passage of light through milk. Skim milk with virtually no fat shows maximum transmittance while homogenized full cream milk exemplifies the minimum
(5) Phase separation (none–extreme)	A continuum exemplified by homogenized and skim milks as minimum and pasteurized full cream milk as maximum
(6) Visual consistency (thin/watery–thick/creamy	The resistance observed when the sample is swirled. The maximum and minimum of the continuum exemplified by pasteurized high-fat milks and skim milks, respectively
Olfactory characteristics (by nose)	
(7) Odor intensity (nil–strong)	Overall odor intensity
(8) Characteristic milkiness (nil–strong)	Characteristic milk aroma exemplified by pasteurized full-cream milks
(9) Sweet smell (nil–strong)	An odor sensation that is analogous to the sweet taste
(10) Degree of freshness (not fresh–very fresh)	Odor characteristic normally associated with fresh, pasteurized, full-cream milk. (Assessors could not define freshness without further associations.)
(11) Cooked odor (nil–strong)	A continuum exemplified by pasteurized milk at the lower end and in-bottle sterilized milk at the other extreme
(12) Stale/old (nil–strong)	A negative odor attribute that is different from "not fresh." Again the assessors could not define staleness by simple association.
(13) Cardboard/oxidized (nil–strong)	Exemplified by a cardboard smell or oxidized fat in milk
(14) Sourness (nil–strong)	An odor sensation that is analogous to sour taste
(15) Cheesiness (nil–strong)	Exemplified by mature Cheddar cheese
(16) Synthetic/artificial (nil–strong)	A foreign odor not normally associated with milk
(17) Antiseptic/phenolic (nil–strong)	Exemplified by TCP (a commercial preparation of trichlorophenol)
(18) Metallic (nil–strong)	An odor associated with metal, particularly copper and iron
(19) Nutty (nil–strong)	Fatty, burnt, roasted, or sulfury odor note, characteristic of certain nuts

Table 11.5 (Continued)

Term	Definition
Oral characteristics (by mouth)	
Flavor	
(20) Flavor intensity (nil–strong)	Overall flavor intensity
(21) Characteristic milkiness (nil–strong)	Characteristic milk flavor exemplified by fresh, good, pasteurized full cream milk
(22) Sweetness (nil–strong)	A taste similar to sucrose. Perceived mainly at the front of the tongue
(23) Degree of freshness (not fresh–very fresh)	Flavor characteristic normally associated with fresh, pasteurized, full-cream milk. (Assessors could not define freshness without further associations.)
(24) Cooked flavor (nil–strong)	A continuum exemplified by pasteurised milk at the lower end and in-bottle sterilized milk at the other extreme
(25) Stale/old (nil–strong)	A negative flavor attribute that is different from "not fresh." Again the assessors could not define staleness by simple association.
(26) Cardboard/oxidized (nil–strong)	Exemplified by a cardboard smell or oxidized fat in milk
(27) Sourness (nil–strong)	A taste sensation that is analogous to sour milk
(28) Cheesiness (nil–strong)	Exemplified by mature Cheddar cheese
(29) Synthetic/artificial (nil–strong)	A foreign flavor not normally associated with milk
(30) Antiseptic/phenolic (nil–strong)	Exemplified by TCP (a commercial preparation of trichlorophenol)
(31) Metallic (nil–strong)	A flavor associated with metals, characterized by the sensation of sucking a copper alloy
(32) Nutty (nil–strong)	Fatty, burnt, roasted, or sulfury odor note, characteristic of certain nuts
(33) Bitter (nil–strong)	A basic taste sensation characterized by a solution of caffeine or quinine sulfate
Mouthfeel	
(34) Oral consistency (thin/watery–thick/creamy)	A textural continuum exemplified by skim milk at the lower end and high-fat milk at the other extreme
(35) Greasiness (not greasy–very greasy)	A sensation associated with a film or particles of fat on the tongue and or teeth. Characterized by high-fat milks
(36) Graininess (smooth–grainy)	The sensation of fine particles or granules. Characterized by a suspension of ground chalk
Aftertaste	
(37) Persistent aftertaste (nil–very persistent)	Duration for which the flavor remains after sample is consumed
(38) Drying effect (nil–very dry)	The feeling of shrivelling or puckering on the tongue or lining of the mouth

11.2.24 Separation Methods

In the course of analyzing components in milk, it is often necessary to fractionate it to facilitate the analysis. The most common fractionation is removing the fat and casein from the soluble components. Centrifugation at relatively low-speed can be used to separate the fat but removal of the casein is more challenging. Several methods exist for this operation but two major ones, the membrane-based techniques of dialysis and ultrafiltration, and centrifugation are discussed below. Other methods include precipitation of the casein with acid or rennet but these cause some changes in the casein and the soluble components.

11.2.24.1 Dialysis and Ultrafiltration

Dialysis and UF are two procedures that can be used to measure components that are of low molecular weight and which are not associated with casein or whey proteins. Rose and Tessier (1959) conducted some pioneering UF experiments of milk at high temperature. Total Ca and Ca^{2+} levels in permeate decreased as UF temperature increased. Similar trends have been found by On-Nom *et al.* (2010) and Chandrapala *et al.* (2010).

A pioneering study of milk partitioning by Davies and White (1960) used dialysis and UF to investigate mineral partitioning between the soluble and colloidal (casein micellar) phases in milk and especially how this was affected by pH and, to a lesser extent, temperature. Dialysis is performed by placing dialysis tubing containing a small amount of water into the milk and allowing sufficient time for equilibrium to take place. Note that this will take longer for components of a larger molecular weight. The milk can be held at different temperatures. At the end of the period the tubing contents are analysed. Davies and White (1960) compared dialysates taken at 4 °C and 20 °C and found that total calcium increased at the lower temperature. On-nom *et al.* (2010, 2012) used this procedure for partitioning milk at temperatures up to 120° C. It was important to remove the dialysis bag from the milk as quickly as possible after the heating process. Visking membranes with a molecular weight cut-off (MWCO) of 12,000 were used, as they were found to have good heat resistance. As milk temperature increased, pH, and total and ionic calcium decreased and there was a linear relationship between ionic calcium and temperature (Figure 11.7). It was also shown experimentally, that the dialysate collected at any temperature would show hardly any change in pH and Ca^{2+} with temperature below that temperature. This was demonstrated by subjecting dialysates collected at 20, 50 and 80 °C to further dialysis at these three temperatures. (Kaombe *et al.*, 2012; On-nom, 2012). This procedure allowed pH and ionic calcium to be determined in milk at high temperatures.

Figure 11.8 shows the variations in pH and ionic calcium for individual reconstituted skim milk powders, with increasing levels of DSHP, which caused a decrease in Ca^{2+} and an increase in pH. Dialysis was used to determine these values at 115 °C and it can be seen that both pH and ionic calcium were much lower at 115 °C. Note that although dialysis was performed at 115 °C, measurements were made at 20 °C.

On-Nom *et al.* (2012) measured pH and Ca^{2+} in milk with small incremental additions of calcium chloride, both at 20 °C and at 110 °C, where some of the milk samples coagulated. At both temperatures, the addition of calcium chloride decreased pH and increased Ca^{2+}. Using this dialysis procedure, it was possible to establish the pH and Ca^{2+} at which coagulation took place, measured at the coagulation temperature (Figure 11.9).

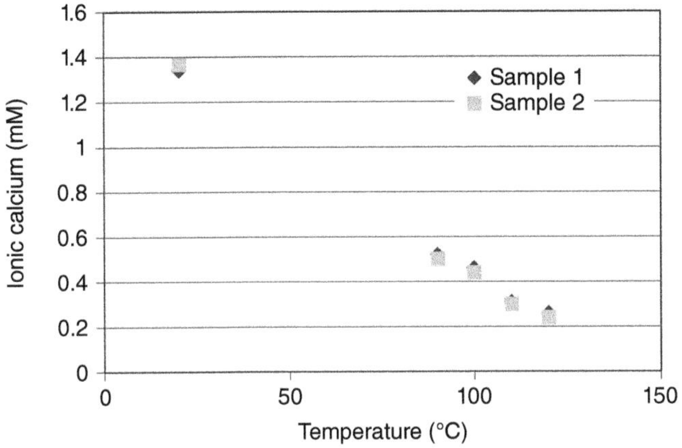

Figure 11.7 Ionic calcium levels in dialysates obtained at various temperatures. (Source: On-Nom 2012. Reproduced with permission.)

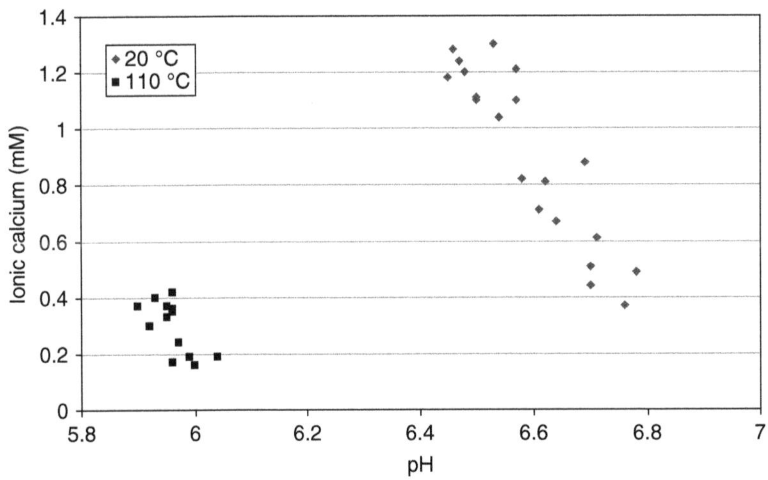

Figure 11.8 Ionic calcium and pH of dialyses of milk obtained at 20 °C and 115 °C. (Source: On-Nom, 2012. Reproduced with permission.)

Dialysis can also be used for removing low-molecular-weight components from milk. In this case the milk is placed in a dialysis bag and dialysed against a large volume of water. It removes lactose and soluble minerals, but if the milk is dialysed against a 4.5% lactose solution, then only soluble minerals are removed. In both of these cases, other low molecular weight compounds such as vitamins, organic acids and free amino acids are also removed.

On-Nom (2012) investigated using a dialysis membrane with MWCO of 300,000 and 1 million daltons. The aim was to examine the transfer of casein from the micelle to the soluble phase. It was shown that addition of TSC and to a lesser extent DSHP resulted in considerable dissociation of casein from the micelle. This could also be done by using an ultracentrifuge, but not at high temperatures.

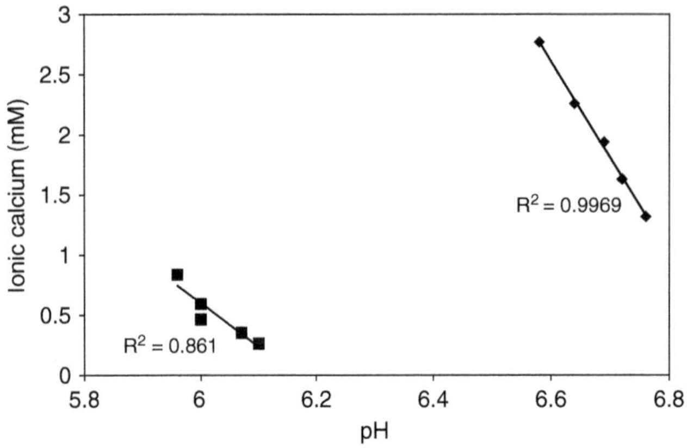

Figure 11.9 Ionic calcium and pH of dialysates of milk, containing added calcium chloride, obtained at 20 °C and 110 °C. (Source: On-Nom *et al.*, 2012. Reproduced with permission of Elsevier.)

Table 11.6 Composition of UF permeates taken at different temperatures. (Source: On-Nom, 2012. Reproduced with permission.)

Samples	pH	Ca^{2+} (mM)
Milk	6.70±0.04	1.76±0.55
Permeate at 80 °C	6.41±0.04	0.57±0.19
Permeate at 120 °C	5.91±0.03	0.29±0.03
Permeate at 140 °C	5.65±0.04	0.19±0.04

Thus dialysis is a useful analytical tool for determining what happens to various components of milk when it is heated, acidified or subjected to addition of stabilisers, mineral supplements or other additives (e.g., phosphates, citrates, EDTA).

Milk has been subjected to ultrafiltration (UF) at high temperature, up to 140 °C, by placing the UF module in the holding tube of a UHT plant (On-Nom, 2012). Some data for pH and Ca^{2+} for permeates collected at different UF temperatures are shown in Table 11.6. One conclusion was that the composition of UF permeate varies according to the temperature at which UF is performed. This is relevant to those milk processors who use UF permeate to standardise the protein content of milk.

11.2.24.2 Centrifugation

Centrifugation is an extremely useful procedure for a number of applications. It allows separation by the normal process of gravity to be considerably accelerated.

The G value is the ratio of the centrifugal force to the gravitational force. It can be determined from Equation 11.8:

$$G = \omega^2 r / g \qquad (11.8)$$

where ω = angular velocity (rad/s) = 2 π N (where N = rotational speed, s^{-1}); r = radius of the centrifuge arm or head and g = acceleration due to gravity (9.81 m s^{-2}). Thus if the rotational speed and the radius of the centrifuge arm or head are known, G can be calculated from Equation 11.8.

Centrifugation can be used for problem solving in a number of ways: for detecting excessive sediment; for detecting fat separation and improper homogenisation; for detecting other instabilities at an early stage, before they become noticeable by the consumer. For example in flavoured milk drinks, reducing the pH of the milk by too much during the preparation may cause the product to separate during storage. Photographs of some chocolate milk samples show that up to four distinct phases can be observed (see Figure 9.4).

The most common use of centrifugation in the dairy industry is centrifugal separation of milk into skim milk and cream. This can be done by either cold separation (~4-15 °C) or by hot separation (~50-55 °C). The viscosity of milk decreases by a factor of three as temperature increases from 4 °C to 50 °C, so separation processes take place three times faster at the higher temperature. This is a major reason for use of hot separation rather than cold separation. Also, all the fat is in the liquid form at 50 °C.

Separation of cream from milk occurs at a relatively low centrifugal force. By contrast, high-speed centrifugation or ultracentifugation (up to 100,000 g) is required to separate casein micelles from the serum phase. This produces a sediment fraction which consists largely of casein micelles and a soluble fraction which contains soluble proteins and peptides, most minerals and salts, lactose and water-soluble vitamins. Note that this soluble fraction differs from the permeate from ultrafiltration in that it contains the soluble proteins which include whey proteins and soluble caseins. One drawback is that ultracentrifugation cannot easily be used to study mineral and protein partitioning at high temperature.

11.2.25 Stability Tests

11.2.25.1 Ethanol Stability Test

The ethanol stability test is also known as the alcohol stability test. It is a simple test which is performed by mixing equal volumes of milk and diluted ethanol solutions. Different strength ethanol solutions are used to determine which concentrations cause the milk to flocculate. Alcohol concentration intervals used are at 2% or 5% increments. The highest level which just fails to cause flocculation is deemed to be the ethanol stability of that milk.

Shew (1981) reported that milk should be stable in 74% alcohol to be suitable for UHT processing. A stability value of 68% was proposed for milk being suitable for pasteurisation. An ethanol stability of 74% provides the simplest procedure for evaluating the suitability of milk for UHT processing. For example, it is much easier than establishing whether milk has a heat coagulation time of greater than 9 min at 140 C, which was considered a criterion for its suitability for UHT processing (von Bockelmann & von Bockelmann, 1998) (see Section 6.2.1). If a milk-based formulation is unstable in 74% alcohol, then it should be UHT processed with due care. As mentioned in Section 6.2.1.4, this criterion may not be a good indicator for milk destined for in-container sterilisation.

In a pioneering study, White and Davis (1958) measured variations in Ca^{2+} in milk from 132 individual cows, each of which was subject to detailed mineral analysis. They found a correlation coefficient of -0.76 between ethanol stability and ionic calcium, indicating that approximately 60% of the variation in ethanol stability could be accounted for by variation in Ca^{2+} concentration. Ca^{2+} as a proportion of the total calcium ranged from 8.8 to 13.0%. Lin (2002) studied variations in Ca^{2+} and ethanol stability in milk for individual cows and goats. Figure 11.10a shows the distribution of ethanol stability values found for milk from individual cows and goats, illustrating the lower levels found in goat's milk compared to cow's milk. Figure 11.10b shows the distribution of ionic calcium concentrations for these same milk samples. Figure 11.11 shows the relationship between ethanol stability and ionic calcium for these milk samples. It is interesting that there is hardly any overlap of the results for the goat's and cow's milk, suggesting that other factors also influence the relationship.

In the context of heat stability, the data shown in Figure 11.10a are interesting because according the criterion that milk should be stable in 74% ethanol to be suitable for UHT processing, the majority of goat's milk samples would be unstable to UHT processing, whereas the majority of cow's milk samples would be stable to UHT processing; this is what is found in practice.

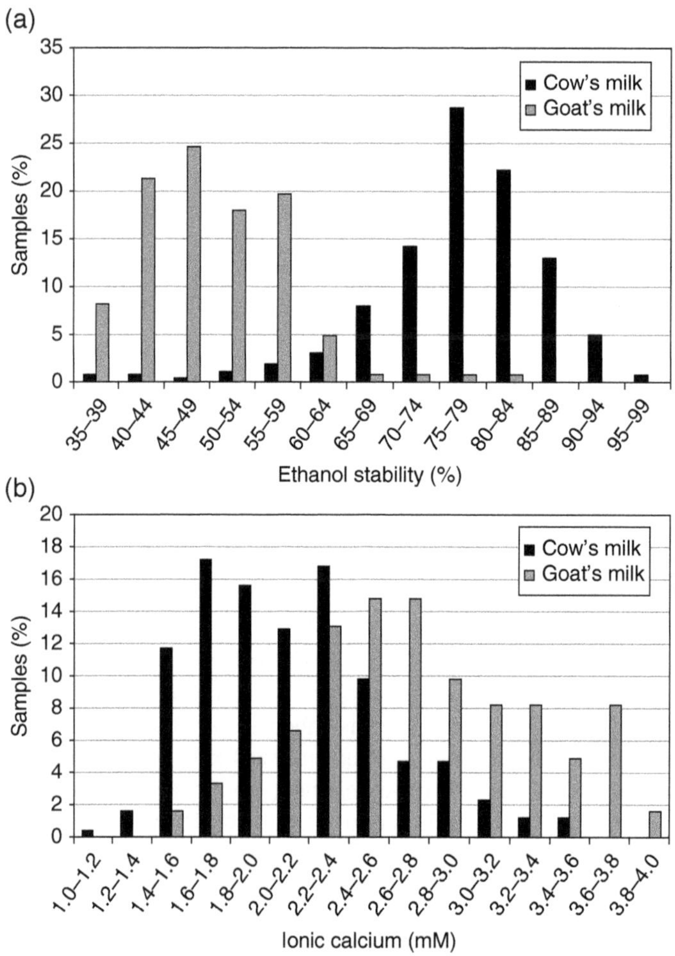

Figure 11.10 Relationship between (a) ethanol stability and (b) ionic calcium of milk from individual cows and goats. (Source: Lin, 2002. Reproduced with permission of University of Reading.)

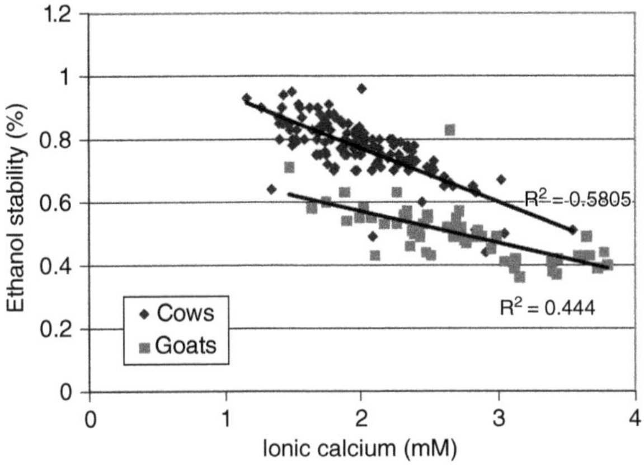

Figure 11.11 Relationship between ethanol stability and ionic calcium for milk from individual cows and goats. (Source: Lin, 2002. Reproduced with permission.)

The two main uses of the ethanol stability test are for determining the suitability of milk for thermal processing and as a general indicator of the quality of raw milk. This is discussed in more detail in Section 6.2.1.4. Although it is simple to perform the results should never be used in isolation, especially for rejecting raw milk destined for processing.

11.2.25.2 Other Heat Stability Tests

Several other tests for the heat stability of milk and milk products, particularly milk powders, evaporated milk, cream and milk creamers, have been reported. Most of these are concerned with the stability of the product in hot coffee and are often referred to as feathering tests, as proposed by Whittaker (1931). While these have not been designed for UHT milk they are suitable for it. The tests commonly involve addition of the milk product to a hot coffee solution and measurement of the amount of sediment formed, for example, the New Zealand coffee-stability test used by Teehan *et al.* (1997) for instant whole milk powder. The causes of instability, as shown by high levels of sediment in the tests, have been reported to include pH [the acidity of coffee is~4.9 (Burgwald, 1923)], the heat treatment conditions used during processing of the product (Oldfield *et al.*, 2000), and mineral imbalances, particularly high ionic calcium levels (Teehan *et al.*, 1997) which are manifested in early and late season milk (McKinnon *et al.*, 2000). Because of the significance of ionic calcium and acidity, tests have also been developed to test the stability of milk containing different levels of calcium chloride (or citrate if the milk is unstable without addition of calcium chloride) (Whitaker, 1931; Ranjith, 1995; Smyth *et al.*, 1999), and at different pH values (Teehan *et al.*, 1997; Ranjith, 1995). Ranjith (1995) recommended adjusting cream or evaporated milk with acetate buffers in the pH range of 4.70 to 4.92 and heating to 80 °C. A related test is the phosphate test proposed by Ramsdell *et al.* (1931) which was used more recently to assess the heat stability of UHT milk (Gaucher *et al.*, 2008a). It involves adding different quantities of 0.5 M potassium dihydrogen phosphate to 10 mL milk samples and heating at 100 °C for 10 min. The test result is the smallest amount of phosphate to cause destabilisation of the milk. Low values signify low heat stability. The ranges found were: for raw milk 0.1 to 1.0 mL, for pasteurised milk, 0.5 to 1.0 mL and for freshly processed UHT milk 1.2 to 2.0 mL. The values remained constant or increased slightly (from 1.6 to ~2 mL) in UHT milk during storage at 4 and 20 °C for >100 days but increased markedly at 40 °C (from 1.6 to ~3.3 mL) after 89 days, before decreasing. The results were surprising as a decrease rather than an increase in heat stability during storage would be expected. The significance of such testing of UHT milk remains unclear.

11.2.25.3 Accelerated Physical Stability

Centrifugation can be combined with light scattering for measuring emulsion stability and sedimentation using a LUMiSizer® (Lum, 2013). This allows accelerated stability testing results to be obtained. Optical properties are measured along the length of the sample in a polycarbonate tube throughout the centrifugation process using a near-infrared light source. The movement of the phase boundary can be used to calculate sedimentation rate which can be converted into a sedimentation rate under normal gravity conditions. The instrument can be used in a reflectance (back scattering) or transmission mode. Crowley *et al.* (2014) used a LUMisizer to observe stability changes when reconstituted skim milk powder was fortified with different calcium salts. These results are described in more detail in Section 9.5.1.

Other instruments for measuring the dispersion state of emulsions and suspensions are the Turbiscan (Formulaction, undated) and the Coulter QuickSCAN™ (Coulter, 1996). These instruments are also equipped with a near-infra-red light source (860 nm) which scans the length of a sample held vertically in a flat bottomed tube and detects the degree of creaming and sedimentation at a point in time or changes over time in the light scattering profiles. Blecker *et al.* (2012) used a Turbiscan to detect rennet-induced coagulation in milk and found it could detect coagulation at a much earlier stage than could the human eye. The QuickSCAN has been used to monitor changes in UHT soymilk over time; in this case, the sample was transferred to the tube aseptically and maintained in that condition for the duration of the analysis, 18 months (Durand, 2003).

11.2.26 Viscosity

Measurement of viscosity is important for milks, creams and desserts. It is particularly significant in UHT milk as an increase in viscosity during storage is an indicator of impending gelation (see Section 7.2.2). Different items of equipment are available for measuring viscosity; these include capillary flow viscometers, rotational viscometers and rheometers.

For low-viscosity fluids such as milk and milk-based drinks, capillary flow viscometers are cheap, and very accurate, whereas for more viscous products and those exhibiting non-Newtonian behaviour, rotational viscometers should be used. Where rotational viscometers are used for more viscous products, which are likely to be non-Newtonian (Chapter 5.2.1.8.3), it is recommended that a standard measurement procedure is followed and that the shearing conditions are stipulated to allow results to be replicated. Alternatively, controlled stress or strain rheometers, can be used to measure the viscous and elastic components of the material. More information on viscosity measurement is provided by Lewis (1990). Some data on viscosity changes for milk during storage are given in Section 7.2.2 and for evaporated milk in Section 9.3.1.

11.2.27 Vitamins

The levels of *water-soluble vitamins* in milk are reduced by heat treatment and in some cases during storage, especially in the presence of dissolved oxygen (see Sections 6.1.5 and 7.1.10). Hence there is often a need to monitor the water-soluble vitamin content of milk and milk products, particularly infant formulae. Several methods have been published for the analyses whereby individual vitamins or mixtures of vitamins are determined per analysis. Methods which simultaneously measure all water-soluble vitamins are available and have been used for ESL and UHT milks and infant formula mixes. For example, Lorenzen *et al.* (2011) combined the methods of Albalá-Hurtado *et al.* (1997) and Heudi *et al.* (2005). The milk is first treated with TCA to precipitate the proteins and to dissociate any vitamin-protein and vitamin-phosphate complexes. The vitamins in the supernatant are then analysed by reversed-phase HPLC using UV detection. The HPLC analysis of Albalá-Hurtado *et al.* (1997) involves the use of a variable wavelength detector and the vitamins are detected at several different wavelengths; the total time for the analysis of eight vitamins [thiamine (B1), riboflavin (B2), nicotinamide (B3 amide), pyridoxine (B6), pyridoxal, pyridoxamine, folic acid (B9) and cyanocobalamin (B12)] was 55 min. The HPLC analysis of Heudi *et al.* (2005) used two wavelengths, 210 and 275 nm, and separated nine vitamins [B1, B2, B3 amide, calcium pantothenate (B5), B6, biotin (B8), B9, B12 and, ascorbic acid (C)] in a single run of 17 min.

Fat-soluble vitamins in milk products processed at high-temperature are less of a concern than water-soluble vitamins as they are affected very little by the heat. Analysis of the fat-soluble vitamins [retinol (A) and tocopherol (E)] involves denaturation of lipoproteins and alkaline saponification followed by extraction with hexane-toluene or hexane-ether. The vitamins in the extract are estimated by HPLC using a silica column and fluorimetric detection (Bruschi *et al.*, 1992; Lorenzen *et al.*, 2011).

11.2.28 Whey Protein Denaturation

As indicated in Section 6.1.3.1, heating causes the whey proteins to denature, that is, unfold, and subsequently aggregate or interact with caseins, particularly κ-casein. This causes the majority of the whey proteins to become insoluble at pH 4.6 and hence the concentration of whey proteins soluble in a pH 4.6 filtrate after precipitation of caseins and insolubilised whey proteins provides a measure of the amount of undenatured whey protein. If the whey protein concentration in the milk before heating is known, the percentage whey protein denaturation can be determined. However, the pre-heating concentration is seldom known, except in research situations, and hence only approximate whey protein denaturation percentages can be obtained using reported concentrations of whey proteins. In cow's milk the total whey protein content is ~6 g/L (Fox, 2003).

The undenatured whey protein content can be measured in pH 4.6 filtrate using the methods described in Sections 11.2.20 and 11.2.21. However, when using wet chemistry methods (categories 1 and 2 in Section 11.2.21), it needs to be recognised that the pH 4.6 filtrate will also contain some naturally occurring peptides – ~0.82 μmole/mL according to McKellar (1981). The most common analysis method is reversed-phase HPLC on a C18 column with detection at 205 nm. This avoids the complication of the presence of native peptides as these are separated from the whey proteins. Quantification of the individual whey proteins can be achieved by using a calibration curve constructed with pure standards (Elliott *et al.*, 2003).

11.2.28.1 Soluble Tryptophan

Measurement of acid-soluble tryptophan by fluorescence has been reported to provide a rapid estimate of undenatured whey proteins, and hence of heat treatment of milk (Birlouez-Aragon *et al.*, 1998). The fluorescence is measured on a pH 4.6 filtrate with an excitation wavelength of 290 nm and emission wavelength of 340 nm. Bovine serum albumin is used to construct a calibration curve. This fluorimetric procedure should not be confused with the fluorimetric determination of advanced Maillard reaction products which uses an excitation wavelength of 350 nm and emission wavelength of 440 nm (Birlouez-Aragon *et al.*, 1998).

11.2.28.2 Turbidity Test

The turbidity test is a quick procedure for qualitatively assessing the extent of whey protein denaturation in milk (Aschaffensburg, 1950). Milk (20 mL) is mixed with ammonium sulfate (4 g) which causes casein and any associated denatured whey protein to precipitate. The mixture is filtered, producing a clear filtrate, which contains any undenatured whey protein that was present in the milk sample. If the filtrate is not clear at this stage, the filtration step should be repeated. The filtrate is boiled for 5 min which causes any undenatured whey protein to be denatured, thereby producing a turbid solution. The turbidity is proportional to the amount of *undenatured* whey protein in the milk.

Historically, heat treatment regulations required that sterilised milk should give a negative turbidity result, that is, contain no undenatured whey protein, whereas UHT milk should give a positive result (IDF, 1972). Commercially sterile UHT milk with a positive turbidity can be produced by indirect heating and direct heating with those produced by direct heating being the more turbid; this is in line with the levels of whey protein denaturation to be expected in these different UHT situations. However, almost all UHT milk samples purchased in the UK show negative turbidity, that is, show no detectable undenatured whey protein.

The turbidity test does not easily distinguish between raw milk and pasteurised milk, since pasteurisation shows only a small level of whey protein denaturation, but it may be useful for determining whether pasteurised milk has been more severely heated. Such milks would also be negative for lactoperoxidase activity.

Burton (1988) reviewed the literature for distinguishing between sterilised and UHT milk. In the 1970s and 1980s considerable effort was expended in this direction, but it is not now considered to be so important. Tests were based on the premise that a greater amount of chemical damage would be caused by the in-container sterilisation process than by UHT processing. It was suggested that the turbidity test would be useful. However, some sterilised milks were found to give a slight turbidity, whereas some UHT milks, especially those where a pre-heat holding stage was included or those with high regeneration efficiencies, gave milk with a negative turbidity. Some of the subjectivity of the turbidity test can be removed by measuring the turbidity using a nephelometer, but it has been found to be not possible to distinguish between the milks in every case (Moermans & Mottar, 1984). One suggestion for this was the complex nature of the denaturation process, and the break in reaction kinetic conditions at 95 °C (Dannenberg & Kessler, 1988). In its favour, the turbidity test is the simplest test to perform in a factory environment, to gauge the level of whey protein denaturation, without resort to sophisticated equipment and to identify milks where the process has been severe from a chemical standpoint. Due to the wide range of conditions used for both processes, it is likely that some UHT milk products will give a negative turbidity. UHT milk processed with a C* greater than 6 has been found to give a negative turbidity (Browning *et al.*, 2001); however, processes with lower C* with different temperature–time profiles may also produce UHT milk with a negative turbidity.

11.3 Advanced Analytical Techniques

There are several analytical techniques which have been developed comparatively recently. Some of these are being used for milk and milk products but there is considerable potential for their greater use. Table 11.7 lists several of these techniques.

These techniques can provide opportunities for predicting whether a UHT product will still be acceptable for consumption after storage for nine months or more. The UHT processor may also encounter safety issues which are not specifically related to the UHT process itself, especially veterinary residues, allergens, mycotoxins and environmental contaminants such as pesticides, heavy metals, including radionuclides, and dioxins. In addition there may be some less common microbiological hazards that may be encountered, possibly as a result of increased heat resistance for bacterial spores. These issues may require more in-depth analysis than what is commonly available in factory laboratories.

A small selection of what are termed here "advanced techniques" are discussed below with some information on their applications to milk and milk products, particularly heated products. Discussion of all possible relevant advanced techniques is beyond the scope of this book.

11.3.1 Chemometrics

Chemometrics is the science of extracting information from chemical systems by data-driven means. Chemometric methods are useful for analysing large data sets to determine trends, groupings and associations (Chen *et al.*, 2014). Thus they are used extensively on data from spectroscopic analyses, sensory evaluation and data from

Table 11.7 Summary of some advanced instrumental analytical methods.

Chromatography; separation of components from complex matrices, GC, LC (HPLC), ion chromatography, electrophoresis; for example, sugars, peptides, minerals, volatile components, proteins/peptides. Used, for example, for authentication, adulteration, stability and flavour problems, for example, proteolysis/gelation and bitterness, other off-flavours
Mass spectrometry: identification of components, for example, antibiotic residues, environmental contaminants, proteins/peptides
Proteomics: combines LC and electrophoresis with mass spectrometry for analysis of proteins and peptides
NMR: identificaton of sugars, free amino acids and a wide range of non-protein nitrogen compounds; structure, protein–water interactions.
Light and laser scattering: Turbiscan®, microscopy, particle size analysis; used for stability issues such as sediment formation, fat separation and other phase changes
Infrared spectrometry: structural changes in protein
Atomic absorption spectroscopy, inductively coupled plasma-mass spectrometry and inductively coupled plasma atomic emission spectroscopy, ion chromatography: mineral analysis; used in fortification and metal-induced off-flavours

multiple analyses on the same product. Chemometrics includes a range of multivariate analysis techniques including principal component analysis (PCA), partial least squares regression, factorial discriminant analysis, canonical correlation analysis and hierarchical cluster analysis (HCA). A review of how chemometric techniques can be used with analytical data for assessing the quality and identity of dairy products was published by Karoui and De Baerdemaeker (2007).

Some examples of the use of chemometric techniques with heat-treated milk are given below.

- Chemometric techniques were used to monitor the authenticity of Brazilian UHT milk processed in industrial plants located in different regions of the country (Souza *et al.*, 2011). A total of 100 samples was submitted to the analysis of adulterants such as starch, chlorine, formol, hydrogen peroxide and urine. Except for starch, all the samples reported the presence of at least one adulterant. The use of PCA and HCA enabled the verification of the occurrence of certain adulterations in specific geographical regions.
- Chapman *et al.* (2001) used quantitative descriptive analysis in conjunction with PCA to determine the key attributes of samples of ultrapasteurised milks with different fat and lactose contents, produced by two plants. They found that the milks could be located along four scales corresponding to the attributes cooked, drying/lingering sweet and bitter. The results illustrated the usefulness of the techniques for determining milk attributes considered important to consumers.
- Similarly, Iwatsuki *et al.* (1999) used PCA to relate sensory characteristics of UHT milks processed by direct and indirect heating processes with various physicochemical data (whey protein denaturation, rennetability, lactoferrin, lactulose and HMF). They also used the same techniques for evaluating the effect of heating conditions on the sensory characteristics of UHT milk and pasteurised milk (Iwatsuki *et al.*, 2000).
- In a variant of sensory evaluation, an electronic nose was evaluated for its discriminatory power between various UHT milk samples. Using PCA of the data, the instrument was able to discriminate normal samples from samples with abnormal odour (Brambilla *et al.*, 2007).
- Gaucher *et al.* (2008a) used multidimensional statistical analysis to analyse a large amount of data from an investigation of the effects of season, region, and processing and storage conditions on various physicochemical characteristics (composition, micelle properties, ethanol stability and phosphate stability) of UHT milk. The effects of each factor was related to these characteristics through PCA.
- In a classic example of the application of chemometric techniques, Jansson *et al.* (2014) related GC-MS data on 24 volatiles isolated by dynamic headspace sampling of samples of conventional and lactose-hydrolysed UHT milk. While PCA analysis showed higher ketone levels in conventional than lactose-hydrolysed milk, significant differences in the volatiles profiles of the different milks could not be established due to day-to-day variation

11.3.2 Nuclear Magnetic Resonance (NMR)

NMR analysis holds considerable potential for investigating UHT and other heat-treated products, although to date little attention has been paid to this application. It is a non-destructive method which can be used to analyse the components of a product and also investigate structural features.

NMR analysers fall into two major categories: low-resolution bench-top instruments typically working at 23 MHz, and high-resolution instruments working at up to 700 MHz. The main difference between these instruments is the type of analyses performed. The low-resolution machines measure the total content of fat and a few other components in high concentration while the high-resolution instruments give detailed analysis of individual chemical species.

The nuclei most commonly used in NMR analyses of food are ^{1}H, ^{13}C and ^{31}P but ^{15}N and ^{43}Ca have also been used. The most appropriate are chosen for a particular application. One-dimensional ^{1}H-NMR is the most common but when used directly on milk the spectra are very complex and difficult to interpret. A clearer spectrum can be obtained by fractionating the milk, for example, by dialysis or ultrafiltration, to produce the soluble phase, free of protein and fat. An alternative approach is to use a second nucleus, for example, ^{13}C, in a two-dimensional analysis. The latter approach was recently used in a study on differentiation of organic and conventional milk (Erich *et al.*, 2015). ^{1}H and ^{13}C are commonly used for lipids with ^{13}C giving spectra with lower signal-to-noise ratio. ^{31}P NMR has been used to analyse phospholipids, phosphorylated carbohydrates, inorganic phosphate and phosphoserine in phosphorylated proteins (Belloque & Ramos, 1999; Ishii *et al.*, 2001).

In the case of composition of the milk, NMR has been used to analyse a range of low-molecular-weight compounds including organic acids, amino acids, nucleic acids, lipids and carbohydrates. It has also been used to identify miscellaneous compounds such as histidino-alanine formed during heat treatment of milk (Henle *et al.*, 1993). Compounds of interest for UHT milk are low-molecular-weight compounds such as free amino acids, organic acids, sugars (e.g., lactose and lactulose), Maillard reaction compounds, urea and other non-protein compounds, and citrate. This information combined with chemometric techniques can be a powerful tool for differentiating between different milk samples. For example, Erich *et al.* (2015) used a two-dimensional NMR approach in a study on differentiation of organic and conventional milk and Klein *et al.* (2010) used NMR to study milk composition during early and late lactation.

A relatively recent field is NMR-based milk metabolomics which was reviewed by Sundekilde *et al.* (2013). This analyses metabolites in milk and relates them with nutritional and technological qualities of the milk. An aim of the studies in this field is to identify compounds which could be used as biomarkers. For example, Klein *et al.* (2012) identified lipid biomarkers in milk for assessing risk of ketosis in dairy cattle.

Another branch of NMR analyses is relaxation NMR which can be used to determine the state of water in a product and its interaction with components especially proteins. It has been used to characterise milk powders (Davenel *et al.*, 2002) and to monitor changes in protein structure during storage of milk protein concentrate (Haque *et al.*, 2015).

NMR has potential for investigating changes in milk during heating and changes such as gelation during storage of UHT milk. It has previously been used to study calcium and phosphorus equilibria in heated casein micelle suspensions (using ^{31}P and ^{43}Ca) (Wahlgren *et al.*, 1986, 1990), and rennet and whey protein gels (Lelievre & Creamer, 1978; Colsenet *et al.*, 2005).

NMR-based screening methods appear to be more advanced for fruit juices than for milk. Analytical services are available for fruit juice which are able to evaluate a multitude of parameters related to quality and authenticity simultaneously from a single data set acquired within a few minutes. This permits:

- targeted analysis for simultaneous quantification of relevant organic compounds;
- non-targeted analysis where the NMR-profile is compared with the corresponding group of reference spectra to look for deviating trends; and
- classification analysis, for example, for determination of country of origin or determination of fruit content of fruit-based drinks (Bruker, 2015).

Such an approach could readily be adopted for milk and dairy products.

11.3.3 Proteomics

Proteomics can be defined as the analysis of the proteins from a common source such as cell, cell compartment or secretion (O'Donnell *et al.*, 2004).

Thus milk proteomics refers to the analysis of the proteins in milk. It is a very powerful technique as a large number of proteins can be identified from a very small volume of milk (a few microlitres). While the term proteomics was first coined in the 1990s, it is a combination of techniques which have been used for some time.

In broad terms, it combines the two main techniques of separation and characterisation. In some cases a fractionation step may be employed before the separation step to simplify the analysis and to target particular protein groups, for example, caseins and serum proteins. The most common separation techniques used are 2D-electrophoresis, commonly PAGE, and liquid chromatography. The 2D PAGE typically involves separation firstly according to isoelectric point using isoelectric focussing over a defined pH range, such as 3-11, and secondly according to molecular mass (Mr), typically using SDS as a denaturant and mercaptoethanol as a reductant to convert S–S bonds into –SH groups. This results in a series of spots on the gel whose intensity after staining is related to the relative abundance of the proteins in the starting material (see, for example, Figure 7.3). The protein spots are removed from the gel and the protein extracted from the gel to be analysed directly or proteolysed with a protease such as trypsin before MS analysis. 2D-PAGE has some limitations such as a lower Mr limit of ~8,000 Da and an upper Mr limit of ~150,000 Da, and a limited dynamic range. These limitations do not apply to liquid chromatographic techniques. Liquid chromatography techniques include a range of forms such as reversed phase, ion exchange and size exclusion. For added separation power, two or more forms can be combined for in-line MS.

Characterisation of the proteins separated by the above techniques is principally performed by MS. Two major types of analysis are possible: mass-related information including the total mass of the protein and the mass of fragments, and sequence information usually obtained from tandem mass spectrometry (MS/MS). The data obtained are then analysed by comparison with databases and the proteins identified. This can be performed either manually or automatically by the mass spectrometer.

Proteomics has been used for some time on milk and milk products with the first review of milk proteomics being published in 2004 (O'Donnell *et al.*, 2004). Since that time, proteomics has been extensively used for milk proteins and other reviews have been published (Manso *et al.*, 2005; Gagnaire *et al.*, 2009; Cunsolo *et al.*, 2011; Abd El-Salam, 2014). Aspects investigated include detection of adulterations (Calvano *et al.*, 2013; Sassi *et al.*, 2015), characterisation of the milk proteins from non-bovine species including human milk (e.g., Poth *et al.*, 2008; Roncada *et al.*, 2013; Yang *et al.*, 2013), changes during coagulation (Chen *et al.*, 2016) and changes in mastitic milk (Zhang *et al.*, 2015).

Of particular relevance are the several studies on changes in heated milk that have been published. Holland *et al.* (2011) used a proteomic approach to study the changes in UHT milk during storage at 4, 28 and 40 °C. They identified three major changes: deamidation, non-disulfide cross-linking and lactosylation (see Sections 7.1.5-7.1.7). Substantial changes occurred in milk stored at elevated temperatures. Other changes during heat treatment or storage which have been studied include formation of advanced Maillard reaction products (Renzone *et al.*, 2015), lactosylation (Arena *et al.*, 2010; Le *et al.*, 2012), aggregate formation in whey protein drinks (Le *et al.*, 2016) and formation of disulfide-linked polymers (Chevalier *et al.*, 2010). Calvano *et al.* (2013) used a proteomic approach to detect the presence of milk powder in fresh milk. Using six peptides they identified in milk powder which were absent in fresh milk, they we able to detect 1% adulteration with milk powder. The identified peptides were attributed to chemical modification of the proteins brought about by the heat treatment including lactosylation, deamidation, cross-linking, oxidation and formation of carboxymethyllysine.

11.3.4 Ultrasonic Techniques

Non-destructive ultrasonic analysis uses high-frequency (0.1-20 MHz), low-power (<100 mW) ultrasound. An excellent example of its applicability to heat-treated milk was provided by Smyth *et al.* (1999). They used a high-resolution ultrasonic resonator operating at 6.5-7 MHz to monitor coagulation of milk with added calcium chloride at different temperatures. Using the two independent measures of ultrasonic attenuation and ultrasonic velocity they were able to determine the coagulation temperatures for given calcium additions with a resolution of better than 0.1 °C. Both measures yielded the same results. The effects of addition of stabilisers to prevent coagulation by calcium could also be assessed. The technique has also been used to study rennet- and acid-induced gelation of skim milk. It is able to detect the gelation point and can be used to monitor pre-gelation, gelation, and post-gelation processes. It allows analysis of weak gels without any distortion of the gel network (Gunasekaran & Ay, 1994). It was also shown to be effective in detecting gelation caused by proteolytic enzymes (Ahvenainen *et al.*, 1991) and hence is applicable to detection of gelaton in UHT milk.

Another application of ultrasonic analysis relevant to heat-treated products is detection of spoilage. Ahvenainen *et al.* (1989) found that ultrasound imaging was an effective non-invasive method for monitoring microbial growth. They suggested that bacterial loads in UHT-processed foods such as soft ice-cream and processed vanilla sauce could be detected at levels of 10^5 cfu/g. Various bacteria were used including *B. cereus, St. aureus, Cl. perfringens* and *E. coli*. Wirtanen *et al.* (1992) examined the effect of factors such as ultrasound frequency, probe area and other technical factors on the sensitivity of ultrasound imaging with respect to detection of bacterial and enzymic spoilage of UHT milk and UHT soft ice cream base packed in Tetra Brik cartons. They found that the frequency of the ultrasound should not be greater than 5 MHz, and in some cases, greater sensitivity was observed at 3.75 MHz. Frequencies in the range 3.75-7.5 MHz could not be used for detecting spoilage in cardboard containers. Gestrelius (1994) observed a decrease in the ultrasound streaming velocity as the product was spoiled by *St. epidermidis* and *B. subtilis*, suggesting that the ultrasound method was capable of detecting infected packs.

A drawback of these methods is that they cannot detect low levels of contamination and are only useful after the sample has been incubated to promote bacterial growth. Thus it becomes difficult to apply them in a quality control role immediately after production. However, it could be worthwhile applying them after their normal period of incubation at the factory, prior to release of the product.

References

Aalaei, K., Rayner, M., Tareke, E. & Sjöholm, I. (2016) Application of a dye-binding method for the determination of available lysine in skim milk powders. *Food Chemistry* **196**, 815–820.

Abd El-Salam, M.H. (2014) Application of proteomics to the areas of milk production, processing and quality control - A review. *International Journal of Dairy Technology* **67**(2), 153–166.

Abdulghani, A.H., Ali, M.Y., Prakash, S. & Deeth, H.C. (2015) Sensory evaluation and storage stability of UHT milk fortified with iron, magnesium and zinc. *Dairy Science & Technology* **95**(1), 33–46.

Ahvenainen, R., Wirtanen, G. & Manninen, M. (1989) Ultrasound imaging - a non-destructive method for monitoring the microbiological quality of aseptically-packed milk products. *Lebensmittel Wissenschaft & Technologie* **22**(6), 382–386.

Ahvenainen, R., Wirtanen, G. & Mattilasandholm, T. (1991) Ultrasound imaging - a nondestructive method for monitoring changes caused by microbial enzymes in aseptically-packed milk and soft ice-cream base material. *Lebensmittel-Wissenschaft & Technologie* **24**(5), 397–403.

Al-Attabi, Z. (2009) *Volatile Sulphur Compounds in UHT milk*. PhD thesis, University of Queensland, Australia.

Al-Attabi, Z., D'Arcy, B.R. & Deeth, H.C. (2009) Volatile sulphur compounds in UHT milk. *Critical Reviews in Food Science and Nutrition* **49**(1), 28–47.

Al-Attabi, Z., D'Arcy, B.R. & Deeth, H.C. (2014) Volatile sulfur compounds in pasteurised and UHT milk during storage. *Dairy Science and Technology* **94**(3), 241–253.

Albalá-Hurtado, S., Veciana-Nogues, M.T., Izquierdo-Pulido, M. & Mariné-Font, A. (1997) Determination of water-soluble vitamins in infant milk by high-performance liquid chromatography. *Journal of Chromatography A* **778**, 247–253.

AlKanal, H.A. (2000) Suitability of methods used for the measurement of proteolysis in UHT milk. *Australian Journal of Dairy Technology* **55**(3), 148–152.

Al-Saadi, J.S. & Deeth, H.C. (2008) Cross-linking of proteins and other changes in UHT milk during storage at different temperatures. *Australian Journal of Dairy Technology* **63**, 93–99.

Al-Saadi, J.S., Easa, A.M. & Deeth, H.C (2013) Effect of lactose on cross-linking of milk proteins during heat treatment. *Australian Journal of Dairy Technology* **66**, 1–6.

Arena, S., Renzone, G., Novi, G., Paffetti, A., Bemardini, G., Santucci, A. & Scaloni, A. (2010) Modern proteomic methodologies for the characterization of lactosylation protein targets in milk. *Proteomics* **10**, 3414–3434.

Aschaffenburg, R. (1950) A simple test for sterilised milk. *Journal of the Society of the Dairy Technology* **3**, 236–237.

Augustin, M.-A. & Clark, P.T. (1991) Heat stability of recombined concentrated milk: changes in calcium activity and pH on sterilization. *Journal of Dairy Research* **58**, 67–74.

Bansal, N., Piraino, P. & McSweeney, P.L.H. (2010) Determination of proteolysis in cheese. In: *Handbook of Dairy Foods Analysis*. (eds L.M.L. Nollet & F. Toldrá), pp. 405–426. CRC Press, Taylor & Francis Group, Boca Raton, FL.

Beeby, R. (1980) The use of fluorescamine at pH 6.0 to follow the action of chymosin on kappa-casein and to estimate this protein in milk. *New Zealand Journal of Dairy Science and Technology* **15**, 99–108.

Belloque, J. & Ramos, M. (1999) Application of NMR spectroscopy to milk and dairy products. *Trends in Food Science and Technology* **10**, 313–320.

Bernfeld, P. (1955) Amylases, α and β. In: *Methods in Enzymology* Vol 1 (eds S.P. Colowick & N.O. Kaplan), pp. 149–50. Academic Press, New York.

Bhandari, V & Singh, H. (2011) Physical methods. In: *Encyclopedia of Dairy Sciences*. (eds J.W. Fuquay, P.F. Fox & P.L.H. McSweeney) 2nd edn, Volume 1, pp. 248–255. Elsevier Academic Press, Amsterdam.

Bintsis, T., Angelides A.S. & Psoni, L. (2008) Modern Laboratory Practices- Analysis of Dairy Products. In: *Advanced Dairy Science and Technology*. (eds T.J. Britz & R.K. Robinson), Blackwell Publishing Ltd, Oxford.

Birlouez Aragon, I., Nicolas, M., Metais, A., Marchond, N., Grenier, J. & Calvo, D. (1998) A rapid fluorimetric method to estimate the heat treatment of liquid milk. *International Dairy Journal* **8**(9), 771–777.

Blecker, C., Habib-Jiwan, J.M. & Karoui, R. (2012) Effect of heat treatment of rennet skim milk induced coagulation on the rheological properties and molecular structure determined by synchronous fluorescence spectroscopy and turbiscan. *Food Chemistry* **135**(3), 1809–1817.

Boumpa, T., Tsioulpas, A., Grandison, A.S. & Lewis, M.J. (2008) Effects of phosphates and citrate on sediment formation in UHT goats' milk. *Journal of Dairy Research* **75** 160–166.

Bradford, M.M. (1976) A rapid and sensitive method for the quantitation of microgram quantitites of protein utilizing the principle of protein-dye binding. *Analytical Biochemistry* **72**, 248–254.

Brambilla, M., Guarino, M. & Navarotto, P. (2007) Electronic nose approach to monitor UHT milk quality: a case study. *Industrie Alimentari* **46**(469), 540–544.

Bratov, A, Abramova, N., Dominguez, C. & Baldi, A. (2000) Ion selective field effect transistor (ISFET)-based calcium ion sensor with photocured polyurethane membrane suitable for ionised calcium determination in milk. *Analytica Chimica Acta* **408**, 57–64.

British Standards (1992) Methods for chemical analysis of liquid milk and cream – Part 12 – Determination of total phosphorus content of milk. BS 1741–12.

Browning, E., Lewis, M. & MacDougall, D. (2001) Predicting safety and quality parameters for UHT-processed milks. *International Journal of Dairy Technology* **54**, 111–120.

Bruker (2015) Analyzing food in a new dimension. How Bruker's food screening solutions help to analyze quality with regard to authenticity based on 1H-NMR-profiling. https://www.bruker.com/products/mr/nmr/food-screener.html (accessed December 2015).

Bruschi L., Manzi P., Panfili G. & Quattrucci E. (1992) Methods of chemical analysis of new milk products: fat-soluble vitamins. *Latte* **17**, 1156–1157.

Bujard E. & Finot P.A. (1978) Measure of available and blocked lysine in industrial milks, *Annales de la Nutrition et de l'Alimentation* **32**, 291–305.

Burgess, K. (2010) Key requirements for milk quality and safety: a processors's perspective. In: *Improving the Safety and Quality of Milk.* (ed. M.W. Griffiths), pp. 64–84, Woodhead Publishing, Cambridge.

Burgwald, L. (1923) Some factors which influence the feathering of cream in coffee. *Journal of Agricultural Research* **26**, 541–546.

Burton, H. (1988) *Ultra-high Temperature Processing of Milk and Milk Products.* Elsevier Applied Science, London.

Button, P. (2007) *UHT Milk Spoilage owing to Lipase and Protease from Psychrotrophic* Pseudomonas *Species: Characteristics of Enzymatic Breakdown, Improved Assays for Enzyme Detection and Sources of the Bacteria.* PhD thesis, University of Melbourne.

Button, P.D., Roginski, H., Deeth, H.C. & Craven H.M (2011) Improved shelf-life estimation of UHT milk by prediction of proteolysis. *Journal of Food Quality* **34**, 229–235.

Byland, G. (2003) *Dairy Processing Handbook*, 2nd edn, Tetra Pak Publishing Systems AB, Lund, Sweden.

Calvano, C. D., Monopoli, A., Loizzo, P., Faccia, M. & Zambonin, C. (2013) Proteomic approach based on MALDI-TOF MS to detect powdered milk in fresh cow's milk. *Journal of Agricultural and Food Chemistry* **61**, 1609–1617.

Carpenter, K.J. (1960) The estimation of the available lysine in animal protein foods. *Biochemistry Journal* **77**, 604–610.

Cartier, P., Chilliard, Y. & Chazal, M.-P. (1984) Dosage de l'activité lipasique et des acides gras libres du lait par titration automatic colourimétrique. *Lait* **64**, 340–355.

CFR (2003) Indirect food additives: adjuvants, production aids and sanitizer. In: 21 CFR 178:1005. P. 348, Office of the Federal Register, US Government Priniting Office Washington, DC.

Chandrapala, J., McKinnon I, Augustin, M.-A. & Udabage P. (2010) The influence of milk composition on pH and calcium activity measured *in situ* during heat treatment of reconstituted skim milk. *Journal of Dairy Research* **77**, 257–264.

Chapman, K. W., Lawless, H. T. & Boor, K. J. (2001) Quantitative descriptive analysis and principal component analysis for sensory characterization of ultrapasteurized milk. *Journal of Dairy Science* **84**, 12–20.

Chavez, M.S., Livia, M.N., Taverna, M.A. & Cuatrin, A. (2004) Bovine milk composition parameters affecting the ethanol stability. *Journal of Dairy Research* **71** 201–206.

Chavira, R., Burnett, T.J. & Hageman, J. (1984) Assaying proteinases with azocoll. *Analytical Biochemistry* **136**(2), 446–450.

Chen, B.Y. (2013) *Best Use for Milk and Some Associated Issues.* PhD thesis, University of Reading, UK.

Chen, B.Y., Grandison, A.S. & Lewis, M.J. (2012) Comparison of heat stability of goat's milk subjected to UHT and in-container sterilisation. *Journal of Dairy Science* **95**, 1057–1063.

Chen, B., Lewis, M.J. and Grandison, A.S. (2014) Effect of seasonal variation on the composition and properties of raw milk destined for processing in the UK. *Food Chemistry*, **158**, 216–223.

Chen, B.Y., Grandison, A.S. & Lewis, M.J. (2015) Effect of seasonal variation on some physical properties and heat stability of milk subjected to ultra-high temperature and in-container sterilisation. *Food Chemistry* **181**, 227–234.

Chen, Y.C., Chen, C.C., Chen, S.T. & Hsieh, J.F. (2016) Proteomic profiling of the coagulation of milk proteins induced by glucono-delta-lactone. *Food Hydrocolloids* **52**, 137–143.
Chevalier, F. & Kelly, A.L. (2010) Proteomic quantification of disulfide-linked polymers in raw and heated bovine milk. *Journal of Agricultural and Food Chemistry* **58**, 7437–7444.
Chinese Ministry of Health (2010) National food safety standard. Sterilised milk. GB 25190-2010. The National Standard of the People's Republic of China. (eagri.cz/public/web/file/294863/GB25190_2010_sterilized_milk.pdf).
Chism, G.W., Huang, A.E. & Marshall, J.A. (1979) Sensitive assay for proteases in sterile milk. *Journal of Dairy Science* **62**, 1798–1800.
Christen, G.L. & Marshall, R.T. (1984) Selected properties of lipase and protease of *Pseudomonas fluorescens* 27 produced in four media. *Journal of Dairy Science* **67**, 1680–1687.
Christen, G.L. & Senica, M. (1987) Casein fluorescein isothiocyanate as a substrate for measuring protease production by psychrotrophic bacteria. *Journal of Food Protection* **50**, 744–749.
Church, F.C., Swaisgood, H.E., Porter, D.H. & Catignani, L. (1983) Spectrophotometric assay using o-phthaldialdehyde for determination of proteolysis in milk and isolated milk proteins. *Journal of Dairy Science* **66**, 1219–1227.
Clare, D.A., Bang, W.S., Cartwright, G., Drake, M.A., Coronel, P. & Simunovic, J. (2005) Comparison of sensory, microbiological, and biochemical parameters of microwave versus indirect UHT fluid skim milk during storage. *Journal of Dairy Science* **88**, 4172–4182.
Cliffe, A.J. & Law, B.A. (1982) A new method for the detection of microbial proteolytic enzymes in milk. *Journal of Dairy Research* **49**, 209–219.
Colsenet, R., Mariette, F. & Cambert, M. (2005) NMR relaxation and water self-diffusion studies in whey protein solutions and gels. *Journal of Agricultural and Food Chemistry* **53**, 6784–6790.
Couch, J.R. & Thomas, M.C. (1976) A comparison of chemical methods for the determination of available lysine in various proteins. *Journal of Agriculture and Food Chemistry* **24**, 943–946.
Coulter (1996) Coulter™ QuickSCAN™: A new concept in stability analysis of concentrated colloidal dispersions (emulsions, suspension, foams, gels). http://www.cyto.purdue.edu/cdroms/cyto2/6/coulter/ss000098.htm (accessed December 2015).
Cronshaw, H.B. (1947) *Dairy Information*. Dairy Industries Ltd, London.
Crowley, S.V., Kelly, A.L. & O'Mahony, J.A. (2014) Fortification of reconstituted skim milk powder with different calcium salts: Impact of physicochemical changes on stability to processing, *International Journal of Dairy Technology* **67**, 474–482.
Cunsolo, V., Muccilli, V., Saletti, R. & Foti, S. (2011) Applications of mass spectrometry techniques in the investigation of milk proteome. *European Journal of Mass Spectrometry* **17**(4), 305–320.
Dannenberg, F. & Kessler, H. (1988) Reaction kinetics of the denaturation of whey proteins in milk. *Journal of Food Science* **53**, 258–263.
Datta, N. & Deeth, H.C. (2003) Diagnosing the cause of proteolysis in UHT milk. *LWT-Food Science and Technology* **36**(2), 173–182.
Davenel, A., Schuck, P., Mariette, F. & Brule, G. (2002) NMR relaxometry as a non-invasive tool to characterize milk powders. *Lait* **82**(4), 465–473.

Davies, D.T. & White, J.C.D. (1960) The use of ultrafiltration and dialysis in isolating the aqueous phase of milk and determining the partitioning of milk constituents between the aqueous and disperse phases. *Journal of Dairy Research* **27**, 171–190.

Davies, D.T. & White, J.C.D. (1962) The determination of calcium and magnesium in milk and milk diffusate. *Journal of Dairy Research* **29**, 285–296.

de Block, J., Merchiers, M., Van Renterghem, R. & Moermans, R. (1996) Evaluation of two methods for the determination of lactulose in milk. *International Dairy Journal* **6**, 217–222.

de Jong, C. & Badings, H.T. (1990) Determination of free fatty acids in milk and cheese: Procedures for extraction, clean up, and capillary gas chromatographic analysis. *Journal of High Resolution Chromatography* **13**, 94–98.

Deeth, H.C. & Lewis, M.J. (2015) Practical consequences of calcium addition to and removal from milk and milk products. *International Journal of Dairy Technology* **65**(1), 1–10.

Deeth, H.C. & Fitz-Gerald, C.H. (2006) Lipolytic enzymes and hydrolytic rancidity. In: *Advanced Dairy Chemistry, 2 Lipids*, (eds P.F. Fox & P.L.H. McSweeney), 3rd edn, pp. 481–55. Springer, New York.

Deeth, H.C. & Touch V. (2000) Methods for detecting lipase activity in milk and milk products. *Australian Journal of Dairy Technology* **55**(3), 153–168.

Deeth, H.C., Fitz-Gerald, C.H. & Snow, A.J. (1983) A gas chromatographic method for the quantitative determination of free fatty acids in milk and milk products. *New Zealand Journal of Dairy Science and Technology* **18**, 13–20.

Deeth, H.C., Fitz-Gerald, C.H. & Wood, A.F. (1975) A convenient method for determining the extent of lipolysis. *Australian Journal of Dairy Technology* **30**, 109–111.

De la Fuente, M., Belloque, J., Juarez, M. (2004) Mineral contents and distribution between the soluble and micellar phases in calcium-enriched UHT milks. *Journal Science of. Food and Agriculture* **84**, 1708–1714.

Delgado, T., Corzo, N., Santa-Maria, G., Jimeno, M.L. & Olane A. (1992) Determination of furosine in milk samples by ion pair reversed phase liquid chromatography. *Chromatographia* **33**,374–376.

Demott, B.J. (1968) Ionic calcium in milk and whey. *Journal of Dairy Science* **51**, 1008–1012.

Durand, A (2003) *Kinetic Analysis of Factors affecting the Design of Soy Beverage Processing.* PhD thesis, University of Newcastle, Australia.

Durand, A., Franks, G. V. & Hosken, R. W. (2003) Particle sizes and stability of UHT bovine, cereal and grain milks. *Food Hydrocolloids* **17**, 671–678.

Elliott, A.J., Dhakal, A., Datta, N. & Deeth, H.C. (2003) Heat-induced changes in UHT milks – Part 1. *Australian Journal of Dairy Technology* **58**, 3–10.

Erbersbobler, H.F. & Hupe, A. (1991) Determination of lysine damage and calculation of lysine bio-availability in several processed foods. *Zeitschrift fur Ernahrugswissenschaft* **30**, 46–49.

Erbersdobler, H.F. & Anderson, T.R. (1983) Determination of available lysine by various procedures in Maillard-type products. In: *The Maillard Reaction in Foods and Nutrition.* (eds G.R. Waller & M.S. Feather), pp. 419–427. American Chemistry Society, Washington, DC.

Erbersdobler, H.F. & Somoza, V. (2007) Forty years of furosine - Forty years of using Maillard reaction products as indicators of the nutritional quality of foods. *Molecular Nutrition & Food Research* **51**, 423–430.

Erich, S., Schill, S., Annweiler, E., Waiblinger, H-U.,Kuballa, T., Lachenmeier, D.W. & Monakhova, Y.B. (2015) Combined chemometric analysis of ^{1}H NMR, ^{13}C NMR and stable isotope data to differentiate organic and conventional milk. *Food Chemistry* **188**, 1–7.

Faist, V., Drusch, S., Kiesner, C., Elmadfa, I. & Erbersdobler, H. (2000) Determination of lysinoalanine in foods containing milk protein by high-performance chromatography after derivatization with dansyl chloride. *International Dairy Journal* **10**, 339–346.

Feier, U. & Goetsch, P.H. (1990) Inter-laboratory studies on precision characteristics of analytical methods. 1. Determination of lactulose in milk. *Milchwissenschaft* **45**, 440–441.

Fink, R. & Kessler, H.G. (1986) HMF values in heat treated and stored milk. *Milchwissenschaft* **41**, 638–641.

Finot, P.A. & Mauron, J. (1972) Le blockage de la lysine par la reaction de Maillard. II. Proprietechimiques des derives N-(desoxy-1-D-fructosyl-1) et N-(desoxy-l-D-lactulosyl-1) de la lysine. *Helvetica Chimica Acta* **55**, 1153–l164.

Fitz-Gerald, C.H. & Deeth, H.C. (1983) Factors influencing lipolysis by skim milk cultures of some psychrotrophic microorganisms. *Australian Journal of Dairy Technology* **38**, 97–103.

Fojan, P., Jonson, P.H., Petersen, M.T.N. & Petersen, S.B. (2000) What distinguishes an esterase from a lipase: A novel structural approach. *Biochimie* **82**, 1033–1041.

Formulaction (Undated). Turbiscan, the reference for stability analysis. http://www.formulaction.com/stability-turbiscan-classic.html (accessed December 2015).

Foss (2014) *MilkoScan™ FT1 Milk standardisation with in-built abnormality screening* [Onlinme] Available: http://www.foss.com.au/industry-solution/dairy/brochures-and-data-sheets.

Fox, P.F. (2003) Milk proteins: general and historical aspects. In: *Advanced Dairy Chemistry Volume 1, Proteins* (eds P.F. Fox & P.L.H. Mc Sweeney), 3rd edn, pp. 1–48. Kluwer Academic/Plenum Publishers, New York.

Friedman, M., Gumbmann, M. & Masters, P. (1984) Protein alkali reactions: chemistry, toxicology and nutritional consequences. *Advances in Experimental Medicine and Biology* **177**, 367–412.

Fritsch, R., Hoffmann, H. & Klostermeyer, H. (1983) Formation of lysinoalanine during heat treatment of milk. *Zeitschrift für Lebensmittel-Untersuchung und–Forschung* **176**, 341–345.

Fweja, L.W.T., Lewis, M.J. & Grandison, A.S. (2007) Alternative strategies for activation of the natural lactoperoxidase system in cows' milk: trials in Tanzania. *Journal of Dairy Research* **74**, 381–386.

Gagnaire, V., Jardin, J., Jan, G. & Lortal, S. (2009) Proteomics of milk and bacteria used in fermented dairy products: From qualitative to quantitative advances. *Journal of Dairy Science* **92**, 811–825.

Garnsworthy, P.C., Masson, L.L., Lock, A.L., & Mottram, T.T. (2006) Variation of milk citrate with stage of lactation and de novo fatty acid synthesis in dairy cows. *Journal of Dairy Science* **89**(5), 1604–1612.

Gaucher, I., Boubellouta, T., Beaucher, E., Piot, M., Gaucheron, F. & Dufour, E. (2008a) Investigation of the effects of season, milking region, sterilisation process and storage conditions on milk and UHT milk physico-chemical characteristics: a multidimensional statistical approach. *Dairy Science and Technology* **88**(3), 291–312.

Gaucher, I., Molle, D., Gagnaire, V. & Gaucheron, F. (2008b) Effects of storage temperature on physico-chemical characteristics of semi-skimmed UHT milk. *Food Hydrocolloids* **22**, 130–143.

Geerts, J.P., Bekhof, J.J. & Scherjon, JW. (1983) Determination of calcium ion activities in milk with an ion selective electrode. *Netherlands Milk and Dairy Journal* **37**, 197–211.

Geier, H. & Klostermeyer, H. (1980) [Enzymatic determination of lactulose.]. *Zeitschrift für Lebensmittel-Untersuchung und –Forschung* **171**, 443–445.

Gestrelius, H., Hertz, T.G., Nuamu, M., Persson, H.W. & Lindstroem, K. (1993) A non-destructive ultrasound method for microbial quality control of aseptically packaged milk. *Lebensmittel Wissenschaft und Technologie* **26**(4), 334–339.

González-Córdova, A.F. & Vallejo-Cordoba, B. (2001) Quantitative determination of short-chain free fatty acids in milk using solid-phase microextraction and gas chromatography. *Journal of Agriculture and Food Chemistry* **49**, 4603–4608.

Goodno, C., Swaisgood, H. & Catignani, G. (1981) A fluorimetric assay for available lysine in proteins. *Analytical Biochemistry* **115**, 203–211.

Grimley, H, Grandison, A.S. and Lewis, M.J. (2009) Changes in milk composition and processing properties during the spring flush period. *Dairy Science and Technology.* **89**, 405–416.

Gunasekaran, S. & Ay, C. (1994) Evaluating milk coagulation with ultrasonics. *Food Technology* **48**(12), 74–78.

Haque, E., Bhandari, B. R., Gidley, M. J., Deeth, H. C. & Whittaker, A. K. (2015) Change in molecular structure and dynamics of protein in milk protein concentrate powder upon ageing by solid-state carbon NMR. *Food Hydrocolloids* **44**, 66–70.

Hardham, J.F., Imison, B.W. & French, H.M. (2000) Effect of homogenisation and microfluidisation on the extent of fat separation during storage of UHT milk. *Australian Journal of Dairy Technology* **55**(1) 16–22.

Harding, F. (ed) (1995) *Milk Quality*, Blackie Academic and Professional, Glasgow.

Harding, F. (2013) Processed milk. In: *Milk Quality*. (ed. F Harding) pp. 113–114. Springer, New York.

Haryani, S., Datta, N, Elliott, A.J. & Deeth, H.C. (2003) Production of proteinases by psychrotrophic bacteria in raw milk stored at low temperature. *Australian Journal of Dairy Technology* **58**, 16–20.

Hasegawa, K., Mukai, K., Gotoh, M., Honjo, S. & Matoba, T. (1987) Determination of lysinoalanine content in commercial foods by gas chromatography-selected ion monitoring. *Agricultural and Biological Chemistry* **51**, 2889–2894.

Henle, T., Walter, A.W. & Klostermeyer, H. (1993) Detection and identification of the cross-linking amino-acids N^{τ}- and N^{π}-(2'-amino-2'-carboxy-ethyl)-l-histidine (histidinoalanine, HAL) in heated milk-products. *Zeitschrift Fur Lebensmittel-Untersuchung Und-Forschung* **197**(2), 114–117.

Heudi, O., Kilinc, T. & Fontannaz, P. (2005) Separation of water-soluble vitamins by reversed-phase high performance liquid chromatography with ultra-violet detection: Application to polyvitaminated premixes. *Journal of Chromatography A* **1070**, 49–56.

Hillbrick, G.C, McMahon, D.J. & Deeth, H.C. (1998) Electrical impedance particle size method (Coulter counter) detects the large fat globules in poorly homogenised UHT processed milk. *Australian Journal of Dairy Technology* **53**, 17–21.

Hillbrick, G.C., McMahon, D.J. & McManus, W.R. (1999) Microstructure of indirectly and directly heated ultra-high-temperature (UHT) processed milk examined using transmission electron microscopy and immunogold labelling. *LWT- Food Science and Technology* **32**, 486–494.

Holland, J, Gupta, R., Deeth, H.C. & Alewood, P.F. (2011) Proteomic analysis of temperature-dependent changes in UHT milk *Journal of Agriculture and Food Chemistry* **59**, 1837–1846.

Hooi, R., Barbano, D.M., Bradley, R.L., Budde, D., Bulthaus, M., Chettiar, M., Lynch, J. & Reddy, R. (2004), Chemical and Physical methods. In: *Standard Methods for the Examination of Dairy Products*, 17th edn, (eds H.M. Wehr & J.F. Frank), Amercian Public Health Association, Washington, USA.

Holt, C., Dalgleish, D.G. & Jenness, R. (1981) Calculation of the ion equilibria in milk diffusate and comparison with experiment. *Analyitcal. Biochemistry* **113**, 154–163.

Hull, M.E. (1947) Studies on milk proteins II. Colourimetric determination of the partial hydrolysis of the proteins in milk. *Journal of Dairy Science* **30**, 881–884.

Humbert, G., Guingamp, M.-F., Linden, G. & Gaillard, J.-L. (2006) The clarifying reagent, or how to make the analysis of milk and dairy products easier. *Journal of Dairy Research* **73**, 464–471.

Hunterlab (2012) Measuring color using Hunter L, a, b versus CIE 1976 L*a*b*. *Application Note* 1005.00 (www.hunterlab.com/de/an-1005-de.pdf).

Hurrell, R.F. & Carpenter, K.J. (1974) Mechanisms of heat damage in protein.4. The reactive lysine content of heat-damaged material as measured in different ways. *British Journal of Nutrition* **32**, 589.

Hurrell, R.F. & Carpenter, K.J. (1981) The estimation of available lysine in foodstuffs after Maillard reactions. *Progress in Food and Nutrition Science* **5**(1-6), 159–176.

Hurrell, R.F., Lerman, P. & Carpenter, K.J. (1979) Reactive lysine in foodstuffs as measured by a rapid dye-binding procedure. *Journal of Food Science* **44**, 1221–1228.

IDF (1972) *Monograph on UHT milk.* IDF Annual Bulletin, part V International Dairy Federation, Brussels.

IDF (1993) *Milk. Determination of Nitrogen Content (Kjeldahl method)* Provisional standard 20B. International Dairy Federation, Brussels.

IDF (1998) *Heat treated Milk. Determination of Lactulose Content. High-performance Liquid Chromatography (Reference Method).* Standard 147B. International Dairy Federation, Brussels.

IDF (2004) *Determination of Lactulose Content - Enzymatic Method.* Standard 175. International Dairy Federation, Brussels.

Ishii, T., Hiramatsu, K., Ohba, T. & Tsutsumi, A. (2001) The liquid-state ^{31}P nuclear magnetic resonance study on microfiltrated milk. *Journal of Dairy Science* **84**, 2357–2363.

Ismail, B. & Nielsen, S.S. (2010) Invited review: Plasmin protease in milk: Current knowledge and relevance to dairy industry. *Journal of Dairy Science* **93**(11), 4999–5009.

ISO (2004) *Milk. Determination of Lactulose Content. Enzymatic Method.* International Standard: ISO 11285. International Organization for Standardization, Geneva, Switzerland.

Iwatsuki, K., Konno, T., Mizota, Y., Sotoyama, K., Sumi, M. & Tomita, M. (2000) Studies on the sensory characteristics and physicochemical properties of UHT processed milk. IV. Sensory characteristics of UHT milk processed by indirect and direct heating methods. *Journal of the Japanese Society for Food Science and Technology (Nippon Shokuhin Kagaku Kogaku Kaishi)* **47**, 844–850.

Iwatsuki, K., Mizota, Y., Sumi, M., Sotoyama, K. & Tomita, M. (1999) Effect of pasteurization and UHT processing conditions on the sensory characteristics of milk. *Journal of the Japanese Society for Food Science and Technology (Nippon Shokuhin Kagaku Kogaku Kaishi)* **46**, 535–542.

Jansson, T., Jensen, S., Eggers, N., Clausen, M.R., Larsen, L.B., Ray, C., Sundgren, A., Andersen, H.J., & Bertram, H.C. (2014) Volatile component profiles of conventional and lactose-hydrolyzed UHT milk-a dynamic headspace gas chromatography-mass spectrometry study. *Dairy Science & Technology* **94**, 311–325.

Jarrett, W.D., Aston, J.W. & Dulley, J.R. (1982) A simple method for estimating free amino-acids in Cheddar cheese. *Australian Journal of Dairy Technology* **37**(2), 55–58.

Jensen, S., Jansson, T., Eggers, N., Clausen, M.R., Larsen, L.B., Jensen, H.B., Ray, C., Sundgren, A., Andersen, H.J. & Bertram, H.C. (2015) Storage-induced changes in the sensory characteristics and volatiles of conventional and lactose-hydrolyzed UHT processed milk. *European Food Research and Technology* **240**(6), 1247–1257.

Jeurnink, T.J.M. and DeKruif, K.G. (1995) Calcium concentration in milk in relation to heat stability and fouling. *Netherlands Milk and Dairy Journal* **49**, 151–165.

Kakade, M.L.& Liener, I.E. (1969) Determination of available lysine in proteins. *Analytical Biochemistry* **27**, 273–280.

Kalogridou-Vassiliadou, D. & Law, B.A. (1982) Sensitivity of proteinase detection in milk with Hide Powder Azure: comparison with direct observations of casein proteolysis. *Journal of Dairy Research* **49**(3), 511–514.

Kaombe, D.D., Du, Y. & Lewis, M.J. (2012) Mineral partition in milk and milk permeates at high temperature. *Journal of Dairy Research* **79**, 1–6.

Karoui, R. & De Baerdemaeker, J. (2007) A review of the analytical methods coupled with chemometric tools for the determination of the quality and identity of dairy products. *Food Chemistry* **102**, 621–640.

Kastanas, P. (1996) *Studies on the Fouling of Heat Exchangers during the Sterilization of Cow's and Goat's Milk*. Ph.D Thesis, University of Reading, UK.

Kastanas, P., Lewis, M.J. & Grandison, A.S. (1995) Design and development of a miniature UHT plant for fouling studies and its evaluation using milk adjusted to different pH values, goat's milk and milk permeate. *Food and Bioproducts Processing* **73**, 83–91.

Keeney, M. & Bassette, R. (1959) Detection of intermediate compounds in the early stages of browning reaction in milk products. *Journal of Dairy Science* **42**(6), 945–960.

Kelly, A.L. & Foley, J. (1997) Proteolysis and storage stability of UHT milk as influenced by milk plasmin activity, plasmin/beta-lactoglobulin complexation, plasminogen activation and somatic cell count. *International Dairy Journal* **7**(6/7), 411–420.

Khalfan, H., Abuknesha, R. & Robinson, D. (1983) Fluorigenic method for the assay of proteinase activity with the use of 4-methylumbelliferone-casein. *Biochemical Journal* **209**, 265–267.

Klein, M.S., Almstetter, M.F., Schlamberger, G., Nuernberger, N., Dettmer, K., Oefner, P.J., Meyer, H.H.D., Wiedemann, S. & Gronwald, W. (2010) Nuclear magnetic resonance and mass spectrometry-based milk metabolomics in dairy cows during early and late lactation. *Journal of Dairy Science* **93**(4), 1539–1550.

Klein, M.S., Buttchereit, N., Miemczyk, S.P., Immervoll, A.-K., Louis, C., Wiedemann, S., Junge, W., Thaller, G., Oefner, P.J. & Gronwald, W. (2012) NMR metabolomic analysis of dairy cows reveals milk glycerophosphocholine to phosphocholine ratio as prognostic biomarker for risk of ketosis. *Journal of Proteome Research* **11**(2), 1373–1381.

Kocak, H.R. & Zadow, J.G. (1985) Age gelation of UHT whole milk as influenced by storage temperature. *Australian Journal of Dairy Technology*, **40**, 14–21.
Korycka-Dahl, M., Ribadeau-Dumas, B., Chene, N. & Martal, J. (1983) Plasmin activity in milk. *Journal of Dairy Science* **66**, 704–711.
Laemmli, U.K. (1970) Cleavage of structural proteins during the assembly of the head of the bacteriophage T4. *Nature* **227**, 680–685.
Lawless, H.T. & Heymann, H. (eds) (2010) *Sensory Evaluation of Food: Principles and Practices*. 2nd edn, Springer, New York.
Le T.X., Datta, N. & Deeth, H.C. (2006) A sensitive HPLC method for measuring bacterial proteolysis and proteinase activity in UHT milk. *Food Research International* **39**, 823–830.
Le, T.T., Deeth, H.C., Bhandari, B., Alewood, P.F. & Holland, J.W. (2012) A proteomic approach to detect lactosylation and other chemical changes in stored milk protein concentrate. *Food Chemistry* **132**, 655–662.
Le, T.T., Nielsen, S.D., Villumsen, N.S., Kristiansen, G.H., Nielsen, L.R., Nielsen, S. B., Hammershøj, M. & Larsen, L.B. (2016) Using proteomics to characterise storage-induced aggregates in acidic whey protein isolate drinks. *International Dairy Journal* **60**, 39–46.
Lelievre, J. & Creamer, L.K. (1978) NMR-study of formation and syneresis of renneted milk gels. *Milchwissenschaft* **33**(2), 73–76.
Lewis, M.J. (1990) *Physcial Properties of Foods and Food Processing Systems*. Woodhead Publishers, Cambridge, UK.
Lewis, M.J. & Heppell, N. (2000) *Continuous Thermal Processing of Foods: Pasteurization and UHT Sterilization*, Aspen Publishers, Gaithersburg, MD.
Lewis, M.J. (2010) The measurement and significance of ionic calcium in milk- A review. *International Journal of Dairy Technology* **64**, 1–13.
Lewis, M.J., Grandison, A.S. Lin, M.-J. & Tsioulpas, A. (2011) Ionic calcium and pH as predictors of stability of milk to UHT processing. *Milchwissenschaft* **66**, 197–200.
Lieske, B. (1998) Effects of ageing of raw milk on some structural properties of native casein micelles. *Milchwissenschaft* **53** 562–565.
Lin, M.-J. (2002) *Measurement of Ionic Calcium and its Role in Milk Stability*. PhD thesis, University of Reading, UK.
Lin, M-J; Lewis, M. & Grandison, A. (2006) Measurement of ionic calcium in milk. *International Journal of Dairy Technology*, **59** 192–199.
Lin, M-J., Grandison, A.S. & Lewis, M.J. (2015) Partitioning of calcium and magnesium (total divalent cations) during membrane filtration of milk. *Journal of Food Engineering* **149**, 153–158.
Lorenzen P.C., Clawin-Raedecker I., Einhoff K., Hammer P., Hartmann R., Hoffmann W., Martin D., Molkentin J., Walte H.G. & Devrese M. (2011) A survey of the quality of extended shelf life (ESL) milk in relation to HTST and UHT milk. *International Journal of Dairy Technology* **64**, 166–178.
Lowry, O.H., Rosebrough, N.J., Farr, A.L. & Randall, R.J. (1951) Protein measurement with the Folin phenol reagent. *Journal of Biological Chemistry* **193**(1), 265–275.
Lum (2013) LumiSizer- Dispersion analyser http://www.lum-gmbh.com/LUMiSizer_product-information.html (accessed December 2015).
Lynch, J.M. & Barbano, D.M. (1999) Kjeldahl nitrogen analysis as a reference method for protein determination in dairy products. *Journal of AOAC International* **82**(6), 1389–1398.

Malvern Instruments Limited (2012) *A Basic Guide to Particle Characterization.* http://www.malvern.com/en/support/resource-center/Whitepapers/WP120620BasicGuidePartChar.aspx.

Manji, B., Kakuda, Y. & Arnott, D.R. (1986) Effect of storage temperature on age gelation of ultra-high temperature milk processed by direct and indirect heating systems. *Journal of Dairy Science* **69**(12), 2994–3001.

Manso, M.A., Leonil, J., Jan, G. & Gagnaire, V. (2005) Application of proteomics to the characterisation of milk and dairy products. *International Dairy Journal* **15**(6-9), 845–855.

Marks, N.E., Grandison, A.S. & Lewis M.J. (2001) Use of hydrogen peroxide detection strips to determine the extent of pasteurization in whole milk. *International Journal of Dairy Technology* **54**, 20–22.

Martin, N.H., Friedlander, A., Mok, A., Kent, D., Wiedmann, M. & Boor, K.J. (2014) Peroxide test strips detect added hydrogen peroxide in raw milk at levels affecting bacterial load. *Journal of Food Protection* **77**, 1809–1813.

Martinez-Castro, I., Calvo, M.M. & Olano, A. (1987) Chromatographic determination of lactulose. *Chromatographia* **23**, 132–136.

May, R.J. & Smith, D.E. (1998) Effect of storage and various processing conditions on the amount of ionic calcium in milk. *Milchwissenschaft* **53**, 605–608.

McKellar, R.C. (1981) Development of off-flavors in ultra-high temperature and pasteurized milk as a function of proteolysis. *Journal of Dairy Science* **64**, 2138–2145.

McKinnon, I.R., Jackson, E.K. & Fitzpatrick, L. (2000) Instant whole milk powder micelles - stability in instant coffee. *Australian Journal of Dairy Technology* **55**(2), 88.

Mehta, B.M. & Deeth, H.C. (2016) Blocked lysine in dairy products – formation, occurrence, analysis and nutritional implications. *Comprehensive Reviews in Food Science and Food Safety* **15**, 206–218.

Mengesha, A.E. & Bummer, P.M. (2010) Simple chromatographic method for simultaneous analyses of phosphatidylcholine, lysophosphatidylcholine, and free fatty acids. *AAPS PharmSciTech* **11**(3), 1084–1091.

Mittal, S.B, Hourigan, J.A. & Zadow, J.G. (1990) Effect of added sodium hexametaphosphate on certain technological aspects of UHT recombined milk. *Australian Journal of Dairy Technology* **45**, 1–4.

Moermans, R. & Mottar, J. (1984) The turbidity test and the quality of UHT treated milk. *Milchwissenschaft* **39**, 94–95.

Morales, F.J. & van Boekel, M.A J.S. (1998) A study on advanced Maillard reaction in heated casein sugar solutions: Colour formation. *International Dairy Journal* **8**, 907–915.

Morales, F.J., Romero, C. & Jimenez Perez, S. (1997) Application of the clarifying reagent index as a heat-induced parameter in commercial milk. *Journal of Food Protection* **60**(10), 1241–1246.

Morales, F.J., Romero, C. & Jimenez, P.S. (2000) Characterization of industrial processed milk by analysis of heat-induced changes. *International Journal of Food Science & Technology* **35**(2), 193–200.

Mottar, J. & Moermans, R. (1988) Optimization of the forewarming process with respect to deposit formation in indirect ultra high temperature plants and the quality of milk. *Journal of Dairy Research* **55**, 563–568.

Mouillet, L., Luquet, F.M., Nicod, H., Boudier, J.F. & Mahieu, H. (1981) La lipolyse des laits. Étude d'une méthode rapide de mesure. *Lait* **61**, 171–186.

Muldoon, P.J. and Liska, B.J. (1969) Comparison of a resin ion-exchange method and a liquid ion-exchange method for determining ionised calcium in skim milk. *Journal of Dairy Science* **52**, 460–464.

Nair, B.M., Laser, A., Burvalla, A. & Asp, N.G. (1978) Gas chromatographic determination of available lysine. *Food Chemistry* **3**, 283.

Newstead, D.F., Paterson, G., Anema, S.G., Coker, C.J. & Wewala, A. R. (2006) Plasmin activity in direct-steam-injection UHT-processed reconstituted milk: effects of preheat treatment. *International Dairy Journal* **16**, 573–579.

Nieuwenhuijse, J.A, Timmermans, W. & Walstra, P. (1988) Calcium and phosphate partitions during the manufacture of sterilized milk and their relations to the heat stability. *Netherlands Milk and Dairy Journal* **42**, 387–421.

Nieuwenhuijse, J.A. & van Boekel, M.A.J.S. (2003) Protein stability in sterilised milk and milk products. In: *Advanced Dairy Chemistry, Volume 1 Proteins*, 3rd edn. (eds P.F. Fox & P.L.H. McSweeney) pp. 947–974, Kluwer Academic/Plenum Press, New York.

Nollet, L.M.L. & Toldra, F. (2010) *Handbook of Dairy Foods Analysis*, CRC Press, Boca Raton, FL.

O'Donnell, R., Holland, J.W., Deeth, H.C. & Alewood, P.F. (2004) Milk proteomics. *International Dairy Journal* **14**, 1013–1023.

Ohnishi, M., Iwata, K. & Hiromi, K (1990) An enzymatic assay for bacterial alpha-amylase using a glucose determination kit and maltohexaose as the substrate. *Starch/Stärke* **42**(3), 103–106.

Oldfield, D.J., Teehan, C.M. & Kelly, P.M. (2000) The effect of preheat treatment and other process parameters on the coffee stability of instant whole milk powder. *International Dairy Journal* **10**, 659–667.

Omoarukhe, E.D., On-Nom, N., Grandison, A.S. & Lewis, M.J. (2010) Effects of different calcium salts on properties of milk related to heat stability. *International Journal of. Dairy Technology* **63**, 504–511.

On-Nom, N. (2012) *Partitioning of Milk at High Temperature*, PhD thesis, University of Reading, UK.

On-Nom, N., Grandison, A.S. & Lewis, M.J. (2010) Measurement of ionic calcium, pH and soluble divalent cations in milk at high temperature. *Journal of Dairy Science* **93**, 515–523.

On-Nom, N., Grandison, A.S. & Lewis, M.J. (2012) Heat stability of milk supplemented with calcium chloride. *Journal of Dairy Science* **95**, 1623–1631.

Pagliarini, E., Vernille, M. & Peri, C. (1990) Kinetic study on colour changes in milk due to heat. *Journal of Food Science*, **55**, 1766–1767.

Park, Y.H. & Hong Y.H (1989) HMF (5-hydroxymethylfurfural) content in domestic market milks. *Korean Journal of Dairy Science* **11**(4), 265–272.

Pathomrungsiyounggul, P., Grandison, A.S. & Lewis, M.J. (2012) Feasibility of using dialysis for determining calcium ion concentration and pH in calcium-fortified soymilk at high temperature. *Journal of Food Science* **77**, E10–E16.

Pellegrino, L., Resmini, P., De Noni, I. & Masotti, F. (1996) Sensitive determination of lysinoalanine for distinguishing natural from imitation Mozzarella cheese. *Journal of Dairy Science* **79**, 725–734.

Perkins, M.L., D'Arcy, B.R., Lisle, A. & Deeth, H.C. (2005a) Solid phase microextraction of stale flavour volatiles from the headspace of UHT milk. *Journal of the Science of Food and Agriculture* **85**, 2421–2428.

Perkins, M.L., Elliott, A.J., D'Arcy, B.R. & Deeth, H.C. (2005b) Stale flavour volatiles in Australian commercial UHT milk during storage. *Australian Journal of Dairy Technology* **60**, 231–237.

Picard, C., Plard, I. & Collin, J.C. (1996) Application of the inhibition ELISA method to the study of proteolysis caused by heat-resistant *Pseudomonas* proteinases specific towards kappa-casein in heated milk. *Milchwissenschaft* **51**(8), 438–442.

Poth, A.G., Deeth, H.C., Alewood, P.F & Holland, J.W. (2008) Analysis of the human casein phosphoproteome by 2-D electrophoresis and MALDI-TOF/TOF MS reveals new phosphoforms and variable deamidation. *Journal of Proteomic Research* **7**, 5017–5027.

Prakash, S, Kravchuk, O. & Deeth, H.C. (2015) Influence of pre-heat temperature, pre-heat holding time and high-heat temperature on fouling of reconstituted skim milk during UHT processing. *Journal of Food Engineering* **153**, 45–52.

Prakash, S., Datta, N. & Deeth, H.C. (2005) Methods of detecting fouling caused by heating of milk. *Food Reviews International* **21**, 267–293.

Prakash, S., Datta, N., Lewis M. J. & Deeth H.C. (2007) Reducing fouling during UHT treatment of goat's milk. *Milchwissenschaft* **62**, 16–19.

Prasad, K.S. (1989) *The Sensory Characteristics of Heat-treated Milks, with Special Reference to UHT Processing.* PhD Thesis, University of Reading, Reading, UK.

Prasad, S.K, Thomson, D.M.H. & Lewis, M.J. (1990) In: *Trends in Food Product Development.* (eds T.C. Yan & C. Tan). Proceedings of the World Congress of Food Science and Technology, Singapore.

Ramsdell, G.A., Johnson, W.M.T. Jr & Evans, F.R. (1931) A test for the detection of milk unstable to heat. *Journal of Dairy Science* **14**, 93–106.

Ranjith, H.M.P. (1995) *Assessment of some Properties of Calcium-Reduced Milk and Milk Products from Heat Treatment and other Processes.* PhD thesis, University of Reading, UK.

Reed, A.W., Deeth, H.C. & Clegg, D.E. (1984) An HPLC procedure for the quantitative determination of free fatty acids in butter. *Journal of the Association of Official Analytical Chemists* **67**(4), 718–721.

Renzone, G., Arena, S. & Scaloni, A. (2015) Proteomic characterization of intermediate and advanced glycation end-products in commercial milk samples. *Journal of Proteomics*, **117**, 12–23.

Resmini, P., Pellegrino, L. & Battelli, G. (1990) Accurate quantification of furosine in milk and dairy products by a direct HPLC method. *Italian Journal of Food Science* **2**(3), 173–183.

Richardson, B.C. & Pearce, K.N. (1981) The determination of plasmin in dairy products. *New Zealand Journal of Dairy Science and Technology* **16**, 209–220.

Robyt, J.T. & Ackerman, R.J. (1968) Amylases. In: *Starch and its Derivatives.* (ed J.A. Radley), 4th edn, pp. 423–429. Chapman and Hall, London.

Rollema, H.S., McKellar, R., Sorhaug, T., Suhren, G., Zadow, J.G., Law, B.A., Poll, J.K., Stepaniak, L. & Vagias, G. (1989) Comparison of different methods for the detection of bacterial proteolytic enzymes in milk. *Milchwissenschaft* **44**(8), 491–496.

Roncada, P., Stipetic, L.H., Bonizzi, L., Burchmore, R.J.S. & Kennedy, M.W. (2013) Proteomics as a tool to explore human milk in health and disease. *Journal of Proteomics* **88**, 47–57.

Rose, D. & Tessier, H. (1959) Composition of ultrafiltrates from milk heated at 80 to 230 °F in relation to heat stability. *Journal of Dairy Science* **42**, 969–980.
Rutherfurd, S.M., & Moughan, P.J. (2007) Development of a novel bioassay for determining the available lysine contents of foods and feedstuffs. *Nutrition Research Reviews* **20**(1), 3–16.
Saint-Denis. T., Humbert, G. & Gaillard, J.L. (2001) Enzymatic assays for native plasmin, plasminogen and plasminogen activators in bovine milk. *Journal of Dairy Research* **68**(3), 437–449.
Saraiva, J., Oliveira, J. C., Hendrickx, M., Oliveira, F.A.R. & Tobback, P. (1996) Analysis of the inactivation kinetics of freeze-dried alpha-amylase from *Bacillus amyloliquefaciens* at different moisture contents. *LWT- Food Science and Technology* **29**(3), 260–266.
Sassi, M., Arena, S. & Scaloni, A. (2015) MALDI-TOF-MS Platform for integrated proteomic and peptidomic profiling of milk samples allows rapid detection of food adulterations. *Journal of Agricultural and Food Chemistry* **63**(27), 6157–6171.
Scanlan, R.A., Lindsay, R.C., Libbey, L.M. & Day, E.A. (1968) Heat-induced volatile compounds in milk. *Journal of Dairy Science* **51**, 1001–1004.
Schokker, E.P. & van Boekel, M.A.J.S. (1997) Production, purification and partial characterization of the extracellular proteinase from *Pseudomonas fluorescens* 22 F. *International Dairy Journal* **7**, 265–271.
Sharma, G. (2003) Colour fundamentals for digital imaging. In: *Digital Colour Difference Handbook.* (eds G. Sharma& R. Bala), p. 31. CRC Press, Boca Raton, FL.
Sheehan, H. & McCleary, B.V. (1988) A new procedure for the measurement of fungal and bacterial α-amylase. *Biotechnology Techniques* **2**, 289–292.
Shew, D. (1981) Technical aspects of UHT processing. In: *New Monograph on UHT milk* Document 133, pp. 115–121. International Dairy Federation, Brussels.
Silanikove, N, Shapiro, F. & Shamay, A. (2003) Use of an ion-selective electrode to determine free Ca ion concentration in milk of various mammals. *Journal of Dairy Research* **70** 241–243.
Silva, R.A.B. Montes, R.H.O., Richter, E.M. & Munoz, R.A.A. (2012) Rapid and selective determination of hydrogen peroxide residues in milk by batch injection analysis with amperometric detection. *Food Chemistry* **133**, 200–204.
Skanderby, M., Westergaard, V. Partridge, A. & Muir, D.D. (2009) Dried milk products. In: *Dairy Powders and Concentrated Products,* (ed A.Y. Tamime), Wiley-Blackwell, Chichester, UK.
Smyth, C., Dawson, K. & Buckin, V. A. (1999) Ultrasonic analysis of heat-induced coagulation in calcium fortified milk. *Progress in Colloid & Polymer Science* **112**, 221–226.
Snoeren, T.H.M. & Both, P. (1981) Proteolysis during the storage of UHT-sterilized whole milk. II. Experiments with milk heated by the indirect system for 4 s at 142 °C. *Netherlands Milk and Dairy Journal* **35**(2), 113–119.
Sodhi, S., Ahuja, S.P. & Singh, S. (1996) Changes in the levels of ionic calcium and proteins during assembly of casein micelles and transition of colostrums to milk in the buffalo and the cow. *Buffalo Journal* **1**, 101–108.
Somogyi, M. (1952) Notes on sugar determination. *Journal of Biological Chemistry* **195**, 19–23.
Souza, S.S., Cruz, A.G., Walter, E.H.M., Faria, J.A.F., Celeghini, R.M.S., Ferreira, M.M.C., Granato, D. & Sant'Ana, A.D. (2011) Monitoring the authenticity of Brazilian UHT milk: A chemometric approach. *Food Chemistry* **124**, 692–695.

Spadaro, A.C., Draghetta, W., Dellama, S.N., Camargo, A.C.M. & Greene L.J. (1979) A convenient manual trinitrobenzene sulfonic acid method for monitoring amino acids and peptides in chromatographic column effluents. *Analytical Biochemistry* **96**, 317–321.

Spangelo, A., Karijord, Ø., Svensen, A. & Abrahamsen, R.K. (1986) Determination of individual free fatty acids in milk by strong anion-exchange resin and gas chromatography. *Journal of Dairy Science* **69**, 1787–1792.

SPX (2008) *Understanding the Gaulin EQA™ analyser.* [Online], Available: http://www.spx.com/en/assets/pdf/TB63_Understand_The_Gaulin_EQA_3079_US_tcm11-7650.pdf [22 July 2015].

Standards Association of Australia (1980) Methods of chemical and physical testing for the dairying industry. Part 2: Liquid milks. Section 1: Method for determining the efficiency of homogenization of milk. In: *Australian Standard AS 2300.2.1.*

Stead, D. (1984a) Evaluation of a fluorimetric assay on the lipases from strains of milk psychrotrophic bacteria. *Journal of Dairy Research* **51**, 123–130.

Stead, D. (1984b) A fluorimetric method for determination of *Pseudomonas fluorescens* AR11 lipase in skim milk powder, whey powder and whey protein concentrate. *Journal of Dairy Research* **51**, 623–628.

Stead, D. (1984c) A new method for measuring lipase activity from psychrotrophic bacteria in milk. *Dairy Industries International* **49**(7), 29–29.

Stead, D. (1989) Assay methods for lipases and phospholipases. In: *Enzymes of Psychrotrophs in Raw Food* (ed. R.C. McKellar), pp. 173–18. CRC, Boca Raton, FL.

Stepaniak, L., Birkeland, S.E., Vagias, G. & Sorhaug, T. (1987) Enzyme-linked immunosorbent assay (ELISA) for monitoring the production of heat stable proteinases and lipase from *Pseudomonas*. *Milchwissenschaft* **42**(3), 168–172.

Stoscheck, C.M. (1990) Quantitation of protein. *Methods in Enzymology* **182**, 50–68.

Sundekilde, U.K., Larsen, L.B. & Bertram, H.C. (2013) NMR-based milk metabolomics. *Metabolites* **3**(2), 204–222.

Sundekilde, U.K. (2012) Milk metabolite variability and heritability and their association with technological properties of bovine milk elucidated by NMR-based metabolomics. PhD thesis, Aarhus University, Denmark.

Teehan, C. M., Kelly, P. M., Devery, R. & O'Toole, A. (1997) Evaluation of test conditions during the measurement of coffee stability of instant whole milk powder. *International Journal of Dairy Technology* **50**(4), 113–121.

Teshima, S., Mitsuhida, N & Ando, M. (1985) Determination of alpha-amylase in biological fluids using a new substrate (β-2-chloro-4-nitrophenylmaltopentaoside). *Clinica Chimica Acta* **150**(3), 165–174.

Tetra Pak (undated) *The Role of Raw Milk Quality in UHT Production*. Tetra Pak, Lund, Sweden.

Tran, T.T.H., Datta, N & Deeth, H.C. (2003) Amylases in enzyme detergents and their effects on a simulated dairy dessert. *Milchwissenschaft* **58**, 527–531.

Tremblay, L., Laporte, M.F., Léonil, J, Dupont, D. & Paquin, P. (2003) Quantificatioin of proteins in milk and milk products. In: *Advanced Dairy Chemistry* Volume 1 (Eds P.F. Fox & O.L.H. McSweeney), 3rd edn, pp 49–138, Kluwer Academic Plenum Publishers, New York.

Tsioulpas, A., Lewis, M.J. & Grandison, A.S. (2007) Effect of minerals on casein micelle stability of cow's milk. *Journal of Dairy Research* **74**, 167–173.

Tsioulpas, A, Grandison, A.S. and Lewis, M.J. (2008) Changes in physical properties of bovine milk from the colostrum period to early lactation, *Journal of Dairy Science* **90** 5012–5017.

Twining, S.S., (1984) Fluorescein isothiocyanate-labeled casein assay for proteolytic enzymes. *Analytical Biochemistry* **143**, 30–34.

Tyler, G. (undated) ICP-OES, ICP-MS and AAS techniques compared. *ICP Optical Emission Spectroscopy* Technical Note 05 Jobin Yvon S.A.S, Horiba Group (http://www.horiba.com/fileadmin/uploads/Scientific/Downloads/OpticalSchool_CN/TN/ICP/ICP-OES__ICP-MS_and_AAS_Techniques_Compared.pdf).

Udabage, P., McKinnon, I. and Augustin, M-A. (2000) Mineral and casein equilibria in milk: effects of added salts and calcium chelating agents. *Journal of Dairy Research* **67**, 361–370.

Vazquez-Landaverde, P., Torres, J.A. & Qian, M.C. (2006) Quantification of trace volatile sulfur compounds in milk by solid-phase microextraction and gas chromatography-pulsed flame photometric detection. *Journal of Dairy Science* **89**, 2919–2927.

Vazquez-Landaverde, P.A., Velazquez, G., Torres, J. A. & Qian, M.C. (2005) Quantitative determination of thermally derived off-flavor compounds in milk using solid-phase microextraction and gas chromatography. *Journal of Dairy Science* **88**, 3764–3772.

von Boekelman, B. & von Boekelman, I. A. (1998) *Long-life Products: Heat-treated, Aseptically Packed: A Guide to Quality.* Tetra Pak Processing Systems AB, Lund, Sweden.

Wahlgren, M., Drakenberg, T., Vogel, H.J. & Dejmek, P. (1986) ^{31}P nuclear magnetic-resonance study of milk fractions. *Journal of Dairy Research* **53**(4), 539–545.

Wahlgren, N.M., Dejmek, P. & Drakenberg, T. (1990) A ^{43}Ca and ^{31}P NMR-study of the calcium and phosphate equilibria in heated milk solutions. *Journal of Dairy Research* **57**, 355–364.

Walstra, P. & Jenness, R. (1984) *Dairy Chemistry and Physics*, John Wiley and Sons, New York.

Whitaker, R. (1931) The feathering of evaporated milk in hot coffee. *Journal of Dairy Science* **14**, 177–188.

White, J.C.D. & Davies, D.T. (1958) The relation between the chemical composition of milk and the stability of the caseinate complex. 1. General introduction, description of samples, methods and chemical composition of samples. *Journal of Dairy Research* **25**, 236–255.

Wiles, P.G., Gray, I.K. & Kissling, R.C. (1998) Routine analysis of proteins by Kjeldahl and Dumas methods: review and inter-laboratory study using dairy products. *Journal of AOAC International* **81**(3), 620–632.

Williams, R.P, d'Ath, L. & Augustin, M.A. (2005) Production of calcium fortified milk powders using soluble calcium salts. *Lait* **85**, 369–381.

Wirtanen, G., Ahvenainen, R. & Mattila Sandholm, T. (1992) Non-destructive detection of spoilage of aseptically-packed milk products: effect of frequency and imaging parameters on the sensitivity of ultrasound imaging. *Lebensmittel Wissenschaft und Technologie* **25**(2), 126–132.

Yang, Y.X., Bu, D.P., Zhao, X.W., Sun, P., Wang, J.Q. & Zhou, L.Y. (2013) Proteomic analysis of cow, yak, buffalo, goat and camel milk whey proteins: quantitative differential expression patterns. *Journal of Proteome Research* **12**, 1660–1667.

Yousif, A.M., Kato, J. & Deeth, H.C. (2003) Effect of storage time and conditions on the seed coat colour of Australian adzuki beans. *Food Australia* **55**, 479–485.

Zhang, L., Boeren, S., van Hooijdonk, A.C.M., Vervoort, J.M. & Hettinga, K.A. (2015) A proteomic perspective on the changes in milk proteins due to high somatic cell count. *Journal of Dairy Science* **98**, 5339–5351.

12

Concluding Comments

In this concluding chapter we discuss some issues which are not yet fully resolved and require future research. Furthermore, there are some issues which we believe processors should take into account to ensure maximal quality and safety of their products. Also included are some key references on high-temperature processing and products.

12.1 Spore-Forming Bacteria

The destruction or removal of spore-forming bacteria is the fundamental aim of sterilisation processes. There is a considerable amount of knowledge of these organisms but in preparing this book we have encountered several aspects which require attention. Some of these are as follows:

12.1.1 Highly Heat-Resistant Spores

As pointed out in Section 4.2.2 and demonstrated in Table 4.2, a wide variety of bacteria which produce highly heat-resistant spores (defined as surviving heat treatment at 100 °C for 30 min) occur in the dairy environment. Some of these, such as *Bacillus sporothermodurans, Paenibacillus* spp and *Anoxybacillus flavithermus,* have emerged as problems in recent years but it is highly probable that others will be encountered in future years. This may be exacerbated by the adaptation of particular species or strains to the dairy environment as has been observed for strains of *Geobacillus stearothermophilus* (S. Flint, 2016, Pers Com).

12.1.2 Enzymes Produced by Spores

It has been demonstrated that enzymes produced by spores in biofilms on equipment and tankers used for raw milk can enter the milk and, if heat-resistant, can remain active in a sterilised product and cause spoilage during storage (Teh *et al.*, 2014). To date, it has been largely assumed that any heat-resistant enzymes in UHT milk originate only from growth of psychrotrophic bacteria such as pseudomonads in raw milk.

12.1.3 Sources of Spores

It is well known that spores enter the milk from a variety of sources. A major source is the environment of housed animals since the spore counts in milk produced by housed

High Temperature Processing of Milk and Milk Products, First Edition. Hilton C. Deeth and Michael J. Lewis.

cows is generally much higher than that of milk from free-range cows. Silage and concentrates often carry a high spore load and these spores pass through the cow into faeces which can be picked up on the teats and udders and enter the milk during milking (Scheldeman *et al.*, 2005). We have recently encountered a case of UHT milk contaminated by *Bacillus thuringiensis*, a close relative of *B. cereus*. This is of interest as this organism is used as a biological control agent and could be in high concentrations in some agricultural environments and animal feed sources.

12.1.4 Identification of Spores

When contamination of a UHT product with a spore-forming organism occurs, it is important that the source of the contamination be identified and, where possible, eliminated. Given the modern microbiological methods such as 16S rRNA gene sequence analysis (e.g., Coorevits *et al.*, 2011) and DNA high-resolution melt analysis (Chauhan *et al.*, 2013), it is now possible to determine whether isolates from raw material, for example, raw milk and ingredients such as milk powder, or a biofilm in the equipment, and from the final product are identical. If this can be determined, it is possible to answer the most frequently asked question following a contamination incident of a UHT product, which is whether the contaminant survived the heat treatment or entered the product downstream of the steriliser. Once this answer is obtained, appropriate action can be taken; either the heat treatment conditions need to be adjusted or the downstream sterility cleaning, sanitisation and sterilisation procedures need to be improved.

12.1.5 Spore Counts in Raw Milk

Spore count is sometimes used as a quality parameter of raw milk. This would intuitively seem perfectly reasonable, particularly for milk supplies which commonly have high counts, for example, where cows are housed and fed silage. However, we query the value of spore counts where they are always low, for example, ≤100/mL.

Spore counts may have limited value for predicting the quality UHT milk but may be more important for the shelf-life ESL milk (see Section 12.6). More information is required on the relationship between raw milk spore counts and the quality of the final product.

12.1.6 Conditions of Activation and Germination of Spores

Because the vegetative form of spore-forming bacteria is much more sensitive to heat than the spore form, activation and germination of the spores and conversion to the vegetative form before heat treatment is seen as an attractive option for reducing or eliminating spore-formers from milk products and ingredients such as cocoa powder (see Section 4.4.3.3). The most common means of activation to initiate germination is heat shock but, unfortunately, this is not straight forward for several reasons. Firstly, not all spores of a particular spore-former are activated to the same degree by a given heat treatment; secondly, the optimal heat shock temperature–time conditions differ for different spore-formers; thirdly, the temperature–time conditions following activation for germination are generally not known; fourthly, the effect of the activation medium, for example, total solids content, on the germination process is unclear; and fifthly, for powders such as milk and cocoa powders, the optimum temperature–time conditions for rehydration is important since water access to the spores is essential for effective activation. A question which arises is whether commonly used heat treatments such as pasteurisation, ESL and UHT treatments cause activation of spores which allow them to geminate and grow in the product during storage. A thorough study of these aspects would be very helpful for processors of ESL and UHT milk products. In fact, some ESL and UHT milk is produced from milk which has already been heat treated, rather than from raw milk. The beneficial and detrimental effects of these double heat treatments are worthy of further investigation.

12.1.7 Psychrotrophic Spore-Formers

Spores of psychrotrophic spore-formers have assumed considerable significance since the introduction of ESL milk. If ESL milk is not contaminated downstream of the high-temperature section of the plant, for example, if it is packaged aseptically, these are the only organisms which can cause spoilage of the refrigerated product. These can apparently be eliminated by heating at ≥134 °C (Blake *et al.*, 1995) but detailed information on their thermal inactivation in milk is currently not available. Further knowledge of psychrotrophic spore-formers in particular milk supplies is therefore required; this includes the development of convenient testing regimes for these organisms.

A spore-former of particular relevance is *Bacillus cereus* because: it can cause spoilage; some strains are pathogenic; it can be psychrotrophic; its spores vary considerably in thermal stability; and it has the ability to form biofilms. For these reasons, it can be a problem in ESL milk and possibly UHT milk, but to date it has not been possible to properly assess the risk of contamination of a product with this organism. Further information about it is required to allow informed risk analyses to be performed.

12.2 Biofilms

Biofilms have been extensively researched but they still pose considerable problems for the dairy industry as they are a major source of recontamination of product (Marchand *et al.*, 2012). That they influence the microbiology of products is evidenced by the persistence of particular bacteria in the milk processed by individual processing plants. This has been noted by several authors for pasteurised milk (e.g., Griffiths & Phillips, 1988; Craven & Macauley, 1992) and may explain incidences of "tenacious periodical contamination" (Scheldeman *et al.*, 2004) of some UHT products. Unfortunately, the role of biofilms seems to be underestimated or poorly understood in the dairy industry. As indicated in Section 6.2.2.8, removal of biofilms in difficult and standard CIP processes fail to remove them (Flint *et al.*, 1999; Bremer *et al.*, 2006). Some spores are resistant to hot alkali and may even adhere more strongly to stainless steel after such treatment (Seale *et al.*, 2011). Hence, effective strategies need to be developed in conjunction with specialist detergent and sanitiser companies to ensure plants are kept free of biofilm build-up. The reader is referred to a recent publication on biofilms (Teh *et al.*, 2015).

12.3 Age Gelation

12.3.1 Mechanism

As discussed in Section 7.2.2, there are aspects of the proposed mechanisms of age gelation which do not fit with reality. The proposed mechanisms are mostly predicated on the fact that gelation is initiated by proteolysis. While this holds for many cases of gelation, it does not hold for all. Several authors have been unable to find a direct correlation between the time of onset of gelation and the level of proteolysis. Age gelation is inhibited at temperatures >~30 °C but proteolysis increases with increasing temperature, and gelation sometimes occurs in the absence of proteolysis. A more comprehensive explanation of the phenomenon is called for. In particular, the cause of gelation in the absence of proteolysis warrants investigation.

More information on the sensory changes that accompany gelation would also be valuable for elucidating the mechanism of gelation. This includes the incidence of bitterness, acidity and other off-flavours, as well as the lack of any off-flavour, accompanying gelation.

12.3.2 Early Prediction of a Milk's Susceptibility

Several approaches have been examined for this purpose although it would be rare for a company to use them. This is largely because the correlations between the results of such predictive testing and the time to gelation have not generally been determined. An increase in viscosity inevitably precedes gelation and may be used as a predictor, and where proteolysis is the key factor, increases in peptide levels give a reasonable guide.

Since a direct correlation of time to gelation and proteolysis is often not found, it is possible that some milk has an inherent susceptibility to gelation although the cause of this may not be known. As discussed in Section 7.2.2.7, one possible approach to assessing this susceptibility is to add various substances which are known to cause coagulation of milk; these include proteolytic enzymes, salts such as calcium chloride and sodium dihydrogen phosphate, alcohol and enzymes from plant sources (IDF, 2007). If none of these reagents fails to cause gelation to occur, one might argue that the product is not susceptible to gelation and therefore unlikely to gel in the future. However, as indicated above, it is paramount that correlations between the results of such tests and gelation time are established before they can be adopted as predictive tests.

12.4 Predictive Modelling

In several parts of this book, the usefulness of modelling the effects of processing and storage of UHT products has been demonstrated (Browning *et al.*, 2001; Tran *et al.*, 2008; Hotrum *et al.*, 2010). It can be used for calculating essential indices such as F_0, B* and C* and for predicting other changes in UHT milk. A particularly useful application is tailoring UHT pilot plant conditions to match those of commercial UHT plants (Tran *et al.*, 2008).

The modelling is based on reaction kinetics data and uses the temperature–time profile of the processing plant in the calculation of the effects. To date, relevant kinetic data are available for several microbiological, chemical and physical changes in white milk but the data are lacking for other products. Furthermore, accurate temperature–time profiles of commercial plants are seldom available within dairy companies. Generally, equipment manufacturers have these profiles as they supply data such as F_0, B*, C* and predicted lactulose levels when the plant is commissioned. It is our opinion that companies should hold an up-to-date temperature–time profile of each processing line so that they can predict the effects of their processing system and any changes they may contemplate for it.

Predictive modelling has become important in recent times with the increasing popularity and market share of ESL milk. When producing ESL milk, the processor aims to minimise chemical change (chiefly flavour change) and maximise the bactericidal effect of the process. Such optimisation can be achieved with predictive modelling provided the temperature–time profile of the ESL plant is known.

12.5 The Shelf-Life of UHT Milk

Historically, the shelf-life of UHT milk has been six months although shorter times such as 12 weeks are also quoted (e.g., Buckenhuskes, 2015). However some UHT products now have a "best before" period of 1 year. Our advice to a UHT milk producer is to resist the temptation to move in this direction, as it is taking the product out of its "comfort zone" and is likely to lead to an increase in customer complaints. Although some products are still satisfactory after twelve months, there have also been cases where the entire batch was satisfactory after six months but had formed a firm gel after nine months. One cannot totally prevent chemical and physical reactions taking place during storage, although with the current knowledge of processing conditions and milk quality, they can be minimised. The incidence of age-gelation appears to have decreased and this may have induced some producers to extend out the "best before" dates. We speculate that the main reason for this apparent reduction in age gelation is the better understanding by UHT milk producers of the important role of raw milk quality and the need to avoid using raw milk with a high bacterial count. An important challenge remains for the UHT producer to establish the quality of raw milk (its freshness) before it can no longer be satisfactorily processed into UHT milk (or in fact to any other product). With the current availability of rapid bacteriological counting methods such as the Bactoscan™, we would recommend testing milk destined for UHT processing and to be wary of milk with a total count of $>10^5$ cfu/mL and to not use milk with $\geq 10^6$ cfu/mL.

One aspect which does not seem to have been investigated is whether the heat-inactivated bacteria play any role in the quality of UHT milk throughout storage. The number of such heat-inactivated cells will be higher in UHT milk manufactured from raw milk of low quality and hence they may make a contribution to quality deterioration in the UHT milk.

We would predict that as aseptic packaging technology continues to improve, the incidence of microbial spoilage will decrease, or, if they occur, will be detected and the products withdrawn before they become an issue. Some UHT milk producers have reduced consumer complaints below 1 in a million and hence are well placed to put an extended "best before" date on their product.

Another interesting query relates to sterility testing of samples and what constitutes an acceptable standard. Previous EU legislation required that samples should show less than 100 cfu/mL after a suitable incubation period. However, the current legislation is less prescriptive and states that there should be no viable micro-organisms or spores capable of growing in the treated product when kept in an aseptic closed container at ambient temperature (see Table 8.2).

Tetra Pak (undated) in its testing regime mentions a count for a sterile product of less than 10 when a calibrated loop of 10 µL is streaked onto a plate. This equates to a count of less than 1000 cfu/mL, meaning that, products with such a count (after incubation) would most likely be satisfactory.

One suggestion is for a processor to adopt an approach having a lower acceptable quality level (AQL), an action level and an upper AQL. One proposal is to set these as 10 cfu/mL, 100 cfu/mL and 1000 cfu/mL respectively and to monitor how many samples fall into each category for every production batch. This information could be very useful in the longer term, as it could be cross-referenced against numbers of defective items and complaints about any batch over its storage period. This type of approach was suggested by von Bockelmann and von Bockelmann (1998) for monitoring psychrotrophic bacteria in raw milk destined for UHT processing. The three levels were a lower AQL of 10^4 cfu/mL; an action level for 10^5 cfu/mL and an upper AQL of 10^6 cfu/mL. It could also be adapted for other tests used to establish sample sterility, such as ATP readings or pH decrease.

12.6 The Shelf-Life of ESL Milk

As discussed in Chapters 3 and 4, the shelf-life of thermally produced ESL milk varies considerably and has been reported to be unpredictable. This can be largely attributed to post-processing contamination as most ESL milk is currently packaged in very clean, but not aseptic, filling machines.

When milk is processed under conditions which destroy spores of psychrotrophic spore-formers (see Section 12.1.7) and packaged aseptically, ESL milk can be expected to have a very long shelf-life under refrigeration. The shelf-life of such an ESL milk has not been determined. Ideally it would be processed at high temperature for a very short time, for example, 140 °C for 1 s, which would maximise the bactericidal effect while minimising the chemical (cooked flavour-producing) effect on the milk. Such temperatures, however, are considered to be in the UHT range by many jurisdictions.

Hence this is really a new type of milk and may require special regulatory consideration in some countries. It is essentially a "commercially sterile" milk where the definition of a "commercially sterile" product is that it does not contain microorganisms which can grow under the normal conditions of storage, which in this case is under refrigeration, nominally ≤7 °C. Consequently, such a milk may require a new label to distinguish it from both UHT milk and ESL milk not packaged aseptically.

Many consumers would certainly not recognise it as "fresh" if it has a very long shelf-life. It is acknowledged that the term "fresh" is somewhat controversial, particularly in relation to ESL milk, and is not used in some countries. In Germany, for example, it has been agreed to label ESL milk as "longer lasting" to distinguish it from "traditionally manufactured" pasteurised milk which is considered to be "fresh" (Buckenhuskes, 2015).

12.7 Non-Thermal Technologies

Thermal processing for producing extended-shelf-life and sterile milk products is a very mature technology, which, although it poses problems from time to time, has served the dairy industry very well. However, with the development of several other technologies which may be capable of replacing, or at least augmenting, thermal processing, the dairy industry needs to be aware of the possible applications of the new technologies. Obviously, any technology would need to meet several criteria before it could be adopted. These include food safety and quality aspects, economic viability, capacity and regulatory aspects. No one technology discussed in Chapter 10 is currently capable of meeting all of these criteria. However, some technologies such as high pressure processing may be appropriate for some products, particularly low-volume, high-value products, products which are sensitive to heat and specialty products which are impossible to produce by heating.

High-pressure processing is now an established technology for processing a variety of foods although its use for dairy products is very limited. The conditions (pressure, temperature, time) necessary to ensure sterility of the final product are severe and currently only suited to small-scale operations.

Similarly, ultra-high-pressure homogenisation (UHPH) has been demonstrated to be capable of producing sterile products but the current scale of such homogenisers is very small. However, as Amador-Espejo *et al.* (2014) have suggested, UHPH treatments potentially offer a real alternative to heat processing for production of shelf-stable milk.

Gamma irradiation at low doses has been shown to significantly extend the shelf life of pasteurised milk, with little effect on its flavour. Hence it now offers an alternative means of producing ESL milk. Although consumer opposition to this technology exists in some countries, the dairy industry should maintain awareness of the potential of this technology.

12.8 Analytical Methods

In Chapter 11, some modern analytical methods which are suitable for milk and dairy products, but to date have been little used for this purpose, were discussed. Some of these, as well as others not discussed in Chapter 11, offer considerable benefit. For example, proteomics has been used quite extensively in recent research and has demonstrated this benefit. NMR has been used very little but recent developments have shown its considerable potential.

Developments in microbiological methods, as discussed in Chapter 8, are proving very beneficial to the dairy industry. In particular, with the cost of gene sequence analysis decreasing markedly, its use in identifying microorganisms and subsequently the source(s) of microbiologically related problems has become a practical reality.

12.9 Using the Literature

The trend today is to make use of internet searches and cite a large number of internet sources, rather than to visit the library. In the main, we have aimed to minimise the use of internet sources but to use them where it was felt that they made a useful contribution. This applies particularly to commercial equipment and products. Also, there is a lot of wisdom contained in some of the older literature, some of which we have referred to throughout this book. As an example of its value, we have described recent incidents of ropy pasteurised milk (Chapter 2). Today this defect is rare but a most useful account of this phenomenon was found in Davis (1955).

There is also a lot of information on UHT processing and products that is publically unavailable, either in company reports or in unpublished reports of student projects in many universities. In this book, we have called on some of the information generated by students but much remains out of the public domain. Unfortunately, much of this information will never become available. This is because the results of many Masters and Doctorate students' projects are not published, and because many of the projects carried out by students do not produce sufficient data themselves for a research publication. We would therefore encourage all academic supervisors of student projects to ensure that as much as possible of the results of student projects are written up and submitted for publication. It would be particularly useful if all PhD theses were made publically available and searchable.

We are not suggesting supervisors should submit papers to journals based on inadequate data but that they devise means by which sufficient data can be generated on specific topics to warrant publication. In some cases, this may involve the judicious accumulation of data from several small projects.

12.10 Further Reading

The final contribution of this book is an extensive, though not exhaustive, lists of references to books, book chapters and reviews related to the material in this book. These lists appear after the list of references relating directly to this chapter. It is suggested that these lists be consulted on topics not covered, or not covered in detail, in this book to obtain a better general overview of these topics. It is recognised that several of the books referenced contain very relevant chapters, some of which are not separately identified.

References

Amador-Espejo, G.G., Suarez-Berencia, A., Juan, B., Barcenas, M.E. & Trujillo, A.J. (2014). Effect of moderate inlet temperatures in ultra-high-pressure homogenization treatments on physicochemical and sensory characteristics of milk. *Journal of Dairy Science* **97**(2), 659–671.

Blake, M.R., Weimer, B.C., McMahon, D.J. & Savello, P.A. (1995) Sensory and microbial quality of milk processed for extended shelf life by direct steam injection. *Journal of Food Protection* **58**, 1007–1013.

Bremer, P.J., Fillery, S. & McQuillan, A.J. (2006) Laboratory-scale clean-in-place (CIP) studies on the effectiveness of different caustic and acid wash steps on the removal of dairy biofilms. *International Journal of Food Microbiology* **106**(3), 254–62.

Buckenhuskes, H.J. (2015) *ESL Milk Production.* DLG-Expert Report 4/4-14: http://www.dlg.org/fileadmin/downloads/food/Expertenwissen/Ernaehrung/e_2014_4_Expertenwissen_ESL.pdf

Browning, E., Lewis, M. & MacDougall, D. (2001) Predicting safety and quality parameters for UHT-processed milks. *International Journal of Dairy Technology* **54**, 111–120.

Chauhan, K., Dhakal, R., Seale, R.B., Deeth, H.C., Pillidge, C.J., Powell, I.B., Craven, H. & Turner, M.S. (2013) Rapid identification of dairy mesophilic and thermophilic sporeforming bacteria using DNA high resolution melt analysis of variable 16S rDNA regions. *International Journal of Microbiology* **165**, 175–183.

Coorevits, A., Logan, N.A., Dinsdale, A.E., Halket, G., Scheldeman, P., Heyndrickx, M., Schumann, P., Van Landschoot, A. & De Vos, P. (2011) *Bacillus thermolactis* sp nov., isolated from dairy farms, and emended description of *Bacillus thermoamylovorans*. *International Journal of Systematic and Evolutionary Microbiology* **61**, 1954–1961.

Craven, H.M. & Macauley, B.J. (1992) Micro-organisms in pasteurised milk after refrigerated storage. 3. Effect of milk processors. *Australian Journal of Dairy Technology* **47**, 50–55.

Davis, J.G. (1955). *A Dictionary of Dairying*, 2nd edition, Leonard Hill, London.

Flint, S.H., van den Elzen, H., Brooks, J.D. & Bremer, P.J. (1999) Removal and inactivation of thermo-resistant streptococci colonising stainless steel. *International Dairy Journal* **9**(7), 429–436.

Griffiths, M.W. & Phillips, J.D. (1988) Modeling the relation between bacterial growth and storage temperature in pasteurized milks of varying hygienic quality. *Journal of the Society of Dairy Technology* **41**(4), 96–102.

Hotrum, N., Fox, M., van Lieverloo, H., Smit, E. & de Jong, P. (2010) Modelling heat processing of dairy products. In: *Improving the safety and quality of milk*, Volume 1: *Milk production and processing* (ed. M.W. Grifiths), pp. 330–348, Woodhead Publishing, CRC, Oxford.

IDF (2007) *Coagulation of Milk. Processes and Characteristics.* IDF Bulletin 420. International Dairy Federation, Brussels.

Marchand, S., De Block, J, De Jonghe, V., Coorevits, A., Heyndrickx, M. & Herman, L. (2012) Biofilm formation in milk production and processing environments; influence on milk quality and safety. *Comprehensive Reviews of Food Science and Food Safety* **11**, 133–147.

Scheldeman, P., Goossens, K., Rodriguez-Diaz, M., Pil, A., Goris, J., Herman, L., De Vos, P., Logan, N.A. & Heyndrickx, M. (2004) *Paenibacillus lactis* sp. *nov.* isolated from raw and heat-treated milk. *International Journal of Systematic & Evolutionary Microbiology* **54**, 885–891.

Scheldeman, P., Pil, A., Herman, L., De Vos, P. & Heyndrickx, M. (2005) Incidence and diversity of potentially highly heat-resistant spores isolated at dairy farms. *Applied and Environmental Microbiology* **71**(3), 1480–1494.

Seale, R.B., Flint, S.H., McQuillan, A.J. & Bremer, P.J. (2011) Effect of NaOH (caustic wash) on the viability, surface characteristics and adhesion of spores of a Geobacillus sp. isolated from a milk powder production line. *Letters in Applied Microbiology* **52**(2), 104–108.

Teh, K. H., Lindsay, D., Palmer, J., Andrewes, P., Bremer, P. & Flint, S. (2014) Proteolysis in ultra-heat-treated skim milk after exposure to multispecies biofilms under conditions modelling a milk tanker. *International Journal of Dairy Technology* **67**(2), 176–181.

Teh, K.H., Flint, S., Brooks J. & Knight, G. (eds) (2015) *Biofilms in the Dairy Industry.* Society of Dairy Technology Series, Wiley Blackwell, Chichester, UK.

Tetra Pak (undated) *Guideline for Microbiological Evaluation of Commercially Sterile Products. Procedures & Guidelines:* Doc. QAM – 588002-0107.
Tran, H., Datta, N., Lewis, M.J. & Deeth, H.C. (2008) Processing parameters and predicted product properties of industrial UHT milk processing plants in Australia. *International Dairy Journal* **18**, 939–944.
von Bockelmann, B. & von Bockelmann, I.A. (1998) *Long-life Products: Heat-treated, Aseptically Packed: A Guide to Quality.* Dr Bernard von Bockelmann, Åkarp, Sweden.

Further Reading: References to Books, Book Chapters and Reviews Arranged Alphabetically within Publication Type

Books

Alfa-Laval (undated) *UHT treatment - Processing Guidelines.*
APV (1999) *Long life Dairy, Food and Beverage Products* APV Technology Update, APV Nordic Unit Systems, Silkeborg, Denmark.
Burton H (1988) *Ultra-High Temperature Processing of Milk and Milk Products.* Elsevier Applied Science, London.
Cronshaw, H.B. (1947) *Dairy Information.* Dairy Industries Ltd, London.
David, J.R.D., Graves, R.H. & Szemplenski, T. (eds) (2013) *Handbook of Aseptic Processing and Packaging.* 2nd edn, CRC Press, Boca Raton.
Davis, J.G. (1955) *A Dictionary of Dairying.* 2nd edn. Leonard Hill, London.
Griffiths, M.W. (ed) (2010) *Improving the Safety and Quality of Milk.* Vol. 1: *Milk Production and Processing.* Woodhead Publishing, Oxford.
Hsu, D.S. (1970) *Ultra-high-temperature (U.H.T) Processing and Aseptic Packaging (A.P.) of Dairy Products.* Damana Tech Inc, New York.
IDF (1981) *New Monograph on UHT Milk.* IDF Doc. 133. International Dairy Federation, Brussels.
IDF (1993) *Protein & Fat Globule Modifications by Heat Treatment, Homogenisation and other Technological Means for High Quality Dairy Products.* Doc. 9303. International Dairy Federation, Brussels.
IDF (1996) *Heat Treatments and Alternative Methods.* IDF Doc. 9602. International Dairy Federation, Brussels.
IDF (1995) *Heat-induced Changes in Milk.* IDF Doc. 9501. International Dairy Federation, Brussels.
IDF (1996) *UHT Cream.* IDF Doc. 315. International Dairy Federation, Brussels.
Institute for Thermal Processing Specialists (2011) *Guidelines for Microbiological Validation of the Sterilization of Aseptic Filling Machines and Packages, including Containers and Closures.* Document G.005.V1.
Juffs, H. & Deeth, H.C. (2007) *Scientific Evaluation of Pasteurisation for Pathogen Reduction in Milk and Milk Products.* FSANZ, Canberra (ISBN 978-0-642-34555-4).
Kessler, H.G. (1981) *Food Engineering and Dairy Technology.* Verlag A. Kessler, Freising, Germany.
Lewis, M.J. (1990) *Physcial Properties of Foods and Food Processing Systems.* Woodhead Publishers, Cambridge, UK.

Lewis. M.J. & Heppell, N.J. (2000) *Continuous Thermal Processing of Foods: Pasteurization and UHT Sterilization*. Aspen Publishers, Gaithersburg, MD.

Nelson, P.E. (ed) (2010) *Principles of Aseptic Processing and Packaging*. 3rd edn, GMA Science and Education Foundation, USA.

Newstead, D.F. (ed.) (2000) *Methods for Quality Assessment of UHT Milks*, 2nd edn, New Zealand Dairy Research Institute (now Fonterra Research Institute), Palmerston North, New Zealand.

SPX (2008). *APV Technolology Update. Long Life Dairy, Food and Beverage Products*. SPX Corporation, Getzville, NY. http://www.apv-tapflo.ro/userfiles/file/Technology_Update_6750_01_08_2008_US.pdf.

Teh, K.H., Flint, S., Brooks J. & Knight, G. (eds) (2015) *Biofilms in the Dairy Industry*, Society of Dairy Technology Series, Wiley Blackwell, Chichester, UK.

Tetra Pak (1988) *Long Life Products; From Raw Materials to Finished Products*. Tetra Pak, Lund, Sweden.

Tetra Pak (undated) *Guideline for Microbiological Evaluation of Commercially Sterile Products. Procedures & Guidelines:* Doc. QAM – 588002-0107.

Tetra Pak (undated) *The Role of Raw Milk Quality in UHT Production*. Tetra Pak, Lund, Sweden.

Various (1996) *Proceedings of the 2nd Biennial UHT Milk Symposium*. Logan, Utah.

von Bockelmann, B. & von Bockelmann, I.A. (1998) *Long-life Products: Heat-treated, Aseptically Packed: A Guide to Quality*. Dr Bernard von Bockelmann, Akarp, Sweden.

Book Chapters

Andersson, I. & Öste, R. (1995) Sensory quality of UHT milk. In: *Heat-induced changes in milk* (ed. P.F Fox) IDF Doc. 9501, pp. 318–330. International Dairy Federation, Brussels.

Datta, N. & Deeth, H.C. (2005) UHT and aseptic processing of milk. In: *Advances in Thermal and Non-thermal Processing*. (eds G. Tewari & V. Juneja), pp. 63–90. Blackwell Publishing, Ames.

Deeth, H.C. &d Datta, N. (2011) Heat treatment of milk | Ultra-High Temperature Treatment (UHT): Heating systems. In: *Encyclopedia of Dairy Sciences*, 2nd edn, vol. 2 (eds J.W. Fuquay, P.F. Fox & P.L.H. McSweeney), pp. 699–707. Academic Press. San Diego.

Deeth, H.C., Datta, N. & Versteeg, C. (2013) Non thermal technologies In: *Advances in Dairy Ingredients* (eds G. Smithers and M.A. Augustin), pp 161–216. Wiley-Blackwell Publishing, Ames, Iowa, USA.

Deeth, H.C. (2010) Improving UHT processing and UHT milk products. In: *Improving the safety and quality of milk*. Volume 1: *Milk Production and Processing*. (ed. M.W. Griffiths), pp. 302–329. Woodhead Publishing, Oxford.

Deeth, H.C. & Lewis, M.J. (2015) Protein stability in sterilised milk and milk products. In: *Advanced Dairy Chemistry*. Volume 2. *Proteins*. (eds P.L.H. McSweeney & J.A. O'Mahony) pp. 247–286, Springer, New York.

Harwalkar, V.R. (1992). Age gelation of sterilized milks. In: *Advanced Dairy Chemistry*, Volume 1 *Proteins*, 2nd edn, (ed P.F. Fox), pp. 691–734. Elsevier Applied Science Publishers, London.

Hinricks, J. & Atama, Z. (2011) Sterilisation of milk and other products. *Encyclopedia of Dairy Sciences*, 2nd edn, vol. 2 (eds J.W. Fuquay, P.F. Fox & P.L.H. McSweeney), pp. 714–724. Academic Press. San Diego.

Hinricks, J. & Kessler, H.G. (1995) Thermal processing of milk – Processes and equipment. In: *Heat-induced Changes in Milk*. IDF Doc. 9501, pp. 9–21. International Dairy Federation, Brussels.

Kelly, A., Datta, N & Deeth, H.C. (2012) Thermal processing of dairy products. In: *Thermal Food Processing: New Technologies and Quality Issues* 2nd edn (ed. D.-W. Sun) pp. 273–307. CRC Press, Taylor & Francis Group, Boca Raton, FL.

Kessler H G (1989) Effect of thermal processing on milk. In: *Developments in Food Preservation 5*. (ed. S. Thorne), pp. 91–130. Elsevier Applied Science, London.

Lewis, M.J. (2010) Improving pasteurised and extended shelf-life milk. In: *Improving the safety and quality of milk*. Volume 1: *Milk Production and Processing* (ed. M.W. Griffiths), pp. 277–301.Woodhead Publishing, Oxford.

Lewis, M.J. & Deeth, H.C. (2009) Heat treatment of milk. In: *Market Milks – Processing and Quality Management* (ed. A.Y. Tamime), pp. 168–204. Blackwell Publishing Ltd, Oxford.

Manzi, P. & Pizzoferrato, L. (2006) UHT thermal processing of milk. In: *Thermal Food Processing: New Technologies and Quality Issues* (ed. D.-W. Sun), pp. 299–334. CRC Press, Boka Raton, FL.

Robertson, G.L. (2011) Heat treatment of milk | Ultra-High Temperature treatment (UHT): Aseptic packaging. In: *Encyclopedia of Dairy Sciences* (eds J.W. Fuquay, P.F. Fox & P.L.H. McSweeney) 2nd edn, Volume 2, pp. 708–713. Academic Press, San Diego.

Robertson, G.L. (2013) Aseptic packaging of foods. In: *Food Packaging: Principles and Practice*. 3rd edn, pp. 367–382. CRC Press, Boca Raton.

Tetra Pak (2003) Long-life milk. In: *Dairy Processing Handbook*, pp 227–245, Tetra Pak processing Systems AB, Lund, Sweden.

Tetra Pak (2015) Long-life milk. In: *Dairy Processing Handbook*, Tetra Pak processing Systems AB, Lund, Sweden (On-line at http://www.tetrapak.com/au/about/newsarchive/dairy-processing-handbook-free-online)

Touch, V. & Deeth, H.C. (2009) Microbiology of raw and market milks. In: *Market Milks – Processing and Quality Management* (ed. A.Y. Tamime), pp. 48–71. Blackwell Publishing Ltd, Oxford.

Reviews

Al-Attabi, Z., D'Arcy, B.R. & Deeth, H.C. (2009) Volatile sulphur compounds in UHT milk. *Critical Reviews in Food Science & Nutrition* **49**(1), 28–47.

Bansal, B. & Chen, X. (2006). A critical review of milk fouling in heat exchangers. *Comprehensive Reviews in Food Science & Food Safety* **5**, 27–33.

Bremer, P., Seale, B., Flint, S. & Palmer, J. (2009) Biofilms in dairy processing. *Biofilms in the Food and Beverage Industries* **181**, 396–431.

Calvo, M.M. & de la Hoz, L. (1992) Flavour of heated milks. A review. *International Dairy Journal* **2**, 69–81.

Chavan, R.S., Chavan, S.R., Khedkar, C.D. & Jana, A.H. (2011) UHT Milk processing and effect of plasmin activity on shelf life: A review. *Comprehensive Reviews in Food Science & Food Safety* **10**, 251–268.

Cogan, T.M. (1977). A review of heat resistant lipases and proteinases and the quality of dairy products. *Irish Journal of Food Science & Technology* **1**(2), 95–105.

Cousin, M. (1982) Presence and activity of psychrotrophic microorganisms in milk and dairy products: A review. *Journal of Food Protection* **45**(2), 172–207.

Datta, N & Deeth, H.C. (2001) Age gelation of UHT milk - a review. *Transactions of the Institute of Chemical Engineers C. Food and Bioproducts Processing* **79**, 197–210.

Datta, N., Elliott, A.J., Perkins M.L. & Deeth, H.C. (2002) Ultra-High Temperature (UHT) treatment of milk: Comparison of direct and indirect modes of heating. *Australian Journal of Dairy Technology* **57**, 211–227.

Deeth, H.C. & Touch V. (2000). Methods for detecting lipase activity in milk and milk products. *Australian Journal of Dairy Technology* **55**(3), 153–168.

Gleeson, D., O'Connell, A. & Jordan, K. (2013) Review of potential sources and control of thermoduric bacteria in bulk-tank milk. *Irish Journal of Agricultural and Food Research* **52**(2), 217–227.

Harwalkar, V.R., Beckett, D.C., McKellar, R.C., Emmons, D.B. & Doyle, G.E. (1983). Age-thickening and gelation of sterilized evaporated milk. *Journal of Dairy Science* **66**(4), 735–742.

Hill, B.M. & Smythe, B.W. (2012) Endospores of thermophilic bacteria in ingredient milk powders and their significance to the manufacture of sterilized milk products: An industrial perspective. *Food Reviews International* **28**(3), 299–312.

Ismail, B. & Nielsen, S. S. (2010) Invited review: Plasmin protease in milk: Current knowledge and relevance to dairy industry. *Journal of Dairy Science* **93**(11), 4999–5009.

Karoui, R., Mazerolles, G. & Dufour, E. (2003) Spectroscopic techniques coupled with chemometric tools for structure and texture determinations in dairy products. *International Dairy Journal* **13**, 607–620.

Kosaric, N., Kitchen, B., Panchal, C.J., Sheppard, J.D., Kennedy, K. & Sergeant, A. (1981) UHT milk: production, quality and economics. *Critical Reviews in Food Science & Nutrition* **21**, 153–199.

Krasaekoopt, W., Bhandari, B. & Deeth, H.C. (2003) Yogurt from UHT milk: A review. *Australian Journal of Dairy Technology* **58**, 26–29.

Law, B. A. (1979) Review of the progress of dairy science: enzymes of psychrotrophic bacteria and their effects on milk and milk products. *Journal of Dairy Research* **46**(3), 573–588.

Lewis, M.J. (2011) The measurement and significance of ionic calcium in milk - A review. *International Journal of Dairy Technology* **64**, 1–13.

Meer, R.R., Baker, J., Bodyfelt, F.W. & Griffiths, M.W. (1991) Psychrotrophic *Bacillus* spp in fluid milk-products - a review. *Journal of Food Protection* **54**(12), 969–979.

Mehta, R.S. (1980) Milk processed at ultra-high-temperatures - a review. *Journal of Food Protection* **43**(3), 212–225.

Mehta B.M. and Deeth H.C. (2015) Blocked lysine in dairy products – formation, occurrence, analysis and nutritional implications. *Comprehensive Reviews in Food Science and Food Safety* **15**, 206–218.

Mostert, J.F. & Holzapfel, W.H. (1979) [Bacterial spoilage of UHT treated milk - a review]. *South African Journal of Dairy Technology* **11**(3), 93–99.

Muir, D. D. (1984) UHT-sterilized milk concentrate - a review of practical methods of production. *Journal of the Society of Dairy Technology* **37**(4), 135–141.

Mulvihill, D. M. & Donovan, M. (1987) Whey proteins and their thermal-denaturation - a review. *Irish Journal of Food Science & Technology* **11**(1), 43–75.
Perkin, A.G. (1985) Review of UHT processing methods. *Journal of the Society of Dairy Technology* **38**(3), 69–73.
Prakash, S., Datta, N. &d Deeth, H.C. (2005) Methods of detecting fouling caused by heating of milk. *Food Reviews International* **21**, 267–93.
Sadeghinezhad, E., Kazi, S.N., Badarudin, A., Zubair, M.N.M., Dehkordi, B.L. & Oon, C.S. (2013) A review of milk fouling on heat exchanger surfaces. *Reviews in Chemical Engineering* **29**(3), 169–188.
Tetra Pak (undated2) *Procedures & Guidelines.Guideline for Microbiological Evaluation of Commercially Sterile Products.* Document number QAM – 588002-0107.
Wijayanti, H.B., Bansal. N. & Deeth, H.C. (2014) Stability of whey proteins during thermal processing: a review. *Comprehensive Reviews in Food Science & Food Safety* **13**, 1235–1251.
Zabbia, A., Buys, E.M. & De Kock, H.L. (2012). Undesirable sulphur and carbonyl flavor compounds in UHT milk: a review. *Critical Reviews in Food Science and Nutrition* **52**(1–3), 21–30.

Index

Note: Page numbers in **bold** refer to tables and page numbers in *italics* refer to figures.

High Temperature Processing of Milk and Milk Products, First Edition. Hilton C. Deeth and Michael J. Lewis.

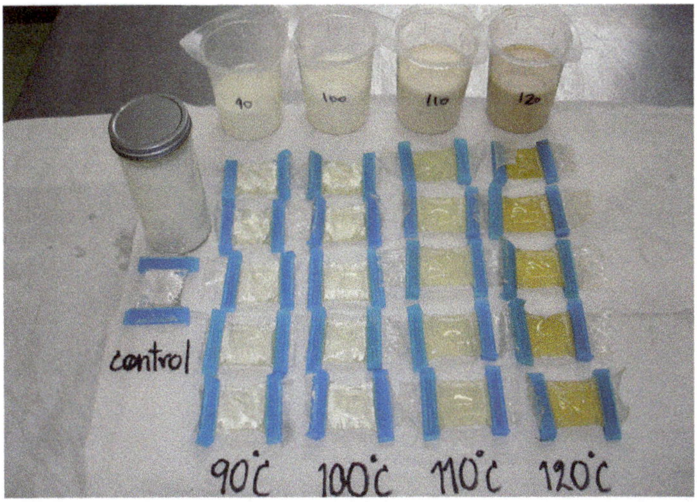

Figure 7.4 Dialysates of milk heated to 90, 100, 110 and 120 °C for 60 min. (Source: On-Nom, 2012. Reproduced with permission.)

Figure 7.5 Lactose-reduced milk samples after storage at 4, 20, 35 and 50 °C for 4 months. (Source: DIAL, 2014. Reproduced with permission.)

Figure 7.6 Milk samples, skim milk, goat's milk, full cream (cow's) milk (X2) and lactose-reduced milk, after storage at 50 °C for 4 months. (Source: DIAL, 2014. Reproduced with permission.)

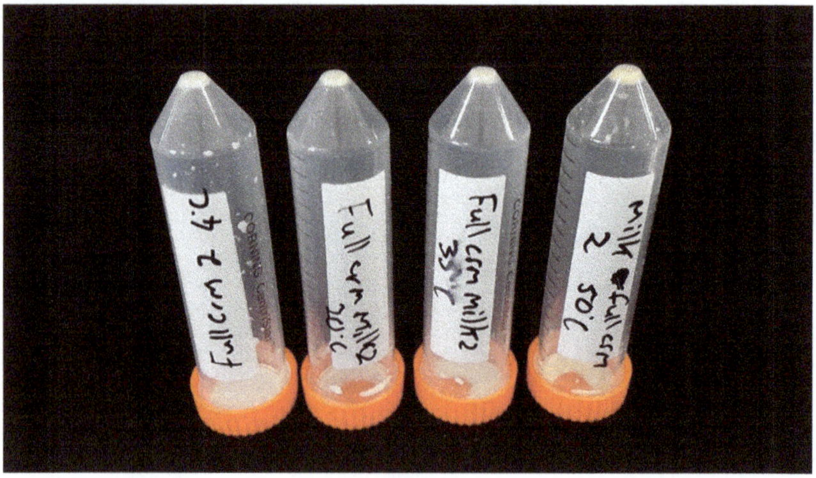

Figure 7.7 Sediment from full cream milk after storage at 4, 20, 35 and 50 °C for 4 months, from approximately 50 mL milk. (Source: DIAL, 2014. Reproduced with permission.)

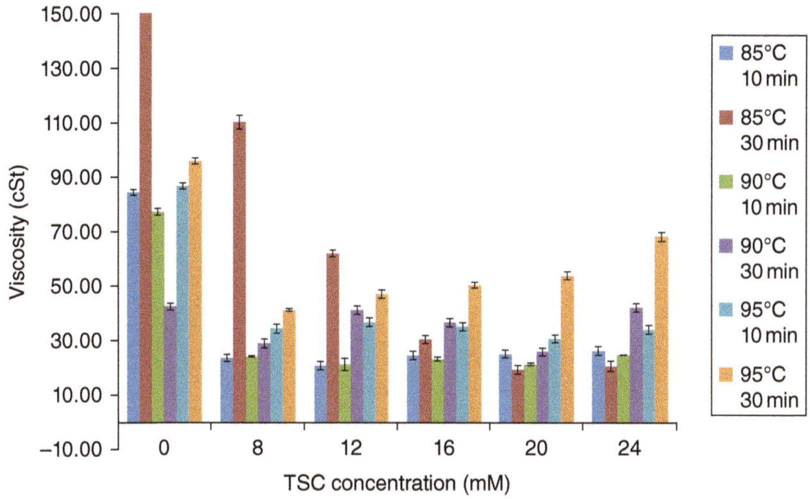

Figure 9.1 The influence of forewarming conditions and stabiliser concentration on the viscosity of evaporated milk for six batches of raw milk. (Source: Chen, 2013. Reproduced with permission.)

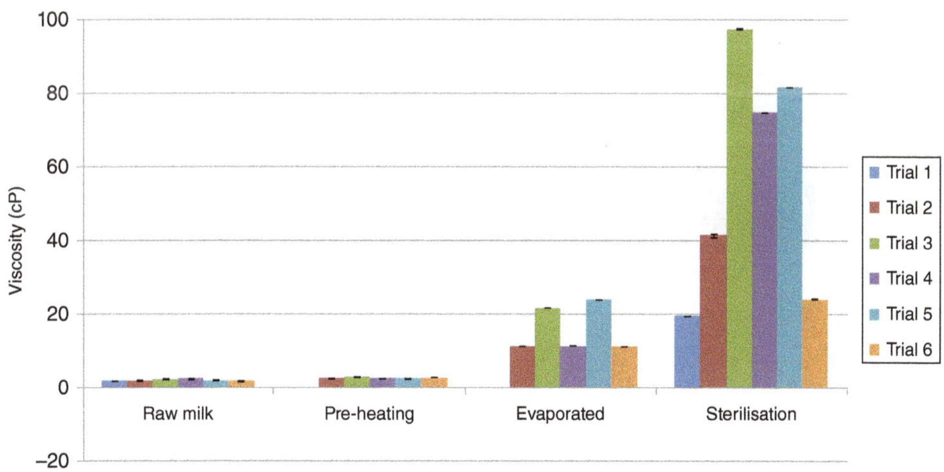

Figure 9.2 Effects of forewarming, evaporation and sterilisation on the viscosity of evaporated milk for six different milk samples. (Source: Chen, 2013. Reproduced with permission.)

Figure 9.4 Some chocolate milk samples after centrifugation.

www.ingramcontent.com/pod-product-compliance
Ingram Content Group UK Ltd.
Pitfield, Milton Keynes, MK11 3LW, UK
UKHW050842101026
15624UKWH00018B/429